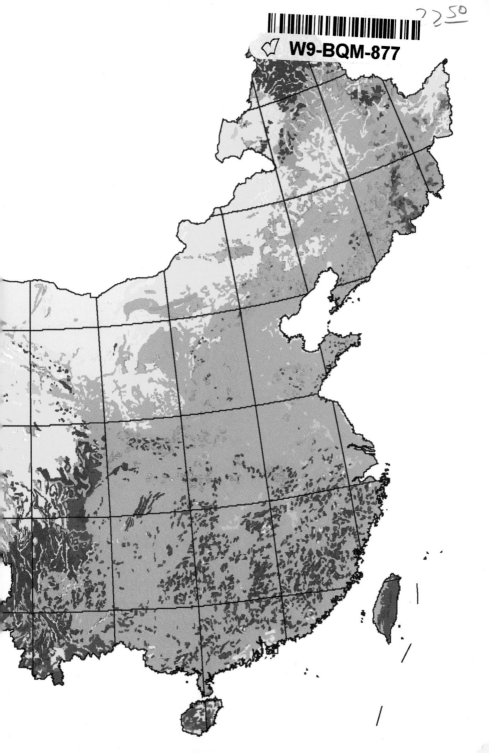

A Field Guide to the Birds of China

A Field Guide to the Birds of China

John MacKinnon

Karen Phillipps

In collaboration with
He Fen-qi

Illustrations by
Karen Phillipps

with 20 plates by **David Showler**

Distribution Maps prepared by
David MacKinnon

OXFORD
UNIVERSITY PRESS

OXFORD

UNIVERSITY PRESS

Great Clarendon Street, Oxford OX2 6DP

Oxford University Press is a department of the University of Oxford.
It furthers the University's objective of excellence in research, scholarship,
and education by publishing worldwide in

Oxford New York Athens Auckland Bangkok Bogotá Buenos Aires Calcutta
Cape Town Chennai Dar es Salaam Delhi Florence
Hong Kong Istanbul Karachi Kuala Lumpur Madrid Melbourne Mexico City
Mumbai Nairobi Paris São Paulo Singapore Taipei Tokyo Toronto Warsaw with
associated companies in Berlin Ibadan

Oxford is a registered trade mark of Oxford University Press
in the UK and in certain other countries

Published in the United States
by Oxford University Press Inc., New York

A catalogue record for this book is available from the British Library

Library of Congress Cataloging in Publication Data
MacKinnon, John Ramsay.
A field guide to the birds of China/John MacKinnon, Karen Phillipps; in collaboration
with He Fen-qi; illustrations by Karen Phillipps; also 20 plates by David Showler;
distribution maps prepared by David MacKinnon.
Includes bibliographical references
1. Birds–China–Identification   I. Phillipps, Karen   II. He, Fen-qi   III. Title.
QL961.C5 M23 2000   598'.0951–dc21   99-050044

ISBN 0 19 854941 5 (Hbk)
ISBN 0 19 854940 7 (Pbk)

Typeset by J&L Composition Ltd, Filey, North Yorkshire
Printed in China

Contents

Warblers, Grassbirds, Laughingthrushes and Babblers—Sylviidae   375
Larks—Alaudidae   465
Flowerpeckers and Sunbirds—Nectariniidae   471
Sparrows, Finches, Pipits and allies—Passeridae   477
Finches and Buntings—Fringillidae   496

APPENDICES

List of Plates

In Memory of
General Nicholas Michilovitch Prjevalsky
Dr Ferdinand Stoliczka
John Whitehead
who died during pioneering expeditions of the Chinese avifauna.

The Authors

John MacKinnon is Professor of Biodiversity Information at the Durrell Institute of Conservation and Ecology at the University of Kent in Canterbury, England. He is currently posted in the Philippines as head of a European Union project to establish an ASEAN Regional Centre for Biodiversity Conservation. He is Chairman of a Special Biodiversity Working Group that advises the Chinese government on biodiversity matters and formerly spent eight years living in China and Hong Kong working on a number of conservation projects in China. His wife is Chinese. Apart from many technical reports on China he has published two books on the country—*Wild China* and *A Photoguide to the birds of China*. Among several others books on natural history of Asia, he is also the senior author of the *Fieldguide to the birds of Borneo, Sumatra, Java and Bali* which remains the standard bird guide to the Greater Sunda region. He has spent several years developing field methods for assessing bird species-richness in forests and computerised species databases and monitoring systems.

Karen Phillipps was born and brought up in Borneo and has spent most of her life in East Asia. She is a professional wildlife illustrator and currently resides in Portugal. She has illustrated a number of other fieldguides including *A field guide to the mammals of Borneo*, *Birds of Hong Kong and south China*, *The birds of Peninsular Malaysia* and *The birds of Sulawesi*. She was illustrator and co-author with John MacKinnon of the *Fieldguide to the birds of Borneo, Sumatra, Java and Bali*. She lived in Hong Kong for over 20 years, where she was a keen member of the Hong Kong Bird Watching Society and visited China a number of times.

Acknowledgements

We would like to thank a large number of people who have helped in the compilation of data, preparation, reviewing and editing of text, making materials available for artwork and commenting on plates and text.

Five collaborators deserve special thanks for reviewing and straightening out many parts of the text: He Fen-qi, David Melville, Craig Robson, Mike Crosby and Dr Ken Searle.

Other experts who reviewed sections of the text and gave valuable comment include Per Alström, Guy Kirwan and Chris Perrins. Thanks are also due to those who have given advice, comment and information to help with the preparation of the book or earlier lists and data on which it is based, namely Guo Rongji, Tim Inskipp, Frank Rozendaal, Phil Round and David Melville.

We are indebted to a number of ornithologists and bird watchers who have provided unpublished notes, species lists, tape recordings and other personal information used in this book: Frank Rozendaal, Paul Andrew, Ben King, Mike Crosby, Jeffrey Boswell, Geoff Carey, Simba Chan, Clive Viney, Dr. Wolfgang Grummt, Jonathan Eames, David Rimlinger, Yang Yuangchan, Rob-Williams, Don Reid, Alan Pearson, Malcolm Coulter, Simon Dowell, Mike Chalmers, Daniel Chamberlain, Martin Williams and Ray Tipper. Thanks are also due to the staff of various nature reserves in China who have kindly supplied reserve bird lists: Wolong, Tangjiahe, Foping, Taibai, Poyang Hu, Fanjingshan, Wuyishan, Niubeiling, Shennongjia, Damingsham, Xishuangbanna, Bawangling, Dongzaigang, Xinghai Hu, Zhalong, Erhai Hu and Yancheng.

In addition, JM would like to thank a number of ornithologists who accompanied him in the field in Sichuan, Yunnan, Jiangxi and Hong Kong (some to their everlasting regret, including three who got lost in the elephant forests of Xishuangbanna and spent a pleasant night huddled together being eaten by mosquitos): Phil Round, Derek Scott, David Melville, Paul Leader, Peter Kennerly, Geoff Carey, Bob Ferguson, Martin Williams, Verity Pickens, Angus Hutton, Tony Galsworthy, Yang Yuangchan, Andrew Laurie, Nigel Watts, Mary Ketterer and Prof. Wei.

We would like to thank the staff of the British Museum at Tring and collections of the Chinese Academy of Sciences in Beijing, Kunming and Guangzhou for their cooperation in allowing us access to examine and photograph specimens. Thanks are due to Dr. Ken Searle for access to, and permission to photograph, live birds in the collections of the Urban Council of Hong Kong.

Prof. Wang Sung of the China Endangered Species Unit and his staff made available over 6000 digital point data records of birds in China gleaned from all available Chinese books and papers.

Particular thanks are due to those who have helped in the physical preparation of the book in typing, editing and other tedious labours of love, especially Lu Hefen who typed most of the original drafts, David MacKinnon who prepared the distribution maps and Dr. Ken Searle who supplied Karen Phillipps with a constant flow of reference material for artwork. Special thanks to Hank and D. D. Tyler for going to so much trouble to provide the cover illustration from an original painting by Karen Phillipps.

INTRODUCTION

Authors' introduction

China is one of the largest countries in the world with a total area of *10 000 000 square kilometres* or 7 per cent of the land surface of the planet. It is a huge country of great extremes from the world's highest peak of Everest, at 8848 m, to one of the world's lowest land points, at −155 m in the Turpan basin, and from the permafrost of the extreme north to the tropical seas of the extreme south. Habitats include forests, grasslands, deserts and wetlands.

This great diversity of habitats, together with China's large size give rise to a great richness of biological diversity. China is ranked as one of the top five richest countries in the world in terms of its vertebrate species—a remarkable fact for a largely temperate landmass. In terms of birds, this richness is demonstrated in over 1300 species recorded or expected from the country, including some of the most spectacular and fascinating birds in the world.

Most families have long lists of members but there are some families and groups that are particularly Chinese. For instance China is the world distribution centre of pheasants with 62 species out of a global total of about 200. Another particularly Chinese group is the laughingthrushes of which China boasts no less than 36 species or more than half the global total. China is particularly rich in parrotbills, leafwarblers, crows and rosefinches as well as in ducks, swans, and geese with a total of 50 species, about a quarter of the world total. Nine of the world's 14 crane species live in China, breeding in the north and migrating in winter to the southern wetlands. Over 100 bird species are endemic to China or only marginally exist outside of China. The bird fauna of China is still poorly-known and new discoveries are reported frequently.

One of the reasons for the incomplete knowledge of Chinese birds is the poor availability of field guides for the region. Until recently there were only keys, lists and poor black and white drawings available for Chinese ornithologists to work with. There is still no complete, taxonomically up-to-date and fully illustrated guide, which is so urgently needed. This guide attempts to fill the gap and will be published in both English and Chinese versions.

The book describes and illustrates over 132 species and describes many hundreds of additional subspecies. Distribution maps are provided, showing summer, winter and resident distributions within China and in neighbouring countries of the region.

Introductory chapters provide a useful background to the region and useful information about the conservation and status of Chinese birds, and hints on how and where to see birds in the region.

The appendices give additional useful information such as lists of endangered and protected species, lists of ornithological clubs and societies for the region and lists of birds endemic to specific regions of China.

The taxonomy of the book will be unfamiliar to most Chinese ornithologists who still follow the outdated taxonomy used by Cheng (1987). In 1990 Sibley and Monroe published a major volume entitled *Distribution and Taxonomy of Birds of the World*. This work is the most comprehensive revision since the 15-volume treatment of Peters *et al.* (1931–86). The volume is based on new DNA-hybridisation studies and provides a far more objective measure of the relationships of genera and families than mere morphology. As a result, the authors produced a rather revised ordering of families. The most dramatic changes involve inclusion of wood-swallows, orioles, fantails, drongos, monarchs, leafbirds and helmet-shrikes as tribes under an enlarged crow family Corvidae, and subjugation of flowerpeckers under the sunbirds.

This revision is gradually being accepted as the new standard. We have followed the new system of families, sub-families and tribes but have included notes under the family and subfamily headings explaining the revisions and relating these to the older arrangement of families.

We have followed the nomenclature of *An Annotated Checklist of the Birds of the Oriental Region* (Inskipp *et al.* 1996) as far as possible. This checklist was published by the Oriental Bird Club (OBC) in an effort to create greater conformity names used within the region. The checklist follows the taxonomic arrangement of Sibley and Monroe but departs in details of nomenclature where it is shown that Sibley and Monroe's names are incorrect or where species splits seem insufficiently justified. In particular, the checklist has reverted to more generally used English names where the Sibley and Mouroe names seem inappropriate such as their use of capitalised hyphenated classifiers. In about a dozen cases we have maintained species splits where lumping by the OBC checklist seems too conservative. The rationale for such divergence are included in the notes section after the relevant species descriptions.

The authors ultimately believe that a phylogenetic classification of species and other taxa should be adopted but in the meantime the necessary data are simply not available, and in most cases the decision as to whether two forms belong to one or two species is the subjective decision of taxonomists. In this respect we accept and expect that where closely related species overlap in space, there will be some hybridisation. Such hybrid zones can be extensive without invalidating the full species status of the two taxa. Sexually aroused birds will generally mate with second best if the right pattern of signals is not available but so long as they prefer to breed with their own then the hybrids will be gradually selected against and species integrity will be maintained. Reduced response to the songs of related forms can be tested in the field as per Alström *et al.* (1992). However, in most cases this has not been done. In the absence of other data the following types of evidence are taken as a strong indication that taxa belong to different species.

- differences in display
- differences in song
- consistent differences in morphology
- breeding sympathy of the two distinct forms
- disjunct distributions which must have been separate for a very long time.

There are many synonyms of English names in current usage. We have therefore listed most currently used common names on the second line of the species entries using the following conventions. Alternatives are separated by a diagonal slash and common general name is given only for the last name of a list. The symbol > indicates that names following are applicable to only part of the whole species. AG: indicates that the alternate general name so introduced can be applied to all preceding names sharing another general name. Thus an entry (Russet/Mountain Bush-Warbler > Javan/Timor Bush-Warbler, AG: Scrub-Warbler) indicates that named Javan and Timor forms are included in this species and that the general name Bush-Warbler can be replaced by Scrub-Warbler. All names are also traceable through the index. Each family and sub-family is introduced with a short description followed by entries for each species.

Species entries cover nomenclature; description of plumage and soft parts; voice; range (global); distribution and status (within region); habits, including diet only where this differs from the general dietary habits of the group which are covered in the family introduction. Chinese names in pinyin romanised from are given in the species descriptions.

For sake of brevity the compass points north, east etc. are abbreviated to capitals without full stops. Thus SE Asia for South-east Asia, N China and SE Sichuan refer to northern China and south-east Sichuan. Species reported or expected from the region but still lacking confirmation are described in the text but placed within square brackets [].

Data on vegetation and altitude assist in explaining distribution and is given under the Status and Distribution headings. Data on microhabitat use, such as forest strata used by a species, are given under the Habits section.

John MacKinnon

Canterbury
January 1999

Introduction to the region

1. Geographical limits of the book

The region covered by this field guide includes the entire land and sea area claimed as Chinese territory on official government maps of China, including Taiwan, Macao and Hong Kong. This territory also includes some areas such as Arunachal Pradesh and parts of the South China Sea that remain regions of international dispute, and this poses problems of inclusion and exclusion of species. The following criteria have been applied to determine how birds from these disputed regions will be treated. All birds recorded within territory currently administered by the government of China or Taiwan are described in full. Where races of these species have not been recorded from SE Xizang but are known from areas of Arunachal Pradesh claimed by China and are also likely occur within SE Xizang because their altitudinal range is greater than 1000 m, they are mentioned in the text with a qualifier such as 'probably occurs in SE Xizang' or 'can be expected in SE Xizang'. Similarly, species not yet recorded in territory actually administered by China but known from areas claimed by China, are considered likely to occur in areas administered by China on the grounds of altitude or ranging behaviour, and are also described in full but enclosed within square brackets. Birds known only from areas claimed by China but likely to occur anywhere else in territory administered by China, such as the lowlands of Arunachal Pradesh and the extreme south of the South China Sea, are either referred to only in footnotes under similar species or are listed in Appendices 3 and 4.

2. Physical description of the region

In order to understand the distribution and movements of birds in China, it is necessary to have a feeling for the country's physical geography and climate.

Physically, China can be viewed as composing three major regions. The highest of these consists of the raised Tibetan (Qinghai–Xizang) Plateau which slopes slightly from a base elevation of about 5000 m in the west to about 4000 m in the east. Along the south edge of this plateau rise the Greater Himalayas with many peaks over 7000 m. On the northern flank of the plateau are, from west to east, the Kunlun, Altun and Qilian mountain ranges.

Due to its high altitude, temperatures over the entire plateau are low, permafrost is widespread, solar radiation is intense and winds are very strong. Rates of desiccation and erosion are high and the water that comes from snow and glacier melt-off gathers in salty lakes scattered over the entire plateau. Soils are weak and generally infertile and the Tibetan people are mostly pastoral.

Draining from the plateau are five of the world's major rivers. In the south is the Yarlung Zangbo or upper stretches of the Brahmaputra that drain east then cut south through the Himalayas to northern India. Next is the Nujiang or Salween which curves around the eastern end of the Himalayas before draining south through Burma, paralleled by the Lancang or Mekong which drains through Indochina and finally the famous Yangtze (Changjiang) which turns north and east again to drain through central China to emerge at Shanghai. To the north the great Yellow River or Huanghe drains across northern China to the Bohai Sea.

At the northern end of the plateau is the graben depression known as the Qaidam basin which drops to altitudes of only 2600 m with salty lakes and mineral concentrations at the bottom.

Lake Qinghai (Koko Nor) at an elevation of 3200 m in the north-eastern part of the plateau is China's largest lake with a surface of 4400 km². However, like other lakes in the region, its water level is dropping.

The second major region of physical China is the arid north-western region. In fact this is the eastern end of the great Eurasian desert and grassland zone. It accounts for about 30 per cent of the land area of China but only supports about five per cent of the human population.

This region consists of series of low-lying basins mostly between 500 and 1000 m above sea-level. As a result of the great distance from any seas, and the surrounding ranges of mountains, the influence of summer monsoons is very weak and conditions range from arid in the west to semi-arid in the east.

China's sandy deserts lie in this realm. Largest and harshest is the huge Taklimakan desert, immediately to the north of the Tibetan Plateau and with an area of 327 000 km² lying in the bottom of the Tarim basin. Most of the desert is loose drifting dunes of yellow sand. These sands are stabilised around the deserts' edges and along some river beds by belts of tamarisk *Tamarix* bushes. This is the only warm temperate desert in China.

To the north of the Tianshan mountains lies another large basin, the Junggar. At 47 000 km², this is China's second largest desert. It is quite different from Taklimakan— cooler, with lower rates of evaporation and more rainfall which allows for more vegetation. Most of the basin is estabilised stony desert with enough plant life to support winter grazing.

Smaller deserts include the Alxa desert in western Inner Mongolia between the Helan mountains and Qilian mountains, and the Ordos desert south of the great bend of the Yellow River. The Alxa includes the Badain Jaran, Tengger and Ulan Buh desert areas. Much of these is mobile dunes but there are some saline lakes and springs. These deserts are spreading toward the Yellow River but major efforts to afforest and stabilise dunes with grasses are having some effect in checking the desertification process.

To the north, the great stony Gobi desert stretches into the heart of Mongolia mostly outside Chinese territory. To the east the deserts give way to semi-desert and temperate steppe grasslands as the climate becomes semi-humid.

Two great mountain ranges rise in the north-west region—the Tianshan and Altai. Both are glaciated on their highest peaks and the streams that drain off these mountains provide an important but narrow fringe of fertility around the edges of the great deserts they separate.

The third major physical region of China is Eastern Monsoon China. This realm, comprising about 45 per cent of the country also contains 95 per cent of the human population. Almost all potentially arable land has already been developed, often overused, degraded or even abandoned. Almost all natural vegetation has been modified by man. However, there are some important mountain ranges around and within this unit that still supports forest and many of China's most interesting birds.

This entire unit enjoys seasonally humid weather but there are major seasonal variations in temperature and wind direction, and a gradual cline of climate from the colder temperature north, through the subtropical south of the country and a narrow tropical fringe in the south.

The unit consists of wide alluvial valleys of the main rivers—particularly the lower reaches of the great Yellow (Huang He) River and Yangtze (Changjiang), coastal plains and a few rather ancient and stable ranges of mountains, rarely exceeding 2000 m.

3. Climate of China

Climate varies greatly over the face of China. The greatest influence on climate is exerted by the pattern of monsoon winds over the Asiatic landmass. In the winter months Asia is much cooler than the Pacific Ocean and a high pressure area is formed. This results in cold dry winds that sweep from the interior landmass south-east towards the ocean. Most of eastern China is swept by this dust-laden monsoon and the weather is very cold. Only a little rainfall comes from the eastward-moving depressions and falls in the Central parts of China. In summer, however, Central Asia heats up and the reverse wind pattern results. The summer monsoon blows from the Pacific and Indian Oceans and brings with it most of China's annual rainfall.

Climatically the north-west and west of China are characterised by low rainfall and extreme summer high and winter low temperatures. Eastern China is more humid, equable and more suitable for agriculture. In Harbin in north-east China the average temperature in January is −18 °C whilst that in July is 22 °C—a difference of 40 °C. By contrast, in Guangzhou, in south China the mean January temperature is 13 °C and in July is 22 °C—a difference of only 9 °C. In north-east China the plains are snowbound all winter whilst in the south frost only occurs on the higher hills. Between these extremes the Changjiang (Yangtze) valley averages about 3 °C in midwinter.

4. Natural vegetation

China exhibits a wide range of bird habitats from the highest peaks in the Himalayas, the largest high altitude plateaux, hot and cold deserts, spectacular wetlands, great grassy steppes, seashores and a range of forest types from permafrost tundra in the north-east, through temperate coniferous and broadleaf forests in eastern China to tropical forests in the far south. Marine areas of China range from cold northern seas to tropical coral beaches and deep oceanic troughs. Birds use all of these habitats, but each species only uses some of the wide range of habitats available. Recognising habitat is a key to correct species identification.

Many birds in China show seasonal ranges in habitat: occurring further north or higher in mountains during the summer months and migrating south or descending in altitude in winter. Indeed the east coast of China is an important migrant flyway for birds that breed in northern latitudes but migrate south in winter. Some of these stay to winter in southern parts of China; whilst others fly further south to equatorial regions of Indonesia and beyond.

The endpaper shows the main habitat types used in the descriptions in this book. The map is somewhat generalised and land relief is also significant. Where there are mountains in an area which at sea-level may support subtropical evergreen forest, the vegetation at higher altitudes will vary through montage formations which often resemble vegetation of lower altitudes much further north. Thus we find subalpine and alpine vegetation in tropical mountains as well as in the northern latitudes of the country.

The major vegetation types mapped are:

1. Coniferous forests—dominated by fir, spruce and hemlock at higher altitudes or latitudes or pines and larches at lower altitudes or latitudes, but often mixed with birches and other broadleaf species. In moist coniferous forest on southern mountains there may be dense bamboo undergrowth which provides a rather special habitat for birds.

2. Deciduous broadleaf forests. These are dominated by oaks, beech, chestnuts and familiar northern genera but are generally much richer in species than their European or American counterparts. There is considerable variation in species composition from north to south.

3. Subtropical mixed forests. These are composed of hardleaved evergreen oaks and chestnuts mixed with some tropical pines and the Chinese fir *Cunninghamia*.

4. Tropical evergreen and semi-evergreen forests. These rich formations are confined to the moist south-west and south of the country, southern Taiwan and on Hainan island. These forests are rich in composition and sometimes very tall and complex in structure with several palm species. This provides a great range of habitat niches for birds and indeed this is where we find the greatest variety of birds in China. Visibility is not always good, however, and rain is frequent.

5. Montane forests show similar changes in altitude as the lowland forests show with latitude, passing successively from evergreen broadleaf forests, through hardleaved oak forests into mixed conifer formations of hemlock and spruce, topped by a narrow fir zone and scrub of rhododendrons and junipers along the alpine treeline. In wetter parts of China the conifer forests have a dense bamboo understorey. Pines dominate on drier ridges or on sandier rocks and soils.

6. Wetlands include a wide range of types from northern swampy marshes and lakes to the great lakes of the Yangtze valley, small alpine marshes of the Tibetan Plateau, salt pans of Xinjiang in north-west China and mangroves and shorelines of the tropical south. In total area wetlands are not extensive and appear relatively insignificant on a vegetation map but these are some of the most important areas for birds, and many of the most spectacular and most endangered species in China are wetland specialists.

7. Grassy steppes form a wide swathe across northern China being wetter and lusher in the east and becoming gradually drier to the west. Somewhat specialised grasslands grow on some of the loose Loess soils in the Yellow River valley.

8. Deserts include three main types—cold deserts on the Tibetan Plateau at altitudes of over 4000 m where cold dry conditions limit the growth of plants. Hot deserts fall into two main types, sandy and stony. Each has its own flora and some spring

into life on those rare occasions when it rains. The desert arifauna is limited but some species specialise on desert conditions.

9. Bare lands are areas of bare rock or generally great height or steepness, or areas of mobile sand dunes where there is no vegetation at all.

10. Alpine meadows are extensive areas on the Tibetan Plateau and smaller areas on some of the major mountain ranges of south-west and north-west China. They have a rather rich, though stunted vegetation, and a distinctive bird fauna including some rare mountain species.

11. Scrub includes a variety of secondary formations of low stature resulting from deforestation, shifting agriculture, fire or general degradation of vegetation as a result of grazing and cutting. In the east of China where most of the original forest has been cleared, these scrub areas provide the best cover for many bird species.

12. Cultivated areas are extensive in the east and south of the country. Such lands are mostly farmland but may contain woods, orchards and plantations as well as land under fallow. All these provide habitat for some bird species and in spring and autumn many migrants pass though these lands on passage. Some birds have become commensals of Man and are mostly to be found around towns and villages.

5. Human population and current land-use

China has a huge human population of over one billion persons mostly in the eastern half of the country, and pressure on natural habitats is very severe. Many wild species are threatened with extinction, others are rare or restricted to very small distributions. Conservation of the remaining habitats of these species is urgent but difficult.

Most of the population belong to the Han Chinese ethnic group but in the north-east, north, north-west and south-west of the country are many ethnic minorities with Mongolians, Uygurs, Tibetans, Dai, Yi and Li among the most important groups.

The country applies a firm birth control policy with Han Chinese restricted to one child per family. Other minority groups are given more freedom to reproduce depending on the size of the minority.

China is comprised of 31 political divisions. These include three autonomous regions, three municipalities, 23 provinces plus the Special Administrative Unit of Hong Kong and the currently self-governing island of Taiwan. These are shown on the rear endpaper using the modern Pinyin system of Anglicised names.

Land-use varies across the landscape. The great plains of the east are used for agriculture. North of the Huai River the stable crops are wheat, Kaoliang and soya beans and for much of the year the fields lie dry and fallow. To the south of the Huai the main crop is rice and crops can be grown year-round. In the south, tropical crops such as sugar cane, pineapples, bananas and edible bamboos are planted, whilst tea is grown in the hills. Shifting cultivation is a problem in some southern areas.

In the northern steppes and western plateaux efforts to develop agriculture are limited by harsh natural conditions and most people are pastoralists and seasonal nomads.

6. Birds in the local economy and culture

There is a saying about the people of southern China that they eat anything with four limbs except tables, anything that flies except aeroplanes and anything that swims except ships. The saying is not entirely true and, in any case, only applicable to some areas of the country but most people certainly see birds firstly as food, secondly as pests, thirdly a possible medicine and fourthly a valuable pet.

Enjoyment and appreciation of birds free and wild is rather uncommon in urban China today, although through China's long history birds have been greatly admired in poems and paintings.

Pheasants are admired for their great beauty but the birds that have attracted most admiration in Chinese culture have been the cranes. Cranes are long-lived and pair for life. They meet the Chinese ideal of faithfulness and do so with elegance and controlled romance in their elaborate courtship rituals.

Cranes, pheasants and the mythical phoenix are the commonest birds in Chinese paintings and in Yunnan, young girls dance the seductive dance of the peacock.

There is a new movement to restore appreciation of birds in Chinese society and culture. The State Environment Protection Agency (SEPA) promote an annual 'bird-loving' week in which most regional governments organise special events to popularise birds among the public and school children with educational displays, painting contests, staged bird releases and other events. In Hong Kong, the Worldwide Fund for Nature (WWF) has for many years organised an annual 24-hour 'Big Bird Race' which generates a lot of local interest in birds.

March 1958 witnessed one of the most bizarre events in Chinese bird relations. 'Sparrows' which effectively included all passerines were declared pests and a three-day purge was waged in Beijing and other large cities to eliminate them. According to witness Han Suyin (1959) 800 000 birds were destroyed by hysterical crowds banging drums and gongs to prevent the birds from settling or resting. Huge piles of dead and exhausted birds were swept up at the end of this victorious war but the subsequent plague of insects forced officials to admit the operation had been a mistake. To put this number into perspective it is interesting to note that it has been calculated that in the USA 4 000 000 birds are killed every day by domestic cats!

The pet trade is huge in China. Every town has its own bird market and one can find a large array of species available for sale. Of special interest are the Peking Robins, Laughingthrushes especially Hwamei, Magpie Robin, shamas, white-eyes, munias, mynas and parakeets. Saddest of all are the robins, flycatchers and other insectivores that will certainly die quickly in their tiny, although exquisitely made cages. The Chinese word Hwamei literally means 'beautiful eyebrow' in recognition that the elegant white brow of the bird conforms with the arched perfection of the classic Chinese beauty. But it is the lusty clear song of the Hwamei that is most admired by bird fanciers.

Many towns have special parks or even quiet streets where proud owners bring their birds each day to hang up in trees among other caged birds to sing to each other.

The great increase in pet birds, due largely to rising economic standards, has exhausted local supply of wild birds in most of eastern and southern China and many of the birds for sale are either bred commercially or imported from Indonesia or Indochina. There are long internal trade routes from the forested areas where there are still tappable wild populations. One very serious and specialised part of the bird trade concerns the illegal smuggling of falcons out of north-west China through Pakistan to the Middle East. Very high prices are paid for these birds, especially the Saker Falcon, and many thousands of birds are taken out each year with severe consequences for the wild population.

In some parts of southern and eastern China you can still see fishermen catching fish using tame cormorants, but even cormorants are becoming rare in China today.

Traditional Beijing Opera requires fantastic costumes and head decorations which put particular pressure on the long tail feathers of Reeve's and Lady Amherst's Pheasants.

Other curious uses of birds in China include the 'sky burials' practiced by some Tibetan Buddhist groups. These people believe that by helping other animals they gain merit for their next life. When they die their bodies are dismembered and left on rocks or flimsy wooden platforms for ravens and vultures to eat.

Hunting for food is widespread in rural areas. Virtually all birds are eaten but pheasants, pigeons, ducks and geese are especially endangered. Small boys catch rails and herons in crude mist nets in the rice fields and sell these in bundles to passing motorists. In some minority areas, hunting is particularly intense. Almost every household in the Xishuangbanna prefecture of southern Yunnan has about three long-barrelled muskets and gangs of young men wander the roads shooting all the small birds they can find. In the cities it is boys with slingshots or airguns that do the most damage. Domestic cats are rare in most parts of China.

Some birds have the misfortune of being regarded as medicinal. Some species are just regarded as a good tonic, others such as small owls are used to treat specific ailments.

In rural villages in the south, men rear fighting cocks and hold illicit gaming fights. Cross-breeding domestic chickens with wild cocks is found to combine the larger size of domestic fowl with the fierceness of the wild junglefowl.

The Plates

Resident 留鸟

Migrant 候鸟

Winter 冬候鸟

Summer Breeders 繁殖鸟

• Vagrant/accidental 迷鸟

↓ On passage 过境鸟

? Uncertain status 不详

PLATE 1 Gamebirds 1 鸡形目一

1. Snow Partridge
Lerwa lerwa
雪鹑

2. Tibetan Snowcock
Tetraogallus tibetanus
藏雪鸡

3. Altai Snowcock
Tetraogallus altaicus
阿尔泰雪鸡

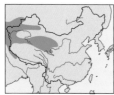

4. Himalayan Snowcock
Tetraogallus himalayensis
暗腹雪鸡

5. Chestnut-throated Partridge
Tetraophasis obscurus
雉鹑

6. Buff-throated Partridge
Tetraophasis szechenyii
四川雉鹑

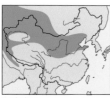

7. Chukar
Alectoris chukar
石鸡

8. Rusty-necklaced Partridge
Alectoris magna
大石鸡

9. Chinese Francolin
Francolinus pintadeanus
中华鹧鸪

10. Grey Partridge
Perdix perdix
灰山鹑

11. Daurian Partridge
Perdix dauurica
斑翅山鹑

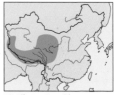

12. Tibetan Partridge
Perdix hodgsoniae
高原山鹑

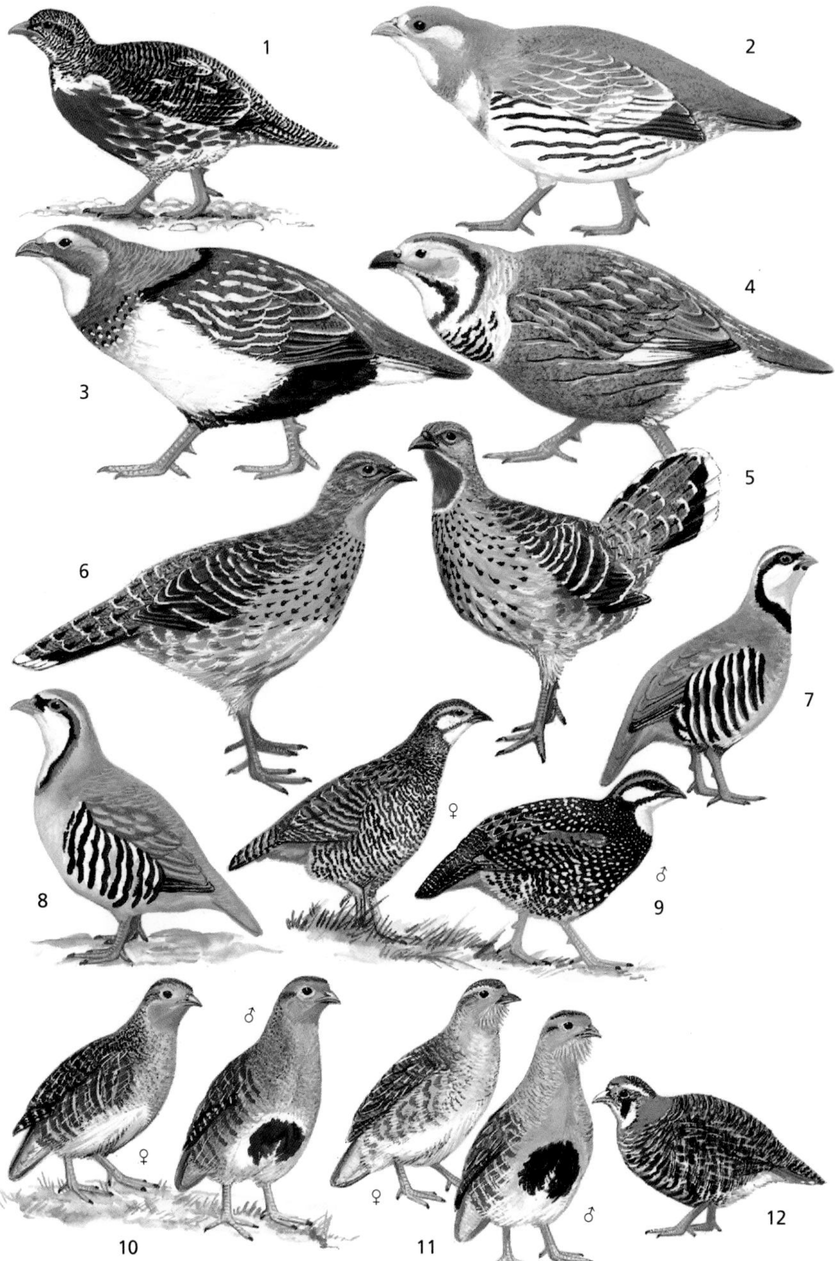
1
2
3
4
5
6
7
8
♀
♂
9
♂
♀
10
♀
♂
11
12

PLATE 2 Gamebirds 2   鸡形目二

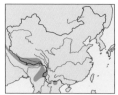

16. Hill Partridge
Arborophila torqueola
环颈山鹧鸪

17. Rufous-throated Partridge
Arborophila rufogularis
红喉山鹧鸪

18. White-cheeked Partridge
Arborophila atrogularis
白颊山鹧鸪

19. Taiwan Partridge
Arborophila crudigularis
台湾山鹧鸪

20. Chestnut-breasted Partridge
Arborophila mandellii
红胸山鹧鸪

21. Bar-backed Partridge
Arborophila brunneopectus
褐胸山鹧鸪

22. Sichuan Partridge
Arborophila rufipectus
四川山鹧鸪

23. White-necklaced Partridge
Arborophila gingica
白眉山鹧鸪

24. Hainan Partridge
Arborophila ardens
海南山鹧鸪

25. Scaly-breasted Partridge
Arborophila charltonii
绿脚山鹧鸪

26. Mountain Bamboo Partridge
Bambusicola fytchii
棕胸竹鸡

27. Chinese Bamboo Partridge
Bambusicola thoracica
灰胸竹鸡

16 *batemani* ♀

intermedia *rufogularis*

17 18

torqueola

19 20 21

22 23 24

25 26

thorocica

27

sonorivox

PLATE 3   **Gamebirds 3**   鸡形目三

28. Blood Pheasant
Ithaginis cruentus
血雉

29. Western Tragopan
Tragopan melanocephalus
黑头角雉

30. Satyr Tragopan
Tragopan satyra
红胸角雉

31. Blyth's Tragopan
Tragopan blythii
灰腹角雉

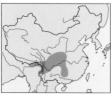

32. Temminck's Tragopan
Tragopan temminckii
红腹角雉

33. Cabot's Tragopan
Tragopan caboti
黄腹角雉

34. Koklass Pheasant
Pucrasia macrolopha
勺鸡

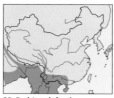

38. Red Junglefowl
Gallus gallus
原鸡

28 sinensis ♀ ♂

29 ♂ ♀

30 ♂ ♀

31 ♀ ♂

32 ♂ ♀

33 ♀ ♂

34 xanthospila ♀ ♂

38 ♂ ♀ jabouillei

PLATE 4 **Gamebirds 4** 鸡形目四

35. Himalayan Monal
Lophophorus impejanus
棕尾虹雉

36. Sclater's Monal
Lophophorus sclateri
白尾梢虹雉

37. Chinese Monal
Lophophorus lhuysii
绿尾虹雉

39. Kalij Pheasant
Lophura leucomelanos
黑鹇

40. Silver Pheasant
Lophura nycthemera
白鹇

41. Swinhoe's Pheasant
Lophura swinhoii
蓝鹇

42. Tibetan Eared Pheasant
Crossoptilon harmani
哈曼马鸡

43. White Eared Pheasant
Crossoptilon crossoptilon
藏马鸡

44. Brown Eared Pheasant
Crossoptilon mantchuricum
褐马鸡

45. Blue Eared Pheasant
Crossoptilon auritum
蓝马鸡

PLATE 5 Gamebirds 5 鸡形目五

46. Elliot's Pheasant
Syrmaticus ellioti
白颈长尾雉

47. Mrs Hume's Pheasant
Syrmaticus humiae
黑颈长尾雉

48. Mikado Pheasant
Syrmaticus mikado
黑长尾雉

49. Reeves's Pheasant
Syrmaticus reevesii
白冠长尾雉

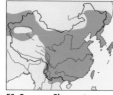

50. Common Pheasant
Phasianus colchicus
雉鸡

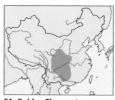

51. Golden Pheasant
Chrysolophus pictus
红腹锦鸡

52. Lady Amherst's Pheasant
Chrysolophus amherstiae
白腹锦鸡

53. Grey Peacock Pheasant
Polyplectron bicalcaratum
灰孔雀雉

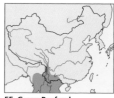

55. Green Peafowl
Pavo muticus
绿孔雀

burmannicus

pallasi *suehschanensis* *torquatus*

47
46
48
53
49
50
51
52
55

PLATE 6 Gamebirds 6 鸡形目六

56. Siberian Grouse
Dendragapus falcipennis
镰翅鸡

57. Willow Ptarmigan
Lagopus mutus
柳雷鸟

58. Rock Ptarmigan
Lagopus mutus
岩雷鸟

59. Black Grouse
Tetrao tetrix
黑琴鸡

60. Western Capercaillie
Tetrao urogallus
西方松鸡

61. Spotted Capercaillie
Tetrao parvirostris
黑嘴松鸡

62. Hazel Grouse
Tetrastes bonasia
花尾榛鸡

63. Chinese Grouse
Tetrastes sewerzowi
斑尾榛鸡

56 ♂ ♀

57 ♂ winter ♀

58 ♂ winter ♀

63 ♂ ♀

62 ♂ ♀

59 ♂ ♀

60 ♀ ♂

61 ♀ ♂

PLATE 7 Swans and Geese 雁及天鹅

66. Mute Swan
Cygnus olor
疣鼻天鹅

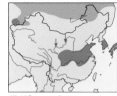

67. Whooper Swan
Cygnus cygnus
大天鹅

68. Tundra Swan
Cygnus columbianus
小天鹅

69. Swan Goose
Anser cygnoides
鸿雁

70. Bean Goose
Anser fabalis
豆雁

71. Greater White-fronted Goose
Anser albifrons
白额雁

72. Lesser White-fronted Goose
Anser erythropus
小白额雁

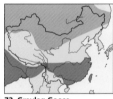

73. Greylag Goose
Anser anser
灰雁

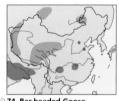

74. Bar-headed Goose
Anser indicus
斑头雁

75. Snow Goose
Anser caerulescens
雪雁

76. Canada Goose
Branta canadensis
加拿大雁

77. Brent Goose
Branta bernicla
黑雁

78. Red-breasted Goose
Branta ruficollis
红胸黑雁

imm. ad. 69

70

73

71 ad. imm.

72

imm. ad. 74

ad.
blue phase

imm.
pale phase

75

78 77 76

ad. white
phase

imm. ad. ♂

66

imm. ad.

67

ad.

imm.

68

PLATE 8　Ducks 1　鸭一

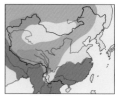

79. Ruddy Shelduck
Tadorna ferruginea
赤麻鸭

80. Crested Shelduck
Tadorna cristata
冠麻鸭

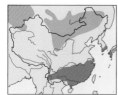

81. Common Shelduck
Tadorna tadorna
翘鼻麻鸭

82. Comb Duck
Sarkidiornis melanotos
瘤鸭

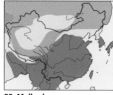

89. Mallard
Anas platyrhynchos
绿头鸭

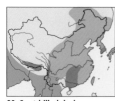

90. Spot-billed duck
Anas poecilorhyncha
斑嘴鸭

91. Philippine Duck
Anas luzonica
棕颈鸭

92. Northern Shoveler
Anas clypeata
琵嘴鸭

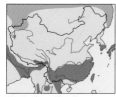

93. Northern Pintail
Anas acuta
针尾鸭

PLATE 9 Ducks 2 鸭二

64. Lesser Whistling-duck
Dendrocygna javanica
栗树鸭

83. Cotton Pygmy-goose
Nettapus coromandelianus
棉凫

84. Mandarin Duck
Aix galericulata
鸳鸯

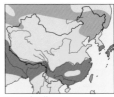

85. Gadwall
Anas strepera
赤膀鸭

86. Falcated Duck
Anas falcata
罗纹鸭

87. Eurasian Wigeon
Anas penelope
赤颈鸭

88. American Wigeon
Anas americana
葡萄胸鸭

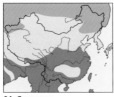

94. Garganey
Anas querquedula
白眉鸭

95. Baikal Teal
Anas formosa
花脸鸭

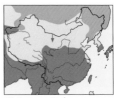

96. Common Teal
Anas crecca
绿翅鸭

ALL BUT = 88"

64

83 ♀

♂

84 ♀

87 ♀

♂

88 ♂

84 ♂

86 ♀

♂

85 ♀

♂

95 ♀

♂

96 ♀

♂

94 ♀

♂

PLATE 10 Ducks 3 鸭三

65. White-headed duck
Oxyura leucocephala
白头硬尾鸭

97. Marbled duck
Marmaronetta angustirostris
云石斑鸭

98. Red-crested Pochard
Rhodonessa rufina
赤嘴潜鸭

99. Common Pochard
Aythya ferina
红头潜鸭

100. Canvasback
Aythya valisneria
帆背潜鸭

101. Ferruginous Pochard
Aythya nyroca
白眼潜鸭

102. Baer's Pochard
Aythya baeri
青头潜鸭

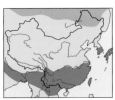

103. Tufted Duck
Aythya fuligula
凤头潜鸭

104. Greater Scaup
Aythya marila
斑背潜鸭

PLATE 11 Ducks 4 鸭四

105. Steller's Eider
Polysticta stelleri
小绒鸭

106. Harlequin Duck
Histrionicus histrionicus
丑鸭

107. Long-tailed Duck
Clangula hyemalis
长尾鸭

108. Black Scoter
Melanitta nigra
黑海番鸭

109. White-winged Scoter
Melanitta fusca
斑脸海番鸭

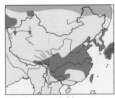

110. Common Goldeneye
Bucephala clangula
鹊鸭

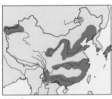

111. Smew
Mergellus albellus
白秋沙鸭

112. Red-breasted Merganser
Mergus serrator
红胸秋沙鸭

113. Scaly-sided Merganser
Mergus squamatus
中华秋沙鸭

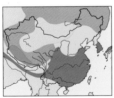

114. Common Merganser
Mergus merganser
普通秋沙鸭

ALL BUT
105 & 106

105

106

107

111

108

109

110

112

113

114

PLATE 12 **Woodpeckers 1** 啄木鸟一

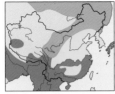

119. Eurasian Wryneck
Jynx torquilla
蚁䴕

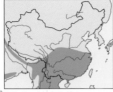

120. Speckled Piculet
Picumnus innominatus
斑姬啄木鸟

121. White-browed Piculet
Sasia ochracea
白眉棕啄木鸟

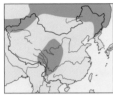

133. Three-toed Woodpecker
Picoides tridactylus
三趾啄木鸟

134. Rufous Woodpecker
Celeus brachyurus
栗啄木鸟

144. Himalayan Flameback
Dinopium shorii
西山金背三趾啄木鸟

145. Common Flameback
Dinopium javanense
金背三趾啄木鸟

146. Greater Flameback
Chrysocolaptes lucidus
大金背啄木鸟

147. Pale-headed Woodpecker
Gecinulus grantia
竹啄木鸟

148. Bay Woodpecker
Blythipicus pyrrhotis
黄嘴噪啄木鸟

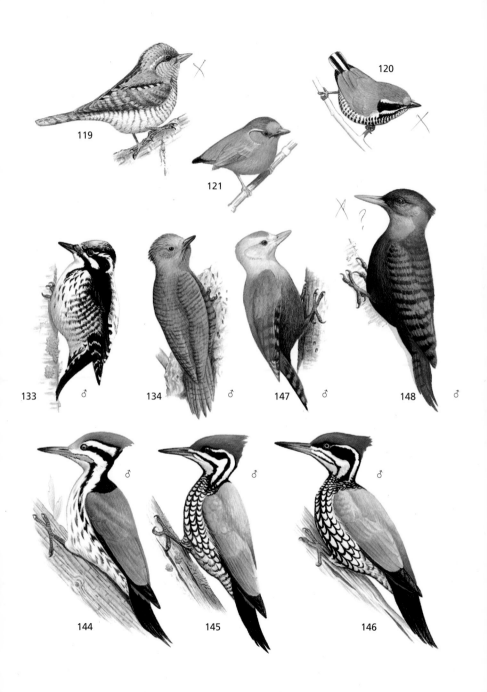

119

120

121

133 ♂

134 ♂

147 ♂

148 ♂

144 ♂

145 ♂

146 ♂

PLATE 13 Woodpeckers 2 啄木鸟二

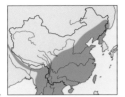

122. Grey-capped Pygmy Woodpecker 星头啄木鸟
Dendrocopos canicapillus

123. Japanese Pygmy Woodpecker
Dendrocopos kizuki
小星头啄木鸟

124. Lesser Spotted Woodpecker
Dendrocopos minor
小斑啄木鸟

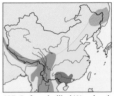

125. Fulvous-breasted Woodpecker
Dendrocopos macei
茶胸斑啄木鸟

126. Stripe-breasted Woodpecker
Dendrocopos atratus
纹胸啄木鸟

127. Rufous-bellied Woodpecker
Dendrocopos hyperythrus
棕腹啄木鸟

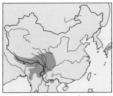

128. Crimson-breasted Woodpecker
Dendrocopos cathpharius
赤胸啄木鸟

129. Darjeeling Woodpecker
Dendrocopos darjellensis
黄颈啄木鸟

130. White-backed Woodpecker
Dendrocopos leucotos
白背啄木鸟

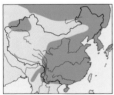

131. Great Spotted Woodpecker
Dendrocopos major
大斑啄木鸟

132. White-winged Woodpecker
Dendrocopos leucopterus
白翅啄木鸟

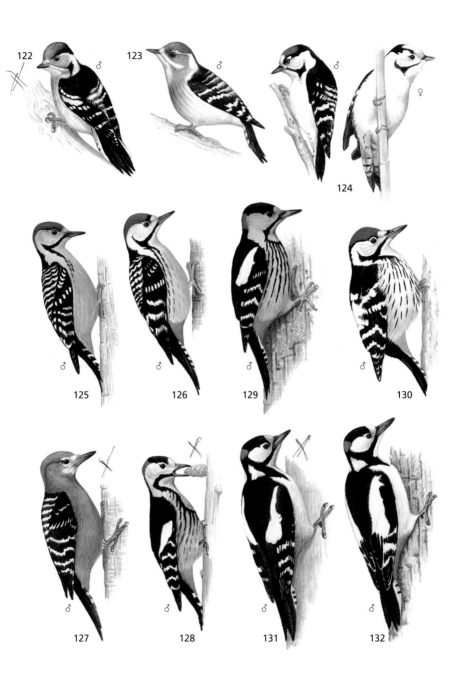

122 ♂

123 ♂

124 ♀

125 ♂

126 ♂

129 ♂

130 ♂

127 ♂

128 ♂

131 ♂

132 ♂

PLATE 14 Woodpeckers 3 啄木鸟三

135. White-bellied Woodpecker
Dryocopus javensis
白腹黑啄木鸟

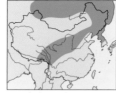

136. Black Woodpecker
Dryocopus martius
黑啄木鸟

137. Lesser Yellownape
Picus chlorolophus
黄冠啄木鸟

138. Greater Yellownape
Picus flavinucha
大黄冠啄木鸟

139. Laced Woodpecker
Picus vittatus
花腹啄木鸟

140. Streak-throated Woodpecker
Picus xanthopygaeus
鳞喉啄木鸟

141. Scaly-bellied Woodpecker
Picus squamatus
鳞腹啄木鸟

142. Red-collared Woodpecker
Picus rabieri
红颈啄木鸟

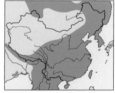

143. Grey-headed Woodpecker
Picus canus
灰头啄木鸟

149. Great Slaty Woodpecker
Mulleripicus pulverulentus
大灰啄木鸟

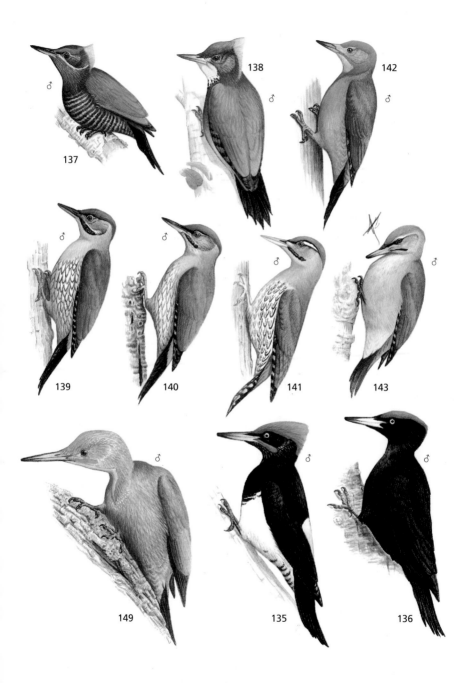

PLATE 15 Barbets and Treecreepers 沈鴷及旋木雀

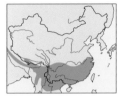

150. Great Barbet
Megalaima virens
大拟啄木鸟

151. Lineated Barbet
Megalaima lineata
[斑头]绿拟啄木鸟

152. Green-eared Barbet
Megalaima faiostricta
黄纹拟啄木鸟

153. Golden-throated Barbet
Megalaima franklinii
金喉拟啄木鸟

154. Black-browed Barbet
Megalaima oorti
黑眉拟啄木鸟

155. Blue-throated Barbet
Megalaima asiatica
蓝喉拟啄木鸟

156. Blue-eared Barbet
Megalaima australis
蓝耳拟啄木鸟

157. Coppersmith Barbet
Megalaima haemacephala
赤胸拟啄木鸟

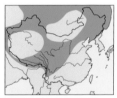

844. Eurasian Treecreeper
Certhia familiaris
旋木雀

845. Bar-tailed Treecreeper
Certhia himalayana
高山旋木雀

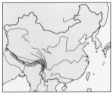

846. Rusty-flanked Treecreeper
Certhia nipalensis
锈红腹旋木雀

847. Brown-throated Treecreeper
Certhia discolor
褐喉旋木雀

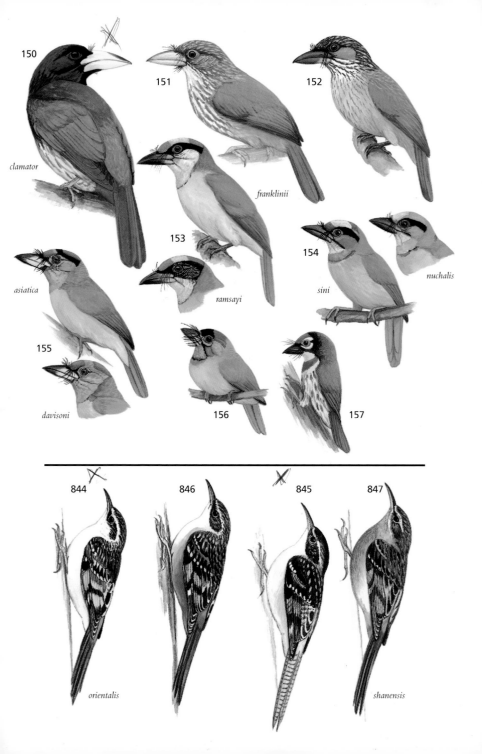

150 clamator

151

152

franklinii

153

154 sini

nuchalis

asiatica

ramsayi

155 davisoni

156

157

844

846

845

847

orientalis

shanensis

PLATE 16   Hornbills   犀鸟

158. Oriental Pied Hornbill
Anthracoceros albirostris
冠斑犀鸟

159. Great Hornbill
Buceros bicornis
双角犀鸟

160. Brown Hornbill
Anorrhinus tickelli
白喉[小盔]犀鸟

161. Rufous-necked Hornbill
Aceros nipalensis
棕颈[无盔]犀鸟

162. Wreathed Hornbill
Aceros undulatus
花冠皱盔犀鸟

NONE

PLATE 17 Kingfishers 翠鸟

170. Blyth's Kingfisher
Alcedo hercules
斑头大翠鸟

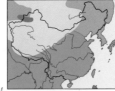

171. Common Kingfisher
Alcedo atthis
普通翠鸟

172. Blue-eared Kingfisher
Alcedo meninting
蓝耳翠鸟

173. Oriental Dwarf Kingfisher
Ceyx erithacus
三趾翠鸟

174. Stork-billed Kingfisher
Halcyon capensis
鹳嘴翡翠

175. Ruddy Kingfisher
Halcyon coromanda
赤翡翠

176. White-throated Kingfisher
Halcyon smyrnensis
白胸翡翠

177. Black-capped Kingfisher
Halcyon pileata
蓝翡翠

178. Collared Kingfisher
Todirhamphus chloris
白领翡翠

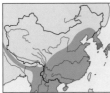

179. Crested Kingfisher
Megaceryle lugubris
冠鱼狗

180. Pied Kingfisher
Ceryle rudis
斑鱼狗

170

171

172

173

174

175

176

rufous morph

177

179 ♀

178

180 ♂

♀

PLATE 18 Bee-eaters, Rollers and Hoopoe 蜂虎，佛法僧及戴胜

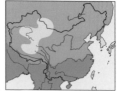

163. Common Hoopoe
Upupa epops
戴胜

167. European Roller
Coracias garrulus
蓝胸佛法僧

168. Indian Roller
Coracias benghalensis
棕胸佛法僧

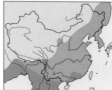

169. Dollarbird
Eurystomus orientalis
三宝鸟

181. Blue-bearded Bee-eater
Nyctyornis athertoni
〔蓝须〕夜蜂虎

182. Green Bee-eater
Merops orientalis
绿喉蜂虎

183. Blue-throated Bee-eater
Merops viridis
蓝喉蜂虎

184. Blue-tailed Bee-eater
Merops philippinus
栗喉蜂虎

185. European Bee-eater
Merops apiaster
黄喉蜂虎

186. Chestnut-headed Bee-eater
Merops leschenaulti
黑胸蜂虎

181

182

imm.

186

183

imm.

184

185

167

168

169

163

PLATE 19 Cuckoos 1 鹃一

187. Pied Cuckoo
Clamator jacobinus
斑翅凤头鹃

188. Chestnut-winged Cuckoo
Clamator coromandus
红翅凤头鹃

200. Drongo Cuckoo
Surniculus lugubris
乌鹃

201. Asian Koel
Eudynamys scolopacea
噪鹃

202. Green-billed Malkoha
Phaenicophaeus tristis
绿嘴地鹃

203. Greater Coucal
Centropus sinensis
褐翅鸦鹃

204. Lesser Coucal
Centropus bengalensis
小鸦鹃

187

juv.

188

201

200

chinensis

♀

♂

juv.

202

saliens

imm.

203   sinensis

imm.

204

PLATE 20 Cuckoos 2 鹃二

189. Large Hawk Cuckoo
Hierococcyx sparverioides
鹰鹃

190. Common Hawk Cuckoo
Hierococcyx varius
普通鹰鹃

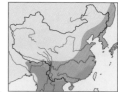

191. Hodgson's Hawk Cuckoo
Hierococcyx fugax
棕腹杜鹃

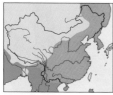

192. Indian Cuckoo
Cuculus micropterus
四声杜鹃

193. Eurasian Cuckoo
Cuculus canorus
大杜鹃

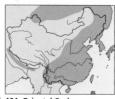

194. Oriental Cuckoo
Cuculus saturatus
中杜鹃

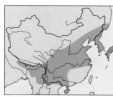

195. Lesser Cuckoo
Cuculus poliocephalus
小杜鹃

196. Banded Bay Cuckoo
Cacomantis sonneratii
栗斑杜鹃

197. Plaintive Cuckoo
Cacomantis merulinus
八声杜鹃

198. Asian Emerald Cuckoo
Chrysococcyx maculatus
翠金鹃

199. Violet Cuckoo
Chrysococcyx xanthorhynchus
紫金鹃

189

imm.

190

imm.

193
canorus

hepatic
♀

191   imm.

hyperythrus

192

♀

♂

194

hepatic ♀

195

♀

196   197

imm.

198

♂

♀

199

♂

PLATE 21   **Trogons and Parrots**   咬鹃及鹦鹉

164. Orange-breasted Trogon
Harpactes oreskios
橙胸咬鹃

165. Red-headed Trogon
Harpactes erythrocephalus
红头咬鹃

166. Ward's Trogon
Harpactes wardi
红腹咬鹃

207. Vernal Hanging Parrot
Loriculus vernalis
短尾鹦鹉

208. Rose-ringed Parakeet
Psittacula krameri
红领绿鹦鹉

209. Grey-headed Parakeet
Psittacula finschii
灰头鹦鹉

210. Blossom-headed Parakeet
Psittacula roseata
花头鹦鹉

211. Derbyan Parakeet
Psittacula derbiana
大紫胸鹦鹉

212. Red-breasted Parakeet
Psittacula alexandri
绯胸鹦鹉

213. Long-tailed Parakeet
Psittacula longicauda
长尾鹦鹉

NONE

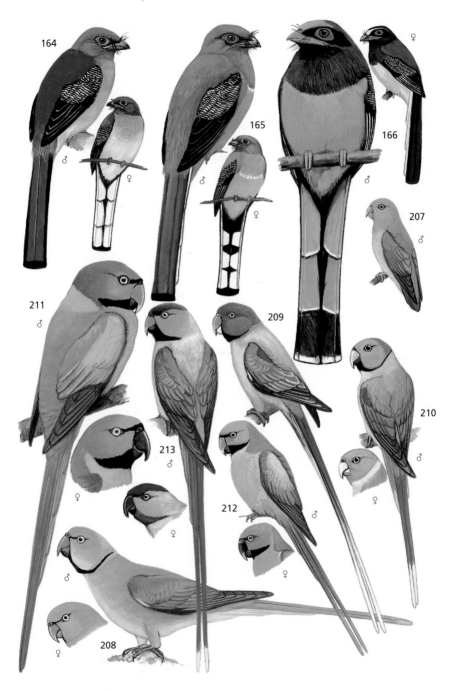

164

165

166 ♀

♂ ♀

207 ♂

211 ♂

209

210

213 ♂

212 ♂

210 ♀

♀

♀

♀

208 ♀

PLATE 22 Swifts 雨燕目

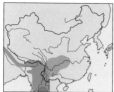

214. Himalayan Swiftlet
Collocalia brevirostris
短嘴金丝燕

215. Germain's Swiftlet
Collocalia germani
爪哇金丝燕

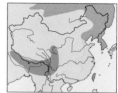

216. White-throated Needletail
Hirundapus caudacutus
白喉针尾雨燕

217. Silver-backed Needletail
Hirundapus cochinchinensis
灰喉针尾雨燕

218. Asian Palm Swift
Cypsiurus balasiensis
棕雨燕

219. Alpine Swift
Tachymarptis melba
高山雨燕

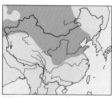

220. Common Swift
Apus apus
普通楼燕

221. Fork-tailed Swift
Apus pacificus
白腰雨燕

222. House Swift
Apus affinis
小白腰雨燕

223. Crested Treeswift
Hemiprocne coronata
凤头树燕

214 *brevirostris*

214 *innominata*

215

218

216

217

223

220

221

219

222

PLATE 23　Owls 1　鸮形目一

224. Barn Owl
Tyto alba
仓鸮

225. Grass Owl
Tyto capensis
草鸮

226. Oriental Bay Owl
Phodilus badius
栗鸮

245. Northern Hawk Owl
Surnia ulula
猛鸮

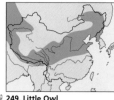

249. Little Owl
Athene noctua
纵纹腹小鸮

250. Spotted Owlet
Athene brama
横斑腹小鸮

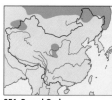

251. Boreal Owl
Aegolius funereus
鬼鸮

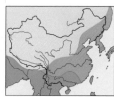

252. Brown Hawk Owl
Ninox scutulata
鹰鸮

224

225

chinensis

226

252

burmanica

245

251

sibiricus

250

249

ludlowi

PLATE 24 Owls 2 鸮形目二

227. Mountain Scops Owl
Otus spilocephalus
黄嘴角鸮

228. Pallid Scops Owl
Otus brucei
纵纹角鸮

229. Eurasian Scops Owl
Otus scops
红角鸮

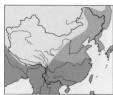

230. Oriental Scops Owl
Otus sunia
东方角鸮

231. Elegant Scops Owl
Otus elegans
琉球角鸮

232. Collared Scops Owl
Otus bakkamoena
领角鸮

246. Eurasian Pygmy Owl
Glaucidium passerinum
花头鸺鹠

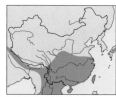

247. Collared Owlet
Glaucidium brodiei
领鸺鹠

248. Asian Barred Owlet
Glaucidium cuculoides
斑头鸺鹠

latouchei

rufous morph

227

228

229

232
glabripes

231
botelensis

rufous
morph

230

246

247

248
whiteleyi

posterior head
pattern

PLATE 25 Owls 3 鸮形目三

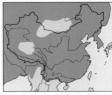

233. Eurasian Eagle Owl
Bubo Bubo
鹏鸮

234. Spot-bellied Eagle Owl
Bubo nipalensis
林鹏鸮

235. Dusky Eagle Owl
Bubo coromandus
乌鹏鸮

236. Blakiston's Fish Owl
Ketupa blakistoni
毛腿渔鸮

237. Brown Fish Owl
Ketupa zeylonensis
褐渔鸮

238. Tawny Fish Owl
Ketupa flavipes
黄脚渔鸮

239. Snowy Owl
Nyctea scandiaca
雪鸮

233
hemachalana

234

235

236

237
leschenaulti

238

239

♂

♀

PLATE 26   **Owls 4**   鸮形目四

240. Brown Wood Owl
Strix leptogrammica
褐林鸮

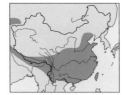

241. Tawny Owl
Strix aluco
灰林鸮

242. Ural Owl
Strix uralensis
长尾林鸮

243. Sichuan Wood Owl
Strix davidi
四川林鸮

244. Great Grey Owl
Strix nebulosa
乌林鸮

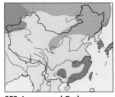

253. Long-eared Owl
Asio otus
长耳鸮

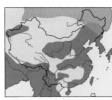

254. Short-eared Owl
Asio flammeus
短耳鸮

240 *ticehursti*

241 *nivicola*
rufous phase

241 *ma*

nikolskii

242

243

244

253

254

PLATE 27 Nightjars and Frogmouth 夜鷹目

255. Hodgson's Frogmouth
Batrachostomus hodgsoni
黑顶蛙口鸱

256. Great Eared Nightjar
Eurostopodus macrotis
毛腿夜鹰

257. Grey Nightjar
Caprimulgus indicus
普通夜鹰

258. Eurasian Nightjar
Caprimulgus europaeus
欧夜鹰

259. Egyptian Nightjar
Caprimulgus aegyptius
埃及夜鹰

260. Vaurie's Nightjar
Caprimulgus centralasicus
中亚夜鹰

261. Large-tailed Nightjar
Caprimulgus macrurus
长尾夜鹰

262. Savanna Nightjar
Caprimulgus affinis
林夜鹰

ONLY ONE

255

256

257 ♂

258 ♂

259

260 ♀

261 ♂

262 ♂

unwini

PLATE 28 Pigeons 1 鸠鸽一

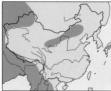

263. Rock Pigeon
Columba livia
原鸽

264. Hill Pigeon
Columba rupestris
岩鸽

265. Snow Pigeon
Columba leuconota
雪鸽

266. Stock Pigeon
Columba oenas
欧鸽

267. Yellow-eyed Pigeon
Columba eversmanni
中亚鸽

268. Common Wood Pigeon
Columba palumbus
斑尾林鸽

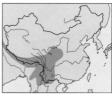

269. Speckled Wood Pigeon
Columba hodgsonii
点斑林鸽

270. Ashy Wood Pigeon
Columba pulchricollis
灰林鸽

271. Pale-capped Pigeon
Columba punicea
紫林鸽

272. Japanese Wood Pigeon
Columba janthina
黑林鸽

NONE EXCEPT 264

263

264

265

267

266

269

268

270

271

272

PLATE 29   **Pigeons 2**   鸠鸽二

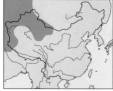

273. European Turtle Dove
Streptopelia turtur
欧斑鸠

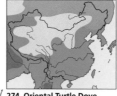

274. Oriental Turtle Dove
Streptopelia orientalis
山斑鸠

275. Laughing Dove
Streptopelia senegalensis
棕斑鸠

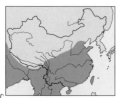

276. Spotted Dove
Streptopelia chinensis
珠颈斑鸠

277. Red Collared Dove
Streptopelia tranquebarica
火斑鸠

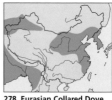

278. Eurasian Collared Dove
Streptopelia decaocto
灰斑鸠

279. Barred Cuckoo Dove
Macropygia unchall
斑尾鹃鸠

280. Brown Cuckoo Dove
Macropygia amboinensis
栗褐鹃鸠

281. Little Cuckoo Dove
Macropygia ruficeps
棕头鹃鸠

282. Emerald Dove
Chalcophaps indica
绿翅金鸠

PLATE 30   Pigeons 3   鸠鸽三

283. Orange-breasted Green Pigeon
Treron bicincta
橙胸绿鸠

284. Pompadour Green Pigeon
Treron pompadora
灰头绿鸠

285. Thick-billed Green Pigeon
Treron curvirostra
厚嘴绿鸠

286. Yellow-footed Green Pigeon
Treron phoenicoptera
黄脚绿鸠

287. Pin-tailed Green Pigeon
Treron apicauda
针尾绿鸠

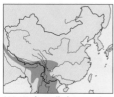

288. Wedge-tailed Green Pigeon
Treron sphenura
楔尾绿鸠

289. White-bellied Green Pigeon
Treron sieboldii
红翅绿鸠

290. Whistling Green Pigeon
Treron formosae
红顶绿鸠

291. Black-chinned Fruit Dove
Ptilinopus leclancheri
黑额果鸠

292. Green Imperial Pigeon
Ducula aenea
绿皇鸠

293. Mountain Imperial Pigeon
Ducula badia
山皇鸠

JUST 289

PLATE 31 Cranes 鹤

297. Siberian Crane
Crus leucogeranus
白鹤

298. Sarus Crane
Crus antigone
赤颈鹤

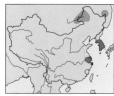

299. White-naped Crane
Crus vipio
白枕鹤

300. Sandhill Crane
Crus canadensis
沙丘鹤

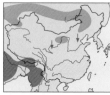

301. Demoiselle Crane
Crus virgo
蓑羽鹤

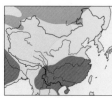

302. Common Crane
Crus grus
灰鹤

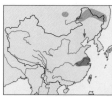

303. Hooded Crane
Crus monacha
白头鹤

304. Black-necked Crane
Crus nigricollis
黑颈鹤

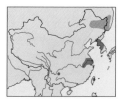

305. Red-crowned Crane
Crus japonensis
丹顶鹤

PLATE 32 **Quails and Rails** 鹌鹑及秧鸡

13. Common Quail
Coturnix coturnix
鹌鹑

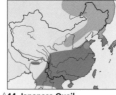

14. Japanese Quail
Coturnix japonica
日本鹌鹑

15. Blue-breasted Quail
Coturnix chinensis
蓝胸鹑

115. Small Buttonquail
Turnix sylvatica
林三趾鹑

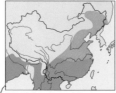

116. Yellow-legged Buttonquail
Turnix tanki
黄脚三趾鹑

117. Barred Buttonquail
Turnix suscitator
棕三趾鹑

306. Swinhoe's Crake
Coturnicops exquisitus
花田鸡

307. Red-legged Crake
Rallina fasciata
红腿斑秧鸡

308. Slaty-legged Crake
Rallina eurizonoides
白喉斑秧鸡

309. Slaty-breasted Rail
Gallirallus striatus
蓝胸秧鸡

311. Corn Crake
Crex crex
长脚秧鸡

PLATE 33   **Rails 2**   秧鸡二

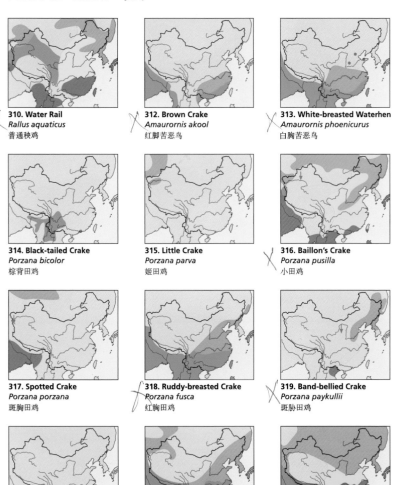

310. Water Rail
Rallus aquaticus
普通秧鸡

312. Brown Crake
Amaurornis akool
红脚苦恶鸟

313. White-breasted Waterhen
Amaurornis phoenicurus
白胸苦恶鸟

314. Black-tailed Crake
Porzana bicolor
棕背田鸡

315. Little Crake
Porzana parva
姬田鸡

316. Baillon's Crake
Porzana pusilla
小田鸡

317. Spotted Crake
Porzana porzana
斑胸田鸡

318. Ruddy-breasted Crake
Porzana fusca
红胸田鸡

319. Band-bellied Crake
Porzana paykullii
斑胁田鸡

320. White-browed Crake
Porzana cinerea
白眉秧鸡

323. Common Moorhen
Gallinula chloropus
黑水鸡

324. Common Coot
Fulica atra
骨顶鸡

PLATE 34   **Rails 3, Jacanas, Painted-snipe and Bustards**   秧鸡至鸨

294. Little Bustard
Tetrax tetrax
小鸨

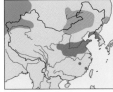

295. Great Bustard
Otis tarda
大鸨

296. McQueen's Bustard
Chlamydotis macqueeni
波斑鸨

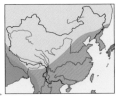

321. Watercock
Gallicrex cinerea
董鸡

322. Purple Swamphen
Porphyrio porphyrio
紫水鸡

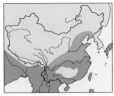

379. Greater Painted-snipe
Rostratula benghalensis
彩鹬

380. Pheasant-tailed Jacana
Hydrophasianus chirurgus
水雉

381. Bronze-winged Jacana
Metopidius indicus
铜翅水雉

321

322

br. ♂

br.

380

381

379   ♀   ♂

294   ♀

br. ♂

295   ♀   ♂

♀   296

♂

PLATE 35   Sandgrouse, Phalaropes and Pratincoles   沙鸡，瓣蹼鹬及燕鸻

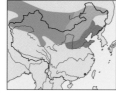

325. Tibetan Sandgrouse
Syrrhaptes tibetanus
西藏毛腿沙鸡

326. Pallas's Sandgrouse
Syrrhaptes paradoxus
毛腿沙鸡

327. Black-bellied Sandgrouse
Pterocles orientalis
黑腹沙鸡

377. Red-necked Phalarope
Phalaropus lobatus
红颈瓣蹼鹬

378. Red-necked Phalarope
Phalaropus lobatus
灰瓣蹼鹬

404. Collared Pratincole
Glareola pratincola
领燕鸻

405. Oriental Pratincole
Glareola maldivarum
普通燕鸻

406. Small Pratincole
Glareola lactea
灰燕鸻

juv.

ad.
non-br.

377

br. ♀

378

br. ♀

ad. non-br.

405

br.

404

406

juv.

br.

325

♀

♂

326

♀

♂

327

♀

♂

PLATE 36 **Waders 1** 鸻形目一

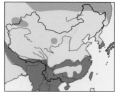

328. Eurasian Woodcock
Scolopax rusticola
丘鹬

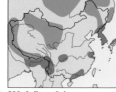

329. Solitary Snipe
Gallinago solitaria
孤沙锥

330. Latham's Snipe
Gallinago hardwickii
澳南沙锥

331. Wood Snipe
Gallinago nemoricola
林沙锥

332. Pintail Snipe
Gallinago stenura
针尾沙锥

333. Swinhoe's Snipe
Gallinago megala
大沙锥

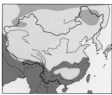

334. Common Snipe
Gallinago gallinago
扇尾沙锥

335. Jack Snipe
Lymnocryptes minimus
姬鹬

376. Ruff
Philomachus pugnax
流苏鹬

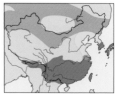

400. Northern Lapwing
Vanellus vanellus
凤头麦鸡

401. River Lapwing
Vanellus duvaucelii
距翅麦鸡

402. Grey-headed Lapwing
Vanellus cinereus
灰头麦鸡

403. Red-wattled Lapwing
Vanellus indicus
肉垂麦鸡

ALL EXCEPT
330, 331, 401 ℤ 403

400

403

402

401

328

331

332

334

330

333

329

♂ br.

376

335

PLATE 37   **Waders 2**   鸻形目二

336. Black-tailed Godwit
Limosa limosa
黑尾塍鹬

337. Bar-tailed Godwit
Limosa lapponica
斑尾塍鹬

338. Little Curlew
Numenius minutus
小杓鹬

339. Whimbrel
Numenius phaeopus
中杓鹬

340. Eurasian Curlew
Numenius arquata
白腰杓鹬

341. Eastern Curlew
Numenius madagascariensis
大杓鹬

352. Grey-tailed Tattler
Heteroscelus brevipes
灰尾〔漂〕鹬

353. Wandering Tattler
Heteroscelus incanus
漂鹬

355. Long-billed Dowitcher
Limnodromus scolopaceus
长嘴鹬

356. Asian Dowitcher
Limnodromus semipalmatus
半蹼鹬

340 ♂

339

338

341 ♀

352

br.

non-br.

336

br.

melanuroides

limosa

337

br.

non-br.

baveri

353

non-br.

lapponica

355

356

355 x

PLATE 38 **Waders 3** 鸻形目三

342. Spotted Redshank
Tringa erythropus
鹤鹬

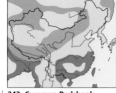

343. Common Redshank
Tringa totanus
红脚鹬

344. Marsh Sandpiper
Tringa stagnatilis
泽鹬

345. Common Greenshank
Tringa nebularia
青脚鹬

346. Nordmann's Greenshank
Tringa guttifer
小青脚鹬

347. Lesser Yellowlegs
Tringa flavipes
小黄脚鹬

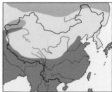

348. Green Sandpiper
Tringa ochropus
白腰草鹬

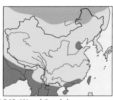

349. Wood Sandpiper
Tringa glareola
林鹬

350. Terek Sandpiper
Xenus cinereus
翘嘴鹬

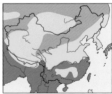

351. Common Sandpiper
Actitis hypoleucos
矶鹬

354. Ruddy Turnstone
Arenaria interpres
翻石鹬

ALL EXCEPT 346 & 347

342 br.

non-br.

343 br.

non-br.

344

345 br.

non-br.

354 br.

non-br.

346 br.

non-br.

347 x

348 non-br.

347

350

349 non-br.

351

PLATE 39   **Waders 4**   鸻形目四

357. Great Knot
Calidris tenuirostris
大滨鹬

358. Red Knot
Calidris canutus
红腹滨鹬

359. Sanderling
Calidris alba
三趾鹬

368. Pectoral Sandpiper
Calidris melanotos
斑胸滨鹬

369. Sharp-tailed Sandpiper
Calidris acuminata
尖尾滨鹬

370. Rock Sandpiper
Calidris ptilocnemis
岩滨鹬

371. Dunlin
Calidris alpina
黑腹滨鹬

372. Curlew Sandpiper
Calidris ferruginea
弯嘴滨鹬

373. Stilt Sandpiper
Micropalama himantopus
高跷鹬

375. Broad-billed Sandpiper
Limicola falcinellus
阔嘴鹬

PLATE 40　Waders 5　鸻形目五

360. Western Sandpiper
Calidris mauri
西方滨鹬

361. Spoon-billed Sandpiper
Calidris pygmeus
勺嘴鹬

362. Little Stint
Calidris minuta
小滨鹬

363. Red-necked Stint
Calidris ruficollis
红颈滨鹬

364. Temminck's Stint
Calidris temminckii
青脚滨鹬

365. Long-toed Stint
Calidris subminuta
长趾滨鹬

366. Least Sandpiper
Calidris minutilla
姬滨鹬

366a. White-rumped Sandpiper
Calidris fuscicollis
白腰滨鹬

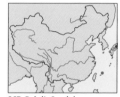

367. Baird's Sandpiper
Calidris bairdii
黑腰滨鹬

374. Buff-breasted Sandpiper
Tryngites subruficollis
黄胸鹬

363   br.   non-br.   juv.

362   br.   juv.   br.   361

364   br.   non-br.   non-br.

365   br.   non-br.   juv.   374

360   juv.   juv.   juv.   juv.

non-br.   367   non-br.   366a   juv.   366   non-br.

PLATE 41　**Plovers**　鸻科

388. Pacific Golden Plover
Pluvialis fulva
金斑鸻

388a. American Golden Plover
Pluvialis dominicus
美国金鸻

389. Grey Plover
Pluvialis squatarola
灰斑鸻

390. Common Ringed Plover
Charadrius hiaticula
剑鸻

391. Long-billed Plover
Charadrius placidus
长嘴剑鸻

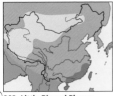

392. Little Ringed Plover
Charadrius dubius
金眶鸻

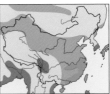

393. Kentish Plover
Charadrius alexandrinus
环颈鸻

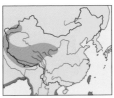

395. Lesser Sand Plover
Charadrius mongolus
蒙古沙鸻

396. Greater Sand Plover
Charadrius leschenaultii
铁嘴沙鸻

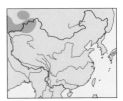

397. Caspian Plover
Charadrius asiaticus
红胸鸻

398. Oriental Plover
Charadrius veredus
东方鸻

399. Eurasian Dotterel
Charadrius morinellus
小嘴鸻

ALL BUT 388, 390, 397 & 399

388

juv.

388a

dominca

juv.

389

juv.

390

juv.

391

393

juv.

392

juv.

395

juv.

397

juv.

atrifrons

ad.br.

398

juv.

399

juv.

ad.br.

♀ br.

396

PLATE 42　**Skuas**　贼鸥

407. Brown Skua
Catharacta antarctica
大贼鸥

408. South-Polar Skua
Catharacta maccormicki
南极贼鸥

409. Pomarine Jaeger
Stercorarius pomarinus
中贼鸥

410. Parasitic Jaeger
Stercorarius parasiticus
短尾贼鸥

411. Long-tailed Jaeger
Stercorarius longicaudus
长尾贼鸥

411

br.

moulting

light morph
juv.

ad. dark morph

juv. intermediate

409

br.
dark

br.
light

non-br.

juv. light

410

br.
light

non-br. light

juv.
intermediate

br.
dark

408

ad. light

ad. intermediate

juv. light

407

ad.

juv.

PLATE 43 **Gulls 1** 鸥科一

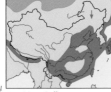

413. Black-tailed Gull
Larus crassirostris
黑尾鸥

414. Mew Gull
Larus canus
海鸥

425. Slender-billed Gull
Larus canus
细嘴鸥

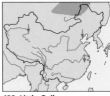

428. Little Gull
Larus minutus
小鸥

429. Ross's Gull
Rhodostethia rosea
楔尾鸥

430. Sabine's Gull
Xema sabini
叉尾鸥

413
1st winter
br.
winter
br.
1st winter.

414
winter
1st winter
br.
1st winter.

425
br.
1st winter
br.
1st winter

br.
non-br.
428
1st winter
br.
br.
1st winter
429
non-br.
br.
non-br.
430
1st winter

PLATE 44 Gulls 2 鸥科二

415. Glaucous-winged Gull
Larus glaucescens
灰翅鸥

416. Glaucous Gull
Larus hyperboreus
北极鸥

417. Slaty-backed Gull
Larus schistisagus
灰背鸥

418. Herring Gull
Larus argentatus
银鸥

419. Heuglin's Gull
Larus heuglini
灰林银鸥

420. Vega Gull
Larus vegae
织女银鸥

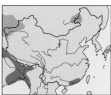

421. Yellow-legged Gull
Larus cachinnans
黄脚银鸥

l. vegae birulai

br.

420

winter

l. vegae vegae

420

winter

winter

421

c. mongolicus

br.

419

h. taimyrensis

419

heuglini

winter

winter

br.

418

winter

br.

a. smithonianus

417

winter

1st winter

br.

1st winter

1st winter

br.

416

winter

1st winter

br.

415

winter

PLATE 45   Gulls 3   鸥科三

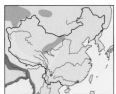

422. Pallas's Gull
Larus ichthyaetus
渔鸥

423. Brown-headed Gull
Larus brunnicephalus
棕头鸥

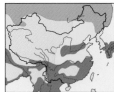

424. Black-headed Gull
Larus ridibundus
红嘴鸥

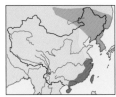

426. Saunders's Gull
Larus saundersi
黑嘴鸥

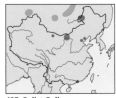

427. Relict Gull
Larus relictus
遗鸥

431. Black-legged Kittiwake
Rissa tridactyla
三趾鸥

422

1st winter

1st winter

br.

non-br.

br.

non-br.

427

1st winter

423

non-br.

non-br.

1st winter

br.

br.

1st winter

424

non-br.

431

1st winter

426

br.

non-br.

1st winter

1st winter

non-br.

br.

PLATE 46　Terns 1　燕鸥一

438. Roseate Tern
Sterna dougallii
粉红燕鸥

439. Black-naped Tern
Sterna sumatrana
黑枕燕鸥

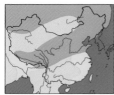

440. Common Tern
Sterna hirundo
普通燕鸥

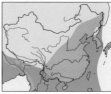

441. Little Tern
Sterna albifrons
白额燕鸥

444. Bridled Tern
Sterna anaethetus
褐翅燕鸥

445. Sooty Tern
Sterna fuscata
乌燕鸥

440 non-br.

br.

br.

juv.

438

br.

br.

juv.

juv.

non-br.

439

juv.

444

br.

juv.

445

br.

juv.

imm.

441

br.

non-br.

juv.

PLATE 47 **Terns 2** 燕鸥二

  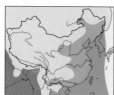

432. Gull-billed Tern
Gelochelidon nilotica
鸥嘴噪鸥

433. Caspian Tern
Sterna caspia
红嘴巨鸥

446. Whiskered Tern
Chlidonias hybridus
须浮鸥

447. White-winged Tern
Chlidonias leucopterus
白翅浮鸥

450. White Tern
Gygis alba
白玄鸥

446 br.

non-br.

br.

non-br.

br.

1st winter

juv.

447 br.

non-br.

br.

non-br.

1st winter

juv.

432 br.

juv.

non-br.

1st winter

433 non-br.

br.

juv.

br.

450 juv.

PLATE 48 Terns 3 燕鸥三

434. River Tern
Sterna aurantia
黄嘴河燕鸥

435. Lesser Crested Tern
Sterna bengalensis
小凤头燕鸥

436. Great Crested Tern
Sterna bergii
大凤头燕鸥

437. Chinese Crested Tern
Sterna bernsteini
黑嘴端凤头燕鸥

442. Black-bellied Tern
Sterna acuticauda
黑腹燕鸥

443. Aleutian Tern
Sterna aleutica
白腰燕鸥

448. Black Tern
Chlidonias niger
黑浮鸥

449. Brown Noddy
Anous stolidus
白顶玄鸥

NONE

434 br. non-br. br.

435 juv. non-br. br.

436 br. juv. non-br.

437 non-br. br. non-br.

442 non-br. br.

443 juv. non-br. br.

448 non-br. br. juv.

449

PLATE 49 **Kites, Honey-buzzard and Bazas** 鹰科一

456. Jerdon's Baza
Aviceda jerdoni
褐冠鹃隼

457. Black Baza
Aviceda leuphotes
黑冠鹃隼

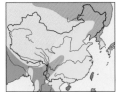

458. Oriental Honey-buzzard
Pernis ptilorhynchus
凤头蜂鹰

459. Black-shouldered Kite
Elanus caeruleus
黑翅鸢

460. Black Kite
Milvus migrans
黑鸢

461. Black-eared Kite
Milvus lineatus
黑耳鸢

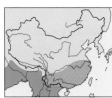

462. Brahminy Kite
Haliastur indus
栗鸢

456

457

459

juv.

juv.

458

orientalis

ruficollis

juv.

460

m. lineatus

m. migrans

juv

461

462

juv.

juv.

PLATE 50 Sea Eagles, other eagles and Vultures 鷹科二

463. White-bellied Sea Eagle
Haliaeetus leucogaster
白腹海鵰

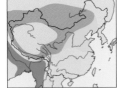

464. Pallas's Fish Eagle
Haliaeetus leucoryphus
玉带海鵰

465. White-tailed Eagle
Haliaeetus albicilla
白尾海鵰

466. Steller's Sea Eagle
Haliaeetus pelagicus
虎头海鵰

467. Lesser Fish Eagle
Ichthyophaga humilis
渔鵰

468. Lammergeier
Gypaetus barbatus
胡兀鹫

469. White-rumped Vulture
Gyps bengalensis
白背兀鹫

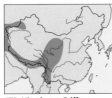

470. Himalayan Griffon
Gyps himalayensis
高山兀鹫

471. Eurasian Griffon
Gyps fulvus
兀鹫

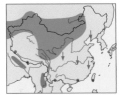

472. Cinereous Vulture
Aegypius monachus
秃鹫

473. Red-headed Vulture
Sarcogyps calvus
黑兀鹫

474. Short-toed Snake Eagle
Circaetus gallicus
短趾鵰

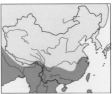

475. Crested Serpent Eagle
Spilornis cheela
蛇鵰

PLATE 51   Harriers   鷹科三

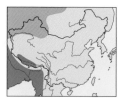

476. Eurasian Marsh Harrier
Circus aeruginosus
白头鹞

477. Eastern Marsh Harrier
Circus spilonotus
白腹鹞

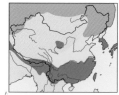

478. Hen Harrier
Circus cyaneus
白尾鹞

479. Pallid Harrier
Circus macrourus
草原鹞

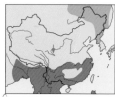

480. Pied Harrier
Circus melanoleucos
鹊鹞

481. Montagu's Harrier
Circus pygargus
乌灰鹞

♂ varient

477

476

478

♀

♂

479

♀

♂

480

481

PLATE 52 **Accipiters** 鹰科四

482. Crested Goshawk
Accipiter trivirgatus
凤头鹰

483. Shikra
Accipiter badius
褐耳鹰

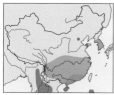

484. Chinese Sparrowhawk
Accipiter soloensis
赤腹鹰

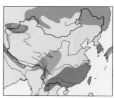

485. Japanese Sparrowhawk
Accipiter gularis
日本松雀鹰

486. Besra
Accipiter virgatus
松雀鹰

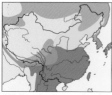

487. Eurasian Sparrowhawk
Accipiter nisus
雀鹰

488. Northern Goshawk
Accipiter gentilis
苍鹰

ALL BUT 482 & 483

482
indicus

ad.

juv.

juv.

juv.

ad.

483

484

juv.

♀

juv.

485

♀

juv.

♂

487

♀

nisosimilis

juv.

486

♂

juv.
488

albidus

♀

juv.
♂

affinis

juv.

♂

schvedowi

PLATE 53 **Buzzards** 鹰科五

489. White-eyed Buzzard
Butastur teesa
白眼鸶鹰

490. Rufous-winged Buzzard
Butastur liventer
棕翅鸶鹰

491. Grey-faced Buzzard
Butastur indicus
灰脸鸶鹰

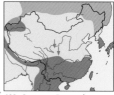

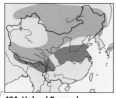

492. Common Buzzard
Buteo buteo
普通鸶

493. Long-legged Buzzard
Buteo rufinus
棕尾鸶

494. Upland Buzzard
Buteo hemilasius
大鸶

495. Rough-legged Buzzard
Buteo lagopus
毛脚鸶

489   ad.   juv.

490   ad.   ad.

491   ad.   juv.

492   *japonicus*   *vulpinus*

493   ad.   rufous morph   juv.

ad.

494

495   ad.   juv.

PLATE 54 **Eagles** 鹰科六

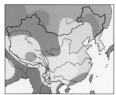

455. Osprey
Pandion haliatus
鱼鹰

496. Black Eagle
Ictinaetus malayensis
林雕

497. Greater Spotted Eagle
Aquila clanga
乌雕

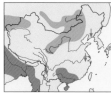

498. Steppe Eagle
Aquila nipalensis
草原雕

499. Imperial Eagle
Aquila heliaca
白肩雕

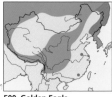

500. Golden Eagle
Aquila chrysaetos
金雕

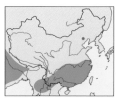

501. Bonelli's Eagle
Hieraaetus fasciatus
白腹隼雕

502. Booted Eagle
Hieraaetus pennatus
靴隼雕

503. Rufous-bellied Eagle
Hieraaetus kienerii
棕腹隼雕

504. Mountain Hawk Eagle
Spizaetus nipalensis
鹰雕

ALL BUT 496, 502 & 503

455

497

pale morph

ad. juv. juv. ad.

498

ad. juv.

496

ad.

juv.

499

ad.

juv.

dark phase

pale phase

502

imm.

ad.

500

ad.

501

juv.

ad.

juv.

orientalis

ad.

503

504

juv.

nipalensis

formosus

PLATE 55 **Falcons** 隼

505. Collared Falconet
Microhierax caerulescens
红腿小隼

506. Pied Falconet
Microhierax melanoleucos
白腿小隼

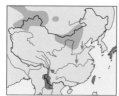

507. Lesser Kestrel
Falco naumanni
黄爪隼

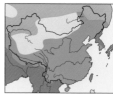

508. Common Kestrel
Falco tinnunculus
红隼

509. Red-footed Falcon
Falco vespertinus
红脚隼

510. Amur Falcon
Falco amurensis
阿穆尔隼

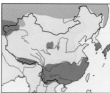

511. Merlin
Falco columbarius
灰背隼

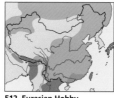

512. Eurasian Hobby
Falco subbuteo
燕隼

513. Oriental Hobby
Falco severus
猛隼

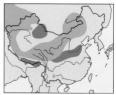

514. Saker Falcon
Falco cherrug
猎隼

515. Gyrfalcon
Falco rusticolus
矛隼

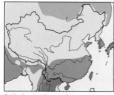

516. Peregrine Falcon
Falco peregrinus
游隼

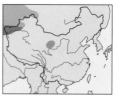

517. Barbary Falcon
Falco pelegrinoides
北非隼

505

506

507 ♂

♀

♂

508 ♂

♂

509

♀

♂

510 ♂

♀

511

♂

palidus

♀

512

513

514

514
altaicus

white
phase

516
peregrinator

515
obsoletus

517

516
calidus

PLATE 56　Grebes and Loons　鸊鷉及潜鸟

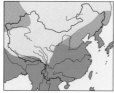

518. Little Grebe
Tachybaptus ruficollis
小鸊鷉

519. Red-necked Grebe
Podiceps grisegena
赤颈鸊鷉

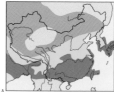

520. Great Crested Grebe
Podiceps cristatus
凤头鸊鷉

521. Horned Grebe
Podiceps auritus
角鸊鷉

522. Black-necked Grebe
Podiceps nigricollis
黑颈鸊鷉

575. Red-throated Loon
Gavia stellata
红喉潜鸟

576. Black-throated Loon
Gavia arctica
黑喉潜鸟

577. Pacific Diver
Gavia pacifica
太平洋潜鸟

578. Common Loon
Gavia immer
白嘴潜鸟

579. Yellow-billed Loon
Gavia adamsii
黄嘴潜鸟

ALL BUT
519, 577 578
& 579

578
br.

520
br.

579
br.

519
br.

578
br.

521

522
br.

576
br.

518
br.

577
br.

juv.

PLATE 57   **Tropicbirds, Albatrosses and Frigatebirds**   其它海鸟

523. Red-billed Tropicbird
Phaethon aethereus
短尾鹲

524. Red-tailed Tropicbird
Phaethon rubricauda
红尾鹲

525. White-tailed Tropicbird
Phaethon lepturus
白尾鹲

526. Masked Booby
Sula dactylatra
兰脸鲣鸟

527. Red-footed Booby
Sula sula
红脚鲣鸟

528. Brown Booby
Sula leucogaster
褐鲣鸟

572. Great Frigatebird
Fregata minor
小军舰鸟

573. Lesser Frigatebird
Fregata ariel
白斑军舰鸟

574. Christmas Island Frigatebird
Fregata ariel
白腹军舰鸟

589. Short-tailed Albatross
Diomedea albatrus
短尾信天翁

590. Black-footed Albatross
Diomedea nigripes
黑脚信天翁

591. Laysan Albatross
Diomedea immutabilis
黑背信天翁

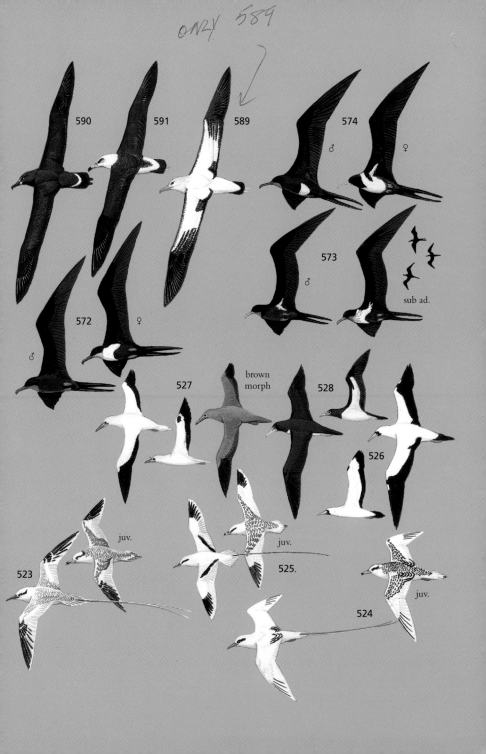

ONLY 589

PLATE 58 **Cormorants and Pelicans** 鸬鹚及鹈鹕

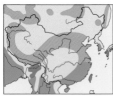

529. Darter
Anhinga melanogaster
黑腹蛇鹈鹕

530. Little Cormorant
Phalacrocorax niger
黑颈鸬鹚

531. Great Cormorant
Phalacrocorax carbo
普通鸬鹚

532. Japanese Cormorant
Phalacrocorax capillatus
暗绿背鸬鹚

533. Red-faced Cormorant
Phalacrocorax urile
红脸鸬鹚

534. Pelagic Cormorant
Phalacrocorax pelagicus
海鸬鹚

564. Great White Pelican
Pelecanus onocrotalus
白鹈鹕

565. Dalmatian Pelican
Pelecanus crispus
卷羽鹈鹕

566. Spot-billed Pelican
Pelecanus philippensis
斑嘴鹈鹕

530
br.

534
br.

532
br.

531
br.

533
br.

529

juv.

531
non-br.

564

juv.

565

565

566

PLATE 59   Herons 1   鷺一

535. Little Egret
Egretta garzetta
白鷺

536. Chinese Egret
Egretta eulophotes
黃嘴白鷺

537. Pacific Reef Egret
Egretta sacra
岩鷺

538. Pied Heron
Egretta picata
白頸黑鷺

542. Great Egret
Casmerodius albus
大白鷺

543. Intermediate Egret
Mesophoyx intermedia
中白鷺

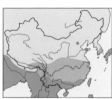

544. Cattle Egret
Bubulcus ibis
牛背鷺

br.

544

538

imm.

537

br.

536

br.

white morph

br.

542

dark
morph

543

br.

535

br.

PLATE 60 Herons 2 鷺二

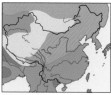

539. Grey Heron
Ardea cinerea
蒼鷺

540. White-bellied Heron
Ardea insignis
白腹鷺

541. Purple Heron
Ardea purpurea
草鷺

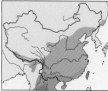

545. Chinese Pond Heron
Ardeola bacchus
池鷺

546. Little Heron
Butorides striatus
綠鷺

547. Black-crowned Night Heron
Nycticorax nycticorax
夜鷺

ALL BUT 540

541

539

imm.

imm.

540

imm.

545

546

547

juv.

juv.

br.

br.

PLATE 61 **Herons 3** 鹭三

548. White-eared Night Heron
Gorsachius magnificus
海南鸦

549. Japanese Night Heron
Gorsachius goisagi
栗头鸦

550. Malayan Night Heron
Gorsachius melanolophus
黑冠鸦

551. Little Bittern
Ixobrychus minutus
小苇鸦

552. Yellow Bittern
Ixobrychus sinensis
黄苇鸦

553. Von Schrenck's Bittern
Ixobrychus eurhythmus
紫背苇鸦

554. Cinnamon Bittern
Ixobrychus cinnamomeus
栗苇鸦

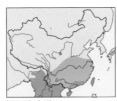

555. Black Bittern
Dupetor flavicollis
黑鸦

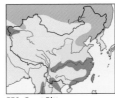

556. Great Bittern
Botaurus stellaris
大麻鸦

PLATE 62 **Storks and Ibises** 鹳及鹮

558. Glossy Ibis
Plegadis falcinellus
彩鹮

559. Black-headed Ibis
Threskiornis melanocephalus
黑头白鹮

560. White-shouldered Ibis
Pseudibis davisoni
黑鹮

561. Crested Ibis
Nipponia nippon
朱鹮

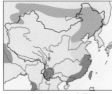

562. Eurasian Spoonbill
Platalea leucorodia
白琵鹭

563. Black-faced Spoonbill
Platalea minor
黑脸琵鹭

567. Painted Stork
Mycteria leucocephala
白头彩鹳

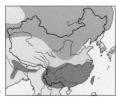

568. Black Stork
Ciconia nigra
黑鹳

569. White Stork
Ciconia ciconia
白鹳

570. Oriental Stork
Ciconia boyciana
东方白鹳

571. Lesser Adjulane
Leptopilos javanicus
秃鹳

(5)

568

570

567

569

571

558

560

562

563

561

br.

559

imm.

PLATE 63 **Petrels and Shearwaters** 鹱形目

580. Northern Fulmar
Fulmarus glacialis
暴雪鹱

581. Tahiti Petrel
Pterodroma rostrata
钩嘴圆尾鹱

582. Bonin Petrel
Pterodroma hypoleuca
点额圆尾鹱

583. Bulwer's Petrel
Bulweria bulwerii
纯褐鹱

584. Streaked Shearwater
Calonectris leucomelas
白额鹱

585. Wedge-tailed Shearwater
Puffinus pacificus
曳尾鹱

586. Flesh-footed Shearwater
Puffinus carneipes
肉足鹱

587. Sooty Shearwater
Puffinus griseus
灰鹱

588. Short-tailed Shearwater
Puffinus tenuirostris
短尾鹱

592. Wilson's Storm-petrel
Oceanites oceanicus
烟黑叉尾海燕

593. Leach's Storm-petrel
Oceanites oceanicus
白腰叉尾海燕

594. Swinhoe's Storm-petrel
Oceanodroma monorhis
黑叉尾海燕

595. Matsudaira's Storm-petrel
Oceanodroma matsudairae
日本叉尾海燕

584

585
pale
phase

587

588

586

581

583

582

594

593

595

592

580

light
phase

dark
phase

PLATE 64 Pittas and Broadbills 八色鸫及阔嘴鸟

596. Eared Pitta
Pitta phayrei
双辫八色鸫

597. Blue-naped Pitta
Pitta nipalensis
蓝枕八色鸫

598. Blue-rumped Pitta
Pitta soror
蓝背八色鸫

599. Rusty-naped Pitta
Pitta oatesi
栗头八色鸫

600. Blue Pitta
Pitta cyanea
蓝八色鸫

601. Hooded Pitta
Pitta sordida
绿胸八色鸫

602. Fairy Pitta
Pitta nympha
仙八色鸫

603. Blue-winged Pitta
Pitta moluccensis
蓝翅八色鸫

604. Silver-breasted Broadbill
Serilophus lunatus
银胸丝冠鸟

605. Long-tailed Broadbill
Psarisomus dalhousiae
长尾阔嘴鸟

605

604 ♂

601

600

603

602

596 ♂

599 ♂

597 ♂

598 ♂

PLATE 65 Leafbirds, Honeyguide, Wallcreper and Ioras 叶鹎，雀鹎及其它

607. Blue-winged Leafbird
Chloropsis cochinchinensis
蓝翅叶鹎

608. Golden-fronted Leafbird
Chloropsis aurifrons
金额叶鹎

118. Yellow-rumped Honeyguide
Indicator xanthonotus
黄腰响密鴷

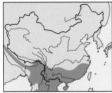

609. Orange-bellied Leafbird
Chloropsis hardwickii
橙腹叶鹎

682. Common Iora
Aegithina tiphia
黑翅雀鹎

683. Great Iora
Aegithina lafresnayei
大绿雀鹎

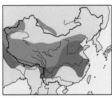

843. Wallcreeper
Tichodroma muraria
红翅旋壁雀

ONLY 843

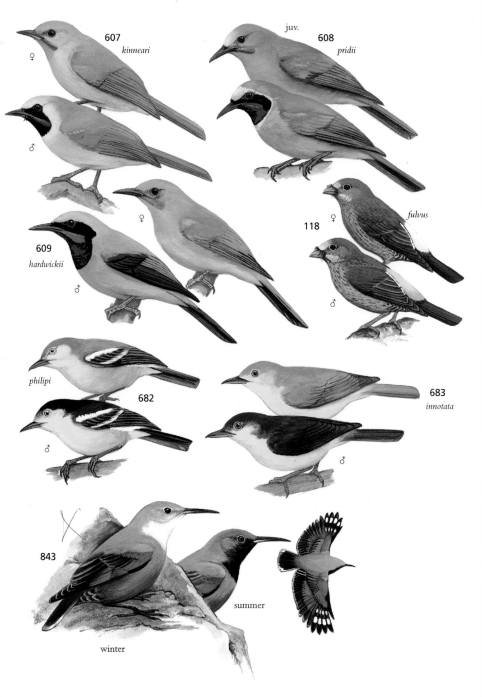

607 kinneari
♀

608 pridii
juv.

609 hardwickii
♂
♀

118 fulvus
♀
♂

philipi
682
♂

683 innotata
♂

843
winter
summer

PLATE 66   Shrikes and Woodshrike   伯劳及林鵙

610. Tiger Shrike
Lanius tigrinus
虎纹伯劳

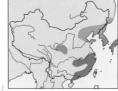

611. Bull-headed Shrike
Lanius bucephalus
牛头伯劳

612. Red-backed Shrike
Lanius collurio
红背伯劳

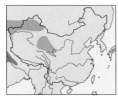

613. Rufous-tailed Shrike
Lanius isabellinus
棕尾伯劳

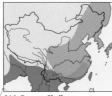

614. Brown Shrike
Lanius cristatus
红尾伯劳

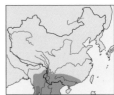

615. Burmese Shrike
Lanius collurioides
栗背伯劳

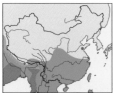

616. Long-tailed Shrike
Lanius schach
棕背伯劳

617. Grey-backed Shrike
Lanius tephronotus
灰背伯劳

618. Lesser Grey Shrike
Lanius minor
黑额伯劳

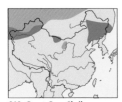

619. Great Grey Shrike
Lanius excubitor
灰伯劳

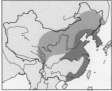

620. Chinese Grey Shrike
Lanius sphenocercus
楔尾伯劳

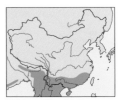

684. Large Woodshrike
Tephrodornis gularis
钩嘴林鵙

juv.

610

♀

611

juv.

612

pallidifrons

♀

winter

♂

summer

613

♂

614

cristatus

♂

615

♀

♂

ioenicuroides

ibellinus

♂

and
imm.

lucionensis

♂

formosae

tricolor

617

♂

imm.

imm.

618

dark morph

616

imm.

619

684

620

latouchei

PLATE 67 **Crows 1** 鸦科一

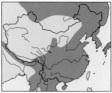

621. Eurasian Jay
Garrulus glandarius
松鸦

622. Siberian Jay
Perisoreus infaustus
北噪鸦

623. Sichuan Jay
Perisoreus internigrans
黑头噪鸦

624. Taiwan Blue Magpie
Urocissa caerulea
台湾蓝鹊

625. Yellow-billed Blue Magpie
Urocissa flavirostris
黄嘴蓝鹊

626. Red-billed Blue Magpie
Urocissa erythrorhyncha
红嘴蓝鹊

627. White-winged Magpie
Urocissa whiteheadi
白翅蓝鹊

628. Common Green Magpie
Cissa chinensis
蓝绿鹊

629. Indochinese Green Magpie
Cissa hypoleuca
短尾绿鹊

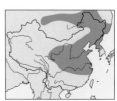

630. Azure-winged Magpie
Cyanopica cyanus
灰喜鹊

631. Rufous Treepie
Dendrocitta vagabunda
棕腹树鹊

632. Grey Treepie
Dendrocitta formosae
灰树鹊

633. Collared Treepie
Dendrocitta frontalis
黑额树鹊

635. Ratchet-tailed Treepie
Temnurus temnurus
塔尾树鹊

636. Black-billed Magpie
Pica pica
喜鹊

621

622

sinensis

623

624

brandtii

626

625

627

628

630

631

629

632

633

635

hemileucoptera

636

PLATE 68   **Crows 2**   鸦科二

637. Mongolian Ground-Jay
Podoces hendersoni
黑尾地鸦

638. Xinjiang Ground-Jay
Podoces biddulphi
白尾地鸦

639. Hume's Groundpecker
Pseudopodoces humilis
褐背拟地鸦

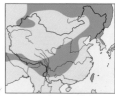

640. Spotted Nutcracker
Nucifraga caryocatactes
星鸦

641. Red-billed Chough
Pyrrhocorax pyrrhocorax
红嘴山鸦

642. Yellow-billed Chough
Pyrrhocorax graculus
黄嘴山鸦

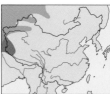

643. Eurasian Jakdaw
Corvus monedula
寒鸦

644. Daurian Jakdaw
Corvus dauuricus
达乌里寒鸦

645. House Crow
Corvus splendens
家鸦

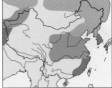

646. Rook
Corvus frugilegus
秃鼻乌鸦

647. Carrion Crow
Corvus corone
小嘴乌鸦

648. Large-billed Crow
Corvus macrorhynchos
大嘴乌鸦

649. Jungle Crow
Corvus levaillantii
丛林鸦

PLATE 69　Orioles and Fairy Bluebird　黄鹂及和平鸟

606. Asian Fairy Bluebird
Irena puella
和平鸟

653. Eurasian Golden Oriole
Oriolus oriolus
金黄鹂

654. Black-naped Oriole
Oriolus chinensis
黑枕黄鹂

655. Slender-billed Oriole
Oriolus tenuirostris
细嘴黄鹂

656. Black-hooded Oriole
Oriolus xanthornus
黑头黄鹂

657. Maroon Oriole
Oriolus traillii
朱鹂

658. Silver Oriole
Oriolus mellianus
鹊色鹂

ONLY 654

653

♀

♂

imm.

656

imm.

654

655

imm.

♀

657

♀

♂

♀

606

658

♂

♀

♂

PLATE 70 **Minivets and Cuckooshrikes** 山椒鸟科

659. Large Cuckooshrike
Coracina macei
大鹃鵙

660. Black-winged Cuckooshrike
Coracina melaschistos
暗灰鹃鵙

661. Rosy Minivet
Pericrocotus roseus
粉红山椒鸟

662. Swinhoe's Minivet
Pericrocotus cantonensis
小灰山椒鸟

663. Ashy Minivet
Pericrocotus divaricatus
灰山椒鸟

664. Grey-chinned Minivet
Pericrocotus solaris
灰喉山椒鸟

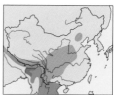

665. Long-tailed Minivet
Pericrocotus ethologus
长尾山椒鸟

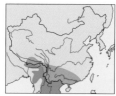

666. Short-billed Minivet
Pericrocotus brevirostris
短嘴山椒鸟

667. Scarlet Minivet
Pericrocotus flammeus
赤红山椒鸟

668. Bar-winged Flycatcher-shrike
Hemipus picatus
褐背鹃鵙

♀ and imm.

♀

♂

659

rexpineti

660

♀

avensis

♂

668

♀

♂

capitalis

♀

661

♂

662

663

♀

♂

♀

664

♂

665

♀

♂

666

667

♀

♂

PLATE 71 Drongos 卷尾及盘尾

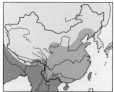

672. Black Drongo
Dicrurus macrocercus
黑卷尾

673. Ashy Drongo
Dicrurus leucophaeus
灰卷尾

674. Crow-billed Drongo
Dicrurus annectans
鸦嘴卷尾

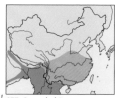

675. Bronzed Drongo
Dicrurus aeneus
古铜色卷尾

676. Lesser Racket-tailed Drongo
Dicrurus remifer
小盘尾

677. Spangled Drongo
Dicrurus hottentottus
发冠卷尾

678. Greater Racket-tailed Drongo
Dicrurus paradiseus
大盘尾

672

imm.

hopwoodi

salangensis

674

imm.

673

675

676
tectirostris

678
grandis

imm.

677

PLATE 72 Monarchs and Paradise-flycatchers 扇尾鹟至方尾鹟

669. Yellow-bellied Fantail
Rhipidura hypoxantha
黄腹扇尾鹟

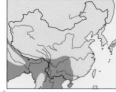

670. White-throated Fantail
Rhipidura albicollis
白喉扇尾鹟

671. White-browed Fantail
Rhipidura aureola
白眉扇尾鹟

679. Black-naped Monarch
Hypothymis azurea
黑枕王鹟

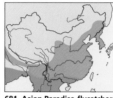

681. Asian Paradise-flycatcher
Terpsiphone paradisi
寿带〔鸟〕

680. Japanese Paradise-flycatcher
Terpsiphone atrocaudata
紫寿带〔鸟〕

726. Brown-chested Jungle Flycatcher
Rhinomyias brunneata 白喉林鹟

759. Grey-headed Canary Flycatcher
Culicicapa ceylonensis 方尾鹟

white morph

681

rufous morph

680

♂

♀

♂

♀

726

759

679

♀

♂

669

670

671

PLATE 73 **Thrushes 1** 鸫一

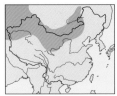

689. Rufous-tailed Rock Thrush
Monticola saxatilis
白背矶鸫

690. Blue-capped Rock Thrush
Monticola cinclorhynchus
蓝头矶鸫

691. White-throated Rock Thrush
Monticola gularis
白喉矶鸫

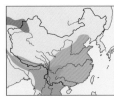

692. Chestnut-bellied Rock Thrush
Monticola rufiventris
栗腹矶鸫

693. Blue Rock Thrush
Monticola solitarius
蓝矶鸫

694. Blue Whistling Thrush
Myophonus caeruleus
紫啸鸫

695. Taiwan Whistling Thrush
Myophonus insularis
台湾紫啸鸫

696. Orange-headed Thrush
Zoothera citrina
橙头地鸫

697. Siberian Thrush
Zoothera sibirica
白眉地鸫

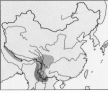

698. Plain-backed Thrush
Zoothera mollissima
光背地鸫

699. Long-tailed Thrush
Zoothera dixoni
长尾地鸫

700. Scaly Thrush
Zoothera dauma
虎斑地鸫

701. Dark-sided Thrush
Zoothera marginata
长嘴地鸫

689 ♀ ♂

690 ♀ ♂

691 ♀ ♂

692 ♀ ♂

pandoo ♂

philippensis 693 ♂

694 *eugenei*

695

innotata

caeruleus

melli 696

697 ♀

caeruleus

698

sibirica ♂

699

701

700

PLATE 74 Thrushes 2 鸫二

702. Grey-backed Thrush
Turdus hortulorum
灰背鸫

703. Black-breasted Thrush
Turdus dissimilis
黑胸鸫

704. Japanese Thrush
Turdus cardis
乌灰鸫

705. White-collared Blackbird
Turdus albocinctus
白颈鸫

706. Grey-winged Blackbird
Turdus boulboul
灰翅鸫

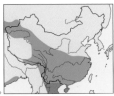

707. Eurasian Blackbird
Turdus merula
乌鸫

708. Island Thrush
Turdus poliocephalus
岛鸫

709. Chestnut Thrush
Turdus rubrocanus
灰头鸫

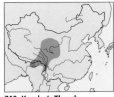

710. Kessler's Thrush
Turdus kessleri
棕背黑头鸫

702

704

703

705

706

707

mandarinus

708
niveiceps

709

gouldi

710

PLATE 75 **Thrushes 3** 鸫三

711. Grey-sided Thrush
Turdus feae
褐头鸫

712. Eyebrowed Thrush
Turdus obscurus
白眉鸫

713. Mistle Thrush
Turdus viscivorus
槲鸫

714. Pale Thrush
Turdus pallidus
白腹鸫

715. Brown-headed Thrush
Turdus chrysolaus
赤胸鸫

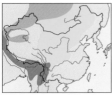

716. Dark-throated Thrush
Turdus ruficollis
赤颈鸫

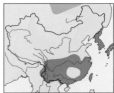

717. Dusky Thrush
Turdus naumanni
斑鸫

718. Fieldfare
Turdus pilaris
田鸫

719. Redwing
Turdus iliacus
白眉歌鸫

720. Song Thrush
Turdus philomelos
欧歌鸫

721. Chinese Thrush
Turdus mupinensis
宝兴歌鸫

711

712 ♀

712 ♂

714 ♀

714 ♂

715 ♀

715 ♂

716
ruficollis ♀

716
ruficollis ♂

716
atrogularis ♀

716
atrogularis ♂

intermediate
1st winter

717
naumanni

717
eunomus

720

718

721

719

713

PLATE 76 **Shortwings and Robins** 鸲四

722. Gould's Shortwing
Brachypteryx stellata
栗背短翅鸫

723. Rusty-bellied Shortwing
Brachypteryx hyperythra
锈腹短翅鸫

724. Lesser Shortwing
Brachypteryx leucophrys
白喉短翅鸫

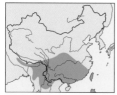

725. White-browed Shortwing
Brachypteryx montana
蓝短翅鸫

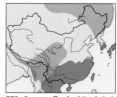

773. Orange-flanked Bush Robin
Tarsiger cyanurus
红胁蓝尾鸲

774. Golden Bush Robin
Tarsiger chrysaeus
金色林鸲

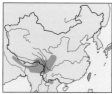

775. White-browed Bush Robin
Tarsiger indicus
白眉林鸲

776. Rufous-breasted Bush Robin
Tarsiger hyperythrus
棕腹林鸲

777. Collared Bush Robin
Tarsiger johnstoniae
台湾林鸲

793. White-tailed Robin
Myiomela leucura
白尾蓝〔地〕鸲

794. Blue-fronted Robin
Cinclidium frontale
蓝额〔长脚〕地鸲

722

723 ♀ ♂

724 ♀ ♂

725 ♀ ♂
cruralis

773
cyanurus ♂ ♂

774 ♂

775 ♀ ♂
indicus

776 ♀ ♂

777 ♀ ♂

793
leucura ♀ ♂

794 ♀ ♂

♀ ♂ indicus

PLATE 77 Robins 2 鸫五

760. Eurasian Robin
Erithacus rubecula
欧亚鸲

761. Japanese Robin
Erithacus akahige
日本歌鸲

762. Ryukyu Robin
Erithacus komadori
琉球歌鸲

763. Rufous-tailed Robin
Luscinia sibilans
红尾歌鸲

764. Common Nightingale
Luscinia megarhynchos
新疆歌鸲

765. Siberian Rubythroat
Luscinia calliope
红点颏，红喉歌鸲

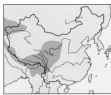

766. White-tailed Rubythroat
Luscinia pectoralis
黑胸歌鸲

767. Bluethroat
Luscinia svecica
蓝点颏

768. Rufous-headed Robin
Luscinia ruficeps
棕头歌鸲

769. Blackthroat
Luscinia obscura
黑喉歌鸲

770. Firethroat
Luscinia pectardens
金胸歌鸲

771. Indian Blue Robin
Luscinia brunnea
栗腹歌鸲

772. Siberian Blue Robin
Luscinia cyane
蓝歌鸲(蓝喉歌鸲)

760

juv.

761 ♀

♂

762 ♀

♂

763

764

766 ♀

♂

ballioni

765

♂

♀

tschebaiewi

767

svecica

♀

♂

♂

768

♀

769

♂

770

771 ♀

♂

772

♀

♂

♂

PLATE 78 Redstarts 红尾鸲

780. Ala Shan Redstart
Phoenicurus alaschanicus
贺兰山红尾鸲

781. Rufous-backed Redstart
Phoenicurus erythronota
红背红尾鸲

782. Blue-capped Redstart
Phoenicurus coeruleocephalus
蓝头红尾鸲

783. Black Redstart
Phoenicurus ochruros
赭红尾鸲

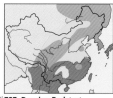

784. Common Redstart
Phoenicurus phoenicurus
欧亚红尾鸲

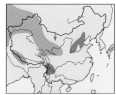

785. Hodgson's Redstart
Phoenicurus hodgsoni
黑喉红尾鸲

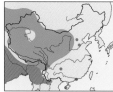

786. White-throated Redstart
Phoenicurus schisticeps
白喉红尾鸲

787. Daurian Redstart
Phoenicurus auroreus
北红尾鸲

788. White-winged Redstart
Phoenicurus erythrogaster
红腹红尾鸲

789. Blue-fronted Redstart
Phoenicurus frontalis
蓝额红尾鸲

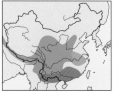

790. White-capped Water Redstart
Chaimarrornis leucocephalus
白顶溪鸲

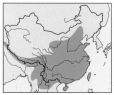

791. Plumbeous Water Redstart
Rhyacornis fuliginosus
红尾水鸲

792. White-bellied Redstart
Hodgsonius phaenicuroides
白腹短翅鸲

PLATE 79   Flycatchers 1   鹟一

727. Spotted Flycatcher
Muscicapa striata
斑鹟

728. Grey-streaked Flycatcher
Muscicapa griseisticta
灰斑鹟

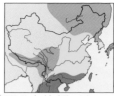

729. Dark-sided Flycatcher
Muscicapa sibirica
乌鹟

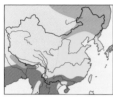

730. Asian Brown Flycatcher
Muscicapa dauurica
北灰鹟

731. Brown-breasted Flycatcher
Muscicapa muttui
褐胸鹟

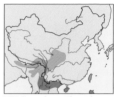

732. Ferruginous Flycatcher
Muscicapa ferruginea
棕尾褐鹟

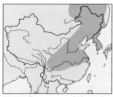

733. Yellow-rumped Flycatcher
Ficedula zanthopygia
白眉〔姬〕鹟

734. Narcissus Flycatcher
Ficedula narcissina
黄眉〔姬〕鹟

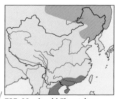

735. Mugimaki Flycatcher
Ficedula mugimaki
鸲〔姬〕鹟

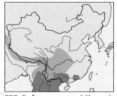

737. Rufous-gorgeted Flycatcher
Ficedula strophiata
橙胸〔姬〕鹟

738. Red-throated Flycatcher
Ficedula parva
红喉〔姬〕鹟

739. White-gorgeted Flycatcher
Ficedula monileger
白喉〔姬〕鹟

ALL BUT 727, 732, OR 739

728
727
729
730
731

732
739

non-br.
738
737
imm. ♂
br.
♂
735
♀

733
♀
♂
734
♀
♂
elisae ♂

PLATE 80   **Flycatchers 2**   鹟二

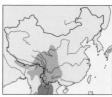

736. Slaty-backed Flycatcher
Ficedula hodgsonii
锈胸蓝〔姬〕鹟

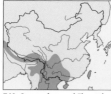

740. Snowy-browed Flycatcher
Ficedula hyperythra
棕胸蓝〔姬〕鹟

741. Little Pied Flycatcher
Ficedula westermanni
灰蓝〔姬〕鹟

742. Ultramarine Flycatcher
Ficedula superciliaris
白眉蓝〔姬〕鹟

743. Slaty-blue Flycatcher
Ficedula tricolor
灰蓝〔姬〕鹟

744. Sapphire Flycatcher
Ficedula sapphira
玉头〔姬〕鹟

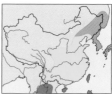

745. Blue-and-white Flycatcher
Cyanoptila cyanomelana
白腹〔姬〕鹟

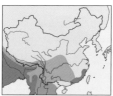

746. Verditer Flycatcher
Eumyias thalassina
铜蓝鹟

755. Pale Blue Flycatcher
Cyornis unicolor
纯蓝仙鹟

758. Pygmy Blue Flycatcher
Muscicapella hodgsoni
侏蓝仙鹟

ONLY 745 & 746

736

741

740

♀

♀

♀

♂

♂

744

742
aestigma

743

♀

♀

♂

♂

minuta   1st year

♂

sapphira

758

746

745

♀

♂

♂

cyanomelana

♂

755
unicolor

♀

juv.

♂

745

cumatalis   ♂

♀

PLATE 81 **Flycatchers 3** 鹟三

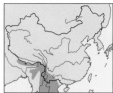

747. Large Niltava
Niltava grandis
大仙鹟

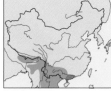

748. Small Niltava
Niltava macgrigoriae
小仙鹟

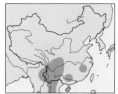

749. Fujian Niltava
Niltava davidi
棕腹大仙鹟

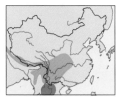

750. Rufous-bellied Niltava
Niltava sundara
棕腹仙鹟

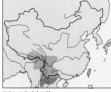

751. Vivid Niltava
Niltava vivida
棕腹蓝仙鹟

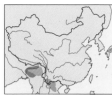

752. White-tailed Flycatcher
Cyornis concretus
白尾蓝仙鹟

753. Hainan Blue Flycatcher
Cyornis hainanus
海南蓝仙鹟

754. Pale-chinned Flycatcher
Cyornis poliogenys
灰颊仙鹟

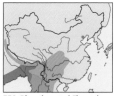

756. Blue-throated Flycatcher
Cyornis rubeculoides
蓝喉仙鹟

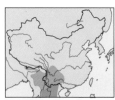

757. Hill Blue Flycatcher
Cyornis banyumas
山蓝仙鹟

ONLY 750 & 756

747
griseiventris
♀
♂

749
♀
♂

748
♀
♂

750
♂

751
oatesi
♀
♂

752
♀
♂

753
juv.
♀
♂

754
cachariensis

756
laucicomans
♀
♂

756
rubeculoides
♀
♂

757
♀
♂

KP

PLATE 82 Forktails, Magpie Robins and Cochoas 其它鸫

778. Oriental Magpie Robin
Copsychus saularis
鹊鸲

779. White-rumped Shama
Copsychus malabaricus
白腰鹊鸲

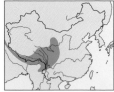

795. Grandala
Grandala coelicolor
蓝大翅鸲

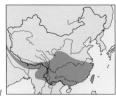

796. Little Forktail
Enicurus scouleri
小燕尾

797. Black-backed Forktail
Enicurus immaculatus
黑背燕尾

798. Slaty-backed Forktail
Enicurus schistaceus
灰背燕尾

799. White-crowned Forktail
Enicurus leschenaulti
黑背燕尾

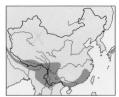

800. Spotted Forktail
Enicurus maculatus
斑背燕尾

801. Purple Cochoa
Cochoa purpurea
紫宽嘴鸫

802. Green Cochoa
Cochoa viridis
绿宽嘴鸫

PLATE 83 **Chats and Wheatears** 鹛

803. Hodgson's Bushchat
Saxicola insignis
白喉石鹛

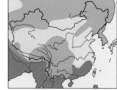

804. Common Stonechat
Saxicola torquata
黑喉石鹛

805. Pied Bushchat
Saxicola caprata
白斑黑石鹛

806. Jerdon's Bushchat
Saxicola jerdoni
黑白林鹛

807. Grey Bushchat
Saxicola ferrea
灰林鹛

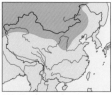

808. Northern Wheatear
Oenanthe oenanthe
穗鹛

809. Variable Wheatear
Oenanthe picata
东方斑鹛

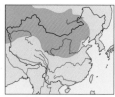

810. Pied Wheatear
Oenenthe pleschanka
白顶鹛

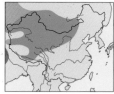

811. Desert Wheatear
Oenanthe deserti
漠鹛

812. Isabelline Wheatear
Oenanthe isabellina
沙鹛

ONLY
804 & 807

803

804

805

806

807

♀ spring

1st winter
810

♂ spring

808
♂ spring
spring

809
picata

capistrata

809
opistholeuca

811

812

PLATE 84   Starlings 1   椋鸟一

813. Chestnut-tailed Starling
Sturnus malabaricus
灰头椋鸟

814. Brahminy Starling
Sturnus pagodarum
黑冠椋鸟

815. Red-billed Starling
Sturnus sericeus
丝光椋鸟

816. Purple-backed Starling
Sturnus sturninus
北椋鸟

817. Chestnut-cheeked Starling
Sturnus philippensis
紫背椋鸟

818. White-shouldered Starling
Sturnus sinensis
灰背椋鸟

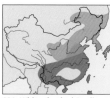

821. White-cheeked Starling
Sturnus cineraceus
灰椋鸟

822. Asian Pied Starling
Sturnus contra
斑椋鸟

823. Black-collared Starling
Sturnus nigricollis
黑领椋鸟

824. Vinous-breasted Starling
Sturnus burmannicus
红嘴椋鸟

813
nemoricolus

814

815
♂

juv.
816

♀

817
♀

♂

818
♀

♂

821

juv.

juv.
perciliaris

822

823

824

juv.

KP

PLATE 85 **Starlings 2** 椋鸟二

819. Rosy Starling
Sturnus roseus
粉红椋鸟

820. Common Starling
Sturnus vulgaris
紫翅椋鸟

825. Common Myna
Acridotheres tristis
家八哥

826. Bank Myna
Acridotheres ginginianus
灰背岸八哥

827. White-vented Myna
Acridotheres cinereus
林八哥

828. Collared Myna
Acridotheres albocinctus
白领八哥

829. Crested Myna
Acridotheres cristatellus
八哥

830. Golden-crested Myna
Ampeliceps coronatus
金冠树八哥

831. Hill Myna
Gracula religiosa
鹩哥

ONLY
829

827

829

825

826

820
juv.

819

juv.

831

828

830

♀

♂

PLATE 86 **Nuthatches** 鳾

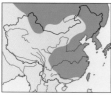

832. Eurasian Nuthatch
Sitta europaea
普通鳾

833. Chestnut-vented Nuthatch
Sitta nagaensis
栗臀鳾

834. Chestnut-bellied Nuthatch
Sitta castanea
栗腹鳾

835. White-tailed Nuthatch
Sitta himalayensis
白尾鳾

836. Chinese Nuthatch
Sitta villosa
黑头鳾

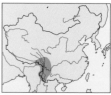

837. Yunnan Nuthatch
Sitta yunnanensis
滇鳾

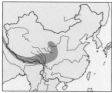

838. White-cheeked Nuthatch
Sitta leucopsis
白脸鳾

839. Velvet-fronted Nuthatch
Sitta frontalis
绒额鳾

840. Yellow-billed Nuthatch
Sitta solangiae
淡紫鳾

841. Giant Nuthatch
Sitta magna
巨鳾

842. Beautiful Nuthatch
Sitta formosa
丽鳾

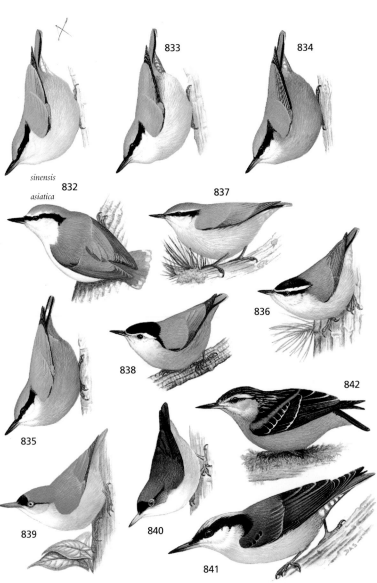

ONLY 532

X

833

834

sinensis 832

asiatica

837

836

838

842

835

839

840

841

PLATE 87 **Tits 1** 山雀一

851. Fire-capped Tit
Cephalopyrus flammiceps
火冠雀

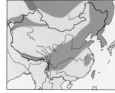

859. Coal Tit
Parus ater
煤山雀

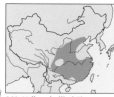

860. Yellow-bellied Tit
Parus venustulus
黄腹山雀

861. Grey-crested Tit
Parus dichrous
褐冠山雀

862. Great Tit
Parus major
大山雀

863. Turkestan Tit
Parus bok
西域山雀

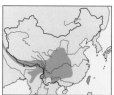

864. Green-backed Tit
Parus monticolus
绿背山雀

865. Yellow-cheeked Tit
Parus spilonotus
黄颊山雀

866. Yellow Tit
Parus holsti
台湾黄山雀

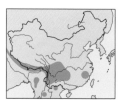

870. Yellow-browed Tit
Sylviparus modestus
黄眉林雀

859
ater

859
aemodius

870

♂
non-br.

♂ br. 860

♀

juv.

861
wellsi

862

862

863

commixtus

tibetanus

864

865
spilonotus
♂

♂

865
rex

juv.

866

♂

juv.

♂ br.

851
olivaceus

♀

PLATE 88 Tits 2 山雀二

849. White-crowned Penduline Tit
Remiz coronatus
攀雀

850. Chinese Penduline Tit
Remiz consobrinus
中华攀雀

852. Marsh Tit
Parus palustris
沼泽山雀

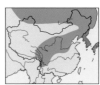

853. Willow Tit
Parus montanus
褐头山雀

854. White-browed Tit
Parus superciliosus
白眉山雀

855. Rusty-breasted Tit
Parus davidi
红腹山雀

856. Siberian Tit
Parus cinctus
西伯利亚山雀

857. Rufous-naped Tit
Parus rufonuchalis
棕枕山雀

858. Rufous-vented Tit
Parus rubidiventris
黑冠山雀

867. Azure Tit
Parus cyanus
灰蓝山雀

868. Yellow-breasted Tit
Parus flavipectus
黄胸山雀

869. Varied Tit
Parus varius
杂色山雀

871. Sultan Tit
Melanochlora sultanea
冕雀

872. Long-tailed Tit
Aegithalos caudatus
银喉〔长尾〕山雀

873. Black-throated Tit
Aegithalos concinnus
红头〔长尾〕山雀

874. Rufous-fronted Tit
Aegithalos iouschistos
黑头〔长尾〕山雀

875. Black-browed Tit
Aegithalos bonvaloti
黑眉〔长尾〕山雀

876. Sooty Tit
Aegithalos fuliginosus
银脸〔长尾〕山雀

caudatus 872

vinaceus

concinnus 873

iouschistos 874

875

♂ 849

850 ♂

♀

juv. 876

852
hellmayri 853 853
songarus

852
ypermelaena *baicalensis*

854

855 856 857

867
juv. 868
berezowskii

867
tianschanicus 858
beavani

♀ ♂

858
varius

871

PLATE 89 **Swallows** 燕

877. White-eyed River-Martin
Pseudochelidon sirintarae
白眼河燕

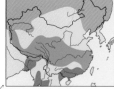

878. Sand Martin
Riparia riparia
崖沙燕

879. Plain Martin
Riparia paludicola
褐喉沙燕

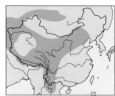

880. Eurasian Crag Martin
Hirundo rupestris
岩燕

881. Dusky Crag Martin
Hirundo concolor
纯色岩燕

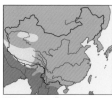

882. Barn Swallow
Hirundo rustica
家燕

883. Pacific Swallow
Hirundo tahitica
洋斑燕

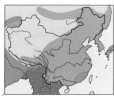

884. Red-rumped Swallow
Hirundo daurica
金腰燕

885. Striated Swallow
Hirundo striolata
斑腰燕

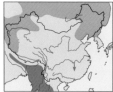

886. Northern House Martin
Delichon urbica
〔白腹〕毛脚燕

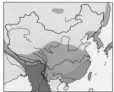

887. Asian House Martin
Delichon dasypus
烟脸毛脚燕

888. Nepal House Martin
Delichon nipalensis
黑喉毛脚燕

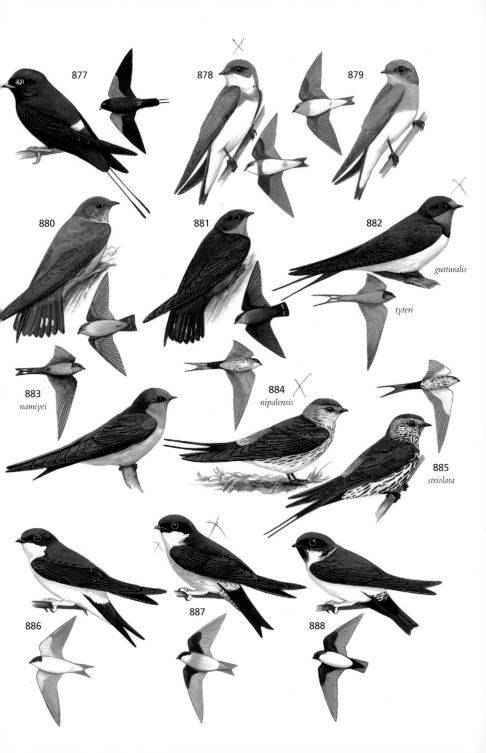

877

878

879

880

881

882

gutturalis

tyteri

883
namiyei

884
nipalensis

885
striolata

886

887

888

PLATE 90　Bulbuls 1　鹎一

893. Striated Bulbul
Pycnonotus striatus
纵纹绿鹎

894. Black-headed Bulbul
Pycnonotus atriceps
黑头鹎

895. Black-crested Bulbul
Pycnonotus melanicterus
黑冠黄鹎

896. Red-whiskered Bulbul
Pycnonotus jocosus
红耳鹎

897. Brown-breasted Bulbul
Pycnonotus xanthorrhous
黄臀鹎

898. Light-vented Bulbul
Pycnonotus sinensis
白头鹎

899. Styan's Bulbul
Pycnonotus taivanus
台湾鹎

900. Himalayan Bulbul
Pycnonotus leucogenys
白颊鹎

901. Red-vented Bulbul
Pycnonotus cafer
黑喉红臀鹎

902. Sooty-headed Bulbul
Pycnonotus aurigaster
白喉红臀鹎

903. Stripe-throated Bulbul
Pycnonotus finlaysoni
纹喉鹎

904. Flavescent Bulbul
Pycnonotus flavescens
黄绿鹎

PLATE 91 **Bulbuls 2** 鹎二

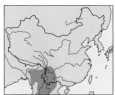

891. Crested Finchbill
Spizixos canifrons
风头雀嘴鹎

892. Collared Finchbill
Spizixos semitorques
领雀嘴鹎

905. White-throated Bulbul
Alophoixus flaveolus
黄腹冠鹎

906. Puff-throated Bulbul
Alophoixus flaveolus
白喉冠鹎

907. Grey-eyed Bulbul
Iole propinqua
灰眼短脚鹎

908. Brown-eared Bulbul
Ixos amaurotis
栗耳〔短脚〕鹎

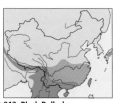

909. Ashy Bulbul
Hemixos flavala
灰短脚鹎

910. Chestnut Bulbul
Hemixos castanonotus
栗背短脚鹎

911. Mountain Bulbul
Hypsipetes mcclellandii
绿翅短脚鹎

912. Black Bulbul
Hypsipetes leucocephalus
黑〔短脚〕鹎

891 892 905 906

908 910 909

912 911 907

white-headed form

PLATE 92   **Prinias and Cisticolas**   鹪莺及扇尾莺

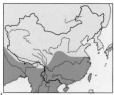

913. Zitting Cisticola
Cisticola juncidis
棕扇尾莺

914. Bright-headed Cisticola
Cisticola exilis
金头扇尾莺

915. White-browed Chinese Warbler
Rhopophilus pekinensis 山鹛

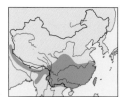

916. Striated Prinia
Prinia criniger
山鹪莺

917. Brown Prinia
Prinia polychroa
褐山鹪莺

918. Hill Prinia
Prinia atrogularis
黑喉山鹪莺

919. Rufescent Prinia
Prinia rufescens
暗冕鹪莺

920. Grey-breasted Prinia
Prinia hodgsonii
灰胸鹪莺

921. Yellow-bellied Prinia
Prinia flaviventris
黄腹鹪莺

922. Plain Prinia
Prinia inornata
褐头鹪莺

913

winter

summer

914

courtoisi

summer

winter

915

916

br.

917

918

br.

920

br.

919

br.

921

br.

922

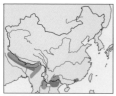

930. Pale-footed Bush Warbler
Cettia pallidipes
淡脚树莺

933. Brownish-flanked Bush Warbler
Cettia fortipes 强脚树莺

936. Yellowish-bellied Bush Warbler
Cettia acanthizoides 黄腹树莺

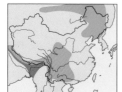

939. Spotted Bush Warbler
Bradypterus thoracicus
斑胸短翅莺

942. Brown Bush Warbler
Bradypterus luteoventris
棕褐短翅莺

943. Russet Bush Warbler
Bradypterus seebohmi
高山短翅莺

964. Booted Warbler
Hippolais caligata
靴篱莺

966. Olivaceous Warbler
Hippolais pallida
草绿篱莺

972. Common Chiffchaff
Phylloscopus collybita
叽咋柳莺

973. Mountain Chiffchaff
Phylloscopus sindianus
东方叽咋柳莺

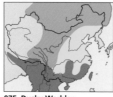

975. Dusky Warbler
Phylloscopus fuscatus
褐柳莺

978. Buff-throated Warbler
Phylloscopus subaffinis
棕腹柳莺

PLATE 94   **Bush Warblers 2 and Tesias**   莺二

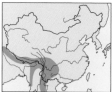

926. Chestnut-headed Tesia
Tesia castaneocoronata
栗头地莺

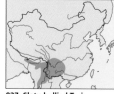

927. Slaty-bellied Tesia
Tesia olivea
金冠地莺

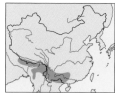

928. Grey-bellied Tesia
Tesia cyaniventer
灰腹地莺

929. Asian Stubtail
Urosphena squamiceps
鳞头树莺

931. Manchurian Bush Warbler
Cettia canturians
满洲树莺

932. Japanese Bush Warbler
Cettia diphone
日本树莺

934. Chestnut-crowned Bush Warbler
Cettia major 大树莺

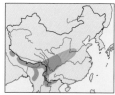

935. Aberrant Bush Warbler
Cettia flavolivacea
异色树莺

937. Grey-sided Bush Warbler
Cettia brunnifrons
棕顶树莺

938. Cetti's Bush Warbler
Cettia cetti
宽尾树莺

940. Long-billed Bush Warbler
Bradypterus major
巨嘴短翅莺

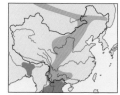

941. Chinese Bush Warbler
Bradypterus tacsanowskius
中华短翅莺

945. Grasshopper Warbler
Locustella naevia
黑斑蝗莺

929

945

938

937

934

935

♂

♀

931

932

borealis

♂

940

major   innae

ad. winter

941

927   928   926

ad.
pale morph

DAS

PLATE 95   **Grasshopper Warblers**   莺三

944. Lanceolated Warbler
Locustella lanceolata
矛斑蝗莺

946. Rusty-rumped Warbler
Locustella certhiola
小蝗莺

947. Middendorff's Warbler
Locustella ochotensis
北蝗莺

948. Pleske's Warbler
Locustella pleskei
史氏蝗莺

949. Savi's Warbler
Locustella luscinioides
鸲蝗莺

950. Gray's Warbler
Locustella fasciolata
苍眉蝗莺

952. Sedge Warbler
Acrocephalus schoenobaenus
水蒲苇莺

953. Streaked Reed Warbler
Acrocephalus sorghophilus
细纹苇莺

954. Black-browed Reed Warbler
Acrocephalus bistrigiceps
黑眉苇莺

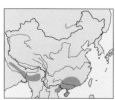

1012. Rufous-rumped Grassbird
Graminicola bengalensis
大草莺

ALL BUT 949, 952 & 1012

1012

946

1st winter

944

1st winter

948

1st winter

949

947

950

954

953

1st winter

952

PLATE 96 **Reed Warblers** 莺四

955. Paddyfield Warbler
Acrocephalus agricola
稻田苇莺

956. Manchurian Reed Warbler
Acrocephalus tangorum
满洲苇莺

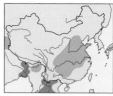

957. Blunt-winged Warbler
Acrocephalus concinens
钝翅〔稻田〕苇莺

958. Eurasian Reed Warbler
Acrocephalus scirpaceus
芦苇莺

959. Blyth's Reed Warbler
Acrocephalus dumetorum
布氏苇莺

960. Great Reed Warbler
Acrocephalus arundinaceus
大苇莺

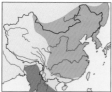

961. Oriental Reed Warbler
Acrocephalus orientalis
东方大苇莺

962. Clamorous Reed Warbler
Acrocephalus stentoreus
噪大苇莺

963. Thick-billed Warbler
Acrocephalus aedon
厚嘴芦莺

970. White-browed Tit Warbler
Leptopoecile sophiae
花彩雀莺

971. Crested Tit Warbler
Leptopoecile elegans
凤头雀莺

956

fresh

957

959

fresh

worn

fresh

juv.

961

955

963

fresh

worn

fresh

fresh

958

♀

♂

960
zarudnyi

971

962
amyae

♀

♂

970

PLATE 97 **LeafWarblers 1** 柳莺一

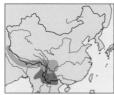

983. Ashy-throated Warbler
Phylloscopus maculipennis
灰喉柳莺

984. Pallas's Leaf Warbler
Phylloscopus proregulus
黄腰柳莺

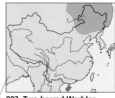

988. Yellow-browed Warbler
Phylloscopus inornatus
黄眉柳莺

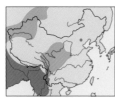

989. Hume's Warbler
Phylloscopus humei
锈眉柳莺

990. Arctic Warbler
Phylloscopus borealis
极北柳莺

992. Two-barred Warbler
Phylloscopus plumbeitarsus
双斑绿柳莺

993. Pale-legged Leaf Warbler
Phylloscopus tenellipes
灰脚柳莺

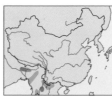

1000. Yellow-vented Warbler
Phylloscopus cantator
黄胸柳莺

1001. Sulphur-breasted Warbler
Phylloscopus ricketti
黑眉柳莺

ALL BUT 983, 989 & 1000

988

fresh

worn

989

fresh

984

worn

983

990

fresh

990
xanthrodryas

fresh

worn

992

993

fresh

1000

worn

1001

PLATE 98   **LeafWarblers 2**   柳莺二

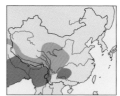

974. Wood Warbler
Phylloscopus sibilatrix
林柳莺

976. Smoky Warbler
Phylloscopus fuligiventer
烟柳莺

977. Tickell's Leaf Warbler
Phylloscopus affinis
黄腹柳莺

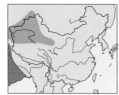

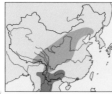

979. Sulphur-bellied Warbler
Phylloscopus griseolus
灰柳莺

980. Yellow-streaked Warbler
Phylloscopus armandii
棕眉柳莺

981. Radde's Warbler
Phylloscopus schwarzi
巨嘴柳莺

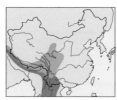

982. Buff-barred Warbler
Phylloscopus pulcher
橙斑翅柳莺

985. Lemon-rumped Warbler
Phylloscopus chloronotus
yan黄腰柳莺

986. Gansu Leaf Warbler
Phylloscopus kansuensis
甘肃黄腰柳莺

987. Chinese Leaf Warbler
Phylloscopus sichuanensis
四川柳莺

991. Greenish Warbler
Phylloscopus trochiloides
暗绿柳莺

994. Large-billed Leaf Warbler
Phylloscopus magnirostris
乌嘴柳莺

996. Emei Leaf Warbler
Phylloscopus emeiensis
娥眉柳莺

999. Hainan Leaf Warbler
Phylloscopus hainanus
海南柳莺

974

worn

fresh

982

976

fuligiventer

tibetanus

fresh

worn

977

fresh

worn

979

981

980

viridianus

worn

fresh

994

trochiloides

991

999

996

985

987

986

DAS

PLATE 99 LeafWarblers 3, other warblers, Tailorbirds and Crests 柳莺三

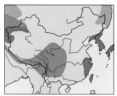

889. Goldcrest
Regulus regulus
戴菊

890. Flamecrest
Regulus goodfellowi
台湾戴菊

967. Mountain Tailorbird
Orthotomus cuculatus
金头缝叶莺

968. Common Tailorbird
Orthotomus sutorius
长尾缝叶莺

969. Dark-necked Tailorbird
Orthotomus atrogularis
黑喉缝叶莺

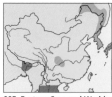

995. Eastern Crowned Warbler
Phylloscopus coronatus
冕柳莺

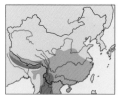

997. Blyth's Leaf Warbler
Phylloscopus reguloides
冠纹柳莺

998. White-tailed Leaf Warbler
Phylloscopus davisoni
白斑尾柳莺

1002. Golden-spectacled Warbler
Seicercus burkii
金眶鹟莺

1006. Chestnut-crowned Warbler
Seicercus castaniceps
栗头鹟莺

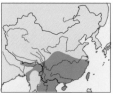

1008. Rufous-faced Warbler
Abroscopus albogularis
棕脸鹟莺

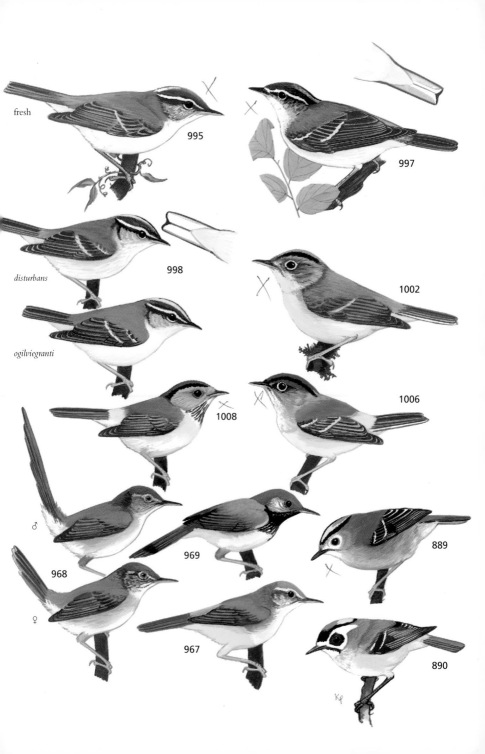

fresh

995

997

disturbans

998

1002

ogilviegranti

1006

1008

♂

968

969

889

♀

967

890

PLATE 100 **Sylvia Warblers** 莺五

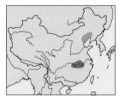

951. Japanese Swamp Warbler
Locustella pryeri
斑背大尾莺

1003. Grey-hooded Warbler
Seicercus xanthoschistos
灰头鹟莺

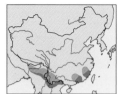

1004. White-spectacled Warbler
Seicercus affinis
白眶鹟莺

1005. Grey-cheeked Warbler
Seicercus poliogenys
灰脸鹟莺

1007. Broad-billed Warbler
Tickellia hodgsoni
宽嘴鹟莺

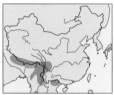

1009. Black-faced Warbler
Abroscopus schisticeps
黑脸鹟莺

1010. Yellow-bellied Warbler
Abroscopus superciliaris
黄腹鹟莺

1011. Striated Grassbird
Megalurus palustris
沼泽大尾莺

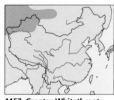

1157. Greater Whitethroat
Sylvia communis
灰〔白喉〕林莺

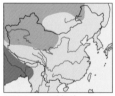

1158. Lesser Whitethroat
Sylvia curruca
白喉林莺

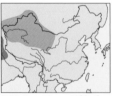

1159. Desert Lesser Whitethroat
Sylvia minula
沙白喉林莺

1160. Desert Warbler
Sylvia nana
漠地林莺

1161. Barred Warbler
Sylvia nisoria
横斑林莺

ONLY
951

1003

1004

1005

1007

1009

1010

1011

ad.
blythi

1158
minuta

1st winter
blythi

1157

spring

1st winter

1159

ad. winter

1160

951

ad. summer

1158

margelanica

1st winter

1161

ad.

DAS

PLATE 101 Laughingthrushes 1 噪鹛一

1013. Masked Laughingthrush
Garrulax perspicillatus
黑脸噪鹛

1014. White-throated Laughingthrush
Garrulax albogularis 白喉噪鹛

1015. White-crested Laughingthrush
Garrulax leucolophus 白冠噪鹛

1016. Lesser Necklaced Laughingthrush
Garrulax monileger 小黑领噪鹛

1017. Greater Necklaced Laughingthrush
Garrulax pectoralis 黑领噪鹛

1018. Striated Laughingthrush
Garrulax striatus
条纹噪鹛

1019. White-necked Laughingthrush
Garrulax strepitans 白颈噪鹛

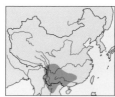

1020. Grey Laughingthrush
Garrulax maesi
褐胸噪鹛

1021. Rufous-necked Laughingthrush
Garrulax ruficollis 栗颈噪鹛

1022. Black-throated Laughingthrush
Garrulax chinensis 黑喉噪鹛

1023. Yellow-throated Laughingthrush
Garrulax galbanus 黄喉噪鹛

1024. Rufous-vented Laughingthrush
Garrulax gularis 栗肛噪鹛

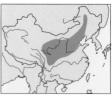

1025. Plain Laughingthrush
Garrulax davidi
山噪鹛

1013

1024

1014

1015

diardi

1016

picticollis

1017

pectoralis

1018

1019

castanotis

1021

1020

maesi

1023

1022

chinensis

1025

monachus

PLATE 102 Laughingthrushes 2 噪鹛二

1026. Snowy-cheeked Laughingthrush
Garrulax sukatschewi 黑额山噪鹛

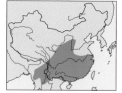

1027. Moustached Laughingthrush
Garrulax cineraceus
灰翅噪鹛

1028. Rufous-chinned Laughingthrush
Garrulax rufogularis 棕颏噪鹛

1029. Barred Laughingthrush
Garrulax lunulatus
斑背噪鹛

1030. White-speckled Laughingthrush
Garrulax bieti 白点鹛

1031. Giant Laughingthrush
Garrulax maximus
大噪鹛

1032. Spotted Laughingthrush
Garrulax ocellatus
眼纹噪鹛

1033. Grey-sided Laughingthrush
Garrulax caerulatus
灰胁噪鹛

1034. Rusty Laughingthrush
Garrulax poecilorhynchus
棕噪鹛

1035. Spot-breasted Laughingthrush
Garrulax merulinus 斑胸噪鹛

1036. Hwamei
Garrulax canorus
画眉

1037. White-browed Laughingthrush
Garrulax sannio 白颊噪鹛

1038. Streaked Laughingthrush
Garrulax lineatus
细纹噪鹛

1026

1027

1028

1029

1030

1031

1035

1032

1033

1034

taewanus

1036

canorus

1037

1038

KP

PLATE 103 **Laughingthrushes 3** 噪鹛三

1039. Blue-winged Laughingthrush
Garrulax squamatus
蓝翅噪鹛

1040. Scaly Laughingthrush
Garrulax subunicolor
纯色噪鹛

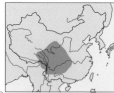

1041. Elliot's Laughingthrush
Garrulax elliotii
橙翅噪鹛

1042. Variegated Laughingthrush
Garrulax variegatus
杂色噪鹛

**1043. Brown-cheeked
Laughingthrush**
Garrulax henrici 灰腹噪鹛

1044. Black-faced Laughingthrush
Garrulax affinis
黑顶噪鹛

**1045. White-whiskered
Laughingthrush**
Garrulax morrisonianus 玉山噪鹛

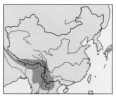

**1046. Chestnut-crowned
Laughingthrush**
Garrulax erythrocephalus 红头噪鹛

**1047. Red-winged
Laughingthrush**
Garrulax formosus 丽色噪鹛

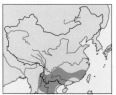

1048. Red-tailed Laughingthrush
Garrulax milnei
赤尾噪鹛

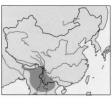

1049. Red-faced Liocichla
Liocichla phoenicea
红翅薮鹛

1050. Emei Shan Liocichla
Liocichla omeinsis
灰胸薮鹛

1051. Steere's Liocichla
Liocichla steerii
黄痣薮鹛

ONLY 1041

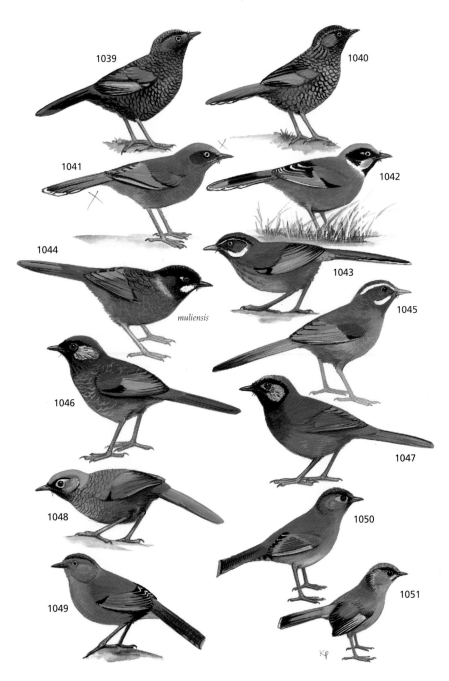

1039

1040

1041

1042

1044

1043

muliensis

1045

1046

1047

1048

1050

1049

1051

KP

PLATE 104　**Scimitar Babblers**　钩嘴鹛

1052. Buff-breasted Babbler
Pellorneum tickelli
棕胸雅鹛

1053. Spot-throated Babbler
Pellorneum albiventre
白腹幽鹛

1054. Puff-throated Babbler
Pellorneum ruficeps
棕头幽鹛

1055. Large Scimitar Babbler
Pomatorhinus hypoleucos
长嘴钩嘴鹛

1057. Rusty-cheeked Scimitar Babbler 锈脸钩嘴鹛
Pomatorhinus erythrogenys

1056. Spot-breasted Scimitar Babbler 斑胸钩嘴鹛
Pomatorhinus erythrocnemis

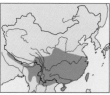

1059. Streak-breasted Scimitar Babbler
Pomatorhinus ruficollis 棕颈钩嘴鹛

1060. Red-billed Scimitar Babbler
Pomatorhinus ochraceiceps
棕头钩嘴鹛

1061. Coral-billed Scimitar Babbler
Pomatorhinus ferruginosus
红嘴钩嘴鹛

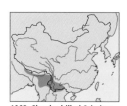

1062. Slender-billed Scimitar Babbler
Xiphirhynchus superciliaris 剑嘴鹛

1055

1053

1052

1056

1054

cowensae

erythrocnemis

1057

godwini

1059

usicus

1060

styani

albipectus

1062

1061

PLATE 105 **Wren Babblers and Wren** 鹩鹛

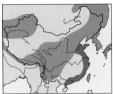

848. Winter Wren
Troglodytes troglodytes
鹪鹩

1063. Long-billed Wren Babbler
Rimator malacoptilus
长嘴鹩鹛

1064. Limestone Wren Babbler
Napothera crispifrons
灰岩鹩鹛

1065. Streaked Wren Babbler
Napothera brevicaudata
短尾鹩鹛

1066. Eyebrowed Wren Babbler
Napothera epilepidota
纹胸鹩鹛

1067. Scaly-brested Wren Babbler
Pnoepyga albiventer
鳞胸鹩鹛

1068. Pygmy Wren Babbler
Pnoepyga pusilla
小鳞〔胸〕鹩鹛

**1069. Rufous-throated Wren
Babbler**
Spelaeornis caudatus 短尾鹩鹛

1071. Bar-winged Wren Babbler
Spelaeornis troglodytoides
斑翅鹩鹛

1072. Spotted Wren Babbler
Spelaeornis formosus
丽星鹩鹛

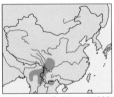

1073. Long-tailed Wren Babbler
Spelaeornis chocolatinus
长尾鹩鹛

1074. Wedge-billed Wren Babbler
Sphenocichla humei
楔头鹩鹛

only 848 & 1068

nipaleasis

848

troglolytes

pusilla

tormosana

1068

troglodytes

soulei

1071

1073

1067

1063

1072

1069

1065

1066

1074

1064

PLATE 106 **Tree Babblers** 其它鹛一

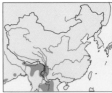

1075. Rufous-fronted Babbler
Stachyris rufifrons
黄喉穗鹛

1076. Rufous-capped Babbler
Stachyris ruficeps
红头穗鹛

1077. Golden Babbler
Stachyris chrysaea
金头穗鹛

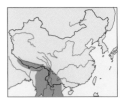

1078. Grey-throated Babbler
Stachyris nigriceps
黑头穗鹛

1079. Spot-necked Babbler
Stachyris striolata
斑颈穗鹛

1080. Striped Tit Babbler
Macronous gularis
纹胸鹛

1081. Chestnut-capped Babbler
Timalia pileata
红顶鹛

1082. Yellow-eyed Babbler
Chrysomma sinense
金眼鹛雀

1083. Rufous-tailed Babbler
Chrysomma poecilotis
宝兴鹛雀

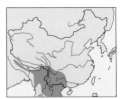

1087. Silver-eared Mesia
Leiothrix argentauris
银耳相思鸟

1088. Red-billed Leiothrix
Leiothrix lutea
红嘴相思鸟

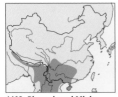

1102. Blue-winged Minla
Minla cyanouroptera
蓝翅希鹛

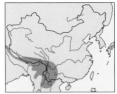

1103. Chestnut-tailed Minla
Minla strigula
斑喉希鹛

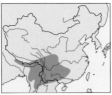

1104. Red-tailed Minla
Minla ignotincta
火尾希鹛

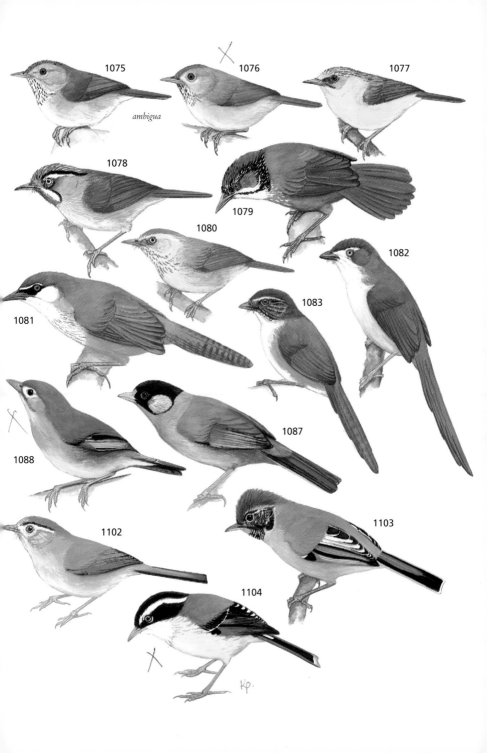

1075

1076

1077

ambigua

1078

1079

1080

1081

1082

1083

1088

1087

1102

1103

1104

KP.

PLATE 107   **Shrike Babblers and Babaxes**   其它鹛二

1084. Chinese Babax
Babax lanceolatus
矛纹草鹛

1085. Giant Babax
Babax waddelli
大草鹛

1086. Tibetan Babax
Babax koslowi
棕草鹛

1089. Cutia
Cutia nipalensis
斑胁姬鹛

1090. Black-headed Shrike Babbler
Pteruthius rufiventer
棕腹鹛鹛

1091. White-browed Shrike Babbler
Pteruthius flaviscapis
红翅鹛鹛

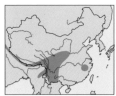

1092. Green Shrike Babbler
Pteruthius xanthochlorus
淡绿鹛鹛

1093. Black-eared Shrike Babbler
Pteruthius melanotis
栗喉鹛鹛

**1094. Chestnut-fronted Shrike
Babbler**
Pteruthius aenobarbus 栗额鹛鹛

1095. White-hooded Babbler
Gampsorhynchus rufulus
白头鹛鹛

ONLY 1084

1084

1085

1086

1089
♀

1090
♀

♂

1089
♂

1091
♀

♂

1092

xanthochlorus

1095

1093
♀

♂

1094
♀

♂

rufulus

KP

PLATE 108 **Fulvettas** 雀鹛

1105. Golden-breasted Fulvetta
Alcippe chrysotis
金胸雀鹛

1106. Gold-fronted Fulvetta
Alcippe variegaticeps
金额雀鹛

1107. Yellow-throated Fulvetta
Alcippe cinerea
黄喉雀鹛

1108. Rufous-winged Fulvetta
Alcippe castaneceps
栗头雀鹛

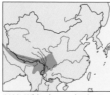

1109. White-browed Fulvetta
Alcippe vinipectus
白眉雀鹛

1110. Chinese Fulvetta
Alcippe striaticollis
高山雀鹛

1111. Spectacled Fulvetta
Alcippe ruficapilla
棕头雀鹛

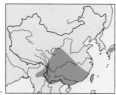

1112. Streak-throated Fulvetta
Alcippe cinereiceps
褐头雀鹛

1113. Ludlow's Fulvetta
Alcippe ludlowi
路德雀鹛

1114. Rufous-throated Fulvetta
Alcippe rufogularis
棕喉雀鹛

1115. Dusky Fulvetta
Alcippe brunnea
褐顶雀鹛

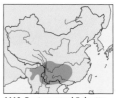

1116. Rusty-capped Fulvetta
Alcippe dubia
褐胁雀鹛

1117. Brown-cheeked Fulvetta
Alcippe poioicephala
褐脸雀鹛

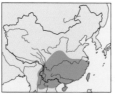

1118. Grey-cheeked Fulvetta
Alcippe morrisonia
灰眶雀鹛

1119. Nepal Fulvetta
Alcippe nipalensis
白眶雀鹛

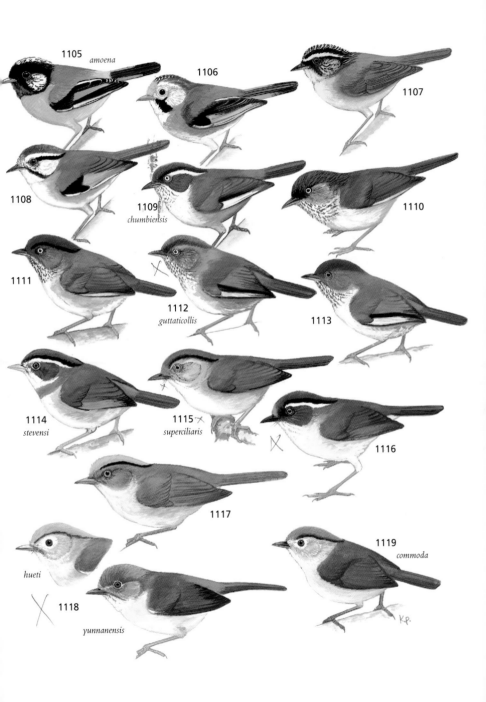

1105 *amoena*
1106
1107
1108
1109 *chumbiensis*
1110
1111
1112 *guttaticollis*
1113
1114 *stevensi*
1115 *superciliaris*
1116
1117
hueti
1118 *yunnanensis*
1119 *commoda*

K.P.

PLATE 109 Sibias and Barwings 斑翅鹛及奇鹛

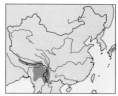

1096. Rusty-fronted Barwing
Actinodura egertoni
锈额斑翅鹛

1097. Spectacled Barwing
Actinodura ramsayi
白眶斑翅鹛

1098. Hoary-throated Barwing
Actinodura nipalensis
纹头斑翅鹛

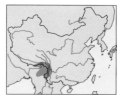

1099. Streak-throated Barwing
Actinodura waldeni
纹胸斑翅鹛

1100. Streaked Barwing
Actinodura souliei
灰头斑翅鹛

1101. Taiwan Barwing
Actinodura morrisoniana
台湾斑翅鹛

1120. Rufous-backed Sibia
Heterophasia annectens
栗背奇鹛

1121. Rufous Sibia
Heterophasia capistrata
黑头奇鹛

1122. Grey Sibia
Heterophasia gracilis
灰奇鹛

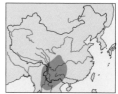

1123. Black-headed Sibia
Heterophasia melanoleuca
黑顶奇鹛

1124. White-eared Sibia
Heterophasia auricularis
白耳奇鹛

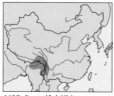

1125. Beautiful Sibia
Heterophasia pulchella
丽色奇鹛

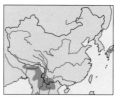

1126. Long-tailed Sibia
Heterophasia picaoides
长尾奇鹛

NONE

1096
ripponi

1097

1098

1099
saturatior

1100

1101

1120

1121

1126

1123

1122

1124

1125

PLATE 110 **Yuhinas** 凤鹛

1127. Striated Yuhina
Yuhina castaniceps
栗耳凤鹛

1128. White-naped Yuhina
Yuhina bakeri
白项凤鹛

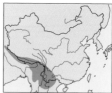

1129. Whiskered Yuhina
Yuhina flavicollis
黄颈凤鹛

1130. Stripe-throated Yuhina
Yuhina gularis
纹喉凤鹛

1131. White-collared Yuhina
Yuhina diademata
白领凤鹛

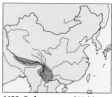

1132. Rufous-vented Yuhina
Yuhina occipitalis
棕肛凤鹛

1133. Taiwan Yuhina
Yuhina brunneiceps
褐头凤鹛

1134. Black-chinned Yuhina
Yuhina nigrimenta
黑颏凤鹛

1135. White-bellied Yuhina
Yuhina zantholeuca
白腹凤鹛

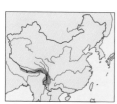

1136. Fire-tailed Myzornis
Myzornis pyrrhoura
火尾绿鹛

ONLY

AND 1131

1127

1134

1129

1132

1128

1133

1130

1131

1135

1136

PLATE 111 **Parrotbills 1** 鸦雀一

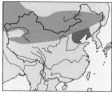

1137. Bearded Parrotbill
Panurus biarmicus
文须雀

1138. Great Parrotbill
Conostoma oemodium
红嘴鸦雀

1139. Three-toed Parrotbill
Paradoxornis paradoxus
三趾鸦雀

1140. Brown Parrotbill
Paradoxornis unicolor
褐鸦雀

1141. Grey-headed Parrotbill
Paradoxornis gularis
灰头鸦雀

1142. Black-breasted Parrotbill
Paradoxornis flavirostris
斑胸鸦雀

1143. Spot-breasted Parrotbill
Paradoxornis guttaticollis
点胸鸦雀

**1154. Lesser Rufous-headed
Parrotbill** 黑眉鸦雀
Paradoxornis atrosuperciliaris

**1155. Greater Rufous-headed
Parrotbill**
Paradoxornis ruficeps 红头鸦雀

1156. Reed Parrotbill
Paradoxornis heudei
震旦鸦雀

ONLY 1138 & 1141

1137

juv. ♂

♀

♂

1139

1138

1140

1143

1141

1142

1155

1154

1156

PLATE 112 **Parrotbills 2** 鸦雀二

1144. Spectacled Parrotbill
Paradoxornis conspicillatus
白眶鸦雀

1145. Vinous-throated Parrotbill
Paradoxornis webbianus
棕头鸦雀

1146. Brown-winged Parrotbill
Paradoxornis brunneus
褐翅鸦雀

1147. Ashy-throated Parrotbill
Paradoxornis alphonsianus
灰喉鸦雀

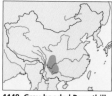

1148. Grey-hooded Parrotbill
Paradoxornis zappeyi
暗色鸦雀

1149. Rusty-throated Parrotbill
Paradoxornis przewalskii
灰冠鸦雀

1150. Fulvous Parrotbill
Paradoxornis fulvifrons
黄额鸦雀

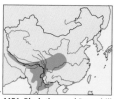

1151. Black-throated Parrotbill
Paradoxornis nipalensis
橙额鸦雀

1152. Golden Parrotbill
Paradoxornis verreauxi
金色鸦雀

1153. Short-tailed Parrotbill
Paradoxornis davidianus
短尾鸦雀

1144

1145
webbianus

1146

mantschuricus

1147

1148

1149

1150
cyanophrys

verrauxi

1152

albifacies

davidiarius

pallidus

1151

1153

thompsoni

PLATE 113 **Larks** 百灵

1162. Singing Bushlark
Mirafra cantillans
歌百灵

1163. Bimaculated Lark
Melanocorypha bimaculata
二斑百灵

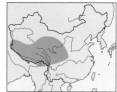

1164. Tibetan Lark
Melanocorypha maxima
长嘴百灵

1165. Mongolian Lark
Melanocorypha mongolica
〔蒙古〕百灵

1166. White-winged Lark
Melanocorypha leucoptera
白翅百灵

1167. Black Lark
Melanocorypha yeltoniensis
黑百灵

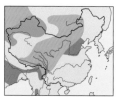

1168. Greater Short-toed Lark
Calandrella brachydactyla
短趾百灵

1169. Hume's Short-toed Lark
Calandrella acutirostris
细嘴短趾百灵

1170. Asian Short-toed Lark
Calandrella cheleensis
〔亚洲〕短趾百灵

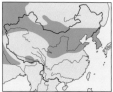

1171. Crested Lark
Galerida cristata
凤头百灵

1172. Eurasian Skylark
Alauda arvensis
云雀

1173. Japanese Skylark
Alauda japonica
日本云雀

1174. Oriental Skylark
Alauda gulgula
小云雀

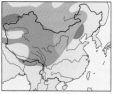

1175. Horned Skylark
Eremophila alpestris
角百灵

1163

1164

1167 ♀

1167 ♂

1165

1166 ♀

1166 ♂

1170

1172
dulcivox

1174

1173

1168

1169

albigula

1175
brandti

1162

1171

PLATE 114 **Sunbirds and Spiderhunters** 太阳鸟

1182. Ruby-cheeked Sunbird
Anthreptes singalensis
紫颊直嘴太阳鸟

1183. Purple-naped Sunbird
Hypogramma hypogrammicum
蓝枕花蜜鸟

1184. Olive-backed Sunbird
Nectarinia jugularis
黄腹花蜜鸟

1185. Purple Sunbird
Nectarinia asiatica
紫色蜜鸟

1186. Mrs Gould's Sunbird
Aethopyga gouldiae
蓝喉太阳鸟

1187. Green-tailed Sunbird
Aethopyga nipalensis
绿喉太阳鸟

1188. Fork-tailed Sunbird
Aethopyga christinae
叉尾太阳鸟

1189. Black-throated Sunbird
Aethopyga saturata
黑胸太阳鸟

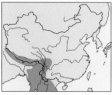

1190. Crimson Sunbird
Aethopyga siparaja
黄腰太阳鸟

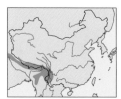

1191. Fire-tailed Sunbird
Aethopyga ignicauda
火尾太阳鸟

1192. Little Spiderhunter
Arachnothera longirostra
长嘴捕蛛鸟

1193. Streaked Spiderhunter
Arachnothera magna
纹背捕蛛鸟

1182

♀

♂

1184

♀

♂

1185

♀

♂

eclipse
♂

1186

♀

♂

1187

♀

♂

1188

✗ ♀

♂

1189

♀

♂

1190

♀

♂

1191

♀

♂

1183

1193

1192

PLATE 115 **Flowerpeckers and Munias** 啄花鸟及文鸟

923. Chestnut-flanked White-eye
Zosterops erythropleurus
红胁绣眼鸟

924. Oriental White-eye
Zosterops palpebrosus
灰腹绣眼鸟

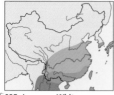

925. Japanese White-eye
Zosterops japonicus
暗绿绣眼鸟

1176. Thick-billed Flowerpecker
Dicaeum agile
厚嘴啄花鸟

1177. Yellow-vented Flowerpecker
Dicaeum chrysorrheum
黄肛啄花鸟

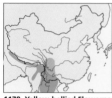

1178. Yellow-bellied Flowerpecker
Dicaeum melanoxanthum
黄腹啄花鸟

1179. Plain Flowerpecker
Dicaeum concolor
纯色啄花鸟

1180. Fire-breasted Flowerpecker
Dicaeum ignipectus
红胸啄花鸟

1181. Scarlet-backed Flowerpecker
Dicaeum cruentatum
朱背啄花鸟

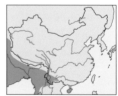

1238. Red Avadavat
Amandava amandava
〔红〕梅花雀

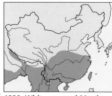

1239. White-rumped Munia
Lonchura striata
白腰文鸟

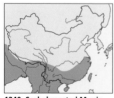

1240. Scaly-breasted Munia
Lonchura punctulata
斑文鸟

1241. Black-headed Munia
Lonchura malacca
栗腹文鸟

PLATE 116 **Sparrows, Snowfinches and Weavers** 麻雀，雪雀及织布鸟

1194. Saxaul Sparrow
Passer ammodendri
黑顶麻雀

1195. House Sparrow
Passer domesticus
家麻雀

1196. Spanish Sparrow
Passer hispaniolensis
黑胸麻雀

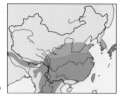

1197. Russet Sparrow
Passer rutilans
山麻雀

1198. Eurasian Tree Sparrow
Passer montanus
〔树〕麻雀

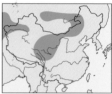

1199. Rock Sparrow
Petronia petronia
石雀

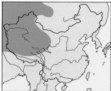

1200. White-winged Snowfinch
Montifringilla nivalis
白斑翅雪雀

1201. Tibetan Snowfinch
Montifringilla adamsi
褐翅雪雀

1202. White-rumped Snowfinch
Pyrgilauda taczanowskii
白腰雪雀

1203. Small Snowfinch
Pyrgilauda davidiana
黑喉雪雀

1204. Rufous-necked Snowfinch
Pyrgilauda ruficollis
棕颈雪雀

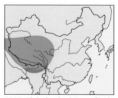

1205. Plain-backed Snowfinch
Pyrgilauda blanfordi
棕背雪雀

1235. Black-breasted Weaver
Ploceus benghalensis
黑喉织布鸟

1236. Streaked Weaver
Ploceus manyar
纹胸织布鸟

1237. Baya Weaver
Ploceus philippinus
黄胸织布鸟

ONLY 1197 Z 1198

1194
stoliczkae
♀
♂

1195
♀
♂

1196
♀
♂

1197
♀
♂

1198
juv.
ad.

1199

1200
1201
1203

1202
1205
1204

1237
♀
♂

1236
♀
♂

1235
♀
♂

KP

PLATE 117 **Wagtails** 鹡鸰

1206. Forest Wagtail
Dendronanthus indicus
山鹡鸰

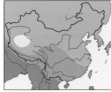

1207. White Wagtail
Motacilla alba
白鹡鸰

1208. Black-backed Wagtail
Motacilla lugens
黑背鹡鸰

1209. Japanese Wagtail
Motacilla grandis
日本鹡鸰

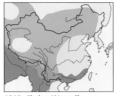

1210. Citrine Wagtail
Motacilla citreola
黄头鹡鸰

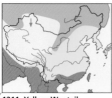

1211. Yellow Wagtail
Motacilla flava
黄鹡鸰

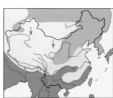

1212. Grey Wagtail
Motacilla cinerea
灰鹡鸰

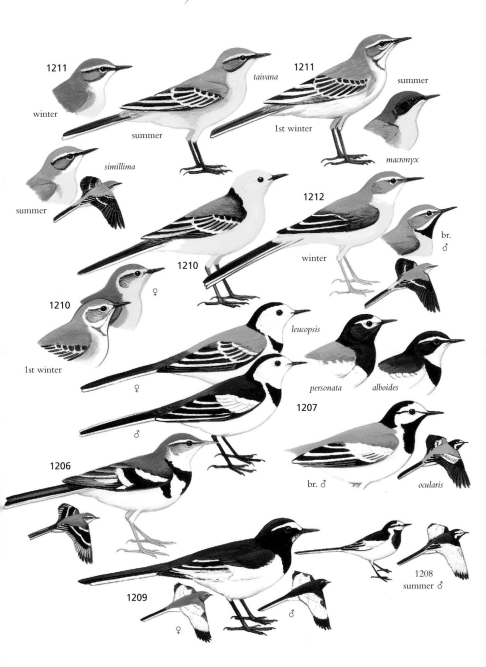

ALL BUT 1209

1211

winter

summer

simillima

summer

taivana

summer

1st winter

1211

summer

macronyx

1212

1210

1210

1st winter

♀

♀

♀

♂

winter

br.
♂

leucopsis

personata

alboides

1207

1206

br. ♂

ocularis

1209

♀

♂

1208
summer ♂

PLATE 118 **Pipits** 鹨

1213. Richard's Pipit
Anthus richardi
田鹨

1214. Paddyfield Pipit
Anthus rufulus
稻田鹨

1215. Tawny Pipit
Anthus campestris
平原鹨

1216. Blyth's Pipit
Anthus godlewskii
布莱氏原鹨

1217. Tree Pipit
Anthus trivialis
林鹨

1218. Olive-backed Pipit
Anthus hodgsoni
树鹨

1219. Pechora Pipit
Anthus gustavi
北鹨

1220. Meadow Pipit
Anthus pratensis
草地鹨

1221. Red-throated Pipit
Anthus cervinus
红喉鹨

1222. Rosy Pipit
Anthus roseatus
粉红胸鹨

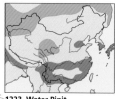

1223. Water Pipit
Anthus spinoletta
水鹨

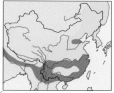

1224. Buff-bellied Pipit
Anthus rubescens
黄腹鹨

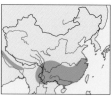

1225. Upland Pipit
Anthus sylvanus
山鹨

ALL BUT 1214, 1215, 1216 & 1217

1213

1214

1215

1220

1216

1221

br.

1218

1217
haringtoni

1225

1223

1219

coutellii

1222

br.

1224

japonicus

PLATE 119 Accentors, Waxwings and Woodswallow 岩鹨及太平鸟

652. Ashy Woodswallow
Artamus fuscus
灰燕鵙

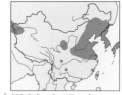

685. Bohemian Waxwing
Bombycilla garrulus
太平鸟

686. Japanese Waxwing
Bombycilla japonica
小太平鸟

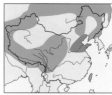

1226. Alpine Accentor
Prunella collaris
领岩鹨

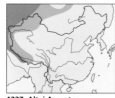

1227. Altai Accentor
Prunella himalayana
高原岩鹨

1228. Robin Accentor
Prunella rubeculoides
鸲岩鹨

1229. Rufous-breasted Accentor
Prunella strophiata
棕胸岩鹨

1230. Siberian Accentor
Prunella montanella
棕眉山岩鹨

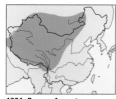

1231. Brown Accentor
Prunella fulvescens
褐岩鹨

1232. Black-throated Accentor
Prunella atrogularis
黑喉岩鹨

1233. Mongolian Accentor
Prunella koslowi
贺兰山岩鹨

1234. Maroon-backed Accentor
Prunella immaculata
栗背岩鹨

PLATE 120　**Finches 1**　雀一

1245. Fire-fronted Serin
Serinus pusillus
金额丝雀

1246. Grey-capped Greenfinch
Carduelis sinica
金翅〔雀〕

1247. Yellow-breasted Greenfinch
Carduelis spinoides
高山金翅〔雀〕

1248. Black-headed Greenfinch
Carduelis ambigua
黑头金翅〔雀〕

1249. Eurasian Siskin
Carduelis spinus
黄雀

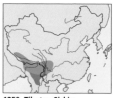

1250. Tibetan Siskin
Carduelis thibetana
藏黄雀

1251. European Goldfinch
Carduelis carduelis
红额金翅〔雀〕

1252. Hoary Redpoll
Carduelis hornemanni
极北朱顶雀

1253. Common Redpoll
Carduelis flammea
白腰朱顶雀

1254. Twite
Carduelis flavirostris
黄嘴朱顶雀

1255. Eurasian Linnet
Carduelis cannabina
赤胸朱顶雀

1245
ad.   juv.

1250
♂   ♀

1246
♂   juv.

caniceps
♂

1251
juv.

♂
paropansi

1249
♂
♀

♂
juv.

1247

1248
juv.

♂

♂
♀

1253
♂

1252
♀

korejevi

1255
♀

rufostrigata

1254

♂

montanella

PLATE 121 **Finches 2** 雀二

1243. Chaffinch
Fringilla coelebs
苍头燕雀

1244. Brambling
Fringilla montifringilla
燕雀

1256. Plain Mountain Finch
Leucosticte nemoricola
林岭雀

1257. Brandt's Mountain Finch
Leucosticte brandti
高山岭雀

1259. Asian Rosy Finch
Leucosticte arctoa
粉红腹岭雀

1260. Spectacled Finch
Callacanthis burtoni
红眉金翅雀

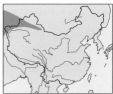

1261. Crimson-winged Finch
Rhodopechys sanguinea
赤翅沙雀

1262. Mongolian Finch
Bucanetes mongolicus
蒙古沙雀

1263. Desert Finch
Rhodospiza obsoleta
巨嘴沙雀

1283. Pine Grosbeak
Pinicola enucleator
松雀

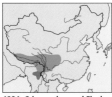

1284. Crimson-browed Finch
Propyrrhula subhimachala
红眉松雀

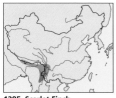

1285. Scarlet Finch
Haematospiza sipahi
血雀

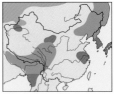

1286. Red Crossbill
Loxia curvirostra
红交嘴雀

1287. White-winged Crossbill
Loxia leucoptera
白翅交嘴雀

ONLY 124<del>f</del> X

1243 ♂ ♀

1244 ♂ ♀

1260 ♂ ♀

1256 ♂ ♀ juv. 1257 ♂

1263 ♂ juv. 1261 ♂ 1262 ♂ imm.

1283 ♂ ♀ 1285 ♂ ♀ 1259 ♂

1286 imm. 1287 ♂ ♀ ♀

PLATE 122 **Rosefinches 1** 朱雀一

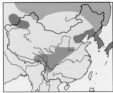

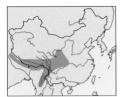

1264. Long-tailed Rosefinch
Uragus sibiricus
长尾雀

1265. Blandford's Rosefinch
Carpodacus rubescens
赤朱雀

1266. Dark-breasted Rosefinch
Carpodacus nipalensis
暗胸朱雀

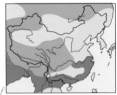
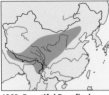
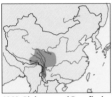

1267. Common Rosefinch
Carpodacus erythrinus
普通朱雀

1268. Beautiful Rosefinch
Carpodacus pulcherrimus
红眉朱雀

1269. Pink-rumped Rosefinch
Carpodacus eos
曙红朱雀

1270. Pink-browed Rosefinch
Carpodacus rhodochrous
玫红眉朱雀

1271. Vinaceous Rosefinch
Carpodacus vinaceus
酒红朱雀

1272. Dark-rumped Rosefinch
Carpodacus edwardsii
棕朱雀

1299. Pink-tailed Bunting
Urocynchramus pylzowi
朱鹀

ONLY 1267 ± 1271

1264
1299
1265
1266
1267
1268
1270
1269
1271
1272

PLATE 123 **Rosefinches 2** 朱雀二

1273. Pale Rosefinch
Carpodacus synoicus
沙色朱雀

1274. Pallas's Rosefinch
Carpodacus roseus
北朱雀

1275. Three-banded Rosefinch
Carpodacus trifasciatus
斑翅朱雀

1276. Spot-winged Rosefinch
Carpodacus rhodopeplus
点翅朱雀

1277. White-browed Rosefinch
Carpodacus thura
白眉朱雀

1278. Red-mantled Rosefinch
Carpodacus rhodochlamys
红腰朱雀

1279. Streaked Rosefinch
Carpodacus rubicilloides
拟大朱雀

1280. Great Rosefinch
Carpodacus rubicilla
大朱雀

1281. Red-faced Rosefinch
Carpodacus puniceus
红胸朱雀

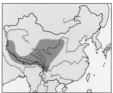

1282. Tibetan Rosefinch
Carpodacus roborowskii
藏雀

NOTE ONLY
1274.

synoleus

beicki ♂

♀

1273

synoicus ♂

1274 ♂

verreauxi

♀

♀

1276

♂

thura ♀

♀ feminius

1281

1277

1278

thura ♂

♀

♂

1280

♂

1279 ♂

♀

1284

1275

♀

1282

♂

PLATE 124 Finches 3 雀三

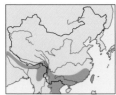

1288. Brown Bullfinch
Pyrrhula nipalensis
褐灰雀

1289. Red-headed Bullfinch
Pyrrhula erythrocephala
红头灰雀

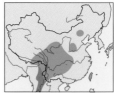

1290. Grey-headed Bullfinch
Pyrrhula erythaca
灰头灰雀

1291. Eurasian Bullfinch
Pyrrhula pyrrhula
灰腹灰雀

1292. Hawfinch
Coccothraustes coccothraustes
锡嘴雀

1293. Yellow-billed Grosbeak
Eophona migratoria
黑尾蜡嘴雀

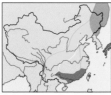

1294. Japanese Grosbeak
Eophona personata
黑头蜡嘴雀

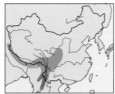

1295. Collared Grosbeak
Mycerobas affinis
黄颈拟蜡嘴雀

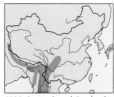

1296. Spot-winged Grosbeak
Mycerobas melanozanthos
白点翅拟蜡嘴雀

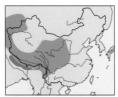

1297. White-winged Grosbeak
Mycerobas carnipes
白斑翅拟蜡嘴雀

1298. Gold-naped Finch
Pyrrhoplectes epauletta
金枕黑雀

juv.　　　♀　　　♂

1291 *griseiventris*

juv.　　　　　ad.

1288

♀　　　♂

juv.

1290

juv.

1289

juv.　　　♂

1292

♀　　　♂

1298

juv.　　　♂

1294

juv.　　　　♂

♀　　　♂

1293

♀　　　♂

1297

♂

♀

1296

♀　　　♂

juv.

1295

PLATE 125 **Buntings 1** 鹀一

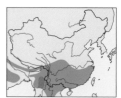

1300. Crested Bunting
Melophus lathami
凤头鹀

1311. Tristram's Bunting
Emberiza tristrami
白眉鹀

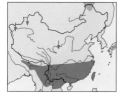

1313. Little Bunting
Emberiza pusilla
小鹀

1314. Yellow-browed Bunting
Emberiza chrysophrys
黄眉鹀

1316. Yellow-throated Bunting
Emberiza elegans
黄喉鹀

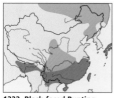

1322. Black-faced Bunting
Emberiza spodocephala
灰头鹀

ALL!

1300 ♂

♀

sordida ♂

♀

♀

1322 personata ♂

spodocephala ♂

♀

1313

♂ 1311 ♀

1st winter ♂

♂

1314

1316

♀

summer ♂

PLATE 126 Buntings 2 鹀二

1301. Slaty Bunting
Latoucheornis siemsseni
蓝鹀

1303. Pine Bunting
Emberiza leucocephalos
白头鹀

1306. Godlewski's Bunting
Emberiza godlewskii
戈氏岩鹀

1307. Meadow Bunting
Emberiza cioides
三道眉草鹀

1309. Grey-necked Bunting
Emberiza buchanani
灰颈鹀

1315. Rustic Bunting
Emberiza rustica
田鹀

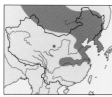

1328. Lapland Longspur
Calcarius lapponicus
铁爪鹀

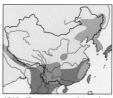

1312. Chestnut-eared Bunting
Emberiza fucata
栗耳鹀

ALL BUT 1303 & 1309

1307

1303

♀

winter ♂

summer ♂

♂

winter ♂

1306

1309

♀

♂

winter ♂

1315

1328

summer ♂

summer ♂

winter ♂

1312

1301

♀

♀

♂

♂

PLATE 127 Buntings 3 鹀三

1317. Yellow-breasted Bunting
Emberiza aureola
黄胸鹀

1318. Chestnut Bunting
Emberiza rutila
栗鹀

1319. Black-headed Bunting
Emberiza melanocephala
黑头鹀

1321. Japanese Yellow Bunting
Emberiza sulphurata
硫黄鹀

1324. Pallas's Bunting
Emberiza pallasi
苇鹀

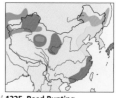

1325. Reed Bunting
Emberiza schoeniclus
芦鹀

1326. Ochre-rumped Bunting
Emberiza yessoensis
红颈苇鹀

ALL BUT 1319 & 1321

1321
♀
♂

1317
♀
♂
1st year ♂

1318
♀
winter ♂
♂

1319
♂
♀
(not to scale)

autumn ♂
1325
♀
♂

winter ♀

1326
summer ♂

1324
♀
summer ♂
winter ♂

♂
1st winter

PLATE 128 Buntings 4 鹀四

1302. Yellowhammer
Emberiza citrinella
黄鹀

1304. Tibetan Bunting
Emberiza koslowi
藏鹀

1305. Rock Bunting
Emberiza cia
灰眉岩鹀

1308. Jankowski's Bunting
Emberiza jankowskii
栗斑腹鹀

1310. Ortolan Bunting
Emberiza hortulana
圃鹀

1320. Red-headed Bunting
Emberiza bruniceps
褐头鹀

1323. Grey Bunting
Emberiza variabilis
灰鹀

1327. Corn Bunting
Miliaria calandra
黍鹀

1329. Snow Bunting
Plectrophenax nivalis
雪鹀

1302

♀ br.

♀

♂

1304

br. ♂

1st winter
♂

♀
1305
stracheyi

♂

♀

♂

1310

♀

1308

♂

♂
non-br.

1320

br. ♂

♀
1323

1st winter
♂

♂

1327

1st winter

br. ♂

1329

♀ br.

History of ornithology in China

Almost nothing was known of the birds of China until after the First Opium War of the 1840s when European traders around the ports of Xiamen and Fuzhou started to send back a trickle of trade skins from the nearby provinces. Knowledge further improved after the Second Opium War in the 1860s when traders started to open up trade routes with the interior of China.

The first significant ornithologist to work in China was the British Consul Robert Swinhoe who from 1854–1875 combined a successful role as both a diplomat and ornithologist, scouring some 2000 miles of the Chinese coastline from Hainan to Beijing with most detailed work on the birds of Taiwan, Hainan and Fujian provinces. Swinhoe was a brilliant linguist and meticulous zoologist, who wrote 120 zoological papers on his various travels and collected 3700 bird skins of 650 species, containing no less than 200 type specimens. Following his 1863 publication of a rather incomplete checklist, in 1871 he published a revised checklist of the birds of China listing 675 species. Swinhoe returned to England in poor health and died in 1877 but his work is commemorated by the brilliant Swinhoe's Pheasant and Swinhoe's Crake.

No less extraordinary was the work of the French missionary Pére Armand David who arrived in Beijing in 1862 and made three major collecting trips through China before returning to France in 1874 to spend the rest of his life teaching other missionaries and writing up his collections. David is famous for discovering the Pere David's Deer and Giant Panda but his trips to Fujian, Qinling Mts, northern China and especially his sojourn in Baoxing county in Sichuan yielded a fantastic number of new finds among plants, insects, mammals and birds. David stated that he saw 772 species of birds in China and his collection of 1300 skins of 470 species yielded 65 new to science. In 1877 he published *Les Oiseaux de la Chine* together with Emile Oustalet, describing a total of 807 species. Eight birds still bear his name, although all were described by other authors.

Whilst Swinhoe and David collected the cream of birds in eastern and central China it was the Russian military explorer, General Nicholas Michailovitch Prjevalsky, who first discovered the bird riches of north-west China in four incredible expeditions between 1871 and 1885 across some of the harshest regions of the world, through the Gobi, Alashan and Taklimakan deserts, Qaidam basin, Lop Nor depression, mountains of eastern Tibet, Altyn Tagh and the Tianshan and Altai mountains of north-west Xinjiang. Three times he visited the great salt lake of Qinghai (Koko Nor) and three times he set out for Lhasa to be finally turned away by the Tibetan army despite pleas to the Dalai Lama that his mission was peaceful. Prjevalsky's fifth expedition was halted at the lake of Issuk Kul on the western border of China by typhus and an early death in 1888, but his Lieutenants, Roborovsky and Koslov, continued the surveys and zoological collections, filling the gaps of knowledge on the birds of the Tibetan plateau. Prjevalsky had collected 5000 bird specimens of 430 species out of total collections of 20000 zoological and 16000 botanical specimens. More than 20 birds were new to science and nine of these still bear his name as first author. Many intrepid and dedicated collectors continued to fill in the gaps of Chinese avifaunal distribution. John La Touche, Charles Rickett and Frederick Styan collected much new material in south-east China. The famous John Whitehead died collecting birds on Hainan. Botanists such as Ernest Wilson, George Forrest, Frank Ludlow and Joseph Rock added birds to their plant collections of west and south-west China. Dr. George Henderson, Captain John Biddulph and Dr. Ferdinand Stoliczka penetrated the scientifically unknown regions of Turkestan. Stoliczka died on returning through Karakoram.

The recent history of ornithology in China has been well documented by Boswell (1986, 1989). From 1949 to 1980 the country was totally closed to Western ornithologists but ornithology was continued by several Western-trained Chinese biologists of whom Cheng Tso Hsin was the most active and prolific. Russia gave China assistance in the early years of communism in building up the Chinese Academy of Sciences and training more ornithologists. Russian aid was withdrawn in 1960 and for two decades, ornithology continued in total isolation. Many new surveys were undertaken in northeast and south-west China and also on the Tibetan Plateau. Books of birds of individual provinces were published with an emphasis on economic fauna which seems to include all birds. Cheng published no less than 202 books and papers in 11 years, and major collections were built up by the Institute of Zoology in Beijing, Institute of Zoology in Kunming and Dept. of Entomology Rare Species Division in Guangzhou as well as several normal colleges and universities.

The Cultural Revolution of 1967–77 was a disaster for ornithology as for many other branches of academia in China. Scientists were sent into the country to undertake menial tasks. Many died but Cheng remained in the Zoology Institute in Beijing although his publication output was much reduced until 1977 when he was able to resume his former level of productivity. He continued to publish right up to his death in 1998 at age 92 years.

Since 1976 Cheng and several other ornithologists have been able to visit other countries and the international links have been gradually restored. Since 1981 foreign biologists have again been active in China.

In 1980 the Ornithological Society of China was established and has grown in size and stature ever since. The Beidaihe Birdwatching Society was launched in 1988. Other local birdwatcher groups have been formed in Kunming, Harbin and at universities and institutes in several other provincial capitals. A national bird ringing programme has been operational since 1983. There are now hundreds of active Chinese ornithologists and species lists are published for many protected areas in the country.

The Hong Kong Bird Watching Society has been active since 1956 and has extended its area of interest into southern China since 1983. The Chinese Wild Bird Federation in Taiwan has been active since 1988. A list of relevant clubs and societies is given in Appendix 5.

Foreign ornithologists visiting China have brought much needed new approaches and methodology—more focus on vocalisations, more emphasis on conservation and trade studies, revised taxonomy and improved field criteria for identification. Many individual ornithologists have published lists and observations from their visits to China and some international organisations such as The Crane Foundation (CF), World Conservation Society (WCS), World-Wide Fund for Nature (WWF) and Japan Birdwatching Society (JBS) have all been active in sponsoring both field studies of Chinese birds and conservation programmes.

The lack of a clear identification guide has restrained the progress of field ornithology in China for decades. Most birds could not be properly identified until they were dead and back in a comparative museum. Even then a host of out-dated, eroneous or synonymous names could be applied.

Cheng's major work *A Synopsis of the Avifauna of China* which only appeared in English in 1987 but had been available in Chinese in various stages of revision since 1949, has provided some conformity of names used within China but provides no figures or aids to identification. Meyer de Schauensee's book *The Birds of China* describes most species but only figures one third of these, uses out-dated Chinese place names and is not available in Chinese. The two-volume *Les Oiseaux de Chine, de Mongolie et de Coree* by Étchéclopar and Hüe (1978, 1982) does figure almost all the Chinese species but is almost unavailable in China and is only available in French. The first pictorial field guide for the whole country was not published until 1995 as *Atlas of Birds of China*, but comprised only the species of Cheng's 1987 list (1189 species) with brief descriptions, distribution maps and very poor quality paintings of sometimes completely unrecognisable birds. Meanwhile Cheng had both revised the taxonomy and updated the completeness of his own checklist publishing *A complete checklist of species and subspecies of the Chinese birds* in 1994 which added more than 50 new species to his 1987 book, bringing the total of Chinese species to 1244. *A Field Guide to the Chinese Birds of China* was published the following year (Yan Chong Wei *et al.* 1995) describing and illustrating 1253 species and was a big improvement on the Atlas. Neither of the two Chinese lanaguage guides have adopted the revised taxonomy of Sibley and Monroe (1990) and neither is available in English. The present work, hopefully, fills this gap and when published in Chinese should help domestic ornithologists to adapt to the latest taxonomic revisions.

Avian biogeography of the region

The avian regions of China

Cheng's excellent book *A Synopsis of the Avifauna of China* analyses the distribution of bird species and subspecies across China and divides the country into 16 avian subregions classed in seven major regions. Nine of the subregions belong to the Palearctic realm and the other seven to the Oriental realm. These realms can in turn be divided into four subrelams as follows.

Realm	Subrealm	Region	Subregion
Palearctic	East Asia	North-eastern	Da Hinggan/Altai Mts Changhai Mt
		North China	Huang–Huai Plain Loess Plateau
	Eremian	Mongolo–Xinjiang	East Meadow West Desert Tianshan Hilly
	Central Asian	Qinghai–Xizang	Qiantang Plateau Qinghal–Zangnan
Oriental	Sino-Indian	South-west Mid-China	South-west Mountainous Eastern HillockPlain Western Mountainous Plateau
		South China	Min–Guan coastal Diannan Hilly Hainan Island Taiwan

Using a different approach BirdLife International have described a global system of Endemic Bird Areas or EBAs (Stattersfield *et al.* 1998). An EBA consists of an area where the ranges of two or more birds of restricted range overlap. For the purposes of the review, limited distribution was taken to be a species whose total known global range was less than 50 000 km². Thirteen such EBAs are identified in China. Their names and the number of restricted-range Chinese species involved are listed in the table below. A full list of the species involved is given in Appendix 2.

EBA code	EBA name	Species
D01	Taklimakan desert	3
D06	Eastern Tibet	2
D07	Southern Tibet	2
D08	Eastern Himalayas	21
D11	Qinghai mountains	4

D12	Central Sichuan mountains	9
D13	West Sichuan mountains	3
D14	Chinese subtropical forests	5
D15	Yannan mountains	4
D20	Hainan	5
D23	Shanxi mountains	2
D24	South-east Chinese mountains	6
D25	Taiwan	12

North-east China. Here one can find some important wetland sites that are important for breeding waterfowl. White Storks, several species of cranes and many ducks, geese and swans breed in these areas. Best localities are Zhalong and Momoge but Lake Kankha on the Russian border is also a wonderful place. These areas should be visited in summer. Forest birds are best seen in Changbaishan reserve or for the real tundra birds, such as Spotted Capercaillie and Hazel Grouse, up in the conifer forests north of Harbin and the Huzhong–Hanma reserve.

North China. Some excellent places to see the birds of northern China include Beidahe on the Hebei coast. This is a great place to watch passage migrants, whilst the mountain reserves of Shennongjia in Hubei and several reserves in the Qinling mountains just south of Xian in Shaanxi are excellent places to visit. The Pangquangou Nature Reserve in Shanxi was established mainly for the protection of Brown Eared Pheasant. This unit contains the EBA D23.

Mongolo–Xinjiang. For the birds of the great northern steppes you should travel up the Yellow River through the Loess and Ordos plateau to the Helanshan and Qilianshan mountains and the Xilingele reserve of Inner Mongolia. In north-west China, the mountain ranges of both Tianshan and Altai provide a range of habitats from alpine and conifer forests to desert conditions. There are some important wetland in Tianshan where geese, swans and some cranes nest. Desert birds can be seen in the Tarim and Turpan depressions but travel is difficult from the main town of Urumqi. The great reserve of the Arjin Mountains has salt lakes and montane habitats and is about as remote as you can get in China. This unit contains EBA D01.

Qinghai–Xizang Plateau. The plateau region of China is full of potential but most birdwatchers do not get far from Lhasa. You can travel into the Chumbi valley between Sikkim and Bhutan to see wetland and some Himalayan birds or into the Yalong Zangpo (Brahmaputra) valley into the moist forests of South-east Xizang. This unit contains EBAs D06, D07 and D11.

South-west China. In south-west China, which includes the subtropical parts of south-east Xizang, you can find the bulk of China's restricted range endemic birds in Sichuan, in the Wolong Panda reserve or Mt Emei sacred mountain reserve. Wolong, in the Qionglai mountains, boasts nine different species of pheasants. Gaoligongshan and Baimazueshan in Yunnan are other excellent reserves to visit in this unit. The unit contains EBAs D12, D13, D14, D15 and parts of D08.

Mid-China. Central and south-east China contain important lakes such as Dongting and Poyang in the Yangtze valley where hundreds of thousands of waterbirds winter including cranes, swans, geese and storks. Visitors should take plenty of clothes in winter and a snorkel in summer. Fanjingshan and Cao Hai in Guizhou, typify the western plateau birds, whilst the endemic mountain birds of the south-east are best seen in Wuyishan reserve in Fujian and Jiangxi provinces where excellent facilities have been created for visitors. The unit contains EBA D24.

South China. Tropical south China has many good places to see birds. Most famous are Xishuangbanna in the tropics of south Yunnan, Dinghushan and Babaoshan in Guangdong province, Jianfenling and Bawangling on Hainan Island and such excellent reserves as Kenting, Taroga and Dawushan in Taiwan. For forest birds on Hong Kong, the visitor should go to Tai Po Kau, Whilst for waders and waterbirds one should visit Maipo marshes reserve in Hong Kong or Dongzaigang in Hainan. The unit contains EBAs D20 and D25.

Some special sites to look for birds.

Knowledge of China's birds is far from complete. The country is huge, some regions are very remote and there have been few active ornithologists in the country. The following section mentions some special regions that deserve more attention. Instead of going to the familiar places in China to clock up new additions to life lists, why not get to some remote areas where you have a good chance of finding something new?

1. Jinping in south-east Yannan. Montane forests straddle the border with Vietnam directly connecting with the Fan Si Pan mountains of the Hoang Lien Son mountain range in north-west Tonkin. Studies by Delacour and others on the Vietnamese side reveal many interesting species such as Ward's Trogon, Slender-billed Scimitar Babbler and other birds that have disjunct populations in those mountains. Many of these species probably also occur on the Chinese side of the border but have not been thoroughly searched for.

2. Gaoligong mountains and Yannan to the west of the Nujiang (Salween) River. This narrow strip of quite wild territory is faunally part of north-east Burma and has many species and subspecies not found elsewhere in China. Recent studies by the Kunming Institute of Zoology and visiting overseas birdwatchers have already made some exciting and surprising finds in this area and there are surly more discoveries to be made here.

3. South-east Xizang. China's territorial claims in this region overlap with most of the Indian state of Arunachal Pradesh. The region has therefore only been accessible to ornithologists from the Indian side and is not well understood by Chinese ornithologists. Several species are known from within China's officially mapped boundaries only here. For instance, the Mishmi hills are famous for two endemic species and several endemic subspecies. These hills are bisected by the current boundary of control between India and China and surely many of these endemic forms would be found on the Chinese side of the border if a thorough search was mounted.

4. The Nansha archipelago or Spratly Islands. China claims almost all of the South China Sea as far south as close to the coasts of Borneo and the Philippines. These claims are disputed by several other south-east Asian countries. Nevertheless the islands are of great interest ornithologically and due to their disputed sovereignty are neither well-studied nor well-protected. Appendix 4 of this book lists some of the species that can be expected there that are not currently listed for China.

5. North-west Xinjiang. The Altai and Tianshan and Kashi regions of Xinjiang have already yielded a fair list of species not otherwise recorded in China. This is a moist meeting point between the Turkestan (Central Asian) and Transbaikal (Eremian) faunal subregions—a verdant moist oasis in a generally arid part of the world. Several more species are known from just outside China's borders in this sector and a diligent search would undoubtedly extend some of their known ranges into the country.

6. The Panda zone. There is an important region, very much corresponding with the distribution of such important Chinese mammals as the Giant Panda *Ailuropoda melanoleuca*, Golden Monkey *Pygathrix roxellanae*, Lesser Panda *Ailurus fulgens* and Takin *Budorcas taxicolor*, that extends from north-west Yunnan through southern to north-west Sichuan, extreme south Gansu, extending through South Shaanxi as far as Shennongjia in West Hubei. This scythe-shaped region to the west and north of the Sichuan basin is not only important for rare mammals but is fantastically rich botanically and encompassed the distribution of a large number of China's endemic birds. The region contains EBA's D12, D13 and D14 and includes the endemic hot spot of Mt Emei. Two of the recent new leaf warbler discoveries were also made in this area.

7. South-east China mountains. Many endemics of south-east China were originally collected at Guadun in north-west Fujian. This locality lies close to what is now the headquarters of the Wuyishan National Nature Reserve. However the forests continue over into Jiangxi, so most of the special species of this locality may also be found in south Jiangxi which is poorly surveyed even in the adjacent Jiangxi Wuyishan Nature Reserve. Surveys of other mountain forests in south and south-east China such as Mangshan (South Human) and Babaoshan (North Guangdong) reveal that many species though to be confined to north-west Fujian have much wider distributions and there are many more areas where significant range extensions and new provincial records could be made. This region contains EBA D24.

8. Hulun Nur. A series of lakes in the north-east corner of Nei Mongol on the west side of the Greater Hinggan Mountains is an oasis for many birds otherwise not recorded in China, as well as an important breeding site for waterfowl. There are probably new species records to be made in this region and certainly new breeding records to be made.

9. North-east China Wetlands. The Three Rivers Plain is a large swampy region with extensive reedbeds and hidden lakes. It is a special breeding area for many of China's most spectacular wetland birds especially cranes and storks but many parts are completely unexplored and many new discoveries must await the keen ornithologist.

10. South-flowing valleys of the Himalayas. Several forested, south-flowing valleys cross into south Xizang which have never been properly explored on the Chinese side. Examples are Chumbi (Torsa) valley and Kulong Chu flowing into Bhutan and the Arun valley flowing into Nepal. There are probably some Himalayan bird species in these valleys that have not yet been recorded in China.

The avian year

China lies north of the equator, ranging from permafrost tundra in the extreme north, through the temperate and subtropical zones to a narrow tropical fringe in the south. Most tropical and subtropical birds are resident, with little seasonality of breeding or movement. Montane species are sometimes altitudinal migrants, breeding in the spring at high altitudes, rearing young through the summer and descending to lower altitudes in winter. Species that are territorial in summer are often flock-living in winter.

Some birds living in the north of the country such as crows and pheasants are hardy enough to live through the winter but many raptors and passerines are summer visitors, breeding in the long days of the northern summer but migrating to the south of China and beyond in winter. Many waders breed far north of China's borders in Siberia or along the Arctic coasts and only occur in China as passage migrants in spring and autumn on their long journeys to and from Indonesia and in some cases even New Guinea and Australia. China's east coast acts as in important flyway for these migrants.

Conservation

China's bird fauna is seriously threatened by loss and fragmentation of habitat, forest fires, hunting and the pet trade, pollution of waterways and inland seas and use of insecticides on agricultural lands. However, the Chinese government is taking the conservation of biodiversity very seriously. China has ratified many international conventions such as the Convention on International Trade in Endangered Species (CITES), Wetlands Convention (Ramsar), Migratory Bird Convention (Bonn Agreement), Convention on Biological Diversity (CBD) and Agenda 21. China hosted a special international workshop in Beijing to try to control illegal wildlife trade across its borders.

The Wildlife Protection laws have been strengthened by additional notices of the State Council to improve the management of nature reserves. There are now severe penalties for poaching or trading in rare animals, including the death penalty. Lists of protected animals have been published as have official Red Data Books for most taxonomic groups, including birds.

The system of nature reserves grows very fast. By the end of 1997 there were 926 official nature reserves totalling 77 000 000 hectares or 7.65 per cent of the country. Plans are in place to extend this to over 10 per cent in the next five-year period. Most of the reserves are managed by the State Forestry Bureau but several other agencies such as the Ministry of Agriculture, SEPA and provincial and municipal governments also manage some reserves.

Following the disastrous floods of 1998, the Prime Minister has ordered a logging ban on all natural forests in the middle and upper catchments of the Huang He (Yellow River) and Changjiang (Yangtze River). This single decision will help protect most of China's endemic bird species.

After decades of over-cutting of forests, which has seen the forest cover of the country fall from about 35 per cent to about 12 per cent the State Forestry Bureau is now able to show that the forest and stock density are recovering as a result of active replanting and recovery of secondary forests. Forest cover is currently at 14 per cent.

However, despite all these commendable efforts, conditions on the ground do not always match up to plans and regulations. China has very high population pressure combined with a large poor sector, willing to put up with considerable hardship and take risks to take anything valuable out of forest lands, lakes and other natural areas.

Moreover, the standards of management of many of the protected areas also leaves much to be desired. Staff are generally under-paid, under-equipped, poorly motivated and poorly trained. Reserves have been abused by agricultural incursion, forest fire, poaching, cutting of timber and firewood, collection of medicinal plants, littering and scarring from visitors etc. Wetlands have been polluted or drained.

Conditions are slowly improving and in particular in those reserves that benefit from support from international conservation agencies or programmes.

Field techniques for birdwatching

Birdwatching is an absorbing pastime. A knowledge of birds and their habits gives a greater sense of purpose to any walk in the country, or even a stroll in a city park or garden. Moreover, the broader the birdwatcher's experiences the better and more acute his observations become. Most birdwatchers are not scientists or ecologists. They come from all walks of life—doctors, lawyers, school teachers, factory workers and youngsters. But there are lots of them and they are the scientists' eyes and ears for monitoring the state of our planet. They collect birdlists and observations that are very important for scientists to notice where the environment is deteriorating. This book is designed to help birdwatchers make their observations more accurate and their records more meaningful.

The following section gives some hints on how to go about watching birds in a more meaningful way and what sorts of equipment to use. As much of China is open terrain and the birds are shy and wary of humans, the birdwatcher will need optical assistance. A powerful spotting scope mounted on a sturdy tripod that is stable in wind is a rewarding investment and proves particularly valuable for watching birds in open mountains, on lakes, other waterways and along the coast. However, such equipment is too slow and cumbersome to use in forest, where binoculars prove more effective and where speed of locating a bird with width of view and brightness of image more valuable than magnification power. Forests in China vary from conifer forests to deciduous woodlands but in the south are some very dense moist tropical and subtropical forests with added problems of leeches, moisture and poor visibility.

1. Hints on birdwatching in forests

Watching birds in tall forest is not easy. You may walk for an hour without seeing anything then suddenly be surrounded by so many twittering birds you can't focus on any. A bird is so high up and so obscured by foliage that you cannot get a good view or anyway it looks like six other species. In the rain you have special problems of water on lenses blurring your vision.

The newcomer to the forest should first get to know the common forest edge species by walking along roads and good trails through forest where observation conditions are better and birds appear more numerous because the light falling sideways on the forest edge creates a very rich feeding zone. Some of the true forest canopy birds can be seen along roads and where forest roads cut through mountains you may have excellent sideways views into the canopy that you will never get when walking on narrow trails.

Leeches in the southern parts of China are an accepted irritation to the hardened birder but may be quite distressing to the newcomer to the forest. There are several ways to minimise damage. First, wear a pair of fine tight socks inside your normal thick walking socks. Tuck your outer socks over your trouser bottoms. This will greatly reduce the number of leeches reaching your feet. You can wear special canvas leech socks outside you normal socks which tie around your trouser legs. You can use a variety of chemicals for protection. Spraying your socks and the inside of your boots with insecticide (Shelltox etc.) will give good protection until you wade through a stream. Salt, tobacco and soap also work. Rubbing a little benzyl benzoate on your skin or clothes will protect you from leeches but this is a hot chemical so don't put it on sensitive areas!

Rain forests are terribly damp if not actually dripping or pouring water. This causes problems with optical devices such as binoculars, cameras and spectacles. Make sure your kitbag is fairly waterproof, keep equipment in polythene bags and wrapped with dry absorbent cloth, paper or dessicant packets. If you wear spectacles try to find a small piece of soft (chamois) leather so that you can wipe your lenses dry whenever you are watching something. If you get moisture inside your binoculars, dry them out as quickly as possible. Quick ways to remove moisture include lying them in the sun with the lenses facing directly at the sun, putting them overnight in a bag of dessicant or sleeping with the binoculars held close for warmth. More drastic measures such as cooking them near a campfire will work but shorten the life of your binoculars considerably! When buying binoculars, select waterproof varieties.

Spotting scopes are very useful for birding in open areas and are almost essential to get good views of waders and waterbirds but need a strong tripod and are awkward and of limited use in the rain forest.

Most birders walk very slowly through the forest trying not to miss anything. This works quite well for canopy birds but you will in fact miss a lot of ground birds—pheasants, pittas and thrushes that are very wary hear you coming and have already slipped away before you arrive. If you can learn to move quickly but silently you will have more success with these shy birds. Also, there are great rewards for sitting very quietly on a log waiting for birds to come past you. You may have to wait for some time but you will sometimes be rewarded with great views of birds you would never otherwise get. Ideally you should alternate between bouts of slow careful searching, faster silent walking and resting periods of silent waiting. Particularly rewarding places for waiting are at fruiting fig trees, red flowering trees, large mistletoe clumps or by pools and streams.

If you have company, limit conversation to a minimum and if you have guides or porters try to persuade them to keep quiet or keep at a distance from yourself. Wear drab clothes and never white.

Two methods are increasingly used by birders to call up birds to aid observation. The first, 'pishing', involves making sibilant, squeaking or rasping sounds which roughly imitate the alarm calls of some small birds. This can have the effect of exciting small birds, especially skulking babblers to call back and through curiosity come close or pop up out of the understorey to investigate the source of alarm. A similar effect can be achieved by imitating the call of the Collared Owlet or other small raptors which small birds come to mob.

The second method is the use of playback of tape recorded songs which can cause a territorial reaction and attract birds for a better view.

In the depths of the forest both methods are pretty harmless, but problems can arise in areas regularly visited by birders such as the main trails of national parks. Here it may be necessary to ban such techniques which can disturb the territorial and nesting behaviour of local birdlife and reduce the natural alarm responses of wild birds. This is particularly the case with taped calls.

Birds are most active in the early morning and this is the best time to see them. Their activity drops off towards midday. The afternoon activity peak is never as energetic as the morning, although there is sometimes a 'false dawn' peak of activity after prolonged rain.

Midday and late afternoon are, however, quite good times to wait for birds to visit water sources in the drier times of year.

2. Keeping field notes

The bird fauna of China is still rather poorly known. Locality birdlists are incomplete, lists for some of the border areas of the north-east and south-west and of islands in the South China Sea are very incomplete; most habitat and feeding generalisations are based on very few records. Knowledge of migration patterns and breeding seasonality are also poorly documented. There are few ornithologists in a region which is geographically extensive and complicated. You may watch birds for your own fun but your observations are important. You may be the only source of information from a given locality in months or years. Try to keep good notes and share your observations with other bird-watchers, or through clubs and journals to help fill some of the gaps in our knowledge. A few pointers are pertinent.

- Always carry and use a notebook. Keep it in a waterproof bag.
- Clearly indicate the locality and date of all notes or lists recorded.
- Use a pencil or ballpoint, not ink or felt-tip (which run when wet).
- Keep records of all birds seen at a locality.
- Make especially detailed notes of any rare or unusual bird sightings and follow these up later by sending records to a bird club or journal.
- Make notes of the dates of migrant species seen in the locality and note the condition of their plumage—breeding, eclipse, non-breeding etc., and numbers.
- Keep notes of any breeding records observed.
- Keep notes of any unusual feeding or behaviour observed.

3. Identification hints: what to look for.

Birdwatching is more fun when you know what you are looking at. Recognition is based on a combination of several characters, including a bird's appearance, voice, behaviour and location. Field identifications may be classed as follows:

1. definite: complete diagnostic features or a combination of features recognised.
2. probable: the features recognised are typical of a certain species, even though not diagnostic, and that species can be expected in that place at that time.
3. unconfirmed: as for 'probable' except that the species is not expected at the locality at the time, i.e. the record contains new information but, without supporting evidence, is inconclusive.

It is important to take notes of the birds you see at the time and not wait until you get home. Without notes the observer will soon forget which birds were seen in a given area or on what dates (important in building up a picture of breeding season and migration periods). Moreover, notes are essential for later identification of those birds not immediately recognised in the field. Make an effort to check for diagnostic features. The most conspicuous character, such as a white bar on the tail, is remembered vividly but other features are overlooked, and when the observer later checks in the field guide he finds two similar birds with white tail bars and cannot remember if his had a brown or grey head.

With a new or unfamiliar bird it is best to make a sketch in your notebook. This does not need to be a work of art, simply enough to record the pertinent details of size; shape; length of bill; presence of crest or other features; colour of plumage; length of wings and tail; colour of any bare facial skin; colour of bill, eyes and feet; any other unusual visual features. Additional notes about the call, behaviour and locality all help in later identification.

Often a bird is half-recognised. It is a 'sort of sunbird' or like 'a Little Pied Flycatcher but with yellow on its breast'. In such instances, list the features that differ from the nearest species you know well. For example, a note entry might read: 'small pinkish dove, similar to Spotted Dove but no white spots on black neck patch, greyer head and more uniform red-brown on back'. Such a description can be readily identified later, in this case as the Island Collared Dove. As it is impossible to remember all the diagnostic characters of every species it is useful to have a field guide close at hand, or with you in the field, for quick reference to re-check characters you may have overlooked on first encounter.

Example of field sketch from notebook (Dollarbird, *Eurystomus orientalis*)

Some species are less easy to identify in the field; some are virtually impossible without close examination and measurement. With experience you will learn which are the 'difficult' birds and what distinguishing characters to look for to separate species of similar general appearance. These diagnostic features are given in the individual species descriptions in this book.

4. Making bird lists

Many birdwatchers keep lists of the birds they see—day lists of all birds seen on a given day, locality lists of all birds ever seen at a given locality, island lists, trip lists, year lists or life lists and so on. Such lists are fun to compile and give an added sense of purpose and achievement to the activity of birdwatching. They are also valuable sources of scientific information both for the study of birds for their own sake and also for conservationists and wildlife managers whose job it is to select, protect and manage reserves and other areas to ensure the survival of important and valuable wild species.

Records for a given reserve or locality make up the locality species list, which is useful for the evaluation of the area and selection of appropriate management.

Obviously, to be useful such lists must be accurate. Do not dilute valuable accurate information by the inclusion of doubtful records. By all means make a note of uncertain species, but separate these from the really confident or definite observations. A small gap of ignorance is better than a false piece of information. Even the experts do not expect to identify every bird they see. Do take great care that all lists are clearly identified with the location and dates or period in which they were collected. Records of unusual sightings should be sent to local bird clubs, museums or journals. Appendix 5 gives addresses of the relevant authorities for reviewing such records in the China region.

Various methods are employed by birdwatchers to compile birdlists. Probably the most common are day lists i.e. all species seen on a given day, or locality or trip lists. The drawback of such lists is their incompleteness. It takes a long time to locate and identify all resident species in a given area and new visitors can be added to the list year after year. A partial list for an area does not give a full picture; moreover it does not indicate the relative abundance of different species nor the richness of the locality. A short list may indicate a rather impoverished fauna or may only reflect the shortness of the visit or adverse weather conditions. Birdlists become far more useful if they indicate the rate of discovery of 'new' species, which is an index of overall species richness in that locality. This is done by keeping notes on both the abundance of species found and the intensity of searching effort upon which these are based.

Ideally, one should record every single bird seen and heard, but in practice this is generally an impossible task. I recommend the following method, however, which has been found to be very useful in China and other parts of South-east Asia.

Make a list by recording each new species seen until you reach 20 species; then start again with a new list. Any one species will only be recorded once in your first list of 20, but may be recorded again in subsequent lists. When you have collected ten or more lists for a given area you will have a very good picture of its avifauna. If the accumulating total number of species recorded is plotted graphically against the number of lists made this gives a species discovery curve whose steepness reflects species richness and indicates how many more species are likely still to be found in that locality.

Species which occur on a high proportion of lists are clearly the most abundant or conspicuous species of the local avifauna. The graphs below show examples of field data of this sort for four sites in Xishuangbanna in south Yunnan. Small isolated forest blocks show impoverished bird faunae compared to larger forest blocks.

The great advantage of this method is that it is relatively independent of observer experience and expertise, and also independent of birdwatching intensity, weather conditions or other factors.

The only rules to keep in mind are that:

(a) All species seen should be included in the lists even if they are not recognised i.e. give your own name or code to unknown species, so that you can at least recognise them as the same species when they are seen again.

(b) The results reflect where the data were collected and will show any bias in the survey pattern. To accurately reflect the bird fauna of the whole area of a reserve or park, all habitat types should be surveyed in similar proportions to their abundance. Do not survey the same trail too many times.

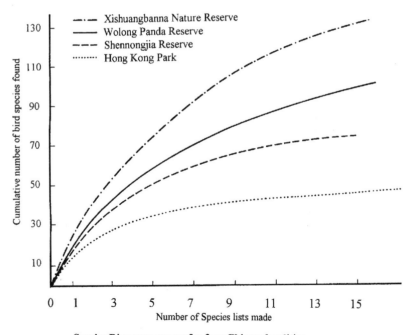

Species Discovery curves for four Chinese localities.

5. Submitting records

If you get into the habit of making good notes, you will be well placed to make a good report when something unusual turns up and your record will have a much greater chance of being accepted by the relevant records screening body. A good description should include:

- Name of observer and witness companions
- Name of species recorded
- Date of observation
- Time of day
- Location (include coordinates of remote areas)
- Habitat (include estimated elevation)
- Weather and lighting conditions
- Distance and observation conditions (include details of optics used)
- Description of bird (size, shape, plumage and bare parts)
- Behavioural notes (flight, gait, posture, feeding and associated species)
- Additional supporting information (the prior familiarity of observer with this species, unusual weather conditions that may explain the unexpected observation etc.)

Appendix 5 lists the most suitable societies, journals etc. to which records should be submitted.

6. Recording bird sounds

The calls of most birds are as distinctive as their appearance. Indeed in some species, such as the babblers, the call pattern may be the only really diagnostic field character. A bird-watcher wandering through a forest will hear far more species of birds than he or she will see. The keen ornithologist is therefore neglecting a great deal of information if he does not learn to recognise the calls of different birds.

To describe bird calls and record their form and structure is not easy. A standard method in many bird books, including this one, is to use human syllable nemonics which approximate the form of the bird call. Thus sharp clicks are represented by consonants: t, ts, ch, j, sh, soft notes by b, l, m, k, the duration of notes by vowels duplicated where necessary to indicate prolonged duration; r, s, z, f, are used to simulate different forms of harsh, reverberating notes. This method has many drawbacks, however, as different people make quite different interpretations of the same birdsong and phrases, also letters that are meaningful or have one sound in one language, such as Chinese, may be unrecognisable in another, such as English.

More exact records of birdsong can of course be made with a tape recorder. This used to involve taking a lot of special equipment into the field, but electronic equipment has improved so fast that very small, cheap recorders are now adequate. Most mini-cassette recorders can be easily adapted for field use but a directional microphone is necessary to amplify the sound of the bird and mask out other irrelevant noises. Two types of directional microphones are available—amplifying unidirectional microphones and parabolic reflectors, where an ordinary microphone is set facing backwards at the focal point of a metal or fibreglass parabola. The parabola gives better quality results but is larger and more clumsy to carry. Another improvement is to record in stereo, using two microphones set well apart or facing in divergent directions. Stereo recordings help to separate the subject sound spatially from any irrelevant sounds recorded at the same time. The secret of all methods, however, is to get as close as possible to the subject. Share your recordings by depositing copies in such bird-sound libraries as the Wildlife Section of the British Library National Sound Archive or the history of National Sounds, Cornell Laboratory of Ornithology, and Bioacous Archives and Library, Florida State Museum University of Florida.

Taped sound has the disadvantage that to listen to it one needs special equipment and it cannot be examined visually. This can be resolved with a sonogram or melogram machine, however, which can produce visual sonographs displaying the intensity and pitch of sound against a time scale. Such sonograms are very useful in the scientific analysis and comparison of bird sounds. Few people can look at a sonogram and imagine what the sound is like.

Anatomy and plumage of a bird

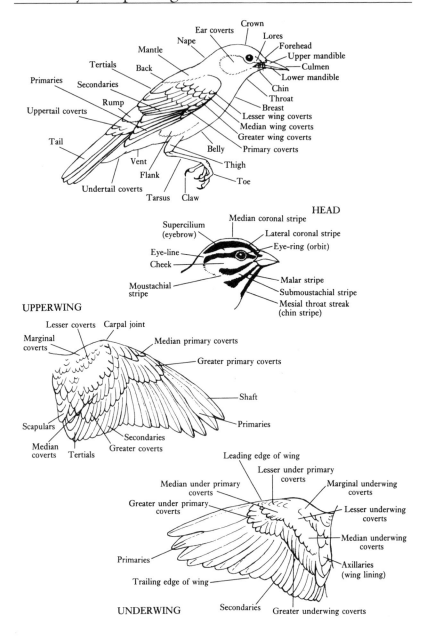

Crown
Ear coverts
Nape
Mantle
Tertials
Back
Primaries
Secondaries
Rump
Uppertail coverts
Tail
Vent
Flank
Undertail coverts
Tarsus
Claw
Lores
Forehead
Upper mandible
Culmen
Lower mandible
Chin
Throat
Breast
Lesser wing coverts
Median wing coverts
Greater wing coverts
Belly
Primary coverts
Thigh
Toe

HEAD
Median coronal stripe
Supercilium (eyebrow)
Lateral coronal stripe
Eye-ring (orbit)
Eye-line
Cheek
Moustachial stripe
Malar stripe
Submoustachial stripe
Mesial throat streak (chin stripe)

UPPERWING
Lesser coverts Carpal joint
Marginal coverts
Median primary coverts
Greater primary coverts
Shaft
Scapulars
Primaries
Median coverts
Tertials Greater coverts
Secondaries

Leading edge of wing
Lesser under primary coverts
Median under primary coverts
Marginal underwing coverts
Greater under primary coverts
Lesser underwing coverts
Median underwing coverts
Primaries
Axillaries (wing lining)
Trailing edge of wing
UNDERWING
Secondaries Greater underwing coverts

Glossary

accidental: a stray or vagrant.

adult: birds sufficiently mature to breed.

apical: terminal, outer end of something.

aquatic: water-living.

arboreal: tree-living.

axillaries: the feathers in the axil or wing lining.

basal: refers to the base.

cap: usually used to denote a larger area than the crown.

casque: an enlargement of the upper part of the bill.

cere: bare, wax-like or fleshy structure at the base of the upper beak containing the nostrils.

cock: male galliform bird. Used as verb to describe jerky raising of head or tail.

collar: a band or bar of contrasting colour passing around either the front or back of the neck.

coronal stripe: a streak on the crown running front to back.

cosmopolitan: a species which is widely distributed, having a near worldwide distribution.

crepuscular: active in twilight and just before dawn.

crest: tuft of elongate feathers usually on the head which, in some species, may be raised or lowered.

crown: top of a bird's head or top branches and foliage of a tree.

cryptic: having protective colouring or camouflage with associated behaviour.

deciduous: of a tree (or forest) which is leafless for part of the year.

diagnostic: character of sufficient distinctiveness to allow identification to be made.

dimorphic: existing in two well-defined, genetically determined plumage types.

diurnal: active in daytime.

duetting: male or female of a pair singing together, in response to each other.

echolocation: emission of high-frequency sounds to locate objects.

eclipse plumage: post-breeding plumage in which distinctive breeding features are obscured; found in ducks, sunbirds, etc.

endemic: indigenous: restricted to a particular area.

face: the lores, orbital area, cheeks, and malar area combined.

feral: domesticated species released and living wild.

ferruginous: rusty-brown colour with orange tinge.

fledgling: a young bird partially or wholly feathered, flightless or partially flighted, but before flight.

flight feathers: the primaries, secondaries and tail feathers which give buoyancy in flight.

flush: frighten out of concealing cover.

foot: claws, toes and tarsus combined.

foreneck: the lower part of the throat.

frontal shield: bare, horny or fleshy skin on forehead, which extends down to the base of the upper bill.

fulvous: brownish-yellow.

gait: manner of walking.

gape: the fleshy interior of the bill.

gliding: straight, level flight with wings outstretched or slightly swept back, without flapping.

gorget: a necklace or distinctively coloured patch on the throat or upper breast.

grace note: soft introductory note given just before main song.

gregarious: living in groups.

gular pouch: an expandable patch of bare skin on the throat of pelicans, cormorants, etc.

hackles: long, slender feathers found on the neck of some birds.

head: forehead, crown, nape and sides of head combined; does not include chin and throat.

hepatic: usually used as a label for the brown colour morph found in some cuckoos.

Holarctic: Palearctic and Nearctic regions combined.

hood: a dark-coloured head and (usually) throat.

immature: used to denote all plumage phases except the adult plumage, i.e. includes juvenile and subadult.

juvenile: fledging to free-flying birds, with the feathers which first replaced the natal down.

leading edge: the forward edge of the wing.

lobe: a fleshy, rounded protuberance (usually on the feet, as an aid to swimming).

local: discontinuous or uneven in distribution; found in only certain areas.

malar area: the area bounded by the base of the bill, the throat and the orbit.

mantle: the back, upperwing coverts and scapulars combined.

median: pertaining to the middle.

melanistic: a blackish morph.

mesial: median; running down the middle, usually of the throat.

migratory: relating to regular geographical movement.

morph: a distinct, genetically determined colour form.

nocturnal: active at night.

non-passerine: refers to those species having a different foot structure from the passerines.

notch: an indentation in outline of feather, wing, tail, etc.

occiput: the rear part of the crown.

ochraceous: dark brownish-yellow.

Old World: the Palearctic, Afrotropical, and Oriental zoogeographical regions.

Oriental: referring to the zoogeographical region, from the Himalayas, India, South-east Asia east to Wallacea.

Palearctic: North Africa, Greenland and Eurasia.

pan-tropical: distributed around the world's tropics.

passerine: refers to the families of 'perching birds'. They differ in foot structure from the non-passerines with three toes always porating forward and one always back.

pelagic: ocean-living.

pied: patterned in black and white.

piratic: stealing food from other birds or species.

'pishing': making a harsh squeaking noise in imitation of bird alarm calls to attract attention of birds.

plume: a greatly elongated feather, usually used in display.

primary forest: original or virgin forest.

proximal: the near, or basal portion.

race: a colloquial term for subspecies.

raptor: a bird of prey.

rictal bristles: bare feather shafts arising around the base of the bill.

roost: a resting or sleeping place/perch for birds.

secondary forest: new forest growing where the original or primary forest has been removed.

shaft streak: a contrasting line of pigment along the midline of a feather which forms a pale or dark streak.

shank: bare part of leg.

skin: a study skin or specimen.

skulk: action of creeping or flitting about in unobtrusive and furtive manner close to the ground.

soaring: rising flight, riding updraughts to gain height without flapping.

song: call used in courtship and territorial context.

spangles: shimmering points in a bird's plumage.

speculum: iridescent dorsal patch on a duck's wing which contrasts with the rest of the wing.

storey: a level of the forest.

streamers: greatly elongated, ribbon-like tail feather or projections of the tail feathers.

subterminal: near the end.

subadult: a later immature plumage phase.

submontane: lower elevations and foothills of mountains.

subspecies: a population which is morphologically distinguishable from other populations of the same species.

terminal: at the end.

terrestrial: ground-living,

trailing edge: the hind edge of the wing.

undergrowth: herbs, saplings and bushes in a forest.

underparts: undersurface of body from throat to undertail coverts.

underwing: the entire undersurface of the wing, including coverts and flight feathers.

upperparts: upper surface of the body.

vagrant: rare and irregular in occurrence.

vent: the area surrounding the cloaca, sometimes used as shorthand for undertail coverts.

washed: suffused with a particular colour.

wattle: a patch of bare skin, often brightly coloured, usually hanging from some part of the head or neck.

web: a fold of skin stretched between two toes; or a vane of a feather.

wing bars: bands formed on the wings when the tips of wing coverts are a different colour from their bases.

wing coverts: a term used for all or part of the lesser, median and greater coverts of the upperwing or underwing.

wing lining: a term loosely applied to the underwing coverts.

FAMILY AND SPECIES DESCRIPTIONS

PHEASANTS AND ALLIES—Family: Phasianidae

This is a worldwide family of ground-living birds with short round wings but often long tails. Males are usually very decorative whilst females are drably camouflaged. They nest on the ground but roost in trees. Some species have loud clear calls. Many species have wing-whirring or shaking displays. Most species have fighting spurs on the legs of males. Flight is flurried and usually only for short distances but the birds can run well. China is a major radiation centre of the pheasant family and boasts 62 species. None are migratory.

1. Snow Partridge *Lerwa lerwa* Xuechun   PLATE 1

Description: Medium-sized (35 cm) greyish partridge with finely barred black and white upperparts, head, neck and tail and rufous brown wash on back and wings. White breast with broad lanceolate chestnut streaks diagnostic. Races not distinguishable in field. Iris—reddish-brown; bill—crimson; feet—orange-red.

Voice: Breeding call similar to call of Grey Partridge. Alarm call is low whistle, becoming shriller with increasing danger.

Range: Himalayas, Tibetan Plateau to C China.

Distribution and status: Common from 2900–5000 m on alpine pastures and screes above the treeline. Nominate race in S Xizang; *major* in Sichuan and *callipygia* in S Gansu and N Sichuan.

Habits: Lives in small to large coveys. When flushed, rise with great clapping of wings and scattering in different directions. Shares same habitat as Tibetan Snowcock but has smaller home ranges. Can be quite tame.

2. Tibetan Snowcock *Tetraogallus tibetanus* Zang xueji   PLATE 1

Description: Medium-sized (53 cm) grey, white and buff snowcock. Head, breast and nape grey with white throat, whitish forebrow, white ear coverts sometimes washed buff and circular white patch on each side of breast. Bare skin around eye orange. Wings streaked grey and white, tail grey with rufous edges. Underparts whitish streaked with black. Races vary from darker *tibetanus* in west with black spotting on throat to paler and less buffy races *przevalskii* and *henrici* in east. Iris—dark brown; bill—yellow; feet—red.

Voice: Breeding birds call day and night with repeated croaking *gu-gu-gu-gu*.

Range: Himalayas from Pamirs to Mishmi Hills and entire Tibetan Plateau.

Distribution and status: Uncommon resident of rocky alpine meadows and stony screes up to 4500 m in summer, some down to 2500 m in winter. Race *tibetanus* in SW Xinjiang and C and W Xizang; *tschimenensis* in Altun Mts, E Kunlun; *aquilonifer* in S Xizang; *przevalskii* from N Xizang to Qinghai, S and W Gansu and N Sichuan; *henrici* in W Sichuan and *yunnanensis* in NW Yunnan.

Habits: As Himalayan Snowcock. Shares habitat with Snow Partridge but has larger home range.

3. Altai Snowcock *Tetraogallus altaicus* A'ertai xueji PLATE 1

Description: Large (60 cm) greyish-brown snowcock. Crown, nape and sides of upper breast uniform grey; sides of head and throat buffish, streaked grey; bare skin around eye yellow. Back, wings and tail brown with white streaking on wing coverts. Lower breast and abdomen dirty white. Iris—dark brown; bill—horn; feet—yellow.

Voice: Harsh, half whistled *geuk-geuk-geuk* shortening to *guk-guk-guk* and ending with rolling *rrruuuuu*.

Range: Altai Mts. to montane NW Mongolia.

Distribution and status: Race *altaicus* and maybe also *orientalis*, rare near snowfields of Altai Mts in NW Xinjiang.

Habits: As Himalayan Snowcock.

4. Himalayan Snowcock *Tetraogallus himalayensis* Anfu xueji PLATE 1

Description: Large (60 cm) buffy-grey snowcock with bold white and deep chestnut pattern on head and neck. White crescent on side of head and neck separates two chestnut lines which join to form a band across upper breast. Upperparts mostly grey, boldly streaked buff. Upper breast otherwise grey spotted black; lower breast grey streaked black and chestnut with white-edged plume-like feathers at sides. Undertail coverts whitish. Bare area around eye yellow. Eastern races are darkest. Iris—dark brown; bill—grey; feet—red or orange.

Voice: High-pitched *shi-er, shi-er* and deeper *wai-wain-guar-guar* call. Softer contact call *ger-u, ger-u*.

Range: Afghanistan, Turkestan to Nepal and NW China.

Distribution and status: Uncommon on rocky alpine meadows and screes near snowline of NW China from 2500–5500 m. Dark race *himalayensis* in Kashi and Tianshan regions of NW Xinjiang; similar *grambszewskii* in W Kunlun Mts; paler race *koslowi* in E Kunlun, Altai Mts, Qaidam basin, most of Qinghai and marginally into the Wuwei region of Gansu.

Habits: Lives in small groups on open ground at high altitude.

5. Chestnut-throated Partridge *Tetraophasis obscurus* PLATE 1
[Verreaux's Monal Partridge] Zhichun

Description: Large (48 cm) greyish-brown partridge with grey breast streaked black. Distinguished from similar Buff-throated Partridge by chestnut throat patch outlined whitish. Has scarlet orbital skin. Iris—brown; bill—grey; feet—pink.

Voice: Harsh, loud group calls are started by one bird and audible far across valleys. Calls often heard before bad weather.

Range: C China.

Distribution and status: Globally Near-threatened (Collar *et al.* 1994). Locally not uncommon. Found in rocky mountains from treeline up to 4600 m in Min and Qionglai Mts of Sichuan and Qilian Mts along Qinghai–Gansu border.

Habits: Lives in small parties on alpine meadows and screes close to treeline and in rhododendron scrub.

6. Buff-throated Partridge *Tetraophasis szechenyii* [Szechenyi's Monal Partridge] Sichuan zhichun   PLATE 1

Description: Large (48 cm) partridge. Distinguished from very similar Chestnut-throated Partridge by buff throat patch with no white border. Orbital skin scarlet. Iris—brown; bill—blackish; feet—pink.

Voice: Loud harsh series of notes, audible over long range.

Range: E Tibetan Plateau to C China.

Distribution and status: Globally Near-threatened (Collar *et al.* 1994). Locally not uncommon in SE and E Xizang, S Qinghai, SW Sichuan and NW Yunnan from 3350–4600 m.

Habits: Lives in small parties in rocky ravines with extensive cover of fir forest and rhododendron scrub and nearby alpine meadows. Freezes or flies downhill to cover when alarmed.

7. Chukar *Alectoris chukar* Shiji   PLATE 1

Description: Medium-sized (38 cm) boldly marked partridge with white throat, and lower face that is bordered by black band which runs through eye and across lower throat and contrasts with bright red bill and fleshy eye-ring. Upperparts pinkish grey; breast orange-buff; flanks barred with black, chestnut and white stripes. Several races differ in minor details but desert birds are palest. Iris—brown; bill—red; feet—red.

Voice: Male gives accelerating series *ka-ka-kaka-kaka-kaka..* followed by nasal chuckles *chukara-chukara-chukar*.

Range: S Europe, Asia Minor, Himalayas, C Asia to Mongolia and N China. Introduced United Kingdom and elsewhere.

Distribution and status: Widespread in northern China and locally common.

Habits: Lives in pairs or coveys in open mountainous areas, high plateaux, steppes and dry grassland.

8. Rusty-necklaced Partridge *Alectoris magna*   PLATE 1
[Przevalski's Partridge] Da shiji

Description: Medium-sized (38 cm). Very similar to Chukar but slightly larger and more yellow in hue. White patch on lower face, chin and throat is outlined with black line, like Chukar, and with a diagnostic chestnut line. Undertail coverts yellower. Orbital skin crimson. Iris—yellowish-brown; bill—red; feet—red.

Voice: Long series of loud hard *kak* notes, speeding up and getting louder then tailing off. Often some double notes, *kabble*, in the middle.

Range: NE Tibetan Plateau to N China.

Distribution and status: Uncommon in hills and mountains of E Qinghai into Gansu in Qilian Mts from 1800–3500 m.

Habits: Inhabits barren plateaux and rocky mountains and gorges. Lives in small parties.

Note: Treated by some authors as a race of Chukar.

9. Chinese Francolin *Francolinus pintadeanus* [Zhonghua] Zhegu   PLATE 1

Description: Male: medium-sized (30 cm) unmistakable black partridge decorated with bold white spots on nape, mantle, underparts and wings, and white bars on back and tail. Head is black with chestnut supercilium, broad white band beneath eye to ear coverts and white chin and throat. Female similar to male but underparts buff, barred black and more rufous on upperparts. Iris—reddish-brown; bill—blackish; feet—yellow.

Voice: Characteristic resonant, harsh, grating call variously transliterated as '*do-be-quick-papa*' or '*come to the peak ha-ha*'. Several birds may call simultaneously at dusk and dawn.

Range: NE India to S China and SE Asia. Introduced to Philippines.

Distribution and status: Common resident in W and S Yunnan, SW Guizhou, Guangxi, Hainan, Guangdong, Fujian, Jiangxi, Zhejiang and Anhui.

Habits: Inhabits dry forest, grass and secondary scrub from lowlands up to 1600 m.

10. Grey Partridge *Perdix perdix* Hui shanchun   PLATE 1

Description: Medium-sized (30 cm) greyish-brown partridge with yellowish-orange supercilium, face and throat. Underparts grey becoming white ventrally. Male has conspicuous chestnut, inverted U-shaped patch on lower breast. Flanks are barred broadly chestnut. Distinguished from Daurian Partridge by having chestnut rather than black breast patch and lacking plumose throat feathers. Rusty-red tail feathers obvious in flight. Iris—brown; bill—yellowish; feet—yellow.

Voice: Male call is creaky *ki-errr-ik, ki-errr-ick* with emphasis on *errr*. On being flushed gives low *grrree-grrree* call.

Range: Eurasia; introduced N America.

Distribution and status: Uncommon resident in Junggar Basin and Altai foothills in NW Xinjiang.

Habits: Pairs raise chicks in tight family coveys. All rise together when flushed. Favours open country with short grass, especially farmland.

11. Daurian Partridge *Perdix dauurica* Banchi shanchun   PLATE 1

Description: Male: smallish (28 cm) greyish-brown partridge with orange face, centre of throat and belly, and a large black inverted U-shaped patch on centre of belly. Distinguished from Grey Partridge by black rather than chestnut breast patch, orange continuing from throat to belly and having plumose whiskers on throat. Female lacks orange and black on breast but also has 'whiskers'. Iris—brown; bill—yellowish; feet—yellow.

Voice: Creaking calls typical of genus.

Range: C Asia to Siberia, Mongolia and N China.

Distribution and status: Fairly common. Nominate race in NW Xinjiang; *przewalskii* in Qinghai and Gansu and *suschkini* in Nei Mongol, Shaanxi, Ningxia, Shanxi, Hebei and NE China.

Habits: As Grey Partridge.

12. Tibetan Partridge *Perdix hodgsoniae* Gaoyuan shanchun   PLATE 1

Description: Smallish (28 cm) greyish-brown partridge with prominent white eyebrow, diagnostic chestnut nuchal collar and black spot on side of face beneath eye. Upperparts heavily banded black. Outer tail feathers rufous. Underparts buffy white with broad black scaling on breast and black scaling on sides. Iris—red-brown; bill—horn-green; feet—pale greenish-brown.

Voice: Rattling creaking call *scherrrrrreck-scherrrrrreck.* Gives shrill *chee, chee, chee, chee, chee* when flushed.

Range: Himalayas and Tibetan Plateau.

Distribution and status: Fairly common resident on rocky slopes with scattered scrub from 2700–5200 m. Race *caraganae* in W and S Xizang; nominate race in SE Xizang; *sifanica* in E Xizang, NW Sichuan, S Qinghai and Gansu; and *koslowi* in N Qinghai.

Habits: Lives in coveys of 10–15 birds. Reluctant to fly, preferring to run to safety but scattering downhill in twos and threes if flushed.

13. Common Quail *Coturnix coturnix* Anchun PLATE 32

Description: Small (18 cm) plump, squat, brown quail with conspicuous buff, spear-shaped streaks and irregular mottling and barring of reddish-brown and black on upperparts of both sexes. Male has dark brown chin and line down centre of throat dividing to curve up to ear coverts with buff gorget immediately below this. Buff supercilium contrasts with brown crown and eye-stripe. Female has similar patterning but with much less contrast. Iris—reddish-brown; bill—grey; feet—fleshy-brown.

Voice: Loud, clear trisyllabic whistle like dripping water and often rendered as '*wet my lips*', uttered mostly at dawn, dusk or at night. Squeaky whistle when flushed.

Range: Europe to W Asia and India, Africa, Madagascar and NE Asia.

Distribution and status: Uncommon. Breeds in Kashi, Tianshan and Lop Nur areas of Xinjiang; on passage in S and SE Xizang.

Habits: Generally in pairs rather than coveys. Favours grain crops in agricultural areas or grasslands.

Note: Some authors place Japanese Quail within this species (e.g. Cheng 1987, 1996).

14. Japanese Quail *Coturnix japonica* Riben anchun PLATE 32

Description: Small (20 cm) plump, squat, greyish-brown quail. Upperparts barred brown and black with long, spear-shaped buff streaks. Underparts buff with black streaks on breast and flanks. Has striped head with long whitish supercilium. In summer, male has chestnut on face, throat and upper breast. Two dark brown bands, on side of neck, separate this species from buttonquails. Not readily separated from Common Quail in winter. Iris – reddish—brown; bill—grey; feet—fleshy-brown.

Voice: Distinctive whistle *gwa kuro* or *guku kr-r-r-r.*

Range: E Asia, NE India, China, Taiwan, SE Asia, Philippines. Introduced Hawaii.

Distribution and status: Locally common. Breeds in NE provinces, Hebei, Shandong and Heshui region of E Gansu, probably SW and S China. Winters over most of C, SW, E and SE China, Taiwan and Hainan.

Habits: As Common Quail. Lives in short grasslands and farmland.

Note: Treated by some authors as a race of Common Quail (e.g. Cheng 1987, 1994; Wild Bird Society of Japan 1982).

15. Blue-breasted Quail *Coturnix chinensis* Lanxiong chun PLATE 32

Description: Male: very small (14 cm) unmistakable quail with bold black and white pattern on throat. Breast, rump, forecrown and eye-stripe blue-grey. Upperparts otherwise rich olive-brown, barred black and streaked white. Rump and centre of abdomen dark chestnut. Female is reddish-brown above mottled with black and streaked white; abdomen is buff finely barred with black. Female is easily confused with buttonquail. Iris—reddish; bill—black; feet—yellow.

Voice: Call is sweet double whistle *ti-yu, ti-yu.*

Range: India, Nicobars, S China, Hainan, Taiwan, SE Asia to Australia.

Distribution and status: Uncommon in lowland grassy areas, scrub and rice stubble in S and E China.

Habits: Typical quail, living in small coveys.

16. Hill Partridge *Arborophila torqueola* [Common Hill Partridge] PLATE 2
Huanjing shanzhegu

Description: Male: medium-sized (29 cm) olive-brown partridge with chestnut cap and nape. Has rufous-buff ear coverts; black lores and eyebrow, topped with white line; white malar stripe and white band between foreneck and breast. Race *batemani* has chestnut sides of neck streaked black. Female has brown breast and chestnut chin and throat with olive-brown crown spotted white. Iris—brown to crimson; bill—brown to black; feet—brown to grey.

Voice: Full call is mournful whistled *poor* or *pheaw* repeated several times followed by 3–6 double whistles *do-eat, do-eat* ... rising in pitch like a hawk cuckoo. In duet, presumed female gives *kwik kwik kwik kwik kwik* call and male responds with series of *do-eat* calls suddenly ending after climax.

Range: Himalayas to W China, Burma and NW Tonkin (Vietnam).

Distribution and status: Local resident in forests from 1800–3000 m. Nominate race in S Xizang and *batemani* in W Yunnan to west of Nujiang (Salween) River.

Habits: Lives in small parties scouring forest floor, scrabbling among leaf litter for food. Scurry away quietly on ground when disturbed.

17. Rufous-throated Partridge *Arborophila rufogularis* PLATE 2
Honghou shanzhegu

Description: Medium-sized (27 cm) greyish partridge with orange-rufous throat and foreneck with large black spots. Underparts grey with bold silver and rufous streaking on flanks. Broad black and buff bars across folded rufous wing. Race *intermedia* has black chin and throat. Iris—brown; bill—grey; feet—pink.

Voice: Loud, clear monotone whistle leading to series of disyllabic whistles *hu-hu, hu-hu* ... rising gradually in pitch. Partner calls in duet with rapid *kew-kew-kew* ...

Range: N India to SW China and SE Asia.

Distribution and status: Rare resident in China. Nominate race in SE Xizang; *euroa* in S and SE Yunnan and *intermedia* in Yingjiang area of W Yunnan to west of Nujiang (Salween) River.

Habits: Lives in broadleaf evergreen forests from 1200–2500 m.

18. White-cheeked Partridge *Arborophila atrogularis*   PLATE 2
Baijia shanzhegu

Description: Medium-sized (28 cm) olive-brown partridge with black face and conspicuous white cheeks. Similar to nominate Rufous-throated but narrower black bars on upperparts, different face pattern, lack of rufous throat, streaked upper breast and lack of rufous on flanks. Iris—reddish-brown; bill—black (male) or brown (female); feet—orange-red.

Voice: Rolling whistle *whew, whew* repeated many times ending with sharper, quicker *whew*; also loud clear double whistle. Soft mellow contact whistles when covey is scattered.

Range: NE India and Burma.

Distribution and status: Local in China. Resident in Yingjiang area of W Yunnan to west of Nujiang (Salween) River.

Habits: Typical of genus. Lives in small coveys in scrub, close to tall forests up to 1300 m.

19. Taiwan Partridge *Arborophila crudigularis* [Formosan Partridge]   PLATE 2
Taiwan shanzhegu

Description: Smallish (24 cm) grey partridge with conspicuous black and white face markings. Crown grey; sides of head black with long white eyebrow, white chin and throat extending into bold white patch below eye. Broad black half collar underlined by whitish and buff. Bare orbital skin red; back and tail olive barred with black; wing rufous with three grey bars; breast blue-grey, streaked with white on flanks; abdomen whitish. Iris—brown; bill—blackish-grey; feet—red.

Voice: Repeated *guru, guru* . . . rising in pitch then falling back suddenly after climax.

Range: Endemic to Taiwan.

Distribution and status: Globally Near-threatened (Collar *et al.* 1994). Confined to central mountains and eastern slopes where it is not uncommon in broadleaf forests between 700 and 2300 m.

Habits: Typical of genus.

20. Chestnut-breasted Partridge *Arborophila mandellii*   PLATE 2
Hongxiong shanzhegu

Description: Medium-sized (28 cm) greyish partridge with orange-brown head and diagnostic broad chestnut band around upper breast and nape. Has long narrow grey supercilium, white moustachial and gorget of white above black bands across throat. Lower breast and flanks grey with bold white and rufous streaks on flanks. Iris—brown to red-brown; bill—black; feet—reddish.

Voice: Loud, rich, long-drawn *quick* followed by a series of ascending double notes leading to climax. Similar to Rufous-throated Partridge.

Range: E Himalayas.

Distribution and status: Globally Vulnerable (Collar *et al.* 1994). Very rare in China. Resident in upper reaches of Danba River in SE Xizang; also Mishmi Hills area of India claimed by China.

Habits: Typical of genus.

21. Bar-backed Partridge *Arborophila brunneopectus*   PLATE 2
[Brown-breasted/Bare-throated Partridge] Hexiong shanzhegu

Description: Medium-sized (28 cm) olive-brown partridge with bold creamy supercilium extending onto neck; black eye-stripe and creamy white throat and cheeks. Has band of black speckles between throat and breast, linking with eye-stripe. Flanks with large bold black and white scales. Wings with bold striped pattern. Iris—reddish-brown; bill—blackish; feet—pink.

Voice: Throaty monotone, buzzing whistles leading into see-sawing whistled couplets, *ti-hu, ti-hu, ti-hu* . . . of steady pitch and volume. Partner duets by adding *kew, kew, kew* . . . call as in Rufous-throated Partridge.

Range: SW China, SE Asia.

Distribution and status: Nominate race is local resident in SW and S Yunnan and Guangxi.

Habits: Lives in evergreen forests from 500–1300 m. Habits typical of genus.

22. Sichuan Partridge *Arborophila rufipectus* [Sichuan Hill Partridge]   PLATE 2
Sichuan shanzhegu

Description: Medium-sized (30 cm) colourful partridge with brown cap and white supercilium combined with broad chestnut band across breast and whitish throat diagnostic. Has red orbital skin and yellowish-rufous ear coverts. Iris—dark brown; bill—grey; feet—pinkish.

Voice: Breeding males make calls in mornings and evenings or occasionally at noon. Call is series of loud ascending whistles, repeated over and over a few seconds apart.

Range: Endemic to C China.

Distribution and status: Globally Critical (Collar *et al.* 1994). Very rare in the wild due to loss of habitat. Confined to lowland subtropical broadleaf forests of S Sichuan (Ganluo, Pingshan and Mabian) from c. 1000–2200 m; also NE Yunnan (Dai Bo *et al.* 1998).

Habits: Typical of genus. Forages in pairs or small parties, scraping at leaf litter of forest floor in search of food.

23. White-necklaced Partridge *Arborophila gingica*   PLATE 2
Baimei shanzhegu

Description: Medium-sized (30 cm) greyish-brown partridge with red legs, white eyebrow and diffuse supercilium, yellow throat and diagnostic, elaborate gorget of black, white then chocolate bands. Iris—brown; bill—grey; feet—red.

Voice: Far-carrying, plaintive two-toned whistle.

Range: SE China endemic with restricted range.

Distribution and status: Globally Vulnerable (Collar *et al.* 1994). Rare resident in N and C Fujian, N Guangdong, Guangxi (Yaoshan). Threatened by habitat loss and hunting.

Habits: Typical of genus. Found in dense forest on low hills from 500–1900 m.

24. Hainan Partridge *Arborophila ardens* Hainan shanzhegu   PLATE 2

Description: Smallish (24 cm) colourful hill partridge with blackish head, white patch over ear coverts and diffuse salmon pink upper breast diagnostic. Upperparts greyish, scaled with black; belly yellow with grey wash on breast and white-streaked flanks. Iris—dark brown; bill—grey; feet—deep pink.

Voice: Breeding birds repeat double-noted calls such as *ju-gu, ju-gu, ju-gu*, falling on first note and rising on second.

Range: Endemic to Hainan.

Distribution and status: Globally Vulnerable (Collar *et al.* 1994). Rare. Found only in a few patches of remaining evergreen hill forest from 900–1200 m.

Habits: Typical of genus.

25. Scaly-breasted Partridge *Arborophila charltonii* PLATE 2
Lüjiao shanzhegu

Description: Medium-sized (29 cm) olive-brown partridge with dull green to pale green legs, whitish supercilium and throat but no black on head. Distinguished from Bar-backed Partridge by finer black barring, broad brown band across breast and lack of bold white marks on flanks. Has dark plumbeous orbital skin. Iris—reddish-brown; bill—yellow-horn; feet—green.

Voice: Loud whistled call consisting of a series of monotone notes repeated with increasing tempo; then a series of shallow descending and ascending phrases; and finally a series of wildly descending and ascending notes ending suddenly.

Range: SE Asia.

Distribution and status: Globally Vulnerable (Collar *et al.* 1994). Local in China. Resident in S Yunnan (Xishuangbanna).

Habits: Typical of genus. Lives in forest and secondary forests and bamboo up to 1000 m.

26. Mountain Bamboo Partridge *Bambusicola fytchii* PLATE 2
Zongxiong zhuji

Description: Medium-sized (34 cm) greyish-brown partridge with rufous outer feathers to long tail, buffy white face and throat with diagnostic, long black line extending only behind eye and onto neck, contrasting with white eyebrow. Upper breast chestnut, spotted or streaked white and grey; lower breast and vent buffy-white with large, black, heart-shaped spots. Sexes similar. Iris—reddish-brown; bill—blackish; feet—grey.

Voice: Male crows with chattering *che-chirree-che-chirree, chiree*. Female also has discordant squawking call. Screams when flushed.

Range: NE India to SW China and SE Asia.

Distribution and status: Nominate race is uncommon resident in Yunnan and S Sichuan in tall grass and bamboo thickets from lowlands up to 2100 m.

Habits: Lives in small coveys. Males in spring very noisy, flying up to a perch to call. Seldom far from streams. On flushing, flies only a few metres before tumbling back into long grass.

27. Chinese Bamboo Partridge *Bambusicola thoracica* PLATE 2
Huixiong zhuji

Description: Medium-sized (33 cm) reddish-brown partridge with diagnostic blue-grey forehead, supercilium and gorget contrasting with rufous face, throat and upper breast. Has distinctive large brown crescent markings on mantle, sides of breast and flanks. Race *sonorivox* has entire face, sides of neck and upper breast greyish-blue with only chin and throat chestnut. Outer tail feathers chestnut. Has two white bars on underwing in flight. Sexes similar. Iris—reddish-brown; bill—brown; feet—greenish-grey.

Voice: Piercing loud *people pray, people pray, people pray* call.

Range: S China endemic. Introduced Japan.

Distribution and status: Nominate race is quite common resident through most of C, S, E and SE China. Race *sonorivox* is resident on Taiwan.

Habits: Lives in family coveys. Flight is heavy and direct. Occurs in dry bush country and bamboo thickets up to 1100 m.

28. Blood Pheasant *Ithaginis cruentus* Xuezhi   PLATE 3

Description: Smallish (46 cm) partridge-like pheasant with long lanceolate feathers, floppy crest and scarlet skin on face, scarlet legs and red in wing and tail. Head blackish with whitish crest and long white streaks. Upperparts generally grey with white streaks and underparts tinged greenish. Varying amounts of red on breast. Female is duller and more uniform than male with buffy breast. Many races differ in plumage details. Nominate race has red streaks on breast and only outer pair of tail feathers lack red; *beicki* has no red on head and tertials are chestnut with green centres; *berezowskii* has no red on head but tertials all chestnut; *sinensis* is like *berezowskii* male but female has coarser markings; *affinis* has no red on outer two tail feathers and has black forehead; *tibetanus* has yellow ear coverts, red on breast but no black gorget; *kuseri* has black gorget and anterior ear coverts; *marionae* has black eyebrow and broken gorget; *rocki* has little red on breast but throat is red with white streaks; *clarkei* has very little red on throat or cheeks and *geoffroyi* has no red on head at all. Iris—yellow-brown; bill—blackish with red cere; feet—red.

Voice: Male makes short *si* call or cackles when alarmed. Sometimes several whistling *si* calls are rolled together *sisisi*. Both sexes give kite-like squeal, *chiu-chiu* to rally scattered flock.

Range: Himalayas, Tibetan Plateau and C China.

Distribution and status: Locally common from 3200–4700 m. Many races in China—*beicki* in Qilian Mts of Qinghai and Gansu; *sinensis* in SE Gansu (Baishuijiang), S Shaanxi (Qinling Mts) and SW Shanxi; *berezowskii* in S Gansu and N Sichuan; *geoffroyi* in E Xizang, S Sichaun and S Qinghai; *affinis* and *cruentus* in S Xizang; *tibetanus* in SE Xizang; *kusneri* in SE Xizang to Cigu area of NW Yunnan; *rocki* in NW Yunnan between Nujiang and Jinsha rivers; *marionae* in NW Yunnan between Lancang and Nujiang rivers and *clarkei* in NW Yunnan in Lijiang Mts.

Habits: Forms small to large flocks that forage over mossy forest floor among rhododendrons in subalpine conifer forests.

29. Western Tragopan *Tragopan melanocephalus* Heitou jiaozhi   PLATE 3

Description: Male: large (71 cm) brilliantly coloured, white-spotted, red and black tragopan. Distinguished from other tragopans by largely black plumage. Female also darker than other female tragopans with large white-centred black oval spots on underparts. Iris—brown; bill—blackish; feet—pink to grey with season.

Voice: Alarm call is wailing cries like a lamb, *waa, waa, waa*. In spring, male calls at intervals throughout day with a loud single *waa* like the bleat of a goat.

Range: W Himalayas.

Distribution and status: Globally Vulnerable (Collar *et al.* 1994). Rare in restricted range in Upper Indus valley (Gar Zanbo) in W Xizang. Breeds from 2400–3600 m, some down to 1350 m in winter.

Habits: As other tragopans. Male has vigorous display with horns and wattles expanded and one wing lowered and the other raised to face the female to full effect.

30. Satyr Tragopan *Tragopan satyra* Hongxiong jiaozhi   PLATE 3

Description: Male: large (70 cm) beautiful crimson tragopan with black head and throat. Crest is black tipped red. Plumage is mostly decorated with circular white or pearl dots outlined with black. Wings and tail bluish barred with buff. Erectile 'horns' blue and lappets blue, revealing green and red patches when fully extended by displaying male. Similar to more widespread Temminck's Tragopan but more crimson and with black-edged white spots rather than scaled pattern on red underparts. Female is duller and mottled black and rufous brown; orbital skin bluish. Iris—brown; bill—black; feet—flesh-coloured.

Voice: Loud, piercing *wak* repeated several times. Loud *kya, kya, kya, kya* like bleating of young goat.

Range: C Himalayas.

Distribution and status: Globally Near-threatened (Collar *et al.* 1994). Local in China with limited range in S and SE Xizang from 2300–4250 m but down to 1800 m in winter.

Habits: As other tragopans. Lives in rhododendron thickets. Male gives spectacular displays.

31. Blyth's Tragopan *Tragopan blythii* Huifu jiaozhi   PLATE 3

Description: Male: large (68 cm) tragopan with scarlet neck and supercilium contrasting with black head. Bare facial skin diagnostic yellow. Lappet and fleshy horns blue. Scaly grey breast and belly distinguishes from other tragopans. Male of race *molesworthi* has darker upperparts and red of underparts confined to narrow gorget; rest of underparts paler grey. Smaller female is brown and mottled, similar to female Satyr and Temminck's Tragopans but paler. Iris—brown; bill—brown; feet—pink.

Voice: Very loud, challenging *gnau, gnau* by male. Displaying male utters *gock ... gock ... gock* call rather like a Great Hornbill.

Range: E Himalayas.

Distribution and status: Globally Near-threatened (Collar *et al.* 1994). Rare in China with restricted range in SE Xizang (nominate race) and NW Yunnan (*molesworthi*) from 1800–4000 m.

Habits: As other tragopans. Lives in undergrowth of rhododendron forest in subalpine conifer forests.

32. Temminck's Tragopan *Tragopan temmminckii* PLATE 3
Hongfu jiaozhi

Description: Large (58 cm) short-tailed pheasant. Male, crimson with small round white spots outlined black over upperparts and grey-white oblong spots on underparts. Head black with golden stripe behind eye and blue facial skin and inflatable throat lappet and 'horns'. Distinguished from Satyr Tragopan by larger grey-white spots on underparts, not outlined black. Female is smaller and speckled brown with large white spots on underparts. Iris—brown; bill—black with pink tip; feet—pink to red.

Voice: Male gives noise like a baby crying, *wu, wa . . ., ga, ga* or *nyear-ni* to mark territory. Female in breeding season gives *wa, wa* calls. Alarm call is duck-like *quack-quack-quack.*

Range: E Himalayas to N Burma, NW Tonkin (Vietnam) and C China.

Distribution and status: Globally Near-threatened (Collar *et al.* 1994). Widespread and locally common from 2000–3500 m (down to 1000 m in winter) in SE Xizang, W and N Yunnan, Sichuan, S Gansu, S Shaanxi, Guizhou, Hubei, Hunan and N Guangxi.

Habits: Lives singly or in family parties on floor of subalpine forests. Not shy. Roosts in low branches. Male inflates throat lappet in display and erects bluish 'horn' wattles. When fully inflated the throat lappet is patterned blue and red.

33. Cabot's Tragopan *Tragopan caboti* Huangfu jiaozhi PLATE 3

Description: Large (61 cm) short-tailed pheasant. Male is rich brown above with large buff spots and straw-buff below. head black with scarlet front collar and patch on side of neck; golden stripe behind eye and orange bare facial skin, including inflatable throat lappet and 'horn' wattles. When inflated the throat lappet is brilliant blue and red. Female is smaller, peppery-grey with white arrow-shaped streaks, outlined black, on underparts. Iris—brown; bill—grey; feet—pink.

Voice: In early breeding season both sexes make regular harsh *ga-ra, ga-ra* or *ga-ga-ga* calls. Breeding birds make *wear-wear-ar-ga-ga-ga* calls. Displaying male makes softer *chi . . . chi* call.

Range: Endemic to SE China.

Distribution and status: Rare in evergreen subtropical hills from 800–1500 m. Nominate race in Fujian (Wuyishan), S Jiangxi, W and S Zhejiang (Wuyanling), and N Guangdong (Babaoshan) and *guangxiensis* in NE Guangxi (Gongcheng) and S Hunan.

Habits: Typical of genus.

34. Koklass Pheasant *Pucrasia macrolopha* Shaoji PLATE 3

Description: Large (61 cm) relatively short-tailed pheasant with distinctive long trailing 'ear-tufts'. Male: crown and crest greyish; throat, broad eye-stripe, nape and 'ear tufts' metallic green; side of neck white; mantle buff; breast chocolate and rest of plumage composed of long white feathers with black shaft streaks. Female: smaller with crest but no long 'ear tufts'; recognisably similar patterning to male. Races differ in details—*joretiana* has short crest and no yellow on breast; *ruficollis* has more rufous on upper breast; *darwini* has buffy underparts and *xanthospila* has more stripes on feathers of upperparts and female is greyer with reduced black markings. Iris—brown; bill—brownish; feet—purplish-grey.

Voice: Calls easily distinguished from other pheasants. Loud, melodious, harsh calls *khwa-kha-kaak* or *kok-kok-kok . . . ko-kras* can be heard at long range. The second last note is high-pitched with stress on the last note.

Range: Himalayas to C and E China.

Distribution and status: Rare and range becoming more restricted in both north and south. In west and north occurs as altitudinal migrant from 1200–4600 m but in east at only 600–1500 m. Several races occur in China—*xanthospila* in N Hebei, W Liaoning and N Shanxi; *ruficollis* in S Gansu, S Shaanxi, Ningxia, N and W Sichuan; *joretiana* in W Anhui; *darwini* in Hubei, E Sichuan, Guizhou, S Anhui, Zhejiang, N Fujian, Jiangxi and N Guangdong; *meyeri* in SE Xizang and NW Yunnan.

Habits: Usually alone or in pairs. Sits tight when alarmed and difficult to flush. Several cocks will explode into calling triggered by sudden loud noise such as gunshot or tree-fall. Male erects 'ear tufts' in display. Prefers open, rocky forests, usually with pines and rhododendrons.

35. Himalayan Monal *Lophophorus impejanus* Zongwei hongzhi PLATE 4

Description: Male: large (70 cm) shimmering purple and green pheasant with white upper back and black belly. Distinguished from Sclater's Monal by long green crest of peacock-like racquet feathers, lack of white tip to tail, purple lower back and green uppertail coverts. Distinguished from Chinese Monal by green crest and cinnamon tail. Smaller female distinguished from other female monals by colour of back being the same as the rest of upperparts; also white uppertail coverts. Iris—brown; bill—blackish-grey; feet—olive-brown.

Voice: Call and alarm is ringing whistle reminiscent of Curlew and generally given from a rock perch.

Range: Himalayas.

Distribution and status: Rare in limited areas of S and SE Xizang from 3000–4100 m.

Habits: Lives in small parties along treeline, feeding in alpine meadows. In courtship, male has elaborate display dance with fanned tail flicked up and down over back, wings raised and occasionally leaping into the air.

36. Sclater's Monal *Lophophorus sclateri* Baiweishao hongzhi PLATE 4

Description: Large (70 cm) shimmering purple-green pheasant with white back and uppertail coverts and black underparts. Distinguished from Himalayan Monal by lack of trailing crest, having white uppertail coverts and white tip to tail. Distinguished from Chinese Monal by lack of crest and tail colour. Smaller female distinguished from other female monals by having back and uppertail coverts pale buff contrasting with rest of brown upperparts. Iris—brown; bill—flesh-coloured; feet—brown.

Voice: Alarm call is shrill, plaintive cry. Call a wild ringing whistle, *go-li . . .*

Range: E Himalayas.

Distribution and status: Globally Vulnerable (Collar *et al.* 1994). Uncommon from 3000 to over 4000 m; descending in winter. Narrow range in SW China in SE Xizang (nominate) and NW Yunnan (*orientalis*).

Habits: As other monals, keeping to tight territories and rather reluctant to fly. Very noisy in evenings.

37. Chinese Monal *Lophophorus lhuysii* Lüwei hongzhi   PLATE 4

Description: Male: large (76 cm) shimmering purple pheasant with green head and golden nape. Underparts black with green metallic sheen. Has elongated trailing crest of bushy purple feathers. Distinguished from Sclater's Monal by having crest, tail colour and only having upper back white. Distinguished from Himalayan Monal by having purple fluffy crest and blue-green tail. Female distinguished from other female monals by having white back. Iris—brown; bill—grey; feet—dark horn (male) paler (female).

Voice: Calls mainly in spring and summer. Male, standing on a rock, makes repeated *guli ... * calls. Female sometimes makes same call. Male gives short *guo-guo-guo* call in display. Alarm call is low *geee*. Quiet in winter except dull *a ... awu, a ... awu*.

Range: Endemic to C China.

Distribution and status: Globally Vulnerable (Collar *et al.* 1994). Formerly common but now rather rare from 2900–4950 m in mountains of W Sichuan extending marginally to NW Yunnan, E Xizang, SE Qinghai and S Gansu.

Habits: Lives in upper conifer zone of sub-alpine forests and bushes in alpine zone above treeline. Found singly or in small groups scratching about for food in alpine meadows. Habits like Himalayan Monal.

38. Red Junglefowl *Gallus gallus* Yuanji   PLATE 3

Description: Familiar, largish (male 70 cm, female 42 cm) wild ancestor of domestic chicken. Male: comb, wattles and face red; hackles, tail coverts, and primaries bronze; mantle chestnut; elongated tail feathers and wing coverts glossy greenish-black. Female drab brown with black streaks on nape and neck. Iris—red; bill-horn; feet—bluish-grey.

Voice: Calls only in spring breeding season. Male call *ge ge-ge ge* with a clear pause in the middle and the last *ge* very short. Like truncated call of domestic cock.

Range: N, NE and E Indian subcontinent, S China, SE Asia, Sumatra and Java. Introduced elsewhere.

Distribution and status: Common in scrub and secondary forest in tropical evergreen belt of SW China (*spadiceus*) and S China and Hainan (*jabouillei*).

Habits: Male solitary or associating with female harem or other males. Feeds on ground but flies well and roosts in trees.

39. Kalij Pheasant *Lophura leucomelanos* Heixian   PLATE 4

Description: Large (70 cm) glossy blue-black pheasant with long trailing crest and red facial skin. Back and rump shiny black, scaled by white feather tips. Nominate race has white sides but *lathami* has black sides. Distinguished from Silver Pheasant by grey or brown legs, shorter tail and little white in plumage. Female is brown and speckled with white chin. Distinguished from female Silver Pheasant by dark brown outer tail feathers. Iris—brown; bill—greenish-horn; feet—grey or brown.

Voice: Low clicking *kurr-kurr-kurrchi-kurr* as contact call when group is dispersed. Male call is loud whistling chuckle with 'drumming' of wings on body. Agonistic call is threatening *koor koor* followed by sharper *waak, waak*. In alarm, a repeated *koorchi koorchi koorchi* call or guttural *whoop-keet-keet*.

Range: Himalayas, NE India, N and W Burma.

Distribution and status: Rather uncommon in China from 2100–3200 m in temperate forests. Nominate race in S and SE Xizang and *lathami* in W Yunnan to west of Nujiang (Salween) River.

Habits: Similar to Silver Pheasant with which it can interbreed. Lives in large parties.

40. Silver Pheasant *Lophura nycthemera* Baixian PLATE 4

Description: Male: large (94–110 cm) bluish-black pheasant with long white tail, white back, black crown and full long black crest. Central tail feathers pure white, other feathers of back and tail finely barred and patterned black. Underparts black. The bare facial skin bright red. Distinguished from Kalij by longer, whiter tail and red legs. Female: olive-brown to chestnut above and brown, streaked and mottled with white or buff, below. Also has dark crest and red facial skin. Distinguished from female Kalij by pink feet and black and white or pale chestnut, rather than dark brown, outer tail feathers. Several races differ in details of plumage especially in female. Nominate female has rather plain brown underparts; female *fokiensis* has brown underparts and outer tail feathers boldly vermiculated black and white; female *rufipes* has no black tip to crest; *whiteheadi* female has black upper back with white shaft streaks and black underparts spotted white; *occidentalis* female has outer tail feathers heavily vermiculated black and white; male *omeiensis* has bold black streaks on inner web of tail feathers. Iris—brown; bill—yellow; feet—bright red.

Voice: Generally quiet. Makes harsh *ji-go, ji-go* or shrill whistle when alarmed. Courting male gives light *lu,lu,lu,lu* . . . calls. Soft mewing sound when feeding.

Range: S China, Hainan and SE Asia.

Distribution and status: Common resident in evergreen forest, bamboo thickets and scrub at moderate altitudes over much of S and E China. Race *fokiensis* in Fujian and E Guangdong; nominate in Guangxi; *whiteheadi* on Hainan; *rongjiangensis* in Guizhou; *omeiensis* in S Sichuan; *beaulieui* in S and SE Yunnan; *occidentalis* in Yunnan west of Nujiang (Salween) River; *rufipes* in SW Yunnan to east of Nujiang River and *jonesi* in Xishuangbanna between Nujiang and Lancang (Mekong) rivers to south of *rufipes*. Some authors do not recognise all these races.

Habits: Lives in small groups. Feeds on figs under feeding monkeys and doves. Inhabits open forests and secondary evergreen forests up to 2150 m.

Note: Interbreeds with Kalij Pheasant in areas of overlap.

41. Swinhoe's Pheasant *Lophura swinhoii* Lanxian PLATE 4

Description: Male: large (72 cm) dark pheasant with short occipital crest, upper back and elongated central tail feathers silvery white. Mantle dark reddish-brown, scapulars maroon. Plumage otherwise black, glossed with iridescent blue-green scaling on upperparts and streaks on underparts. Fleshy face wattles scarlet. Female: smaller (55 cm); general plumage finely patterned greyish-brown with fine barring on wing. Wings and tail dark chestnut. Lacks crest and has reduced red facial wattles. Underparts are cinnamon with black barring. Iris—brown; bill—yellowish-horn; feet—red.

Voice: Deep harsh *ge ge*.

Range: Endemic to Taiwan.
Distribution and status: Globally Near-threatened (Collar *et al.* 1994). Rare in moist lowland forests from 800–2800 m. Population has been boosted by reintroduction.
Habits: Shy and alert. Often active at dusk and dawn. Male gives wing-whirring displays.

42. Tibetan Eared Pheasant *Crossoptilon harmani*   PLATE 4
[Elwes's Eared Pheasant] Haman maji

Description: Large (86 cm) greyish pheasant with white 'ears'. Similar to White Eared Pheasant but general colour darker and ear tufts longer. Throat, 'ears' and nape white, contrasting with black cap and grey body; wings blackish. Uppertail coverts paler grey; centre of breast whitish; arched filamentous tail blackish with bronze and purple gloss. Iris—pale orange; bill—horn-coloured; feet—red.
Voice: Loud, harsh distinctive calls like loud guineafowl in early morning. Also heron-like squawk.
Range: Endemic to SE Tibetan Plateau and marginally into adjacent NE India.
Distribution and status: Globally Vulnerable (Collar *et al.* 1994). Locally common. Recorded in SE Xizang in lower Yarlung Zanbo valley from 91 °E–95 ° 30'E.
Habits: As others of genus. Lives in small flocks in juniper and rhododendron scrub and alpine scrub and meadows from 2400–5000 m.
Note: Formerly treated as a race of White Eared Pheasant (e.g. Cheng 1987).

43. White Eared Pheasant *Crossoptilon crossoptilon* Zang maji   PLATE 4

Description: Large (80 cm) white pheasant with black floppy plumous tail. Flight feathers black and wings grey in some races. Crown is black and bare facial skin scarlet. Fluffy white moustachial does not extend into such long 'ears' as in other members of the genus. Iris—orange; bill—pale pink; feet—red.
Voice: Loud, harsh *ge ge ge ge* calls given from tree perch at dusk can be heard from far away.
Range: SE Tibetan Plateau to C China and possibly NE Burma.
Distribution and status: Quite common but declining over wide distribution. Lives in scrub at treeline from 3000–4000 m. Nominate race in SE Qinghai, Sichuan and E Xizang; *dolani* in S Qinghai (Yushan); *drouynii* from S Qinghai to NW Sichuan and SE Xizang between Jinsha and Nujiang; *lichiangense* in S Sichuan and NW Yunnan (Lijiang). Some intergrading in overlap zones.
Habits: Lives in small flocks, feeding in grassy clearings. Reluctant to fly and scurries into nearby bushes when disturbed.

44. Brown Eared Pheasant *Crossoptilon mantchuricum* He maji   PLATE 4

Description: Large (100 cm) brownish 'eared' pheasant. Very like Blue Eared Pheasant but general plumage greyish-brown rather than bluish-grey and with longer plumose tail. Iris—orange; bill—pink; feet—red.
Voice: Makes *gu-ji gu-ji* calls when foraging. Courting males make deep *gu-gu gu-gu* call.
Range: Endemic to N China.

Distribution and status: Globally Vulnerable (Collar *et al*. 1994). Very rare. Survives only in a few localities in Shanxi, Beijing Municipality and NW Hebei above 1300 m.

Habits: As other eared pheasants. Lives in stunted montane forest, feeding in scrub and grassy clearings.

45. Blue Eared Pheasant *Crossoptilon auritum* Lan maji PLATE 4

Description: Large (95 cm) bluish-grey pheasant with black velvety cap, scarlet bare orbital area and white moustachial stripe extending into long 'ear' tufts. There is a whitish bar behind nape. The tail is arched with filamentous grey central feathers contrasting with purplish-blue outer feathers. This species is the most north-westerly in distribution of the eared pheasants. Iris—orange; bill—pink; feet—red.

Voice: Alerted male gives *gela-gelage* call. Courting male calls *wu, wu, wu* or *ga, ga*. Both sexes make *ziwo-ge, ziwo-ge* call in alarm.

Range: Endemic to C China.

Distribution and status: Globally Near-threatened (Collar *et al*. 1994). Locally common in NE and E Qinghai, Gansu, Ningxia, NE Xizang (Tanggula Mts.) and N Sichuan from 2000–4000 m. The Helanshan population of Ningxia Nei Mongol is disjunct.

Habits: Like other eared pheasants, lives in small parties in open alpine meadows and juniper–rhododendron scrub at high altitudes.

46. Elliot's Pheasant *Syrmaticus ellioti* Baijing changweizhi PLATE 5

Description: Male: large (81 cm) brownish pheasant with pale head and brown pointed elongated tail barred with silvery-grey. Boldly patterned with white sides of neck, bars on wing and white abdomen and vent. Black chin and throat with white belly diagnostic. Bare facial skin is scarlet. Rump is black scalloped with narrow white edges. Female (45 cm) has reddish-brown crown with grey nape and mantle. Upperparts are otherwise mottled and vermiculated chestnut, grey and black. Throat and foreneck are black and rest of underparts are white with cinnamon barring. Iris—orange-brown; bill—yellow; feet—bluish-grey.

Voice: Deep voice. Usually calls in morning. Male more than female with *gu-gu-gu, ge-ge-ge* or *ji-ji-ji, ju-ju-ju*.

Range: Endemic to SE China.

Distribution and status: Globally Vulnerable (Collar *et al*. 1994). Relatively common in forested hills of E Jiangxi, S Anhui, W Zhejiang, N Fujian, Hunan, E Guizhou and N Guangdong from 200–500 m in coastal areas or 1000–1500 m inland.

Habits: Keeps to thickets and bamboo in mixed woodlands. Rather shy and alert. Lives in small groups.

47. Mrs Hume's Pheasant *Syrmaticus humiae* [Black-necked PLATE 5
Bar-tailed/Hume's Pheasant] Heijing changweizhi

Description: Male: large (92 cm) brown pheasant with long white tail, narrowly and sparsely banded black or black and brown. Wing has two white bars and bluish lesser coverts. Lower back and rump white with black scaling. Head and neck glossy purplish; facial skin red. Female is smaller (50 cm) with rather scaly olive-brown mantle and back; banded brown tail and buffy underparts. Wing is mottled brown and black with two whitish bars and pale lesser coverts. Iris—orange-brown; bill—greenish-horn; feet—dull light grey.

Voice: Male makes various calls; low grunting *ge-ge-ge* when foraging with first note loud and clear; rapid *guk-guk-guk-guk-guk* in alarm and clucking notes when hurrying away.

Range: NE India, W, N and E Burma, NW Thailand to SW and S China.

Distribution and status: Globally Vulnerable (Collar *et al.* 1994). Race *burmanicus* is rare in Yunnan and W Guangxi.

Habits: Lives in small parties in higher hills in scrubby and rugged terrain from 780–1800 m (higher in NE India).

48. Mikado Pheasant *Syrmaticus mikado* Hei changweizhi PLATE 5

Description: Male is large (86 cm) elegant blackish, long-tailed pheasant with distinct glossy purple-blue edges to feathers of mantle, breast and rump, forming conspicuous scalloped pattern against sooty-black feather centres. Pointed tail is barred black and mottled white. Wings are black with a conspicuous white bar and white tips to secondaries and tertials. Bare orbital skin is crimson. Smaller female is mottled grey below, brown with mottling of reddish and black with white streaks above. Iris—brown; bill—greyish; feet—greenish.

Voice: Rather high *wok, wok, wok* alarm call. Shrill whistle by male in breeding season.

Range: Endemic to Taiwan.

Distribution and status: Globally Near-threatened (Collar *et al.* 1994). Uncommon resident of the central mountains of Taiwan from 1800–3000 m.

Habits: Keeps to dense forest in lower conifer and mixed forest zone, also bamboo forests and steep scrub. Shy and secretive.

49. Reeves's Pheasant *Syrmaticus reevesii* Baiguan changweizhi PLATE 5

Description: Male is large (180 cm) unmistakable pheasant with extraordinarily long, barred tail (up to 1.5 m). Head is boldly patterned black and white. Upperparts are golden-yellow scaled with black feather edges. Centre of belly and thighs are black. Female is mottled reddish-brown, with a scaled breast and much shorter tail. Iris—brown; bill—horn; feet—grey.

Voice: Rarely calls. Alarmed male gives fast *gu-gu-gu-gu* call. Breeding male calls similar *gu-gu-gu*, female response is *ge-ge-ge*. Lost juveniles make plaintive *xia yiyo, xia yiyo* call.

Range: Endemic to C and E China.

Distribution and status: Globally Vulnerable (Collar *et al.* 1994). Rare and becoming increasingly restricted over the past 50 years as a result of habitat loss and collection for long tail feathers. Formerly found as far north-east as Hubei. Today limited to three populations—a tiny area of S Shandong; a small area of E Hubei and W Anhui; and a larger area of Guizhou, E Sichuan, SE Gansu, S Shaanxi, W Hubei and NE Yunnan. Lives in wooded hills from 300–1800 m in steep valleys and canyons of deciduous oak forest and mixed conifers.

Habits: Typical of genus. The long tail feathers are commonly used in the flamboyant head-dresses of Beijing Opera.

50. Common Pheasant *Phasianus colchicus* Zhiji PLATE 5

Description: Male: large (85 cm) familiar pheasant of Europe and N America, imported originally from China. Male has glossy black head with prominent 'ear tufts' and broad fleshy-red orbital skin. Some races have white neck-ring. Body is gloriously coloured by shimmering decorative feathers ranging from dark green to copper to gold; wings greyish and tail is elongated, pointed and brown with black bands. Much smaller female (60 cm) is cryptic, banded pale brown. Flight is fast with noisy rapid take-off when flushed. Great variation in plumage detail shown by 19 regional races in China. Eastern races have light greyish-green lower back and rump. Of these *formosanus, kiangsuensis, torquatus, karpowi* and *pallasi* have a white neck-ring; *rotschildi, suehschanensis* and *elegans* have no ring or only part of a ring; other races always have a broken ring. Races *pallasi* and *elegans* have green rather than purple breast. Western races have white wing coverts and chestnut lower back and rump with broken white neck-ring. Of these *shawii* has green breast and *mongolicus* has a purple breast. Intermediate races have yellowish-brown wing coverts and buffy lower back and rump with white neck-ring absent or insignificant. Of these *tarimensis* has purplish-red breast, the rest are green-breasted. Iris—yellow; bill—horn; feet—greyish.

Voice: Male's call is two-noted explosive hacking followed by whirr of wings.

Range: SE Western Palearctic, C Asia, SE Siberia, Ussuriland, China, Taiwan, Korea, Japan and Tonkin. Introduced to Europe, Australia, New Zealand, Hawaii and N America.

Distribution and status: Widespread in scrubby open areas. Formerly common but now locally persecuted to low levels. Many regional races occur—*pallasi* in Heilongjiang and NE Nei Mongol; *karpowi* in Liaoning, Jilin, E Hebei and SE Nei Mongol; *torquatus* in E and SE China; *kiangsuensis* in N Shaanxi, C Nei Mongol, Shanxi and W Hebei; *decollatus* in Guizhou, E Sichuan and W Hubei; *rothschildi* in SE Yunnan (Mengzi); *tatatsukasae* in SW Guangxi; *formosanus* in C Taiwan; *elegans* in SW China; *suehschanensis* in NE Xizang, N Sichuan and SE Qinghai; *strauchi* in N and NE Sichuan, S Shaanxi, C and S Gansu, E Qinghai; *satscheuensis* in N Qinghai and NW Gansu; *shawii* in SW Xinjiang; *tamirensis* in the Tarim and Turpan basins of Xinjiang; *mongolicus* in NW Xinjiang; *vlangalii* in the Qaidam basin of Qinghai; *sohokhensis* in the Tengger desert of W Nei Mongol; *alaschanicus* from Helan Mts of Ningxia; *edzinensis* around Sogo Lake in W Nei Mongol.

Habits: Males alone or in small groups, females with their own young and occasionally mixed groups. Lives in open woods, scrub, semi-desert and farmlands over a wide altitudinal range.

51. Golden Pheasant *Crysolophus pictus* Hongfu jinji PLATE 5

Description: Male: smallish but elongate (98 cm) dazzling pheasant with golden-plumed crown and back; gold and black barred nape ruff; metallic green mantle and crimson underparts. Wing is metallic blue and tail is elongated, arched and central feathers are blackish with small buff spots, otherwise cinnamon. Smaller female is buffy-brown, finely banded with black, above and paler buff below. Iris—yellow; bill—greenish-yellow; feet—yellowish-horn.

Voice: Female gives *cha-cha* calls in spring. Other females respond to the calls. Male reply is *gui-gui, gui* or *gui-gu, gu, gu* or melodious short *gu gu gu* . . . In flight, male gives quick *zi zi zi* . . . call.

Range: Endemic to C China.

Distribution and status: Globally Near-threatened (Collar *et al.* 1994). Common from 800–1600 m, rarely up to 2800 m. Resident in SE Qinghai, S Gansu, Sichuan, S Shaanxi, W Hubei, Guizhou, N Guangxi and W Hunan.

Habits: Singly or in small parties, favouring scrubby hillsides and secondary forest in subtropical broadleaf forest and deciduous broadleaf forest. Often kept in captivity.

52. Lady Amherst's Pheasant *Chrysolophus amherstiae* Baifu jinji   PLATE 5

Description: Male: medium-sized (150 cm) distinctive colourful pheasant with glossy dark green crown, throat and upper breast, short scarlet crest and white mane scalloped with black edges. Back and wings are glossy dark green and belly is white. Rump is yellow and tail is extremely long, slightly arched and composed of white feathers narrowly banded black. There are several orange-tipped elongated tail coverts. Smaller female (60 cm) has upperparts barred black and cinnamon with white throat and chestnut breast finely edged black. Flanks and undertail coverts buff barred with black. Iris—brown; bill—bluish-grey; feet—bluish-grey.

Voice: Breeding male gives loud, harsh, far-carrying *ga-ga-ga* call or harsh *gua*; group call is soft *shu-shu-shu-sss*. Alarm is piercing *shi-ya*. Female call is *guo-guo-guo* to summon chicks. Male in threat calls *ja-ja-ja-ja*.

Range: NE Burma and SW China. Introduced Europe.

Distribution and status: Globally Near-threatened (Collar *et al.* 1994). This is an uncommon bird of forested hills from 1800–4600 m in SE Xizang, Yunnan, S Sichuan, W Guizhou and W Guangxi to the west of the distribution of Golden Pheasant.

Habits: Typical of genus. Lives in scrubby and secondary forest on slopes of wooded mountainsides.

53. Grey Peacock Pheasant *Polyplectron bicalcaratum*   PLATE 5
Hui kongquezhi

Description: Male: smallish (75 cm) brownish-grey pheasant with whitish throat and large violet-green ocelli on mantle and tail. Brush-like recurved crest. Bare facial skin pinkish. Underparts barred buffy white and dark brown. Female smaller, lacks crest, ocelli smaller and tail shorter. Iris—pale grey; bill—dark grey; feet—bluish-grey.

Voice: Male territorial call is loud, explosive *trew-tree*. Female gives loud *ga-ga* call or faster *ok-kok-kok-kok* in alarm.

Range: SW China and SE Asia.

Distribution and status: Rare and diminishing due to hunting and habitat loss. Nominate race is found in W and S Yunnan and SE Xizang. Lives in montane evergreen forest up to 2000 m.

Habits: Male has spectacular courtship display, crouched on ground with tail fanned and wings spread and raised. Maintains cleared calling spots and is highly territorial.

54. Hainan Peacock Pheasant *Polyplectron katsumatae*   NOT ILLUSTRATED
Hainan kongquezhi

Description: Very similar to Grey Peacock Pheasant but slightly smaller (70 cm), darker and browner in colour and with purple but no green sheen in the ocelli. Iris—grey; bill—bluish-grey; feet—bluish-grey.

Voice: Male has melodious, loud call *guang-gui, guang-gui* with first note of each couplet longer. Female gives faster *ga, ga, ga* call.

Range: Endemic to Hainan.

Distribution and status: Very rare in the remaining hill forest of SW Hainan. Should be added to Globally Threatened lists.

Habits: As Grey Peacock Pheasant.

Note: Formerly treated as a race of Grey Peacock Pheasant (e.g. Cheng 1987, 1996). (See Johnsgard 1986.)

55. Green Peafowl *Pavo muticus* Lü kongque   PLATE 5

Description: Unmistakable, huge (male 240 cm, female 110 cm) pheasant with elongated tail (male only) and vertical plume crest on head. Male has iridescent green mantle, neck and breast, and enormously elongated 'fan' tail of ocellated iridescent feathers. Female lacks long tail and is less finely coloured with whitish underparts. Iris—reddish-brown; bill—horn; feet—dark grey.

Voice: Loud, resonant trumpeting *kay-yaw, kay-yaw* given at dawn and dusk from tree perch.

Range: NE India to SW China, SE Asia and Java.

Distribution and status: Globally Vulnerable (Collar *et al.* 1994). Race *imperator* formerly widespread in Yunnan up to 1500 m but now confined to only two or three counties due to hunting for decorative feathers and food.

Habits: Lives along rivers in lowland forest and scrub. Male has fantastic display, fluttering fanned elongated uppertail coverts at female.

56. Siberian Grouse *Dendragapus falcipennis* [Sharp-winged Grouse,   PLATE 6
AG: Spruce Grouse] Lian chiji

Description: Medium-sized (41 cm) grouse with black throat patch outlined with white. Upperparts olive-brown, barred black; belly white barred black. Central tail feathers brown; outer tail feathers black with broad white tips. Red eyebrow wattles prominent. Distinguished from smaller Hazel Grouse by blackish breast and mantle, shorter crest and diagnostic narrow pointed primaries. Iris—dark brown; bill—black; feet—blackish.

Voice: Courting male makes loud, harsh *ka cha, ka cha* calls; rolling *u-u-u-rrr*; mellow *koo* when mating.

Range: SE Siberia and Sakhalin Island.

Distribution and status: Globally Near-threatened (Collar *et al.* 1994). Rare vagrant in Lesser Hinggan Mts and Heilong (Amur) valley of Heilongjiang, NE China. Not yet recorded to breed in China but may do so.

Habits: Occurs in mossy conifer forest with dense undergrowth between 200 and 1500 m.

Note: Has also been placed in the genera, *Falcipennis* and *Canachites*.

57. Willow Ptarmigan *Lagopus lagopus* [Liu] Leiniao   PLATE 6

Description: Stocky, medium-sized (38 cm) grouse. Summer: rufous-buff, barred, spotted and vermiculated with black. Tail black, mostly concealed by brownish uppertail coverts. Abdomen, wing coverts and tips of outer tail feathers white. Winter: white with black tail, mostly concealed by white uppertail coverts. Prominent red eyebrow wattles at all seasons. Iris—dark brown; bill—horn to black; feet—feathered white.

Voice: Various harsh calls—*krrrow; ko-bek; keh-uk* and accelerating *ka, ke ke-ke-ke-ke kekekeke-rrr*. Female has contact call *nyow*.

Range: N Eurasia to Mongolia, SE Siberia and Sakhalin.

Distribution and status: Race *okadai* is very rare in willow scrub along the Heilong (Amur) River in extreme NE China.

Habits: Males gather in courtship display at night in spring to call and jump in the air. Can be attracted to imitation of female's contact call.

58. Rock Ptarmigan *Lagopus mutus* Yan leiniao   PLATE 6

Description: Stocky, medium-sized (38 cm) grouse. Similar to Willow Ptarmigan but much greyer in summer and distinguished in winter by black instead of white lores. Also bill smaller and more slender. Species are geographically far separated in China so cannot be confused in the field. Iris—dark brown; bill—horn to black; feet—feathered white.

Voice: Harsh *kuh, kuh, kwa-kwa-kwa* or male's belching *arr, arr*.

Range: Tundra and high mountain regions of N Holarctic.

Distribution and status: Race *nadezdae* is a rare bird of stony tundra screes, juniper scrub and moorland of Altai Mts from 1300–2000 m in extreme NW Xinjiang.

Habits: Lives in small parties. Quite tame. Very hardy, mostly above treeline.

59. Black Grouse *Tetrao tetrix* Hei qinji   PLATE 6

Description: Male: large (54 cm) black grouse with bluish-green gloss. Wing black with bold white bar formed by base of coverts and inner remiges. Black tail is lyre-shaped curving apart to allow white undertail coverts to be fanned and erected during 'lek' displays. Red comb-like wattles form eyebrow patch. Female: smaller (41 cm) dark brown, barred by buffy feather tips and with rounded tail. Iris—dark brown; bill—black; feet—leaden grey.

Voice: Male gives gurgling *gurururu* calls.

Range: W and N Europe to Siberia and N Korea.

Distribution and status: Locally common in pine and larch forests and forested steppes of NE China (*ussuriensis*); around Hulun Nur in NE Nei Mongol (*baikalensis*) and in the Kashi, Tianshan and Altai Mts of Xinjiang in NW China (*mongolicus*).

Habits: Has elaborate courtship displays when several males gather at 'lek' in front of females.

60. Western Capercaillie *Tetrao urogallus* [Xifang] Songji PLATE 6

Description: Male: very large (86 cm) black, stocky grouse with blunt rounded tail which can be fanned like a turkey. Purplish sheen on upperparts; greenish sheen on breast. Red wattles form eyebrow patch. Female: smaller (61 cm) brown and strongly mottled black grey and white. Iris—dark brown; bill—yellow; feet—feathered grey.
Voice: Deep rattling *ge-ge-ge*. Ringing *peng-peng-peng*.
Range: Scotland, N Europe, N Asia.
Distribution and status: Race *taczanowskii* is a rare bird in the conifer forests of Altai Mts in extreme NW Xinjiang from 3200–4400 m.
Habits: Feeds on pine shoots in trees. Male erects tail and throat feathers when calling.

61. Spotted Capercaillie *Tetrao parvirostris* [Black-billed Capercaillie] PLATE 6
Heizui songji

Description: Male: very large (86 cm) black, stocky grouse with blunt rounded tail which can be fanned like a turkey. Purplish sheen on upperparts; greenish sheen on breast. Red wattles form eyebrow patch. Uppertail feathers elongated, black with white tips forming arc of large white spots when tail is spread. Scapulars and wing coverts tipped white. Underparts black with white spots. Female: smaller (61 cm) dark brown finely vermiculated with buff and white barring; whitish wing bars and white tips to central tail feathers. Iris—dark brown; bill—black; feet—feathered grey.
Voice: Very loud rolling calls.
Range: E Siberia to Sakhalin.
Distribution and status: Rare in larch and pine forests of NE China in Greater and Lesser Hinggan and Changbai Mt ranges from 300–1000 m. Vagrants reach Liaoning and Hebei borders.
Habits: As Western Capercaillie.

62. Hazel Grouse *Tetrastes bonasia* Huawei zhenji PLATE 6

Description: Small (36 cm) grouse with prominent crest and black throat bordered broadly with white. Upperparts ashy grey-brown, finely vermiculated. Wings mottled with black and brown; white tips to scapulars and coverts forming bands. Tail feathers brownish, outer feathers subterminally black with white tips. Underparts buff with rufous and black crescent-shaped spots at feather centres. Flanks scaled rufous. Red eyebrow wattles not prominent. Iris—dark brown; bill—black; feet—horn-coloured.
Voice: Audible wingbeats *boorr boorr*. Advertising call is drawn-out sucking *tseeuu-eee tititi*. In alarm, a rapid twitter *pyittittittitt-ett-ett*.
Range: N Eurasia to N Altai Mts and Sakhalin.
Distribution and status: Fairly common in conifer forests and wooded plains of NE China from 800–2100 m. Race *sibiricus* found in taiga forests of Greater Hinggan Mts; race *amurensis* in Lesser Hinggan Mts and at south end of Greater Hinggan Mts and Heilong valley with vagrants reported from Liaoning and NE Hebei. Vagrants can also be expected in Altai Mts of NW China.
Habits: Can be called up by imitation of its call. Lives in pairs. Chicks can fly up into trees when a few days old. Prefers dense tangles of birch and alders near streams.
Note: Has been placed in the genus *Bonasa*.

63. Chinese Grouse *Tetrastes sewerzowi* [Severtzov's Hazel Grouse]   PLATE 6
Banwei zhenji

Description: Small (33 cm) rich brown barred grouse with prominent crest and black throat patch outlined in white. Upperparts brown barred with black. Outer tail feathers subterminally black with white tip. There is a white line behind eye, whitish patch on shoulder and white tips to wing coverts. Underparts rufous on breast becoming whiter ventrally and all barred with black. Female is duller with white streaking on throat and more buffy underparts. Race *secunda* is generally more rufous. Iris—brown; bill—black; feet—grey.

Voice: Rarely calls. When feeding, makes harsh *gir gir gir* noise; soft *gu, gu, gu* calls in alarm and *dir, dir, dir* call to summon the young.

Range: Endemic to C China.

Distribution and status: Globally Near-threatened (Collar *et al.* 1994). Rather uncommon in conifer forests and thickets in open areas from 2500–4000 m. Nominate race from Qiling Mts of Qinghai, C Gansu and N Sichuan; race *secunda* from W Sichuan and E Xizang.

Habits: As Hazel Grouse. Uses *Salix* thickets along streams.

Note: Has been placed in the genus *Bonasa*.

WATER FOWL—Order: Anseriformes

A familiar large, worldwide order of swimming waterbirds with webbed feet and rather specialised, characteristic broad, flat bills. They have rather short legs, narrow pointed wings set well back on the body and generally short tails. They fly fast with a continuous audible whistling flap.

Ducks are divided taxonomically into two families and several tribes. Chinese representatives can be classed into five groups: Tree-ducks or Whistling-ducks (*Dendrocygna*) which have clear whistling calls and are now classed in their own family Dendrocygnidae; stiff-tailed ducks (*Oxyura*); Swans (*Cygnus*); Geese (*Anser, Branta, Tadorna, Sarkidiornis, Nettapus*); true ducks— Anatini, which includes dabbling ducks (*Aix, Anas, Marmaronetta, Rhodessa*) which swim high in the water; diving ducks (*Aythya, Polysticta, Histrionicus, Clangula, Melanitta, Bucephala*) which dive for food and to escape when alarmed; and Mergansers (*Mergus*) with serrated fish-catching bills.

There are a total of 45 species in China of which seven are only winter visitors.

WHISTLING DUCKS—Family: Dendrocygnidae

64. Lesser Whistling-duck *Dendrocygna javanica* [Lesser Tree-Duck]   PLATE 9
[Li] Shuya

Description: Medium-sized (41 cm) reddish-brown duck. Crown dark brown; head and neck buffy; back dark brown, scalloped rufous; underparts pale reddish-brown. Iris—brown; bill—grey; feet—dark grey.

Voice: Shrill, musical whistle *seasick seasick* in flight.

Range: India, S China, SE Asia and Greater Sundas.

Distribution and status: Breeds in S Yunnan and SW Guangxi. Occasionally summers in lower Changjiang, S Guangdong, Hainan and Taiwan. Winters south to tropical zone. Locally common.

Habits: Found in small parties on lakes, swamps, mangroves and paddy fields. Partly nocturnal.

DUCKS, SWANS AND GEESE—Family: Anatidae

65. White-Headed Duck *Oxyura leucocephala* Baitou yingweiya PLATE 10

Description: Squat (46 cm) unmistakable stubby brown duck with stiff tail held erect or flat on water. Head black and white. Head of male white with black crown and collar. In breeding plumage, bill is blue. Head of female and juvenile dark greyish. Occasionally first spring male may have all black head. Bill is distinctively swollen at base. Iris—yellowish; bill—blue male, grey female; feet—greyish.

Voice: Usually silent. In breeding display males swim in groups with erect tails and neck and make low rattling noise and piping calls. Females give low, harsh notes.

Range: Mediterranean and western Asia to NW China.

Distribution and status: Globally Vulnerable (Collar *et al.* 1994). Extremely rare and status uncertain. Recorded breeding in Junggar Basin and Tianshan in W Xinjiang and once wintering in Hubei (Honghu).

Habits: Lives on fresh water lakes.

66. Mute Swan *Cygnus olor* Youbi tian'e PLATE 7

Description: Large (150 cm) elegant white swan with orange bill and characteristic black nob at base of forehead (in male). Holds neck in graceful S-shape while swimming and often holds wings in high arched posture. Young are downy grey or dirty white with grey-mauve bill. Adult is aggressive in defence of nesting area. Iris—brown; bill—orange; feet—black.

Voice: Despite name it hisses in threat and emits deep explosive *heeorr* call.

Range: Europe to C Asia, wintering to N Africa and India.

Distribution and status: Breeds on a few lakes in N and C China including Hulun Nur area in NE China, with occasional wintering birds reaching southern China.

Habits: Nest is made on pile of reeds. Flies with powerful slow beat and loud whistling noise. Wintering birds gather in large flocks on lakes or rivers.

67. Whooper Swan *Cygnus cygnus* Da tian'e PLATE 7

Description: Tall (155 cm) white swan with black bill and extensive yellow basal area. The yellow extends along the side of the upper mandible to form a point. While swimming, the neck is held straighter than in Mute Swan. Immature is more uniformly coloured than Mute Swan, and has paler bill. Much larger than Tundra Swan. Iris—brown; bill—black with yellow base; feet—black.

Voice: Flight call is characteristic *klo-klo-klo* but contact call is a loud, melancholy, bugle-like call.

Range: Greenland, N Europe, N Asia, wintering to C Europe, C Asia and China.
Distribution and status: Breeds among reed beds on a few northern lakes but flocks migrate south in winter. Scarce; generally much rarer than Tundra Swan. One record Hong Kong.
Habits: Much quieter in flight than Mute Swan.

68. Tundra Swan *Cygnus columbianus* Xiao tian'e PLATE 7

Description: Tallish (142 cm) white swan with black bill but much less extensive yellow on bill base than Whooper. Yellow on side of upper mandible does not form such a sharp forward point and culmen of bill is black. Smaller than Whooper Swan with which it can easily be confused. Iris—brown; bill—black with yellow base; feet—black.
Voice: Calls are similar to Whooper but higher-pitched. Groups chorus with crane-like drawn-out *klah* notes.
Range: N Europe and N Asia, wintering to Europe, C Asia, China and Japan.
Distribution and status: Race *jankowskii* breeds in Siberian tundra but migrates in winter through north-east China to lakes in Changjiang valley where it is uncommon but more common than Whooper Swan.
Habits: Flocks fly in V-formation like other swans.

69. Swan Goose *Anser cygnoides* Hong yan PLATE 7

Description: Large (88 cm) long-necked goose with rather long black bill forming straight line with forehead, and narrow white line around base of bill. Upperparts ashy-brown with feathers edged buff. Foreneck white, crown and back of neck reddish-brown. Sharp contrast line between front and back of neck. Legs pink and vent whitish. Flight feathers black. Distinguished from Lesser and Greater White-fronted Geese by black bill, less white on forehead and white front of neck. Iris—brown; bill—black; feet—deep orange.
Voice: Honking and cackling in flight. Typical prolonged honk rises in pitch.
Range: Breeds Mongolia, NE China and Siberia, wintering in C and E China, Taiwan and Korea.
Distribution and status: Globally Vulnerable (Collar *et al.* 1994). Breeds in NE China but migrates across E and C China to mostly winter in the lower Changjiang valley, rarely to SE coasts. Vagrants reach Taiwan. Up to 50 000 birds winter on Poyang Lake—a major part of the world population of the species.
Habits: Forms flocks on fresh water lakes, feeding on surrounding grasslands and fields.

70. Bean Goose *Anser fabalis* Dou yan PLATE 7

Description: Large (80 cm) grey goose with orange feet. Similar to Pink-footed Goose but feet orange; neck darker and bill black with orange subterminal band. In flight looks darker and longer-necked than other grey geese. Both upperwing and underwing lack pale grey tones of Pink-footed or Greylag. Iris—dark brown; bill—orange, yellow and black; feet—orange.
Voice: Similar to Pink-footed Goose including similar *wink-wink* phrase but deeper in Bean more like *hank-hank*.

Range: Breeds in taiga of Europe and Asia wintering in temperate zone.
Distribution and status: Fairly common winter visitor. Birds of races *rossicus* and *johanseni* winter in the Kashi region of W Xinjiang and in Shaanxi; *serrirostris* and *sibiricus* are seen on passage through NE and N China and winter in the lower Changjiang valley, SE coastal provinces, Hainan and Taiwan. Irregular vagrant to Hong Kong.
Habits: Forms flocks on marshes and rice stubble near lakes.

71. Greater White-fronted Goose *Anser albifrons* PLATE 7

[White-fronted Goose] Bai'e yan

Description: Large (70–85 cm) grey goose with orange legs, white patch surrounding base of bill and large black belly patches, narrower in juvenile. Very similar to Lesser White-fronted Goose with which it often associates in winter. For separation see under that species. In flight looks heavy and underwing darker than Greylag or Pink-footed but paler than Bean Goose. Iris—dark brown; bill—pink with yellow base; feet—orange.
Voice: Noisy cackling. In flight gives musical *lyo-lyok* call of varying pitch. Calls higher in pitch than Bean or Greylag but lower than Pink-footed.
Range: Breeds in tundra of Northern Hemisphere; winters in farmlands of temperate zone.
Distribution and status: Locally common on wintering grounds. Race *frontalis* recorded on passage through NE China, Shandong and Hebei. Wintering in Changjiang valley and eastern provinces to Hubei and Hunan, and Taiwan. Also in S Xizang (Qamdo). Up to 20 000 birds winter on Poyang Lake.
Habits: Forms large concentrations in winter at favoured wintering grounds. Rather shy.

72. Lesser White-fronted Goose *Anser erythropus* Xiao bai'e yan PLATE 7

Description: Medium-sized (62 cm) grey goose with orange legs, white patch a round base of bill and blackish patches on belly. Very similar to Greater White-fronted Goose with which it associates in winter. Distinguished by smaller size, shorter bill, shorter neck, white patch around bill extending further up forehead, yellow eye-ring and smaller dark patches on belly. In flight has proportionally longer wings and faster wing beat. Iris—dark brown; bill—pink; feet—orange.
Voice: On breeding grounds alarm call is loud *queue-oop*. Flight call is squeaky and higher-pitched than Greater White-fronted usually with repeated *kyu-yu-yu*.
Range: Breeds in Asian and European Arctic, winters in temperate steppe and farmland of Balkans, Middle East and E China.
Distribution and status: Globally Vulnerable (Collar *et al.* 1994). Uncommon in China but several thousand concentrate at Poyang Lake to winter, constituting a large proportion of the world population. May breed in NE China.
Habits: Winters besides larger rivers and lakes in China, mixing with Greater White-fronted Goose and feeding on farmlands and cut reed beds. Rather agile and sometimes runs on land.

73. Greylag Goose *Anser anser* Hui yan PLATE 7

Description: Large (76 cm) grey-brown goose with diagnostic combination of pink bill and pink feet. There is no white at base of bill. Feathers of upperparts are grey with white edges giving the bird a scalloped pattern. The breast is pale ashy-brown and upper- and lowertail coverts are white. In flight pale forewing contrasts with dark flight feathers. Iris—brown; bill—pink; feet—pink.

Voice: The call is a deep clanging, honking.

Range: N Europe and N Asia, wintering to N Africa, India, China and SE Asia.

Distribution and status: Breeds across N China, wintering on lakes of S and C China in small flocks. A few birds winter on Poyang Lake in Jiangxi. Probably feeds on seagrass in coastal areas of Hainan.

Habits: Inhabits steppes, moors and lakes; feeding on short grasslands and agricultural fields.

74. Bar-headed Goose *Anser indicus* Bantou yan   PLATE 7

Description: Smallish (70 cm) goose with diagnostic head pattern of white crown with two black transverse bars on back of head. White of throat extends in a lateral bar down side of neck. Young birds lack black patterning which is light grey. In flight the bird has uniformly pale upperparts with only narrow trailing edge of wings dark. Underparts largely white. Iris—brown; bill—yellow with black tip; feet—orange.

Voice: Typical goose honking in flight. Honks low and nasal.

Range: Breeds in C Asia, wintering to N India and Burma.

Distribution and status: Breeds in marshes and highland moors of extreme N China, Qinghai and Xizang and migrates in winter to C China and S Xizang.

Habits: A hardy goose of barren cold deserts and salt lakes. Winters on fresh water lakes.

75. Snow Goose *Anser caerulescens* Xue yan   PLATE 7

Description: Unmistakable smallish (80 cm) white goose with black wing tips. Juvenile has greyish crown, back of neck and upperparts. Blue phase birds may occur in which head and upper neck are white but rest of plumage is mostly black with a blue shoulder patch on wing. Iris—brown; bill—reddish-pink; feet—pink.

Voice: Flocking birds in flight give pleasant, nasal, high-pitched *la-luk* call sounding like a small dog yapping at long range.

Range: Breeds Arctic tundra of N America and in small numbers on Wrangel Island of Siberia; winters in subtropical and temperate zone of N America and occasionally in Japan and E China.

Distribution and status: Occasional birds winter in Hebei and E China; now very rare due to decline in Siberian population.

Habits: Winters on cultivated fields and stubble in coastal regions.

[76. Canada Goose *Branta canadensis* Jianada yan   PLATE 7

Description: Large (100 cm) grey goose with black neck and head and diagnostic white patch from behind eye down to throat. In flight back and tail contrast with white vent and uppertail coverts. Iris—brown; bill—black; feet—black.

Voice: Very vocal especially during flight.

Range: Widespread in N America. Introduced Europe and New Zealand. Spreading into NE Asia.

Distribution and status: Not originally native in China but feral population in Beijing and elsewhere, and spreading globally. Winters occasionally in Japan.
Habits: Social goose of lakes, rivers and farmland.]

77. Brent Goose *Branta bernicla* Hei yan   PLATE 7

Description: Medium-sized (62 cm) dark grey goose with black bill and feet and white undertail. Has diagnostic white pattern on side of grey neck, sometimes forming half collar on foreneck. Sides of breast finely barred whitish. Juvenile lacks white patch on neck and has fine white wing bars. Iris—brown; bill—black; feet—black.
Voice: Call is low, rolling *raunk, raunk* given on land and in flight.
Range: Breeds in Arctic tundra of N America and Siberia; Winters south along coastal grasslands and estuaries.
Distribution and status: Rare winter visitor. Small number of race *nigricans* winter around coasts of Yellow Sea. Vagrant to Shanxi.
Habits: Does not mix significantly with other species. Flies low over water. Roosts in coastal bays at high tide. Feeds on maritime pastures and *Zostera* eelgrass.

78. Red-breasted Goose *Branta ruficollis* Hongxiong heiyan   PLATE 7

Description: Small (54 cm) unmistakable gaudy goose with round head and tiny bill. Body plumage black and white with diagnostic red breast, foreneck and patch on side of head. Distinctive white patch at base of bill. In flight, small size, short neck and very black plumage contrast with white ventral region. More extensive black in tail than Brent Goose. Iris—brown; bill—black; feet—black.
Voice: Call is jerky, staccato, repeated *kik-yoik, kik yoik*. Uttered constantly by feeding flocks.
Range: Breeds in Arctic tundra of Siberia on Taymyr Peninsula, wintering to SE Europe.
Distribution and status: Globally Vulnerable (Collar *et al.* 1994). Uncommon. Vagrants have been recorded in Hubei and Hunan (Dongting Lake).
Habits: Associates in winter with other geese. Flies in tight flocks rather than V-formation. Roosts on lakes or lagoons. Flocks very noisy.

79. Ruddy Shelduck *Tadorna ferruginea* Chi maya   PLATE 8

Description: Large (63 cm) orange-chestnut duck with buff head. Shape is rather goose-like. Male in summer has narrow black collar. In flight the white wing coverts are conspicuous and speculum is bronze-green. Iris—brown; bill—blackish; feet—black.
Voice: Rolling, honked *aakh*, sometimes repeated as *pok-pok-pok-pok*. Female call is deeper than male.
Range: SE Europe and C Asia, wintering to India and S China.
Distribution and status: This is a hardy bird breeding in NE and NW China and the Qinghai–Xizang Plateau up to 4600 m but migrating for the winter to S and C China.
Habits: Nests in holes in the banks of small streams. Keeps to inland lakes and rivers. Rarely visits the coast.

[80. Crested Shelduck *Tadorna cristata* Guan maya PLATE 8

Description: Large (61 cm) distinctive duck with white foreneck and forewing. Sexually dimorphic. Male: black cap, hindneck and breast. Female with white patch around eye and white forehead and foreneck but grey-brown breast and belly, finely barred white. Iris—brown; bill—red (male) pink (female); feet—red.
Voice: Unknown.
Range: Borderlands between Siberia, China and Korea.
Distribution and status: Globally Critical (Collar *et al.* 1994). Almost certainly extinct. Not seen (globally) since 1971. Wintered along coasts of Yellow Sea. Shot in NE China in 1936 (unverified). However, in response to distributed questionaires, several fishermen and hunters in NE China claim to have seen the bird (Zhao 1993). There are Russian claims of sightings in 1985.
Habits: Mixed with flocks of other ducks.]

81. Common Shelduck *Tadorna tadorna* Qiaobi maya PLATE 8

Description: Large (60 cm) strikingly patterned black and white duck with glossy greenish-black head contrasting sharply with bright red bill and pronounced nob (male) at base of forehead. There is a chestnut band across the breast. Female is similar to male but slightly duller and with reduced or lacks bill nob. Juvenile is mottled brown with dull red bill and white patch on side of face. Iris—pale brown; bill—red; feet—red.
Voice: Vocal in spring. Male gives low whistles; female gives chattering *gag-ag-ag-ag-ag*.
Range: W Europe to E Asia, wintering to N Africa, India and S China.
Distribution and status: Breeds in N and NE China but migrates in winter to SE China where it is quite common.
Habits: Nests in holes in banks of salty or brackish lakes and rarely fresh water lakes.

82. Comb Duck *Sarkidiornis melanotos* Liuya PLATE 8

Description: Male is unmistakable, very large (76 cm), black and white duck with prominent black fleshy comb above bill. White head and neck are finely spotted black. Black upperparts are glossed with metallic green and bronze. Female is like male but much smaller and lacks comb. The female nests in natural tree holes and lives on wooded pools and rivers. Iris—brown; bill—black; feet—grey.
Voice: Low croaks. In breeding season gives wheezy whistles, grunts and hisses during display.
Range: S America, Africa, India, Burma and SW China.
Distribution and status: Nominate race is recorded in wetlands of SE Xizang and extreme S Yunnan where it is becoming rare. Stragglers sometimes found in SE China.
Habits: Lives in flocks in swamps and wooded waterways.
Note: White-winged Wood Duck *Cairina scutulata* could occur in lowlands of SE Xizang. Similar to female Comb Duck but with black neck and white wing patch.

83. Cotton Pygmy-goose *Nettapus coromandelianus* PLATE 9
[White Pygmy-goose/Cotton Teal] Mianfu

Description: Small (30 cm) dark green and white duck. Male: crown, neck band, back, wings and tail black with green iridescence; plumage otherwise whitish. In flight white wing patch conspicuous. Female: drabber with brown replacing the glossy black and buff replacing white; dark stripe through eye; no white wing patch. Iris—red (male) dark (female); bill—greyish; feet—grey.

Voice: In flight utters a soft musical laughing note *kar kar kar wark* several times in succession, also a soft *kwak*.

Range: India, S China, SE Asia and locally to New Guinea and Australia.

Distribution and status: Breeds Changjiang and Xijiang valleys, along S and SE coasts including Hainan and in SW Yunnan. Strays recorded on Taiwan and as far north as Hebei.

Habits: Usually on ponds, canals, grassy backwaters or paddy fields. Nests in tree holes and regularly perches in tall trees.

84. Mandarin Duck *Aix galericulata* Yuanyang　　　　PLATE 9

Description: Small (40 cm) decorative duck. Male has sweeping bold white brow, golden mane of hackles and extraordinary cinnamon display 'sails' that are held erect concealing wings. Female is less ostentatious—smart grey with elegant white eye-ring and rear eye-stripe. In eclipse plumage male resembles female but has red bill. Iris—brown; bill—red (male) grey (female); feet—yellowish.

Voice: Rather silent. Male gives short whistled *hwick* in flight. Female makes low clucking notes.

Range: NE Asia, E China and Japan. Introduced elsewhere.

Distribution and status: Globally Near-threatened (Collar *et al.* 1994). Breeds in north-east China but migrates to southern China in winter. Widely recorded but generally rare. Commonly trapped and kept in captivity.

Habits: Nests in holes in trees or banks and lives on wooded streams.

85. Gadwall *Anas strepera* Chibang ya　　　　PLATE 9

Description: Male: medium-sized (50 cm) grey duck with black bill, brownish head, black rear end, white patch on secondaries and orange legs distinctive. Slightly smaller and more slender-billed than Mallard. Female: like female Mallard but flatter head, orange sides to bill, white belly and white on secondaries. Iris—brown; bill—grey in breeding male otherwise orange with grey centre; feet—orange.

Voice: Rather silent except in courtship. Male utters short *nheck* and low whistle, female gives repeated *gag-ag-ag-ag-ag* higher in pitch than Mallard.

Range: Holarctic to Mediterranean, N Africa, N India to S China and S Japan. Breeds in temperate zone; winters south.

Distribution and status: Uncommon seasonal migrant. Nominate race breeds in NE China and W Xinjiang. Recorded on passage over N China and winters widely across S Xizang and most of China south of Changjiang River.

Habits: Lives on fresh water lakes and marshes in open country; rarely coastal estuaries.

86. Falcated Duck *Anas falcata* Luowen ya　　　　PLATE 9

Description: Male: large (50 cm) with distinctive chestnut crown, shiny sweeping green side of head extending into long drooping nuchal crest and long-plumed black and white tertials. White throat and spot at base of bill distinguish from much smaller Common Teal. Female is dull brown with dark mottling. Similar to female Gadwall but with dark grey bill and legs, plain head and neck, weaker scalloping on flanks and buffish line on sides of uppertail coverts; has bronzy-brown speculum. Iris—brown; bill—black; feet—dark grey.

Voice: Fairly quiet. In breeding season, male gives low whistle followed by wavering note *uit-trr*. Female replies with gruff *quack*.

Range: Breeds NE Asia, migrates to E and S China.

Distribution and status: Breeds on lakes and wetlands of NE China but migrates in winter over much of the country as far west as NW Yunnan. Regularly winters in Hong Kong.

Habits: Forms large flocks, resting on water and often mixed with other species.

87. Eurasian Wigeon *Anas penelope* [Wigeon] Chijing ya   PLATE 9

Description: Medium-sized (47 cm) rather compact big-headed duck. Male has diagnostic chestnut head and buff crest. Rest of plumage grey with white patch on flanks, white belly and black undertail coverts. In flight white wing coverts contrast against dark flight feathers and green speculum. Female is rich uniform rufous-brown or greyish-brown with white belly. In flight, pale grey wing coverts contrast with darker flight feathers. Underwing grey—darker than American Wigeon. Iris—brown; bill—blue-green; feet—grey.

Voice: Male—musical piping whistle *whee-oo*, female—short growled *quack*.

Range: Palearctic; winters south.

Distribution and status: Breeds in NE and probably NW China. Migrates south to winter over most of China south of 35 °N including Taiwan and Hainan. Locally common.

Habits: Mixes with other waterbirds on lakes, marshes and estuaries.

[88. American Wigeon *Anas americana* [Baldpate] Putaoxiong ya   PLATE 9

Description: Medium-sized (52 cm) duck, very much like Eurasian Wigeon but slightly larger. Breeding male distinctive with broad green eye-stripe across whitish head but in winter both sexes difficult to separate from Eurasian Wigeon. American Wigeon is longer-billed, longer-necked and longer-tailed. It has more contrast between grey of neck and rufous of flanks; some young males have black band around base of bill; axillaries and central underwing are pure white rather than greyish-white; greater wing coverts are whiter forming pale panel on wing. Iris—brown; bill—grey; feet—blue-grey.

Voice: As Eurasian Wigeon, but male call more throaty and less shrill.

Range: N and C North America wintering to S America. Some birds occasionally winter to Japan.

Distribution and status: Rare. Not yet documented for China but mixes with Eurasian Wigeon wintering in coastal bays off Honshu and Kyushu, so must stray into E China Sea. Hybrids of intermediate plumage are recorded in Japan and also Hong Kong.

Habits: As Eurasian Wigeon but with preference for fresh water marshes.]

89. Mallard *Anas platyrhynchos* Lütou ya   PLATE 8

Description: Medium-sized (58 cm) wild form of domestic duck. Male has distinctive dark shiny green head and neck separated from chestnut breast by white collar. Female is mottled brown with dark eye-stripe. Distinguished from female Pintail by shorter and blunter tail and from female Gadwall by larger size and different wing pattern. Iris—brown; bill—yellow; feet—orange.

Voice: Male has soft rasping *kreep*. Female gives familiar *quack quack quack* of domestic duck.

Range: Holarctic; winters south.

Distribution and status: Breeds in NW and NE China. Winters in SW Xizang and C and S China south of 40 °N including Taiwan. Locally common.

Habits: Frequents lakes, ponds and river estuaries.

90. Spot-billed Duck *Anas poecilorhyncha* Banzui ya   PLATE 8

Description: Large (60 cm) dark brown duck with pale head, dark crown and eye-stripe and diagnostic yellow-tipped black bill with black terminal spot in breeding season. Throat and cheeks are buffy. Race *zonorhyncha* has dark stripe across cheeks and blacker body. Plumage heavily scalloped by pale-edged dark feathers. Speculum is bluish (*zonorhyncha*) or greenish-purple (*haringtoni*) usually with white band on trailing edge. White tertials sometimes visible at rest and diagnostic in flight. Sexes are alike but female is duller. Iris—brown; bill—black tipped with yellow; feet—coral-red.

Voice: Female call is *quack* like domestic duck; often in descending series. Male utters rasping *kreep*.

Range: India, Burma, NE Asia and China.

Distribution and status: Race *zonorhyncha* breeds across the eastern half of China with northern birds wintering south of Changjiang river. Race *haringtoni* is resident in S and SW Yunnan, Guangdong and Hong Kong. Widespread and quite common.

Habits: Inhabits lakes, rivers and coastal mangroves and lagoons.

91. Philippine Duck *Anas luzonica* Zongjing ya   PLATE 8

Description: Large (63 cm) unmistakable grey-mottled dabbling duck with distinctive head pattern. Head and neck rufous contrasting with black crown and eye-stripe. In flight looks dark with whitish underwing and green speculum, narrowly bordered before and after with white. Sexes alike. Iris—dark brown; bill—bluish-grey; feet—brownish-grey.

Voice: Like Mallard but harsher.

Range: Endemic to Philippines.

Distribution and status: Globally Near-threatened (Collar *et al.* 1994). Accidental to S Taiwan.

Habits: Inhabits marshes, rivers, lakes, ponds and tidal creeks.

92. Northern Shoveler *Anas clypeata* [Common Shoveler] Pizui ya   PLATE 8

Description: Unmistakable large (50 cm) duck with extremely long, broad spatulate bill. Male: chestnut belly, white breast and dark glossy green head distinctive. Female is mottled brown with whitish tail and dark eye-stripe. Coloration similar to female Mallard but bill unmistakable. In flight pale grey-blue upper wing coverts contrast with dark flight feathers and green speculum. Iris—brown; bill—blackish (breeding male) orange-brown (female); feet-orange.

Voice: Similar to Mallard but softer and lower, also chuckling *quack*.

Range: Breeds through Holarctic; winters south.

Distribution and status: Breeds in NE and NW China migrating to winter over most of China south of 35 °N including Taiwan. Locally common.

Habits: Favours coastal lagoons, ponds, lakes and mangrove swamps.

93. Northern Pintail *Anas acuta* [Pintail] Zhenwei ya PLATE 8

Description: Medium-sized (55 cm) duck with long pointed tail. Male has brown head, white throat, scalloped grey flanks, black tail with extremely elongated central feathers, grey wings with a green-bronze coloured speculum and white underparts. Female is more drab—brown, speckled with black above; below buff with black spots on breast; wings grey with brown speculum; grey bill and feet. Distinguished from other female *Anas* by elegant shape, plain brown head and pointed tail. Iris—brown; bill—blue/grey; feet—grey.

Voice: Rather silent. Female gives low guttural *kuuk-kuuk*.

Range: Breeds throughout Holarctic; winters south.

Distribution and status: Breeding recorded in Tianshan in NW Xinjiang and S Xizang. Migrates across most of China to winter south of 30 °N including Taiwan.

Habits: Frequents marshes, lakes, large rivers and sea coasts. On water it is a surface feeder but sometimes up-ends in shallow water.

94. Garganey *Anas querquedula* [Garganey Teal] Baimei ya PLATE 9

Description: Medium-sized (40 cm) dabbling duck. Male with distinctive chocolate head and broad white eyebrow. Brown back and breast contrast with white belly. Scapulars are long, black and white. Speculum glossy green with white margins. Female is brown with distinctive strongly patterned head, white belly and dull olive speculum with white trailing edge. Eclipse male is like female except wing pattern. In flight the blue-grey wing coverts of the drake are a good characteristic. Iris—hazel; bill—black; feet—bluish-grey.

Voice: Generally silent. Drake gives a rattle-like croak. Female gives slight *kwak*.

Range: Breeds through Palearctic; winters south.

Distribution and status: Breeds in NE and NW China. Migrates south to winter over most of China south of 35 °N including Taiwan and Hainan. Rather uncommon.

Habits: Gregarious in large flocks in winter. Sometimes frequents coastal lagoons. Sleeps on water during day, feeding at night.

95. Baikal Teal *Anas formosa* Hualian ya PLATE 9

Description: Male: medium-sized (42 cm) with dark crown and characteristic yellowish crescentic patches on face, patterned with glossy green. Breast is spotted with rufous wash, flanks are scaled as in Common Teal. Scapulars are elongated with black centres and white upper edges. Speculum is bronze-green and vent is black. Female: like Garganey and Common Teal but larger and with white spot at base of bill; also white crescent on side of face. Iris—brown; bill—grey; feet—grey.

Voice: Male gives deep chuckling *wot-wot-wot* call in spring. Female gives low *quack*.

Range: Breeds NE Asia, wintering in China, Korea and Japan.

Distribution and status: Globally Vulnerable (Collar *et al.* 1994). Breeds on small lakes in NE China. Winters in small numbers in parts of C and S China and rarely Hong Kong. The population of the bird has been dramatically declining over the past 20 years.

Habits: Gregarious, forming large flocks and mixing freely with other species. Feeds by dabbling and also on paddy fields. Roosts on lakes and estuaries.

96. Common Teal *Anas crecca* [Green-winged Teal] Lüchi ya   PLATE 9

Description: Small (37 cm) fast-flying duck with conspicuous green speculum in flight. Male has distinct metallic green, buff-bordered eye-stripe through chestnut head, a long white stripe on scapulars and buff patch bordering dark undertail; rest of plumage generally greyish. Female mottled brown with pale belly. Distinguished from female Garganey by bright green speculum, dark forewing and plain head. American race *carolinensis* is distinguished by vertical white breast stripe and no white wing stripe. Iris—brown; bill—grey; feet—grey.

Voice: Male call is a metallic rattling *kirrik*; female a thinner, higher, short *quack*.

Range: Breeds throughout Palearctic; winters south.

Distribution and status: Nominate race breeds in Tianshan in NW Xinjiang and in NE provinces. Migrates over most of China to winter in all non-desert parts south of 40 °N. Locally common. American race *carolinensis* has been recorded in Japan and Hong Kong and may be overlooked elsewhere.

Habits: Generally in pairs or flocks on small lakes or ponds often mixing with other waterfowl. Flies with very fast wing beats.

97. Marbled Duck *Marmaronetta augustirostris* [Marbled Teal]   PLATE 10
Yunshiban ya

Description: Small (40 cm) slender, pale, sandy-brown duck with diagnostic dusky bill and mask. Looks large-headed. Plumage blotched above and below with buffish-white spots. Underwing pale in flight. No speculum. Iris—dark brown; bill—bluish-grey; feet—olive-green to dull yellow.

Voice: Rather silent. Both sexes utter nasal squeaks in display.

Range: Mediterranean and W Asia.

Distribution and status: Globally Vulnerable (Collar *et al.* 1994). Rare in China and throughout range. Status uncertain. Recorded breeding only in W Xinjiang on Kekamkyi Lake (Harvey 1986).

Habits: Forms small flocks even in breeding season. Inhabits small shallow lakes and ponds. Flies low and slow.

98. Red-crested Pochard *Rhodonessa rufina* Chizui qianya   PLATE 10

Description: Large (55 cm) buffy duck. Breeding male unmistakable with rusty head and orange bill contrasting with black foreparts. White flanks, black rear, and white

underwing and flight feathers conspicuous in flight. Female: brown without white flanks but with white lower face, throat and side of neck. Forehead, cap and nape dark brown, darkest around eye. Eclipse male as female but with red bill. Iris—reddish-brown; bill—orange-red (male) black with yellow tip (female) feet—pink (male) grey (female).
Voice: Rather silent. In courtship display male gives rasping wheeze and female a grating chatter.
Range: Breeds in E Europe and W Asia; winters to Mediterranean, Middle East, India and Burma.
Distribution and status: Locally common seasonal migrant. Breeds in NW China as far east as Wuliangsuhai and Ulan Suhai Lake in Nei Mongol. Winters in scattered localities across C, SE and SW China.
Habits: Lives on lakes or slow rivers with good fringing vegetation or reeds.
Note: Formerly assigned to the genus *Netta*.

99. Common Pochard *Aythya ferina* Hongtou qianya PLATE 10

Description: Medium (46 cm) smart-looking duck with chestnut-red head contrasting with clear grey bill and black breast and mantle. Rump is black but back and sides appear grey. At close range they can be seen to be white with fine black vermiculations. In flight the grey wing band contrasts little with darker rest of wing. Female has grey back but brownish head, breast, tail and buffy 'spectacles'. Iris—red (male) brown (female); bill—grey with black tip; feet—grey.
Voice: Male gives soft wheezy whistle. Female gives harsh, growled *krrr* when flushed.
Range: W Europe to C Asia, winters in N Africa, India and S China.
Distribution and status: Breeds in NW China but migrates in winter to E and S China.
Habits: Lives on ponds and lakes with abundant water vegetation.

100. Canvasback *Aythya valisneria* Fanbei qianya PLATE 10

Description: Largish (56 cm) diving duck with rufous head, black breast and rear and greyish-white 'saddle'. Colour pattern as in Pochard but larger and looks more long-necked with much longer bill and characteristic high-domed head profile. Head becomes blacker at front and crown. Female duller and browner than male, but more contrast between breast and sides than in female Pochard. Iris—red (male) or dark brown (female); bill—black; feet—bluish-grey.
Voice: Rather silent. In display male gives soft cooing notes and female responds with harsh *krrr*.
Range: Breeds Alaska, Canada and N USA: winters in temperate N America and sometimes Japan.
Distribution and status: Accidental recorded off N Taiwan.
Habits: Winters on open lakes, coastal lagoons, sheltering bays, and estuaries. Flies high in V-formation. Shy but sociable. Forms rafts.

101. Ferruginous Pochard *Aythya nyroca* [White-eyed Pochard/ PLATE 10
Ferruginous Duck] Baiyan qianya

Description: Medium-sized (41 cm) all dark duck with white undertail. Male has rich chestnut head, neck, breast and flanks and has white eye; female is dark sooty-brown with dark eye. In profile head has high raised crown. In flight, flight feathers are seen to be white with narrow black trailing edge. Both sexes distinguished from female Tufted Duck by white undertail coverts (sometimes also shown by female Tufted), head shape, lack of tuft and lack of subterminal black band on bill. Distinguished from Baer's Pochard by less extensive white on sides. Iris—white (male) brown (female); bill—bluish-grey; feet—grey.

Voice: Male gives whistled *wheeoo* in courtship; female utters harsh *gaaa*. Otherwise rather silent.

Range: Palearctic wintering in Africa, Middle East, India and SE Asia.

Distribution and status: Globally Vulnerable (Collar *et al.* 1994). In China locally common to rare. Breeds in W Xinjiang; Wuliangsuhai (30 000 birds) and Ulan Suhai Lake of Nei Mongol; scattered lakes in S Xinjiang and probably other parts of W China. Winters in mid-Changjiang valley, also NW Yunnan and seen on passage elsewhere. Vagrant to Hebei and Shandong.

Habits: Inhabits marshes and fresh water lakes. In winter also uses estuaries and coastal lagoons. Shy and wary; usually in pairs or small parties.

102. Baer's Pochard *Aythya baeri* Qingtou qianya PLATE 10

Description: Compact (45 cm) blackish duck with dark brown breast and white belly and sides white; underwing and white secondaries with black leading edge on wing in flight. Breeding male has glossy green sheen to head. Distinguished from male Tufted Duck by lack of tuft, smaller and less sharply delineated white patch on sides and white undertail (*n.b.* Tufted is occasionally found with white undertail). From Ferruginous Pochard by less rufous shade of brown and white of belly extending onto sides. Iris— white (male) brown (female); bill—bluish-grey; feet—grey.

Voice: Harsh *graaaak* calls during courtship by both sexes; otherwise rather silent.

Range: Siberia and NE China wintering to SE Asia.

Distribution and status: Globally Vulnerable (Collar *et al.* 1994). Formerly common, now uncommon seasonal migrant in China. Breeds in NE China; on passage across E China and wintering over most of southern China. Occasional birds recorded at Mai Po, Hong Kong.

Habits: Shy, pair-living duck. Mixes with other ducks. Inhabits ponds, lakes and slow rivers.

103. Tufted Duck *Aythya fuligula* Fengtou qianya PLATE 10

Description: Medium-sized (42 cm) squat chunky duck with distinctive long crest. Male is black with white belly and sides. Female is dark brown with brown flanks and short crest. In flight shows a white band on secondaries. Undertail occasionally white. Female can show pale cheek patches. Juvenile like female but with brown eye. Head profile on top, flatter and with more prominent brow than Ferruginous Pochard. Iris—yellow; bill and feet—grey.

Voice: Generally silent in winter. Harsh, low *kur-r-r, kur-r-r* given in flight.

Range: Breeds throughout N Palearctic; winters south.

Distribution and status: Breeds in NE China, migrating over most of China to winter in SE including Taiwan. Locally common.

Habits: Frequents lakes and deep ponds, diving for food. Flight is fast.

104. Greater Scaup *Aythya marila* Banbei qianya — PLATE 10

Description: Medium-sized (48 cm) squat duck. Male is longer than Tufted Duck with grey back and no head tuft. Female differs from female Tufted in having broad white ring around base of bill. Larger than very similar Lesser Scaup but lacks slight crest and in flight can be separated by white base to primaries. Iris—whitish-yellow; bill—bluish-grey; feet—grey.

Voice: In courtship display male gives soft *coos* and whistles; female responds with harsh gruff notes. Otherwise rather silent.

Range: Holarctic. Breeds N Asia and winters in temperate coastal waters including SE Asia and Philippines.

Distribution and status: Uncommon winter visitor. Recorded on passage around Yellow Sea and wintering in coastal provinces of SE and S China and Taiwan.

Habits: Marine duck mostly in coastal waters or estuaries; sometimes on fresh water lakes. Sociable and lives in flocks.

Note: Lesser Scaup *Aythya affinis* has been recorded as occasional vagrant to Japan and some reported sightings suggest it probably also reaches China.

105. Steller's Eider *Polysticta stelleri* Xiao rongya — PLATE 11

Description: Smallish (45 cm) blackish marine duck. Breeding male unmistakable with white head; black collar, eye-ring, chin and nob on back of head; breast creamy with dark spot on side of breast. Long black and white scapular plumes. Females and eclipse male dark brown with pale eye-ring; green speculum edged white fore and aft; elongated glossy wing coverts. Male has white upper wing coverts forming a stripe when folded. In flight underwing is white. Iris—reddish-brown; bill—blue-grey; feet—blue-grey.

Voice: In display, male gives low growled *croons* and short barks. Female gives low barks, growls and whistles. Otherwise fairly silent.

Range: Breeds Siberia and Alaskan Arctic. Winters to N Europe, NW North America and through Aleutians to Kamchatka Peninsula and N Japan.

Distribution and status: Globally Vulnerable (Collar *et al.* 1994). Rare winter visitor. Winters in estuaries of Wusuli River in Heilongjiang; also recorded Beidaihe (Hebei) in 1985.

Habits: Breeds on fresh water pools but mostly lives in coastal waters near stream outlets. Sociable and forms flocks. Often cocks tail whilst swimming.

106. Harlequin Duck *Histrionicus histrionicus* Chou ya — PLATE 11

Description: Small (42 cm) compact dark marine duck with white spots on face and ear coverts, high head and small bill. Breeding male is grey with chestnut sides and white stripes on back of head, upper and lower breast and wing coverts. Scapulars are elongated and black and white. Non-breeding male is dark brown but white of scapulars and lower breast stripe usually still visible. Female is similar but without white on scapulars or breast stripe. In flight underwing is black. Iris—dark brown; bill—grey; feet—grey.

Voice: In courtship male utters high-pitched whistle and female gives short, harsh calls. Otherwise silent.

Range: E Asia and N America, Greenland and Iceland.

Distribution and status: Very rare winter migrant. Recorded in Heilongjiang (Harbin), Liaoning (Lüshun), Hebei (Beidaihe), and Shandong (Qingdao).

Habits: Lives in pairs or flocks. Winters in marine waters in rocky bays. Swims with tail cocked. Flies fast and low, pattering along surface before take-off.

107. Long-tailed Duck *Clangula hyemalis* Changwei ya PLATE 11

Description: Winter male: medium-sized (58 cm) grey, black and white duck with elongated central tail feathers, black breast and large black patch on side of neck. Winter female is brown with white head and belly. Black cap and black patch on side of neck. In flight, combination of black underwing and white belly distinctive. Male's black breast bar diagnostic. Iris—dull yellow; bill—male grey with pink band near tip, female grey; feet—grey.

Voice: Male rather vocal especially in display, uttering loud yodelling *ow-ow-ow-lee . . . caloo caloo*. Female gives variety of low weak quacking notes.

Range: Holarctic.

Distribution and status: Very rare winter visitor. Wintering in Hebei, mid-Changjiang valley and Fujian.

Habits: In winter stays in shallow coastal waters, rarely in fresh water. Feeds by diving. Flies in long straggling lines low over water.

108. Black Scoter *Melanitta nigra* Hei haifanya PLATE 11

Description: Smallish (50 cm) all dark, squat marine duck. Male entirely black with large yellow lump on base of bill. Female dusky-brown with black crown and nape but buffy-grey face and foreneck. In flight, wings are blackish with dark underwing. Iris—dark brown; bill—grey; feet—dark grey.

Voice: Rather silent. Displaying male utters piping, high whistle; female a harsh, grating rasp.

Range: Breeds in Eurasian tundra and Alaska. Winters to coasts of N America and Europe and through Aleutians to Japan and east coast of Korea.

Distribution and status: Very rare winter visitor to China. Recorded wintering in Jiangsu (Zhenjiong) and Fujian (Liangjiang).

Habits: Gregarious in tight rafts at sea. Migrating birds may temporarily rest on fresh water.

109. White-winged Scoter *Melanitta fusca* [Velvet Scoter] PLATE 11
Banlian haifanya

Description: Medium-sized (56 cm) dark, squat marine duck. Adult male all black with white spot beneath and behind eye. Female dusky brown with whitish spot between eye and bill and another on ear coverts. In flight, white secondaries distinguish it easily from Black Scoter; also pale feet. Iris—white (male) brown (female); bill—male black knobbed at base, otherwise pink and yellow, female greyish; feet—pinkish.

Voice: Rather silent. Displaying male gives loud piping and female a harsh *karrr*.
Range: Northern Hemisphere. Asian population winters off coast of Korea and Japan.
Distribution and status: Race *stejnegeri* is uncommon winter visitor to east coast provinces and Nanchang area of Sichuan. Seen on passage over NE China especially Beidaihe. Two records from Hong Kong.
Habits: Breeds inland, winters at sea. Habits as Black Scoter.
Note: Two species may be involved: nominate *fusca* (Velvet Scoter) and *deglandi*, including *stejnegeri* (White-winged Scoter), see Stepanyan (1990).

110. Common Goldeneye *Bucephala clangula* Que ya   PLATE 11

Description: Medium-sized (48 cm) dark diving duck with large high-domed head and golden eye. Breeding male has white breast and belly, white patch on secondaries and upperwing coverts. Large round white spot at base of bill; otherwise head black with green gloss. Female is smoky-grey with whitish scalloping; brown head without white spot or purple gloss; usually narrow white forecollar. Non-breeding male like female but spot near base of bill remains pale. Iris—yellow; bill—blackish; feet—yellow.
Voice: Rather silent. In flight has whistling wing beat. In display, male utters series of strange whistling and grating notes. Female gives harsh *graa*; also given when flushed.
Range: Holarctic. Breeds N Asia wintering to C and SE China.
Distribution and status: Rare seasonal migrant. Breeds in N Heilongjiang and NW China. Recorded on passage in northern part of China and wintering widely over southern China and Taiwan.
Habits: Sociable forming large rafts on lakes, coastal waters; only loosely mixing with other species. Feeds by diving. Swims with tail cocked. Sometimes rests on land.

111. Smew *Mergellus albellus* Bai qiushaya   PLATE 11

Description: Small (40 cm) elegant pied duck. Breeding male is white with black mask, nape stripe, mantle, primaries and pattern of narrow lines at side of breast. Sides of body finely vermiculated with grey lines. Female and non-breeding male are grey above with two white wing bars, white underparts, blackish ocular region and chestnut forehead, crown and nape. Distinguished from mergansers by white throat. Iris—brown; bill—blackish; feet—grey.
Voice: Generally silent. Male gives low croaks and whistles in display. Female gives low growling note.
Range: N Europe and N Asia, wintering to N India, China and Japan.
Distribution and status: Breeds in swampy areas of NE Nei Mongol, migrating south over much of country in winter. Widespread but generally uncommon.
Habits: Lives on small ponds and rivers and breeds in tree holes.

112. Red-breasted Merganser *Mergus serrator*   PLATE 11
Hongxiong qiushaya

Description: Medium-sized (53 cm) dark fishing duck with long, slender, hooked bill and long silky, spiky crest. Male is black and white with fine vermiculated pattern on sides. Distinguished from Scaly-sided Merganser by rufous on breast, streaked dark and from Common Merganser by dark breast and longer crest. Female and non-breeding male duller and brown with reddish head grading into greyish-white neck. Iris—red; bill—red; feet—orange.

Voice: Rather silent. Displaying male utters various soft, catlike mews. Female gives grating calls in display and in flight.

Range: Holarctic, India, China; winters SE Asia.

Distribution and status: Breeds in N Heilongjiang; migrates across much of China and winters in coastal provinces of SE China and Taiwan.

Habits: As other mergansers.

113. Scaly-sided Merganser *Mergus squamatus* [Chinese Merganser]   PLATE 11
Zhonghua qiushaya

Description: Male: large (58 cm) greenish-black and white duck with long narrow reddish bill with hooked tip. Black head has shaggy crest. Flanks are patterned with characteristic white feathers edged and shafted black which give scaly pattern. Distinguished from Red-breasted Merganser by white breast and from Common Merganser by scaly sides. Female is duller and greyer and differs from Red-breasted Merganser in having sides patterned with broad grey and narrow black concentric bands. Iris—brown; bill—orange; feet—orange.

Voice: As Red-breasted Merganser.

Range: Breeds Siberia, N Korea and NE China; wintering to S and C China, Japan and Korea; accidental in SE Asia.

Distribution and status: Globally Vulnerable (Collar *et al.* 1994). Rare and decreasing in China. Breeds in NE China; on passage over north-east coast and occasionally wintering in C, SW, E and S China and Taiwan.

Habits: Found on fast-flowing rivers and sometimes open lakes. Lives in pairs or family groups. Dives after fish.

114. Common Merganser *Mergus Merganser* [Goosander]   PLATE 11
Putong qiushaya

Description: Largish (68 cm) fish-catching duck with long, slender hooked bill. Breeding male unmistakable with greenish-black head and back contrasting with clean creamy breast and underparts. In flight wing is white with black outer third. Female and non-breeding male dark grey above, pale grey below with rufous-brown head and white chin. All plumages have shaggy mane much shorter than Scaly-sided Merganser and thicker than smaller Red-breasted Merganser. In flight secondaries and secondary coverts all-white not divided by black bar as in Red-breasted. Iris—brown; bill—red; feet—red.

Voice: Rather silent. Displaying male gives twanging *uig-a* call and female has several harsh calls.

Range: Northern Hemisphere.

Distribution and status: Fairly common resident and seasonal migrant. Nominate race breeds in NW and NE China migrating to winter over most of China south of Huanghe (Yellow River). Vagrant on Taiwan. Race *comatus* is an altitudinal migrant resident on the Qinghai–Xizang Plateau lakes with some birds wintering to SW China.

Habits: Sociable, forms flocks on lakes and fast-flowing rivers. Dives for fish.

BUTTONQUAILS—Order: Turniciformes Family: Turnicidae

The buttonquails are tiny, short-tailed, compact birds similar in general appearance to the true quails of the pheasant family but lacking the hind toe. They also show reversed sexual roles with the female more brightly coloured and more aggressively territorial than the male. She often mates with several males and leaves them to incubate the eggs and raise the chicks. There are only three species in China.

115. Small Buttonquail *Turnix sylvatica* [Little/Common   PLATE 32
Buttonquail] Lin sanzhichun

Description: Very small (14 cm) rufous-brown buttonquail Distinguished by rufous chest, white streaks on upperparts and reddish and black blotches on flanks. Larger female is darker and redder than male. Iris—yellow; bill—grey; feet—whitish.

Voice: Low mooing or crooning notes.

Range: Africa, S Eurasia, India, SE Asia, SE China, Philippines and Java.

Distribution and status: Race *mikado* is a rare lowland resident in Guangdong, Taiwan and Hainan in open grasslands.

Habits: Similar to the commoner Barred Buttonquail.

116. Yellow-legged Buttonquail *Turnix tanki*   PLATE 32
Huangjiao sanzhichun

Description: Small (16 cm) rufous-brown buttonquail with bold black spots on upperparts and sides of breast. In flight wing coverts are pale buff, contrasting with dark brown flight feathers. Distinguished from other buttonquails by yellow legs. Female has more chestnut on nape and back than male. Iris—yellow; bill—pinkish-horn; feet—yellow.

Voice: Loud booming calls.

Range: E Asia, India, China and SE Asia.

Distribution and status: Fairly common up to 2000 m. Race *blandfordii* breeds across most of SW, S, C, E and NE China with northern populations migrating to southern China in winter.

Habits: Lives in small parties in scrub, grass, marshland, and on cultivated fields, especially rice stubble.

117. Barred Buttonquail *Turnix suscitator* [Lesser Sunda/sunda   PLATE 32
Buttonquail] Zong sanzhichun

Description: Small (16 cm) russet-brown quail-like bird. Larger female has black chin and throat, blackish crown with grey and white mottled head. Male has brown mottled crown, brown and white streaked face and chin, and black barring on chest and flanks. Both sexes have brown mottled upperparts and rufous chest and flanks. Iris—brown; bill—grey; feet—grey.

Voice: Female courting call is purring *krrrr* maintained for several seconds at a time, often at night.

Range: India, Japan, S China, SE Asia, Philippines, Sulawesi, Sumatra, Java, Bali, and Lesser Sundas.

Distribution and status: Race *blakistoni* is resident in tropical S China and Hainan; *rostrata* on Taiwan; and *plumbipes* in SW Yunnan to west of Nujiang (Salween) River and SE Xizang. Locally common in suitable habitats from sea-level to about 1500 m.
Habits: Lives singly or in pairs in open grassy habitats. When flushed these birds jump up, fly low over the ground for 20 m or so then drop down into the herbs and hide.

HONEYGUIDES—Order: Piciformes Family: Indicatoridae

This small family is largely African in distribution with just two rare species in Asia, one marginal to China.

The honeyguides have two toes pointing backwards like barbets but in colour and shape they resemble finches with short strong bills and no rictal bristles.

They nest in tree holes and feed largely on bees and wasps and their wax combs. Their name is derived from some African species that have learned to lead humans and honey badgers to wild bee hives to encourage them to open up the hive.

118. Yellow-rumped Honeyguide *Indicator xanthonotus*   PLATE 65
Huangyao xiangmilie

Description: Smallish (15 cm) dull olive-grey finch-like bird. Male: yellow brow, crown and cheeks, bright golden back and rump and white stripe on tertials diagnostic. Underparts whitish with dark streaks. Female duller with less yellow on head. Iris—brown; bill—yellowish-brown; feet—grey-green.
Voice: Quiet *weet* call in flight.
Range: Pakistan (formerly), N and NE Indian subcontinent, NE Burma.
Distribution and status: Rare in temperate forest zone from 1450–3500 m along S face of Himalayas. Recently recorded by staff of Kunming Institute of Zoology in SE Xizang.
Habits: Lives close to honeybee combs on which it makes regular raids.

WOODPECKERS—Family: Picidae

Woodpeckers are a large family of birds with long, powerful beaks for chiselling into wood. They are distributed almost worldwide except for the Australasian region. All species work over the trunks and branches of trees, chipping or poking under bark or into rotten wood for insects and grubs which are picked up with a long sticky tongue. The feet are adapted for clinging to trees with only two forward-pointing toes and one or two pointing backwards. The stiffened tail feathers are used as a stabilising prop when drilling. Woodpeckers excavate tree holes for nesting, fly with an erratic dipping flight and have harsh discordant calls. Thirty-one species occur in China. They vary in size from the tiny (9 cm) White-browed Piculet to the large (47 cm) Great Slaty Woodpecker.

119. Eurasian Wryneck *Jynx torquilla* Yilie   PLATE 12
Description: Small (17 cm) grey-brown woodpecker with distinctive cryptic mottled plumage and barred underparts. Bill relatively short, conical and pointed. Tail long for a woodpecker and indistinctly banded. Iris—pale brown; bill—horn; feet—brown.

Voice: Series of loud, nasal, moaning *teee-teee-teee-teee* like kestrel. Young beg with high ticking *tixixixixix* . . . call.

Range: Africa, Eurasia, India, China, Taiwan and SE Asia.

Distribution and status: Locally common. Race *chinensis* breeds in C, N and NE China, wintering to S and SE China, Hainan and Taiwan. Nominate race passes through NW China on migration (maybe winters in Tianshan) and race *himalayana* winters in SE Xizang.

Habits: Perches across tree limbs unlike other woodpeckers and does not drum or chip away at wood to feed. When cornered it gives display twisting head from side-to-side (hence name). Generally single. Feeds mostly on ground on ants. Prefers thickets.

120. Speckled Piculet *Picumnus innominatus* Banji zhuomuniao   PLATE 12

Description: Tiny (10 cm) olive-backed, tit-like, woodpecker with diagnostic boldly spotted underparts and black and white striped face and tail. Forecrown tipped orange in male. Iris—red; bill—blackish; feet—grey.

Voice: Sharp-noted *tsit* repeated; or alarm rattle.

Range: Himalayas to S China, SE Asia, Borneo and Sumatra.

Distribution and status: Nominate race occurs in SE Xizang; *chinensis* is resident across C, E, S and SE China; race *malayorum* occurs in W and S Yunnan. Uncommon in evergreen broadleaf forests up to 1200 m.

Habits: Found in mixed tropical submontane forest, on dead trees or branches of trees, particularly in areas of bamboo. Makes a slight, persistent tapping noise when looking for food.

121. White-browed Piculet *Sasia ochracea*   PLATE 12
Baimei zongzhuomuniao

Description: Tiny (9 cm) green and orange, tit-like, short-tailed woodpecker. Forehead of male is yellow, female rufous; upperparts olive-green; eyebrow white; underparts rufous; has only three toes. Immature: duller. Iris—red; bill—blackish; feet—yellow.

Voice: Single, sharp noted *tsit* repeated several times; or in alarm a rapid, insistent *kih-kih-kih-kih-kih*.

Range: Himalayas to S China.

Distribution and status: This is an uncommon bird of lowlands and hills up to 2000 m. Race *reichenowi* in W and SW Yunnan; nominate race in SE Xizang; race *kinneari* in SE Yunnan, Guangxi and S Guizhou.

Habits: Found in lower and middle storeys of broadleaf and secondary forests and particularly bamboo. Makes a slight, persistent tapping noise on tree trunks and fine stems when looking for food.

122. Grey-capped Pygmy Woodpecker *Dendrocopus canicapillus*   PLATE 13
[Grey-headed Pygmy Woodpecker] Xingtou zhuomuniao

Description: Small (15 cm) black and white striped woodpecker with no red on underparts and grey crown; male has a red streak above and behind the eye. Orange-buff wash on belly is streaked blackish. Race *nagamichii* lacks white humeral patch and *omissus*, *nagamichii* and *scintilliceps* have black barring on white back. Iris—whitish-brown; bill—grey; feet—greenish-grey.

Voice: Shrill squeaking trill *ki-ki ki ki rrr...*

Range: Pakistan, China, SE Asia, Borneo and Sumatra.

Distribution and status: Widespread but generally uncommon in most forest types up to 2000 m. Race *doerriesi* is resident in NE China; *scintilliceps* from Liaoning through E China; *nagamichii* in S and SE China; *szetschuanensis* in C China; *omissus* in SW China; *obscurus* in S Yunnan; *semicoronatus* in SE Xizang; *swinhoei* on Hainan; and *kaleensis* on Taiwan.

Habits: As other small woodpeckers.

123. Japanese Pygmy Woodpecker *Dendrocopos kizuki* PLATE 13
[Pygmy Woodpecker] Xiao xingtou zhuomuniao

Description: Small (14 cm) pied woodpecker. Upperparts black with white spots on back and in rows across wings. Edges of outer tail feathers white. Has white patch behind ear coverts; short white supercilium and moustachial and faint red stripe at rear above supercilium. Underparts buff streaked black with greyish bar across breast and white bib. Iris—brown; bill—grey; feet—grey.

Voice: Noisy. Sharp *khit* or *khit-khit-khit* and buzzing *kzz, kzz* and drumming.

Range: NE China, SE Siberia, Korea, Japan, Ryukyu Islands.

Distribution and status: Uncommon resident in NE Heilongjiang through Liaoning (*permutatus*) and from Hubei as far as Huanghe river in Shandong (*wilderi*) from lowlands to over 2000 m.

Habits: Lives singly or in pairs. Sometimes joins mixed bird flocks. Uses all kinds of woodlands, parks and gardens.

124. Lesser Spotted Woodpecker *Dendrocopos minor* PLATE 13
Xiaoban zhuomuniao

Description: Small (15 cm) woodpecker with black upperparts marked with rows of white spots and whitish underparts streaked narrowly black on sides. Male has red cap but black nape and whitish forecrown. Race *amurensis* has clean white unstreaked underparts. Iris—reddish-brown; bill—black; feet—grey.

Voice: Shrill, feeble notes, *peet peet peet peet peet peet*. Also weak *kik* call. Drums more slowly and weakly than Greater Spotted Woodpecker.

Range: Europe, N Africa, Asia Minor to Mongolia, Siberia, and Korea.

Distribution and status: Locally common. Race *kamtschakensis* breeds in the Altai Mts and N Junggar basin of NW Xinjiang and recorded wintering in N Heilongjiang. Race *amurensis* is resident in NE China.

Habits: Deeply undulating flight. Favours deciduous and mixed woods, subalpine birch groves and orchards.

[125. Fulvous-breasted Woodpecker *Dendrocopus macei* PLATE 13
[Streak-bellied Woodpecker] Chaxiongban zhuomuniao

Description: Rather small (18 cm) black and white barred woodpecker. Crown red in male, black in female; sides of face white with black malar stripe and collar; upperparts

barred black and white; underparts buff streaked with black and red undertail coverts. Iris—brown; bill—bluish-black above, bluish-grey below; feet—olive.
Voice: Ringing call *tuk-tuk* and trill *tirri-tierrier-tierrierie.*
Range: Himalayas, India, SE Asia and Greater Sundas.
Distribution and status: Occurs in E Himalayas as a common resident up to 2000 m. Expected in SE Xizang.
Habits: Prefers open and secondary forest, plantations, gardens.]

126. Stripe-breasted Woodpecker *Dendrocopos atratus* PLATE 13
Wenxiong zhuomuniao

Description: Medium-sized (22 cm) boldly marked black, white and red woodpecker. Upperparts black with rows of white spots. Underparts whitish-fulvous with rufous vent and black moustachial stripe extending onto neck. Black streaks on breast. Forecrown white. Male has red crown extending to nape with black band at front of crown. Distinguuished from Fulvous-breasted by heavier streaking on breast, blacker tail and whiter face. Iris—reddish-brown; bill—greenish-horn with blackish tip; feet—grey-green.
Voice: Explosive *tchick* and shrill descending whinny. Curious creaking, grating sound.
Range: NE India to SW China and SE Asia.
Distribution and status: Rare resident in China. Recorded in W, NW and S Yunnan.
Habits: Favours tropical evergreen forests from 800–2200 m.

127. Rufous-bellied Woodpecker *Dendrocopos hyperythrus* PLATE 13
Zongfu zhuomuniao

Description: Medium-sized (20 cm) colourful woodpecker. Back, wings and tail black with rows of white dots; sides of head and underparts diagnostic rich rufous; vent red. Male has red crown and nape. Female has black crown, spotted white. Race *marshalli* has red of nape extending to rear of ear coverts; nominate race has more cinnamon underparts than other two races. Iris—brown; bill—grey with black tip; feet—grey.
Voice: Long rattling, staccato cry *kii-i-i-i-i-i-i*; trailing off at end; like Greater Flameback but weaker. Both sexes drum on trees.
Range: Himalayas, China and SE Asia.
Distribution and status: Uncommon. Race *marshalli* breeds in W Xizang; nominate race is altitudinal migrant from 1500–4300 m. through SE Xizang to Sichuan and NW, W and S Yunnan. Race *subrufinus* breeds in Heilongjiang at moderate altitudes and migrates across E China to winter in S China.
Habits: Favours conifer or mixed forests.

128. Crimson-breasted Woodpecker *Dendrocopos cathpharius* PLATE 13
Chixiong zhuomuniao

Description: Smallish (18 cm) pied woodpecker. Has broad white wing panel and broad black malar stripe that extends to form band across lower breast. Diagnostic crimson breast patch and red vent. Male has red nape. Female has black nape but may have red patch on side of neck (*ludlowi*). Juvenile has entire crown red but lacks red on breast.

Race *ludlowi* has scarlet patch behind ear coverts and well-marked crimson on breast. Nominate race has little or no crimson on breast but crimson of male's hindcrown extends to sides of neck. Iris—reddish; bill—dark grey; feet—greenish.

Voice: Loud monotonous *chip* and shrill *kee-kee-kee*.

Range: Nepal to SW and C China, N Burma and N Indochina.

Distribution and status: Rare. Nominate race occurs in SE Xizang south of Tsangpo (Brahmaputra); *ludlowi* is resident in SE Xizang resident in W and C Yunnan, *pernyii* from S Gansu, NW and S Sichuan to NW Yunnan; and *innixus* in Qinling Mts of S Shaanxi, Shennongjia area of W Hubei and NE Sichuan. Found in broadleaf oak and rhododendron forests from 1500–2750 m.

Habits: Feeds low down and often on dead trees. Feeds on nectar as well as insects.

129. Darjeeling Woodpecker *Dendrocopos darjellensis*   PLATE 13
Huangjing zhuomuniao

Description: Medium-sized (25 cm) pied woodpecker with very fulvous face and breast with heavy black streaking and light crimson vent. Entire back black; broad white shoulder patch and rows of white spots on wings and outer tail feathers. Male has crimson nape, black in female. Iris—red; bill—grey with black tip; feet—greenish.

Voice: Low *puk puk* call. Drumming in breeding season.

Range: Nepal to SW China, Burma and N Indochina.

Distribution and status: Uncommon in humid forests above 1200 m up to 4000 m. Nominate race resident in S Xizang (Nyalam, Zham); *desmursi* is an altitudinal migrant in SE Xizang, W and NW Yunnan and NW and S Sichuan.

Habits: Feeds at all heights. Sometimes joins mixed-species flocks.

130. White-backed Woodpecker *Dendrocopos leucotos*   PLATE 13
Baibei zhuomuniao

Description: Medium-sized (25 cm) pied woodpecker with diagnostic white lower back. Entire crown crimson in male (black in female) with white forehead. Underparts white with black streaks and pale crimson vent. Wings and outer tail feathers barred with white spots. Race *tangi* has centre of belly buffy yellow; *insularis* has centre of belly brownish. Distinguished from Three-toed Woodpecker by lack of yellow forecrown and more boldly banded wings. Iris—brown; bill—black; feet—grey.

Voice: Quiet *kik* call rather like Blackbird alarm. Drums powerfully for c. 1.7 second bursts accelerating but getting fainter at end.

Range: E Europe to Japan and China.

Distribution and status: Disjunct distribution but fairly common where present. Nominate race in NE China and extreme N Xinjiang; *sinicus* in Hebei and SE Nei Mongol; *fohkiensis* on Wuyishan in NW Fujian and Guanshan in N Jiangxi; *tangi* in Qinling Mts of S Shaanxi to C Sichuan and *insularis* on Taiwan. Inhabits deciduous and mixed forests on mountains from 1200–2000 m.

Habits: Requires old dying trees. Rather tame.

131. Great Spotted Woodpecker *Dendocopus major* PLATE 13
Daban zhuomuniao

Description: Medium-sized (24 cm) familiar black and white, spotted woodpecker. The boldly pied male has a narrow red nape band lacking in the female. Both sexes have a red vent but lack of red or orange on the black-streaked white breast distinguish from related Crimson-breasted and Rufous-bellied Woodpeckers. Iris—reddish; bill—grey; feet—grey.

Voice: Drums loudly and has harsh shrieking calls.

Range: Eurasian temperate forest zone, NE India, W, N and E Burma and N Indochina.

Distribution and status: The most widespread woodpecker in China; found throughout temperate woodlands, farmlands and city parks. Nine races are recognised: *tianshanicus* in NW China; *brevirostris* breeds in Greater Hinggan Mts and winters in Lesser Hinggan Mts and plains of NE China; *wulashanicus* in Helan and Wula Mts of Ningxia and N Shaanxi; *japonicus* in Liaoning, Jilin and E Nei Mongol; *cabanisi* in eastern N China; *beicki* in NC China; *stresemanni* in SC and SW China; *mandarinus* in S and SE China; and *hainanus* on Hainan.

Habits: Typical of the genus, nesting in excavated tree holes and feeding on insects and grubs under tree bark.

132. White-winged Woodpecker *Dendrocopos leucopterus* PLATE 13
[AG: Spotted/Pied Woodpecker] Baichi zhuomuniao

Description: Medium-sized (23 cm) pied woodpecker. Similar to Great Spotted Woodpecker with large area of white on closed wing instead of black and white spots. Iris—brown; bill—grey with black tip; feet—grey.

Voice: As genus. Rattling calls and drumming.

Range: Afganistan, Aral Sea, Transcapia through W Xinjiang.

Distribution and status: Globally Near-threatened (Collar *et al.* 1994). An uncommon bird of the Kashi and Junggar basins through Tianshan foothills east to Lop Nor, up to 2500 m.

Habits: Typical woodpecker of poplar forests along streambeds and in foothills of Tianshan.

133. Three-toed Woodpecker *Picoides tridactylus* PLATE 12
Sanzhi zhuomuniao

Description: Medium-sized (23 cm) black and white woodpecker with yellow forecrown (white in female) and only three toes. No red in plumage and centre of mantle and back white. Rump black. Race *funebris* has brown rump, white of back confined to patches on mantle and browner underparts. Iris—brown; bill—black; feet—grey.

Voice: Soft *kik* call. Powerful drumming in 1.3 second bursts similar to Black Woodpecker.

Range: Holarctic.

Distribution and status: Locally common. Nominate race is resident in NE China; race *tianschanicus* in W and NW Xinjiang; and *funebris* from SE Xizang and NW Yunnan to Sichuan, NE Qinghai and Gansu. Inhabits conifer and mixed forests from 2000–4300 m in Himalayas but lowlands in the north.

Habits: Favours old spruce and subalpine birch forest. Makes holes in rings around trunks to collect sap.

134. Rufous Woodpecker *Celeus brachyurus* Li zhuomuniao   PLATE 12

Description: Medium-sized (21 cm) dark rufescent woodpecker. Entire body rufescent with black barring on wings and upperparts, and to a lesser extent on the underparts. Male has red patch just below and behind the eye. Iris—red; bill—black; feet—brown.

Voice: Short, hurried, high-pitched laughing call *kwee-kwee-kwee-kwee* . . . of 5–10 notes on a descending scale. Drums in short decelerating bursts.

Range: S Asia, SE Asia to Greater Sundas.

Distribution and status: A common bird up to 1500 m. Race *phaioceps* is resident in SE Xizang and W and S Yunnan; *fokiensis* in S and SE China; and *holroydi* in Hainan.

Habits: Prefers rather open forest, secondary forest, forest edge, gardens and plantations at low altitudes. Pecks rarely audible.

135. White-bellied Woodpecker *Dryocopus javensis* [Great Black   PLATE 14
Woodpecker] Baifu heizhuomuniao

Description: Unmistakable, large (42 cm) black and white woodpecker. Upperparts and breast black, belly white. Male has red crest and cheek patches; female is largely black with red restricted to the hindcrown. Iris—yellow; bill—horn-slate; feet—grey-blue.

Voice: A loud, sharp, rising yelp *kiyow*; also a loud laugh in flight *kiau kiau kiau* . . . Hammers loudly.

Range: India, China, S. E. Asia, Philippines, and Greater Sundas.

Distribution and status: Race *forresti* is an uncommon resident in SW Sichuan and Yunnan.

Habits: Prefers open lowland forest including mangroves. It is a noisy and conspicuous bird, usually solitary. It forages at all levels.

136. Black Woodpecker *Dryocopus martius* Heizhuomuniao   PLATE 14

Description: Very large (46 cm) diagnostically all-black woodpecker with yellowish bill and red crown. Female has red only on hindcrown. Unmistakable. Race *khamensis* has greenish gloss on head and neck. Iris—whitish; bill—ivory with darker tip; feet—grey.

Voice: Loud *krri-krri-krri* in alarm or in flight. Loud clear *klee-ay* and in spring a loud laughing *kwee-kwee-kwee-kwee-kwee-kwee-kwee* without drop in pitch at end. Also nasal clucking squeaks. Loud burst of drumming of 2–3 seconds duration.

Range: Europe to Asia Minor, Siberia, China and Japan.

Distribution and status: Generally uncommon. Nominate race is resident in Altai Mts of NW China and boreal conifer forests of NE China; *khamensis* is resident in subalpine conifer forest on eastern side of Tibetan Plateau in Qinghai, E Xizang, Gansu, Sichuan and NW Yunnan. Lives from lowlands in north to 2400 m in south of range.

Habits: Flight uneven but not undulating like other woodpeckers. Feeds mostly on ants, making huge feeding craters.

137. Lesser Yellownape *Picus chlorolophus* [Lesser Yellow-naped   PLATE 14
Woodpecker] Huangguan zhuomuniao

Description: Medium-sized (26 cm) bright green woodpecker with fluffy yellow-edged nuchal crest, red markings on face and white malar stripe. Male has distinctive red eyebrow and malar stripe with white upper malar stripe. Female has red only on side of crown. Flanks are barred with white and flight feathers are black. Iris—red; bill—grey; feet—greenish-grey.

Voice: Loud, shrill, descending alarm *kwee-kwee-kwee* or single *pee-ui* call.

Range: Himalayas, S China, SE Asia, Malay Peninsula and Sumatra.

Distribution and status: Uncommon in subtropical forests from 800–2000 m. Nominate race is resident in SE Xizang; *chlorolophoides* in W and S Yunnan; *citrinocristatus* in Fujian; and *longipennis* in Hainan.

Habits: A noisy and conspicuous woodpecker, sometimes travelling in small groups of following mixed-species bird waves.

138. Greater Yellownape *Picus flavinucha* [Greater Yellow-naped   PLATE 14
Woodpecker] Da huangguanzhuomuniao

Description: Large (34 cm) green woodpecker with yellow throat and long yellow crest. Tail black; flight feathers are barred black and brown; rest of plumage green. Female has rufous brown throat. Distinguished from Lesser Yellownape by lack of red on head. Iris—reddish; bill—greenish-grey; feet—greenish-grey.

Voice: Slow *chup* or *chup-chup* followed by staccato roll.

Range: Himalayas, S China, SE Asia and Sumatra.

Distribution and status: Rare in mixed subtropical forest, pine forest and secondary growth from 800–2000 m. Nominate race occurs in W Yunnan; *lylei* in SE Xizang and S Yunnan; *styani* in S Guangxi and Hainan; and *ricketti* in S Sichuan and C Fujian.

Habits: A conspicuous and noisy woodpecker sometimes seen in small family groups.

139. Laced Woodpecker *Picus vittatus* [Laced/Scaly-bellied Green   PLATE 14
Woodpecker] Huafu zhuomuniao

Description: Medium-sized (30 cm) green woodpecker. Crown red in male, black in female; back green; rump yellow; tail black; primaries black with white stripes; throat buff; breast buff with bold green lacing of dark feather edges; black eye-stripe and malar stripe flecked with white; cheeks bluish-grey. Iris—red; bill—black; feet-greenish.

Voice: Plaintive shrill *kweep* call with falling tone.

Range: Bangladesh, SE Asia to Sumatra and Java.

Distribution and status: Race *eisenhoferi* is locally common in suitable habitat up to 200 m in S Yunnan (Xishuangbanna).

Habits: Found in open forest and plantations. Feeds on the ground as well as on fallen trees and among bamboo.

140. Streak-throated Woodpecker *Picus xanthopygaeus*   PLATE 14
Linhou zhuomuniao

Description: Medium-sized (29 cm) green woodpecker with yellowish rump, red crown and grey cheeks. Distinguished from Grey-headed Woodpecker by white malar stripe and green scaly belly. Female has black crown. Iris—pinkish-white with red inner ring; bill—grey with yellow on sides; feet—greyish-green.

Voice: Rather quiet. Gives single falsetto note and drums on trees.

Range: Himalayas, India, SW China and SE Asia.

Distribution and status: Very rare resident in open lowland forests of W Yunnan.

Habits: More terrestrial than most woodpeckers and feeds mostly on ants and termites.

141. Scaly-bellied Woodpecker *Picus squamatus*　　　　PLATE 14
Linfu zhuomuniao

Description: Large (35 cm) green woodpecker with diagnostic pale underparts boldly scalloped with black. Male has crimson crown and crest; broad whitish supercilium bordered above and below with black; whitish cheeks; black moustachial streak; yellow rump. Female duller with black crown speckled grey. Distinguished from Laced and Streak-throated Woodpeckers by barred black and white tail. Iris—red to pink; bill—horn-yellow with grey tip; feet—yellow-green.

Voice: Wild, ringing, melodious *klee-gu*; occasional long drawn-out nasal *cheenk* repeated every 10–15 seconds. Loud vibrating hammering.

Range: Pakistan, N India and Himamalayas.

Distribution and status: Local in S Xizang (Gyirong).

Habits: Flight undulating. Feeds on trees and ground. Lives in pairs or family groups.

142. Red-collared Woodpecker *Picus rabieri* [Rabier's Woodpecker]　　PLATE 14
Hongjing zhuomuniao

Description: Largish (30 cm) green woodpecker with diagnostic red crown, nape, collar and moustachial stripe. Female has blackish-green crown. Distinguished from female Laced Woodpecker by dull rump and plain underparts. Iris—pale brown; bill—grey with dark tip; feet—greyish.

Voice: Shrill cries and drumming.

Range: Laos and N Vietnam.

Distribution and status: Local resident in SE Yunnan (Hekou).

Habits: Found in primary and secondary lowland evergreen forests.

143. Grey-headed Woodpecker *Picus canus* [Grey-faced/　　PLATE 14
Black-naped Green Woodpecker] Huitou zhuomuniao

Description: Medium-sized (27 cm) green woodpecker with diagnostic uniform grey underparts and grey cheeks and throat. Male has scarlet patch on forecrown, loral stripe and narrow black moustachial streak. Nape and tail black. Female lacks red crown patch and has grey crown. Bill is relatively short and bluntish. Many races vary in size and colour tone. Female *sobrinus* has crown and nape black. Female *tancolo* and *kogo* have black streaks on hindcrown and nape. Iris—reddish-brown; bill—greyish; feet—bluish-grey.

Voice: Similar to Green Woodpecker *Picus viridis* 'laughing' but finer, clearer and slowing down at end. Alarm is repeated, agitated *kya*. Drums frequently in rapid loud bursts of at least one second.

Range: Eurasia, India, China, Taiwan, SE Asia and Sumatra.
Distribution and status: Uncommon but widespread in a wide range of forest types and even urban parks. Ten races recognised in China. Race *biedermanni* recorded for Altai Mts of NW Xinjiang; *jessoensis* in NE China; *zimmermanni* in eastern N China; *guerini* over the rest of N China; *sobrinus* in SE China; *hainanus* on Hainan; *tancolo* on Taiwan; *sordidor* in SE Xizang and SW China; *hessei* in S Xishuangbanna in S Yunnan; and *kogo* in E Xizang and Qinghai.
Habits: Shy and wary. Uses small forest blocks and woodland fringes as well as larger forests and woodlands. Sometimes comes to ground to find ants.

[144. Himalayan Flameback *Dinopium shorii* Xishan   PLATE 12
jinbeisanzhizhuomu

Description: Largish (30 cm) woodpecker with golden upperparts, red rump and long red crest. Face is striped black and white. Distinguished with difficulty from Common Flameback by two black lines down sides of throat separated by brown rather than single central black stripe. Underparts are boldly streaked and scalloped with black. Iris—dark brown or crimson; bill—blackish; feet—plumbeous or greenish-brown.
Voice: As Common Flameback.
Range: Himalayas, W Burma.
Distribution and status: Probably resident in SE Xizang. Not yet recorded in China but occurs in disputed territory of NE India up to 700 m.
Habits: As Common Flameback.]

145. Common Flameback *Dinopium javanense*   PLATE 12
[Common Golden-backed (Three-toed) Woodpecker/Common Goldenback] Jinbei sanzhizhuomuniao

Description: Medium-sized (30 cm) colourful woodpecker. Face striped black and white. Male has red crown and crest; crown of female black streaked with white; back and rump red; mantle and wing coverts golden; breast scaly, white feathers edged black. Differs from Greater Flameback by having only one broad black malar stripe and only one hind toe. Iris—red; bill and feet—black with only three toes.
Voice: A harsh prolonged trill between partners *churrrr* and a short *chee chee* call or a hard *kluuk-kluuk-kluuk* in flight.
Range: India, S. E. Asia, Philippines and Greater Sundas.
Distribution and status: Race *intermedium* is an uncommon woodpecker in rather open lowland forest and cultivated areas up to 1000 m in SE Xizang and S Yunnan (Xishuangbanna).
Habits: Lives in pairs which call to each other regularly; prefers secondary and open forests, plantations and gardens.

146. Greater Flameback *Chrysocolaptes lucidus* [Greater   PLATE 12
Goldenback/Greater Golden-backed Woodpecker] Dajingbei zhuomuniao

Description: Largish (31 cm) colourful woodpecker. Very similar to Common Flameback but slightly larger, has two black malar stripes which fuse on the cheek and has four toes instead of three. Female crown black with white spots. Iris—pale yellow; bill—grey; feet—black.

Voice: Harsh, loud, strident, explosive burst of stuttered shrieks similar to a large cicada.

Range: India, China, Philippines, and Greater Sundas.

Distribution and status: Race *guttacristatus* is a local resident in S Yunnan (Xishuangbanna) and SE Xizang up to 1200 m.

Habits: Prefers rather open forest and forest edge. Lives in pairs and sometimes drums loudly.

147. Pale-headed Woodpecker *Gecinulus grantia*   PLATE 12
Zhu zhuomuniao

Description: Medium-sized (25 cm) reddish-brown woodpecker with paler buffish head. Upperparts diagnostic plain reddish-brown. Distinguished from Bay Woodpecker by lack of black banding on upperparts. Male has red forecrown. Race *indochinensis* has head golden buffy-olive, upperparts yellow-olive-green, tail black with olive base, underparts brownish-olive and male crown and nape crimson. Race *viridanus* has head pale buff, upperparts olive-green with red wash, tail crimson-brown with pale chestnut bars, underparts dark olive-green and male crown and nape rose, tinged orange. Iris—brown; bill—bluish-white; feet—olive.

Voice: Rather noisy. Repeated nasal contact call *chaik-chaik-chaik.* . . . Harsh, rattling, jay like *cheereker-chereker-chereker*.

Range: Nepal to S China, N Burma and Indochina.

Distribution and status: Uncommon. Three races in China—*viridanus* in SE China; *indochinensis* in SW and S Yunnan; nominate in W Yunnan (Yingjiang).

Habits: Prefers bamboo and secondary forests up to 1000 m.

Note: Sometimes treated as race of Bamboo Woodpecker *G. viridis*.

148. Bay Woodpecker *Blythipicus pyrrhotis* Huangzui   PLATE 12
zaozhuomuniao

Description: Largish (30 cm) woodpecker. Rufous plumage with black barring and long pale yellow bill diagnostic. Distinguished from Pale-headed Woodpecker by black barring on plumage. Male has crimson patch on sides of neck and nape. Distinguished from Rufous Woodpecker by bolder dark barring and pale yellow bill. Race *hainanus* lacks black barring on back and vent. Iris—reddish—brown; bill—pale greenish-yellow; feet—brownish-black.

Voice: Harsh cackle. Loud penetrating 'laughter' which increases in tempo but descends in pitch *keek, keek-keek-keek, keek, keek* rather like Plaintive Cuckoo.

Range: Nepal to S China, Hainan and SE Asia.

Distribution and status: Generally uncommon in evergreen forests from 500–2200 m. Nominate race is resident in subtropical Yunnan and SE Xizang; *hainanus* is confined to Hainan; and *sinensis* occurs in S and SE China.

Habits: Does not drum.

149. Great Slaty Woodpecker *Mulleripicus pulverulentus*   PLATE 14
Dahuizhuomuniao

Description: Unmistakable, very large (50 cm) lanky, grey woodpecker. Entire plumage grey with buff throat; male has red malar patch and reddish wash on throat and neck. Iris—dark brown; bill—dirty white with grey base and tip; feet—dark grey.

Voice: A loud braying or whinnying cackle, typically *woik woik.*

Range: N India to Indochina, Malay Peninsula and Greater Sundas

Distribution and status: Race *harterti* is rare in S Yunnan (Xishuangbanna) and SE Xizang in lowland forest up to 1000 m.

Habits: A noisy and conspicuous bird where present. Prefers semi-open habitats. Sometimes travels in noisy family parties. Birds forage in emergent trees and sometimes drum noisily.

BARBETS—Family: Megalaimidae

Barbets are medium or small-sized, colourful birds with large, powerful beaks. They are closely related to woodpeckers and share their habit of excavating tree holes for nesting. They also have the same unusual arrangement of toes, with two pointing forward and two backward for clinging to vertical tree trunks. However, their diet is comprised of fruits and flowers and they are particularly fond of small figs. Almost all species have the habit of sitting motionless in a tree-top for long periods, emitting monotonous, loud, repetitive calls. Since the general colour of most species is bright green they are very inconspicuous but all species can be identified by call. Eight species occur in China.

150. Great Barbet *Megalaima virens* Da nizhuomuniao   PLATE 15

Description: Very large (30 cm) big-headed barbet with dark inky-blue head and massive straw-coloured bill. Upperparts mostly green, belly yellow streaked with dark green and undertail coverts bright red. Iris—brown; bill—pale yellow/brown with black tip; feet—grey.

Voice: The usual call is a constantly-repeated, far carrying, drawn-out *piho piho* but the bird makes other calls including a loud rasp and repetitious *tuk, tuk, tuk* in duet.

Range: Himalayas to S China and N Indochina.

Distribution and status: Resident and quite common in evergreen forests of S China up to moderate altitudes of over 2000 m. Nominate race is resident in China S of 30 °N except *marshallorum* in S Xizang; *clamator* in Yunnan west of Nujiang (Salween) and *magnifica* in Yunnan between Nujiang and Lancang (Mekong) Rivers.

Habits: Several birds sometimes gather together in an open tree crown to call. Has heavy rising and falling flight like a woodpecker.

151. Lineated Barbet *Megalaima lineata* [Bantou]   PLATE 15
Lü nizhuomuniao

Description: Largish (29 cm) pale-headed, streaky barbet. Plumage generally green with a pale yellowish-brown head and neck, and diagnostic streaking on head and underparts. Iris—buff; bill—pale yellow; feet—yellow.

Voice: A low-pitched irregular ringing *bul-tok* – *bul-tok* at about 1 second intervals; also a loud *kuerr-kuerr* and a rare 'counting call' consisting of a long trill followed by a series of four-syllable trilled notes.

Range: W Himalayas to NE India, SE Asia, Java and Bali.

Distribution and status: Race *hodgsoni* recorded in lowlands of S and SW Yunnan. Locally common, may occur in SE Xizang.

Habits: Similar to other barbets but most common in relatively dry, open, wooded habitats.

Note: Included in *M. zeylanica* by Cheng (1987).

152. Green-eared Barbet *Megalaima faiostricta* PLATE 15

Huangwen nizhuomuniao

Description: Large (24 cm) green barbet. Distinguished from Lineated Barbet by blacker bill; darker streaking on head; green ear coverts and lack of yellow orbital skin. Iris—brown; bill—blackish; feet—black.

Voice: Throaty *per-roo-roo-rook*; also mellow, fluty *pooouk* with rising inflection.

Range: SE China, Thailand and Indochina.

Distribution and status: Race *praetermisa* is rare in China, recorded only in Guangzou Bay and on Naozhou Islands in SE Guangdong.

Habits: Lives in open and deciduous woodlands. Habits as other barbets.

153. Golden-throated Barbet *Megalaima franklinii* PLATE 15

Jinhou nizhuomuniao

Description: Largish (23 cm) colourful barbet with diagnostic yellow chin and upper throat, and pale grey lower throat. Crown is patterned red, yellow, red and has broad black eyestripe. Iris—reddish; bill—black; feet—black.

Voice: Monotonous ringing *ki-ti-yook* or *pukwowk*; also higher wailing *peeyu, peeyu*.

Range: Nepal to S China and SE Asia.

Distribution and status: Nominate race is occasional in hilly evergreen forest from 1200–2200 m in SE Xizang, W, S and SE Yunnan and SW Guangxi. Race *ramsayi* occurs in SW Yunnan.

Habits: More montane than other barbets.

154. Black-browed Barbet *Megalaima oorti* [Mueller's Barbet] PLATE 15

Heimei nizhuomuniao

Description: Smallish (20 cm) green barbet with gaudy blue, red, yellow and black head. Distinguished from other barbets by smallish size, black eyebrow, blue cheeks, yellow throat and red spot on sides of neck. Immature is duller. Iris—brown; bill—black; feet—greyish-green.

Voice: Hollow *tok-tr-trrrrrt* call about 20 times per minute with emphasis on third syllable.

Range: S China, SE Asia, Malay Peninsula and Sumatra.

Distribution and status: Race *sini* is common resident in Guangxi (Yaoshan); *nuchalis* on Taiwan; and *faber* on Hainan. Lives in subtropical forests from 1000–2000 m.

Habits: Typical barbet of upper and middle canopy.

155. Blue-throated Barbet *Megalaima asiatica*   PLATE 15
Lanhou nizhuomuniao

Description: Medium-sized (20 cm) green barbet with diagnostic crimson forecrown and hindcrown with black or bluish midcrown. Ocular region, face, throat and side of neck bright blue. There is a red spot on each side of breast. Iris—brown; bill—grey with black culmen; feet—grey.

Voice: The call is a continuous fast repeated *took-a-rook, took-a-rook* given by a bird sitting motionless in the treetops.

Range: India to S China and SE Asia.

Distribution and status: Common resident of tropical evergreen forest and secondary forest at lower altitudes in SE Xizang, S and SW Yunnan. The commonest barbet in Xishuangbanna (S Yunnan). Nominate race probably occurs in SE Xizang and is found in Yunnan west of Lancang (Mekong) River; *davisoni* is resident east of Lancang River.

Habits: Often found feeding in small parties in fruit trees, especially figs.

156. Blue-eared Barbet *Megalaima australis*   PLATE 15
Lan'er nizhuomuniao

Description: Small (18 cm) barbet with blue crown and chin and black forecrown, malar patch and throat band. Iris—brown; bill—black; feet—greenish-grey.

Voice: Fast, repeated, endless rattle *tu-trruk* about 100 times a minute with continuous head turning, or shrill, trilled notes repeated more slowly (like a pea whistle) with the head held very still.

Range: NE Indian subcontinent to S. China, SE Asia and Greater Sundas.

Distribution and status: Race *cyanotis* is a locally common resident of primary forests, plantations and secondary forest up to about 1600 m in SW and S Yunnan and SE Xizang.

Habits: A familiar call in wooded countryside although the bird itself is not so easily seen. Quietly joins pigeons and other birds feeding in fig trees.

157. Coppersmith Barbet *Megalaima haemacephala*   PLATE 15
[Crimson-breasted Barbet] Chixiong nizhuomuniao

Description: Small (17 cm) red-crowned, barbet. Back, wings and tail bluish-green; underparts dirty white with heavy black streaks. Immature lacks red and black on head but has yellow spot under eye and under chin. Iris—brown; bill—black; feet—red.

Voice: A resonant, monotonous, metallic *took-took-took* continued for several minutes at steady rate of about 110 notes per minute. Tail flicks forwards with each *took*. Another slower, less regular call when head bobs but tail is held still.

Range: Pakistan to S China, Philippines, Sumatra, Java and Bali.

Distribution and status: Race *indica* is an uncommon resident of open lowland forests up to 1000 m in W and S Yunnan (Xishuangbanna) and probably SE Xizang.

Habits: Similar to other barbets but prefers open habitat such as woods, gardens and plantations. In the early morning several birds may congregate to call from the top of a bare branch.

HORNBILLS—Order: Bucerotiformes Family: Bucerotidae

Hornbills are large, black or brown and white, mainly arboreal birds, with long, heavy bills. Many species have hollow casques on top of the bill which give a deep resonant tone to their harsh calls. Hornbills are found throughout Africa and tropical Asia. They eat fruit, insects and small vertebrates.

The nesting habits of the family are interesting. The incubating females are usually sealed into a tree hole with mud by their mates, leaving only a small aperture through which food can be passed. When the young are hatched the female breaks out but reseals the nest entrance again until the young are ready to leave. Five species of hornbill occur in China.

158. Oriental Pied Hornbill *Anthracoceros albirostris*   PLATE 16
[Malaysian/Southern Pied Hornbill] Guanban xiniao

Description: Small (75 cm) black and white hornbill with a large yellow/white casque sometimes marked with black. Plumage black except for a white patch under the eye, white lower belly, thighs and undertail coverts, white tips to flight feathers of wing and white outer tail feathers. Iris—dark brown; naked orbital and gular skin—white; bill and casque—yellow/white with black spot on base of lower mandible and front of casque; feet—black.
Voice: Incessant strident cackle *ayak-yak-yak-yak-yak*.
Range: N India, S China, SE Asia and Greater Sundas.
Distribution and status: Formerly a conspicuous bird but now rare in lowland primary and secondary forests of SE Xizang, S Yunnan and S Guangxi.
Habits: Prefers more open forest and forest edge. Pairs or noisy parties, flap or glide from tree to tree. Eats insects more than fruit.
Note: Includes both Northern (*albirostris*) and Southern (*convexus*) Pied Hornbills but not the Malabar Hornbill (*A. coronatus*). *A. coronatus* has precedence if lumped together.

159. Great Hornbill *Buceros bicornis* [Great Indian Hornbill]   PLATE 16
Shuangjiao xiniao

Description: Very large (125 cm) black and cream hornbill with black subterminal bar on white tail and broad yellow-stained white bar on black wing. Bill and concave-topped casque yellow; face black. White plumage of head and chest often stained yellow. Iris—red (male) whitish (female); bill—yellow; feet—black.
Voice: Loud, harsh barked *gok* or *wer gok*.
Range: India, SE Asia, Malay Peninsula and Sumatra.
Distribution and status: Race *homrai* is a rare resident of lowland evergreen forest in W and SW Yunnan and SE Xizang. Absent from many former haunts as a result of over-hunting.
Habits: Generally in pairs. Flies noisily over forest. Feeds and roosts in upper canopy of primary forest.

160. Brown Hornbill *Anorrhinus tickelli* Baihou [xiaokui] xiniao   PLATE 16
Description: Small (74 cm) brownish hornbill with blue orbital skin and small casque on dull yellow bill. White-tipped primaries conspicuous in flight; also white tips to outer

tail feathers. Male has whitish throat and rufous underparts. Iris—reddish-brown; bill—dull yellow; feet—black.

Voice: Eerie, piercing screams and high-pitched yelp, usually inflected upwards, *klee-ah*.

Range: NE India, SW China and SE Asia.

Distribution and status: Rare. Recorded in Southern Xishuangbanna, S Yunnan. Also occurs in SE Xizang.

Habits: Inhabits evergreen forests up to 1500 m. Lives in noisy parties.

161. Rufous-necked Hornbill *Aceros nipalensis*　　　　　PLATE 16
Zongjing [Wu Kui] xiniao

Description: Medium-sized (117 cm) dark hornbill with minimal casque and dull yellow bill with blue orbital skin and red gular skin. Male has rufous head and underparts; female blackish. Both sexes have white tips to primaries and terminal half of tail white. Iris—reddish; bill—yellow; feet—blackish.

Voice: Soft, barking *Kup*, given by both sexes. Less deep than call of Great Hornbill.

Range: Himalayas from Nepal to SW China, N Burma and Indochina.

Distribution and status: Rare. Recorded Southern Xishuangbanna in S Yunnan and in SE Xizang.

Habits: More montane than other hornbills, living in evergreen forest from 600–1800 m.

162. Wreathed Hornbill *Aceros undulatus*　　　　　PLATE 16
Huaguan zhoukuixiniao

Description: Large (105 cm, female smaller) white-tailed hornbill. Both sexes have black back, wings and belly, but male has creamy head with reddish plume from the nape, and naked gular pouch with a distinct black stripe. Female has black head and neck, and blue gular pouch. Iris—red; bill—yellow with corrugated casque; feet—black.

Voice: A repeated, short, hoarse, dog-like yelp *koe-guk*.

Range: NE Indian subcontinent, SW China, SE Asia and Greater Sundas.

Distribution and status: Rare. Heads of male and female found in village hut in SW Yunnan (Tengchong). Probably occurs in SE Xizang. Status in China uncertain.

Habits: Flies in pairs or small flocks over forest with heavy wing beat.

HOOPOE—Order: Upupiformes Family: Upupidae

Hoopoes are a small family of two species distributed through Africa, Europe, Madagascar and Asia. Hoopoes are characterised by colourful plumage, an erectile crest and a long curved bill. One species occurs in China.

163. Common Hoopoe *Upupa epops* [>Eurasian Hoopoe]　　　PLATE 18
Daisheng

Description: Unmistakable medium-sized (30 cm) brightly coloured bird with long erectile crest of black-tipped, pinkish-rufous plumes. Head, mantle, shoulders and underparts pinkish-rufous with wings and tail striped black and white. The bill is long and decurved. Nominate race has subterminal white bar below black tips of crest. Iris—brown; bill—black; feet—black.

Voice: Low, soft *hoop-hoop hoop* on a monotone accompanied by head-bobbing display.
Range: Africa, Eurasia, Indochina.
Distribution and status: Common resident and migrant over most of China up to
3000 m. Nominate race is migrant and possibly breeds in W Xinjiang; *longirostris* is resident in S Yunnan, Guangxi and Hainan. Race *saturata* breeds over the rest of China and
S Xinjiang with northern birds wintering south of the Changjiang Valley; accidental in
Taiwan.
Habits: An active bird of open, often moist, ground where it feeds by probing the
ground with its long bill. Erects crest when alarmed and on alighting after flight.
Note: African and Madagascar forms are sometimes treated as 2–3 separate species.

Trogons—Order: Trogoniformes Family: Trogonidae

Trogons are medium-sized, brightly coloured birds with short bills, short legs and feet,
short wings but long broad tails and soft fluffy plumage. The family is pan-tropical. Two
toes point backwards. Nesting is in tree holes where buff-coloured eggs are laid. Trogons
are insectivorous, hunting from sometimes quite low perches in thick forest. They have
harsh distinct calls. Three species occur in China.

164. Orange-breasted Trogon *Harpactes oreskios* PLATE 21
Chengxiong yaojuan

Description: Medium-sized (29 cm) brown and orange trogon. Head, neck and breast
greenish-grey (greyer in female, brown in immature); back and tail reddish-brown; primaries black and wing coverts barred black; yellowish to orange lower breast and belly.
Edges and underside of graduated tail white. Iris—orbital skin blue; bill—bluish-black;
feet—grey.
Voice: Male song is a five-note cadence *kek tau-tau-tau-tau*. Repeated harsh *kek-kek*.
Range: S China, SE Asia, Borneo, Sumatra and Java.
Distribution and status: Race *stellae* is uncommon resident in S Yunnan
(Xishuangbanna) up to 1500 m.
Habits: A solitary but noisy and quite conspicuous bird of the lower levels of the forest. Hunts from a perch and not so shy of humans.

165. Red-headed Trogon *Harpactes erythrocephalus* PLATE 21
Hongtou yaojuan

Description: Large (33 cm) red-headed trogon. Red head of male diagnostic. Lacks
nuchal collar and has narrow white crescent on red breast. Female differs from other
female trogons by having red belly and white breast crescent and differs from all male
trogons in having cinnamon-brown head. Iris—brown; bare skin round eye—blue;
bill—bluish; feet—pinkish.
Voice: Mellow repeated *tiaup* and also a rattled note *tewirrr*.
Range: Himalayas to S China, SE Asia and Sumatra.
Distribution and status: Uncommon resident in tropical and subtropical forests up to
2400 m. Race *yamakanensis* from S Sichuan to Guizhou, N Guangdong and Fujian; *helenae*

from SE Xizang and extreme W Yunnan; nominate in S Yunnan between Nujiang (Salween) and Lancang (Mekong) rivers; *intermedius* in SE Yunnan; and *hainanus* on Hainan.

Habits: Hunts from low perch in dense forest.

166. Ward's Trogon *Harpactes wardi* Hongfu yaojuan   PLATE 21

Description: Large (38 cm) crimson and maroon trogon. Male: head, upper breast, upperparts and central tail feathers maroon-brown. Wings blackish with white edges to primaries; forehead and crown pinkish-red. Lower breast to undertail coverts pinkish-red; undertail of lateral graduated tail feathers diagnostic pink. Bare skin around eye blue. Female similar but dark parts greyer dusky and pale parts diagnostic primrose-yellow instead of pink. Iris—brown; bill—pink; feet—pinkish-brown.

Voice: Loud, rapidly-delivered, mellow *klew* notes speeding up but dropping in pitch, very different from call of Red-headed Trogon. Alarm is *whirr-ur*.

Range: E Himalayas to NE India, NE Burma and NW Tonkin (Vietnam).

Distribution and status: Globally Vulnerable (Collar *et al.* 1994). Rare resident in Gongshan area of NW Yunnan, Gaoligong Mts of W Yunnan and in SE Xizang from 1600–3000 m. May occur in Jinping area of SE Yunnan, part of the Fan-si-pan population of NW Tonkin.

Habits: Lives in montane evergreen forests. Typical trogon behaviour.

ROLLERS—Order: Coraciformes Family: Coraciidae

Rollers are medium-sized, brightly coloured, long-winged birds found in Europe, Asia, Africa and Australia. They have powerful, sharp beaks and feed mostly on large insects. Like the kingfishers and bee-eaters they have short legs, with the front three toes fused at the base; lay roundish white eggs in burrows and tree holes and the young retain their feather sheaths until they are nearly fully grown. Three species occur in China including one somewhat atypical of the family—a broad-billed roller.

167. European Roller *Coracias garrulus* Lanxiong fofaseng   PLATE 18

Description: Largish (30 cm) roller with bright azure blue head, underparts and forewings, with black flight feathers and pinkish-brown mantle, back and tertials. Iris—dark brown; bill—black; feet—dull yellow.

Voice: Call is harsh *chack-ack* with emphasis on first syllable, like Eurasian Jackdaw or Black-billed Magpie. Also jay-like *rrak-rrak, rrak-rehhh*.

Range: Europe to C Asia and Transcapia migrating to Africa and India.

Distribution and status: Rare. Race *semenowi* is recorded on migration in NW and W Xinjiang and W Xizang. May breed in Xinjiang.

Habits: Hunts from perch to dive down on insects. In display flight, pitches side-to-side like Northern Lapwing.

168. Indian Roller *Coracias benghalensis* Zongxiong fofaseng   PLATE 18

Description: Largish (33 cm) bluish-grey bird with slender decurved black bill. At rest looks dull but at close range it is a beautiful mixture of dull turquoise-blue crown, tail coverts and wings; lilac throat, upper mantle and part of flight feathers; dull greenish back and central tail feathers. In flight the brilliant blue in wings and tail are very conspicuous. Distinguished from Dollarbird by black bill and from European Roller by less blue on head and breast. Iris—brown; bill—grey; feet—dull yellowish.

Voice: The call is a harsh crow-like *chak chak*.

Range: S Asia, India to SW China.

Distribution and status: An occasional bird of open country and farmland in S and SW China and S Xizang.

Habits: As European Roller.

169. Dollarbird *Eurystomus orientalis* [Broad-billed Roller]   PLATE 18
Sanbaoniao

Description: Medium-sized (30 cm) dark roller with broad red bill (black in immature). The overall colour is dark bluish-grey but the throat is bright blue and in flight there are contrasting circular light blue patches in the centre of each wing which give the bird its name. Iris—brown; bill—coral-red with black tip; feet—orange/red.

Voice: Hoarse rasping croaks *kreck-kreck* in flight or from perch.

Range: Widely distributed from E Asia, SE Asia, Japan, Philippines, Indonesia to New Guinea and Australia.

Distribution and status: Widespread, but never very common, at edge of forests up to 1200 m. Race *calonyx* breeds from NE China through N, C, E, S, SE and SW China and Hainan; accidental in Taiwan. Northern birds migrate south in winter; southern birds are resident. Race *cyanicollis* occurs in SE Xizang.

Habits: Sits in dead trees in open country near forest, making occasional flights after passing insects or diving onto insects on the ground. Has a curious, heavy, twisting and flapping flight similar to a nightjar. Two or three sometimes fly and dive together at dusk, especially during courtship. Sometimes mobbed by small birds because its head and beak give it a predatory appearance.

KINGFISHERS—Family: Alcedinidae

Kingfishers are a group of brightly coloured birds (many with metallic blue feathers), with short legs and tail, big heads and long powerful beaks. They eat insects, small vertebrates and several species catch fish. They nest in burrows in the ground, in tree trunks, river banks or termite mounds. The eggs are whitish and almost spherical. The family has a worldwide distribution and there are 11 species recorded for China including two which occur only as accidentals. Some kingfishers give rather loud, harsh calls. The three front toes are partly fused at the base.

170. Blyth's Kingfisher *Alcedo hercules* Bantou dacuiniao   PLATE 17

Description: Medium-sized (23 cm) brilliant blue and rufous kingfisher. Similar to Common Kingfisher but noticeably larger with darker blackish crown, nape and sides of head. Distinguished from Common Kingfisher by buff spots in front of and below eye, buff stripe on sides of neck and blackish ear coverts streaked silvery-blue. Iris—brown; bill—black with reddish base in male; feet—black.

Voice: Similar squeak to Common Kingfisher but deeper-pitched.

Range: NE India, Burma, S China and Indochina.

Distribution and status: Globally Vulnerable (Collar *et al.* 1994). Resident in SE Xizang, S Yunnan and Hainan. Rare up to 900 m.

Habits: Keeps to larger shady forest streams.

171. Common Kingfisher *Alcedo atthis* [River/European/   PLATE 17
The Kingfisher] Putong cuiniao

Description: Small (15 cm) shimmering blue and rufous kingfisher. Upperparts metallic pale greenish-blue with white spot on side of neck; underparts rufous-orange with white chin. Juvenile is duller with dark breast band. Orange stripe through eye and across ear coverts is diagnostic and distinguishes from the darker Blue-eared and Blyth's Kingfishers. Iris—brown; bill—black (male) and with orange lower mandible (female); feet—red.

Voice: Shrill, drawn-out squeak *tea-cher*.

Range: Widespread in Eurasia, SE Asia, Indonesia to New Guinea.

Distribution and status: Nominate race breeds in Tianshan and winters at lower altitudes of W Xizang. Race *bengalensis* is a common resident up to 1500 m over NE, E, C, S and SW China including Hainan and Taiwan.

Habits: Frequents fresh water lakes, streams, canals, fish ponds and mangroves in open country. Perches on rocks or overhanging branches; bobs head whilst looking for fish then plunges into water to catch prey.

172. Blue-eared Kingfisher *Alcedo meninting* [Deep-blue Kingfisher]   PLATE 17
Lan'er cuiniao

Description: Small (15 cm) bright blue-backed kingfisher. Back darker metallic blue than Common Kingfisher; underparts brighter orange/red; blue ear coverts distinctive. White patch on side of neck. Iris—brown; bill—blackish; feet—red.

Voice: A high-pitched *chiet* usually uttered in flight, rapid chatter when perched.

Range: India to China and SE Asia, Philippines and Indonesia.

Distribution and status: Rare resident in Xishuangbanna, S Yunnan up to 1000 m.

Habits: As Common Kingfisher but prefers more wooded terrain.

173. Oriental Dwarf Kingfisher *Ceyx erithacus* [Three-toed/   PLATE 17
Malay Kingfisher, AG: Pygmy-Kingfisher] Sanzhi cuiniao

Description: Very small (14 cm) red and yellow kingfisher. Bright yellow underparts and bluish-black back and wing coverts diagnostic. Iris—brown; bill—red; feet—red.

Voice: During flight a high-pitched whistle *tsriet-siet* or *tsie-tsie*.

Range: India, Burma to Sundas and Philippines.

Distribution and status: An uncommon bird of lowland forests up to 1500 m in S Yunnan, Hainan and possibly SE Xizang.

Habits: Confined to forest, usually close to streams. Flies at great speed from one low perch to another; hunts for insects or other small prey.

Note: Reddish-winged form attributed to Rufous-backed Kingfisher *C. rufidorsa* has been recorded from Sikkim and can be expected in both SE Xizang and S Yunnan. Some authors treat this form as a subspecies or even morph of *C. erithacus*.

174. Stork-billed Kingfisher *Halcyon capensis* Guanzui feicui   PLATE 17

Description: Very large (35 cm) blue-backed kingfisher with diagnostic massive red beak; crown, sides of face and back of neck grey/brown. Underparts pinkish-orange. Iris—brown; bill—red; feet—red.

Voice: Loud, sharp, laughing call *wiak-wiak* and harsh alarm cry.

Range: India, SE Asia, Philippines and Sundas.

Distribution and status: A rare stray recorded in Xishuangbanna, S Yunnan.

Habits: Frequents edges of large rivers.

Note: Sometimes placed in genus *Pelargopsis*.

175. Ruddy Kingfisher *Halcyon coromanda* Chi feicui   PLATE 17

Description: Medium-sized (25 cm) violet and rufous kingfisher. Upperparts bright violet rufous, except for contrasting pale blue rump; underparts rufous. Iris—brown; bill—red/orange; feet—red/orange.

Voice: Rapid, mellow, dissyllabic or trisyllabic note, slowing down.

Range: Widespread from India to Japan, China, SE Asia to Philippines and Indonesia.

Distribution and status: *H. c. coromanda* is a rare resident in Xishuangbanna, S Yunnan, also expected in SE Xizang. *H. c. major* is a rare bird recorded breeding in Changbaishan, Jilin and wintering along the east coast S of 33 °N. *H. c. bangsi* is a rare resident of Taiwan and Lanyu Island.

Habits: Inhabits coastal forest, swamp forest and mangrove or streams and pools in forest.

176. White-throated Kingfisher *Halcyon smyrnensis*   PLATE 17
[White-breasted/Smyrna Kingfisher] Baixiong feicui

Description: Largish (27 cm) blue and brown kingfisher with a white chin, throat and breast. Head, neck, and rest of underparts brown; mantle, wings and tail bright iridescent blue (can appear greenish-turquoise in early morning light); upperwing coverts and wing tips black. Iris—dark brown; bill—deep red; feet—red.

Voice: Loud screaming, tittering call *kee kee kee kee* given in flight or from perch and harsh *chewer chewer chewer*.

Range: Middle East, India, China, SE Asia, Philippines, Andamans and Sumatra.

Distribution and status: A fairly common resident up to 1200 m over most of China S of 28 °N including Hainan. Vagrant to Taiwan.

Habits: Active, noisy hunter of open fields, rivers, ponds and coast.

177. Black-capped Kingfisher *Halcyon pileata* Lanfeicui   PLATE 17

Description: Large (30 cm) blue, white and black kingfisher. Black head diagnostic. Wing coverts black but upperparts otherwise metallic bright royal blue/purple. Flanks and vent washed rufous. White wing patch conspicuous in flight. Iris—dark brown; bill—red; feet—red.

Voice: Shrill loud call when alarmed.

Range: Breeds in China and Korea, migrating south in winter as far as Indonesia.
Distribution and status: Breeds and summers over most of E, C and S China from Liaoning to Gansu and SE including Hainan. Vagrant to Taiwan. Not rare along cleaner rivers up to 600 m. Northern populations migrate south in winter.
Habits: Prefers the banks of large rivers, estuaries and mangroves. Perches over rivers on overhanging branches. More of a river bird than White-throated Kingfisher.

178. Collared Kingfisher *Todiramphus chloris* [White-collared/ PLATE 17
Mangrove Kingfisher] Bailing feicui

Description: Medium-sized (24 cm) blue and white kingfisher. Crown, wings, back and tail bright iridescent greenish-blue; black stripe through eye; white spot above beak; white collar and underparts diagnostic. Iris—brown; bill—dark grey above, paler below; feet—grey.
Voice: Harsh scream *chew-chew-chew-chew-chew* with downward inflection or double notes *chek-chek, chek-chek, chek-chek.*
Range: S and SE Asia, Indonesia to New Guinea and Australia.
Distribution and status: A common coastal kingfisher but only marginally reaches China. Recorded on Jiangsu coast; expected on Nansha archipelago.
Habits: Perches on a rock or tree and hunts along beach or in any open space near water including gardens, towns and plantations.

PIED KINGFISHERS—Family: Cerylidae

179. Crested Kingfisher *Megaceryle lugubris* [Greater Pied/ PLATE 17
Mournful Kingfisher] Guan yugou

Description: Very large (41 cm) pied kingfisher with bushy crest. Upperparts slaty-black spotted and barred with white including pied bushy crest. Large white patch extends from malar area to sides of neck underlined by black moustachial line. Underparts white with black streaky chest bar and buff barring on flanks. Wing lining white in male, cinnamon in female. Iris—brown; bill—black; feet—black.
Voice: Harsh loud screech, squeaky flight call *aeek.*
Range: Himalayas and foothills in N India, N Indochina, S and E China.
Distribution and status: *M. l. guttulata* is an occasional resident over the whole of C, E and S China including Hainan up to 2000 m. *M. l. lugubris* is an uncommon bird in Liaoning, NE China.
Habits: Frequents fast-flowing, boulder-strewn, clear rivers and streams. Perches on large rocks. Flies with a slow, powerful beat and does not hover.

180. Pied Kingfisher *Ceryle rudis* [Lesser Pied Kingfisher] PLATE 17
Ban yugou

Description: Medium-sized (27 cm) black and white kingfisher. Distinguished from Crested Kingfisher by smaller size, smaller crest and prominent white eyebrow. Upperparts black spotted with white. Primaries and tail basally white, distally black. Underparts white with broad band of black streaks on upper breast and narrow black bar below it. Breast band of female is somewhat reduced. Iris—brown; bill—black; feet—black.

Voice: Shrill whistling calls.

Range: NE India, Sri Lanka, Burma, China, Indochina and Philippines.

Distribution and status: Race *insignis* is a fairly common resident on lakes and ponds of SE China and Hainan. Race *leucomelanura* is an occasional resident in W and S Yunnan. Expected in SE Xizang.

Habits: Lives in noisy pairs or groups on larger water bodies and in mangroves. The only kingfisher to regularly hover over water in search of prey.

BEE-EATERS—Family: Meropidae

Bee-eaters are a small family found throughout the Old World. They are colourful birds with green plumage predominating. They have short legs; long, graceful outlines with long, slender, curved bills; long pointed wings and, in many species, elongated streamer-like central tail feathers. Most species are gregarious and prefer open country. Parties of birds sit on bare perches and make sweeping flights after insects which they carry back to the perch and hit against a hard surface to break and soften the prey before eating. The three front toes are partly fused. Bee-eaters nest in burrows in the ground where they lay white eggs. Six species occur in China. Some of these are migratory.

181. Blue-bearded Bee-eater *Nyctyornis athertoni* PLATE 18
[Lanxu] Ye fenghu

Description: Medium-sized (30 cm) green forest bee-eater with blue puffy breast and thick, heavy decurved bill. Adult has bluish crown and yellow-buff belly with green streaks. Under tail buffy-yellow. Immature birds are all green. Iris—orange; bill—blackish; feet—dull green.

Voice: Deep, guttural, croaking chuckle and even-pitched rolling *kirrr-r-r-r*.

Range: Himalayas, N India to S China, Hainan and SE Asia.

Distribution and status: An uncommon resident of primary and logged forest up to 1800 m. Nominate race occurs in SE Xizang and W and S Yunnan; *brevicaudata* on Hainan.

Habits: Lives in upper and middle canopy of tall forest. Hunts quietly from high perches. Fans and flicks tail at intervals. More of a forest bird than other bee-eaters.

182. Green Bee-eater *Merops orientalis* [Little Green Bee-eater] PLATE 18
Lühou fenghu

Description: Small (20 cm) green bee-eater with long tail streamers, coppery crown and nape, bluish throat and sides of face and black gorget. Distinguished from Chestnut-headed by blue throat and tail streamers and from Blue-throated by green tail and belly and black gorget. Iris—reddish-crimson; bill—brownish-black; feet—yellowish-brown.

Voice: Higher-pitched and more metallic trills than other bee-eaters. Pleasant jingling *tree-tree-tree* and also *tit-tit-tit* from perch and on wing.

Range: Africa, Middle East to SW China and SE Asia.

Distribution and status: Uncommon. Breeds in lowlands of W and S Yunnan in SW China.

Habits: Inhabits dry open country up to 1500 m. Hunts from dead tree perch in small flocks. Fond of dust bathing.

183. Blue-throated Bee-eater *Merops viridis* Lanhou fenghu   PLATE 18

Description: Medium-sized (28 cm, including elongated central tail streamers) bluish bee-eater. Adult: crown and mantle chocolate; eye-stripe black; wings bluish-green; rump and streamered tail pale blue; underparts pale green with diagnostic blue throat. Immature lacks elongated tail feathers and head and mantle are green. Iris—red or brown; bill—black; feet—grey or brown.

Voice: Fast trilling notes given in flight *kerik-kerik-kerik*.

Range: S China, SE Asia, Greater Sundas and Philippines.

Distribution and status: Nominate race is uncommon summer breeder in Hubei and China south of Changjiang river. Resident on Hainan.

Habits: Favours open country and woodlands in low-lying areas, often close to the sea. Birds congregate at breeding sites in sandy areas. Less prone to hunt in gliding flight than Blue-tailed Bee-eater, preferring to hunt by waiting on a perch for an insect to fly past. Occasionally picks insects off the water surface or off the ground.

184. Blue-tailed Bee-eater *Merops philippinus* Lihou fenghu   PLATE 18

Description: Largish (30 cm including elongated tail streamers), elegant bee-eater. Black stripe through eye is edged above and below with blue. Head and mantle green, rump and tail blue. Chin is yellow, throat chestnut and abdomen pale green. Underwing is orange.

Voice: Plaintive trill *kwink-kwink, kwink-kwink, kwink-kwink-kwink* given in flight.

Range: Breeds in S Asia, Philippines, Sulawesi and New Guinea. Visits the Sundas in winter.

Distribution and status: Nominate race is summer breeder in SE Xizang, S Sichuan, Yunnan, Guangxi and Guangdong; resident on Hainan. Common in open habitats up to 1200 m.

Habits: Gregarious parties gather in open areas to hunt. They settle on bare twigs and telegraph wires and make lazy, circular, swallow-like gliding flights after insects. More of an aerial feeder than other bee-eaters. Calling groups of these bee-eaters sometimes pass high overhead.

Note: Closely related Blue-cheeked Bee-Eater *M. persicus* which may occur in SW Xinjiang.

185. European Bee-eater *Merops apiaster* Huanghou fenghu   PLATE 18

Description: Medium-sized (28 cm) brilliantly coloured bee-eater with distinctive golden back. Throat yellow with narrow black gorget and rest of underparts blue. Neck, crown and nape chestnut. Juvenile lacks central elongated tail feathers and has green back. Iris—red; bill—black; feet—grey.

Voice: Far-carrying *kruuht* repeated frequently. Similar to cry of Black-headed Gull.

Range: S Europe, N Africa, Middle East, C Asia and NW Indian subcontinent. Migrates to south of range in winter.

Distribution and status: Rare in W Tianshan and extreme N Xinjiang.

Habits: Gregarious; glides elegantly over open country, hawking for insects. Flutters rapidly.

186. Chestnut-headed Bee-eater *Merops leschenaulti* [Bay-headed   PLATE **18** Bee-Eater] Litou fenghu

Description: Smallish (20 cm) green and brown bee-eater lacking elongated central tail streamers. Crown, nape and mantle bright chestnut; wings and tail green; rump bright blue; throat yellow, bordered with chestnut; a black gorget across upper chest; abdomen pale green; eye-stripe black. Orange underwing visible in flight. Iris—reddish-brown; bill—black; feet—dark brown.

Voice: A liquid ringing trill given in flight, *kree-kree-weet-weet-weet* and variations.

Range: S and SE Asia and Greater Sundas.

Distribution and status: Nominate race is summer breeder in SE Xizang and W Yunnan. Common in open and wooded areas up to 1200 m.

Habits: Typical of the group.

CUCKOOS—Order: Cuculiformes Family: Cuculidae

A large, worldwide family of insectivorous birds with slender bodies and long wings and tails. The two outer toes point backward and the two inner toes forward. Cuckoos have strong curved bills which are used for catching large insects. Some species specialise on caterpillars. Two main groups of cuckoos occur in China. True cuckoos are arboreal, have pointed wings, often barred or streaked plumage and breed parasitically by laying eggs in the nests of other birds who then raise the foster nestling. Malkohas have colourful bills, very long tails and long strong legs. They creep about in thick tangles of vines and low bushes and have low, clucking calls. Sixteen species occur in China; some are very difficult to identify. Many species are best recognised by voice.

187. Pied Cuckoo *Clamator jacobinus* [Black-and-white Cuckoo]   PLATE **19** Banchi fengtoujuan

Description: Large (34 cm) black and white cuckoo with a diagnostic crest. Like Chestnut-winged but lacks black of head and wings black with white band at base of primaries; tail black with broad white tip. Juvenile exhibits less contrast with pattern of brown on buff. Iris—brown; bill—black with yellow base; feet—grey.

Voice: Loud metallic and arresting *ple-ue; pee-pee-piu* etc. Also harsh *chu-chu-chu-chu* in alarm.

Range: Africa, Iran to India and Burma; winters to Africa.

Distribution and status: Nominate race is very rare in China. Only recorded in S Xizang.

Habits: Inhabits deciduous forest and open scrub. Highly migratory. Lives in small parties.

188. Chestnut-winged Cuckoo *Clamator coromandus*   PLATE 19
[Red-winged Cuckoo, AG: Crested-Cuckoo] Hongchi fengtoujuan

Description: Large (45 cm) black, white and rufous long-tailed cuckoo with promi-
nent erectile crest. Crown and crest black; back and tail black with blue gloss; wings
chestnut; throat and breast orange-brown; nuchal collar white; abdomen whitish.
Immature: upperparts with rufous scaling; throat and breast whitish. Iris—red-brown;
bill—black; feet—black.
Voice: Loud harsh screech *chee-ke-kek* and a hoarse whistle.
Range: Breeds India, S China and SE Asia, migrates to Philippines and Indonesia.
Distribution and status: Occasionally breeds in suitable habitat up to 1500 m in E,
C, SW, S and SE China, SE Xizang, Hainan and accidental on Taiwan.
Habits: Similar to a malkoha, scrambling about in low vegetation, hunting for insects.
In flappy and coucal-like flight the crest is lowered.

189. Large Hawk-Cuckoo *Hierococcyx sparverioides* Yingjuan   PLATE 20

Description: Largish (40 cm) greyish-brown, hawk-like cuckoo with a rufescent subterminal bar to white-tipped tail; chest rufous, mottled white and grey; abdomen barred
white and brown, washed rufous; chin black. Immature: upperparts brown with rufous
barring; underparts buff with blackish streaks. Differs from hawk in posture and bill
shape. Iris—orange; bill—black above, yellow-green below; feet—pale yellow.
Voice: In breeding season calls *pi-peea* or *brain-fever* with increasing speed and shriller
pitch to a frantic climax.
Range: Resident in Himalayas, S China, Philippines, Borneo and Sumatra, visits
Sulawesi and Java in winter.
Distribution and status: Nominate race is uncommon summer breeder in S Xizang,
C, E, SE and SW China and Hainan; some birds resident in S Yunnan and Hainan.
Accidental to Taiwan and Hebei.
Habits: Prefers open forests up to 1600 m. Typical secretive tree-top cuckoo.

190. Common Hawk Cuckoo *Hierococcyx varius*   PLATE 20
Putong yingjuan

Description: Medium-sized (34 cm) cuckoo with 'Shikra' plumage of grey upperparts
with barred tail, rufous breast and banded belly and thighs. Throat white with black chin
and band down centre of throat. Distinguished from Hodgson's Hawk Cuckoo by buff
tip to tail and barring on underparts. Female is brown above with dark brown scaling
and whitish underparts with heavy rufous black streaks. Iris—yellow; bill—yellowish-
green; feet—yellow.
Voice: Calls monotonously in spring—loud, shrieking, high-pitched *wee-piwhit* (*brain
fever*) accent on *pi* in series of rising pitch then stopping suddenly and after a pause
starting again. Female gives harsh grating call.
Range: India and Sri Lanka.
Distribution and status: Common up to 1200 m in SE Xizang.
Habits: Arboreal cuckoo of woods, gardens and semi-evergreen forests.

191. Hodgson's Hawk Cuckoo *Hierococcyx fugax* [Hodgson's/   PLATE 20
Horsfield's/Fugitive Hawk Cuckoo] Zongfu dujuan

Description: Medium-sized (28 cm) slaty-grey cuckoo with barred tail and rufous breast. Smaller than Large Hawk Cuckoo and separated from other hawk cuckoos by slaty-grey upperparts; grey sides of head; lack of moustachial line (except in juvenile) and white belly. Has white band on nape. Chin is black and throat whitish. Tail is narrowly edged rufous. Race *nisicolar* has white streaks on rufous breast. Lacks white band on nape. Lacks narrow rufous on tail, is smaller and has different call. Iris—red or yellow; bill—black with yellow base and tip; feet—yellow.

Voice: Call is distinct sibilant, insistent *gee-whizz*, or *fe-ver* repeated about 20 times (*nisicolar*) or three-noted call *ju-ichi, ju-ichi* of Northern Hawk Cuckoo (*hyperythrus*).

Range: Breeds SE Siberia, Korea, Japan and NE China, wintering to S China and southern SE Asia; also resident in S Thailand, Peninsular Malaysia.

Distribution and status: Race *hyperythrus* breeds in NE provinces wintering in S and SE China. Race *nisicolar* breeds in China south of 32 °N and winters to SE Asia.

Habits: Lives in woods and deciduous forests, winters in evergreen forests.

Note: Some authors treat Northern Hawk Cuckoo (*hyperythrus*) as separate species

192. Indian Cuckoo *Cuculus micropterus* [Short-winged Cuckoo]   PLATE 20
Sisheng dujuan

Description: Medium-sized (30 cm) greyish cuckoo. Similar to Common Cuckoo but distinguished by subterminal black tail bar and grey, darker iris. Grey head contrasts with dark slaty back. Female is browner than male; immature distinguished by whitish-buff scaling on head and upper back. Iris—reddish-brown; eye-ring—yellow; bill—black above, greenish below; feet—yellow.

Voice: Loud, clear, deliberately enunciated four-note whistle *one more bottle* persistently repeated, often at night. Fourth note is lower.

Range: S Asia, SE Asia, Philippines, Borneo, Sumatra with offshore islands and W Java.

Distribution and status: Nominate race is a common summer breeder of lowland forests up to 1000 m. Occurs from NE to SW and SE China. Resident on Hainan.

Habits: Generally keeps high to the canopy of forest and secondary forest. More frequently heard than seen.

193. Common Cuckoo *Cuculus canorus* [Eurasian/Grey Cuckoo]   PLATE 20
Da dujuan

Description: Medium-sized (32 cm) cuckoo. Upperparts grey with blackish tail; black bars on whitish abdomen. Hepatic female form is rufous, barred black on the back. Distinguished from Indian Cuckoo by yellow iris and lack of subterminal tail bar and from female Oriental Cuckoo by unbarred rump. Juvenile has white patch on nape. Iris and eye-ring—yellow; bill—dark above, yellow below; feet—yellow.

Voice: Classic, loud, clear *kuk-oo* generally heard only in breeding area.

Range: Breeds in Eurasia, migrating to Africa and SE Asia.

Distribution and status: Common summer breeder through most of China. Race *subtelephonus* in Xinjiang to C Nei Mongol; nominate in Altai Mts of N Xinjiang, NE

China, Shaanxi and Hebei; *fallax* in E and SE China; and *bakeri* from Qinghai and Sichuan to S Xizang and Yunnan.

Habits: Favours open wooded areas and large reed beds where it sometimes sits on telephone wires looking for Great Reed Warbler nests.

194. Oriental Cuckoo *Cuculus saturatus* [Himalayan Cuckoo]   PLATE 20
Zhong dujuan

Description: Smallish (26 cm) grey cuckoo with broadly barred abdomen and flanks. Male and grey female have grey breast and upperparts, blackish-grey unbarred tail and buffy underparts barred black. Immature and hepatic female have upperparts rufous-brown heavily barred black and underparts whitish barred black up to chin. Distinguished from Common and Indian Cuckoos by bolder broader breast bars and by song. Hepatic female differs from that of Common Cuckoo in having barred rump. Iris—reddish-brown; eye-ring—yellow; bill—horn; feet— orange-yellow.

Voice: A four-note hoot lacking a grace note. Similar to Common Hoopoe but often preceded by guttural *kkukh* and sometimes followed by series of quiet *bu bu* notes.

Range: Breeds in N Eurasia and Himalayas migrating in winter to SE Asia and Greater Sundas. Resident races occur on the Greater Sundas.

Distribution and status: Fairly common summer breeder of hills and mountains from 1300–2700 m. Race *horsfieldi* breeds in NW and NE China south to 32 °N; race *saturatus* breeds south of 32 °N including Taiwan and Hainan.

Habits: Secretive bird of forest canopy. Rarely seen except in spring breeding season when calling very frequently.

195. Lesser Cuckoo *Cuculus poliocephalus* [Small Cuckoo]   PLATE 20
Xiao dujuan

Description: Small (26 cm) grey cuckoo with banded belly. Upperparts grey with head, neck and upper breast paler grey. Lower breast and rest of underaprts white with clean blackish bars and buff wash ventrally. Tail unbanded grey with narrow white tip. Female like male but also has hepatic rufous morph, banded black all over. Has yellow eye-ring. Like Common Cuckoo but smaller and best identified by call. Iris—brown; bill—yellow with black tip; feet—yellow.

Voice: Husky, chattering call *that's your choky pepper* with emphasis on *choky* and rising in pitch, followed after slight pause by *choky pepper* descending in pitch.

Range: Himalayas to India, C China and Japan; winters to Africa, S India and Burma.

Distribution and status: Uncommon. Nominate race breeds in S Jilin, Liaoning and Hebei across to Sichuan, S Xizang, Yunnan, Hainan, Guangxi and eastern provinces. Migrates across SE China, Hainan and Taiwan. Ranges from 1500–3000 m in Himalayas, but lower in north of range.

Habits: Similar to Common Cuckoo. Keeps to well-wooded country.

196. Banded Bay Cuckoo *Cacomantis sonneratii* [Banded Cuckoo]   PLATE 20
Liban dujuan

Description: Small (22 cm) brown, finely-barred cuckoo. Adult: upperparts rich brown, underparts whitish, all finely barred black; has a conspicuous pale eyebrow.

Immature: brown with black streaks and blotches rather than barring. Iris—yellow-red; bill—above blackish, below yellowish; feet—greyish-green.

Voice: A shrill, rhythmic four-note call *smoke-yer-pepper* distinguished from four-note call of Indian Cuckoo by being quicker, more plaintive and less deliberately enunciated. In the breeding season a rising call of four slow notes followed by three to six faster notes of two or three syllables rising in pitch to a sudden stop; also *tay-ta-tee* call.

Range: India, China, Borneo, Sumatra with offshore islands, Java and Philippines.

Distribution and status: Nominate race is a rare lowland bird up to 900 m and rarely up to 1200 m in SW Sichuan to S Yunnan (Xishuangbanna).

Habits: Prefers open forest, forest edge, secondary scrub and cultivated areas. Regularly heard but rarely seen.

197. Plaintive Cuckoo *Cacomantis merulinus* Basheng dujuan   PLATE 20

Description: Small (21 cm) greyish-brown and rufous cuckoo. Adult: head grey; back and tail brown; breast and belly orange-rufous. Immature: upperparts brown, barred black; underparts whitish with fine barring; resembles adult Banded Bay but without eye-stripe. Iris—crimson; bill—blackish above, yellow below; feet—yellow.

Voice: Mournful whistle *tay-ta-tee, tay-ta-tee*, increasing in speed and rising in pitch. Sometimes heard at night. A second call of two or three whistles breaking into a descending series *pwee, pwee, pwee, pee-pee-pee-pee*.

Range: E India, S China, Greater Sundas, Sulawesi and Philippines.

Distribution and status: A common resident and seasonal migrant up to 2000 m. Race *querulus* breeds in SE Xizang, S Sichuan, Yunnan, Guangxi, Guangdong and Fujian; resident on Hainan.

Habits: Prefers open woodland, secondary forest and cultivated areas, including towns and villages. Regularly mobbed by small birds. Has a very familiar call but is difficult to see.

198. Asian Emerald Cuckoo *Chrysococcyx maculatus*   PLATE 20
Cui jinjuan

Description: Small (17 cm) shining green cuckoo. Male: head, upperparts and breast shining green; belly white with green banding. Female: rufous crown and nape, coppery-green upperparts and white underparts barred dark buff. Immature has rufous head and streaked crown. In flight a broad white band on underwing at base of flight feathers. Iris—reddish-brown; bare eye-ring—orange; bill—orange yellow; feet—black.

Voice: Loud whistled twitters.

Range: Breeds in northern SE Asia, migrating south in winter as far as Malay Peninsula and Sumatra.

Distribution and status: An uncommon resident and summer breeder of lowland forests and secondary growth up to 1200 m. Breeds in S Sichuan, Hubei and Guizhou; resident in SE Xizang, Yunnan and Hainan.

Habits: Recognised by calls but otherwise easily overlooked, feeding quietly in tree crowns.

199. Violet Cuckoo *Chrysococcyx xanthorhynchus* Zi jinjuan　　　　PLATE 20

Description: Small (16 cm) violet (male) or bronze-green (female) cuckoo. Male: head, chest and upperparts violet; abdomen white with violet banding. Female: eyebrow, cheeks and underparts white, banded bronze; crown brownish; upperparts otherwise bronze-green. Iris—red; bill—yellow with red base (male), upper mandible black with red base (female); feet—grey.

Voice: High-pitched *kie-vik, kie-vik,* usually given in dipping flight; also a shrill musical descending accelerating trill.

Range: E Asia, SE Asia, Greater Sundas and Philippines.

Distribution and status: Very rare lowland resident. Race *linborgi* is found in W (Weixi) and S Yunnan (Xishuangbanna); nominate race occurs in SE Xizang.

Habits: Prefers forest edge, gardens and plantations rather than primary forest. Generally secretive, creeping about branches catching insects or perched motionless at the top of a tall tree in an exposed position to call.

200. Drongo Cuckoo *Surniculus lugubris* Wu juan　　　　PLATE 19

Description: Medium-sized (23 cm) black cuckoo. Plumage all over glossy black, except for white thighs and white barring on undertail coverts and underside of outer tail feathers; also rarely visible white nuchal patch. Juveniles are irregularly spotted with white. Tail forked like a drongo. Iris—brown (male), yellow (female); bill—black; feet—blue grey.

Voice: Loud clear call of four to seven even-spaced *pi* notes on steadily ascending scale preceded by higher introductory note; also a rapidly trilled series of rising notes ending with about three descending notes.

Range: India, China, SE Asia, Indonesia and Philippines.

Distribution and status: An uncommon lowland bird with resident and migrant populations up to 900 m. Race *dicruroides* breeds in S Sichuan, Yunnan, SE Xizang, Guizhou, Guangdong and Fujian. Resident on Hainan.

Habits: Inhabits forest, forest edge and secondary scrub. Secretive. Similar to drongos in appearance but not in posture, movements or flight.

201. Asian Koel *Eudynamys scolopacea* [Indian/Common Koel]　　　　PLATE 19
Zao juan

Description: Large (42 cm) entirely black (male) or white-speckled grey-brown (female) cuckoo with green bill. Iris—red; bill—pale green; feet—blue-grey.

Voice: Loud *kow-wow* with stress on second syllable, repeated up to 12 times with increasing tempo and pitch by day and at night. Also a shriller, faster *kuil, kuil, kuil, kuil* call.

Range: India, China, SE Asia and Indonesia.

Distribution and status: Race *chinensis* is summer breeder in most of China south of 35 °N; *harterti* is resident on Hainan.

Habits: A maddening bird whose loud calls endlessly mock the bird watcher who rarely gets a glimpse of this shy bird which keeps to dense cover in mangroves, secondary forest, forest, gardens and plantations. Brood parasite of crows, drongos and orioles.

202. Green-billed Malkoha *Phaenicophaeus tristis* Lüzui dijuan   PLATE 19

Description: Large (55 cm), very long-tailed malkoha. Head and mantle grey; underparts brownish-grey with dark shaft lines on throat and breast; back, wing and tail dark metallic green; rectrices tipped white. Iris—brown; bare skin round eye—red; bill—green; feet—grey-green.

Voice: A clucking and croaking call, rather frog-like.

Range: Himalayas, China, SE Asia and Sumatra.

Distribution and status: Locally common resident. Race *saliens* is found in Yunnan and S Guangxi; *hainanus* on Hainan and adjacent Leizhou Peninsula of Guangdong.

Habits: As other malkohas. Prefers dense thickets and tangles in middle canopy of primary and secondary forests and plantations.

COUCALS—Family: Centropodidae

Coucals have black and chestnut plumage with long tails and live in secondary scrub habitats and long grass. They are poor fliers and hop about in a characteristic manner. They were formerly included within the Cuckoo family but are not brood parasites and differ in DNA structure.

203. Greater Coucal *Centropus sinensis* [Crow-Pheasant]   PLATE 19
Hechi yajuan

Description: Large (52 cm) long-tailed coucal. Plumage entirely black except for clean chestnut red-mantle, wings and wing coverts. Iris—red; bill—black; feet—black.

Voice: Series of deep *boop* notes beginning slowly, increasing in tempo and falling in pitch; then pitch rises and tempo slows to a drawn-out series at constant pitch or an abbreviated call of four *boop* notes at the same pitch. Also a sudden *plunk* sound.

Range: India, China, SE Asia, Greater Sundas and Philippines.

Distribution and status: A common resident up to 800 m in south of country. Race *intermedius* on Hainan and in S and W Yunnan; *sinensis* in E Yunnan to Fujian.

Habits: Frequents forest edge, secondary scrub, reedy river banks and mangroves. Often comes to the ground but also hops about in small bushes and trees. Prefers thicker vegetation than Lesser Coucal.

204. Lesser Coucal *Centropus bengalensis* Xiao yajuan   PLATE 19

Description: Largish (42 cm) rufous and black, long-tailed cuckoo. Similar to Greater Coucal but smaller and duller, almost dirty coloured. The chestnut mantle and wings are pale and suffused with black. Immature is streaked brown. Intermediate plumage is common. Iris—red; bill—black; feet—black.

Voice: Several deep hollow *hoop* notes increasing in tempo and descending in pitch, like water pouring from a bottle; more rapid than Greater Coucal. A second call of three *hup* notes breaking into a series *logokok, logokok, logokok*.

Range: India, China, SE Asia, Philippines and Indonesia.

Distribution and status: Race *lignator* is a common resident up to 1000 m over China south of 27 °N and in Anhui, Taiwan and Hainan.

Habits: Prefers hillside scrub, marshes and open grassy areas including tall grassland. Often on the ground or making short flappy flights, low over vegetation.

PARROTS—Order: Psittaciformes Family: Psittacidae

Parrots are a large, diverse family of colourful birds found throughout the world's tropics and Australasia. They have large heads, powerful, hooked beaks and strong flexible feet with two toes pointing backwards. They nest in tree holes and feed mostly on fruit, seeds and pollen. Their flight is fast and their calls are harsh and piercing. Nine species have been recorded in China. Two are based on feral populations and one species is dubious. Other species may be seen as free-flying escapes from the cage-bird trade.

[205. Rainbow Lorikeet *Trichoglossus haematodus*   NOT ILLUSTRATED
[Coconut Lorikeet] Caihong yingwu

Description: Medium-sized (24 cm) colourful parrot. Head blackish-brown, streaked grey; yellow collar; green back; red chest and underwing; purple-black abdomen; thighs banded green and yellow; yellow band under wing conspicuous in flight. Iris—red; bill—red; feet—grey.

Voice: Harsh repeated screech during flight *keek, keek, keek, keek*; chattering and twittering calls when settled.

Range: Australia, Pacific, New Guinea, Moluccas, Lesser Sundas to Bali.

Distribution and status: Feral population established on Hong Kong Island but shows some decline.

Habits: A gregarious bird, flying over towns and parks in noisy screeching parties.]

[206. Yellow-crested Cockatoo *Cacatua sulphurea*   NOT ILLUSTRATED
[Lesser Sulphur-crested Cockatoo] Xiaokuihua fengtouyingwu

Description: Large (33 cm), noisy, conspicuous white parrot with a long, erectile yellow crest and yellow cheeks. Iris—dark brown; bill—black; feet—dark grey.

Voice: A loud raucous screeching *kerk-kerk-kerk* and assorted whistles.

Range: Endemic to Sulawesi and Lesser Sundas. Introduced to Singapore and Hong Kong.

Distribution and status: Globally Endangered (Collar *et al.* 1994). Feral colony firmly established on Hong Kong Island.

Habits: Lives in pairs and congregates in small groups. Conspicuous in flight with heavy, fast flapping interspersed with glides, screeching at each other. When calling from a perch the crest is erected and lowered.]

207. Vernal Hanging Parrot *Loriculus vernalis* Duanwei yingwu   PLATE 21

Description: Tiny (13 cm) short-tailed parrot with red bill and rump diagnostic, underside of wing is turquoise with green wing lining. Male has blue throat. Iris—yellow; bill—red; feet—yellow.

Voice: High-pitched, squeaky *tsee-sip* or *pee-zeez-eet*, given in flight.

Range: India, S China, SE Asia and Java.

Distribution and status: Uncommon resident in tropical and subtropical evergreen forests. Nominate race is only known from SW Yunnan. Other records are probably escapes. Threatened by bird trade in China.

Habits: Flies over forest with 'whirring' wings. Clamber about in comical manner. Often hangs upside-down.

[208. Rose-ringed Parakeet *Psittacula krameri* PLATE 21
Hongling lüyingwu

Description: Medium-sized (38 cm) long-tailed green parrot with red bill. Tail blue with yellow tip. Lacks maroon patch on wing 'shoulder'. Male has green head with bluish nape, narrow moustachial line extending around sides of neck immediately above narrow pink collar. Female has entire head uniform green. Iris—yellow; bill—red with blue cere; feet—greenish.

Voice: Harsh shrill screams.

Range: E Africa, India, SE China and SE Asia.

Distribution and status: Introduced and established in Hong Kong and adjacent Guangdong. Also in Macao. Could occur in extreme W Yunnan.

Habits: Typical of genus.]

209. Grey-headed Parakeet *Psittacula finschii* Huitou yingwu PLATE 21

Description: Medium-sized (35 cm) long-tailed green parrot with slaty-grey head and black throat with chestnut patch on 'shoulder' of wing diagnostic. Yellow tips to tail streamers. Derbyan lacks chestnut wing patch and has black loral stripe and grey breast. Female Blossom-headed lacks black throat and head paler grey. Iris—yellow; bill—upper mandible vermilion tipped yellow, lower mandible yellow; feet—grey.

Voice: Soft melodious but high-pitched calls *tooi-tooi* less harsh than most of genus. Shrill whistle *sweet*, with rising inflection like Long-tailed Broadbill.

Range: E Himalayas to SW China and SE Asia.

Distribution and status: Fairly common resident in SE Xizang, Yunnan and SW Sichuan up to 2700 m.

Habits: Lives in small flocks in subtropical broadleaf forests and comes into cultivated lands to raid maize fields.

Note: Placed by some authors as a race of Slaty-headed Parakeet *p. himalayana* (e.g. Cheng 1987, 1994).

210. Blossom-headed Parakeet *Psittacula roseata* PLATE 21
Huatou yingwu

Description: Medium-sized (30 cm) green parrot with very long tail. Male has rose-pink head with violet tinge on nape; Black throat extending to form narrow black collar. Has small maroon patch on 'shoulder' of wing and bluish tail with pale yellow tips. Tail less long than Grey-headed Parakeet. Female has grey head and lacks black throat and collar; distinguished from juvenile Grey-headed Parakeet by bill colour and maroon shoulder patch. Iris—yellow; bill—yellow above, dark grey below; feet—grey.

Voice: Softer and less piercing whistle than other parakeets.

Range: N and E India, SE Asia and S China.

Distribution and status: Nominate race is rare resident in S Guangxi, W Guangdong and probably S Yunnan and SE Xizang.

Habits: Lives in deciduous forests and open woodlands and secondary growth.

211. Derbyan Parakeet *Psittacula derbiana* Da zixiong yingwu PLATE 21

Description: Large (43 cm) long-tailed parrot with bright red upper bill (male), violet blue-grey head and breast with broad black moustachial stripe. Male has greenish wash around eyes and forehead, and a narrow black frontal band extending into narrow eyeline. Tail is graduated with bluish central feathers. Distinguished from other parakeets by underpart of neck, breast and upper abdomen being purplish-violet and lack of maroon on 'shoulder' of wing. Female all-black bill, no blue on forecrown. Iris—yellow; bill—male: red upper mandible, black lower mandible, female: black; feet—grey.

Voice: The call is a high-pitched shrill whistle.

Range: NE India and SE Tibetan Plateau to SW China.

Distribution and status: Globally Near-threatened (Collar *et al.* 1994). Common in hill and montane forests of SE Xizang, SW Sichuan and W and NW Yunnan up to 4000 m.

Habits: Flies fast over forest in screeching flocks. Commonly caught at nest for pet trade and locally endangered by such collecting.

212. Red-breasted Parakeet *Psittacula alexandri* [Moustached PLATE 21
Parakeet] Feixiong yingwu

Description: Medium-sized (34 cm) colourful parrot with diagnostic pink breast. Adult: crown and cheeks violet-grey with black lores; nape, back, wings and tail green; pronounced black 'moustache'; thighs and vent pale green. Immature birds have buffy-brown head with less prominent black moustachial. Iris—yellow; bill—male upper mandible red, lower black, female upper mandible black, lower dark brown; feet—grey.

Voice: Repeated piercing *kekekek* particularly of young birds and a raucous trumpet-like scream.

Range: India, S China, SE Asia and Greater Sundas.

Distribution and status: Occasionally found flying free in Hong Kong and other cities of S Guangdong, probably feral. Race *fasciata* is uncommon resident in SE Xizang, Yunnan, SW Guangxi and Hainan.

Habits: A communal bird, travelling, roosting and nesting in parties. Noisy and conspicuous in flight with birds flashing low over open spaces to settle with a clatter of wings in trees to feed or rest, screeching frequently.

[213. Long-tailed Parakeet *Psittacula longicauda* [Pink-cheeked PLATE 21
Parakeet] Changwei yingwu

Description: Largish (40 cm) parrot with very long tapering tail and green breast. Male: green crown, red sides of head and bold black malar stripe; mantle washed pale blue; yellow-tipped tail and wings bluish. Female: duller, with greenish moustachial and no blue on back. In flight the wing lining is yellow. Distinguished from Red-breasted

Parakeet by green underparts and red sides of head. Iris—greenish-yellow; bill—male upper mandible red with horn tip, female black; feet—grey.

Voice: Harsh strident screeches given from tree-tops and in flight.

Range: Andamans, Nicobars, Malay Peninsula, Sumatra and Borneo.

Distribution and status: Race *modesta* has been recorded from Sichuan and nominate race recorded from Guangxi. Both records are considered dubious (see Schauensee 1984) being either misidentified Blossom-headed Parakeets or escapes.

Habits: Flies fast in large flocks between feeding sites and roosts. Congregates in huge numbers at coastal roost sites.]

SWIFTS—Order: Apodiformes Family: Apodidae

A worldwide family of fast-flying, insectivorous birds; swifts superficially look rather like swallows but are in fact most closely related to hummingbirds. Swifts have long, pointed backswept wings and either short, squarish or long, pointed tails and tiny legs. They rarely perch in trees but usually rest by clinging to cliffs with their sharp claws. They nest in caves, hollow trees and under house roofs in cup-shaped nests made of mud or, in some species, saliva. They feed on the wing using the wide mouth to catch insects. Some of the cave-nesting swiftlets use a form of sonar echolocation with clicking calls to find their way in the dark. There are nine species of swifts in China, although some authors split the Himalayan Swiftlet into two or three species. Some swifts are very difficult to identify on the wing. Another dark swift has been reported from Beidaihe (Hebei) but never collected or named.

214. Himalayan Swiftlet *Collocalia brevirostris* Duanzui jinsiyan   PLATE 22

Description: Smallish (14 cm) blackish swiftlet with long blunt wings and slightly forked tail. The rump varies from brownish to greyish and underparts paler brown with slight darker streaks. Legs only slightly feathered. Race *innominata* has greyer rump than nominate. Race *inopina* is the darkest race. Race *rogersi* is slightly smaller than nominate with rump only slightly darker than back; underparts whiter; legs unfeathered. Iris—dark; bill—black; feet—black.

Voice: Twittering *chit chit* and low rattle like knitting needle being drawn across a comb.

Range: Himalayas to C China, SE Asia and W Java.

Distribution and status: Nominate race occurs in SE Xizang; *innominata* breeds in C China and winters to Thailand; *inopina* breeds in E Yunnan, Sichuan and C China. Race *rogersi* is rare resident only in Yongde area of SW Yunnan. Cheng (1987, 1994) merges *inopia* with *innominata*.

Habits: Fast-flying in flocks about the open peaks and ridges of high mountains. Nests in rock crevices and makes an inedible mossy nest.

Note: Some authors treat Indochinese Swiftlet *C. rogersi* and Chinese Swiftlet *C. innominata* as separate species. Birds listed as Black-nest Swiftlet *C. maxima maxima* by Meyer de Schauensee (1984) are misidentified specimens of this species. This has led to confusion in the literature (e.g. Yan *et al.* 1996). Many authors place this species in the genus *Aerodramus*.

215. Germain's Swiftlet *Collocalia germani* [Oustalet's/German's   PLATE 22
Swiftlet] Zhuawa jinsiyan

Description: Smallish (12 cm) dark swiftlet with slightly forked tail. Upperparts black-ish-brown with paler, whitish/greyish rump contrasting with dark tail. Underparts greyish-brown with paler band across belly. Iris—dark brown; bill—black; feet—purplish-red.

Voice: High-pitched *tscheerrr* is commonly uttered near breeding sites.

Range: Indochina, Malaysia and N Borneo.

Distribution and status: Rare. Nominate race breeds on Dazhou islet off SE Hainan where there are maximum of 200 nests in three caves. Probably more common on other islands in the South China Sea.

Habits: Breeds in coastal rock crevices. Whitish-yellow transluscent nests of saliva are harvested for making soup. Swifts can echo-locate.

Note: Some authors place this species within Edible-nest Swiftlet *C. fuciphaga* (e.g. Cheng 1987). Has been placed in genus *Aerodramus*.

216. White-throated Needletail *Hirundapus caudacutus*   PLATE 22
[Spine-tailed Swift/Northern Needletail] Baihou zhenweiyuyan

Description: Large (20 cm) blackish swift with white chin and throat, white undertail coverts and small white patch on tertials; back brown with silvery-whitish saddle. Distinguished from other needletails by white throat. Iris—dark brown; bill—black; feet—black.

Voice: High-pitched chitters when chasing each other.

Range: Breeds in N. Asia, China, Himalayas but migrates in winter to Australia and New Zealand.

Distribution and status: Race *nudipes* breeds in S Qinghai, SE and E Xizang, Sichuan and N and W Yunnan; nominate race breeds in NE China, recorded on passage through E and S China and Hainan; *formosana* is resident on Taiwan.

Habits: Similar to other needletails. Flies at very high speed over forest and ridges. Sometimes flies low over water to drink.

217. Silver-backed Needletail *Hirundapus cochinchinensis*   PLATE 22
[White-vented/Grey-throated Needletail, AG: Needle-tailed Swift] Huihou zhenweiyuyan

Description: Largish (18 cm) blackish swift with a pale brown saddle on back and rump and a short squarish needle-tail. Chin and throat is greyish and the undertail coverts are white. Distinguished from White-throated by greyish throat and lack of white on tertials. Lores never white. Iris—dark brown; bill—black; feet—dark purple.

Voice: Soft rippling *trp-trp-trp-trp-trp*.

Range: NW India, SE Asia, Sumatra and W Java.

Distribution and status: Nominate race breeds on Hainan and islands of South China Sea; also Himalayas to SE Xizang.

Habits: Similar to White-throated Needletail.

218. Asian Palm Swift *Cypsiurus balasiensis* Zong yuyan   PLATE 22

Description: Small (11 cm) entirely dark brown, slender swift. Distinguished from swiftlets by its larger, narrower wings and very deeply-forked tail. Iris—dark brown; bill—black; feet—purplish.

Voice: High-pitched chattering *cheereecheet* is regularly uttered.

Range: India, China, SE Asia, Greater Sundas, Sulawesi and Philippines.

Distribution and status: Race *infumatus* is locally common in tropics of Yunnan and Hainan up to 1500 m where there are palm trees.

Habitat: Distribution is determined by the presence of fan-palms, such as *Livistona*, used as nest sites and resting places. The nest is cemented under a palm leaf.

[219. Alpine Swift *Tachymarptis melba* Gaoshan yuyan   PLATE 22

Description: Large (21 cm) swift with slightly forked tail and diagnostic white throat and white breast separated by dark brown band across upper breast. Has rather broad wings. Iris—brown; bill—black; feet—black.

Voice: Less piercing than Common Swift, *chit rit rit rit rit it it itititit chet et et et et . . .*

Range: SE Europe, N Africa, Middle East, C Asia, Himalayas and India. Winters in tropical Africa.

Distribution and status: Rare seasonal migrant. Race *nubifuga* occurs in Himalayas up to 2500 m. Recorded in Arunachal Pradesh and should occur in SE Xizang. Could also occur in NW Xinjiang.

Habits: Lives in mountainous regions. Has relatively slow wing beat.]

220. Common Swift *Apus apus* Putong louyan   PLATE 22

Description: Large (17 cm) all dark swift with moderately forked tail and pale throat. Forehead paler than crown and outer wing paler than inner wing. Iris—brown; bill—black; feet—black.

Voice: Screaming cries *srrreeee*.

Range: Palearctic; wintering to southern Africa.

Distribution and status: Very common in breeding areas. Race *pekinensis* breeds over most of northern China and as far south as Chengdu (Sichuan). Migrants pass through both east and west of country.

Habits: Colonial nester, building under eaves of houses, also on cliffs. Nests made of mud. Lives in noisy flocks, swerving around at high speed and calling excitedly.

221. Fork-tailed Swift *Apus pacificus* [White-rumped/Pacific Swift]   PLATE 22
Baiyao yuyan

Description: Unmistakable, largish (18 cm), dusky-brown swift with long, deeply-forked tail, whitish chin, and white patch on rump. Distinguished from House Swift by larger size, paler colour, darker throat, narrower white rump saddle, slimmer shape, and forked tail. Iris—dark brown; bill—black; feet—purplish.

Voice: Buzzing and twittering sound, and long, high-pitched squeaks *skree-ee-ee*.

Range: Breeds in Siberia and E Asia, but migrates in winter through SE Asia to Indonesia, New Guinea and Australia.

Distribution and status: Common summer breeder. Nominate race breeds in NE, N and E China and E Xizang and Qinghai; recorded on passage across S China, Taiwan and Hainan, also NW Xinjiang. Race *kanoi* breeds in C, SW, S and SE China and Taiwan.

Habits: Generally found in flocks over open places and often mixed with other species of swifts. Flies more slowly than needletails and makes erratic flutters and turns when feeding.

222. House Swift *Apus affinis* [Little Swift] Xiao baiyao yuyan   PLATE 22

Description: Medium-sized (15 cm) blackish swift with a white throat and rump and notched rather than forked tail. Distinguished from the larger Fork-tailed Swift by darker colour, whiter throat and rump and almost square-cut tail. Iris—dark brown; bill—black; feet—brown-black.

Voice: Very vocal. Trill of fast-repeated, loud, shrill, whickering screams given in flight, *der-der-der-dit-derdiddiddoo*, especially before roosting in the evening.

Range: Africa, Middle East, India, Himalayas, S China, Japan, SE Asia, Philippines, Sulawesi and Greater Sundas.

Distribution and status: Common resident and seasonal migrant up to 1500 m. Race *nipalensis* breeds across most of S China and Hainan. Race *kuntzi* is resident on Taiwan.

Habits: Lives in large flocks, hunting with a steady even flight over open areas. Nests under house eaves, cliff overhangs or cave-mouths.

Note: East Asian races sometimes separated as *A. nipalensis* (e.g. Chantler and Driessens 1995; Viney *et al.* 1994). All Chinese specimens formerly assigned to the southern race *subfurcatus* (e.g. Cheng 1994).

TREESWIFTS—Family: Hemiprocnidae

Treeswifts are a small family confined to SE Asia. Very similar to true swifts but members of this family do perch in trees and are distinguished by more elongated wings and tails. Treeswifts hawk in wide circling flight from high vantage points on tree perches; often in shrill, calling flocks. Their tiny nests are also attached to branches where they glue a single white egg. There is only one species in China.

223. Crested Treeswift *Hemiprocne coronata* Fengtou shuyan   PLATE 22

Description: Grey swift-like bird with long tail and long curved wings and diagnostic erectile crest. Upperparts dark grey with light grey band on tertials; black mask around eye; underparts grey. Male has rufous patch on side of face and ear coverts. Immature is browner, with minimal crest and heavily scaled white and dark brown. Iris—brown; bill—black; feet—red.

Voice: Loud, harsh, high-pitched cries *cher-tee-too-cher-tee-too-cher-tee-too*.

Range: Indian subcontinent and SE Asia.

Distribution and status: Rare. Resident in W and S Yunnan (Xishuangbanna) and SE Xizang up to 1000 m.

Habits: Prefers edge or gaps of evergreen rain forest. Perches on bare tree perches and makes circular hawking flights like a bee-eater or wood-swallow.

Owls—Order: Strigiformes

Owls are a familiar worldwide order of large-eyed, nocturnal birds of prey with haunting calls. They have broad round heads, flat faces and forwardly-directed eyes. Most species have distinct facial discs around the eyes. There are two families—barn owls and true owls. All owls lay white eggs and most species nest in tree holes or even holes in buildings. At night, when they are active, they are of course difficult to see so the best means of identification is the call.

Barn and Grass Owls—Family: Tytonidae

Nocturnal raptors with very round, heartshaped faces, dark eyes and broad facial discs which amplify sound to the ears. They hunt largely by ear. The wing feathers are soft for silent flight and their calls are harsh screeches. There are three species in China.

224. Barn Owl *Tyto alba* [Common Barn Owl] Cangxiao PLATE 23

Description: Unmistakable medium-sized (34 cm) whitish owl. Broad, white heart-shaped facial disc diagnostic. Upperparts ochraceous-buff with fine markings; underside white with fine black spots. General colour variable and immature birds darker buff. Iris—dark brown; bill—dirty yellow; feet—dirty yellow.

Voice: Harsh, hoarse, high-pitched screech *wheech* or *se-rak*; also a high *ke ke ke ke ke* call.

Range: Americas, W Palearctic, Africa, Middle East, Indian subcontinent, SE Asia, Sundas, New Guinea, Australia.

Distribution and status: Occasional stragglers of race *javanicus* recorded in Yunnan.

Habits: Hides during the day in dark holes in houses, trees, caves, cliffs or dense vegetation. Emerges at dusk over open ground, flying low on silent wings, nests in tree holes or usually in buildings.

225. Grass Owl *Tyto capensis* [>Eastern Grass Owl] Caoxiao PLATE 23

Description: Medium-sized (35 cm) owl with heart-shaped facial disc. Like Barn Owl but much darker with buffy face and breast and dark brown upperparts. Spotted, mottled and vermiculated like Barn Owl. Iris—brown; bill—creamy; feet—whitish.

Voice: Loud screech.

Range: Africa, New Guinea, Japan, Australia, Indian subcontinent to SW and S China, SE Asia and Philippines.

Distribution and status: Rare resident and winter visitor. Race *chinensis* is found in SE Yunnan (Mengzi), Guizhou, Guangxi, Guangdong, Hong Kong, Fujian north to Anhui. Race *pithecus* is resident on S Taiwan.

Habits: Inhabits open and tall grassland.

Note: Sometimes treated as race of Eastern Grass Owl *T. longimembris*.

226. Oriental Bay Owl *Phodilus badius* [Asian Bay/Bay Owl]   PLATE 23
Lixiao

Description: Medium-sized (27 cm) reddish-brown owl, rather similar in shape to Barn Owl with heart-shaped facial mask and sometimes erect 'ears'. Upperparts reddish-brown with black and white spots; underparts pinkish-buff with black spots; face pinkish. Iris—dark; bill—yellow; feet—feathered.

Voice: A soft hoot and ringing *hooh-weeyoo* also mournful musical whistles described as *kwankwit-kwankwit-kek-kek-kek* given in flight in darkness.

Range: Indian subcontinent, S China, SE Asia, Greater Sundas.

Distribution and status: Race *saturatus* is a rare forest resident up to 1500 m in S Yunnan, SW Guangxi and Hainan.

Habits: Poorly-known, shy, nocturnal forest owl. Sits rather horizontally in the daytime.

TRUE OWLS—Family: Strigidae

True owls are similar to barn owls but have generally shorter legs and smaller facial discs. Several species have prominent erectile 'ear tufts'. The plumage of all species is elaborately patterned with grey, brown, white and black giving them good camouflage when resting during the day. There are a total of 28 true owl species in China.

227. Mountain Scops Owl *Otus spilocephalus*   PLATE 24
Huangzui jiaoxiao

Description: Small (18 cm) tawny-rufous owl with small eartufts, yellow eyes, cream bill, lack of bold streaks or bars and a row of very large triangular white spots on scapulars diagnostic. Iris—greenish-yellow; bill—creamy-buff; feet—whitish-grey.

Voice: Soft, far-carrying, double-noted metallic whistled hoot *plew plew* at about 12-second intervals given most of the year.

Range: Himalayas, NE Indian subcontinent, S China, SE Asia, Sumatra and N Borneo.

Distribution and status: A rare resident from 1000–2500 m in moist tropical montane forests; race *latouchei* from SW Yunnan to SE China and Hainan; race *hambroecki* on Taiwan.

Habits: As other scops owls. Replies to imitations of its call.

228. Pallid Scops Owl *Otus brucei* [Striated/Bruce's Scops Owl]   PLATE 24
Zongwen jiaoxiao

Description: Small (21 cm) pale sandy-grey scops owl with yellow eyes. Similar to grey morph of Eurasian Scops Owl but paler and more uniform sandy-grey above without white spots on crown or hindneck, rufous-buff rather than pale or cream-coloured scapulor spots, wings fall short of tailt plainer facial disk, and greyer below with well-defined black streaks without white crosbars. Juvenile has underparts entirely barred. Best identified by calls at night. The only local scops owl with feathered toes. Iris—yellow; bill—blackish; feet—grey.

Voice: Call is soft and dove-like. Male gives hollow, resonant, low-pitched *whoop* repeated about eight times in 5 seconds; also longer *whaoo* at 3–5-second intervals, or *ooo-ooo—ooo-ooo*.

Range: Middle East to Pakistan and W China, winters NW and W India.

Distribution and status: Very rare in China. Recorded in W Xinjiang in Kunlun Mts and Kashi region. Presumed to breed.

Habits: As other scops owls. Lives in arid and semi-arid areas with tree cover, fruit groves and irrigated agricultural lands.

229. Eurasian Scops Owl *Otus scops* Hong jiaoxiao   PLATE 24

Description: Small (20 cm) 'eared' owl with yellow eyes and heavily streaked plumage. Occurs in rufous and grey colour forms. No overlap with Oriental Scops Owl which has a different call. Iris—yellow; bill—horn; feet—brownish-grey.

Voice: Call is deep monotonous whistle repeated about every 3 seconds, *chook* rather like a toad. Female call slightly higher than that of male.

Range: W Palearctic to Middle East and C Asia.

Distribution and status: Very restricted range in China. Race *pulchellus* breeds in Tianshan and Kashi area of W Xinjiang.

Habits: Migratory. Truly nocturnal small owl of open country where there are clumps of trees.

Note: Some authors have included races of Oriental Scops Owl *Otus sunia* and Riukiu Scops Owl *Otus elegans* in this species (e.g. Cheng 1987).

230. Oriental Scops Owl *Otus sunia* [Asian Scops Owl]   PLATE 24
Dongfang jiaoxiao

Description: Small (19 cm) mottled brown owl with short eartufts, yellow eyes and heavily black-streaked breast. Grey and rufous forms occur. Distinguished from Collared Scops Owl by smaller size, pale eyes and lack of pale nuchal collar; from Mountain Scops and White-fronted Scops by black-streaked breast, smaller size and greyer colour. Darker and smaller than Pallid Scops Owl with more streaking on underparts and less on upperparts. Iris—orange-yellow; bill—grey-horn; feet—greyish.

Voice: Rough guttural *toik-toitoink* or *toik toik tatoink* with last note emphasised.

Range: Breeds Himalayas, Indian subcontinent, E Asia, Japan, China, SE Asia and Philippines; some winter south.

Distribution and status: Fairly common resident. Race *stictonotus* is resident in NE China and E China north of Changjiang River; nominate race is found in SE Xizang; *malayanus* in S, SW and SE China, Hainan and Taiwan; *japonensis* is accidental in Taiwan.

Habits: Hunts in smaller trees of forest edge, clearings and secondary growth.

Note: Placed by some authors as races of Eurasian Scops Owl *O. scops* (e.g. Cheng 1987).

231. Elegant Scops Owl *Otus elegans* [Ryukyu Scops Owl]   PLATE 24
Liuqiu jiaoxiao

Description: Small (22 cm) rufous-brown scops owl with yellow eyes. Plumage liberally spotted with white dots. Best distinguished from Oriental Scops Owl by call but lacks heavy black streaks on crown. Distinguished from Collared Scops Owl by yellow eyes and lack of collar. Iris—yellow; bill—dark grey; feet—grey with patterned legs.

Voice: Vocal. Call is hoarse coughing *uhu* or *kuru*. Can be called in by playback or imitation of call. Calls repeated 15–30 times per minute.

Range: Nansei Islands (Ryukyu) and Taiwan.

Distribution and status: Race *botelensis* is an endangered resident on Lanyu islet off SE Taiwan.

Habits: Inhabits subtropical lowland forest.

Note: Until recently, treated as a race of Eurasian Scops Owl (e.g. Cheng 1987).

232. Collard Scops Owl *Otus bakkamoena* [>Japanese/Indian Scops Owl] Ling jiaoxiao   PLATE 24

Description: Largish (24 cm) greyish or brownish scops owl with conspicuous ear tufts and diagnostic pale, sandy nuchal collar. Upperparts greyish or sandy-brown, mottled and blotched with black and buff; underparts buff, streaked black. Iris—dark brown; bill—greenish-grey; feet—dirty yellow.

Voice: Male gives a soft upward-inflected hoot *woop* also a steady series of gruff notes at one-second intervals. Female call is higher-pitched quavering, downward-inflected *wheoo* or *pwok* about five times a minute, also a gentle twitter. Pairs often call in concert.

Range: Indian subcontinent, E Asia, China, Japan, SE Asia, Greater Sundas and Philippines.

Distribution and status: A quite common owl up to 1600 m, including tree-lined suburban streets of towns. Race *ussuriensis* in NE China to S Shaanxi; *erythrocampe* in China south of 32 °N; *lettia* in SE Xizang; *umbratilis* on Hainan; and *glabripes* on Taiwan.

Habits: Sits on a low perch for much of the night, giving its mournful call in season. Hunts from its perch and pounces on prey on the ground.

Note: Some authors split this species into Indian Scops Owl *O. bakkamoena* and Collared Scops Owl *O. lempiji*. In this case all Chinese forms, except *lettia*, belong to *O. lempiji*. We prefer to follow Inskipp *et al.* (1996) in retaining the combined species.

233. Eurasian Eagle Owl *Bubo bubo* Diaoxiao   PLATE 25

Description: Huge (69 cm) owl with long 'ear' tufts and enormous orange eyes. Colour is mottled streaky brown. Breast is yellowish, boldly streaked dark brown with each feather finely barred brown. Feathers extend onto toes. Iris—orange; bill—grey; feet—yellow.

Voice: The call is a hollow *poop*. Gives beak-clicking sounds.

Range: Palearctic, Middle East, Indian subcontinent.

Distribution and status: Widespread but generally rare. Inhabits mountainous terrain with woods, generally nesting on cliffs, rarely on ground. Race *ussuriensis* is resident in NE and eastern N China; *kiautschensis* through C, E, S and SE China; *tibetanus* from S, SE and E Xizang, NW Yunnan, W Sichuan, Qinghai and S Gansu; *tarimensis* is resident in Tarim basin of S Xinjiang; *yenisseensis* occurs in Altai Mts and *auspicabillis* in Tianshan Mts of NW China; *hemachalana* is resident in W Xinjiang and W Xizang, also N Qinghai and W Nei Mongol.

Habits: When found in daytime it is aggressively mobbed by crows and gulls. Alarmed birds assume a wide display posture with wings arched and head lowered. Flies fast with shallow wing beats.

234. Spot-bellied Eagle Owl *Bubo nipalensis* [Forest Eagle Owl]   PLATE 25
Lin Diaoxiao

Description: Huge (63 cm) brownish owl with long, thick 'ear' tufts and greyish underparts, scalloped by dark brown barred feather tips rather than streaked, is diagnostic. Upperparts barred brown and buff, not streaked. Brown eyes and yellow bill also distinctive. Iris—brown; bill—yellow; feet—feathered buff.

Voice: Soft but resounding moaning, *boom* repeated at about three second intervals and audible from great distance. Apparently also kite-like whistle.

Range: Indian subcontinent to SW China and SE Asia.

Distribution and status: Globally Near-threatened (Collar *et al.* 1994). Rare in China. Nominate race is resident up to 1200 m, rarely up to 2100 m in C and SW Sichuan, SE and SW Yunnan. Probably also occurs in SE Xizang.

Habits: Typical eagle owl of evergreen subtropical broadleaf and moist deciduous broadleaf forests. Altitudinal migrant. Hunts around clearings and streams.

235. Dusky Eagle Owl *Bubo coromandus* [AG: Horned Owl]   PLATE 25
Wu diaoxiao

Description: Large (56 cm) light brownish-grey owl with long blackish 'ear' tufts. Upperparts sandy-grey with black streaking. Underparts buffy-grey with narrow black streaks; each feather with dusky cross bars. Greyer than any other large horned owls in China. Iris—yellow; bill—grey; feet—feathered grey.

Voice: Deep, resonant hollow call, *wo, wo, wo-o-o-o-* ... getting faster but quieter like a bouncing ping pong ball; given both day and night.

Range: Indian subcontinent, Burma and China.

Distribution and status: Rare in China. Nominate race is resident in Jiangxi (Nanchang) and Zhejiang (Jiande).

Habits: Lives in woods and open forests in moist areas with groves of large leafy trees.

236. Blakiston's Fish Owl *Ketupa blakistoni* Maotui yuxiao   PLATE 25

Description: Huge (70 cm) dark brown 'eared' owl with yellow eyes and heavy black streaking and black bars on primaries of folded wing. Black streaks of breast have many cross bars. Distinguished from eagle owls by greyer colour, yellow eyes and grey toes. Lacks broad chest streaks of Eurasian Eagle Owl and 'ear' tufts broader and less black. Brown facial disc diagnostic. Iris—yellow; bill—grey-horn; feet—grey with feathered legs.

Voice: Short, deep *boo boo, uoo*; distinct from two-noted call of eagle owls.

Range: NE Asia, Korea, Sakhalin Island and N Japan.

Distribution and status: Globally Endangered (Collar *et al.* 1994). Rare and local in NE China. Race *doerriesi* is resident in Hulun Nur area of NE Nei Mongol and through Greater Hinggan Mts to Harbin in Heilongjiang.

Habits: Lives along forested rivers.

237. Brown Fish Owl *Ketupa zeylonensis* He yuxiao   PLATE 25

Description: Large (53 cm) rufous-brown eared owl, streaked black and white above and finely streaked dark brown on underparts. Has whitish-buffy chin with dark streaks and fulvous underparts. Each streak has many fine cross bars. Distinguished from Eurasian Eagle Owl by yellowish underparts, finer streaking on breast, yellow eyes, naked legs and lack of pale eyebrows. Distinguished from Tawny Fish Owl by less orange and less heavily streaked underparts and call. Iris—yellow; bill—grey; feet—grey.

Voice: Deep, booming *oomp-ooo-oo* or *boom-o-boom* with emphasis on middle note; repeated at intervals; also a cat-like mewing call.

Range: Middle East to Indian subcontinent, Burma, Indochina and S China.

Distribution and status: Rare resident in extreme south of China. Race *orientalis* is found in S and SE Yunnan, Guangxi, Hainan and Guangdong and Hong Kong; race *leschenaulti* occurs up to 1500 m in SE Xizang and extreme SW Yunnan (Tengchong).

Habits: Slow flight with legs dangling. Lives in tropical forests along shady rivers. Semi-diurnal fishing habits.

238. Tawny Fish Owl *Ketupa flavipes* Huangjiao yuxiao   PLATE 25

Description: Very large (61 cm) rufous eared owl with yellow eyes and buffy-white throat patch. Upperparts: buffy-rufous boldly streaked dark brown with unbranched streaks. Distinguished from Eurasian Eagle Owl by yellow eyes and naked feet. Distinguished from Brown Fish Owl by brighter rufous underparts, bolder streaked upperparts and by call. Iris—yellow; bill—horn-black with greenish grey cere; feet—grey.

Voice: Deep *whoo-hoo* call and cat-like mewing.

Range: Himalayas to S China and Indochina.

Distribution and status: Globally Near-threatened (Collar *et al.* 1994). Rather rare resident up to 1500 m over much of C and S China from SW Gansu, S Shaanxi, Sichuan, Guizhou, Anhui, Jiangsu, Zhejiang, Taiwan and Guangdong. May occur in SE Xizang.

Habits: Lives in dense forests along streams in hilly terrain.

239. Snowy Owl *Nyctea scandiaca* Xuexiao   PLATE 25

Description: Unmistakable huge (61 cm) white owl with yellow eyes and generally spotted with black feather tips on crown, back, wings and lower breast making the bird appear grey against snow. Iris—yellow; bill—grey; feet—yellow.

Voice: Male gives duck-like alarm call *krek-krek-krek*. Female has a barked alarm call and also gives whistled *seeuee* calls.

Range: N Holarctic.

Distribution and status: A rare winter visitor to open country at northern latitudes in both NE and NW China.

Habits: Partially diurnal but also active after dusk hunting voles and picas. Sometimes sits on prominent rock or mound. Nests on the ground.

240. Brown Wood Owl *Strix leptogrammica* He Linxiao   PLATE 26

Description: Large (50 cm) heavily-barred brown owl without 'ear' tufts. Conspicuous facial disc of rufous 'spectacles', black eye-ring and white eyebrows; underparts buff with fine dark brown barring, washed chocolate on chest in some races upperparts dark brown, strongly barred buff and white. Race *ticehursti* (figured) is palest. Iris—dark brown; bill—whitish; feet—bluish-grey.

Voice: Distinctive deep *boo-boo* or four-syllable *goke-galoo, huhu-hooo* and others.

Range: Indian subcontinent to S China, SE Asia and Greater Sundas.

Distribution and status: Rare and secretive in subtropical montane forest. Race *ticehursti* in southern China; *caligrata* on Hainan and Taiwan; race *newarensis* is probably to be found in SE Xizang.

Habits: Rarely seen nocturnal owl. If disturbed by day it compresses its plumage to look like a dead piece of wood, watching with half-closed eyes. Pairs call to each other at dusk prior to hunting.

241. Tawny Owl *Strix aluco* Hui Linxiao PLATE 26

Description: Medium-sized (43 cm) brownish 'earless' owl. Colour generally mottled and streaked rich reddish-brown but greyish individuals are sometimes found. Underparts are streaked with transverse barring. Upperparts have some pale bars and there is a whitish V-shape on the facial disc. No ear tufts are visible. Iris—dark brown; bill—pale yellow; feet—yellow feathered but toes plumbeans at ends.

Voice: Very loud resonant *HU-HU* repeated at intervals (*nivicola*).

Range: W Palearctic, Middle East, Himalayas, China, Korea and Taiwan.

Distribution and status: Commonest owl in temperate woodlands. Race *nivicola* is resident in S and SE Xizang, and most of southern and C China; *yamadae* is resident on Taiwan; and *ma* in Hebei and Shandong.

Habits: Nocturnal in habits, generally sleeping in shaded spot during daytime. Sometimes discovered and mobbed by small songbirds. Nests in tree holes.

242. Ural Owl *Strix uralensis* Changwei linxiao PLATE 26

Description: Large (54 cm) pale greyish-brown owl with dark eyes and wide pale facial disc. Underparts greyish-buff boldly streaked dark brown. Flanks indistinctly barred. Upperparts dark brown with blackish streaks and rufescent and white spots. Eyebrow whitish. Wings and tail barred. Much larger than Tawny Owl but much smaller than Great Grey Owl. Looks like a buzzard in flight. Iris—brown; bill—orange; feet—feathered with buff and grey bars.

Voice: Courtship call is deep, far-carrying *whoohoo*, then after four-second pause *whoohoo owhoohoo*. Also gives series of about eight *poo* notes rising at end of call. Female has harsher versions of these calls, also croaking, heron-like *kuveh*. Alarm is dog-like bark, *khau khau*.

Range: Conifer forest zone of Palearctic and Japan.

Distribution and status: Rare resident. Race *nikolskii* is resident in Greater and Lesser Hinggan Mts of NE China and race *coreensis* occurs in Changbai mountains of Jilin, Liaoning and maybe to Hebei.

Habits: Lives in conifer forests. Aggressive near nest site. Mainly nocturnal.

243. Sichuan Wood Owl *Strix davidi* Sichuan linxiao PLATE 26

Description: Large (54 cm) greyish-brown, earless owl with distinctive grey facial disc and brown eyes. Looks like a large Tawny Owl with simpler streaks on underparts. Very similar to allopatric Ural Owl. Iris—brown; bill—yellow; feet—feathered with grey and brown bands.

Voice: As Ural Owl

Range: Endemic to C China.

Distribution and status: Globally Vulnerable (Collar *et al.* 1994). Rare resident in SE Qinghai and N, C and W Sichuan from 2700–4200 m in open conifer and mixed subalpine forests.

Habits: As Ural Owl.

Note: Sometimes classed as race of Ural Owl (e.g. Cheng 1987).

244. Great Grey Owl *Strix nebulosa* [Dark Wood Owl] PLATE 26
Wu linxiao

Description: Very large (65 cm) earless, grey owl with unmistakable face pattern of concentric pale and dark rings forming facial disk, with bright yellow eyes separated by opposing C-shaped white fringes. Centre of throat is black but white partial collars extend laterally underlining the facial disc. Overall plumage pale grey with heavy dark brown streaking on underparts and upperparts. Wings and tail barred grey and dark brown. Much larger than either Tawny Owl or Ural Owl. Iris—Yellow; bill—yellow; feet—densely feathered with orange toes.

Voice: Courtship call is series of 10 or more hoots delivered at half-second intervals falling in pitch and volume at end. Female responds with feeble *chi eop-chiepp-chiepp.* Also growls and grunts. Alarm is low, penetrating *grrroooo.*

Range: Boreal forests of Palearctic.

Distribution and status: Very rare in China. Race *lapponica* is resident in Hulun Nur area of Nei Mongol in Greater Hinggan Mts.

Habits: Lives in conifer and mixed forests or deciduous forests. Calm or aggressive but generally fearless at nest.

245. Northern Hawk Owl *Surnia ulula* Mengxiao PLATE 23

Description: Medium-sized (38 cm) brown owl with hawk-like tail and striking dark brown and white face pattern. Has puffy speckled forehead, overshadowing eye discs which are centrally white, fringed by broad dark brown arc, in turn backed by white arc and broad dark spot on sides of neck. Upper breast whitish; lower breast whitish, narrowly barred brown. Upperparts rufous-brown with large whitish spots. Wings and tail barred. Like large, thick-headed sparrowhawk in flight. Iris—yellow; bill—yellowish; feet—feathered pale.

Voice: Courtship call, mostly at night, is tremulous bubbly trill audible up to 1 km. Female responds with *kshuulip.* Alarm call is shrill, falcon-like *quiquiquiqui.*

Range: N Holorctic.

Distribution and status: Rare in China. Race *tianshanica* breeds in Tianshan Mts of NW Xinjiang and also seen on migration in W Xinjiang. Nominate race winters in Hulun Nur region of NE Nei Mongol and in Greater Hinggan Mts.

Habits: Lives in conifer and mixed forests and thickets of birch and larch. Rather diurnal. Alights from steep descent.

246. Eurasian Pygmy Owl *Glaucidium passerinum*   PLATE 24
Huatou xiuxiao

Description: Very small (18 cm) fluffy, plump owl with grey head covered in small white spots; small orange eyes and whitish underparts streaked lightly with greyish-brown. Upperparts are greyish-brown with white spots. Wings and tail barred. Has short indistinct white supercilium. Iris—orange-yellow; bill—grey-horn; feet—yellow with feathered legs.

Voice: Territorial call is soft, whistled *hjunk* repeated about every two seconds at dusk and dawn. Sometimes this call is interspersed with high, stammering hoot, *hjuuk . . . huhuhu . . . hjuuk . . . huhuhu . . . hjuuk . . .* Female call is similar but more nasal. Has different call in autumn—a series of shrill notes of increasing volume and pitch.

Range: Palearctic in conifer forest zone.

Distribution and status: Very rare in China. Race *orientale* recorded in Lesser Hinggan Mts in Heilongjiang, and Hebei (Dongling). Status uncertain.

Habits: Rather diurnal. Flight undulating like a woodpecker.

247. Collared Owlet *Glaucidium brodiei* [Collared Pygmy Owl]   PLATE 24
Ling xiuxiao

Description: Tiny (16 cm) barred owl with yellow eyes, a pale nuchal collar and no 'ear tufts'. Upperparts pale brown barred reddish-buff; crown grey with small white or reddish-buff 'eye-spots', brown bar across white throat; breast and belly buff, barred with black; thighs and vent white streaked with brown. Has orange and black false eye pattern on nape. Iris—yellow; bill—horn; feet—grey.

Voice: Mellow whistled monotone *pho, pho-pho pho* given by day or night. Imitation of this call is very effective in attracting this owl and also small mobbing songbirds.

Range: Himalayas to S China, SE Asia, Sumatra and Borneo.

Distribution and status: Common in all types of forest from 800–3500 m. Nominate race resident in SE Xizang, C, E, SW, S and SE China and Hainan; *paradalotum* occurs on Taiwan.

Habits: Seen in day in tall trees when calling or mobbed by other birds. At night it hunts from prominent perches keeping to taller trees. Flies with very fast wing beat.

248. Asian Barred Owlet *Glaucidium cuculoides* [Barred/Cuckoo   PLATE 24
Owlet, AG: Pygmy-Owl] Bantou xiuxiao

Description: Small (22 cm) finely barred rufous-brown owl without 'ear' tufts. Upperparts rufous-chestnut, barred ochraceous with broken white line at edge of scapulars; underparts mostly brown, barred ochraceous; ventrally whitish with chestnut flanks; bold white chin stripe conspicuous, underlined with brown and buff gorget. Iris—yellow-brown; bill—greenish with yellow tip; feet—greenish-yellow.

Voice: Unlike most owls: a rapid trill descending in pitch but increasing in volume given at dawn and dusk. Also barking two-note whistle repeated in series of increasing volume and speed to a crescendo.

Range: Himalayas, NE India to S China and SE Asia.

Distribution and status: Not uncommon in fragments of forest in lowlands and hills. Five races recognised—*austerum* in SE Xizang; *rufescens* in W Yunnan; *brugeli* in S Yunnan; *persimilie* on Hainan; and *whitelyi* over rest of C, S and SE China. Accidental in Shandong. **Habits:** Frequents gardens, villages, primary and secondary forest. It is principally nocturnal but sometimes active by day. Calls mostly in evening and early morning.

249. Little Owl *Athene noctua* Zongwenfu xiaoxiao PLATE 23

Description: Small (23 cm) earless owl with flat-topped head and intensely staring bright yellow eyes. Flat brow and pale eyebrows and broad white moustachial give it a menacing expression. Upperparts brown with white streaks and spots. Underparts white, mottled and streaked brown. Two white or buff bars on scapulars. Iris—bright yellow; bill—yellow-horn; feet—feathered white.

Voice: Territorial call given by day and night is drawn-out, rising *goooek*. Female responds with same call in falsetto. Also gives loud piercing *KEEoo* or *piu*. Alarm is shrill *Kyitt, Kyitt*.

Range: W Palearctic, Middle East, NE Africa, C Asia to NE China.

Distribution and status: Common and widespread resident up to 4600 m over most of western and northern China. Race *orientalis* is found in Kashi and Tianshan regions of W Xinjiang; *ludlowi* in W, S and E Xizang; *impasta* in Qinghai, Gansu and Sichuan; and *plumipes* from SW Gansu east to Shandong and north to Greater Hinggan Mts of NE China.

Habits: Partly diurnal. Dumpy and curious, bobbing and twisting head about nervously. Sometimes stands up tall on long legs. Quick, flappy undulating flight. Sits on fence posts and wires. Can hover.

250. Spotted Owlet *Athene brama* Hengbanfu xiaoxiao PLATE 23

Description: Small (20 cm) brown, earless owl with yellow eyes. Upperparts greyish-brown with small white spots on crown and larger spots on wings and back. Has broken whitish-buff nuchal collar. Has whitish eyebrows and throat. Underparts whitish with grey barring on breast and sides. No streaking on underparts. Shorter tail and flatter head than Asian Barred Owlet. Iris—yellow; bill—grey; feet—feathered white.

Voice: Harsh, screechy *chirurrr-chirurrr-chirurrr* followed by or alternating with *cheevak, cheevak, cheevak*. Also noisy, discordant screeches and chuckles.

Range: S Iran to Indian subcontinent, SW China and SE Asia.

Distribution and status: Race *pulchra* (*poikila*) is rare resident in Xishuangbanna of S Yunnan. Race *ultra* occurs in SE Xizang.

Habits: Lives in small parties in lightly wooded open areas and cultivation and scrub.

251. Boreal Owl *Aegolius funereus* [Tengmalm's Owl] Guixiao PLATE 23

Description: Small (25 cm) spotted owlet with high, squarish head and large white 'spectacles'. Raised eyebrow gives astonished expression. Has black spots just below eye. Underparts white with blotchy brown streaks. Large white patch on scapulars. White facial disc separate from Little Owl and much smaller Furasian Pygmy Owl. Iris—bright yellow; bill—grey-horn; feet—yellow feathered white.

Voice: Territorial call is rapid series of about 7–8 deep whistles rising toward end, *popopopoppopapa*; audible over great distance. Also gives nasal *ku-weeuk* call or shrill *chee-AK*. Begging young emit explosive, harsh calls.

Range: Boreal Holarctic.

Distribution and status: Rare in China. Race *pallens* is breeder or resident in Tianshan of W Xinjiang; *sibiricus* breeds or is resident in Hulun Nur area of NE Nei Mongol in Greater Hinggan Mts and race *beickianus* is resident in C Gansu, N Sichuan and E Qinghai. Probably much overlooked.

Habits: Nests in woodpecker holes and sometimes polygamous. Nocturnal. Inhabits dense conifer forests.

252. Brown Hawk Owl *Ninox scutulata* [Oriental Hawk-Owl]   PLATE 23
Yingxiao

Description: Medium-sized (30 cm) large-eyed dark hawk-like owl. Lack of facial disc characteristic. Upperparts dark brown; underparts buff, broadly streaked reddish-brown; vent, chin and a spot at base of bill white. Iris—bright yellow; bill—bluish-grey with green cere; feet—yellow.

Voice: Mellow, rising falsetto whistle *pung-ok*, the second note short with rising inflection, repeated about every one or two seconds, sometimes for long periods, usually at dawn and dusk.

Range: Indian subcontinent, NE Asia, China, SE Asia, Sulawesi, Borneo, Sumatra and W Java.

Distribution and status: Both resident and summer breeding migrant races are uncommon at low to moderate altitudes up to 1500 m. Race *ussuriensis* breeds from NE China to C and E China, migrating south in winter; *burmanica* is resident in south parts of China and on Hainan; *totogo* is resident on Taiwan including Lanyu islet and Penghu Island; *lugubris* is resident in SE Xizang.

Habits: Active shortly before dusk at the edge of the forest and chases insects which it catches in mid-air. Sometimes a family group will hunt around a clearing together. Calls irregularly, particularly when the moon is up.

253. Long-eared Owl *Asio otus* Chang'er xiao   PLATE 26

Description: Medium-sized (36 cm) owl with round buff facial disc outlined brown and white and two long erectile 'ears' (usually not visible). Eyes are red-yellow with glazed look. Has prominent white X-pattern in centre of face above bill. Upperparts: brown, blotched dark and spotted buff and white. Underparts buff with rufous mottling and broad brown streaks or blotches. Distinguished from Short-eared Owl by longer ears; more prominent white X on face; less fine streaking on lower breast and belly; finer and denser brown of wing tip in flight and less white underwing. Iris—orange-yellow; bill—grey-horn; feet—pinkish.

Voice: Male gives muffled *ooh* calls at about two-second intervals. Female responds with relaxed, nasal *paah*. Alarm is *kwek kwek*. Begging young give far-carrying, mournful *peee-e* call.

Range: Holarctic.

Distribution and status: Fairly common resident and seasonal migrant in north of China. Nominate race is resident in Kashi and Tianshan region of W Xinjiang; breeds in E and NE Nei Mongol and S Qinghai, S Gansu and NE China. Migrates across most of

China and winters in coastal provinces of S and SE China and Taiwan; also along major rivers.

Habits: Nests in crows nests in conifer forests. Nocturnal in habits. Wings long and narrow and flies with leisurely gull-like beat.

254. Short-eared Owl *Asio flammeus* Duan'er xiao   PLATE 26

Description: Medium-sized (38 cm) long-winged, tawny owl. Conspicuous facial disc with short 'ear' tufts, not visible in the field, and piercing bright yellow eyes in dark eyerings. Upperparts tawny, heavily streaked black and buff; underparts buff streaked dark brown. Black carpal patch conspicuous in flight. Iris—yellow; bill—dark grey; feet—whitish.

Voice: Sneezing bark *kee-aw* given in flight.

Range: Holarctic and S America; winter migrant in SE Asia.

Distribution and status: Uncommon seasonal migrant. Nominate race breeds in NE China and found wintering over most of the moister parts of the country below 1500 m.

Habits: Prefers grassy open areas.

FROGMOUTHS—Family: Batrachostomidae

A family of curious-looking nocturnal birds, related to nightjars but adapted to live inside forest. Frogmouths are aptly named for their enormous wide gape facilitating catching insects on the forest floor and off branches. They are found from SE Asia to New Guinea and Australia. All species have mottled camouflaged plumage. They sit very upright by day on a low perch. Frogmouths lay a single egg on a precarious downy cupshaped nest, balanced on a horizontal twig. Only one species occurs in China.

255. Hodgson's Frogmouth *Batrachostomus hodgsoni*   PLATE 27
Heiding wakouchi

Description: Unmistakable (24 cm) mottled brown, black and white bird with huge gape and staring pale brown eyes. Plumage is camouflaged bark pattern. Female is more rufous than male with large patch of black bordered white feathers on throat and breast, and less complex speckling. The only frogmouth in China. Iris—yellowish-brown; bill—pale horn; feet—brownish-pink.

Voice: Series of soft, buzzing *gwaaa* notes with rising inflection; also mournful inflected whistle, *pheew*, descending in pitch.

Range: Himalayas and SE Asia.

Distribution and status: Uncommon resident in SW Yunnan (Santaishan). Resident in SE Xizang.

Habits: Lives in evergreen forests and scrub up to 1900 m.

NIGHTJARS—Families: Eurostopodidae and Caprimulgidae

Nightjars are short-legged, entirely insectivorous, nocturnal birds which have a net of bristles around the bill for catching insects while on the wing. In the day they rest on the ground. They fly in an erractic, slow, flapping manner and emit monotonous calls. The eggs are laid on the ground in a scrape without any nest material. One eared and six 'non-eared' species are found in China. One of these is only known from a female specimen.

EARED NIGHTJARS—Family: Eurostopodidae

256. Great Eared Nightjar *Eurostopodus macrotis*　　　　PLATE 27
[Giant Eared-Nightjar] Maotui yeying

Description: Very large (40 cm) dark brown-barred nightjar with prominent ear tufts. Has buffy crown paler than rest of head. Iris—brown; bill—horn; feet—brown.

Voice: Loud, 3–4-noted, hollow whistle. The first note is short and sometimes inaudible, the second is longer and dips in pitch, the third note, also long, rises then tails off *pit, pee-wheeoow*.

Range: India to S China, SE Asia, Philippines and Sulawesi.

Distribution and status: Race *cerviniceps* is accidental in W Yunnan.

Habits: Similar to other nightjars. Glides like a harrier. Favours forest edge and open scrub. Often seen at dusk over forest.

NIGHTJARS—Family: Caprimulgidae

257. Grey Nightjar *Caprimulgus indicus* [Jungle/Indian Jungle/　　PLATE 27
Japanese Nightjar] Putong yeying

Description: Medium-sized (28 cm) greyish nightjar. Male: lacks rusty nuchal collar of Large-tailed; white tail markings on outer four pairs of tail feathers. Female similar to male but white patches buffy. Iris—brown; bill—blackish; feet—chocolate.

Voice: Hard, sharp rapidly repeated *chuck* at a steady rate of about six per second, then ending with a *chrrrr*. Wintering birds rarely call.

Range: Indian subcontinent, China, SE Asia and Philippines; migrates to Indonesia and New Guinea.

Distribution and status: Race *jotaka* breeds across most of E, N and S China, wintering to the south; on passage in Hainan; race *hazarae* is resident in SE Xizang up to 3300 m.

Habits: Prefers rather open mountain forest and scrub. Typical nightjar flight, settles on ground or on horizontal branch in daytime.

258. Eurasian Nightjar *Caprimulgus europaeus* Ou yeying　　　PLATE 27

Description: Medium-sized (27 cm) rufous-greyish, heavily mottled and streaked nightjar without ears. Male has small white spot near tip of wing and white tips to outer two tail feathers in flight. Female lacks white. Race *plumipes* is paler and sandy-buff all over with spots on tertials and larger white wing spot than nominate. Race *unwini* is paler, greyer and less marked than nominate. Race *dementievi* is pale and grey but more heavily streaked. Iris—dark brown; bill—dark horn; feet—grey.

Voice: Flight contact call is short *quoik* often repeated several times. Song is sustained churring up to 10 minutes without pause with occasional switches in pitch. Alarm call is *chuck* or *chuck-ek*. Male in display gives *fee oorr-feeoorr-feeoorr* . . . call.

Range: Breeds Europe, N Asia to N China and Mongolia, also NW Africa. Migrates to Africa.

Distribution and status: Rare in China. Nominate race breeds in Altai Mts and *unwini* in the Kashi and W Tianshan regions of W Xinjiang; *plumipes* breeds in E Tianshan to NW Gansu and in deserts near bend of Huanghe in Nei Mongol. Race *dementieri* should occur in Hulun Nur area of NE Nei Mongol.

Habits: Hawks after moths with jerky flight, tumbling about in their pursuit. In display flight, male glides with wings clapped then held in stiff V and tail fanned at an angle. Sometimes mobs raptors and sometimes mobbed as a raptor. Drinks by dipping over water like a swift.

[**259. Egyptian Nightjar** *Caprimulgus aegyptius* Aiji yeying   PLATE 27

Description: Medium-sized (26 cm) rather uniform, sandy-grey nightjar. Upperparts mottled and wing coverts spotted buff. Scapulars marked with star-like blackish-brown spots. Has indistinct buff nuchal collar. Tips of wing darker brown above than upperparts but whitish underwing. Sexes alike and males without white markings. Rarely shows narrow whitish tips to two outer tail feathers. Distinguished from female Eurasian Nightjar of pale race *plumipes* by different scapular pattern, less streaking and absence of buffish line across forewing at rest. Appears longer and heavier in flight. Iris—brown; bill—horn; feet—grey.

Voice: Song is fast series of purring *powrr* notes (3–4 per second) for several minutes. Gives series of croaking *toc* notes interspersed with short churrs. In alarm gives *tuk-l tuk-l* notes. Also *chuc chuc* calls and low grumbling sounds.

Range: Breeds NW Africa, Middle East to C Asia; wintering subtropics of Africa.

Distribution and status: Nominate race (includes *arenicolor*) is recorded as accidental in W Xinjiang. Cheng (1994) questions the validity of these records.

Habits: Flies with deep, powerful wing beats and glides with level wings. Sings from ground at dawn and dusk.]

260. Vaurie's Nightjar *Caprimulgus centralasicus* [Central Asian   PLATE 27
Nightjar] Zhongya yeying

Description: Small (19 cm) pale nightjar. Known only from one female specimen. Upperparts sandy-buff-grey with dark brown freckles and reticulations. Tail buffy-grey with six indistinct brown bars and three indistinct buff bars. Underparts creamy-buff finely barred brown; small buffy-white patch on side of throat. Iris—brown; bill—horn; feet—grey.

Voice: Not recorded

Range: Presumed endemic to SW Xinjiang.

Distribution and status: Globally Vulnerable (Collar *et al.* 1994). Rare and unknown. Probably resident among sandy foothills along the edge of Taklimakan desert below Kunlum Mts but so far recorded only from Pishan (Guma) in 1929.

Habits: Presumed as others of genus.

261. Large-tailed Nightjar *Caprimulgus macrurus* [White-tailed/   PLATE 27
Long-tailed Nightjar/Coffinbird] Changwei yeying

Description: Largish (30 cm) greyish-brown nightjar. Prominent white patch on centre of four outer primaries and broad white tips of two outer pairs of tail feathers diagnostic. In females these patches are buffy; throat bar white. Iris—brown; bill—greyish-brown; feet—greyish-brown.

Voice: Rich, deep *tchoink* like two stones being struck together at a steady rate of about three per second following a purring warm-up; also low growling.

Range: Indian subcontinent, SE Asia, Philippines, Indonesia to New Guinea and Australia.

Distribution and status: A local but common bird of forest edge and wooded country, including mangroves, up to 1200 m. Race *ambiguus* is resident in extreme S and SE Yunnan and *hainanus* lives on Hainan.

Habits: Rests in forest edge or wooded areas during the day in a shady place on the ground. Calls for half an hour each dusk and dawn from a perch or on the wing. Hunts interspersed with rests on ground, often on roads where birds are frequently killed by cars.

262. Savannah Nightjar *Caprimulgus affinis* [Allied Nightjar]   PLATE 27
Lin yeying

Description: Smallish (22 cm) uniform-coloured nightjar. Male has diagnostic white outer tail feathers. White throat band divided into two patches. Female is more rufous and lacks white tail markings. Iris—brown; bill—reddish-brown; feet—dull red.

Voice: Penetrating plaintive *chweep* uttered constantly for half an hour at dusk and dawn, on the wing.

Range: India to S China, SE Asia, Sulawesi, Philippines, Sundas.

Distribution and status: This is a common nightjar of tropical lowlands in dry open coastal areas including large cities. Race *amoyensis* is resident in SE Xizang, S Yunnan, S Guangxi and coastal Guangdong, Hong Kong and Fujian; *stictomus* lives on Taiwan.

Habits: Typical nightjar resting in the day on the ground, or in cities on the tops of tall flat buildings. Hawks for insects attracted by city lights.

PIGEON AND DOVES—Order: Columbiformes Family: Columbidae

The large, worldwide family of pigeons feed predominantly on fruits, seeds and berries and have rather compact plump bodies and short, stout bills. They nest on flimsy twig platform nests, laying white eggs. Calls are repetitive, melodious coos and in flight pigeons make a characteristic flapping noise. Thirty-one species occur in China, comprising three species groups.

1. Fruit doves and green pigeons (*Treron, Ptilinopus*) are smaller, arboreal birds with generally brightly coloured plumage without metallic colours.
2. Imperial pigeons (*Ducula, Columba*) are large and arboreal with metallic sheen in the plumage and generally grey or whitish in colour on the underparts.

3. Ground doves (*Macropygia, Streptopelia and Chalcophaps*) regularly visit the ground and have either highly iridescent, greenish upperparts or drab, reddish-brown colours.

The following pigeons of small islands can be found in the disputed Nansha archipelago (Spratly's) in the southern South China Sea but are not generally regarded as Chinese species—Silvery Pigeon *Columba argentina*, Island Collared Dove *Streptopelia bitorquata*, Peaceful Dove *Geopelia striata*, Nicobar Pigeon *Caloenas nicobarica*, Cinnamon-headed Green Pigeon *Treron fulvicollis*, Little Green Pigeon *Treron olax*, Pink-necked Green Pigeon *Treron vernans*, Grey Imperial Pigeon *Ducula pickeringii* and Pied Imperial Pigeon *Ducula bicolor*.

FRUIT DOVES AND GREEN PIGEONS

263. Rock Pigeon *Columba livia* [Feral Pigeon/Rock Dove]   PLATE 28
Yuan ge

Description: Medium-sized (32 cm) bluish-grey pigeon with black wing bars and terminal tail bar and a greenish-violet gloss on head and chest. This is the wild form of the familiar town and domestic pigeon. Iris—brown; bill—horn; feet—dark red.
Voice: Familiar *oo-roo-coo* of the domestic pigeon.
Range: Parts of Indian subcontinent and S Palearctic but introduced almost worldwide and feral populations are now established in many towns and cities.
Distribution and status: Race *neglecta* is locally common in NW China and the Himalayas; *nigricans* occurs from S Qinghai to E Nei Mongol and Hebei. Elsewhere feral forms may be found.
Habits: Originally a bird of cliffs but easily adapts to life in towns and around temples. It lives in flocks. Wheeling flight is characteristic.

264. Hill Pigeon *Columba rupestris* Yan ge   PLATE 28

Description: Medium-sized (31 cm) grey pigeon with two black wing bars. Very similar to Rock Pigeon but with paler belly and back and whitish subterminal band on tail leaving grey base of tail contrasted between pale back and this band. Iris—pale brown; bill—black with flesh-coloured cere; feet—red.
Voice: Repeated croak like human hiccough. High-pitched, tremulous *coo* on taking flight and landing.
Range: Himalayas, C Asia to NE China.
Distribution and status: Common resident and seasonal migrant up to 6000 m. Race *turkestanica* is resident in W Xinjiang and Xizang. Nominate race breeds across the rest of N and C China to the NE provinces.
Habits: Lives in colonies on crags, hilly terrain with cliffs and caves.

265. Snow Pigeon *Columba leuconota* Xue ge   PLATE 28

Description: Large (35 cm) unmistakable pigeon coloured like black-headed gull. Head dark grey; collar, lower back and underparts white; upper back brownish-grey; rump black. Tail is black with broad white medial band. Wings are grey with two black bars. Iris—yellow; bill—dark grey with magenta cere; feet—reddish-pink.

Voice: Prolonged high *coo-ooo-ooo* and croaking courtship calls.

Range: Himalayas and W China.

Distribution and status: Nominate race is resident in S Xizang. Race *gradaria* is resident in E and SE Xizang and mountains of NW Yunnan, W Sichuan and Qinghai. Absent from dry montane steppes but quite common in suitable habitat from 3000–5200 m, especially in the wetter Himalayas.

Habits: Lives in pairs or small flocks. Glides over alpine pastures, rocky cliffs and snow-fields

266. Stock Pigeon *Columba oenas* Ou ge PLATE 28

Description: Medium-sized (31 cm) grey pigeon with pinkish breast, green metallic patch on side of neck and two vestigial black bars on wing. Distinguished from Rock Pigeon by grey rump and incomplete wing bars; black trailing edge to primaries; less metallic-purple sheen on neck; also darker eye. Iris—brown; bill—yellow; feet—red.

Voice: Male gives disyllabic *oou-o, oou-o, oo-o*.

Range: Europe, N Africa, Asia Minor, Iran, Turkestan to NW China.

Distribution and status: Race *yarkandensis* is rare resident in Kashi and Tianshan region of W Xinjiang.

Habits: Gives display flight like Rock Pigeon.

267. Yellow-eyed Pigeon *Columba eversmanni* [Pale-backed PLATE 28
Pigeon/Eastern Stock Dove] Zhongya ge

Description: Small (26 cm) grey pigeon with white back, two partial black wing bars, and small patch of iridescent green and purple on side of neck. Very similar to Stock Pigeon but separated by white back; larger pale area at base of primaries; pink top of head; whiter underwing; pale eye and broad eye-ring. Iris—yellow; bill—yellow; feet—flesh-coloured.

Voice: Song of three single notes followed by three disyllabic notes *quooh, quooh, quooh-cuu-gooh-cuu-gooh-cuu-gooh*.

Range: Turkestan to NW India.

Distribution and status: Globally Vulnerable (Collar *et al.* 1994). Rare seasonal migrant. Breeds in Kashi and Tianshan regions of W Xinjiang.

Habits: Inhabits lightly wooded areas in cultivation, cliffs and ruins.

268. Common Wood Pigeon *Columba palumbus* PLATE 28
Banwei linge

Description: Large (42 cm) plump grey pigeon with pink breast and iridescent green patch on side of neck above bean-shaped buff yellow patch. In flight has black flight feathers separated from grey coverts by broad white transverse band. Juvenile lacks buff patch on side of neck and has warm rufous breast. Iris—yellow; bill—reddish; feet—red.

Voice: Hoarse cooing of five-syllables, second note stressed and drawn-out *cu-cooh-cu, coo-coo*.

Range: Europe to Russia, N Africa, Iran, N India.

Distribution and status: Rare resident. Race *casiotis* is resident in Kashi and Tianshan regions of W Xinjiang.

Habits: Loud clap of wings on take-off. Display flight is climb with wing clap at top followed by diving descent, gliding on half-closed wings. Lives in flocks. Feeds on agricultural lands.

269. Speckled Wood Pigeon *Columba hodgsonii* PLATE 28
Dianban linge

Description: Medium-sized (38 cm) brownish-grey pigeon with white spots on wing coverts. Distinguished from all other pigeons by tapering, elongated feathers on neck; no metallic gloss on plumage. Head grey; mantle maroon; lower back grey. Iris—greyish-white; bill—black with purple base; feet—yellowish-green with bright yellow claws.
Voice: Very deep *whock-whr-a-o . . . whroo.*
Range: Himalayas to Burma and C China.
Distribution and status: Common resident in S, SE and E Xizang, Yunnan and Sichuan from 1800–3300 m.
Habits: Lives in twos, threes or small flocks. Mostly arboreal. Freezes in alarm. Even hangs upside-down. Lives in subalpine forest zone with rocky crags.

270. Ashy Wood Pigeon *Columba pulchricollis* Hui linge PLATE 28

Description: Medium-sized (35 cm) grey pigeon with diagnostic broad buff collar on nape, stippled with black. Mantle with lilac and green iridescent sheen. Head paler grey than upperparts. Breast grey grading to greyish-white vent. Chin white. Iris—white to yellow; bill—grey-green with purple base; feet—red.
Voice: Deep calls *hu . . . hu . . . hu.*
Range: Himalayas, Tibetan Plateau, N Burma, N Thailand.
Distribution and status: Rare resident in S and SE Xizang and W Yunnan in broadleaf forests at 1200–3200 m. Fairly common on Taiwan.
Habits: Singly, in pairs or in small flocks. Shy. Flies off with clatter of wings when approached. Rather silent.

271. Pale-capped Pigeon *Columba punicea* [Purple Wood Pigeon] PLATE 28
Zi linge

Description: Mediumsized (35 cm) cinnamon and grey wood pigeon with greyish-white crown and nape. Lower face and underparts cinnamon diagnostic. Upperparts: mantle and wing coverts chestnut-brown; slaty rump and blackish-brown tail. Entire plumage with green and amethyst sheen. Green sheen strongest on back and sides of neck. Orbital skin and cere magenta. Iris—creamy-yellow to red; bill—pale with magenta base; feet—crimson.
Voice: Soft mewing similar to Green Imperial Pigeon but not as loud or prolonged.
Range: NE India to SE Asia.
Distribution and status: Globally Vulnerable (Collar *et al.* 1994). Very rare resident in Chumbi valley of S Xizang and Hainan. Must also occur in SE Xizang in subtropical zone.
Habits: Lives in small to large flocks. Eats fruits and, when available, bamboo seeds; sometimes grain in fields. Flight slow like Green Imperial Pigeon.

272. Japanese Wood Pigeon *Columba janthina* Hei linge PLATE 28

Description: Large (43 cm) blackish wood pigeon with green metallic sheen on sides of neck and purplish sheen on rest of plumage. Unmistakable charcoal-grey plumage above and below. Iris—dark brown; bill—dark blue; feet—red.

Voice: Long-drawn *oo-woo, oo-woo* and courtship bleating.

Range: Japan.

Distribution and status: Globally Near-threatened (Collar *et al.* 1994). Nominate race is rare summer visitor to Shandong (Weihai) and islands off Taiwan.

Habits: Inhabits subtropical broadleaf forests on small islands. Lives in flocks.

273. European Turtle Dove *Streptopelia turtur* Ou banjiu PLATE 29

Description: Smallish (27 cm) pinkish-brown dove with patch of fine black and white lines on side of neck and wing coverts dark brown, boldly scaled pale rufous-brown. Distinguished from Oriental Turtle Dove by smaller size, paler colour, lack of white tips to wing coverts, browner back and nape, whiter markings on neck and edge of tail and more vinous-pink breast. Orbital skin red. Iris—yellow; bill—grey; feet—pink.

Voice: Call is deep rumbling *toorrrrr, toorrrrr, toorrrrr.*

Range: Europe, Asia Minor, N Africa and SW Asia.

Distribution and status: Race *arenicola* is locally common resident in W Xizang and Xinjiang.

Habits: Shy dove of open agricultural lands with woods, trees or hedges.

274. Oriental Turtle Dove *Streptopelia orientalis* PLATE 29
[Rufous/Eastern Turtle Dove] Shan banjiu

Description: Medium-sized (32 cm) pinkish dove distinguished from Spotted Dove by bold black and whitish striped patch on neck, rufous-edged dark scalloped plumage on upperparts and grey rump. Tail is blackish with pale grey edge. Underparts generally pinkish. Feet are red. Distinguished from European Turtle Dove by larger size. Iris—yellow; bill—grey; feet—pink.

Voice: The call is a melodious *kroo kroo-kroo kroo.*

Range: Himalayas, India, NE Asia, Japan, China and Taiwan. Northern birds migrate to south of range.

Distribution and status: Widespread and common. Race *meena* is resident in W and NW China; nominate is resident or summer breeder over most of S Xizang to NE China; race *orii* is resident on Taiwan; and *agricola* is found in S and SW Yunnan. Large flocks pass through S China on spring passage. Occurs up to high altitudes in the Himalayas.

Habits: Lives in pairs and feeds mostly on the ground in open agricultural areas and around villages and monasteries.

275. Laughing Dove *Streptopelia senegalensis* Zong banjiu PLATE 29

Description: Small (25 cm) pink-brown dove with long tail and short wings. Distinguished from Eurasian Collared Dove by smaller size, lack of black neck band and darker colour, more similar to European Turtle Dove. Lacks neck pattern and wing pattern of turtle doves and has brown neck band speckled black. Has white terminal tail bar and distinctive blue-grey wing panel. Iris—brown; bill—grey; feet—pink.

Voice: Five-syllable, rapid, subdued cooing with third and fourth notes higher than others, *dododeedeedo*. Several birds may call together.

Range: N Africa, Middle East, Afghanistan and Turkestan area of W Xinjiang.

Distribution and status: Rare resident in Kashi and Tianshan area of W Xinjiang.

Habits: Rather tame. Lives in open farmland with scattered trees. Slow laboured flight.

276. Spotted Dove *Streptopelia chinensis* [Burmese Spotted/Spot-necked Dove] Zhujing banjiu
PLATE 29

Description: Familiar medium-sized (30 cm) pinkish-brown dove with a longish tail. Outer tail feathers broadly tipped white. Flight feathers darker than body; has a diagnostic conspicuous black patch on the side of the neck, finely spotted with white. Iris—orange; bill—black; feet—red.

Voice: A melodious, gently repeated *ter-kuk-kurr* with last note stressed.

Range: Widely distributed and common in SE Asia to Lesser Sundas and introduced elsewhere as far as Australia.

Distribution and status: Common resident, found throughout C, SW, S and E China in open lowland areas and villages. Race *tigrina* west of Salween River in SW Yunnan; *vacillans* over the rest of Yunnan and S Sichuan; *hainana* on Hainan; *formosana* on Taiwan; and nominate race over the rest of its distribution.

Habits: The bird is a commensal of man, living around villages and rice fields, feeding on the ground and frequently encountered in pairs sitting on open roads. When disturbed it flies close to the ground with a distinctive slow wing beat.

277. Red collared Dove *Streptopelia tranquebarica* Huo banjiu
PLATE 29

Description: Small (23 cm) vinaceous-red dove with diagnostic white-fronted black collar at rear of neck. Male has greyish head, pinkish underparts and cinnamon wing coverts. The primaries are blackish and the tail is slaty with white edges and outer tips. The female is paler and duller with dull brown head and less red in plumage. Iris—brown; bill—grey; feet—red.

Voice: The call is a deep *cru-u-u-u-u* repeated several times with emphasis on first note.

Range: Himalayas, India, China to SE Asia and Philippines.

Distribution and status: Resident in S and E China in rather open woodland and drier coastal forest and secondary growth over S and E Qinghai–Xizang Plateau and across most of N, C, E and southern China. Northern populations winter to south. Regular visitor to Hong Kong in winter.

Habits: Feeds on the ground in busy searching walk.

278. Eurasian Collared Dove *Streptopelia decaocto* Hui banjiu
PLATE 29

Description: Medium-sized (32 cm) brownish-grey dove with distinct black and white hindneck collar. Paler and greyer than turtle doves and much smaller, pinker than Red Collared Dove. Iris—brown; bill—grey; feet—pink.

Voice: Loud three-syllable cooing *coo-cooh-coo* with emphasis on second note.

Range: Europe to Turkestan, Burma and China.

Distribution and status: Rather common especially in north of range. Race *stoliczkae* in Kashi and Tianshan area of W Xinjiang; nominate race from Sichuan through N China to India; *xanthocyclus* straggler to Anhui, Fuzhou and Yunnan.

Habits: Rather tame; occupies agricultural farmlands and villages. Roosts on houses, poles and telegraph wires.

279. Barred Cuckoo Dove *Macropygia unchall* Banwei juanjiu   PLATE 29

Description: Large (38 cm) long-tailed brown dove. Back and tail heavily barred black or brown. Head grey with iridescent blue-green sheen on nape. Breast pinkish, grading into white vent. Female lacks green iridescence. Heavier barring on back and barred tail distinguish from other cuckoo doves in the region. Iris—yellow/pale brown; bill—black; feet—red.

Voice: A loud resonant *kro-uum* or *u-wa* with the second note louder and higher than the first, which is only audible at short-range.

Range: Himalayas to SE Asia, Java and Bali.

Distribution and status: Rather uncommon in subtropical forests from 800–3000 m. Race *tusalia* is summer visitor to C Sichuan and resident in S Yunnan (Xishuangbanna); *minor* is resident in Wuyi Mts of N Fujian and Guangdong, also Hainan; stragglers recorded Shanghai.

Habits: Lives in small flocks. Flies swiftly through canopy. When on the ground, raises its tail.

280. Brown Cuckoo Dove *Macropygia amboinensis* [Philippine   PLATE 29
Cuckoo Dove] Lihe juanjiu

Description: Medium-sized (38 cm) long-tailed, brownish pigeon with pinkish-brown head and underparts. Male distinguished from Barred Cuckoo Dove by lack of black barring on upperparts, lack of green nape and cinnamon undertail coverts. Female has distinctive pale cinnamon forehead and crown and fine black barring on underparts. Iris—yellow; bill—pinkish; feet—pink.

Voice: Deep *wua wu* calls in breeding season.

Range: Philippines, Taiwan.

Distribution and status: Race *phaea* is common resident on Lanyu islet off SE Taiwan.

Habits: As other cuckoo doves.

Note: Placed by some authors in Red Cuckoo Dove *M. phasianella* (e.g. Cheng 1994), alternatively in Philippine Cuckoo Dove *M. tenuirostris*, but combined here within super-species *amboinensis* as per Inskipp *et al.* (1996).

281. Little Cuckoo Dove *Macropygia ruficeps* Zongtou juanjiu   PLATE 29

Description: Medium-sized (30 cm) long-tailed reddish dove. Smaller than Brown Cuckoo Dove and with buff chest, black barring and dark subterminal bar on outer tail feathers. The male has a green and lilac iridescent sheen on the nape. Female lacks iridescence and has heavier dark mottling on breast. Iris—grey-white; bill—brown with black tip; feet—coral-red.

Voice: Fast *wup-wup-wup-wup* ... at about two notes per second for up to 40 times. After a short pause the call starts again.

Range: Widespread and common submontane forest bird in SE Asia, Sumatra, Java, Borneo and Lesser Sundas.

Distribution and status: Race *assimilis* is a rare resident in S Yunnan (Xishuangbanna) up to 2000 m.

Habits: Prefers forest edge and flocks often raid adjacent rice fields.

282. Emerald Dove *Chalcophaps indica* [Green-winged Dove/   PLATE 29
Common Emerald-Dove] Lüchi jinjiu

Description: Medium-sized (25 cm) rather short-tailed, ground dove with reddish-pink underparts. Crown grey with white forehead and grey rump and iridescent green wings. Female lacks grey crown. In flight two conspicuous black and white bars can be seen on the back. Iris—brown; bill—red with orange tip; feet—red.

Voice: Deep soft, mournful, drawn-out two-noted *tuk-hoop* with emphasis on second note.

Range: Indian subcontinent to Australia.

Distribution and status: Nominate race is common in tropical zone of S China through S Yunnan, Guangxi, Hainan, Guangdong, S Taiwan and SE Xizang in lowland and submontane primary and secondary forests.

Habits: Usually singly or in pairs on the forest floor in thick cover. Flies very fast and low through the forest with a clap of wings on take off. Drinks at streams and pools.

283. Orange-breasted Green Pigeon *Treron bicincta*   PLATE 30
[AG: Pigeon] Chengxiong lüjiu

Description: Medium-sized (29 cm) green pigeon with conspicuous yellow stripes and edges on black wing feathers. Face green at front, nape and upper back grey. Male: underparts yellowish-green with band of lilac across upper breast followed by orange on lower breast; female has green breast. Tail grey with a black subterminal bar often broken by central all-grey feathers. Iris—blue and red; bill—greenish-blue; feet—dark red.

Voice: Attractive modulated whistle followed by gurgling notes *ko-wrrrook, ko-wrrroook, ko-wrrroook*, harsh croaking alarm and a chuckling call *kreeeew-kreeew-kreeew*.

Range: India, SE Asia, Java and Bali.

Distribution and status: Race *domvilii* is a rare resident of Hainan and straggler to Taiwan.

Habits: Typical of the genus. Lives in pairs or sometimes small parties. Feeds in small, fruit-bearing bushes and trees. Has a typical *Treron* tail-flicking display. Prefers wooded lowlands and plantations.

284. Pompadour Green Pigeon *Treron pompadora*   PLATE 30
Huitou lüjiu

Description: Medium-sized (26 cm) green pigeon with maroon mantle, back and wing coverts in male. Similar to respective sexes of Thick-billed Green Pigeon but with thinner, entirely blue-grey bill and lack of obvious orbital ring. Male has orange wash on breast. Female separated by short streaks rather than bars on undertail coverts. Iris—outer ring pink, inner ring pale blue; bill—blue-grey; feet—red.

Voice: Series of mellow, powerful, musical whistles up and down scale.

Range: India, Sri Lanka, SE Asia, Indonesia, Philippines.

Distribution and status: Race *phayrei* is a very rare resident in Xishuangbanna, S Yunnan. Most available habitat now converted to rubber plantations.

Habits: Lives in small to large flocks in lowland evergreen rainforests. Visits salt licks.

285. Thick-billed Green Pigeon *Treron curvirostra* [AG: Pigeon] PLATE 30
Houzui lüjiu

Description: Medium-sized (27 cm) thickset green pigeon. Male: back, mantle and upper innerwing coverts maroon in male; dark green in female; forehead and crown grey; neck green; underparts yellowish-green; wings blackish with yellow edges to feathers and a bold yellow wing bar; central tail feathers green, others grey with medial black bar; flanks barred green and white; undertail coverts cinnamon. Iris—yellow, orbital skin bright blue-green; bill—green with red base; feet—crimson.

Voice: Loud, throaty guttural notes and modulated whistles.

Range: Nepal and NE Indian subcontinent to SE Asia, Philippines and Greater Sundas.

Distribution and status: Rare resident. Race *nipalensis* in SW and S Yunnan (Xishuangbanna) and *hainana* in SW Hainan; vagrant Hong Kong. Lives in forests in lowlands.

Habits: Feeds in flocks, with much thrashing of wings in low canopy trees.

286. Yellow-footed Green Pigeon *Treron phoenicoptera* PLATE 30
Huangjiao lüjiu

Description: Medium-sized (33 cm) green pigeon with yellow feet and diagnostic yellow-olive band extending across upper breast and round neck contrasting with grey underparts and narrow grey hind collar. Uppertail greenish with broad, dark grey terminal bar. Iris—outer ring pink, inner pale blue; bill—grey with green cere; feet—yellow.

Voice: A series of modulated, mellow, musical whistles, similar to Orange-breasted but louder and in lower key.

Range: India, Sri Lanka, Burma and Indochina.

Distribution and status: Race *viridifrons* is very rare resident in W and S Yunnan (Xishuangbanna). Reduced in numbers by habitat loss and hunting.

Habits: Occupies semi-evergreen forests and secondary forests up to 800 m. Joins other pigeons and hornbills in fruiting fig trees.

287. Pin-tailed Green Pigeon *Treron apicauda* Zhenwei lüjiu PLATE 30

Description: Medium-sized (30 cm, tail projects up to 10 cm more) green pigeon with diagnostic pin-like elongated, central tail feathers. Male: cinnamon undertail coverts and orange wash on breast; female: light green breast and undertail coverts white with dark streaks. Iris—red; bill—green with turquoise base; feet—crimson.

Voice: Duets. First bird gives soft *cuc-coo* and second follows with higher-pitched *huu*, repeated in accelerated sequence. Also low growling.

Range: Himalayas to SE Asia.

Distribution and status: Rare resident. Nominate race in W and SW Yunnan, SW Sichuan and SE Xizang; *laotinus* in Xishuangbanna, S Yunnan east of Lancang (Mekong) River.

Habits: Occupies evergreen forests from 600–1800 m. Often forages in small flocks. Behaviour typical of genus.

288. Wedge-tailed Green Pigeon *Treron sphenura* [>Korthal's PLATE 30 Green-Pigeon, AG: Pigeon] Xiewei lüjiu

Description: Medium-sized (33 cm) green pigeon. Male: head green; crown and breast orange; mantle purplish-grey; wing coverts and upper back purplish-chestnut; rest of wings and tail dark green with greater wing coverts and darker flight feathers edged yellow; vent yellowish with dark streaks; flanks edged yellow; undertail coverts cinnamon. Female: pale yellow undertail coverts and vent with large dark markings; lacks golden and chestnut colouring of male. Iris—pale blue to red; bill—turquoise at base, cream at tip; feet—red.

Voice: A deep whistled *koo* note varied with a curious grunting note.

Range: Himalayas, SW China, SE Asia, Sumatra, Java and Lombok.

Distribution and status: Uncommon resident. Nominate race lives in S Sichuan (Emei region); *yunnanensis* is found in S Xizang and Yunnan. It is a local bird of high mountains from 1400–3000 m.

Habits: Occurs in oak–laurel and montane heath forests and is quite tame and approachable.

289. White-bellied Green Pigeon *Treron sieboldii* [Japanese PLATE 30 Green Pigeon] Hongchi lüjiu

Description: Medium-sized (33 cm) green pigeon with diagnostic whitish belly. Sides of belly and undertail flecked grey. Male has maroon wing coverts, greyish mantle and orange crown. Female is mostly green. Bluish orbital skin. Iris—red; bill—bluish; feet—red.

Voice: Mellow *wu-wua wu, wu-wua wu* or *ah oh ah oh*.

Range: S and C China, Taiwan, Japan and Indochina.

Distribution and status: Globally Near-threatened (Collar *et al.* 1994). Very rare resident and migrant. Nominate race recorded as accidental in Hebei; *sororius* in Taiwan, migrating through Jiangsu and Fujian; *fopingensis* in Qinling Mts of S Shaanxi and E Sichuan, and *murielae* in Guangdong, Guangxi and Hainan, wintering in Hong Kong.

Habits: Typical of genus. Occupies evergreen forest and secondary forest; flocking into fruit trees. Flight very fast.

290. Whistling Green Pigeon *Treron formosae* [Formosan PLATE 30 Red-capped Green Pigeon] Hongding lüjiu

Description: Medium-sized (33 cm) green pigeon with brownish wing 'shoulder' and green and white-scaled vent and undertail. Male has green breast, yellow throat and orange cap. Differs from Wedge-tailed in having greyish-green mantle, tail pattern, and black and red eye. Orbital skin blue. Iris—red; bill—blue; feet—red.

Voice: *Po po peh* with last note at higher pitch.
Range: Nansei (Ryukyu) islands, Taiwan, Philippines.
Distribution and status: Globally Near-threatened (Collar *et al.* 1994). Nominate race is rare resident on S Taiwan and Lanyu islet.
Habits: Typical of genus. Lives in tropical lowland evergreen forests.

291. Black-chinned Fruit Dove *Ptilinopus leclancheri* PLATE 30
Heike guojiu

Description: Male: small (28 cm) green dove with white head and diagnostic black chin and purple breast bar. Lower breast greyish-green, belly cream and undertail coverts light cinnamon. Upperparts green with black flight feathers. Female lacks white head but has pectoral band. Juvenile as female but without pectoral band. Iris—red; bill—yellow with red and yellow cere; feet—pink.
Voice: No information.
Range: Philippines, Palawan.
Distribution and status: Race *taiwanus* is rare resident on Taiwan.
Habits: Rather shy. Lives in small flocks in lowland rain forests.

292. Green Imperial Pigeon *Ducula aenea* Lü huangjiu PLATE 30

Description: Large (43 cm) green and grey pigeon. Head, neck and underparts pale pinkish-grey; undertail coverts chestnut; upperparts dark green with a diagnostic bronze iridescence. Iris—reddish-brown; bill—blue-grey; feet—dark red.
Voice: Loud single *oom*, a reverberant *kruk-kroorr* and loud full call of several chuckling notes ending in a rolling note.
Range: India to S China, SE Asia, Philippines, Sundas and Sulawesi.
Distribution and status: A rather uncommon resident of lowland evergreen forests. Race *sylvatica* lives in S Yunnan (Xishuangbanna) and Hainan; *kwantungensis* in S Guangdong (Luofu Mt).
Habits: A bird of the high tree-tops. In courtship flight it makes spectacular vertical climbs to stalling point then abruptly dives and levels out again.

293. Mountain Imperial Pigeon *Ducula badia* PLATE 30
Shan huangjiu

Description: Large (46 cm) darkish pigeon. Head, neck, chest and abdomen vinaceous-grey; chin and throat white; mantle and wing coverts dark maroon; back and rump dark greyish-brown; tail is brownish-black with a broad pale grey terminal band; undertail coverts buff. Iris—white, grey or red; bill—crimson with white tip; feet—crimson.
Voice: A click followed by two melancholy booming coos *click-broom . . . broom.*
Range: India, SE Asia, Borneo, Sumatra and W Java.
Distribution and status: Race *griseicapilla* is locally common in hill forests from 400–2300 m in SW and S Yunnan (Xishuangbanna) and Hainan; race *insignis* in SE Xizang.
Habits: Montane populations make daily flights to lowland feeding areas.

CRANES, BUSTARDS AND RAILS—Order: Gruiformes

BUSTARDS—Family: Otididae

A small family of largish, terrestial birds of desert and semi-desert and steppes with longish legs, stout bodies and long necks. The wings are long and broad and bustards fly with the neck extended. Most species have crests, ruffs or neck plumes, displayed in the breeding season. The nest is a scraped depression in the ground. Three species occur in China.

294. Little Bustard *Tetrax tetrax* Xiao bao   PLATE **34**

Description: Small (43 cm) brownish bustard with mottled upperparts and whitish underparts. Breeding male unmistakable with black neck ruff and contrasting white 'V' on front of neck joined at back by narrow collar and second broader white ring around base of neck. In flight wings appear almost white; only four front primaries having much black. Iris—yellowish; bill—greenish-horn; feet—greenish-yellow.

Voice: Display call is dry, far-carrying *prrrt*. Whistling noise made by fourth primaries in flight.

Range: Parts of southern W Palearctic, Middle East and C Asia.

Distribution and status: Globally Near-threatened (Collar *et al.* 1994). Very rare. Breeds in Tianshan in steppes, migrating across NW China and straggler to Sichuan.

Habits: Lives in small flocks. Displays by leaping with wings fluttering and puffing-out ruff.

295. Great Bustard *Otis tarda* Da bao   PLATE **34**

Description: Huge (100 cm) bustard with grey head, rufous neck and upperparts broadly barred rufous and black with white underparts and undertail. Breeding male has white plumes down front of neck and rufous plumes at side of neck. In flight wing is whitish with black secondaries and dark tips to primaries. Iris—yellow; bill—yellowish; feet—yellowish-brown.

Voice: Normally silent. Males make deep moans in displays.

Range: Europe, NW Africa to Middle East, C Asia, and N China; vagrant Pakistan

Distribution and status: Globally Vulnerable (Collar *et al.* 1994). Nominate race resident in steppes and semi-deserts of Tianshan, Kashi and Turpan areas of Xinjiang; race *dybowskii* breeds in E Nei Mongol and Heilongjiang, wintering in N China from Gansu to Shandong; stragglers as far south as Fujian. Lives in steppes and semi-desert, wintering in agricultural lands.

Habits: Lives in parties of 5–15 birds. Gait is deliberate; flight is powerful. Male in display puffs out breast feathers.

296. McQueen's Bustard *Chlamydotis macqueeni* Boban bao   PLATE **34**

Description: Medium-sized (70 cm) brown mottled bustard with whitish underparts. Breeding male has grey neck with black stripe of fluffy plumes down sides of neck. In flight has bold black wing bars and large white panel at base of black-tipped primaries. Iris—golden-yellow; bill—dark above, yellow is below; feet—brownish-yellow.

Voice: Virtually silent.
Range: Middle East to C Asia and NW India.
Distribution and status: Rare resident in W Xinjiang and NW Nei Mongol.
Habits: Leaping displays. Lives in semi-deserts, dunes and salt plains. Wild and shy.
Note: Formerly treated as a race of Houbara Bustard *O. undulata* (e.g. Cheng 1987) but considered distinct from *undulata* based on difference in male display and lack of hybrids (Gaucher *et al.* 1996).

CRANES—Family: Gruidae

Familiar, tall, pair-living birds of wetlands and open country. All species are highly migratory. Nine species occur in China.

297. Siberian Crane *Grus leucogeranus* Bai he   PLATE 31

Description: Large (135 cm) unmistakable white crane with orange bill and scarlet bare face and pink legs. In flight black primaries are seen. Young are golden-brown. Iris—yellow; bill—orange; feet—pink.
Voice: Pleasant, soft musical*koonk koonk* cheerfully given in flight.
Range: Breeds SE Russia and Siberia, winters to Iran, NW India and E China.
Distribution and status: Globally Endangered (Collar *et al.* 1994). Fluctuating winter population of over 2000 birds winters on Poyang and nearby lakes in Changjiang valley, passing through NE China on migration.
Habits: Feed on small plant bulbs and roots growing on retreating water level of Poyang Lake.

298. Sarus Crane *Grus antigone* Chijing he   PLATE 31

Description: Very tall (150 cm) unmistakable grey crane with bare head and upper neck mostly red. In flight shows black primaries. Immature has rusty-feathered neck and head. Iris—yellow; bill—greenish-horn; feet—pink.
Voice: Loud, far-carrying trumpeting, usually as a duet by paired birds on ground and in flight, day and night. Neck outstretched with bill pointing to the sky.
Range: N and NE India, Burma, SW China to Indochina and N Australia.
Distribution and status: Globally Near-threatened (Collar *et al.* 1994). Very rare. Race *sharpii* formerly in S Yunnan.
Habits: Inhabits marshes and paddy fields.

299. White-naped Crane *Grus vipio* Baizhen he   PLATE 31

Description: Tall (150 cm) grey and white crane with bare red patch on side of face edged and patterned black. Throat and back of neck white. Grey of nape, breast and front of neck extends up side of neck in a narrow pointed line. Primaries are black. Rest of plumage various shades of grey. Iris—yellow; bill—yellow; feet—crimson.
Voice: Bugling calls.
Range: Siberia, N Mongolia and N China wintering to C and S China, Korea and Japan.
Distribution and status: Globally Vulnerable (Collar *et al.* 1994). Breeds in NE and NW China in marshy swamps and reedy lake shores. Birds migrate south in winter to

lakes and river banks of the lower Changjiang (Yangtze) with vagrants reaching Taiwan and Fujian.

Habits: Inhabits swamps and marshes near lakes and rivers. Feeds on agricultural lands.

300. Sandhill Crane *Grus canadensis* Shaqiu he   PLATE 31

Description: Tall (104 cm) grey crane with whitish face and red forehead and cap, and dark grey flight feathers
Iris—yellow; bill—grey; feet—grey.
Voice: Trumpeting, rattling *gar-oo-oo*, audible from a great distance.
Range: Breeds N America and E Siberia; winters south. Accidental in E China.
Distribution and status: Straggler to Jiangsu (Shuyang) in 1981, Zhejiang and also Poyang Lake (Jiangxi).
Habits: Inhabits grassy tundra and prairies beside rivers, swamps and lakes.

301. Demoiselle Crane *Grus virgo* Suoyu he   PLATE 31

Description: Smallish (105 cm) elegant, pale bluish-grey crane with white crown and long white-plumed ear tufts contrasting against blackish head, neck and elongated breast feathers. Elongated tertials are not bushy and overhang tail. Black of breast extends lower than in Common Crane. Iris—red in male, orange in female; bill—yellowish-green; feet—grey.
Voice: Call is a trumpeting similar to Common Crane but shriller and flatter.
Range: N Africa (probably extinct), Turkey to C Asia and China.
Distribution and status: Breeds in NE China, Ordos plateau in W Nei Mongol and NW China, and winters in S Xizang. A bird of high plateaux, steppes, semi-desert and cold deserts and swamps up to 5000 m.
Habits: Flies in V-formation with neck extended.

302. Common Crane *Grus grus* Hui he   PLATE 31

Description: Medium-sized (125 cm) grey crane. Forecrown is black, centre of crown is red, head and neck dark slaty-grey. There is a broad white curved stripe from behind the eye down to back of neck. Plumage otherwise grey with brownish wash on back and elongated bushy tertials. Iris—brown; bill—dingy green toward yellower tip; feet—black.
Voice: Pairs duet with clear far-carrying horn sound *Kaw-Kaw-Kaw*. Migrating flocks give trumpeting *krraw*.
Range: Palearctic.
Distribution and status: Breeds in NE and NW China but migrates south in winter to S China and Indochina. Favours wetlands, marshes and shallow lakes. Becoming quite rare.
Habits: On migration rests and feeds in arable fields. Has mating dance of high fluttering leaps. Flies with extended neck in V-formation.

303. Hooded Crane *Grus monacha* Baitou he   PLATE 31

Description: Small (97 cm) dark grey crane with white head and neck and black forehead and red forecrown. Flight feathers black. Immature has buff wash on head and neck and black eye patch. Iris—yellow-red; bill—greenish; feet—blackish.

Voice: Loud *kurrk* call.

Range: Breeds in N Siberia and NE China; winters in S Japan and E China.

Distribution and status: Globally Conservation Dependent (Collar *et al.* 1994). Uncommon. Could breed in bogs of Lesser Hinggan Mts in Heilongjiang, but not documented. Breeds at Lake Kanka (Xinghai Hu), Wusuli River and maybe Ussuri (Heilongjiang) valley as well as Hulun Nur area of E Nei Mongol.

Habits: Inhabits swampy wetlands near lakes and rivers.

304. Black-necked Crane *Grus nigricollis* Heijing he — PLATE 31

Description: Tall (150 cm) greyish crane with head, throat and entire neck black except for a white patch below and behind eye. Bare lores and crown are red. Tail, primaries and elongated tertials are black. Iris—yellow; bill—horn-grey/green more yellow toward tip; feet—black.

Voice: Call is a series of loud trumpeting honks.

Range: Breeds Tibetan Plateau, wintering to Bhutan, NE India, SW China and N Indochina.

Distribution and status: Globally Vulnerable (Collar *et al.* 1994). Breeds on high plateaux on Qinghai–Xizang Plateau and as far east as NW Sichuan in swamps and around lakes. Winters in south in wet cultivated areas of S Xizang, Yunnan and Guizhou where it may locally be a pest of agricultural crops. This is now a rare bird.

Habits: Flies like other cranes with neck outstretched and in V-formation or in pairs.

305. Red-crowned Crane *Grus japonensis* Danding he — PLATE 31

Description: Tall (150 cm) elegant white crane with bare red crown. Lores, cheeks, throat and sides of neck black. A broad white band extends from ear coverts down back of neck. Rest of plumage white except for black secondaries and elongated drooping tertials. Iris—brown; bill—greenish-grey; feet—black.

Voice: Trumpeting calls in breeding areas.

Range: Breeds in Japan, NE China and SE Siberia, wintering in Korea and E China.

Distribution and status: Globally Vulnerable (Collar *et al.* 1994). Breeds in NE China but migrates south to winter in eastern provinces and Changjiang lakes, also to Korea and Japan. Accidental in Taiwan. This formerly common bird is now rare and local in wide river valleys, woods and swamps. Large flocks still winter at Tancheng, Jiangsu.

Habits: Dancing display on breeding grounds is much revered in local cultures. Flies like other cranes with neck extended and in V-formation.

RAILS—Family: Rallidae

Rails are a worldwide family of rather secretive, swamp-living birds of medium size. They have strong straight bills and long legs with very long toes. The wings are short and the flight weak and flappy. Rails are good runners, but dash for cover and hide in thick reed clumps rather than try to outrun predators. Most species can swim and some do so regularly, coots having lobed feet for this purpose. Most rails have loud cacophonous calls, sometimes with more than one bird joining in. Rails frequent a range of habitats including swamps, lake edges, reed and cane beds, grasslands, rice fields, secondary forest; few species live in forest. Rails nest on the ground and feed on a mixture of plant shoots, seeds and invertebrates. There are 19 species recorded in China, one only as an accidental visitor.

306. Swinhoe's Crake *Coturnicops exquisitus* [Swinhoe's/   PLATE 32
Asian Yellow Rail] Hua tianji

Description: Very small (13 cm) speckled crake. Upperparts brown with black streaks finely barred white; chin, throat and belly white; breast buffy-brown; flanks and undertail broadly barred dark brown and white. Juvenile darker. Short tail is cocked. In flight, white secondaries and black primaries conspicuous. Iris—brown; bill—dull yellow; feet—yellow.
Voice: Grunts, squeals and cackles; call like tapping stones.
Range: Breeds NE Asia migrating to S Japan and S China in winter.
Distribution and status: Globally Vulnerable (Collar *et al.* 1994). A rare breeder in reedy vegetation of NE Nei Mongol; winter visitor to wet grasslands at Poyang Lake in Jiangxi and rice fields of Fujian and Guangdong; transient through E China.
Habits: Very secretive.
Note: Placed by some authors in *Porzana* (e.g. Cheng 1994).

307. Red-legged Crake *Rallina fasciata* [Malay Banded/   PLATE 32
Malay Crake] Hongtui banyangji

Description: Medium-sized (23 cm) reddish-brown, short-billed rail with red legs. Head, back and chest chestnut; wings and tail reddish-brown; belly and undertail black with white bars; chin white. Conspicuous white spotting on wing coverts and white barring on flight feathers. Similar to Band-bellied Crake but white bars on flank and belly much bolder. Iris—red; bill—brown; feet—red.
Voice: Loud series of nasal *pek* calls at half-second intervals given at dawn and dusk in breeding season; also slow descending trill.
Range: NE Indian subcontinent, SE Asia, Philippines, Sundas and Maluku. Some races winter to south of range.
Distribution and status: Accidental summer visitor to Lanyu Islet, Taiwan.
Habits: A rarely seen inhabitant of open swampy places near woods in low-lying areas; shy and poorly-known.

308. Slaty-legged Crake *Rallina eurizonoides* [Slaty-legged   PLATE 32
Banded/Ryukyu/Philippine/Banded Crake] Baihou banyangji

Description: Medium-sized (27 cm) brownish rail with chestnut head and breast; whitish chin; narrow white bars on blackish abdomen and undertail. White on the wing limited to sparse barring on inner secondaries and primaries. Iris—red; bill—greenish-yellow; feet—grey.
Voice: Double *beep-beep* (like cartoon roadrunner) call given at night.
Range: Breeds Indian subcontinent, SE China, SE Asia, Philippines and Sulawesi. Some winter in Sri lanka, Malay Peninsula, Sumatra and W Java.

Distribution and status: Race *amauroptera* is an uncommon resident, up to 700 m, in S Guangxi and straggler to Hainan; also a summer breeding visitor to Hong Kong. Probably often overlooked and maybe becoming more widespread. Race *formosana* is resident in Taiwan.

Habits: Shy rail in forest, forest edge, scrub and paddy fields.

309. Slaty-breasted Rail *Gallirallus striatus* [Blue-breasted Banded   PLATE 32
Rail] Lanxiong yangji

Description: Medium-sized (29 cm) rufous-crowned rail with fine transverse white barring on back. Crown chestnut; chin white; breast and back grey; wings and tail finely barred white; flanks and undertail more coarsely barred black and white. Iris—red; bill—black above, reddish below; feet—grey.

Voice: Hard, sharp double note *terrek* or buzzing *kech, kech, kech* repeated 10–15 times starting weakly, becoming stronger, then fading again.

Range: Indian subcontinent, S China, SE Asia, Philippines and Greater Sundas. Some winter to south of range, in Sulawesi and Lesser Sundas.

Distribution and status: Race *gularis* is resident in S and SW China; *taiwanus* in Taiwan. Generally uncommon up to 1000 m.

Habits: Found in mangroves, swamps, rice paddies, grassland and even on dry coral islands. Infrequently seen as it is retiring and partially nocturnal. Generally solitary.

310. Water Rail *Rallus aquaticus* Putong yangji   PLATE 33

Description: Medium-sized (29 cm) dark rail with streaked upperparts. Crown brown, face grey with pale grey eyebrow and dark eye-stripe. Chin white; neck and breast grey. Flanks barred black and white. Immature birds have some indistinct white barring on upperwing coverts. Iris—red; bill—red to black; feet—red.

Voice: Soft *chip chip chip* call and strange, pig-like grunts and squeals.

Range: Palearctic; migrants reach SE Asia and Borneo.

Distribution and status: Fairly common in breeding range. Race *korejewi* in NW and NC China and Sichuan; *indicus* breeds in NE China, migrating south to winter in SE China and Taiwan.

Habits: A shy bird of thick waterside vegetation, marshes and mangroves.

311. Corn Crake *Crex crex* Changjiao yangji   PLATE 32

Description: Medium-sized (26.5 cm) mottled, yellowish-brown, short-billed rail. Upperparts greyish-brown, streaked with bold black feather centres; broad rufous wing patch. Broad supercilium grey; eye-stripe rufous; chin whitish and throat and breast greyish. Flanks and undertail barred chestnut, black and white. In flight, long rusty-brown wings characteristic. Flies with limp beat and legs dangling. Iris—brown; bill—yellowish-brown; feet—dull yellow.

Voice: Far-carrying, disyllabic, loud, rasping *crek crek*, generally given at night or early morning. Sometimes more than one bird calling together.

Range: W Palearctic to C Asia and S Russia, introduced E USA.; migrates to sub-Saharan Africa.

Distribution and status: Globally Vulnerable (Collar *et al.* 1994). W Xinjiang as a rare breeder in W Tianshan and winter visitor to Bangong Lake in W Xizang.
Habits: A shy bird of dry grassland and cultivated fields.

312. Brown Crake *Amaurornis akool* [Crimson-legged Crake]   PLATE 33
Hongjiao ku'eniao

Description: Medium-sized (28 cm) dull-coloured crake with red legs. Uniform olive-brown above with slaty-grey face and breast; belly and undertail brown. Juvenile less grey. No barring in plumage. Flight feeble with dangling legs. Iris—red; bill—yellow-green; feet—carmine-red.
Voice: Long-drawn vibrating whistle, descending scale.
Range: Indian subcontinent to China, Arakan (W Burma) and NE Indochina.
Distribution and status: Breeds in reedy or grassy swamps. Locally common in hill rice fields of S China.
Habits: A shy skulking and largely crepuscular bird. Constantly cocks tail.

313. White-breasted Waterhen *Amaurornis phoenicurus*   PLATE 33
Baixiong ku'eniao

Description: Largish (33 cm) unmistakable dark slaty-grey and white rail. Grey crown and upperparts; white face, forehead, breast and upper belly; rufous lower belly and undertail. Iris—red; bill—greenish with red base; feet—yellow.
Voice: Monotonous *uwok-uwok* call. Noisy weird chorusing of several birds with grunts, croaks and chuckles *turr-kroowak, per-per-a-wak-wak-wak* and others for up to 15 minutes at a time, at dawn or at night.
Range: Indian subcontinent, S China, SE Asia, Philippines, Sulawesi, Maluku and Sundas.
Distribution and status: Race *chinensis* breeds in lowland areas of China S of 34 °N. Winters in Yunnan, Guangxi, Hainan, Guangdong, Fujian and Taiwan. Accidental in Shandong, Shanxi and Hebei. Generally common in suitable habitat up to 1500 m.
Habits: Generally single, occasionally in twos and threes, skulking in damp scrub, lake sides, river banks, mangrove and fields. Feeds in open spaces so is seen more often than other rails. It also clambers about in bushes and small trees.

314. Black-tailed Crake *Porzana bicolor* [Elwe's; Rufous-backed   PLATE 33
Crake] Zongbei tianji

Description: Medium-sized (22 cm) distinctive rufous-brown and grey crake. Head and neck dark ashy grey; rest of upperparts rufous-brown. Chin white, tail blackish; rest of underparts dark grey. Sexes alike. Iris—red; bill—greenish with small red base; feet—red.
Voice: Prolonged, descending trill, preceded by harsh rasping notes.
Range: E Himalayas, N Burma, N Thailand, N Indochina and SW China.
Distribution and status: A rare or overlooked resident in swamps and reedy stream banks from foothills to 3600 m in SE Xizang, Yunnan, S Sichuan and E Guizhou.
Habits: Skulking and crepuscular.

315. Little Crake *Porzana parva* Ji tianji　　　　　　PLATE 33

Description: Small (19 cm) crake. Male: brown above and grey below with slight white spotting. Female: underparts paler with buff instead of grey and whitish face, chin and throat. Very similar to slightly smaller Baillon's Crake but male duller brown on upperparts with less white spotting, less banding on flanks, green legs and red base to bill. Juvenile distinguished from juvenile Baillon's by having solid rather than ring-shaped white spots on upperparts. Iris—red; bill—greenish with red base; feet—greenish.

Voice: Male song is leisurely, low-pitched croaking, accelerating to stammering end. Female call is more hurried ending in vibrant trill.

Range: Breeds W Palearctic to C Asia, winters in sub-Saharan Africa, Middle East and Pakistan.

Distribution and status: A rare breeder in the Tarim basin of W Xinjiang, also recorded as a migrant in Tianshan.

Habits: Inhabits swamps, wet meadows and ponds with floating vegetation. Swims freely.

316. Baillon's Crake *Porzana pusilla* [Dwarf Rail] Xiao tianji　　PLATE 33

Description: Tiny (18 cm) greyish-brown, short-billed rail with white streaks on back and fine white barring on flanks and undertail. Male: crown and upperparts red-brown, streaked black and white; chest and face grey. Female duller with brown ear coverts. Juvenile has whitish chin and curious ring-shaped white spots on upperparts. Distinguished from Little Crake by richer brown of upperparts, more white spotting, more barring on flanks, lack of red base to bill and pinkish legs. Iris—red; bill—greenish; feet—yellowish-green to dull green.

Voice: A dry scratchy trill descending the scale, similar to frog or male Garganey.

Range: N Africa and Palearctic, migrating to Indonesia, Philippines, New Guinea and Australia.

Distribution and status: Fairly common in suitable habitat. Breeds in NE China, Hebei, Shaanxi, Henan and Kashi region of W Xinjiang. Migrates across most of China, with wintering population in Guangdong. Straggler to Taiwan.

Habits: Inhabits swampy lakes and grassy marshes. Threads its way fast and delicately through reedbeds, rarely flies.

317. Spotted Crake *Porzana porzana* Banxiong tianji　　　　PLATE 33

Description: Medium-sized (23 cm) dark, short-billed rail with extensive white spotting. General coloration brownish above, streaked with grey, black and white; underparts grey, spotted with white and barred black and white on flanks; undertail buff. Iris—brown; bill—yellow with red base, grey in juvenile; feet—greenish.

Voice: Far-carrying, rhythmically repeated, whiplash whistle *hwitt*, given at dusk and after dark.

Range: Breeds in W Palearctic as far east as Lake Baikal but migrates to Africa, India and W China.

Distribution and status: Rare. May breed in Altai Mts. Recorded on passage in W Xinjiang.

Habits: Inhabits wet grassland and rice fields.

318. Ruddy-breasted Crake *Porzana fusca* [Ruddy Crake]   PLATE 33
Hongxiong tianji

Description: Small (20 cm) reddish-brown, short-billed rail. Hindcrown and upperparts plain brown. Sides of head and breast deep brownish-pink (redder in race *erythrothorax*); chin white; abdomen and undertail blackish with fine white barring. Similar to Red-legged and Band-bellied Crakes but smaller and lacks any white on wings. Iris—red; bill—brownish; feet—red.

Voice: Rather quiet; in breeding season 3–4-second bursts of descending squeaky trills rather like Little Grebe; soft *chuck* when foraging.

Range: Breeds Indian subcontinent, China, E Asia, Philippines, Sulawesi and Sundas. Northern populations winter to south of range; winters Borneo.

Distribution and status: Race *erythrothorax* is a locally common resident on Taiwan; *phaeopyga* in E, C and S China; and race *bakeri* in SW China.

Habits: Inhabits reed beds, rice fields and dry bush land beside lakes but shy and not often seen. Occasionally ventures out at edge of reed beds. Partially nocturnal. Calls at dawn and dusk.

Note: Sometimes placed in genus *Amaurornis*.

319. Band-bellied Crake *Porzana paykullii* [Siberian Ruddy/   PLATE 33
Chinese Banded Crake] Banxie tianji

Description: Medium-sized (22 cm) reddish-brown, short-billed rail with red legs. Crown and upperparts dark brown; chin white; sides of head and breast chestnut; flanks and undertail blackish with fine white bars. Distinguished from Ruddy-breasted Crake by white wing bars and from Red-legged Crake by finer white barring. Less white on wing coverts than Red-legged; no white on flight feathers; nape and neck dark. Juvenile brown instead of chestnut. Iris—red; bill—yellowish; feet—red.

Voice: Strange call given at night, sounds like drumbeats or a wooden rattle.

Range: Breeds in NE Asia; migrates in winter to SE Asia and Greater Sundas.

Distribution and status: Globally Near-threatened (Collar *et al.* 1994). Uncommon breeder in N and NE China in midsummer. Passage migrant over C and E China.

Habits: Inhabits wet tussocky meadows and paddy fields.

Note: Sometimes placed in genus *Rallina*.

320. White-browed Crake *Porzana cinerea* [Ashy/Grey-bellied   PLATE 33
Crake] Baimei yangji

Description: Smallish (20 cm) greyish-brown, short-billed rail with conspicuous head pattern of white stripes above and below black eye-stripe. Crown, back and upperparts dull brown; sides of head and breast grey; abdomen, flanks and undertail buffy-brown. Iris—red; bill—blackish; feet—greenish-yellow.

Voice: In breeding areas, noisy, high-pitched, thin reedy piping *cutchi cutchi cutchi* with two or more birds joining in at once.

Range: Southern SE Asia, Philippines and Greater Sundas to New Guinea, N Australia and Pacific islands.

Distribution and status: Vagrant April 1991, Mai Po, Hong Kong.

Habits: A shy bird of flooded grassland, marshes and paddy fields. Usually lives in pairs.

321. Watercock *Gallicrex cinerea* Dong ji   PLATE 34

Description: Large (40 cm) black or buffy-brown rail with short green bill. Female brown with finely barred underparts. Black breeding plumage of male with pointed red horny frontal plate. Iris—brown; bill—yellow-green; feet—green, red in breeding male.
Voice: Deep booming in its summer nesting grounds but generally silent in winter.
Range: Resident Indian subcontinent, Arakan, southern SE Asia, Sumatra and Philippines; summer breeder in Himalayas, NE Asia, China, north-eastern SE Asia and Taiwan; wintering populations visit Japan, peninsular Malaysia, Borneo, Java, Sulawesi and Lesser Sundas.
Distribution and status: Race *cinerea* is a summer breeder over most of E, C, S and SW China, Hainan and Taiwan. Migrates south in winter.
Habits: A shy skulking and largely nocturnal bird of reedy swamps. Sometimes comes into adjacent rice fields to feed on paddy.

322. Purple Swamphen *Porphyrio porphyrio* [Purple Gallinule/   PLATE 34
Purple Waterhen/Purple Coot/Purple Moorhen] Zi shuiji

Description: Unmistakable large (42 cm) chunky, purplish-blue rail with a massive red bill. Entire plumage blue-black with a purple and green sheen except for white undertail coverts. Has a red frontal shield. Iris—red; bill—red; feet—red.
Voice: Cackling grunts and hoots; trumpeted nasal *wak*.
Range: Palearctic to Africa, Oriental region, Australia and Oceania.
Distribution and status: Race *poliocephalus* is a rare breeding bird of SW Yunnan and possibly extreme SE Xizang. Vagrant wintering birds recorded W Yunnan, Guangxi and Hong Kong.
Habits: Inhabits reed-lined swamps and lakes, walking over floating vegetation and through reed beds. Sometimes comes out into open flooded grasslands, paddy fields or even burnt grass areas in small parties. Constantly flicks tail.
Note: Part of a variable complex that may involve several species.

323. Common Moorhen *Gallinula chloropus* [Moorhen]   PLATE 33
Hei shuiji

Description: Unmistakable medium-sized (31 cm) black and white aquatic rail with bright red frontal shield and short bill. Plumage entirely slaty-black except for line of white streaks along flanks and two white patches under tail, conspicuously displayed when tail is cocked. Iris—red; bill—dull greenish with red base; feet—green.
Voice: Loud harsh croaking calls *pruruk-pruuk-pruuk*.
Range: Almost worldwide except Australasia and Oceania. Northern populations winter south.
Distribution and status: Race *indica* breeds in W Xinjiang, including Tianshan; nominate race breeds over most of E, S and SW China, Hainan and Taiwan and SE Xizang. Populations winter S of 32 °N.
Habits: Frequents lakes, pools and canals. Largely aquatic, swimming slowly about, dabbling at surface vegetation. Also feeds on open grass. Constantly flicks tail on land or in water. Flies rather weakly after a long paddling take-off run over water.

324. Common Coot *Fulica atra* [Black/Eurasian/European/   PLATE 33
The Coot] Guding ji

Description: Unmistakable large (40 cm) black aquatic rail with conspicuous white bill and frontal shield. Plumage entirely slaty-black except for narrow whitish trailing edge to wing visible in flight. Iris—red; bill—white; feet—greyish-green.

Voice: Various loud calls and sharp *kik kik.*

Range: Palearctic, Middle East, Indian subcontinent. Northern populations winter S to Africa, SE Asia and Philippines, rarely reaching Indonesia. Also New Guinea, Australia and New Zealand.

Distribution and status: Race *atra* is a common breeding bird on lakes and streams across N China. Migrates S in winter with large wintering populations S of 32 °N.

Habits: Highly aquatic and gregarious; regularly dives to collect water weeds from the lake bed. Fights and chases occur in breeding season. Birds make long running take-off over water.

STORKS AND ALLIES

SANDGROUSE—Order: Ciconiiformes strokes and allies Family: Pteroclidae

Small family of cryptically coloured, pigeon-like birds of desert and semi–deserts found in Africa, Middle East and Asia. Mostly gregarious and congregating to drink at water sources. Nest is a scrape on the ground and the young leave the nest immediately on hatching. Run and fly strongly. Three species occur in China.

325. Tibetan Sandgrouse *Syrrhaptes tibetanus* Xizang maotuishaji   PLATE 35

Description: Large (40 cm) sandgrouse with black underwings and white-tipped elongated central tail feathers; lacks black patch but with dark speckled breast band separating white belly from orange-buff face and throat. Iris—brown; bill—bluish-horn; feet—bluish under feathered legs.

Voice: Loud double note in flight *guk-guk* or *caga-caga* deeper and more musical than other sandgrouse. Also pleasant *koonk-koonk* on varying pitch.

Range: Ladakh, Pamirs and Tibetan plateau.

Distribution and status: Locally common in suitable habitat of Xizang, SW Xinjiang, NW Sichuan and S and E Qinghai.

Habits: As other sandgrouse species. Lives in flocks on high, barren plateaux and rocky screes. Rather tame.

326. Pallas's Sandgrouse *Syrrhaptes paradoxus* Maotui shaji   PLATE 35

Description: Large (36 cm) sandy-coloured sandgrouse with elongated central tail feathers, heavy black speckling on upperparts and orange-buff pattern on side of face. Orbital ring pale blue. Lacks black throat patch but has diagnostic black patch on belly. Male has unstreaked pale grey breast with small pectoral band of fine flecks. Female has narrow black throat bar and sides of neck are spotted. In flight, wings are pointed and underwing is white with narrow black trailing edge to secondaries. Iris—brown; bill—greenish; feet—bluish but legs feathered.

Voice: Noisy in flocks with *kirik* or *cu-ruu cu-ruu-cu-ou-ruu*; rapid repeated *kukerik* calls and clucking, stifled *cho-ho-ho-ho*. Wings make humming sound in flight.

Range: C Asia and N China.

Distribution and status: Resident across suitable habitats in northern parts of China with birds from NE China wintering south in Hebei, Shaanxi, Liaoning and Shandong.

Habits: Subject to occasional eruptions when it occurs well beyond its normal distribution. Lives in open arid country, steppes and semi-desert and uses cultivated land.

327. Black-bellied Sandgrouse *Pterocles orientalis* Heifu shaji   PLATE 35

Description: Largish (34 cm) sandy-brown sandgrouse. Male has grey head neck and throat with chestnut patch on sides of neck and below face. Wing is boldly banded black and cinnamon. Female is paler, more speckled black. Both sexes have black lower breast and abdomen with buffy breast band outlined anteriorly by fine black gorget. In flight the combination of black underparts with white underwing is diagnostic. Iris—brown; bill—greenish-grey; feet—greenish-grey.

Voice: Flight call is characteristic rolling or bubbling *durrrll*.

Range: Spain and N Africa, Middle East, N and NW India, Afghanistan and S Russia.

Distribution and status: Rare in China. Race *arenarius* breeds in N and NW Xinjiang (Tianshan), migrating through SW Xinjiang (Kashi).

Habits: Lives in dry, sparsely vegetated regions up to edge of cultivation. Partially migratory.

SANDPIPERS—Family: Scolopacidae

The sandpiper family is a large, worldwide family of waders, generally found on the sea shores or in wet, open places often close to the sea. A few species range inland to higher altitudes but of the Chinese species only the Eurasian Woodcock is a regular forest dweller. All members of the family have long legs, long pointed wings and generally elongated slender bills. In some species the bills are extraordinarily long and used for probing deep into mud to search for hidden worms or crustaceans. Most species are migratory. Many of these migratory waders have a different plumage in winter than in the breeding season. Some breed north of China and are only seen on migration but most have breeding populations in China and both plumages are illustrated in the plates. Often large mixed flocks are seen. Waders are a demanding group to recognise as there are so many similar species and they are often seen at rather long range. A spotting scope is recommended and careful note should be taken of the general appearance and whether there are conspicuous white wings bars visible in flight. Calls are valuable in identification as many species have very distinctive flight calls. Fifty-three species occur in China.

WOODCOCKS AND SNIPES—Subfamily: Scolopacinae

Medium to small waders with long straight bills and cryptic plumage. They live in marshy areas and feed by probing deep into mud for worms and other prey. Some species are partially nocturnal. They tend to sit tight when disturbed, relying on their camouflage for protection, but explode from the ground when approached too close.

328. Eurasian Woodcock *Scolopax rusticola* [Woodcock] Qiu yu PLATE 36

Description: Large (35 cm) plump, short-legged wader with long, deep-based, straight bill. Distinguished from snipe by much larger size and transverse bands on crown and nape. Rises with noisy swishing wing beats. Territorial or 'roding' flight is slow, at treetop height, with bill angled down. Looks heavy and broad-winged in flight. Iris—brown; bill—pinkish at base with black tip; feet—greyish-pink.

Voice: Usually silent when flushed but occasional rapid *etsh-etsh-etsh*. Males give grunting *oo-oort* calls followed by explosive squeak in 'roding' flight.

Range: Palearctic; migrant SE Asia.

Distribution and status: Breeds in N Heilongjiang, Tianshan of NW Xinjiang and in Sichuan and S Gansu. Migrants across much of the country and wintering populations can be found over most of China S of 32 °N, also Taiwan and Hainan.

Habits: Nocturnal forest bird. It hides by day, sitting tight on the ground, and flies by night to feeding grounds in open areas.

329. Solitary Snipe *Gallinago solitaria* Gu shazhui PLATE 36

Description: Large (29 cm) dull-coloured snipe. Larger, duller and less boldly marked than Wood Snipe. Lacks blackish lateral crown stripes and base of bill greyer. Feet do not extend beyond tail in flight. Duller and less yellow than Common, Swinhoe's or Pintail Snipes with whitish rather than buffy face stripes. Scapulars edged white. Breast pale gingery-brown; belly barred white and rufous-brown. No white on underwing or on trailing edge of secondaries. Race *japonica* has bill longer and more slender than Latham's and coloration paler. Iris—brown; bill—olive-brown with dark tip; feet—olive.

Voice: Characteristic harsh *pench* when flushed, similar to Common Snipe but deeper and harsher; *chok-achock-a* call in display.

Range: Himalayas and C Asian mountains, wintering in foothills from Pakistan to Japan and Kamchatka Peninsula.

Distribution and status: Uncommon in bogs, marshes and paddy fields. Nominate race breeds in Tianshan of W Xinjiang, Himalayas around eastern edge of Qinghai–Xizang Plateau to NW Sichuan, Qinghai and W Gansu. Winters in Kashi region of W Xinjiang, SE Xizang and Yunnan. Race *japonica* breeds in NE provinces, wintering in Changjiang valley and Guangdong.

Habits: Solitary. Flight is slower than Common Snipe but with similar zig-zag turns.

330. Latham's Snipe *Gallinago hardwickii* [Japanese Snipe] PLATE 36
Aonan shazhui

Description: Large (30 cm) chunky, yellow-brown snipe. Similar to Swinhoe's in colour and proportions but larger. At rest, tail projects further beyond folded wing. Can only be safely identified on voice in breeding season or in the hand by tail details. Iris—dark brown; bill—greenish-brown with dark tip; feet—greenish.

Voice: In courtship display flight *ji, ji, ji, zubiyak, zubiyak, zubiyak* then dives with *ga ga ga.*

Range: Breeds in Japan, migrates in winter to Australia.

Distribution and status: Globally Near-threatened (Collar *et al.* 1994). Passage migrant in Taiwan. Straggler in Hebei and Liaoning. May be overlooked in China due to difficulty in identification but seems to migrate directly from Japan to Australia.

Habits: Inhabits marshes, rice stubble, grassland and bamboo thickets in scrub. Often in parties of up to 20 birds. When flushed, flies heavily with few zig-zags, dropping quickly back into cover.

331. Wood Snipe *Gallinago nemoricola* Lin shazhui PLATE 36

Description: Largish (31 cm) dark-backed snipe with whitish facial stripes. Breast fulvous barred with brown, rest of underparts white finely barred with brown. Distinguished from other snipes by darker colour, slower bat-like flight with bill pointing down and habitat. Smaller, more boldly marked than Solitary Snipe with black lateral crown stripes and less grey on base of bill. Iris—dark brown; bill—greenish-brown, dark at tip; feet—greyish-green.

Voice: Low croaking *tok-tok* in flight.

Range: Breeds in Himalayas, winters in India and SE Asia.

Distribution and status: Globally Vulnerable (Collar *et al.* 1994). Rare. Breeds in E Xizang and W Sichuan. Winters in SE Xizang and in W and NE Yunnan.

Habits: Uses swampy puddles and pools in tall grass and scrub of hills and mountains up to 5000 m.

332. Pintail Snipe *Gallinago stenura* [Pin-tailed Snipe] PLATE 36
Zhenwei shazhui

Description: Small (26–27 cm) plump, short-legged snipe with rounded wings and relatively short blunt bill. Upperparts brown and comparatively plain, narrowly streaked and vermiculated white, yellow and black; underparts white, washed rufous on chest and finely barred black; eye-stripe narrow in front of eye and poorly defined behind. Distinguished from Common and Swinhoe's Snipes with difficulty by smaller size, shorter tail, yellow feet trailing further behind tail in flight and by call. Differs from Common Snipe in greyer, cooler plumage and by lack of white trailing edge to wing and lack of broad white bars on underwing coverts; wings in flight appear broader and less rounded. Flight-slower, more deliberate and less erratic than Common Snipe. Legs appear thinner and less yellow than Swinhoe's Snipe. Iris—brown; bill—brown with dark tip; feet—yellowish.

Voice: Rasped *squak-squak* in alarm with nasal twang.

Range: Breeds in NE Asia but migrates S in winter as far as India, SE Asia and Indonesia.

Distribution and status: Common passage migrant over all China with wintering populations in Taiwan, Hainan, Fujian, Guangdong and Hong Kong.

Habits: Frequents paddy fields and marshes, and damp hollows in forest and mangroves but usually in drier situations than Common Snipe. Habits similar to other snipes, including fast leap and zig-zag flight, giving alarm call when disturbed.

333. Swinhoe's Snipe *Gallinago megala* [Chinese/Marsh Snipe]   PLATE 36
Da shazhui

Description: Largish (28 cm) colourful snipe with long, pointed wings, large squre head and long bill. Distinguished with difficulty in the field from Pintail Snipe by longer tail; thicker, yellower legs and feet less trailing in flight. Distinguished from Common Snipe by more white on sides of tail tip, longer tail projection, lack of broad white bars on underwing and usually a lack of white trailing edge to wing in flight. Distinguished with difficulty from Latham's by primaries extending well beyond tertials. Spring birds rather dusky on breast and neck. Iris—brown; bill—brown; feet—olive-grey.

Voice: Harsh, rasping cry, similar to Common Snipe but higher pitched and slightly slurred. Usually only a single call.

Range: Breeds NE Asia; migrates S in winter to N Borneo, Indonesia and as far as Australia.

Distribution and status: Common passage migrant over E and C China with wintering populations in Hainan, Taiwan, Guangdong, Hong Kong and occasionally Hebei.

Habits: Inhabits swamps and wet grasslands, including paddy fields. Habits as other snipes but reluctant to fly, and rises and flies slower and less erratically.

334. Common Snipe *Gallinago gallinago* [Fantail/The Snipe]   PLATE 36
Shanwei shazhui

Description: Medium-sized (26 cm) brightly coloured snipe with slim pointed wings and long bill. Stripes on face buff with broad clear eye-stripe; upperparts dark brown, streaked and vermiculated white and black, underparts buffish streaked brown. Similar in colour to Swinhoe's, Latham's and Pintail Snipes but distinguished by broader white trailing edge to secondaries and broad white bars on underwing, also faster, higher and more erratic flight with frequent urgent calls. Buffy supercilium contrasts with paler cheeks. Pale outer edges to scapulars broader than inner edges. Scapular line duller than median line. Iris—brown; bill—brown; feet—olive.

Voice: Song is loud rhythmic *tick-a, tich-a, tich-a* ... often from a perch. Loud cry *jett.. jett* on a rising note given in alarm when flushed.

Range: Breeds Palearctic; Winters to Africa, India, SE Asia and Philippines.

Distribution and status: Breeds in NE China and Tianshan of NW China. Common passage migrant over most of China. Winters in S Xizang, Yunnan and most of China S of 32°N.

Habits: Found in marshes and paddy fields, generally keeping under cover of long reeds and grasses but leaping up when flushed and fleeing with erratic zig-zag flight, giving alarm call. Performs aerial display with climbs and dives, outer tail feathers are spread to vibrate with humming sound.

335. Jack Snipe *Lymnocryptes minimus* Ji yu   PLATE 36

Description: Small (18 cm) short-billed, narrow-winged snipe. Distinguished from all other snipe in not having longitudinal stripe down centre of crown; having all-dark, wedge-shaped tail and having green and purple gloss on upperparts. Tail all dark without rufous band. Distinguished from Broad-billed Sandpiper by straighter bill and more prominent scapular stripes. In flight, toes do not extend beyond tail and lacks white leading edge to wing. Iris—brown; bill—yellow with black tip; feet—dull yellow.

Voice: Usually silent when flushed or subdued *gah*.

Range: Breeds boreal Europe to W Siberia; wintering to S Europe, Africa, Middle East, India and SE Asia.

Distribution and status: Rare migrant and winter visitor. Migrants pass through NE China and down E coast with small numbers wintering irregularly in S Guangdong, dubiously in Hong Kong and occasionally Taiwan. Another population winters in the Kashi and Tianshan regions of W Xinjiang.

Habits: Solitary. Has aerial switchback display around breeding territory. Rarely flies by day, even when alarmed. Freezes tight or runs to safety. Lives in marshes and rice fields. Bobs head constantly when feeding.

GODWITS, SANDPIPERS, CURLEWS AND PHALAROPES—Subfamily: Tringinae

A large worldwide subfamily of waders, generally found in wetlands and along coasts. They have long legs and bills and feed on molluscs, worms and crustacea. Phalaropes were formerly regarded as a distinct family of three species of specialised pelagic waders. They are slim, dainty birds with narrow pointed bills. They have dense plumage with duck-like down, rendering them highly buoyant. The feet are lobed rather than webbed. Outside the breeding season, these birds stay almost permanently at sea in flocks, weaving about and feeding on the sea a short distance apart. All breed in the northern hemisphere and visit the tropics only in the winter. They show reversed sexual dimorphism.

336. **Black-tailed Godwit** *Limosa limosa* Heiwei chengyu   PLATE 37

Description: Large (42 cm) long-legged, long-billed wader. Similar to Bar-tailed Godwit but larger, bill less upturned, eye-stripe more pronounced, upperparts less mottled, terminal half of tail blackish with rump and base white. White wing bar distinctive; narrow in race *melanuroides*, broader in rare nominate form. Iris—brown; bill—pink at base; feet—greenish-grey.

Voice: Generally quiet but occasionally a loud *wikka wikka wikka* or *kip-kip-kip* in flight.

Range: Breeds in N Palearctic but migrates S in winter from Africa as far as Australia.

Distribution and status: Race *melanuroides* breeds in Tianshan in NW Xinjiang and in Hulun Nur and Dalai lake areas of NE China, in Nei Mongol. Large flocks of passage migrants pass through most of China with a few birds wintering along S coast and on Taiwan.

Habits: Frequents coastal mudflats and the shores of rivers and lakes. Feeds as Bar-tailed Godwit but in muddier and deeper places with head sometimes almost submerged.

337. Bar-tailed Godwit *Limosa lapponica* Banwei chengyu   PLATE 37

Description: Large (40 cm) long-legged wader with long, slightly upturned bill. Upperparts mottled grey and brown; conspicuous white eyebrow; underparts with some grey on breast. Distinguished from Black-tailed Godwit by narrow pale wing bar and fine brown bars on white tail and rump. Eastern form *baueri* has brownish lower back and whiter underwing. Iris—brown; bill—base pink, black-tipped; feet—dark green or grey.

Voice: Rather quiet but occasional deep nasal *kurrunk* or clear barking double note *kak-kak* and in flight a soft *kit-kit-kit-kit*.

Range: Breeds in N Europe and Asia but migrates S in winter as far as Australia and New Zealand.

Distribution and status: Passage migrants of race *baueri* (*=novaezealandiae*) recorded in Tianshan in NW Xinjiang and NE and E provinces. Winters in small parties along S coast and on Taiwan and Hainan.

Habits: Frequents tidal waters, estuaries, sandflats and shallows. Guzzles food with rapid head movements, probing deep into water.

338. Little Curlew *Numenius minutus* [Little Whimbrel]   PLATE 37
Xiao shaoyu

Description: Tiny (30 cm) curlew with medium-length decurved bill and bold buff eyebrow. Differs from Whimbrel by smaller size and shorter, straighter bill. Rump never white. Raises wings when alighting. Iris—brown; bill—brown with conspicuous pink base; feet—blue-grey.

Voice: Chattering *te-te-te* in flight or when flocks feed, and sharp harsh *chay-chay-chay* in alarm.

Range: Breeds NE Asia; migrates S in winter to Australia.

Distribution and status: Rare but regular passage migrant through E China and Taiwan.

Habits: Prefers dry, open, inland grassy areas, rarely coastal mudflats.

Note: Formerly treated as a race of Eskimo Curlew *N. borealis* (e.g. Cheng 1987).

339. Whimbrel *Numenius phaeopus* [Hudsonian Curlew]   PLATE 37
Zhong shaoyu

Description: Small (43 cm) curlew with pale eyebrow, black coronal stripes and long decurved bill. Similar to Eurasian Curlew but much smaller and bill proportionally less long. Rump brownish in commoner race *variegatus*, but some individuals with white rump and underwing approach nominate *phaeopus* in appearance. Iris—brown; bill—black; feet—blue-grey.

Voice: Distinctive, loud, flat whinnying whistle *he-he-he-he-he-he-he*

Range: Breeds in N Europe and Asia but migrates S in winter to SE Asia, Australia and New Zealand.

Distribution and status: Common passage migrant across most of China, especially coasts and estuaries on E and S coasts. A few birds overwinter in Taiwan and Guangdong.

Habits: Favours mudflats, tidal estuaries, pastures near coast, marshes and rocky beaches, generally in small to large flocks and often mixing with other waders.

340. Eurasian Curlew *Numenius arquata* [Western/Common/ PLATE 37
European/The Curlew] Baiyao shaoyu

Description: Large (55 cm) curlew with very long decurved bill. Rump white, grading into white and brown barred tail. Distinguished from Eastern Curlew by whiter rump and tail, and from Whimbrel by larger size, uniform head pattern and proportionally longer bill. Iris—brown; bill—brown; feet—slate-blue.

Voice: Loud plaintive rising cry *cur-lew.*

Range: Breeds in N Palearctic but migrates S in winter as far as Indonesia and Australia.

Distribution and status: Breeds in NE China. Passage migrants across most of country. Never numerous, but in winter more common that Eastern Curlew. A regular visitor to lower Changjiang valley, S and SE coasts of China, Hainan, Taiwan and Yarlung Zangbo valley of S Xizang.

Habits: Frequents tidal estuaries, river banks and mudflats, rarely far from the sea. Often seen singly, sometimes in small flocks or mixing with other curlew species.

341. Eastern Curlew *Numenius madagascariensis* [Far-Eastern/ PLATE 37
Australian Curlew] Da shaoyu

Description: Very large (63 cm) curlew with very long and heavily decurved bill. Darker and browner than Eurasian Curlew, lower back and tail brown; underparts buff. In flight barred underwing distinguishes from white of Eurasian Curlew. Iris—brown; bill—black with pink base; feet—grey.

Voice: Similar to Eurasian Curlew but flatter, *coor-ee.* Strident *ker ker-ke-ker-ee* when agitated.

Range: Breeds in NE Asia but migrates S in winter as far as Australasia.

Distribution and status: Globally Near-threatened (Collar *et al.* 1994). Uncommon but regular passage migrant through E China and Taiwan.

Habits: As Eurasian Curlew; very shy. Single birds sometimes in flocks of Eurasian Curlew.

342. Spotted Redshank *Tringa erythropus* Heyu PLATE 38

Description: Medium-sized (30 cm) red-legged, grey wader with long straight bill. Breeding plumage unmistakable; black with fine white spotting. In winter, similar to Common Redshank but larger and greyer with longer, slimmer bill, more slender shape and less red on base of bill. Also separated by white spotting on dark wings and more prominent eye-stripe. Differs in flight by lack of white trailing bar and legs trail further beyond the tail. Iris—brown; bill—black with red base; feet—orange.

Voice: Very distinctive sharp, explosive whistle *chee-wik* in flight and at rest; also shorter *chip* in alarm.

Range: Breeds in Europe, migrating to Africa, India and SE Asia in winter.

Distribution and status: Breeding recorded in Tianshan of NW Xinjiang. A common passage migrant across most of China with large flocks wintering in the southern provinces, Hainan and Taiwan.

Habits: Similar to Common Redshank. Prefers fish ponds, mudflats and marshes.

343. Common Redshank *Tringa totanus* [Redshank] PLATE 38
Hongjiao yu

Description: Medium-sized (28 cm) wader with reddish-orange legs and red basal half of bill. Upperparts brownish-grey; underparts white, streaked brown on chest. Smaller, stouter and with shorter, thicker bill than Spotted Redshank, with more red on base of bill. In flight, white rump conspicuous and secondaries give conspicuous white trailing edge to wing. Tail finely barred black and white. Iris—brown; bill—red at base, black at tip; feet—orange-red.

Voice: Very noisy. Musical whistle *teu hu-hu* dropping in pitch, given in flight, or single *teyuu* on ground.

Range: Breeds in Africa and Palearctic; migrates S in winter as far as Sulawesi, Timor and Australia.

Distribution and status: Common. According to Cheng (1994), the nominate race breeds in NW China, the Qinghai–Xizang plateau and E Nei Mongol. Large flocks pass S and E with wintering populations remaining in the Changjiang valley and southern provinces, Hainan and Taiwan. Hale (1971), however, recognises several subspecies in China: *ussuriensis* migrates across China, *terrignotae* occurs in NE and E China, *craggi* breeds in NW Xinjiang and *eurhinus* ranges into W China.

Habits: Frequent mudbanks, beaches, salt pans, dried-up swamps and fish ponds, paddy fields near the sea or occasionally well inland. Generally in small parties associating with other waders.

344. Marsh Sandpiper *Tringa stagnatilis* Ze yu PLATE 38

Description: Medium-sized (23 cm) delicate sandpiper with white forehead and very thin, straight black bill and long greenish legs. Wings and tail blackish and has pale supercilium. Upperparts greyish-brown with white rump and lower back; underparts white. Distinguished from Common Greenshank by smaller size, paler forehead, proportionally longer thinner legs and fine, straighter bill. Iris—brown; bill—black; feet—greenish

Voice: Song is repeated *tu-ee-u*. Commonest calls in winter are repeated *kiu* like Common Greenshank but higher-pitched or repeated *yup-yup-yup* on being flushed.

Range: Breeds in Palearctic but migrates S in winter to Africa, S and SE Asia and as far as Australia and New Zealand.

Distribution and status: Fairly common. Breeds in Hulun Nur area of NE Nei Mongol. Passage migrant through Tianshan and down E coast, Hainan and Taiwan. Occasionally through centre of country.

Habits: Frequents lakes, salt pans, marshes, pools and occasionally coastal mudflats. Generally alone or in twos and threes but can form large flocks in winter. Rather shy.

345. Common Greenshank *Tringa nebularia* [Greater/Eurasian/ PLATE 38
The Greenshank] Qingjiaoyu

Description: Medium-sized (32 cm), greyish wader with stout greenish legs and stout, slightly upturned long grey bill. Standing: upperparts appear greyish-brown with mottled texture, blackish wing tips and barred tail; underparts white with brown streaking on throat, breast and flanks. Distinctive long white back, particularly conspicuous in

flight. Underwing streaked dark (white in Nordmann's Greenshank). Separated from Marsh Sandpiper by larger size, stockier shape, proportionally shorter legs and distinctive call. Iris—brown; bill—grey with black tip; feet—yellowish-green.
Voice: Noisy. Loud, ringing *chew chew chew.*
Range: Breeds in Palearctic from Britain to Siberia; wintering S to Africa, Indian subcontinent, SE Asia, and Malaysia to Australia.
Distribution and status: Common winter visitor. Birds on passage across China with large flocks wintering in S Xizang and all of China S of Changjiang river including Taiwan and Hainan.
Habits: Frequents coasts and inland marshes and mudflats on large rivers. Generally alone or in twos or threes. Forages with sideways sweeps of bill in water. Bobs head nervously up and down.

346. Nordmann's Greenshank *Tringa guttifer* [Spotted   PLATE 38
Greenshank] Xiao qingjiaoyu
Description: Medium-sized (31 cm) grey wader with yellowish legs and bicoloured bill. Very similar to Common Greenshank and separated with difficulty by larger head; shorter thicker neck; yellow base to stouter, blunter bill; paler upperparts with more scaling and less streaking (in winter), paler barring on tail; shorter, thicker and yellower legs; feet less trailing in flight and by call. In the hand, shows webbing between all three toes compared with between two toes only in Common Greenshank. Iris—brown; bill—black with yellow base; legs and feet—yellow/green.
Voice: Harsh *gwark* very different from Common Greenshank.
Range: Breeds N Sakhalin Island in NE Asia and migrates S in winter through Japan and China to Bangladesh and SE Asia; vagrants reach Borneo and Philippines.
Distribution and status: Globally Endangered (Collar *et al.* 1994). Very rare. Passage migrant through E China coastal provinces. Small numbers reported each spring in Hong Kong.
Habits: As Common Greenshank; prefers mudflats.

347. Lesser Yellowlegs *Tringa flavipes* Xiao huangjiaoyu   PLATE 38
Description: Medium-sized (23 cm) greyish-brown-backed wader with straight bill and conspicuous yellow legs; smaller and more slender than Common Redshank. The white rump patch in flight is cut-off square above the tail coverts, not wedge-shaped as in Redshank or Greenshank. Iris—brown; bill—black; legs and feet—yellow.
Voice: Single or double notes like a Common Redshank but quieter. Also has *tuk-tuk-tuk* alarm call and continuous *pill-e-wee, pill-e-wee, pill-e-wee* yodelling flight call.
Range: Breeds in Alaska and Canada, migrating south in winter to S America.
Distribution and status: Regularly recorded in Japan and probably overlooked in China. One Hong Kong record.
Habits: A lively and graceful wader, walking on strongly flexed legs.
Note: Greater Yellowlegs *T. melanoleuca* has reached Japan and could reach China—larger, thicker more upturned bill and heavier streaking on underparts in summer.

348. Green Sandpiper *Tringa ochropus* Baiyao caoyu   PLATE 38

Description: Medium-sized (23 cm), stocky, dark greenish-brown wader with white belly and vent In flight, black underwing and white rump and barred tail characteristic. Upperparts greenish-brown with white flecks; wings and lower back almost black; tail white with black terminal bars. Feet extend beyond tail in flight. Appears very black and white in field. Distinguished from Wood Sandpiper by shorter greenish legs, stockier appearance and less spotted underparts and darker underwing. Iris—brown; bill—dusky-olive; feet—olive-green

Voice: Loud, liquid *tlooeet-ooeet-ooeet* with the second note more drawn out.

Range: Breeds in N Palearctic but migrates S in winter as far as Africa, Indian subcontinent, SE Asia, N Borneo and Philippines.

Distribution and status: Breeding reported only in Kashi and Tianshan regions of W Xinjiang, but a common passage migrant over most of China and wintering populations in Tarim Basin, Yarlung Zangbo valley of S Xizang, most E China provinces, Changjiang valley and all China S of 30 °N. Rarely coastal.

Habits: Usually solitary, frequenting small pools and ponds, marshes and ditches. When surprised it flies off with a zig-zag, snipe-like flight.

349. Wood Sandpiper *Tringa glareola* Lin yu   PLATE 38

Description: Medium-sized (20 cm), slim, brownish-grey wader with whitish belly and vent and white rump. Upperparts greyish-brown, rather spotted; long white eyebrow; white tail barred brown. In flight, barred tail, white rump and underwing and lack of wing bar characteristic; also feet extend well beyond tail. Distinguished from Green Sandpiper by longer, yellower legs, pale underwing, longer supercilium and slimmer appearance. Iris—brown; bill—black; feet—yellowish to olive-green.

Voice: High-pitched whistle *chee-chee-chee* or in alarm *chiff-iff-iff*, less ringing than Common Greenshank.

Range: Breeds in N Palearctic but migrates S in winter to Africa, Indian subcontinent, SE Asia and as far as Australia.

Distribution and status: Breeds in Heilongjiang and E Nei Mongol. Common passage migrant across all of China. Wintering populations remain on Hainan, Taiwan, Guangdong and Hong Kong; occasionally also Hebei and probably E coast.

Habits: Prefers muddy coastal habitats but also occurs well inland in paddy fields and fresh water swamps up to 750 m. Generally in small loose flocks of up to 20 and occasionally mixes with other waders.

350. Terek Sandpiper *Xenus cinereus* [Avocet Sandpiper]   PLATE 38
Qiaozui yu

Description: Medium-sized (23 cm), stocky, grey wader with long, upturned bill. Upperparts grey with indistint white anterior eyebrow; black primaries conspicuous; in breeding plumage has black scapular line; belly and vent white. Narrow white trailing edge to wing conspicuous in flight. Iris—brown; bill—black with yellow base; feet—orange.

Voice: Soft, pleasant whistles, *hu hu hu* or sharper trill *tee-tee-tee* or *tit-ter-tee*.

Range: Breeds in N Palearctic but migrates S in autumn as far as Australia and New Zealand.

Distribution and status: Common passage migrant through E and W China. Some non-breeding birds stay all summer in south China.

Habits: Frequents coastal mudflats, creeks and estuaries, mixing with other waders to feed but flying separately. Generally solitary or in ones and twos, occasionally in larger flocks.

351. Common Sandpiper *Actitis hypoleucos* Ji yu PLATE 38

Description: Smallish (20 cm) brown and white, short-billed, restless wader with tail extending well beyond wings. Upperparts brown, flight feathers blackish; underparts white with a brown-grey patch on side of breast. In flight, white wing bar, lack of white rump and white barred outer tail feathers are distinctive. Underwing barred black and white. Iris—brown; bill—dark grey; feet—pale olive-green.

Voice: Thin, high-pitched piping *twee-wee-wee-wee*.

Range: Breeds across Palearctic and in Himalayas but migrates S in winter to Africa, Indian subcontinent, SE Asia and as far as Australia.

Distribution and status: Common. Breeds in NW, NC and NE China; migrating S with wintering populations along coasts, rivers and wetlands of lowland areas S of 32 °N.

Habits: Frequents a wide range of habitats from coastal mudflats and sand-bars to upland paddy fields up to 1500 m and along streams and riversides. Walks with incessant bobbing gait and also flies in a characteristic manner with stiff-winged glides.

352. Grey-tailed Tattler *Heteroscelus brevipes* [Siberian/Polynesian/ PLATE 37
Grey-rumped Tattler] Huiwei [piao]yu

Description: Medium-sized (25 cm), stocky, cold grey wader with stout, straight bill conspicuous black eye-stripe and white eyebrow and short, bright yellow legs. Chin whitish; plumage unmarked grey on upperparts; pale grey on chest with white abdomen. Rump is finely barred. Underwing dark in flight. Iris—brown; bill—black; feet—yellowish.

Voice: Sharp double whistle *too-weet* or soft trill.

Range: Breeds in Siberia but migrates S in winter to Malaysia, Australia and New Zealand.

Distribution and status: Fairly common passage migrant over E China with some wintering birds remaining on Taiwan and Hainan.

Habits: Frequents rocky beaches, coral banks and sandy or shingle beaches rather than mudflats. Generally solitary or in small groups. Does not mix with other waders. Runs about in a characteristic crouched manner with tail high.

Note: Sometimes treated as a race of Wandering Tattler *T. incana* (e.g. Cheng 1987).

353. Wandering Tattler *Heteroscelus incanus* Piao yu PLATE 37

Description: Medium-sized (28 cm) grey wader with dull yellow legs and straight bill. Plumage in winter rather uniform grey with white supercilium, chin, throat, belly and undertail coverts. Very similar to Grey-tailed Tattler, but with shorter, darker bill, darker underwing, slightly longer wing extends beyond tail. Best distinguished by voice. Iris—brown; bill—grey; feet—dull yellow.

Voice: Rapid series of clear whistles, *pew, tu, tu, tu, tu, tu* on one pitch.

Range: Breeds Alaska and NW USA. Winters S down W coast of USA to Philippines, Australia and S Pacific islands.

Distribution and status: Rare in China. Occasionally winters in Taiwan. Probably often overlooked or confused for Grey-tailed Tattler.

Habits: Uses beaches but mostly found on rocky coasts. Teeters and bobs continuously whilst feeding.

354. Ruddy Turnstone *Arenaria interpres* [Turnstone] Fanshi yu   PLATE 38

Description: Medium-sized (23 cm) unmistakable short-billed wader with distinctive short, bright orange legs and feet. Complex pattern of black, brown and white on head and breast distinctive. Bill shape is diagnostic. In flight from above boldly patterned black and white. Iris—brown; bill—black; feet—orange.

Voice: Metallic staccato rattle *trik-tuk-tuk-tuk* or ringing *kee-oo*.

Range: Breeds Holarctic in northern latitudes; migrating S in winter to S America, Africa, and tropical Asia to Australasia and New Zealand.

Distribution and status: A fairly common passage migrant through E China with some wintering birds remaining in Taiwan, and in Fujian and Guangdong. Some non-breeding birds summer in Hainan.

Habits: Frequents mudflats, sandy shores and rocky reeflets in small flocks. Sometimes feeds inland or close to the sea in open areas. Generally does not mix with other species. Turns over stones and other items on beach to look for crustaceans underneath. Runs actively.

355. Long-billed Dowitcher *Limnodromus scolopaceus*   PLATE 37
Changzui yu

Description: Largish (30 cm) grey wader with long straight bill. Similar to Asian Dowitcher but distinguished by smaller size, paler legs and bill, and in flight by unbarred white wedge on back and more prominent white trailing edge to secondaries, whitish wing lining barred black. Distinguished from Short-billed Dowitcher with difficulty by darker and more barred tail and call. Iris—brown; bill—yellowish with dark tip; legs—greenish-grey.

Voice: Short, high-pitched *kreek* given singly or in short series; similar to Wood Sandpiper.

Range: Breeds NE Siberia and NW Nearctic; winters in N America.

Distribution and status: Single birds have been recorded wintering in Hong Kong.

Habits: A bird of marshes and coastal mudflats.

Note: Short-billed Dowitcher *L. griseus* has been recorded in Japan and could occur in China. Distinguished from Long-billed by darker axillaries and underwing coverts but best separated by call. Call is mellow *too-du-du*.

356. Asian Dowitcher *Limnodromus semipalmatus* [Asiatic/   PLATE 37
Snipe-billed Dowitcher AG: Godwit] Banpu yu

Description: Large (35 cm) grey wader with long straight bill. Back grey; rump, lower back and tail white with fine black bars, underparts pale with buffy-brown on chest. Distinguished from godwits by smaller size and straight all-black bill, swollen at tip.

Larger than other dowitchers with darker legs and darker back in flight. Iris—brown; bill—black; legs—blackish.

Voice: Plaintive yelping *chep chep* and occasionally gives a quiet plaintive *miau*.

Range: Breeds S Russia, Mongolia, NE China and SE Siberia; migrant E Indian subcontinent, SE Asia, Philippines and Indonesia to N Australia.

Distribution and status: Globally Near-threatened (Collar *et al.* 1994). Breeds in Qiqihar region of Heilongjiang, Xianghai in Jilin and Hulun Nur in E Nei Mongol. Passage migrants pass through E and S China.

Habits: Inhabits mudflats where it is recognised by its characteristic feeding action walking forward stiffly, rocking to plunge its bill deep in the mud at each step, like a mechanical toy.

357. Great Knot *Calidris tenuirostris* Da binyu   PLATE 39

Description: Largish (27 cm) longish-billed greyish wader. Larger than Red Knot, with longer, thicker bill slightly decurved at tip; upperparts darker with faint streaking; crown streaked; chest and sides spotted black even in non-breeding plumage (looks like dark breast band at a distance); rump and wing bar white. Spring and summer birds show large black spots on breast and rufous wing bar. Iris—brown; bill—black; feet—greenish-grey.

Voice: Low *chucker-chucker-chucker* call or double-noted low whistle *nyut-nyut*.

Range: Breeds in NE Siberia but migrates S in winter to Indian subcontinent, SE Asia, Philippines and as far as Australia.

Distribution and status: Fairly common migrant on E coast. Wintering birds rarely stay in Hainan, Guangdong and Hong Kong.

Habits: Frequents tidal mudflats and sandbars, often in large flocks.

358. Red Knot *Calidris canutus* [Lesser/The Knot]   PLATE 39
Hongfu binyu

Description: Medium-sized (24 cm), thick-set, short-legged, greyish wader with a shortish, stout, dark, straight bill and pale eyebrow. Upperparts grey with faint scaling; underparts whitish, lightly buff on neck, chest and flanks. In flight shows narrow white wing bar and pale grey rump. Underparts rufous in summer. Iris—dark brown; bill—black; feet—yellowish-green.

Voice: Low-pitched throaty *knutt . . . knutt*, also a musical chatter when feeding.

Range: Breeds in the Arctic and migrates S in winter as far as S America, Africa, Indian subcontinent, Australia and New Zealand.

Distribution and status: Scarce to common migrant across E China. Small populations winter along coasts of Taiwan, Hainan, Guangdong and Hong Kong.

Habits: Frequents sand- and mudflats and estuaries. Generally very social in large compact flocks. Mixes with other waders; feeds with rapid stabbing of the bill, sometimes completely submerging the head to catch food.

359. Sanderling *Calidris alba* Sanzhi yu   PLATE 39

Description: Smallish (20 cm) greyish wader with conspicuous black 'shoulder'. Appears whiter than any other sandpiper with broad white wing bar in flight; tail dark down centre, white at sides. Lack of hind toe distinctive. Summer birds are rufous buff above. Iris—dark brown; bill—black; feet—black.

Voice: Squeaking flight note *cheep cheep cheep* or liquid *plit*.

Range: Holarctic. Breeds in northern latitudes and migrates S in winter as far as Australia and New Zealand.

Distribution and status: Fairly common winter visitor and migrant. Occasional on passage in W Xinjiang, S Xizang, NE China, Guizhou and Hainan with fair numbers wintering along S and SE coasts and S Taiwan.

Habits: Frequents sandy beaches along sea coats, rarely on mud. Generally runs along the water's edge following receding waves, picking at tiny food organisms washed on the beach. Sometimes solitary but normally gregarious.

Note: May be placed in genus *Crocethia*.

360. Western Sandpiper *Calidris mauri* Xifang binyu   PLATE 40

Description: Small (16 cm) stint-like sandpiper with sturdy black, lightly decurved bill and blackish legs. Breeding plumage is rufous with heavily streaked breast. Upperparts brownish-grey in winter with face and underparts white. Has dusky stripe through eye, leaving white supercilium and dusky streaks on side of upper breast. Paler and greyer than Dunlin; bill less decurved than Curlew Sandpiper; colour more uniform and darker than Least Sandpiper. Bill longer than other small dark-legged stints. Iris—brown; bill—black; feet—black.

Voice: Thin, high-pitched raspy *jeet* or *cheet*, longer than similar call of White-rumped Sandpiper.

Range: Breeds in E Siberia and Alaska; winters to Gulf of Mexico and down W coast of USA.

Distribution and status: Accidental on W coast of Taiwan. Recorded once from Beidaihe (Hebei). Recorded more than a dozen times in Japan and could reach China mainland more often.

Habits: Frequents coastal and inland wetlands.

361. Spoon-billed Sandpiper *Calidris pygmeus*   PLATE 40
[Spoonbill Sandpiper] Shaozui yu

Description: Small (15 cm) greyish-brown short-legged wader with streaked upperparts and prominent white eyebrow. The diagnostic spatulate bill is not easily seen in the field in side view. Closely resembles Red-necked Stint in winter but greyer with whiter forehead and breast. In summer upperparts and upper breast are rufous. Iris—brown; bill—black; feet—black.

Voice: Shrill, quiet *preep preep* on take-off or shrill *wheet*.

Range: Breeds in N Europe and Asia but migrates to Burma, S China and Hainan in winter with vagrants reaching SE Asia.

Distribution and status: Globally Vulnerable (Collar *et al.* 1994). Uncommon winter visitor and passage migrant. Recorded on passage along E coast and Taiwan, W Xinjiang and S Xizang. Some birds winter along coasts of Fujian and Guangdong.

Habits: A bird of sandflats feeding with bill almost vertically downwards, worked actively with a diagnostic sideways 'vacuum cleaning' motion.

362. Little Stint *Calidris minuta* Xiao binyu   PLATE 40

Description: Small (14 cm) greyish sandpiper with short stout bill and dark grey legs. Underparts white with grey wash on side of upper breast and indistinct dark band through eye; white supercilium. Very similar to Red-necked Stint but slightly longer legs and blunter tip to longer bill. Spring birds may be acquiring into rufous breeding plumage. Distinguished from breeding Red-necked Stint by white chin and throat, creamy white V on mantle and dark-spotted breast. Iris—brown; bill—black; feet—black.

Voice: Call is short, incisive, distinctive *stit*. Flight call is weak *pi, pi, pi*.

Range: Breeds N Europe and NW Asian tundra. Migrates to Africa, Middle East and Indian subcontinent.

Distribution and status: Rare passage migrant. First noticed in Hong Kong in 1986 but since seen annually in low numbers each spring. Recorded at Beidaihe (Hebei). Status elsewhere in China unknown.

Habits: Rather tame. Feeds with fast pecks or probes. Gregarious and mixes with other small waders.

363. Red-necked Stint *Calidris ruficollis* [Rufous-necked Stint,   PLATE 40
AG: Sandpiper] Hongjing binyu

Description: Small (15 cm) greyish-brown stint with black legs and pale, streaked upperparts. Winter: upperparts greyish-brown, mottled and streaked; eyebrow white; centre of rump and tail dark brown; sides of tail white; underparts white. Distinguished from Long-toed Stint by greyer, more uniform plumage and black legs. In spring and summer plumage, neck, crown and wing coverts are rufous. Distinguished from Little Stint by stouter bill, shorter legs and longer wings. Iris—brown; bill—black; feet—black.

Voice: Weak, whistling *chit-chit-chit*, slightly coarser and lower-pitched than Little Stint.

Range: Breeds in N Siberia but migrates S in winter to SE Asia and as far as Australia.

Distribution and status: Very common passage migrant cross E and C China. Some winter along the coasts of Hainan, Guangdong, Hong Kong and Taiwan.

Habits: Frequents coastal mudflats in large flocks of active birds, walking briskly or running about, picking up tiny food items and, when excited, bobbing up and down with a backward throw of the head. More of a coastal bird than Long-toed Stint.

364. Temminck's Stint *Calidris temminckii* Qingjiao binyu   PLATE 40

Description: Small (14 cm) stocky, short-legged, grey wader. Upperparts (in winter) dull uniform grey. Underparts: grey breast grading to whitish abdomen. Tail extends beyond closed wing. Distinguished from other stints by pure white outer tail feathers most easily seen as bird's alight, also by distinct call and greenish or yellowish legs. Summer plumage: brownish-grey breast and some rufous on wing coverts. Iris—brown; bill—black; legs and feet—greenish or yellowish.

Voice: Distinctive short rapid, cicada-like trill *tirrrrrrit . . .*

Range: Breeds in N Palearctic; migrant in winter to Africa, Middle East, Indian subcontinent, SE Asia, Philippines and Borneo.

Distribution and status: Regular but rare passage migrant over all China. Wintering populations remain in Taiwan, Fujian, Guangdong and Hong Kong.

Habits: As other stints, favouring mudflats and marshy areas, in small to large parties. Primarily a fresh water bird but also visits tidal creeks. Towers jerkily upwards when flushed and flies fast in tight, wheeling flocks Has rather horizontal stance.

365. Long-toed Stint *Calidris subminuta* Changzhi binyu   PLATE 40

Description: Small (14 cm) grey-brown stint with boldly black-streaked upperparts and greenish-yellow legs. Crown brown; has a conspicuous white eyebrow. Breast pale brownish-grey; abdomen white; centre of rump and tail dark brown; outer tail pale brown. Summer birds are more rufous-brown. In winter, separated from similar Red-necked Stint by pale legs and from Temminck's Stint by boldly marked upperparts. Shows indistinct wing bar in flight. Iris—dark brown; bill—black; feet—greenish-yellow.

Voice: Soft *prit* and a purring *chirrup*.

Range: Breeds in Siberia but migrates S in winter to Indian subcontinent, SE Asia, Philippines and as far as Australia.

Distribution and status: Uncommon migrant and winter visitor. Recorded on passage over most of E and C China. Winters on Taiwan and in Guangdong and Hong Kong.

Habits: Frequents coastal mudbanks, small ponds, paddy fields and other muddy areas. Occurs singly or in flocks, often mingles with other waders. Less shy than other waders, usually being the last to fly off when approached. Looks more upright than other stints.

[366. Least Sandpiper *Calidris minutilla* Ji binyu   PLATE 40

Description: Very small (14 cm) wader. Very similar to Long-toed Stint but feet yellow and less long-toed. Upperparts in breeding season less contrasting with narrower white V on scapulars and less distinctly striped crown. In winter, browner than dark-legged stints and more uniform above than Long-toed Stint with narrower dark feather centres on upperparts. In profile, has distinctive hunched posture. In flight shows white sides to rump and uppertail; white wingbar longer and more prominent than in Long-toed Stint. Iris—brown; bill—blackish; feet—dull yellow.

Voice: Variable. Shrill, rising *trreee* delivered in slow irregular series; lower *prrrt*.

Range: Breeds N America; winters S USA to S America.

Distribution and status: Vagrant to region. Several records of accidentals reaching Japan. Not yet recorded in China but can be expected.

Habits: Gregarious and feeds on open coasts and mudflats. Also uses inland wetlands. Rather tame.]

366a. White-rumped Sandpiper *Calidris fuscicollis*   PLATE 40
Baiyao binyu

Description: Small (17 cm) sandpiper with rufous-scaled mantle and rufous-streaked crown. Underparts white with bold arrow-head streaks on breast and flanks. In winter, loses rufous tone and streaks; streaking on underparts reduced and grey wash over upper breast. In flight, has complete white uppertail coverts (although rump is grey). Bill short and slightly decurved. Iris—brown; bill—brown with yellowish base to lower mandible; feet—grey.

Voice: Flight call is distinctive, thin squeaky *jeeeet* or *eeet* and short, clear *tit* or *teep.*

Range: Breeds N Alaska and N Canada; winters to S America.

Distribution and status: Vagrant. One doubtful record Hong Kong. One accepted record from Beidaihe, Hebei.

Habits: Uses coastal and inland wetland. Mixes with other waders. Flight strong.

[**367. Baird's Sandpiper** *Calidris bairdii* Heiyao binyu PLATE 40

Description: Small (19 cm) stint-like sandpiper with dark grey legs and slender, pointed bill. Long wings extend beyond tail when standing. Plumage buffy-brown, always with complete finely-streaked breast band. Juvenile, scalloped with whitish edges to feathers of upperparts. In flight has white wing bar; white uppertail coverts and grey tail with longer, blacker central tail feathers. Most migrants are in juvenile plumage. Wing tips project beyond tail when standing. Iris—brown; bill—black; feet—dark grey.

Voice: Call distinctive, low raspy trilling *preet.*

Range: Breeds in NE Siberia and Arctic N America. Migrates to S America.

Distribution and status: Not recorded in China but accidentals reach Japan and should be looked for in China.

Habits: Uses upper beaches and inland wetlands. Feeds briskly with rotary movement of tail in time with bobbing of head.]

368. Pectoral Sandpiper *Calidris melanotos* Banxiong binyu PLATE 39

Description: Medium-sized (22 cm) mottled, brown sandpiper with yellow legs, slightly decurved two-tone bill and heavily streaked breast. Has indistinct white supercilium and brownish cap. Streaking on breast stops abruptly at white belly. Breeding male has brackish breast. Juvenile has buffy wash on streaked breast. Winter birds less rufous. In flight, wings look dark with faint white wing bar and broad black centre to rump and uppertail. Bill longer than Sharp-tailed Sandpiper. Iris—brown; bill—yellow base black tip; feet—yellow.

Voice: Call is rich, low *churk* or *trrit.*

Range: Breeds Arctic Siberia and N America; wintering to S America, Australia and New Zealand.

Distribution and status: Rare passage migrant. First recorded in Hong Kong in 1986. Single birds seen occasionally since. Likely to be regular but overlooked. Single records from Beidaihe in Hebei, SE China and Taiwan.

Habits: Feeds on wet meadows, marshes and edges of ponds.

369. Sharp-tailed Sandpiper *Calidris acuminata* [Siberian PLATE 39
Pectoral Sandpiper] Jianwei binyu

Description: Smallish (19 cm) short-billed wader with rufous cap, pale eyebrow and buffy breast. Bold black streaks on underparts are characteristic. Belly white; tail centrally black, laterally white. Similar to winter Long-toed Stint but much more rufous cap.

Summer birds more rufous, generally brighter than Pectoral Sandpiper. Juvenile even more colourful. Iris—brown; bill—black; legs and feet—yellowish to green.

Voice: Soft *trrt* or *wheep*, sharp liquid twittering *whit-whit*, *whit-it-it* and soft grunting.

Range: Breeds in Siberia but migrates S in winter to New Guinea, Australia and New Zealand.

Distribution and status: Fairly common passage migrant. Recorded over NE China, coastal provinces and Yunnan. Some birds recorded in winter on Taiwan (including Lanyu islet).

Habits: Frequents marshes and mudflats, swamps, lakes and paddy fields. Mixes freely with other waders.

370. **Rock Sandpiper** *Calidris ptilocnemis* Yan binyu   PLATE 39

Description: Non-breeding adult: smallish (21 cm) dumpy-looking, ash-grey sandpiper with yellow legs and yellow-based and yellow-based bill. Lower back, central tail feathers and flight feathers black. Conspicuous white wing bar in flight. Short but distinct supercilium. Breast whitish mottled grey. In breeding season upperparts are rufous-brown with dark chestnut fringes to scapulars and tertials; blackish wash on breast. Iris—brown; bill—yellow with grey tip; feet—yellow.

Voice: Short *whit* or *tweet* and low twitters from feeding flocks.

Range: Breeds NE Siberia, Alaska and Aleutians; winters down W coast of USA, and regularly to Japan.

Distribution and status: Accidental. Recorded at Beidaihe (Hebei) on spring passage.

Habits: Uses exposed rocky coasts in winter, rarely on beaches. Gregarious and mixes with Ruddy Turnstone and other waders.

371. **Dunlin** *Calidris alpina* [Red-backed Sandpiper]   PLATE 39
Heifu binyu

Description: Small (19 cm) medium-billed, greyish wader with white eyebrow. Bill slightly decurved at tip. Tail black down centre, white at sides. Differs from Curlew Sandpiper by dark rump, shorter legs and darker breast. Distinguished from Broad-billed Sandpiper by longer legs and more uniform head with single coronal stripe. In summer black breast is diagnostic; rufous above. Iris—brown; bill—black; feet—greenish-grey.

Voice: Harsh, nasal whistle *dwee* in flight.

Range: Breeds in N Holarctic; winters S Rare migrant to SE Asia.

Distribution and status: Common passage migrant and winter visitor. Race *centralis* occurs on passage from NW and NE China to SE China. Race *sakhalina* recorded on passage across NE China and winters in S and SE coastal provinces and along major rivers south of the Changjiang, also on Taiwan and Hainan.

Habits: Frequents coastal and inland mudflats, singly or in small parties, often mixes with other waders. Feeds busily with a somewhat hunched posture.

372. **Curlew Sandpiper** *Calidris ferruginea* Wanzui binyu   PLATE 39

Description: Smallish (21 cm) wader with conspicuous white rump and long decurved black bill. Upperparts mostly grey with little streaking; underparts white; eyebrow, wing bar and bar across the uppertail coverts white. In summer breast and general plumage

deep rufous with white chin. White rump less conspicuous in breeding plumage. Iris—
brown; bill—black; feet—black.
Voice: Twittering calls. Plaintive *chew* or *wheep*, sharp *whit-whit*, *whit-it-it* and soft grunting.
Range: Breeds in N Siberia but migrates S in winter to Africa, Middle East, Indian subcontinent and as far as Australia.
Distribution and status: Uncommon passage migrant across all China with a few
wintering on Hainan and in Guangdong and Hong Kong. Uses inland salt lakes with
few coastal records N of Hong Kong.
Habits: Coastal mudflats and close to the sea in paddy fields and fish ponds. It is usually mixed in with other stints and sandpipers. Runs over mud as the tide falls, probing
and picking for food. Rests on sand spits on one leg and flies in fast, close flocks.

373. Stilt Sandpiper *Micropalama himantopus* Gaoqiao yu   PLATE 39

Description: Unmistakable largish (21 cm) sandpiper with very long legs and long,
slightly decurved bill. Legs longer and paler than in Curlew Sandpiper. In flight has
white lower rump and uppertail as in Curlew Sandpiper but almost no wingbar. Feet
extend beyond tail in flight. Breeding plumage distinct with rufous nape and mask and
underparts barred black and white. Winter plumage is rather uniform grey above with
grey streaking on underparts extending well down breast and flanks. Iris—brown; bill—
black; feet—yellowish-green.
Voice: Flight call is rattling trill *Kirrr*, also clearer *whu*.
Range: Breeds Alaska and N Canada; winters to S America; vagrant to Japan.
Distribution and status: Accidental on N coast of Taiwan.
Habits: Feeds belly-deep in water with other long-billed waders, with rapid probing.
Note: Placed in genus *Calidris* by many authors.

374. Buff-Breasted Sandpiper *Tryngites subruficollis*   PLATE 40
Huangxiong yu

Description: Medium-sized (14 cm) brownish sandpiper with buffy head, neck and
breast diagnostic in winter. Head has brownish crown but is otherwise unmarked buff
giving bland expression. Chin and vent white. Virtually no white markings on upperside
in flight. Underwing white with dark leading and trailing edges and diagnostic black
crescent patch on primary coverts. Similar to juvenile Ruff but distinguished by brighter
legs, bland expression, and significant smaller size. Flight very graceful like Ruff. Iris—
brown; bill—dark brown; feet—bright yellow-ochre.
Voice: Usually silent. Sometimes gives low growling flight call *pr-r-r-reet*.
Range: Breeds Alaska and N Canada; wintering to S America.
Distribution and status: Accidental. Recorded on E coast of Taiwan. Reported inter-
mittently on Japan and probably reaches Chinese mainland occasionally.
Habits: Uses inland wetlands.

375. Broad-billed Sandpiper *Limicola falcinellus* Kuozui yu   PLATE 39

Description: Smallish (17 cm) sandpiper with decurved bill, often with conspicuous
black carpal patch and diagnostic double coronal stripe. Bill has slight kink giving it a

rather 'broken' look compared to smoothly decurved bill of Dunlin. Upperparts streaked grey-brown; underparts white with streaking on breast; rump and tail with black centre and white sides. Distinguished in winter from Dunlin by head pattern and short legs. Can be confused with Jack Snipe but bill less straight and has less conspicuous scapular stripes. Iris—brown; bill—black; feet—greenish-brown.

Voice: Dry trill *ch-r-r-reep.*

Range: Breeds in N Europe and N Siberia but migrates S to winter in tropics and as far as Australia.

Distribution and status: Fairly common winter visitor and passage migrant. Nominate race passes through W Xinjiang; *sibirica* passes down E coast to winter on Taiwan, Hainan and Guangdong coasts.

Habits: Frequents intertidal mudflats and sand banks and marshy areas where it is a quiet solitary bird. Probes with bill held vertically. Crouches when alarmed.

376. Ruff *Philomachus pugnax* [Reeve (female only)] Liusu yu   PLATE 36

Description: Winter: largish (male 28 cm, female 23 cm) short-billed, dull brownish, long-legged wader with small head, long neck and straight bill. Upperparts dark brown with pale scaling; throat pale buff; head and neck buff; underparts white, often lightly barred on the flanks. In flight the narrow white wing bar and white oval patches either side of the dark tail base are distinctive. Female noticeably smaller than male. Juvenile buff-coloured. Summer male is rufous or partly white with unmistakablefluffy ruff. Iris—brown; bill—brown with yellowish base or grey in winter; feet—variably yellow or green to orange-brown.

Voice: Low *chuck-chuck* but generally silent in winter grounds.

Range: Breeds in N Europe and Asia but migrates S in winter to Africa and S Asia, and vagrants as far as Australia.

Distribution and status: Uncommon non-breeding visitor. Recorded on passage in W Xinjiang, S Xizang and down E coast and Taiwan. Small numbers winter on Guangdong, Fujian and Hong Kong coasts.

Habits: Frequents marshy areas and mudflats where it mingles with other waders.

377. Red-necked Phalarope *Phalaropus lobatus* [Northern   PLATE 35
Phalarope] Hongjing banpuyu

Description: Very small (18 cm) fine-billed grey and white wader usually seen swimming on the sea. Crown and eye-patch black; upperparts grey with dark feather centres; underparts whitish; dark rump and broad white wing bar conspicuous in flight. Flight is fast and jerk. Summer plumage darker with white throat; rufous eyebrow behind eye extending down neck and forming bib on throat; golden-buff scapulars. Distinguished from stints by finer, longer bill and black eye-patch. Iris—brown; bill—black; feet—grey.

Voice: Single or repeated *chek.*

Range: Breeds Holarctic; winters in seas worldwide.

Distribution and status: Rare passage migrant, recorded inland but sometimes common in winter in coastal waters and bays off Hainan, Taiwan and Hong Kong.

Habits: Wintering flocks remain at sea, feeding on surface plankton. They are rather tame, approachable birds. Sometimes they come inland and feed on ponds or mudflats.

378. Red Phalarope *Phalaropus fulicaria* [Grey Phalarope]   PLATE 35
Hui banpuyu

Description: Small (21 cm) straight-billed grey wader. Very similar to Red-necked Phalarope but forecrown whiter, upperparts paler and more uniform and bill deeper and broader, sometimes with yellow base. In the hand, the lobes of the feet are yellow. Iris—brown; bill—black with yellow base; feet—grey.
Voice: Like Red-necked Phalarope.
Range: Breeds Arctic Ocean; winters mainly to seas off W Africa and Chile.
Distribution and status: Very rare inland in W Xinjiang (Tianshan) and Heilongjiang (Harbin). Uncommon winterer in coastal waters off Hong Kong, Taiwan and Shanghai.
Habits: As Red-necked Phalarope.

PAINTED-SNIPES—Family: Rostratulidae

A family consisting of two species of rather specialised snipe-like waders. Single Asian representative is boldly marked with white flash stripes on the head and shoulders and ornate wings with many bars, stripes and eye-like spots. The bill is elongated and slightly decurved. Females are larger and more colourful than males and are most active in defence of territories and mate with several males—a habit shared only with button-quails, jacanas and a few other birds that show sexually reversed roles. Birds nest on the ground in reed beds. The male incubates.

379. Greater Painted-snipe *Rostratula benghalensis* [Painted Snipe]   PLATE 34
Cai yu

Description: Smallish (25 cm) colourful plump snipe-like wader with short tail. Female: head and chest dark chestnut with white eye-patch and yellow central/median crown stripe; back and wings greenish with a white V-shaped marking on back and white band around shoulder to white underparts. Male: smaller and duller than female, more mottled, less buff, and wing coverts spotted golden; eye-patch yellow. Iris—red; bill—yellow; feet—yellowish.
Voice: Generally silent but female courting call is a deep hollow note, also soft purring notes.
Range: Africa, India to China and Japan, SE Asia, Philippines, Sundas and Australia.
Distribution and status: Locally common resident and seasonal migrant up to 900 m in suitable habitat. Nominate race breeds from S Liaoning and Hebei through E China to all of China S of the Changjiang River, Hainan and Taiwan. Northern populations winter S of Changjiang. Vagrant to S Xizang; must be resident in SE Xizang.
Habits: Inhabits swampy grassland and paddy fields. Bobs its tail up and down as it walks. In flight dangles its legs like a rail.

JACANAS—Family: Jacanidae

A small, pan-tropical family of medium-sized waterbirds, resembling rails in general appearance but having immensely elongated toes which facilitate walking over water lily leaves and other floating plants on fresh water lakes and ponds. Several species are polyandrous or have more than one cock mating with one hen—a generally unusual arrangement among birds. Two species occur in China.

380. Pheasant-tailed Jacana *Hydrophasianus chirurgus* Shuizhi PLATE 34

Description: Largish (33 cm) long-tailed, dark brown and white jacana. White wings distinctive in flight. In non-breeding plumage crown, back and chest bar greyish-brown; chin, foreneck, eyebrow, throat and abdomen white; wings whitish. Has a black stripe through the eye and down side of neck and a golden patch on the lower nape. Has peculiar pointed outer primaries with elongated tips. Iris—yellow; bill—yellow/slaty-blue (breeding); feet—grey-brown/bluish (breeding).
Voice: Loud nasal mewing in alarm.
Range: Indian subcontinent to China, SE Asia; migrating S as far as Philippines and Greater Sundas.
Distribution and status: Formerly common seasonal migrant. Now rather rare due to lack of quiet habitat. Breeds through all southern China S of 32 °N including Taiwan, Hainan and SE Xizang. Some birds winter in Taiwan and Hainan.
Habits: Walks on top of floating vegetation, often on water lily and lotus leaves in small ponds and lakes. Picks for food and makes short fluttering flights to new feeding spots.

381. Bronze-winged Jacana *Metopidius indicus* Tongchi shuizhi PLATE 34

Description: Medium-sized (29 cm) brown and black jacana with a bold white eyebrow. Head, neck and underparts black with green gloss. Upperparts olive bronze; tail chestnut; forehead chestnut; eyebrow white; young birds have brown crown and some white on chest. Iris—brown; bill—green with red base and yellow tip; feet—dull green.
Voice: Loud piping alarm calls and low guttural notes.
Range: India, S China, SE Asia, Sumatra and Java.
Distribution and status: Rare resident in S Yunnan (Xishuangbanna).
Habits: Like other jacanas. Shy and rarely seen.

THICK-KNEES—Family: Burhinidae

A small family of large waders with long legs and cryptic coloration. Live in open, stony land or on beaches. Mainly crepuscular in habits, sitting tight by day. Fly with legs trailing.

382. Eurasian Thick-knee *Burhinus oedicnemus* [Stone Curlew] Shixing

382

Description: Large (41 cm) fulvous plover-like bird with big staring, yellow eyes and rather hunched posture. White wing bar outlined with brown above and black below; flight feathers black on closed wing but with two white panels in flight. Iris—yellow; bill—black with yellow base; feet—yellow.
Voice: Sharp, quick, whistled screams *pick-pick-pick-pick-pick*; ending with slower *pick-wick, pick-wick,*

382. Eurasian Thick-knee
Burhinus oedicnemus
石鸻

◼ Resident   ▨ Winter
• Vagrant/accidental

accented on second note, reminiscent of Curlew. Also single whistling scream, slowly repeated.

Range: S Europe, N Africa and Middle East to C Asia.

Distribution and status: Very rare in China. Nominate race resident up to 1000 m in extreme SE Xizang and vagrant recorded on Guangdong coast.

Habits: Cursorial. Lives in open, dry, stony areas with bushy cover. Sometimes in small groups. Rests through day; active at dusk and by night. Freezes with head flat on ground.

383. Great Thick-knee *Esacus recurvirostris* [Great Stone Plover/ Great Stone Curlew] Da shixing

383

383. Great Thick-knee
Esacus recurvirostris
大石鸻

◼ Resident   ▨ Winter

Description: Large (52 cm) big-headed wader with heavy, slightly upturned bill and staring yellow eyes. Diagnostic black and white pattern on head and bold black and white bar across wing. Primaries and secondaries black with bold white patches in flight. Iris—yellow; bill—black with yellow markings on base; feet—dull yellow.

Voice: Loud hiss in threat; loud creaking alarm note, *see-eek*; wild wailing whistle of two or more syllables with upward inflection, often given at night.

Range: Iran, S Pakistan, India, Sri Lanka, Burma; wintering birds extending to SW China, Hainan and SE Asia.

Distribution and status: Rare in China. Recorded on Hainan and in SW and S Yunnan (Tengchong, Luxi, Xishuangbanna).

Habits: Lives in pairs on sand and gravel bars of large rivers and sea beaches.

Note: Forms species pair with Beach Thick-knee *E. neglectus* and may well be conspecific.

OYSTERCATCHERS, STILTS AND PLOVERS—Family: Charadriidae

OYSTERCATCHERS AND STILTS—Subfamily: Recurvirostrinae

Small subfamily of black and white waders with generally long legs and bills. Four diverse species occur in China. Oystercatchers have special chisel-shaped bills for opening bivalve shells. The stilts have very long legs and bills.

OYSTERCATCHERS—Tribe: Haematopodini

384. Eurasian Oystercatcher *Haematopus ostralegus* Li yu

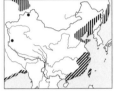

384. Eurasian Oystercatcher
Haematopus ostralegus
蛎鹬

▨ Winter   ⦀ Summer breeder

• Vagrant/accidental

Description: Medium-sized (44 cm) unmistakable black and white wader with straight, blunt-tipped, long red bill and pink legs. Upper back, head and breast black; lower back and uppertail coverts white. Rest of underparts white. Upperwing black with broad white band across base of secondaries. Underwing white with narrow black trailing edge. Iris—red; bill—orange-red; feet—pink.

Voice: Sharp *Kleep* contact call; more drawn-out *Kle-eap*; sharper *Kip*. In display, gives piping call which slows and dies off.

Range: Europe to Siberia; winters to south.

Distribution and status: Uncommon seasonal migrant. Race *osculans* breeds in NE coastal provinces and Shandong with wintering birds along S and SE coasts and Taiwan. Vagrant in Tianshan (W Xinjiang) and W Xizang.

Habits: Slow, deep-flapping flight. Feeds along rocky coasts on molluscs which it detaches with its chisel-like bill. Forms small flocks.

STILTS—Tribe: Recurvirostrini

385. Ibisbill *Ibidorhyncha struthersii* Huan zuiyu

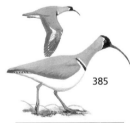

385. Ibisbill
Ibidorhyncha struthersii
鹮嘴鹬

▨ Resident   ▨ Winter

Description: Unmistakable, large (40 cm) grey, black and white wader with diagnostic red legs and red, long, decurved bill. Has black and white gorget separating grey upper breast from white lower parts. Underwing white. Upperwing has large white panel in centre of wing. Juvenile has buffy scaling on upperparts, black pattern less clearly demarcated and pinkish legs and bill. Iris—brown; bill—crimson; feet—crimson.

Voice: Repeated ringing *Klew-klew* like a sandpiper; also rapid loud whimbrel-like *tee-tee-tee-tee*.

Range: Himalayas and SC Asia.

Distribution and status: Rare and local resident and altitudinal migrant. Resident in W Xinjiang, W S and E Xizang, Qinghai, Gansu, Sichuan, Ningxia, Shaanxi, Hebei, Henan and N Yunnan. Vagrant in Xishuangbanna (S Yunnan).

Habits: Lives along stony, fast-flowing rivers at 1700–4400 m. Display posture is crouched with head extended forward and rear of black cap raised.

386. Black-winged Stilt *Himantopus himantopus* [Pied/ Common Stilt] Heichi changjiaoyu

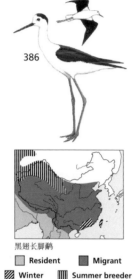

386

黑翅长脚鹬

	Resident		Migrant
	Winter		Summer breeder

Description: Tall, elongate (37 cm) pied wader with distinctive thin black bill, black wings, long red legs and white plumage distinctive. Has black patch on back of neck. Juvenile browner with grey wash on crown and back of neck. Iris—pink; bill—black; legs and feet—pinkish-red.

Voice: High-pitched piping calls and tern-like *kik-kik-kik*.

Range: India, China and SE Asia.

Distribution and status: Nominate race is uncommon seasonal migrant. Breeds in W Xinjiang, E Qinghai and NW Nei Mongol. Passage birds recorded over rest of country and wintering populations in Taiwan, Guangdong and Hong Kong.

Habits: Frequents shallow coastal and fresh water swamps.

387. Pied Avocet *Recurvirostra avosetta* Fan zuiyu

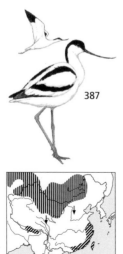

387

反嘴鹬

	On passage		Winter
			Summer breeder

Description: Unmistakable tall (43 cm) black and white wader with long grey legs and long, slender, markedly upturned black bill. From below in flight looks all white with black wing tips. Above has black wing bar and shoulder stripe. Iris—brown; bill—black; feet—black.

Voice: Makes frequent clear flutish calls *kluit, kluit, kluit...*

Range: Europe to China, India and S Africa.

Distribution and status: Breeds in N China but gather in large wintering flocks along the SE coast and through Xizang to India. Accidental in Taiwan.

Habits: Swishes bill from side to side whilst feeding. Swims well and upends. Flies with fast fluttery wingbeat but makes long glides. Adults give 'broken wing' display to distract predators from young.

PLOVERS—Subfamily: Charadriinae

A large worldwide subfamily of waders characterised by shortish, straight beaks with a hard swelling at the tip. The legs are long and powerful and most lack the hind toe. The wings are longish but the tails are short. Most plovers are patterned brown, black and white. They are birds of the water's edge or open spaces. There are 16 species in China, Four of which are only winter visitors or passage migrants.

388. Pacific Golden Plover *Pluvialis fulva* [Eastern/Asiatic/ PLATE 41
Asian Golden Plover] Jin [ban-] xing

Description: Medium-sized (25 cm) stout wader with a large head and a short heavy bill. In winter buffy golden-brown colour with paler eye-stripe, sides of face and underparts. Lacks white wing bar and wing lining shows no contrast in flight. Breeding male develops black face, throat, centre of breast and belly; white border to face and sides of breast. Female also acquires black on underparts but less extensive. Iris—brown; bill—black; legs—grey.

Voice: Abrupt clear shrill single or double whistle *chi-vit* or *tu-ee.*

Range: Breeds N Russia, N Siberia and NW Alaska; winters to E Africa, Indian subcontinent, SE Asia and Malaysia to Australia, New Zealand and Pacific.

Distribution and status: Migrates across all China. A common winter visitor to coastal and open areas of China, Hainan and Taiwan S of 25 °N.

Habits: Feeds singly or in flocks on mudflats, sandbars, open grassy areas, lawns and airports, especially near coast.

Note: Sometimes included in American Golden Plover *P. dominicus* (e.g. Cheng 1987).

388a. American Golden Plover *Pluvialis dominicus* PLATE 41
Meiguo jinxing

Description: Very similar to Pacific Golden Plover. Distinguished by slightly longer primary projection, bolder white supercilium, generally greyer plumage. In flight, feet do not extend beyond tail. Iris—brown; bill—black; legs—grey.

Voice: As Pacific Golden, a clear shrill single or double whistle *chu-weet*, second note higher.

Range: Breeds Alaska and N Canada; winters to S America. Vagrants reach Japan and China.

Distribution and status: Rare vagrant. Recorded Hong Kong. Probably usually overlooked among Pacific Golden Plovers.

Habits: As Pacific Golden Plover.

389. Grey Plover *Pluvialis squatarola* [Black-bellied Plover] PLATE 41
Hui banxing

Description: Medium-sized (28 cm) stout wader with a short heavy bill. Distinguishable in winter from Pacific Golden by larger size and bigger head and bill, by colour—brownish grey upperparts, whitish underparts, and in flight by whitish wing bar, rump and black axillaries appearing as black patch at base of otherwise white underwing. Breeding male has similar black underparts to Pacific Golden but has more silvery upperparts and white undertail. Iris—brown; bill—black; legs—grey.

Voice: Mournful three-noted whistle *chee-woo-ee*, slurred and both dipping and rising in pitch.

Range: Breeds N Holarctic; winters to tropical and subtropical coasts.

Distribution and status: Migrates across NE, E and C China. A common winter visitor to coasts and estuaries of S China, Hainan and Taiwan and the lower Changjiang River.

Habits: Feeds in small flocks on tidal mudflats and sandbars.

390. Common Ringed Plover *Charadrius hiaticula* [Ringed Plover] PLATE 41
Jianxing

Description: Medium-sized (19 cm) plumpish black, brown and white plover. Distinguished from Little Ringed Plover by larger size, lack of white fringe above black forecrown, orange legs and conspicuous white wing bar in flight. Immature has black markings of adult replaced with brown, all black bill and yellow legs. Iris—brown; bill—black with yellow base; feet—orange.

Voice: Mellow whistle *tu-weep*, second note on higher pitch.

Range: Breeds in Arctic Canada, Greenland and Palearctic, migrating to S Europe, Africa and Middle East.

Distribution and status: Straggler to NE China and occasional visitor to Hong Kong. Probably elsewhere but overlooked.

Habits: As other plovers.

391. Long-billed Plover *Charadrius placidus* [AG: Ringed Plover] PLATE 41
Changzui jianxing

Description: Largish (22 cm) robust black, brown and white plover. with longish, entirely black bill and longer tail than Common Ringed and Little Ringed Plovers. White wing bar less bold than in Common Ringed. Breeding plumage distinct with black frontal bar and complete breast band but eye-stripe is greyish-brown, not black. Immature stages as in Common Ringed and Little Ringed Plovers. Iris—brown; bill—black; legs and feet—dull yellow.

Voice: Loud, clear disyllabic piping, *piwee*.

Range: Breeds in NE Asia, E and C China; S in winter to SE Asia.

Distribution and status: Globally Near-threatened (Collar *et al.* 1994). Breeds in NE, C and E China. Wintering population remains S of 32 °N along coasts, rivers and lakes. Generally uncommon.

Habits: Similar to other plovers but preferring gravel areas along river edge and mudflats.

Note: Sometimes treated as a race of Common Ringed Plover (e.g. Cheng 1987).

392. Little Ringed Plover *Charadrius dubius* [Little Dotterel] PLATE 41
Jinkuang xing/Heiling xing

Description: Small (16 cm) black, grey and white, short-billed plover. Distinguished from Kentish and Malaysion Plovers by the full black or brown band across the chest and yellow legs. Distinguished from Common Ringed Plover by more conspicuous yellow eye-ring and lack of wing bar. In immature the black parts of the adults are brown. No white wing bar in flight. Tropical race *jerdoni* is slightly smaller. Iris—brown; bill—grey; legs—yellow.

Voice: Clear, soft, drawn-out, descending whistle *pee-oo* in flight.

Range: N Africa, Palearctic, SE Asia to New Guinea. Northern populations winter south.

Distribution and status: Race *curonicus* breeds over N, C and SE China migrating through eastern provinces and wintering along coasts and estuaries of S Yunnan, Hainan, Guangdong, Fujian and Taiwan. Race *jerdoni* breeds in S Xizang, S Sichuan and Yunnan, migrating further S in winter. Breeds in Hong Kong but race uncertain. Generally common.

Habits: Usually encountered along sandy banks of coastal streams and rivers, also marshes and mudflats; sometimes found far inland.

393. Kentish Plover *Charadrius alexandrinus* Huanjing xing   PLATE 41

Description: Small (15 cm) short-billed, brown and white plover. Distinguished from Little Ringed Plover by black legs, white wing bar in flight and whiter outer tail. Males have a black patch at the side of breast; in females this patch is brown. Race *dealbatus* has longer and heavier bill. Iris—brown; bill—black; legs—black.

Voice: A soft single unmusical rising note *pik*, repeated.

Range: USA, Africa and S Palearctic; winters south.

Distribution and status: Nominate race breeds in NW and NC China, wintering in Sichuan, Guizhou, NW Yunnan and SE Xizang. Race *dealbatus* (including *nihonensis*) breeds along the E and S coasts of China, including Hainan and Taiwan and also in Hebei; wintering in lower Changjiang valley and along coasts S of 32°N. Generally common.

Habits: Feeds singly or in small flocks, often mixed with other waders on beaches or sandy grassland areas near coast, also coastal rivers and marshes.

394. Malaysian Plover *Charadrius peronii* [Malay Plover,   NOT ILLUSTRATED
AG: Sand Plover] Malai shaxing

Description: Small (20 cm) black, brown and white, short-billed wader. Differs from Lesser and Greater Sand Plovers by smaller size and the narrow black (male) or rufous (female) collar. Differs from Kentish Plover by generally complete white rear collar and ear patches being separate rather than a continuous line through eye. Iris—brown; bill—black; legs—grey.

Voice: A quiet soft *kwik* similar to Kentish Plover.

Range: Malay Peninsula, Indochina, Philippines, Sulawesi and Sundas.

Distribution and status: Recorded for China without precise locality (Meyer de Schauensee 1984). The species occurs on many coral islands across the S China Sea area claimed by China.

Habits: Lives in pairs on sandy beaches. Prefers small bays of pure coralline sand. Does not form mixed flocks. Generally rather tame.

395. Lesser Sand Plover *Charadrius mongolus* [Mongolian Plover,   PLATE 41
AG: Sand-Plover/Dotterel] Menggu shaxing

Description: Medium-sized (20 cm) grey, brown and white, shortbilled wader. Very similar to Greater Sand Plover with which it often mixes but distinguishable by smaller size and shorter, more gracile bill; also fainter white wing bar in flight. Early arrivals may show distinctive breeding plumage of broad rufous chest bar and black mask with fully black forehead in race *atrifrons* (including *shaferi*) which is commonest form in China. Iris—brown; bill—black; legs—dark grey.

Voice: Short quiet trill or a sharp *kip-ip*.

Range: Breeds C Asia to NE Asia, migrating to coasts of Africa, Indian subcontinent, SE Asia, Malaysia and Australia.

Distribution and status: Race *pamirensis* breeds on the sandy banks of lakes and rivers in Tianshan and Kashi regions of W Xinjiang. Race *atrifrons* breeds across the Qinghai–Xizang plateau. Both races winter well south. Race *mongolus* breeds in Siberia but migrates through E China. A few birds remain to winter along the S China coast. Race *stegmanni* winters in Taiwan. All races are quite common.

Habits: Found mixed with other waders on coastal mudflats and sands, sometimes in large flocks numbering several hundred.

396. Greater Sand Plover *Charadrius leschenaultii* [Geoffrey's   PLATE 41
Plover/Large/Great/Large-billed Sand Plover, AG: Sand Dotterel] Tiezui shaxing

Description: Medium-sized (23 cm) grey, brown and white, shortbilled plover. Distinguished from Lesser Sand Plover by larger size and longer, thicker bill and longer, yellower legs; from all other wintering plovers except Lesser Sand Plover by lack of chest bar or collar Breeding plumage is distinctive with rufous breast bar and black mask with white forehead. Iris—brown; bill—black; legs—yellowish-grey.

Voice: Low soft trilling *trrrt* on take-off.

Range: Breeds Turkey to Middle East, C Asia to Mongolia; winters to coasts of Africa, Indian subcontinent, SE Asia, Malaysia to Australia.

Distribution and status: Breeds in Tianshan and Kashi regions of W Xinjiang and north of the Yellow River bend in Nei Mongol; migrating across China with a few wintering along the coasts of Taiwan, Guangdong and Hong Kong.

Habits: Frequents coastal mudflats and sandy patches, mixing with other waders especially Lesser Sand Plover.

397. Caspian Plover *Charadrius asiaticus* Hongxiong xing   PLATE 41

Description: Medium-sized (20 cm) brown and white, short-billed wader. As Oriental but smaller, with greyer legs and with no range overlap. Iris—dark brown; bill—blackish; legs—grey-buff.

Voice: A loud sharp *kuwit*.

Range: Breeds from Caspian Sea to Tianshan Mountains of C Asia, migrating to Africa in winter.

Distribution and status: Breeds in Tianshan and Junggar Basin of NW Xinjiang. Rare.

Habits: Breeds in flat arid areas, feeds in short grassland and beside rivers and pools.

Note: Formerly treated as conspecific with Oriental Plover.

398. Oriental Plover *Charadrius veredus* Dongfang xing   PLATE 41

Description: Medium-sized (24 cm) brown and white, short-billed wader. Winter plumage: broad brown chest band, narrow beak and whitish face; Upperparts uniformly brown with no wing bar. In summer plumage, breast band is orange underlined black, with no black face mask. Distinguished from Pacific Golden, Lesser and Greater Sand Plovers by yellow or pinkish legs. Some older birds acquire some white on head. In flight underwing including axillaries light brown. Iris—hazel; bill—olive-brown; legs—yellow to pinkish.

Voice: A sharp piping whistle *kwink* and on rising in flight a loud repeated *chip-chip-chip*.

Range: Breeds in Mongolia and N China, migrating to Malaysia and Australia in winter.

Distribution and status: Breeds in arid steppes and muddy or stony flats in deserts around Hulun Lake in E Nei Mongol and in Liaoning. Migrates through E China but not common.

Habits: Feeds in grassy areas and beside rivers and marshes.

Note: Formerly treated as conspecific with Caspian Plover *C. asiaticus* (e.g. Cheng 1987).

399. Eurasian Dotterel *Charadrius morinellus* Xiaozui xing   PLATE 41

Description: Medium-sized (21 cm) distinctive plover. Broad white supercilium, meeting on nape, and narrow white breast band diagnostic. Breeding plumage with grey throat and upper breast separated by narrow black and white gorget from chestnut lower breast and flanks with black belly distinctive. Female is brighter than male. Winter plumage duller, with buffy abdomen but eyebrow and white gorget still clearly contrasting. Crown and back speckled with white. Profile in flight is deep-chested and long-winged. Iris—dark brown; bill—blackish; legs—yellowish.

Voice: Clear, rhythmic *weet-weeh* calls in alarm and deep rolling *brroot* in flight.

Range: Breeds in heaths of N Palearctic, migrating to Mediterranean and Arabian Gulf and Caspian Sea in winter.

Distribution and status: Rare. Breeds in NW Xinjiang in Dzungarian Ala Tau and Tarbagatai Mts. Recorded as a passage migrant in Tianshan, NE Nei Mongol and N Heilongjiang.

Habits: Inhabits barren mountain tops and mossy tundra.

400. Northern Lapwing *Vanellus vanellus* [Common/The Lapwing/   PLATE 36
Green Plover/Peewit] Fengtou maiji

Description: Largish (30 cm) black and white plover with long narrow black forward-curving crest. Upperparts glossy greenish-black; tail white with broad black subterminal bar; head with dark crown and black ear coverts with dirty white sides of head and throat; breast blackish; belly white. Iris—brown; bill—blackish; legs and feet—orange-brown.

Voice: Nasal, drawn-out *pee-wit*.

Range: Palearctic; S in winter to Indian subcontinent and northern SE Asia.

Distribution and status: Fairly common. Breeding over most of N China and wintering S of 32 °N.

Habits: Favours cultivated fields, rice stubble or short grass.

401. River Lapwing *Vanellus duvaucelli* Jüchi maiji   PLATE 36

Description: Medium-sized (30 cm) noisy black, white and grey plover. Black throat and cap with slender trailing crest contrast with greyish-brown sides of head, back and breast. Belly, rump and undertail white with black primaries, tail and patch in centre of belly. No wattle. Shan black patch at bend of wing in slow, laboured flight. Iris—brown; bill—black; feet—greenish.

Voice: Loud insistent *did, did, did*, sometimes followed by *did-did-do-weet, did-did-do-weet* on rising; resembling call of Red-wattled lapwing but distinctive.

Range: E Himalayas, NE India to SW China and SE Asia.

Distribution and status: Uncommon resident in SE Xizang, W and SW Yunnan and Hainan.

Habits: Inhabits sand banks and shingle beds along rivers. Has elaborate spinning courtship display.

402. Grey-headed Lapwing *Vanellus cinereus* Huitou maiji   PLATE 36

Description: Large (35 cm) noisy black, white and grey plover. Head and breast grey; mantle and back brown; wing tips, band across breast and centre of tail are black and the rest of hindwing, rump, tail and abdomen are white. Immature like adult but browner and lacks black chest band. Iris—brown; bill—yellow, tipped black; feet—yellow.

Voice: Loud plaintive *chee-it, chee-it* alarm call and sharp *kik* given in flight.

Range: Breeds in NE China and Japan; migrating in winter to NE Indian subcontinent, SE Asia and rarely Philippines.

Distribution and status: Globally Near-threatened (Collar *et al.* 1994). Breeds in NE provinces to Jiangsu and Fujian; migrating over E and C China and wintering in Yunnan and Guangdong and sometimes Taiwan. Generally uncommon.

Habits: Inhabits open areas near water, river flats, rice fields and marshes.

403. Red-Wattled Lapwing *Vanellus indicus* Rouchui maiji   PLATE 36

Description: Large (33 cm) black, white and brown plover with black head, throat and centre of breast. White patch on ear coverts. Mantle, wing coverts and back pale brown; wing tips, rear edge of tail and subterminal tail bar black; wing patches, base and tip of tail and rest of underparts white. A red wattle above the base of the bill gives the species its name. Iris—brown; bill—red, tipped black; feet—yellow.

Voice: Loud shrill alarm calls *did-he-do-it, pity-to-do-it* and a *ping* note.

Range: Arabian Gulf and Indian subcontinent to SW China and SE Asia.

Distribution and status: Race *atronuchalis* is locally common in W and SW Yunnan in dry open spaces.

Habits: Inhabits open spaces, farmland, rice fields, marshes and river flats. Conspicuous in flapping alarm display flight.

PRATINCOLES AND COURSERS—Family: Glareolidae

PRATINCOLES—Subfamily: Glareolinae

A small subfamily, found from Africa to Australia, of curious long-winged birds with powerful, arched, pointed bills. They are insectivorous, catching food on the wing like a swallow or running after it on the ground. Most species are migratory. Three species occur in the China region.

404. Collared Pratincole *Glareola pratincola* [Common Pratincole]   **PLATE 35**
Ling yanxing

Description: Medium-sized (25 cm) tern-like wader with short bill and forked white tail with black terminal band. Upperparts olive-brown. Chin and throat buff, bordered by black ring. White lid under eye. Dark underwing with dark chestnut axillaries and underwing coverts in breeding season. Iris—dark brown; bill—black with red base; feet—black.
Voice: Harsh tern-like calls, *kik* and *kirrik* in flight.
Range: Europe, N Africa, Middle East to C Asia; wintering to Africa.
Distribution and status: Strays to Hong Kong (Hopkin 1989).
Habits: Typical pratincole, living in open terrain and hawking after insects at dusk.
Note: Black-winged Pratincole *G. nordmanni* breeds in E Kasakhstan and can be expected in NW Xinjiang. Similar to Collared Pratincole but secondaries black above, underwing uniform's dark and nape washed chestnut.

405. Oriental Pratincole *Glareola maldivarum* [Eastern Collared/   **PLATE 35**
Large Indian Pratincole] Putong yanxing

Description: Medium-sized (23 cm) long-winged plover-like bird with a forked tail and black border around the buffy throat (less clear in winter migrants). Upperparts brown with olive gloss; wings blackish; uppertail coverts white; abdomen grey; undertail white; forked tail black with white base and outer edges. Iris—dark brown; bill—black with scarlet base; feet—dark brown.
Voice: Sharp, raucous, grating *tar-rak*.
Range: Breeds E Asia; migrant S in winter through Indonesia to Australia.
Distribution and status: Locally common. Nominate race breeds through N, NE and E China, Xinjiang and Hainan and resident on Taiwan. Recorded on passage over most of eastern China.
Habits: A graceful wader in small to large, noisy flocks, mixing with other waders in open areas, marshes and rice stubble. Runs well and bobs its head but also flies after insects in the air with graceful swallow-like flight. Often seen on airfields.

406. Small Pratincole *Glareola lactea* Hui yanxing   **PLATE 35**

Description: Small (18 cm) pale, swallow-like pratincole with black underwing coverts. Upperparts sandy grey with white rump; Primaries black; secondaries white with black tips; tail square-cut may look forked because of black wedge-shaped subterminal patch. Underparts white with buffy wash on breast. Iris—brown; bill—black with small red patch at base; feet—brownish-grey.

Voice: High-pitched, rolling *prrit* or *tirrit* in flight.
Range: E Afghanistan, India, Sri Lanka to SE Asia.
Distribution and status: Uncommon. Breeds in S and SW Yunnan, transient through Yunnan. Up to 750 m in Zangbo (Dibang) valley of SE Xizang.
Habits: Frequents sandbars and banks of large rivers. Flies at dusk, hawking with martins and bats.

SKUAS, GULLS, SKIMMERS, TERNS AND AUKS—Family: Laridae

SKUAS, GULLS AND TERNS—Subfamily: Larinae

SKUAS—Tribe: Stercorariini

A small, worldwide tribe of dark-backed seabirds, rather similar in appearance to gulls but some are characterised by elongated central tail feathers. They are notorious for their aggressive harrying of other seabirds, forcing them to drop or regurgitate their food. Formerly regarded as a family; Sibley and Monroe (1990) reassign the skuas as a tribe under Laridae. Five species occur in the seas around China; all are rare.

407. Brown Skua *Catharacta antarctica* Da zei'ou   PLATE 42

Description: Large (63 cm) heavily-built brown skua with dark brown cap and very short protruding central tail feathers. Broad white patch at base of primaries. Tail short and rounded. Differs from south Polar Skua in large size, broader-based wings, larger bill and having little contrast between upperparts and head; some lack dark head; also in overall more reddish tone to brown plumage and more streaky texture. Iris—brown; bill—blackish-grey; feet—blackish-grey.
Voice: Generally silent.
Range: Antarctic and southern oceans.
Distribution and status: Accidental. One record from N Taiwan.
Habits: Aggressive large skua. Raids breeding colonies of other seabirds, eggs and chicks. Harasses other birds to regurgitate food. Attacks intruders to nesting areas with dive-bombing flight.
Note: Sometimes considered conspecific with Great skua *C. skua* of N Atlantic. Great Skua regularly ranges to the Aleutian Is and could occasionally reach Chinese waters.

[408. South-Polar Skua *Catharacta maccormicki* [McCormick's Skua]   PLATE 42
Nanji zei'ou

Description: Largish (53 cm) dark brownish skua with small pointed projections to central tail feathers. Head and abdomen paler brown than wings. White 'flashes' at base of primaries conspicuous from below and above. Smaller and with smaller bill than Brown Skua. Stockier than jaegers with broad rounded wings and heavier body. Light morph lacks black cap. dark morph has at least some white on face. Non-breeding adults and immature darker and more streaky. Iris—dark brown; bill—black; feet—black.

Voice: Generally silent at sea.

Range: Breeds in Antarctic.

Distribution and status: Occasionally passes through S China Sea.

Habits: As other skuas, pirating regurgitated food from other seabirds. Has powerful hawk-like flight. Follows ships and perches on masts.]

409. Pomarine Jaeger *Stercorarius pomarinus* [Pomatorhine Jaeger,   PLATE 42
AG: Skua] Zhong zei'ou

Description: Largish (56 cm) dark skua with elongated spatulate central tail feathers. Occurs in two colour morphs. Light morph: top of head black, sides of head and back of neck yellowish; underparts white with sooty sides and chest bar; upperparts sooty chocolate with paler whitish-grey bases to primaries; central two tail feathers extend 5 cm in blunt, broad, twisted trailers. Dark morph has no white or yellow. Non-breeding adult resembles immature—paler, more mottled and with grey cap. Iris—dark; bill—black; feet—black.

Voice: Generally silent at sea.

Range: Breeds in Arctic; winter migrant to southern seas.

Distribution and status: Regularly seen in S China Sea. Recorded off S China coast, Hong Kong, S Jiangsu and an inland record from S Shanxi.

Habits: As Parasitic Jaeger but keeps more out to sea. Parasitic, pirates food from other seabirds.

410. Parasitic Jaeger *Stercorarius parasiticus* [Arctic Jaeger, AG: Skua]   PLATE 42
Duanwei zei'ou

Description: Small (45 cm) dark skua with elongated, central tail feathers. Light morph: top of head black; sides of head and nape collar yellow; underparts white with or without grey chest bar; upperparts dark, sooty chocolate except whitish bases to primaries giving conspicuous wing flash in flight. Dark morph: sooty brown all over except pale wing flash. Central tail feathers are elongated into sharp, pointed streamers distinct from the shorter blunt ones of the Pomarine Jaeger. Non-breeding adult: paler, more mottled and with grey cap. Smaller than Pomarine Jaeger with slimmer bill and narrower-based wings. Iris—dark; bill—black; feet—black.

Voice: Generally silent at sea.

Range: Breeds in Arctic; winter migrant to southern seas.

Distribution and status: Less common than Pomarine Jaeger; recorded in S China Sea and off Hong Kong; one record from N Taiwan.

Habits: Flies low over sea but attacks other feeding seabirds, twisting and turning with them until they drop or disgorge food. Sometimes also follows ships, feeding on discarded refuse.

411. Long-tailed Jaeger *Stercorarius longicaudus* [AG: Skua]   PLATE 42
Changwei zei'ou

Description: Largish (50 cm) dark skua with elongated, central tail feathers. Similar in both light and dark colour morphs to Parasitic Jaeger but smaller, more slender, more buoyant and with much longer central tail streamers (14–20 cm beyond tip of tail). Light morph does not have grey chest band. Dark morph is rare. Non-breeding adult duller and with tail streamers reduced. Juvenile has bolder black and white barring on vent than other jaegers. Differs from juvenile Parasitic Jaeger in having only two white primary shafts in upperwing. Iris—dark; bill—black; feet—black.

Voice: Generally silent at sea.

Range: Breeds in Arctic but is winter migrant to southern seas.

Distribution and status: Rare visitor to S China Sea. Sight records off Hong Kong. Regular passage migrant off E coast of China and Taiwan.

Habits: As other jaegers.

SKIMMERS—Tribe: Rynchopini

A small tribe of curious tern-like birds with long wings and a long bill with longer lower mandible used for fishing.

412. **Indian Skimmer** *Rynchops albicollis* Jianzui ou

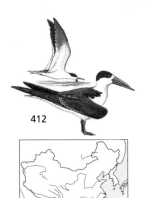

412

412. Indian Skimmer
Rynchops albicollis
剪嘴鸥

■ **Migrant** ▥ **Summer breeder**

Description: Large (42 cm) tern-like bird with large orange bill. Dark brown cap and upperparts contrasting with white underparts, neck collar and wing bar on secondaries, uppertail coverts and tail is diagnostic. Immature is brown, scaled white. Wings very long. Lower bill diagnostically longer than upper bill. Iris—brown; bill—orange-red with yellow tip; feet—red.

Voice: Shrill chattering scream.

Range: India, Burma, Indochina.

Distribution and status: Accidental. Occasionally reported off S China coast. Could occur on Zangbo (Dibang) River in SE Xizang.

Habits: Unique fishing technique, flying fast and low over water with lower bill lowered under surface.

GULLS—Tribe: Larini

A large, worldwide tribe of fish-eating and scavenging seabirds. Most species are white with black wing tips and varying degrees of black, grey and brown on the head and upperparts. Juveniles are mottled brown and usually take several years to acquire full adult plumage. Gulls are larger, more round-winged and heavier in flight than terns. Gulls are most common at temperate latitudes where major sea upwellings support rich pelagic fisheries. Eighteen species are recorded from China. Some are very difficult to identify and field identification is complicated by the immature plumages.

413. Black-tailed Gull *Larus crassirostris* Heiwei ou PLATE 43

Description: Medium-sized (47 cm) gull with long narrow wings; dark grey upperparts; white rump and sharply defined broad black subterminal band on otherwise white tail. Dark streaks on crown and nape in winter. Four small white spots on tips of closed wing. First-winter birds are washed brown with rather pale face and black-tipped pink bill. Has black tail and white uppertail coverts. Second-year birds as adult but with brown wing tips and more black on tail. Iris—yellow; bill—yellow with red tip and subterminal black ring; feet—greenish-yellow.
Voice: Plaintive mewing.
Range: Coasts of Japan and China Seas.
Distribution and status: Common. Breeds along Shandong and Fujian coasts. Winters along S and E coasts and Taiwan, also inland in Yunnan and on Changjiang River. Seasonal migrants recorded from Liaoning, Hebei, Shanxi and N Yunnan.
Habits: Breeds on rocky islets. Loosely colonial.

414. Mew Gull *Larus canus* [Common Gull] Hai'ou PLATE 43

Description: Medium-sized (45 cm) gull with legs and slender unmarked, greenish-yellow bill and white tail. Wing tips have white tips to primaries and large white mirrors. Wintering birds have sparse brown streaking on head and neck, sometimes has black tip to bill. First-winter birds have black subterminal tail band, heavy brown streaks on head, neck, breast and flanks and brown mottling on upperparts. Second-year birds like adult but browner head, blacker wing tips and smaller mirrors. Iris—yellow; bill—greenish-yellow; feet—greenish-yellow.
Voice: Calls are higher and weaker than Herring Gull, including *KaKaKa . . .*, screamed *Kleee-a* and persistent *Klee-uu- Klee-uu-Klee-uu . . .*
Range: Europe and Asia to Alaska and western N America.
Distribution and status: Fairly common. Race *kamtschatschensis* breeds N of China. Seen on migration over NE provinces. Winters on all coasts including Hainan and Taiwan; also on inland lakes and rivers over most of eastern and southern China. Race *heini* may also occur in China.
Habits: Colonial breeder on fresh water.

415. Glaucous-winged Gull *Larus glaucescens* Huichi ou PLATE 44

Description: Large (65 cm) grey-mantled gull with white tail, yellow bill and pink legs. Winter birds have slight brown streaking on back of head and nape. Primaries grey but not black and no obvious wing mirror. First-winter birds are uniformly pale buffy-brown with little contrast on underparts; hindneck noticeably whitish and heavy black bill. Much darker than first-winter Glaucous Gull. Iris—brown; bill—yellow with red spot; feet—pink.
Voice: Call is *kow-kow* or screaming *ka-ka-ako*.
Range: Breeds in Aleutians and coasts of Alaska and W Canada. Winters on W Coast of USA and Siberia.

Distribution and status: Vagrant. Recorded in spring off coast of Fujian. Since 1985 occasional first-winter birds recorded in Hong Kong but hybrids are possible.

Habits: Large aggressive gull, feeds along coasts and on rubbish tips.

416. Glaucous Gull *Larus hyperboreus* Beiji ou PLATE 44

Description: Large (71 cm) white-winged gull with pink legs and yellow bill. Looks robust and fierce. Back and wings pale grey. Much paler than any other Chinese gull. Winter adult has brown streaks on top of head, nape and sides of neck. Takes four years to again adult plumage. First-winter birds pale with milk coffee colour becoming paler by year. Bill is pink with dark tip. Iris—yellow; bill—yellow with red spot; feet—pink.

Voice: Similar to Herring Gull.

Range: Circumpolar. Breeds northern subarctic; wintering S of breeding areas to Florida, California, France, China and Japan.

Distribution and status: Uncommon winter visitor. Race *barrovionus* recorded off NE provinces, Hebei, Shangdong, Jiangsu and Guangdong. Irregular records of first-winter birds in Hong Kong.

Habits: Solitary and colonial breeder. Gregarious. Feeds along coastline and raids rubbish tips.

417. Slaty-backed Gull *Larus schistisagus* Huibei ou PLATE 44

Description: Large (61 cm) dark grey-backed gull with pink legs and yellow bill with red spot. Similar to Herring Gull complex but upperparts darker grey and legs deeper pink. Also has broader white scapular crescent. Winter adults have brown streaks on back of head and neck. First-winter birds are darker than most Herring Gull complex and have completely dark brown tail. Iris—yellow; bill—yellow with red spot; feet—deep pink.

Voice: Similar to Herring Gull.

Range: Breeds on E coast of Siberia and N Japan. Ranges through NW Pacific and winters along coasts of Japan and Korea.

Distribution and status: Quite common along coasts in winter. Found wintering near Harbin and Lushun in NE China, Hebei (Beidaihe), Shandong (Weihai and Yantai), Fujian, Guangdong, Hong Kong, Hainan and Yunnan (Shiping).

Habits: Typical large marine gull.

418. Herring Gull *Larus argentatus* Yin'ou PLATE 44

Description: Large (64 cm) pale grey gull with pale pink legs. Upperparts pale grey. Winter birds show streaking on head and neck. White tertial crescents are broad but scapular crescents are narrower. Looks heavy and deep-breasted with heavy bill and long sloping forehead and flat crown, giving aggressive appearance. Small mirror visible in flight only on outer primary. Up to six small white tips visible on folded wing. Adult plumage reached in fourth year. First-winter birds are mottled brown with black bill. Second-winter birds are slightly paler and greyer with black-tipped yellow bill. Juvenile birds cannot be reliably identified from other species of Herring Gull complex. Iris—pale yellow; bill—yellow with red spot; feet—pale pink.

Voice: Calls include loud *Kleow*, crowing *KlaOW-KlaOW-Kla-OW* . . . and short whinnying *ge-ge-ge*.

Range: Breeds N America, Europe and Mediterranean; wintering in south of range.

Distribution and status: Race *smithsonianus* seems to be a rare winter visitor to Chinese coasts as far south as Hong Kong.

Habits: Loosely colonial and gregarious. An aggressive and opportunistic scavenger of coasts, inland waters, rubbish tips etc.

Note: Formerly included races of Vega, Heuglin's and Yellow-legged Gulls (e.g. Cheng 1987, 1994).

419. Heuglin's Gull *Larus heuglini* Huilin yin'ou   PLATE 44

Description: Large (60 cm) dark grey gull of Herring Gull complex with bright yellow legs. Upperparts are medium to dark grey (*taimyrensis*) or dark grey (*heuglini*) and always darker than any other gull of this complex and darker than Mew Gull. Winter adults have only light to moderate head streaking and no, or only a trace of, black band on bill. Streaking heaviest on nape. White tertial crescent broad but scapular crescent is fine or absent. White tips on outer two primaries are small, being progressively larger toward primaries six and seven. In flight, shows medium-sized mirror on outer primary and smaller one on primary nine. Underside of primaries dark and contrasts with white underwing coverts and white tips of secondaries. Has red orbital ring. Iris—pale yellow; bill—yellow with red spot; feet—yellow.

Voice: Similar to Herring Gull.

Range: Breeds on coasts of NW Russia; winters S and E

Distribution and status: Common winter resident along S China coast. Race *taimyrensis* is by far the commonest 'Herring Gull' at Hong Kong but a few darker *heuglini* also winter each year.

Habits: As Herring Gull.

Note: Formerly placed as races of Herring Gull *L. argentatus*. Alternatively can include other northern forms, i.e. races of Vega Gull, in this species (e.g. Stepanyan 1990, Beaman 1994).

420. Vega Gull *Larus vegae* Zhinü yin'ou   PLATE 44

Description: Large (62 cm) grey gull of Herring Gull complex with pink legs. Winter birds invariably have dark streaks on head and nape, sometimes on breast; upperparts vary from pale to medium grey (*birulae*) or medium to darker grey (*vegae*). Both have bluish cast. Usually has broad white tertial and scapular crescents. Up to five equally-sized prominent white tips can be seen on closed wing. In flight shows medium-sized white mirror on 10th primary and a smaller one on ninth primary. Pale primaries and inner secondaries do not show much contrast with white underwing coverts. Iris—pale yellow to brownish; bill—yellow with red spot; feet—pink.

Voice: Similar to Herring Gull.

Range: Breeds in N Russia and N Siberia; wintering S.

Distribution and status: Quite common in winter. Races *vegae* and *birulae* migrate across NE China and winter in Bohai Sea, along E and S coasts, around Taiwan and inland along major rivers of S China. In Hong Kong *birulae* is the commonest race contrary to earlier literature.

Habits: As Herring Gull.

Note: Formerly treated as race of Herring Gull *L. argentatus* (e.g. Cheng 1987, 1994). Alternatively included as races of Heuglin's Gull (e.g. Stepanyan 1990, Beaman 1994).

421. Yellow-legged Gull *Larus cachinnans* Huangjiao yin'ou   PLATE 44

Description: Large (60 cm) gull of Herring Gull complex with pale to medium grey upperparts and yellow legs. Winter birds lack brown streaking on head and nape. Three races vary in details. Race *cachinnans* has palest grey upperparts with legs in winter bright yellow to flesh. Race *mongolicus* is darker grey than *cachinnans* but paler than *L. vegae vegae*. Winter birds virtually the same shade of grey as *L. vegae birulae* but head is white; has broad white tertial and scapular crescents; usually four small equal-sized white tips visible on closed wing; shows large mirror in flight only on outer primary and larger than in wintering *L. vegae birulae* or *L. heuglini taimyrensis*; legs pale pink. Winter birds ascribed to race *barabensis* are slightly darker and smaller with narrower tertial and scapular crescents and with mirrors on outer two primaries; legs vary from pink to dull yellow but normally bright orange; eye dark yellow; appears small-billed and round-headed. Sometimes has faint black band on bill. Iris—yellow; bill—yellow with red spot; feet—pink to yellow.

Voice: Similar to Herring Gull.

Range: Breeds from Black Sea through Kazakstan, S Russia, NW China and Mongolia to NE China. Winters south to Israel, Arabica Gulf, Indian Ocean and E Asian countries.

Distribution and status: Nominate race breeds in W Tianshan and Kashi areas of W Xinjiang and winters in Indian Ocean. Race *mongolicus* breeds in NE China in Hulun Nur area of E Nei Mongol and passesthrough China to winter in Indian Ocean with rare visitors on S China coast. Rare birds wintering in Hong Kong appear to be race *barabensis* which breeds in N Kazakstan.

Habits: As Herring Gull

Note: Formerly treated as races of Herring Gull *L. argentatus* (e.g. Cheng 1987, 1994). Some authors have split this species further (e.g. Johansen 1960).

422. Pallas's Gull *Larus ichthyaetus* [Great Black-headed Gull]   PLATE 45
Yu ou

Description: Large (68 cm) grey-backed gull with black head and yellowish bill. Shows white lids to eye. Looks like a huge Common Black-headed Gull but bill heavy and differently coloured. Same size or larger than Herring Gull. In winter head is white with dark patch across eye and some dark streaking on crown and lacks most of the red on bill. In flight entire underwing white except small black tip with two mirrors. First-winter birds have white head with grey mottling on head and mantle, with black-tipped yellow bill and black tip to tail. Iris—brown; bill—yellow with black and red bands near tip; feet—greenish-yellow.

Voice: Hoarse calls like a crow.

Range: Breeds discontinuously from Black Sea to Mongolia on suitable lakes; winters from E Mediterranean and Red Sea to coasts of Burma and W Thailand.

Distribution and status: Fairly common on large lakes within range. Breeds Qinghai Lake (Koko Nor) and Gyaring Lake in E Qinghai and Ulansuhai Lake in Nei Mongol. Migrates through W Xinjiang, Sichuan, Gansu, Yunnan, Xizang and estuary of Zhujiang River. Occurs in small numbers as occasional winter visitor to Hong Kong.

Habits: Lives along the sandy shores of deltas and inland seas and steppe lakes. Often rests on water.

423. Brown-headed Gull *Larus brunnicephalus* Zongtou ou PLATE 45

Description: Medium-sized (42 cm) white gull with grey back and large white patch at base of primaries and diagnostic white spots on black wing tips. Winter birds have dark brown patch behind eye. Summer birds have entire head and neck brown. Distinguished from Black-headed Gull by pale iris, heavier bill, larger size and different wing tip pattern. First-winter bird winter adult but lacks white spots of wing tips and has black terminal tail band. Iris—whitish yellow or grey, red orbital ring; bill—red; feet—vermilion.

Voice: Harsh *gek, gek* and loud wailing *ko-yek, ko yek*.

Range: Breeds in C Asia; in winter to India, China and SE Asia.

Distribution and status: Generally rare but locally common at breeding sites (e.g. Koko Nor). Breeds C Xizang and Qinghai; migrants seen on passage in N and SW China. Some birds winter in W Yunnan and occasionally Hong Kong. Records for Xinjiang may be erroneous.

Habits: Joins other gulls in flocks over sea, coasts and river estuaries.

424. Black-headed Gull *Larus ridibundus* [Common Black-headed PLATE 45
Gull] Hongzui ou

Description: Medium-sized (40 cm) grey and white gull with (in winter) black spot behind eye and red bill and feet or with dark chocolate brown hood extending only to hindcrown and white hindneck in breeding plumage. The leading edge of the wing is white and the black of wing tip is not extensive or spotted white. First-winter birds have a black subterminal tail band; rear edge of wing black; plumage mottled with brown. Distinguished from Brown-headed Gull by smaller size, more conspicuous white leading edge to wing, lack of white spots in black wing tip. Iris—brown; bill—red (tipped black in immature); feet—red (paler in immature).

Voice: Harsh *kwar* calls.

Range: Breeds Palearctic; migrant to India, SE Asia, Philippines.

Distribution and status: Very common. Breeds in W Tianshan area of NW China and in NE China wetlands. Winters in large numbers in E China and all China S of 32 °N on lakes, rivers and coasts.

Habits: When at sea, sits on water, floating objects or fish-trap poles or wheels in tern-like flocks with other gulls over fish shoals. Inland, roosts on water or land. In some towns, quite tame and fed by public.

425. Slender-billed Gull *Larus genei* Xizui ou PLATE 43

Description: Small (42 cm) gull with diagnostic slender red bill, red feet and pinkish underparts in adult breeding plumage. In flight has white primaries with black tips. Profile has short thick neck held at sloping forward angle. Non-breeding birds have grey ear spot. First-winter birds have orange bill and small black spots in front of eye and at rear of ear coverts. Slight brown mottling on wing and black subterminal tail band. Distinguished from winter. Black-headed Gull by less pronounced dark ear spot; no black tip to bill; more slender bill; stiff-necked oppearance; and more orange bill and legs. In flight has long-necked, long-tailed, hump-backed jizz. Iris—yellow; bill—red; feet—red.

Voice: Nasal call; deeper than Black-headed Gull.

Range: Breeds N Africa, Mediterranean, Red Sea and Arabian Gulf. Occasional influxes of wintering birds in SE Asia. Perhaps a more easterly breeding population awaits discovery?

Distribution and status: Occasional winter visitor off S coast and Hong Kong.

Habits: Typical small gull.

426. Saunders's Gull *Larus saundersi* Heizui ou   PLATE 45

Description: Smallish (33 cm) gull. Similar to Black-headed Gull in both summer and winter plumages but smaller with stubby black bill. Black of head in summer extends onto back of neck and is darker than Black-headed Gull; also has prominent white eye-ring. Has zebra pattern on folded primaries, clear white trailing edge to wing and black underside of outer primaries in flight. Iris—brown; bill—black; feet—dark red.

Voice: Shrill *eek eek*.

Range: Endemic, as breeding bird, to E China.

Distribution and status: Globally Endangered (Collar *et al.* 1994). Rare. Breeds in only a few sites on E coast—Liaoning, Hebei and Jiangsu as far as Changjiang River mouth. Winters along S coast including Hong Kong.

Habits: Flight very buoyant and tern-like. Mixes within gull flocks. Keeps to tideline. Feeds by flying and then dropping almost vertically with a final twist before landing and then catching crabs and worm prey. If it misses it quickly takes to the air again. Rarely swims.

427. Relict Gull *Larus relictus* Yi ou   PLATE 45

Description: Medium-sized (45 cm) black-headed gull with red bill and feet. Distinguished from Brown-headed and smaller Black-headed Gulls by darker brown on head with blackish hood and by crescent-shaped white tips to folded wing and larger white mirror in black wing tip in flight. Also white eyelids are broader. Winter birds have dusky ear patch but differ from Brown-headed and Black-headed Gulls in having dark streaks on crown and nape. First-winter birds have black bill, black wing tips and black terminal tail band and only moderate brown mottling on neck and wings, giving paler trailing edge in flight than Black-headed or Brown-headed Gulls. Iris—reddish-brown; bill—red; feet—red.

Voice: A laughing *ka-kak, ka-ka kee-a*.

Range: Breeds on lakes of C Asia; wintering range poorly-known. Recorded as accidental or winter visitor in S China Sea off Hong Kong.

Distribution and status: Globally Near-threatened (Collar *et al.* 1994). Locally common but very restricted range with only scattered breeding sites in China—Hulun Nur area of E Nei Mongol and the Ordos highlands, near Ho Hot in bend of Huang He

(Yellow River) and probably further west in W Nei Mongol. Migrants, sometimes in large numbers, recorded off Hebei (Beidaihe and Happy Island), W Nei Mongol and Beijing and Hong Kong.

Habits: Colonial nester.

428. Little Gull *Larus minutus* Xiao ou PLATE 43

Description: Small (26 cm) gull with black head and bill and red legs. Black of head extends further down back of head than Black-headed Gull. In flight, entire underwing is dusky with narrow white trailing edge. In winter, head is white with grey cap, eye patch and crescent on ear coverts. Tail is slightly notched. First-winter birds are marked with black 'W' pattern in flight like a prion and also black tip to tail; distinguished from first-winter Ross's Gull by tail shape and darker crown. Iris—dark brown; bill—dark red; feet—red.

Voice: Call is loud, nasal *kep* repeated in series. Display call is *ke-KAY ke-KAY ke-KAY* . . . uttered with wings beating low down.

Range: Breeds E and W Siberia, Baltic, SE Europe and N America. Winters to Mediterranean and N Africa through Middle East, C Asia to Japan and E Siberia and eastern USA.

Distribution and status: Very rare in China. Breeds on Ergun River of NE Nei Mongol and reported as migrant in Tianshan of W Xinjiang, Hebei (Beidaihe), Jiangsu (Zhenjiang). One record Hong Kong.

Habits: Colonial and gregarious. Flight is buoyant and tern-like, dipping to sea surface with trailing legs.

429. Ross's Gull *Rhodostethia rosea* Xiewei ou PLATE 43

Description: Small (31 cm) distinctive gull with black bill, red legs, black necklace, pale grey upperparts, rosy underparts and wedge-shaped tail diagnostic. Winter adult lacks collar and less or no pink visible. Differs from Black-headed Gull by lack of black in wing, dusky eye patch and wedge-shaped tail. Reaches adult plumage in second year. First-winter like winter adult but with 'W' mark across upperparts in flight and partial tail band. Differs from first-winter Little Gull by lack of dark cap and wedge-shaped tail. Iris—dark brown; bill—black; feet—red.

Voice: Silent out of breeding season.

Range: Breeds in high Arctic and winters within the Arctic Circle but vagrants in most northern seas.

Distribution and status: Rare. Occasional vagrants to China in winter. Recorded at Lushun in Liaoning.

Habits: Flight is light and buoyant.

[430. Sabine's Gull *Xema sabini* Chawei ou PLATE 43

Description: Small (34 cm) gull with forked tail and dark grey head. Wing tricoloured with sharp contrast between black triangle of outer wing with white triangle of secondaries; coverts and mantle uniform grey or, in juvenile, brown scalloped buff; rump, underparts and underwing white. Slightly forked tail is white in adult or distally black in

juvenile. Distinguished from Black-headed Gull by forked tail, very distinctive wing pattern and buoyant tern-like flight. Iris—brown, red orbital ring; bill—black with yellow tip in adult, black in immature; legs and feet—dark grey in adult, pinkish in immature.
Voice: Harsh grating cries.
Range: Breeds in high Arctic, winters in E Atlantic and E Pacific. Vagrants recorded from many parts of the world.
Distribution and status: Rare vagrant to S China Sea.
Habits: Stragglers mix with flocks of other seabirds. More pelagic than other gulls, generally far offshore.]

431. Black-legged Kittiwake *Rissa tridactyle* Sanzhi ou   PLATE 45

Description: Medium-sized (41 cm) gull with shallowly-forked tail. Yellow bill, black legs and wholly black wing tips diagnostic. Winter adults have head and nape mottled with grey. First-winter birds have black bill, dusky cap and hind collar, dark broken 'W' shape across upperparts in flight plus black terminal tail band. Distinguished from first winter Ross's Gull by hind collar and darker 'W' pattern and forked tail and from Little Gull by whiter secondaries and pale crown. Iris—brown; bill—yellow; feet—black.
Voice: Nasal falsetto *Kittiwake*.
Range: Circumpolar. Breeds along Arctic coasts of N America and Asia including Aleutians. Winters S to Mediterranean, W Africa and N Pacific.
Distribution and status: Very rare winter visitor. Reported wintering in Liaoning (Lushun), Hebei and Jiangsu (Shaweishan). Two records Hong Kong.
Habits: Lives in colonies on rock ledges of cliffs and caves. Totally pelagic. Follows ships.

TERNS AND NODDIES—Tribe: Sternini

A tribe of graceful seabirds. Terns have short legs, long pointed wings, forked tails and fine pointed bills. They have a buoyant flight and often hover over the water before diving onto small fish. They congregate in large wheeling flocks wherever the fishing is good and are often found in coastal waters or even inland lagoons and rivers. Many species are migratory, breeding in extreme northern and southern latitudes and only reach the tropics in their respective winter seasons. Nineteen species are recorded in China.

432. Gull-billed Tern *Gelochelidon nilotica* Ouzui zao'ou   PLATE 47

Description: Medium-sized (39 cm) pale tern with shallow forked tail and heavy black bill. Adult in winter has white underparts, grey upperparts and white head with grey mottling on nape and black patch through eye. In summer entire cap is black. Juvenile as first-winter adult but with brown mottling on crown and upperparts. Iris—brown; bill—black; feet—black.
Voice: Repeated *kuwk-wik* or *kik-hik, hik hik hik*.
Range: Almost worldwide, breeding in Americas, Europe, Africa, Asia and Australia and passing through Indonesia and New Guinea as a migrant.
Distribution and status: Uncommon resident and winter visitor. Nominate race breeds in W Xinjiang (Tianshan) and NE Nei Mongol (Hulun Nur); *affinis* breeds

around Bohai Sea and SE China and Taiwan, wintering in SE China, Taiwan and passing through or wintering on Hainan.

Habits: Frequents coastal estuaries, lagoons, and inland fresh water. Often hovers and generally feeds by skimming beetles and other prey off water or mud, rarely plunges into water.

433. Caspian Tern *Hydroprogne caspia* Hongzui ju'ou   PLATE 47

Description: Very large (49 cm) tern with diagnostic massive red bill. Black cap of summer becomes streaked white in winter. Undersides of primaries black. Juvenile has barred brown upperparts. First-winter as adult but with brown speckling on wings and heavy black hood. Iris—brown; bill—red with blackish tip; feet—black.

Voice: Harsh rasping *kraaah*.

Range: N and S America, Africa, Europe, Asia through Indonesia to Australia.

Distribution and status: Nominate race is rather uncommon resident and seasonal migrant. Breeds in coastal provinces from Bohai Sea to Hainan and Changjiang River; northern birds wintering with southern residents in S and SE China, Taiwan and Hainan.

Habits: Favours coasts, lakes, mangroves and estuaries.

434. River Tern *Sterna aurantia* Huanghe heyan'ou   PLATE 48

Description: Distinctive, medium-sized (40 cm), black-capped tern with large yellow bill, red or orange legs and streamered tail. Upperparts, rump and tail dark grey with outer tail feathers white and brackish wing tips in worn plumage. Winter adult has black tip to bill and forehead and crown become whitish. Juvenile as winter adult but with brown on crown and upperparts and grey wash on sides of breast. Like Black-bellied Tern but much larger and with white underparts. Iris—brown; bill—deep yellow; feet—red.

Voice: Aggressive *ping* when swooping at intruders on breeding grounds.

Range: Breeds on large rivers from Iran E through Indian subcontinent, Burma and SE Asia.

Distribution and status: Rare in China. Resident in SE Xizang and W and SW Yunnan.

Habits: Lives on fresh water and occasionally coastal creeks. Has strong slow flight with long wings and tail giving distinct jizz. Mixes with other terns. Rests on sand spits.

435. Lesser Crested Tern *Sterna bengalensis* [Crested Tern]   PLATE 48
Xiao fengtou yan'ou

Description: Medium-sized (40 cm) crested tern. Similar to Great Crested Tern but smaller with black forehead in breeding plumage, and distinctive orange bill. In winter only the forehead is white, the crest remains black. Juvenile like non-breeding adult but brownish mottling on upperparts and with dark grey flight feathers. Iris—brown; bill—orange; feet—black.

Voice: Raucous screams *kirrik*.

Range: Breeds N Africa, Red Sea, Arabian Gulf, Indian subcontinent, SE Asia, Philippines, Malaysia to N Australia. Range through adjacent oceans.

Distribution and status: Very rare and decreasing. Recorded off Guangdong and Hong Kong. More common in S China Sea.

Habits: Highly social in large flocks, often mixed with other species, especially Great Crested Tern. Frequents coastal waters and mud, sand or coral shores, often feeding far out to sea; rests in quarrelsome parties on fishing platforms and buoys. Dives vertically, submerging completely.

436. Great Crested Tern *Sterna bergii* [Greater Crested Tern/   PLATE 48
Swift Tern] Da fengtou yan'ou

Description: Large (45 cm) crested tern. In summer the crown and crest are black, becoming mottled with white in transition to white crown and grey mottled crest in winter. Upperparts grey, underparts white. Young birds darker grey than adults and mottled with brown and white on upperparts; tail grey. Bill colour is the best character to distinguish from other crested terns. Iris—brown; bill—greenish-yellow; feet—black.

Voice: Sharp rasping *kirrik* or clear *chew*.

Range: Distributed throughout coasts and islands of Indian Ocean, Arabian Gulf, tropical seas of Pacific and coasts of Australia and S Africa.

Distribution and status: Race *cristatus* breeds on small islands off the S and SE coasts, Taiwan and Hainan. Decreasing in numbers. Common in S China Sea.

Habits: Fishes in small parties of twos or threes, sometimes with other terns. Rather clumsy plunges. Rests on beaches, buoys and fishing platforms or on the water on floating objects. Often ventures quite far out to sea.

437. Chinese Crested Tern *Sterna bernsteini* Heizuiduan   PLATE 48
fengtou yan'ou

Description: Medium-sized (38 cm) crested tern with diagnostic black-tipped yellow bill. Winter birds have forehead white and black cap centrally streaked white leaving U-shaped black patch around nape. Distinguished from Greater and Lesser Crested Terns by black one-third to tip of yellow bill. Immature like immature Lesser Crested but browner, inner part of wing lining paler, two dark bars on inner wing; back and tail whitish with brown mottling. Iris—brown; bill—yellow with black tip; feet—black.

Voice: Harsh, high-pitched cries.

Range: Thought to breed in E China, migrating south in winter to seas of S China, Philippines and occasionally N Borneo.

Distribution and status: Globally Conservation-dependent (Collar *et al.* 1994). Very rare and probably close to extinction. Formerly thought to breed around Shandong coast.

Habits: Similar to other crested terns. Favours open sea and small islets.

438. Roseate Tern *Sterna dougallii* Fenhong yan'ou   PLATE 46

Description: Medium-sized (39 cm) black-crowned tern with very long, deeply-forked white tail. Adult in summer has black crown, pale grey upperwings and back and white underparts, pinkish on chest. In winter forehead white, crown mottled, pinkish wash absent. Webs of outer primaries blackish. Juvenile: black bill and legs; blackish-brown crown, nape and ear coverts; mantle darker brown than Common Tern; tail white, lacking streamers. Iris—brown; bill—black with a red base in breeding season; feet—reddish in breeding season otherwise black.

Voice: Musical *chew-it* whilst fishing or harsh *aaak* in alarm.

Range: E and W Atlantic, Indian Ocean, China Sea to N Australia and W Pacific.

Distribution and status: Rare seasonal migrant. Race *bangsi* breeds on islets off Fujian, Guangdong and S Taiwan; winters at sea, occasionally in S China Sea.

Habits: Inhabits coral formations, coral and granite islands and sandy beaches but generally uncommon. Often mixes with other terns. Flies in graceful manner, plunging steeply to catch small fish.

439. Black-naped Tern *Sterna sumatrana* Heizhen yan'ou PLATE 46

Description: Smallish (31 cm) very white tern with very long forked tail and distinctive black nape band. Upperparts pale grey; underparts white; head white except for black spot in front of eye and black band over nape. First-winter birds have brown mottling on crown and blackish mottling on nape. Juvenile: side of head and nape greyish-brown; upperparts brownish scalloped buff and grey; rump whitish, rounded unforked tail. Iris—brown; bill—black with yellow tip (adult), dirty yellow (juvenile); feet—black (adult) or yellow (juvenile).

Voice: Sharp *tsii-chee-chi-chip* or in alarm *chit-chit-chitrer*.

Range: Tropical islands and coasts of Indian and W Pacific Oceans and N Australia.

Distribution and status: Regular but uncommon breeder and summer visitor. Nominate race breeds on offshore rocks and islets off SE and S coasts, Hong Kong, Taiwan and Hainan. Also in archipelagos of S China Sea. Some birds winter around Hainan and islands to south.

Habits: Gregarious bird, flocking with other terns along sandy and coral beaches, rarely over mud and never far inland.

440. Common Tern *Sterna hirundo* Putong yan'ou PLATE 46

Description: Smallish (35 cm) black-crowned tern with deeply-forked tail. Breeding: entire crown black, breast grey. Non-breeding: upperwing and back grey; uppertail coverts, rump and tail white; forehead white; crown mottled black and white, blackest on nape; underparts white. In flight, non-breeding adult and immature are characterised by a blackish bar on the forewing and blackish edge to outer tail feathers. First-winter has browner upperparts and scaling on mantle. Iris—brown; bill—black in winter, red base in summer; feet—reddish, darker in winter.

Voice: Harsh *keer-ar* descending, with emphasis on first note.

Range: Breeds in N America and Palearctic but migrating S in winter to S America, Africa, Indian Ocean, Indonesia and Australia.

Distribution and status: Common summer breeder and passage migrant. Nominate race breeds in NW China; *longipennis* in NE China and eastern N China and *tibetana* through central N China, C China, Qinghai and Xizang. The latter two races are recorded as passage migrants through S and SE China, Taiwan and Hainan.

Habits: Frequents coastal waters and occasionally inland fresh water. Rests on elevated perches such as fishing platforms and rocks. Strong flyer and feeds by plunging steeply into the sea.

441. Little Tern *Sterna albiforns* [Least Tern] Bai'e yan'ou PLATE 46

Description: Small (24 cm) pale tern with shallow forked tail. Summer: crown, nape and eye-stripe black; forehead white. Winter: black of crown and nape reduced to crescent. Wing with darker leading edge and white rear edge. Juvenile: similar to nonbreeding adult but mottled with brown on crown and mantle; tail white with brown tip; bill dusky. Iris—brown; bill—yellow with black tip (summer) or black; feet—yellow.

Voice: Rasping high-pitched shrieks.

Range: W coast of USA, Caribbean, W Palearctic, Africa, Indian Ocean, Indian subcontinent, E and SE Asia through Indonesia to Australia.

Distribution and status: Common summer breeder. Nominate race breeds in Kashi area of W Xinjiang; *sinensis* breeds over much of China from NE to SW to S coast and Hainan. Breeds well inland as well as along coasts. Recorded on passage through Taiwan.

Habits: Inhabits sandy seashores, mixing with other terns. Fast wing beat and regular hovering and diving characteristic with the bird climbing back up as soon as it has dived.

442. Black-bellied Tern *Sterna acuticauda* Heifu yan'ou PLATE 48

Description: Medium-sized (33 cm) river-living tern with deeply-forked tail. Black-capped with orange bill and legs. Upperparts, rump and tail pale grey; underparts white, with diagnostic black belly patch. Winter adult has black tip to bill; forehead mottled white; black belly patch much reduced or absent. Distinguished from River Tern in smaller size, paler upperparts, black on belly, more slender bill and more buoyant flight. Iris—dark brown; bill—bright orange-red; feet—orange-red.

Voice: Constant, shrill *krek krek* in flight.

Range: S Asia from Indus valley E through India, Sri Lanka, Burma and Thailand.

Distribution and status: Globally Vulnerable (Collar *et al.* 1994). Very rare in China. Resident on Yingjiang River in SW Yunnan.

Habits: Gregarious; flies up and down slow-flowing rivers and rests on sand banks. Hunts upwind then hurries downwind for another flight over the same stretch of water.

443. Aleutian Tern *Sterna aleutica* Baiyao yan'ou PLATE 48

Description: Medium-sized (34 cm) tern with black bill. Very similar to black-billed race of Common Tern with white facial streak separating black cap from grey underparts, but has diagnostic dark bar on the underside of secondaries in front of white trailing edge. Winter adult has crown and underparts white; differing from Common Tern in blacker bill and legs, greyer plumage and diagnostic underwing pattern. Juvenile: legs and lower mandible dusky-red with buff crown and underparts. Iris—dark brown; bill—black; feet—black.

Voice: Squeaky notes *eek eek* like Saunders's Gull or wader-like *twee-ee-ee* in flight.

Range: Breeds in Siberia, Aleutians and Alaska. Winters southward in oceans

Distribution and status: Rare but probably regular passage migrant. Up to 200 were recorded in Hong Kong in autumn 1992 and annually thereafter.

Habits: Flies strong and deep with slow stiff beats. Gregarious. Wings point upward when perched.

444. Bridled Tern *Sterna anaethetus* [Brown-winged Tern]   PLATE 46
Hechi yan'ou

Description: Medium-sized (37 cm) dark-backed tern with long, deeply-forked tail. Adult: dark brownish-grey on upperwing, back and tail except white leading edge to wing and white outer tail feathers; underparts white. Distinguished from Sooty Tern by narrow white forehead and narrow supercilium extending beyond eye. Juvenile is browner with mottled brown crown, grey breast and back barred buff, less spotted than juvenile Sooty Tern with white neck and breast. Iris—brown; bill—black; feet—black.

Voice: Staccato yapping *wep-wep* and harsh grating alarm calls *kee-errr-krr*.

Range: Widespread throughout Atlantic, Indian and Pacific Oceans as far as Australia.

Distribution and status: Nominate race is an offshore resident in S China Sea; seen inshore mostly during the summer moult, off Fujian, Hong Kong, Taiwan and Hainan.

Habits: Keeps well out to sea, coming inshore only in bad weather or breeding season. Not very social; single or in small parties. The flight is graceful and buoyant. It feeds by scooping insects or fish off the surface. It does not dive. Frequently rests on flotsam or at night on the spars of ships. Breeds in mixed colonies with Black-naped Terns.

445. Sooty Tern *Sterna fuscata* Wu yan'ou   PLATE 46

Description: Medium-sized (44 cm) black-backed tern with deeply-forked tail. Similar to Bridled Tern but upperwings and back darker sooty-brown, lacks grey collar and white forehead does not extend into a supercilium. Immature is sooty-brown with white vent and bars of white spots on back and upperwings. Iris—brown; bill—black; feet—black.

Voice: Nasal *ker-waky-wak*, or *wide-a-wake*.

Range: Widespread throughout tropical Atlantic, Indian and Pacific Oceans.

Distribution and status: Race *nubilosa* lives far out to sea but is occasionally recorded off SE coasts, Hong Kong and Taiwan especially after typhoons.

Habits: Truly oceanic tern staying well out to sea or on small rocky and sandy islets. Follows ships at night. Flight is easy and buoyant, soaring on updraughts.

446. Whiskered Tern *Chlidonias hybridus* Xu fu'ou   PLATE 47

Description: Smallish (25 cm) pale tern with dark belly (in summer) and shallow forked tail. Breeding: distinctive with black forehead and grey breast and belly. Non-breeding plumage: white forehead grades streakily into black hindcrown and nape; underparts white; wings, nape, back and uppertail coverts grey. Juvenile similar to adult but with brown mottling. Distinguished from non-breeding White-winged Tern by blacker crown, grey rump and lack of separate black cheek patch. Iris—dark brown; bill—red (breeding) or black; feet—red.

Voice: Harsh, staccato *kitt* or *ki-kitt*.

Range: Breeds in S Africa, southern W Palearctic, S Asia and Australia.

Distribution and status: Race *swinhoei* is a seasonal migrant, breeding across the eastern half of China; wintering S with some on Taiwan.

Habits: Lives in small or occasionally large flocks, often coming up to 20 km inland to feed over flooded land and paddy fields, taking food in shallow plunges or low skimming flights.

447. White-winged Tern *Chlidonias leucopterus* [White-winged   PLATE **47**
Black Tern] Baichi fu'ou

Description: Small (23 cm) tern with shallow-forked tail. Breeding adult unmistakable
with head, back and chest black, sharply contrasting with white tail and pale grey wings.
Upperwing whitish, underwing coverts distinctively, black. Non-breeding adult: pale
greyish upperparts; back of head mottled greyish-black; underparts white. Distinguished
from non-breeding Whiskered Tern by more complete white collar, crown less black and
more mottled, black ear coverts separate from black crown and paler rump. Iris—dark
brown; bill—red (breeding), black (non-breeding); feet—orange-red.
Voice: Repeated *kweek* or sharp *kwek-kwek*.
Range: Breeds in S Europe and Arabian Gulf across Asia to C Russia and China;
migrates in winter to S Africa and through Indonesia to Australia and occasionally New
Zealand.
Distribution and status: Uncommon seasonal migrant and winter visitor. Breeds in
Tianshan of NW Xinjiang, NE China and bend of Huang He (Yellow River); recorded
on passage across northern China and wintering on S and SE coasts, also on larger rivers
and Taiwan and Hainan. Mostly coastal but may come inland over flooded paddy.
Habits: Frequents coastal areas, estuaries and river mouths in small flocks; ranges well
inland to feed over paddy fields and swamps. Feeds by skimming low over water, working into the wind and chasing insects. Commonly perches on poles.

448. Black Tern *Chlidonias niger* Hei fu'ou   PLATE **48**
Description: Small (24 cm) blackish tern with white underwing. Separated from
White-winged Tern by black bill, white underwing, darker grey wing and darker legs.
Winter adult loses black head and breast but retains black cap extending behind eye;
small black spot in front of eye and small black patch on side of breast in front of wing
in flight. Tail is more deeply notched than White-winged Tern. Iris—brown; bill—black;
feet—dull red.
Voice: Short, nasal, shrill *kyeh* and contact *klit*.
Range: Breeds by fresh water in N America, W Palearctic and C Russia. Winters south
to C America, S and W Africa with strays recorded as far afield as Japan and Australia
Distribution and status: Very rare in China. Breeds in Tianshan in W Xinjiang and
probably near Hulun Nur in E Nei Mongol. Vagrants recorded in Tianjin and Beijing,
Hebei (Beidaihe) and Yancheng in Jiangsu.
Habits: Lives along coasts and around inland waters. More marine than White-winged
Tern. Flight with fast deep wing beat. Holds bill at drooping angle in flight.

449. Brown Noddy *Anous stolidus* [Common Noddy/Noddy Tern]   PLATE **48**
Baiding xuan'ou

Description: Distinctive, medium-sized (42 cm), dark sooty-brown tern with whitish
crown and notched tail. Plumage uniform sooty-brown apart from whitish crown and
white eye-ring. Juvenile has dark forehead and crown, white eye-ring and whitish tips
to feathers of back and wing coverts. Immature as adult but lacking pale crown. Iris—
brown; bill—black; feet—blackish-brown.

Voice: Harsh *karrk* and *kwok-kwok*.

Range: Throughout tropical and subtropical oceans and N Australia, breeding throughout its range.

Distribution and status: Race *pileatus* is a not uncommon bird out at sea but only uncommon inshore. Formerly bred on islets off Taiwan.

Habits: An oceanic bird with a slow, lazy wheeling flight. Rarely dives like other terns. Takes small fry as they skip out of the water when pursued by predatory fish. Sometimes alights on water to collect food. In courtship, partners nod heads in display, hence the English name.

Note: Similar Black Noddy *Anous minutus* is smaller, more slender and with an almost pure white crown which extends further back to cover nape. May be expected in southern S China Sea or off SE Taiwan.

450. White Tern *Gygis alba* [Common White/Fairy Tern/   PLATE 47
White Noddy] Bai xuan'ou

Description: Small (30 cm) pure white tern with black eye-ring. Adult: entire plumage whitish except for black eye-ring. Tail slightly forked with outer feathers shorter than second and third feathers. The bill is very slender, sharp and slightly upturned. Juvenile: dark ear spot, greyish-brown mottled mantle and upperwings, black primary shafts and more rounded wings. Iris—brown; bill—blackish with blue base; feet—bluish-black with whitish webs.

Voice: Soft buzzing calls.

Range: Widely distributed throughout tropical and subtropical oceans of Atlantic and Indo-Pacific.

Distribution and status: Very rare visitor, recorded once off Macao and known from Xisha Is of S China Sea.

Habits: Flies with light rising and sinking manner, occasionally diving for food but never submerges.

AUKS—Subfamily: Alcinae

A small subfamily of northern seabirds, somewhat resembling penguins and swimming underwater with wing propulsion. The birds have fluffy plumage and are good divers, feeding mostly on plankton. Four species occasionally enter Chinese seas.

451. Marbled Murrelet *Brachyramphus marmoratus*
Ban haique

Description: Small (24 cm) blackish-brown and white murrelet with slender bill. In winter chin, throat, neck, nape and underparts and scapulars are white and contrast cleanly against rest of dark plumage. Eye-ring white. In breeding plumage white parts of plumage become barred with grey-black. Young birds are like adults but with less contrast. Iris—brown; bill—brown; feet—pinkish.

451. Marbled Murrelet
Brachyramphus marmoratus
斑海雀

▨ Winter   ↓ On passage

Voice: Call is *meer-meer-meer*.

Range: Breeds Alaska and E Siberia coasts. Ranges through Aleutian chain to Japan and W USA.

Distribution and status: Globally Near-threatened (Collar *et al*. 1994). Very rare migrant. Race *perdix* recorded in Liaoning (Lushun) and off Shandong (Qingdao).

Habits: Carries bill and tail cocked when swimming.

Note: Asian race *perdix* may warrant recognition as a separate species.

452. Ancient Murrelet *Synthliboramphus antiquus*
Bianzui haique

452. Ancient Murrelet
Synthliboramphus antiquus
扁嘴海雀

▦ Resident   ■ Migrant
▨ Winter   ▥ Summer breeder

Description: Smallish (25 cm) black and white murrelet with thick head and short, stout pale bill, giving penguin-like appearance. Breeding birds possess diagnostic combination of uncrested head, blue-grey back and white underparts. Has black throat and diffuse white eyebrow. Non-breeding birds lose eyebrow and black throat. Underwing in flight is white with dark leading and trailing edges and dusky axillaries. Iris—brown; bill—ivory with dark tip; feet—grey.

Voice: Low piping whistle and metallic clinking notes.

Range: Breeds in Aleutians, Alaska and coasts of E Siberia, N Japan and Korea. Winters along coasts of S China and W USA.

Distribution and status: Occasional in Hong Kong waters.

Habits: Sits low in water. Clears water when diving. Flies low and direct for short distance before dropping back on to sea.

453. Japanese Murrelet *Synthliboramphus wumizusume*
[Crested Murrelet] Guan haique

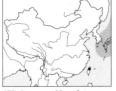

453. Japanese Murrelet
Synthliboramphus wumizusume
冠海雀

■ Resident • Vagrant/accidental

Description: Small (25 cm) black, grey and white murrelet with pointed black crest. Forehead, crown and nape slaty-black; cheeks and upper throat grey. White stripes on side of head above eye extend to meet on upper nape. Upperparts greyish-black; underparts whitish with greyish-black flanks. has crest only in summer. Generally similar to Ancient Murrelet in winter but distinguished by distribution of black and white on head. Bill very short. Iris—brown-black; bill—grey-white; feet—yellowish-grey.
Voice: Shrill whistles.
Range: Japan, ranging into nearby seas. Vagrant Sakhalin.
Distribution and status: Vagrant. Accidental off Taiwan.
Habits: Typical small murrelet.

454. Rhinoceros Auklet *Cerorhinca monocerata*
[Horn-billed Auklet, AG: Auk] Jiaozui haique

454. Rhinoceros Auklet
Cerorhinca monocerata
角嘴海雀

■ Resident ▥ Summer breeder

Description: Largish (32 cm) dusky auk with orange or yellow bill and two diagnostic white head stripes and yellow feet. Upperparts barred and blotched grey. In breeding plumage, white eyebrow behind eye and white moustachial line under eye are prominent; bill and feet are more orange and there is a small pale horn and on base of upper mandible, giving the species its name. Iris—yellow; bill—orange in breeding season, otherwise yellow; feet—yellow.
Voice: Generally silent.
Range: Breeds along coasts of E Siberia, N Japan, Kurils, Aleutians, Alaska and NW USA, ranging through adjacent seas.
Distribution and status: Rare winter visitor. Accidental in Liaoning (Lushun).
Habits: Nests in burrows; feeds far out at sea; rafts in large flocks; flight strong; sits low in water.

HAWKS, EAGLES, VULTURES AND FALCONS—Family: Accipitridae

OSPREY—Subfamily: Pandioninae

Formerly a mono-species family, now reduced to a subfamily. The Osprey is a fishing hawk with characteristic long, narrow, angled wings, specialised for diving quite deep into water to catch fish. Otherwise much like other hawks.

455. Osprey *Pandion haliaetus* [Fish Hawk] Yu'e   PLATE 54

Description: Medium-sized (55 cm) brown, black and white hawk. White head and underparts with black eye-stripe diagnostic. Upperparts mostly dull brown. Has a short dark erectile crest. Races differ in whiteness of head and extent of streaking on underparts. Iris—yellow; bill—black with grey cere; naked tarsus; feet—grey.
Voice: A loud plaintive whistle in breeding season. Young in nest scream loudly when parents sighted.
Range: Cosmopolitan.
Distribution and status: Widely distributed but generally rare resident over most of China and a summer visitor to the NE and NW.
Habits: A dramatic fishing hawk, plunging deep into the water to catch prey from tree perch over water, or circles or hovers slowly over water before diving.

HAWKS, EAGLES AND VULTURES—Subfamily: Accipitrinae

Hawks and eagles are largish to very large, predatory, hook-billed birds with powerful talons or claws, specialised for killing and tearing up vertebrate prey. They differ from the falcons in having generally blunter, more rounded wings and paler eyes (yellow or red). Eagles and vultures are specialised for soaring on air currents and hunt largely from the air, whereas the other hawks hunt from branches but sometimes also soar; some species even hover over intended prey. Vultures feed mostly on carrion and have long barish necks for probing into carcasses. Members of the subfamily make very large, stick pile nests in trees or on cliffs and the young give characteristic screaming calls. There are 48 species in China, including some migrants.

456. Jerdon's Baza *Aviceda jerdoni* [Asian Baza/Lizard Hawk]   PLATE 49
Heguan juansun

Description: Medium-sized (45 cm) brown hawk with long crest often held vertically above head. Upperparts brown, underparts white with black mesial stripe, streaks on chest; dark rufous bars on belly. Distinguished from Crested Goshawk by much longer crest and wing tips almost reaching tip of tail. In flight distinguished by very long broad wings, noticeably broad near tip, and square-cut tail. Iris—yellow-red; bill—black with pale blue-grey cere; feet and legs—yellow.
Voice: Plaintive mewing on wing *pee-weeoh*, second note fading away—similar to Crested Serpent Eagle.

Range: Indian subcontinent, S China, SE Asia and Sulawesi.
Distribution and status: Globally Near-threatened (Collar *et al.* 1994). Few records from SW Yunnan and Hainan. Breeding probable but not confirmed.
Habits: Hunts from tree perches, favouring forest edge, often beneath canopy.

457. Black Baza *Aviceda leuphotes* [Black-crested Baza]   PLATE 49
Heiguan juansun

Description: Smallish (32 cm) unmistakable black and white hawk with long black crest often held vertically above head. General plumage black with broad white breast bar, white wing patches and dark chestnut-banded belly. In flight, short rounded, grey wings are conspicuously patterned with black lining and black wings tips. Has crow-like flapping flight and glides with wings held flat. Iris—red; bill—horn with grey cere; feet—dark grey.
Voice: Weak 1–3 noted airy screams, like a seagull mewing.
Range: Indian subcontiment, S China, SE Asia; winters to Greater Sundas.
Distribution and status: Three races occur: *wolfei* in Sichuan, *syama* in S and SW China and nominate on Hainan. Locally not uncommon in lowlands in open wooded areas.
Habits: Lives in pairs or small flocks making short flapping flights to catch large insects in air or on ground.

458. Oriental Honey-buzzard *Pernis ptilorhyncus* [Crested/   PLATE 49
Asian/Eastern Honey-buzzard] Fengtou fengying

Description: Largish (58 cm) dark hawk with or without crest. Colour very variable with light, normal and dark forms of two distinct races each mimicking hawk eagles and buzzards. Upperparts from white to rufous to dark brown; underparts heavily spotted and barred. Tail has irregular barring. All forms can have a contrasting pale gular patch outlined with heavy black streaks and often a black mesial streak. In flight, the relatively small head, longish neck and long narrow wings and tail are distinctive. Iris—orange; bill—grey; feet—yellow; at close range scalelike feathers in front of eye are diagnostic.
Voice: Loud, high-pitched ringing four syllable note *wee-wey-uho* or *weehey-weehey*.
Range: E Palearctic, Indian subcontinent and SE Asia to Greater Sundas.
Distribution and status: Long-crested race *ruficollis* is sparsely distributed through S Sichuan and Yunnan with some regional migration. E Palearctic short-crested race *orientalis* breeds from Heilongjiang to Liaoning, migrating through C and E China to winter in Taiwan, SE China and Hainan. Not uncommon up to 1200 m in forest.
Habits: Characteristic flight of a few wing beats followed by a long glide. Soars high in the sky with wings held level. Has curious habit of raiding bee and wasp nests.

459. Black-shouldered Kite *Elanus caeruleus* [Black-winged/   PLATE 49
Indonesian Kite] Heichi yuan

Description: Small (30 cm) white, grey and black kite. Black shoulder patch and long black primaries diagnostic.
Adult: grey crown, back, wing coverts and base of tail. Face, neck and underparts white. The only pale hawk which hovers when looking for prey. Immature similar to adults but tinged brown. Iris—red; bill—black with yellow cere; feet—yellow.

Voice: Soft whistle *wheep, wheep.*

Range: Africa, S Eurasia, Indian subcontinent, S China, Philippines and Indonesia to New Guinea.

Distribution and status: Rare resident in Yunnan, Guangxi, Guangdong and Hong Kong in open lowland and hilly habitats up to 2000 m. Formerly recorded in Hubei and Zhejiang.

Habits: Sits on exposed perches such as dead trees or telegraph poles, also hovers like a kestrel.

460. Black Kite *Milvus migrans* [Common/Pariah>Yellow-billed Kite] PLATE 49
Hei yuan

Description: Medium-sized (55 cm) dark brown raptor with diagnostic shallow forked tail. In flight a pale patch at base of primaries contrasts with blackish wing tips. Head sometimes paler than back. Distinguished from Black-eared Kite by rufous fore post of head and cheeks. Immature are streaked buff on head and underparts. Iris—brown; bill—grey; cere—yellow; feet—yellow.

Voice: Shrill whinnying *ewe-wir-r-r-r-r.*

Range: Africa, Indian subcontinent to Australia.

Distribution and status: Race *govinda* is resident in W Yunnan and SE Xizang.

Habits: Frequents open country, towns and villages. Circles gracefully or flies with slow deliberate beat. Perches on poles, wires, trees, buildings or ground, scavenges at rubbish dumps.

461. Black-eared Kite *Milvus lineatus* Hei'er yuan PLATE 49

Description: Largish (65 cm) dark brown raptor with slightly forked tail and conspicuous pale subterminal pattern on base of primaries in flight. As Black Kite but ear coverts black, larger size and whiter pale wing patch. Iris—brown; bill—grey with blue-grey cere; feet—grey.

Voice: As Black Kite.

Range: N Asia to Japan.

Distribution and status: Common and widespread. This is the commonest raptor in China. Resident throughout China, Taiwan and Hainan in suitable habitat up to 5000 m on Qinghai–Xizang plateau.

Habits: As Black Kite. Lives around towns and villages in west of country; rivers and coasts in the east.

Note: Formerly classed as race of Black Kite *Milvus migrans* and synonymous with *Milvus korschun lineatus* (e.g. Cheng 1987).

462. Brahminy Kite *Haliastur indus* [Red-backed/ PLATE 49
White-headed Kite] Li yuan

Description: Medium-sized (45 cm) white and russet-brown kite. Adult: head, neck and chest white; wings, back, tail and belly rich reddish-brown; contrasting black primaries. Immature brownish all over with streaked chest, becoming greyish-white in second year and reaching full adult plumage in third year. Distinguished from Black Kite and Black-eared Kite by rounded tail. Iris—brown; bill and cere—greenish-grey; legs and feet—dull yellow.

Voice: Shrill, querulous, mewing cries *shee-ee-ee* or *kweeaa*.

Range: Indian subcontinent and S China to Australia.

Distribution and status: Uncommon and increasingly rare on large rivers and along coast from lower Changjiang and Xijiang Rivers, SW Yunnan and SE coast.

Habits: Wheels singly or several together over waterways or close to water.

463. White-bellied Sea Eagle *Haliaeetus leucogaster* [White-bellied Fish Eagle] Baifu haidiao   PLATE 50

Description: Large (70 cm) white, grey and black eagle. Adult: head, neck, and underparts white, wings, back and tail grey; primaries black. Immature: white areas of adult are pale brown and grey areas are dark brown. Wedge-shaped tail characteristic. Iris—brown; bill and cere—grey; bare tarsus and feet—pale grey.

Voice: Loud honking cry *ah-ah-ah-ah*.

Range: India, SE Asia, Philippines and Indonesia to Australia.

Distribution and status: A not uncommon resident around S China coasts. Recorded along SE coast, Hainan, Xisha and Nansha islands (Paracels and Spratlys).

Habits: Frequently seen sitting very upright in a waterside tree, or on a cliff. Soars and glides gracefully with wings held in a pronounced dihedral with slow, powerful wing beats. Catches surfacing fish in spectacular dives.

464. Pallas's Fish Eagle *Haliaeetus leucoryphus* [AG: Fishing Eagle] Yudai haidiao   PLATE 50

Description: Large (80 cm) dark brown eagle with buffy-golden head, neck and breast and broad white band across lower part of wedge-shaped tail diagnostic. Immature is tawny brown with black secondaries in flight contrasting with pale median underwing coverts and black, wedge-shaped tail with palish base. Conspicuous pale panel at base of primaries from below. Has dark brown ear coverts and stripe through eye. Ruff of lanceolate feathers on neck. Iris—yellow; bill—grey with grey cere; feet—yellowish-white or grey.

Voice: Loud screams. Very noisy in breeding season.

Range: Iraq and Aralin (winter) to C Asia, Indian subcontinent and Burma.

Distribution and status: Globally Vulnerable (Collar *et al.* 1994). Uncommon to rare resident and seasonal migrant. Breeds in W and C Xinjiang, Qinghai, NE Nei Mongol (Hulun Nur), Heilongjiang and S Xizang. Migrant over much of C and NE China as far south as Jiangsu (Shaweishan Is.).

Habits: Fishes on inland lakes, marshes, rivers on high plateaux and in arid regions. Perches on trees or poles then swoops to pick off fish near surface.

465. White-tailed Eagle *Haliaeetus albicilla* [AG: Sea Eagle] Baiwei haidiao   PLATE 50

Description: Large (85 cm) brown eagle with pale brown head and breast, yellow bill and entirely white tail diagnostic. Underwing of blackish flight feathers contrasting with dark chestnut underwing coverts. Has large bill and short, wedge-shaped tail. Appears vulture-like in flight. Distinguished from Pallas's Fish Eagle by all white tail. Juvenile has lanceolate feathers on breast but not forming a ruff as in Pallas's Fish Eagle. Body brown with irregular rusty or white spots, depending on age. Iris—yellow; bill—yellow with yellow cere; feet—yellow.

Voice: Loud yelping *klee klee-klee-klee*, like a puppy dog or call of Black Woodpecker.

Range: Greenland, Europe, N Asia, China and Japan to India.

Distribution and status: Globally Near-threatened (Collar *et al.* 1994). Nominate race is uncommon seasonal migrant to C and E China, in a variety of habitats along rivers, near lakes and coasts. Breeds around Hulun Nur in NE Nei Mongol and perhaps other sites in NE China. Recorded wintering at Cao Hai (Guizhou), NW Yunnan and Poyang (Jiangxi).

Habits: Appears sluggish, sitting hunched and inert for many hours. Flies with very slow wing beat. Soars on slightly raised bowed wings.

466. Steller's Sea Eagle *Haliaeetus pelagicus* [Kamchatkan Sea Eagle] PLATE 50
Hutou haidiao

Description: Huge (100 cm) black eagle with massive yellow bill, white forewing, rump, vent and wedge-shaped tail diagnostic. Immature is dark grey-brown with grey-edged whitish tail and pale panel on wing. Resembles White-tailed Eagle but with massive yellow bill. Iris—brown; bill—yellow; feet—yellow.

Voice: Call is gruff barking *kyow-kyow-kyow* or strong *kra, kra, kra, kra* when squabbling over food or roost.

Range: Breeds along coasts of E Siberia, Kamchatka, Sakhalin, Korea and Kurile Is and in Amur delta. Winters in Kamchatka Peninsula, Japan and Korea.

Distribution and status: Globally Vulnerable (Collar *et al.* 1994). Rare and declining winter visitor due to pollution of fish in Bohai Sea. Recorded in Liaoning and Hebei.

Habits: Locally winters in flocks. Feeds mostly on fish plucked from sea surface.

467. Lesser Fish Eagle *Ichthyophaga humilis* [AG: Fishing Eagle] PLATE 50
Yu diao

Description: Medium-sized (60 cm) brownish eagle with grey head and neck and white belly. Distinguished from Grey-headed Fish Eagle *I. ichthyaetus* by smaller size and dark tail. Immature paler brown with buffy unstreaked underparts. Iris—yellow or brown; bill—dark grey; feet—grey.

Voice: Occasional querulous *hak hak* cries.

Range: Himalayan foothills, SE Asia, Borneo and Sumatra.

Distribution and status: Globally Near-threatened (Collar *et al.* 1994). Race *plumbea* recorded as rare visitor on Hainan.

Habits: Frequents forested rivers and wooded swamps and catches fish from near surface of water.

468. Lammergeier *Gypaetus barbatus* [Bearded Vulture] Hu wujiu PLATE 50

Description: Large (110 cm) yellowish vulture with a black band through eye contrasting with otherwise whitish head. Underparts are orange-buff and upperparts brown with buff streaks. Has slight beard and adult has bare red eye-ring. In flight, the straight pointed wings and long wedge-shaped tail are diagnostic. Iris—yellow; bill—grey; feet—grey.

Voice: Generally silent. Gives loud whistle in breeding season.

Range: Africa, S Europe, Middle East, E and C Asia.

Distribution and status: High mountains up to 7000 m in W and C China.

Habits: Harries wild sheep and domestic animals waiting for animals to fall and injure themselves on cliffs or freeze in winter. Carries small prey and bones of larger prey to drop them onto rocks to break them for easier eating.

469. White-rumped Vulture *Gyps bengalensis* [White-backed/   PLATE 50
Indian White-backed Vulture] Baibei wujiu

Description: Large (84 cm) vulture with short rounded tail and diagnostic long naked neck. When seen in flight from below, whitish wing lining and white neck collar contrast with blackish flight feathers and dark brown underparts; from above generally blackish with dark grey secondaries and whitish rump. Immature is brownish-grey with no contrasting white. Soars with wings in slight 'V'. Iris—reddish; brown; bill—grey; feet—grey.

Voice: Harsh grunts and shrieks on carcasses.

Range: Indian subcontinent, SW China and SE Asia.

Distribution and status: Globally Near-threatened (Collar *et al.* 1994). Very rare in W and SW Yunnan.

Habits: Circles high in sky searching for carrion. Roosts in trees when fed or drinking at water.

470. Himalayan Griffon *Gyps himalayensis* [AG: Vulture]   PLATE 50
Gaoshan wujiu

Description: Large (120 cm) pale buffy-brown vulture with white streaked underparts. The head and neck are lightly feathered with white down. Has collar of fluffy buff plumes. Primaries black. Immature dark brown with streaking from pale feather shafts. Flight appears in very slow motion. Has long, upturned 'fingers' at wing tips with wings held just above horizontal. Tail shorter than Eurasian Griffon, adults are generally paler with less streaking below and juvenile is darker than Eurasian Griffon. Iris—orange; bill—grey; feet—grey.

Voice: Gives occasional clucking and whistling sounds.

Range: C Asia to Himalayas.

Distribution and status: Common scavenger in some parts of the Himalayas, Qinghai–Xizang Plateau and W and C China in open high altitude habitats.

Habits: Usually seen soaring, sometimes in small parties, or roosting among rocky crags.

[471. Eurasian Griffon *Gyps fulvus* Wujiu   PLATE 50

Description: Large (100 cm) brown vulture with fluffy whitish ruff at base of neck and yellowish-white head and neck. Immature has brown ruff. Very similar to Himalayan Griffon and separated in flight by upperparts cinnamon rather than pale khaki and breast with finer pale shaft streaks. Distinguished from Cinereous Vulture by pale underparts and square or rounded but never wedge-shaped tail. Iris—brown; bill—horn with black cere; feet—dull greenish-yellow.

Voice: Raucous screeching.

Range: S Europe, N Africa, C Asia, Afghanistan, Pakistan, Nepal, Himalayas and N India.

Distribution and status: Occasionally ranges to NE Indian subcontinent up to 3000 m. Not recorded in China but known from disputed areas or Arunachal Pradesh and probably occurs in SE Xizang.

Habits: A large vulture of open, rocky mountains.]

472. Cinereous Vulture *Aegypius monachus* [Black Vulture] PLATE 50
Tu jiu

Description: Huge (100 cm) dark brown vulture with fluffy ruff and bluish neck. Juvenile has blackish face, black bill and pink cere; adult has buffy bare head with blackish throat and area below eye, horn coloured bill and pale blue cere. Juvenile often has fluffy tuff on back of head. In flight, most likely to be confused for dark *Aquila* eagle than another vulture. Wings long and broad with parallel edges, prominently indented trailing edge and with seven deeply splayed 'fingers' at tip. Tail short and wedge-shaped, very powerful head and bill. Iris—dark brown; bill—horn with blue cere; feet—grey.

Voice: Generally silent.

Range: Spain, Balkans, Turkey to C Asia and N China. Stragglers wander outside breeding range.

Distribution and status: Globally Near-threatened (Collar *et al.* 1994). Rare but more common in N of range. Breeds in Kashi and Tianshan in W Xinjiang, S and E Qinghai, Gansu, Ningxia, Nei Mongol and N Sichuan. Sporadic sightings over S Xizang, C, E and SE China and Taiwan.

Habits: Feeds on carrion but will hunt live prey. Dominates other vultures at carcass. Mixes with Himalayan Griffon. Soars for hours.

473. Red-headed Vulture *Sarcogyps calvus* [King Vulture] PLATE 50
Hei wujiu

Description: Large (80 cm) black vulture with red head, neck and legs and white neck ruff diagnostic. Immature is browner with pink rather than red bare parts. In flight has short, wedge-shaped tail, white sides and grey white line on base of secondaries. Juvenile has whitish undertail coverts and may show whitish area on lower back. Glides on level wings with 'hand' slightly lowered; soars on only slightly raised wings. Iris—brown; bill—black with red cere; feet—red.

Voice: Hoarse croaks leading into scream when squabbling. Roaring call in copulation.

Range: Indian subcontinent and SE Asia.

Distribution and status: Globally Near-threatened (Collar *et al.* 1994). Very rare in China. Recorded in extreme SW Yunnan, but probably also occurs in SE Xizang.

Habits: Soars over open and marshy plains up to 1500 m. Feeds with other scavengers.

474. Short-toed Snake Eagle *Circaetus gallicus* [Short-toed Eagle]   PLATE 50
Duanzhi diao

Description: Large (65 cm) pale, heavy-bodied eagle. Upperparts greyish-brown; underparts white with dark streaked or uniform brown throat and breast and indistinct bars on belly; tail with broad indistinct bars. Immature are even paler than adults. In flight, long, broad wings with distinctive longitudinal barring on coverts and flight feathers. Iris—yellow; bill—black; cere—grey; feet—greenish.
Voice: Generally silent in winter, occasional plaintive mewing.
Range: Africa, Palearctic, Indian subcontinent, N China and Lesser Sundas.
Distribution and status: Breeds in Tianshan in NW Xinjiang and probably in NC China. Rare migrants recorded in scattered localities of N China. Probably much overlooked and certainly expected in SE Xizang.
Habits: Inhabits forest edge and secondary scrub. Circles and glides with wings held very straight and level. Often hovers like a giant kestrel.

475. Crested Serpent Eagle *Spilornis cheela* She diao   PLATE 50

Description: Medium-sized (50 cm) dark eagle with very broad, rounded wings and rather short tail. Adult: upperparts dark brown/grey, underparts brown with belly, flanks and vent spotted with white. Tail has broad whitish-grey bar between black bars; short, puffy, broad crest of black and white feathers; yellow naked area between eye and bill characteristic. In flight, broad white tail bar and white trailing edge to wing diagnostic. Immature: similar to adult but browner with more white in plumage. Iris—Yellow, bill—grey-brown; feet—yellow.
Voice: A very vocal eagle, frequently soaring over forest canopy, uttering a loud shrill cry *kiu-liu* or *kwee-kwee, kwee-kwee, kwee-kwee-kwee*.
Range: Indian subcontinent, S China, SE Asia, Palawan and Greater Sundas.
Distribution and status: Resident in SE Xizang and W Yunnan (*burmanicus*); rest of China south of Changjiang (*ricketti*); Hainan (*hoya*) and Taiwan (*rutherfordi*). Probably the commonest eagle over wooded and forested hills up to 1900 m.
Habits: Frequently seen circling over forest or plantations, pairs often calling to each other. Courting pairs perform sluggish aerobatics. Often perches on large branches in shady parts of the forest where it can watch the ground.

476. Eurasian Marsh Harrier *Circus aeruginosus* [Western Marsh   PLATE 51
Harrier] Baitou yao

Description: Medium-sized (50 cm) dark harrier. Male: similar to subadult male Eastern Marsh but head more buff with less dark streaking. Female and immature: similar to Eastern Marsh but back darker brown, tail unbarred and crown lacks bold dark streaking. Female never has pale rump. White patch on under wing primeries (if present) lacks dark mottling. Iris—yellow (male) light brown (juvenile female); bill—grey; feet—yellow.
Voice: Generally silent.
Range: Breeds W and C Palearctic to W China; winters south to Africa, Indian subcontinent and S Burma.

Distribution and status: Breeds in Tianshan and surrounding NW Xinjiang. Migrates S to winter in NE India. Uncommon.

Habits: As Eastern Marsh Harrier.

477. Eastern Marsh Harrier *Circus spilonotus* [Spot-backed   PLATE 51
Harrier/Spotted Marsh Harrier] Baifu yao

Description: Medium-sized (50 cm) dark harrier. Male is similar to male Pied Harrier but black throat and breast heavily streaked white. Female is distinct among all female harriers except Western Marsh in having brown, or sometimes pale, uppertail coverts. Plumage dark brown with buff crown, nape, throat and leading edge of wing. Crown and nape streaked dark brown. Tail is barred; a whitish patch on base of primaries boldly mottled dark when viewed from below. Some individuals have entire head buffy and buff patches on breast. Immature like female but darker, with only crown and nape buff. Iris—yellow (male), light brown (juvenile female); bill—grey; feet—yellow.

Voice: Generally silent.

Range: Breeds in E Asia; winters S to SE Asia and Philippines.

Distribution and status: Breeds in NE China migrating to winter S of 30°N. Fairly common in lowlands.

Habits: Frequents open areas, especially grassy marshes or reedbeds. Glides gracefully low over vegetation and sometimes hovers. Flight heavier and less buoyant than Pied Harrier.

Note: Treated by some authors as a race of Eurasian Marsh Harrier *C. aeruginosus*.

478. Hen Harrier *Circus cyaneus* [Northern/White-rumped Harrier]   PLATE 51
Baiwei yao

Description: Male: largish (50 cm) grey or brown harrier with conspicuous white rump and black wing tips. Larger than Montagu's; larger and darker than Pallid. Lacks black bar across secondaries of Montagu's; black wing tips more extensive than Pallid. Brown female distinguished from Montagu's by pale collar, plain head and lack of rufous bars on underwing coverts. From Pallid by dark trailing edge to 'hand' of wing and paler secondaries and streaking of upper breast. Juvenile from both Pallid and Montagu's by shorter, broader wings with more rounded tip. Iris—pale brown; bill—grey; feet—yellow.

Voice: Generally silent.

Range: Breeds Holarctic; migrates S in winter to N Africa, S China, SE Asia and Borneo.

Distribution and status: Fairly common seasonal migrant. Nominate race breeds in Kashi region of W Xinjiang and Hebei and NE provinces. Found on passage over eastern half of China and wintering in E Qinghai, SE Xizang and rest of China S of Changjiang River.

Habits: Frequents open country, grasslands and cultivation. Flight is slower and more laboured than Pallied or Montagu's Harriers.

479. Pallid Harrier *Circus macrourus* Caoyuan yao   PLATE 51

Description: Male: medium-sized (46 cm) pale grey harrier with small black wedge on wing tip. Very like Montagu's Harrier without clear-cut white rump and with whitish head and distinct wing tip pattern. Lacks black line across base of secondaries. Paler than Hen Harrier. Brown female from female Hen Harrier by lack of dark trailing edge of 'hand' in flight and from Montagu's by pale collar and lack of rufous bars on underwing coverts. Shows paler primaries and darker secondaries; contrast between pale leading underwing coverts and dark rear wing coverts and between dark upper breast and pale lower breast. Juvenile from Hen Harrier by longer, thinner wings and from Montagu's by wing 'fingers' never all dark. Iris—yellow; bill—yellow; feet—yellow.

Voice: Silent in winter.

Range: Breeds from C Palearctic to W China; winters S in Africa, Indian subcontinent, S China and Burma.

Distribution and status: Rare seasonal migrant in China. Breeds in Tianshan area of W Xinjiang. Sporadic records from W Xinjiang, Hebei, Jiangxi, Jiangsu, Guangxi, S Xizang and Hainan.

Habits: Typical of genus. Glides low over open country in search of prey.

480. Pied Harrier *Circus melanoleucos* Que yao PLATE 51

Description: Smallish (42 cm) slender-winged harrier. Male: black, white and grey plumage. Unstreaked black head, throat and breast diagnostic. Female: streaky brown upperparts tinged grey, white rumps, barred tail, buff underparts streaked rufous; underside of flight feathers barred blackish. Immature: dark brown upperparts with whitish band on uppertail coverts. Underparts rufous-chestnut with rufous-buff streaks. Iris—yellow; bill—horn; feet—yellow.

Voice: Generally silent.

Range: Breeds in NE Asia; migrating S in winter to SE Asia, Philippines and N Borneo.

Distribution and status: Breeds in NE migrating to winter in S and SW China. Not uncommon.

Habits: Glides low over open country, marshes, reed beds and rice fields.

481. Montagu's Harrier *Circus pygargus* Wuhui yao PLATE 51

Description: Male: medium-sized (46 cm) grey harrier with black wing tips. Smaller slimmer and more bouyant than Hen Harrier. Separated from Hen and Pallid by single black band across upper secondaries and two bands on underwing. Pale rump not as distinct as in Hen Harrier. Brown female distinguished from Hen and Pallid Harriers by lack of pale collar and by broad gap between dark bars on secondaries in flight from below. Juvenile from Hen Harrier by longer, slender wings and from Pallid by all dark 'fingers' in flight. Iris—yellow; bill—yellow; feet—yellow.

Voice: Male display call is shrill *kek, kek, kek*. Female gives fast *jick-jick-jick* in alarm.

Range: W and C Palearctic, Indian subcontinent and C Asia.

Distribution and status: Rare seasonal migrant. Breeds in Tianshan area of W Xinjiang with sporadic records of wintering birds in Shangdong, Changjiang River, Fujian and Guangdong.

Habits: As others of genus.

482. Crested Goshawk *Accipiter trivirgatus* Fengtou Ying PLATE 52

Description: Large (42 cm) powerful *Accipiter* with small nuchal crest. Adult: upperparts grey-brown with banding on wings and tail; underparts white; chest streaked rufous; bold brown bars on white belly and thighs. The throat is white with black a mesial moustache lines. Immature: as adult but streaks and bars of underside are blackish and upperparts are paler brown. In flight, appears to have shorter, rounder wings than other *Accipiter* hawks. Iris—brown (juresile) or yellow (adult); bill—grey with yellow cere; legs and feet—yellow.

Voice: A shrill scream: *he-he-he-he-he-he* and prolonged yelp.

Range: Indian subcontinent, SW China, Taiwan, SE Asia, Philippines and Greater Sundas.

Distribution and status: Locally not uncommon in lowland forest in SC and SW China including Hainan (*indicus*) and Taiwan (*formosae*). Now common in Hong Kong.

Habits: Keeps to relatively thick forest cover. During breeding season, often soars over forest canopy calling loudly.

483. Shikra *Accipiter badius* [Little Banded Goshawk] He'er ying   PLATE 52

Description: Medium-sized (33 cm) rather pale *Accipiter*. Male: pale blue-grey upperparts with contrasting black primaries; white throat with faint grey mesial stripe; chest and belly narrowly barred rufous and white. Female: like male but back is brown and throat greyer. Immature: grey-brown scalloped rufous, brown-streaked underparts and black mesial line; distinguished from immature Eurasian Sparrowhawk by streaked underparts and from Besra by paler upperparts and narrower tail bars. Iris—yellow to brown; bill—brown; feet—yellow.

Voice: Generally silent. Piping *kyeew* in breeding area.

Range: Africa to Indian subcontinent, S China, SE Asia.

Distribution and status: Race *cenchroides* extreme W Xizang. Race *poliopsis* is a rare lowland resident in Guizhou, Guangxi, Guangdong, Yunnan and Hainan.

Habits: Hunts from tree perches at forest edge, open wooded areas and farmland, chasing birds and sometimes circling in the sky.

484. Chinese Sparrowhawk *Accipiter soloensis*[Horsfield's/Grey   PLATE 52
Sparrowhawk/Frog-Hawk AG: Goshawk] Chifu ying

Description: Medium-sized (33 cm) hawk with very pale underparts. Adult: upperparts pale blue-grey with sparse white tips to back feathers and faint black barring on outer tail feathers; underparts white with faint pinkish wash on breast and flanks and some light grey barring on flanks and slight barring on thigs. Underwing of adult distinctive—almost entirely white except for black tips to primaries. Immature: brown upperparts with dark bars on tail and white underparts, streaked on throat and barred brown on chest and thighs. Iris—red or brown; bill—grey with black tip; cere—orange; feet—bright orange.

Voice: In breeding season shrill nasal piping in rapid accelerating series with descending pitch.

Range: Breeds in NE Asia and China; migrating S in winter to SE Asia, Philippines, Indonesia and New Guinea.

Distribution and status: Not uncommon, breeding throughout the southern half of China up to 900 m. Migrant through Taiwan and Hainan.

Habits: Hawks after small birds in open wooded areas. Also eats frogs. Usually hunts from a perch in a swift dash but sometimes circles overhead.

485. Japanese Sparrowhawk *Accipiter gularis* [Japanese Lesser/   PLATE 52
Asiatic Sparrowhawk] Riben songqueying

Description: Small (27 cm) hawk, very similar in appearance to Chinese Sparrowhawk and Besra but noticeably smaller and more dashing with narrower tail bands. Adult male: upperparts dark grey; tail grey with several dark bands; pale rufous breast and belly with very thin vertical mesial stripe; no pronounced moustachial. Female: upperparts brown instead of grey; lacks rufous on underparts which are heavily barred brown. Immature: streaked rather than barred on chest and more rufous. Iris—yellow (immature) to red (adult); bill—blue-grey, tipped black, with green-yellow care; feet—green-yellow.
Voice: Occasional harsh cries.
Range: Breeds E Palearctic; winters in SE Asia, Philippines and Greater Sundas.
Distribution and status: Race *gularis* breeds in NE provinces and perhaps Altai Mts migrating to winter in SE China S of 32 °N. Not uncommon.
Habits: Typical forest sparrowhawk. Wingbeat is fast and fluttery. Migrates in flocks.
Note: Included within Besra by Cheng (1987, 1994).

486. Besra *Accipiter virgatus* Song queying   PLATE 52
Description: Medium-sized (33 cm), dark hawk. Adult male: upperparts dark grey with strongly barred tail. Underparts white with rufous and brown bars; throat white with black mesial stripe and black moustachial. Female and immature: less rufous on breast and barred with reddish-brown on underside; back brown, tail brown with dark bars. Immature has more strongly streaked breast. Iris—yellow; bill—black with yellow cere; legs and feet—yellow.
Voice: Young birds give repeated *shew-shew-shew* cry when hungry.
Range: Indian subcontinent, S China, SE Asia, Philippines and Greater Sundas.
Distribution and status: Race *affinis* is resident in C and SW China and Hainan. Race *nisoides* is resident in SE China. Race *fuscipectus* is resident on Taiwan. Widespread in wooded and forested hills and mountains from 300–1200 m.
Habits: Sits quietly in the forest watching for reptile or avian prey.

487. Eurasian Sparrowhawk *Accipiter nisus* [Northern/   PLATE 52
Common Sparrowhawk] Que ying

Description: Medium-sized (male 32 cm, female 38 cm) short-winged hawk. Male: brownish-grey above with underparts finely barred rufous on white and with banded tail. Rufous cheeks diagnostic. Female: larger; brown above with white underparts, narrowly barred grey-brown on breast, belly and thighs. No mesial stripe on throat. Cheek patch less rufous. Immature: distinctive from other immature *Accipiter* hawks in having brown barring on breast and no streaks. Iris—bright yellow; bill—horn with black tip and yellow cere feet—yellow.
Voice: Occasional shrill cries.
Range: Breeds in Palearctic; migrants reach Africa, Indian subcontinent, SE Asia.

Distribution and status: Race *nisosimilis* breeds in NE provinces and Tianshan, NW Xinjiang. Migrates to winter in SE and C China, Taiwan and Hainan. Race *melaschistos* breeds from C Gansu S through W Sichuan and S Xizang to N Yunnan. Migrates to winter in SW China. A common forest bird.

Habits: Hunts from tree perch or on the wing in 'ambush-flight' along forest edge or in open wooded areas.

488. Northern Goshawk *Accipiter gentilis* Cang ying   PLATE 52

Description: Large (56 cm) powerful hawk without crest or mesial throat stripe and with distinctive broad white eyebrow. Underparts of adult are white, finely barred pinkish-brown. Upperparts rather uniform grey. Juvenile is browner with pale feather edges giving scaly pattern to upperparts and having bold black streaking on underparts. Iris—red (adult) yellow (juvenile); bill—horn-grey; feet—yellow.

Voice: Begging call of young is melancholy *PEEE-leh*. Alarm call is cackling *kyekyekye...*

Range: N America, Eurasia, N Africa.

Distribution and status: Race *schvedowi* breeds in Greater and Lesser Hinggan Mts of NE China and W Tianshan of NW China, migrating across China to winter S of Changjiang River; *khamensis* breeds along the eastern hills of the Qinghai–Xizang Plateau in SE and E Xizang, NW Yunnan, W Sichuan and S Gansu, wintering in lowlands and S Yunnan. Race *fujiyamae* winters on Taiwan; *albidus* winters in NE China and *buteoides* winters in Tianshan region of NW China. Fairly common in temperate and subalpine forests.

Habits: A hawk of woodlands, able to twist and turn fast on its broad rather rounded wings. It preys largely on pigeons but also can take gamebirds and mammals as large as hares.

489. White-eyed Buzzard *Butastur teesa* Baiyan kuangying   PLATE 53

Description: Medium-sized (43 cm) grey buzzard with white throat, two dark cheek stripes and a central stripe from chin. Has small white patch on nape. Underparts brown and whitish. At rest, wings almost reach end of rufous-tinged tail. Similar to Grey-faced Buzzard but with white eye. Immature is brown with buffy head, grey wings and rufous thighs. In flight, adult's silvery grey-brown underside to broad, blunt wings, contrast with darker body and coverts. Patch of buffy-grey on wing shoulder of upperwing. Juvenile in flight is almost all-creamy white below with narrow black wing tips. Iris—white; bill—bluish-grey with yellow cere; feet—yellow.

Voice: Peculiar, plaintive mewing *pit-weer, pit-weer*, uttered at nest and in flight.

Range: Indian subcontinent and Burma.

Distribution and status: Extremely rare straggler to China. Recorded at Gyangze in S Xizang.

Habits: As Grey-faced Buzzard. Sluggish and rather tame. Perches on pole or tree perch and pounces on small prey including grasshoppers.

490. Rufous-winged Buzzard *Butastur liventer*   PLATE 53
Zongchi kuangying

Description: Medium-sized (40 cm) buzzard with chestnut wings and tail and pale underparts; head and nape brownish-grey; upperparts brown, mottled and streaked with black. Chin, throat and chest grey, abdomen and vent white. Wings are long and rather pointed; tail is long, slender and square-cut. Iris—yellow; bill—yellow with black tip; cere—yellow; feet—yellow.

Voice: Silent except during breeding season when it noisily gives repeated long-drawn shrill mews, *pit-piu* with first note higher.

Range: SW China, SE Asia, Sulawesi and Java.

Distribution and status: Globally Near-threatened (Collar *et al.* 1994). A rare resident below 800 m in S Yunnan.

Habits: Inhabits dry open forest near rivers or swamps. Generally hunts from tree perch.

491. Grey-faced Buzzard *Butastur indicus* Huilian kuangying   PLATE 53

Description: Medium-sized (45 cm) brownish buzzard with prominent white chin and throat and central black stripe and black moustachial. Sides of head blackish; upperparts brown, streaked and barred blackish; breast brown streaked black. Rest of underparts barred rufous, distinguishing it from White-eyed Buzzard. Tail is slender and square-cut. Iris—yellow; cere—yellow; bill—grey; feet—yellow.

Voice: Tremulous *chit-kwee* with rising second note.

Range: Breeds in NE Asia; winters SE Asia, Philippines and Indonesia.

Distribution and status: Breeds in coniferous forest in NE provinces migrating through E China, on passage occurs in Qinghai, S of Changjiang River and on Taiwan.

Habits: Inhabits open wooded areas up to 1500 m. Rather slow and laboured flight, preferring to hunt from a tree perch.

492. Common Buzzard *Buteo buteo* [The/Steppe Buzzard]   PLATE 53
Putong kuang

Description: Largish (55 cm) reddish-brown buzzard. Upperparts dark reddish-brown; sides of face buff streaked reddish with prominent chestnut moustachial; underparts whitish with rufous streaks and washed rufous on flanks and thighs. In flight the broad, rounded wings and white patch at the base of the primaries are diagnostic. There is usually a black subterminal tail bar. Soars with wings in slight. Iris—yellow to brown; bill—grey, tipped black; cere—yellow; feet—yellow.

Voice: Loud mewing *peeioo*.

Range: Breeds Palearctic and Himalayas; some winter to N Africa, Indian subcontinent and SE Asia.

Distribution and status: Race *japonicus* breeds in conifer forest in NE provinces and winters S of 32 N including SE Xizang, Hainan and Taiwan. Race *vulpinus* winters in Tianshan and Kashi regions of W Xizang and also in Sichuan. Quite common up to 3000 m.

Habits: Prefers open country where it circles on thermals high overhead or rests on exposed tree branches. One of the few hawks that regularly hovers.

493. Long-legged Buzzard *Buteo rufinus* [Long-legged Hawk]   PLATE 53
Zongwei kuang

Description: Large (64 cm) rufous buzzard with long wings and tail. Head and breast usually pale, darkening toward belly but several morphs vary in colour from cream to rufous to very dark. Blackish forms have dark bars on flight feathers and tail. Uppertail usually unbarred pale rusty-orange. In flight, similar to rufous form of Common Buzzard but with distinctive large black carpal patch on underwing. Wings kinked when gliding (c.f. straight wings in Common Buzzard) and held at high angle when soaring. Juvenile has finely barred outer tail and dark trailing edge to underwing. Lack of black band at tip of upper tail separates from Rough-legged Buzzard. Iris—yellow; bill—grey; feet—yellow.
Voice: Loud, wailing, kitten-like mewing. Similar to Common Buzzard but less vocal.
Range: Breeds SE Europe to C Palearctic, NW India, E Himalayas and W China; winters S.
Distribution and status: Rare resident and seasonal migrant. Nominate race breeds in Kashi, Urumqi and Tianshan areas of Xinjiang. Migrant or wintering to Gansu, Yunnan and S and SE Xizang.
Habits: Sluggish. Usually hunts from perch. Soars and sometimes hovers. Will follow fires.

494. Upland Buzzard *Buteo hemilasius* Da kuang   PLATE 53
Description: Large (70 cm) rufous buzzard with several morphs. Similar to Long-legged Buzzard but larger and with whitish upperside of tail, usually banded; dark thighs; more distinct dark banding on secondaries; dark rufous wing lining in pale morphs. Dark morph has smaller white patch on underside of primaries than in Long-legged. Tail usually brown rather than rufous. Iris—yellow or whitish; bill—bluish-grey with greenish-yellow cere; feet—yellow.
Voice: Mewing calls, more prolonged and nasal than Common Buzzard.
Range: Tibetan Plateau, Mongolia, C and E China.
Distribution and status: Fairly common in north of range, rare in south. Breeds in N and NE China and E and S parts of Xizang–Qinghai Plateau. Probably also breeds in NW China. Northern birds migrate S to winter in C and E China with accidental reported in Guangxi, Guangdong and Fujian.
Habits: Powerful buzzard able to catch hares and snowcock. Reported to kill lambs.

495. Rough-legged Buzzard *Buteo lagopus* Maojiao kuang   PLATE 53
Description: Medium-sized (54 cm) brown buzzard. Similar to Common Buzzard but has white inner part of tail, black carpal patches and pale head. Some pale morph Common Buzzards have as pale tails but also have pale underwings. In Rough-legged there is a greater contrast between dark wings and pale tail. Base of primaries whiter than Common Buzzard and contrasting with black carpal patch. Pale head of female and

juvenile contrasts with dark breast. Juvenile has less black trailing edge to underwing in flight. Adult male has dark head and paler breast. Tarsus is feathered. Iris—yellow-brown; bill—dark grey with yellow cere; feet—yellow.

Voice: Calls like a Common Buzzard but more powerful.

Range: Holarctic.

Distribution and status: Rare winter visitor and migrant. Nominate race winters in Kashi and Tianshan region of NW China; *kamtschakensis* migrates through or winters in W Xinjiang and NE provinces, Shandong, Shaanxi and Jiangsu and also winters in Yunnan, Fujian, Guangdong and Taiwan.

Habits: Hovers more frequently than Common Buzzard. Flies like a large harrier.

496. Black Eagle *Ictinaetus malayensis* [Indian/Asian Black Eagle]   PLATE 54
Lin diao

Description: Large (70 cm), brownish-black eagle. At rest wings extend beyond tail tip. Distinguished in flight from other dark eagles by long broad tail, long wings with narrow 'arm' broadening to broad 'hand' with pronounced 'fingers'. Has inconspicuous pale patch at the base of the primaries, slight grey barring on the tail and uppertail coverts. Immature: paler with buff streaks and edges to feathers and pale thighs. Iris—brown; bill—black, tipped grey; cere—yellow; feet—yellow.

Voice: Repeated plaintive *kleeee-kee* or *hee-lee-leeeuw*.

Range: Indian subcontinent, SE China, SE Asia, Sulawesi, Moluccas and Greater Sundas.

Distribution and status: Rare resident in Taiwan, Fujian and N Guangdong. Occasional in SW Yunnan and SE Xizang up to 3000 m.

Habits: Inhabits forests where it is usually seen circling low over the canopy. Regularly raids the nests of other birds.

497. Greater Spotted Eagle *Aquila clanga* Wu diao   PLATE 54

Description: Large (70 cm) uniform dark brown eagle with short tail. Plumage colour varies with age and race. Juvenile has conspicuous white spots and bars on upperwings and back. All plumages show white U shape on uppertail coverts in flight seen from above. Tail much shorter than Golden or Imperial Eagles. Iris—brown; bill—grey; feet—yellow.

Voice: Generally silent.

Range: Breeds E Europe, S Russia, S Siberia, Turkestan, NW and N Indian subcontinent, N China; winters to NE Africa, S India, S China and SE Asia through to Indonesia.

Distribution and status: Globally Vulnerable (Collar *et al.* 1994). Breeds across northern China wintering in or migrating through S of the country. Never common but rather regular.

Habits: Inhabits open swampy areas near lakes and marshes or open areas during migration. Feeds mostly on frogs, snakes, fish and birds.

498. Steppe Eagle *Aquila nipalensis* Caoyuan diao   PLATE 54

Description: Large (65 cm) uniformly dark brown rugged-looking eagle with squarish tail. Adult is difficult to separate from other all-dark eagles but has grey remiges below and sparse barring and dark trailing edge to wings. Sometimes shows pale wing band on greater underwing coverts like juvenile. Head is small and slim and protruding compared to Spotted Eagle. Wings longer and fingers more splayed than Spotted Eagle. In flight holds wings straight and level. Glides with wings slightly bowed. Juvenile distinctive milky coffee colour with white bar across underwing, white tip to black tail and white band along trailing edge of wing contrasting with black flight feathers. Upperwing with two buffy wing bars and buffy V on uppertail coverts. Tail sometimes wedge-shaped. Iris—pale brown; bill—grey with yellow cere; feet—yellow.

Voice: Grating harsh calls and cackles.

Range: Breeds Altai, Mongolia and SE Siberia; winters in N Indian subcontinent, S China and SE Asia.

Distribution and status: Fairly common on northern steppes. Breeding or summer visitor in Kashi and Tianshan regions of W Xinjiang east to Qinghai, Nei Mongol and Hebei. Seen on passage over most of the country and wintering in Guizhou, Guangdong and Hainan. Some records dubious—all Hong Kong records now regarded as doubtful.

Habits: Sluggish. Sometimes forms large flocks on migration.

Note: Formerly treated by many authors as a race of Tawny Eagle *A. rapax.* (e.g. Cheng 1987).

499. Imperial Eagle *Aquila heliaca* Baijian diao PLATE 54

Description: Large (75 cm) dark brown eagle with buff crown and nape and white-tipped feathers on side of mantle. Base of tail is banded black and grey, contrasting with rest of dark brown plumage. In flight all-black body and underwing coverts are diagnostic. Glides with bent wing. Juvenile is buff with dark streaks over body and coverts. In flight has narrow white trailing edge to wings and dark tail, dark flight feathers except for pale primary wedge. Has large creamy area on lower back and rump. From above in flight, has two pale bars on coverts. Iris—pale brown; bill—grey with yellow cere; feet—yellow.

Voice: Fast, barking *owk, owk, owk.*

Range: Palearctic, NW Indian subcontinent (former breeder now winter visitor), China.

Distribution and status: Globally Vulnerable (Collar *et al.* 1994). Uncommon seasonal migrant declining and threatened. Nominate race breeds in Tianshan region of NW Xinjiang. Birds sometimes recorded on passage across NE coastal provinces and wintering around Qinghai Lake (Koko Nor), NW Yunnan, Gansu, Shaanxi, middle reaches of Changjiang River and in Fujian and Guangdong. Small numbers annually visit Hong Kong.

Habits: Lives in open country. Heavy and sluggish, perches for hours on stump or pole. Pirates food from other raptors. Slow flight like a vulture.

500. Golden Eagle *Aquila chrysaetos* Jin diao PLATE 54

Description: Large (85 cm) rich brown eagle with golden-crested head and conspicuous white rump in flight. Bill is massive. In flight distinguished by long, evenly rounded tail, wings held in shallow V. Separated from Imperial by lack of white on scapulars.

Immature is distinctive with white wing patches and white base of tail. Iris—brown; bill—grey; feet—yellow.

Voice: Generally silent.

Range: N America, Europe, Middle East, E and W Asia and N and NE Africa.

Distribution and status: Race *daphanea* is widespread but uncommon in mountainous regions over much of the country and to high altitudes in the Himalayas. Race *canadensis* breeds in NE Nei Mongol and winters in Changbai range of NE China. Vagrants seen occasionally in E and S coastal provinces.

Habits: Lives in rugged steppes and mountains with cliffs and open country where it feeds on game birds, marmots and other mammals. Soars majestically on thermals.

501. Bonelli's Eagle *Hieraaetus fasciatus* [Slender Hawk-eagle] PLATE 54
Baifu sundiao

Description: Large (59 cm) raptor with dark wing tips, narrowly barred wings and tail and distinctive silhouette with shortish, broad, rounded wings and longish tail. Adult has black terminal band to pale tail; dark underwing coverts with pale leading edge of wing; pale breast with dark streaks. Adult in flight from above has white patch on upper back. Juvenile has black trailing edge to wing, dark bar along greater coverts but coverts otherwise pale. Upperparts generally brown; head buff with dark streaks and darkish sides to face. In flight, soars on flat level wings. Iris—yellow-brown; bill—grey with yellow cere; feet—yellow.

Voice: Shrill creaking cry; chattering *kie, kie, kikiki.*

Range: N Africa; Eurasia, Indian subcontinent, E China; winters to Lesser Sundas.

Distribution and status: Uncommon resident. Nominate race breeds in SW Guangxi, Guangdong, Guizhou, Hubei, middle reaches of Changjiang River, Fujian and Zhejiang. Stray recorded in Hebei in summer.

Habits: Lives in open mountain regions; often soaring in pairs. Quick wing beat.

502. Booted Eagle *Hieraaetus pennatus* Xue sundiao PLATE 54

Description: Smallish (50 cm), rufous-breasted (dark morph) or buffy-white-breasted (light morph), crestless eagle with feathered legs. Upperparts brown mottled with black and buff, darker brown on wings and tail. In flight dark primaries contrast strongly with buff (light morph) or rufous (dark morph) underwing coverts; undertail pale. Iris—brown; bill—blackish with yellow cere; feet—yellow.

Voice: Thin high *keee.*

Range: Breeds Africa, SW Eurasia, NW Indian subcontinent and N China; migrates S in winter to Africa, Indian subcontinent; vagrant SE Asia.

Distribution and status: Breeds in Tianshan and Korla in W Xinjiang and N Xizang. Rare migrants seen in several areas of NE and E China.

Habits: As Rufous-bellied Eagle

503. Rufous-bellied Eagle *Hieraaetus kienerii* PLATE 54
[Chestnut-bellied Eagle] Zongfu sundiao

Description: Smallish (50 cm) rufous, black and white eagle with short crest. Adult: blackish crown, cheeks, and upperparts; tail dark brown with black bars and white tip; chin, throat and breast white, streaked black; flanks, abdomen, thighs and undertail rufous with black streaks on abdomen. In flight shows conspicuous, rounded pale patch at base of primaries. Immature: upperparts blackish-brown with blackish eye patch and whitish eyebrow; underparts whitish. Iris—red; bill—blackish, cere—yellow; feet—yellow.

Voice: High-pitched scream *chirrup* preceded by several ascending preliminary notes.

Range: S Indian subcontinent, Himalayas, Hainan, SE Asia, Philippines, Sulawesi and Greater Sundas.

Distribution and status: Recorded only as a rare resident on Hainan in forests up to 1500 m. Could occur in S Yunnan and SE Xizang.

Habits: Generally seen circling or gliding fairly low over the trees.

504. Mountain Hawk Eagle *Spizaetus nipalensis* Ying diao   PLATE 54

Description: Large (74 cm), slender eagle with feathered legs, very broad wings and long, rounded tail. Has long erectile crest. Occurs in dark and light morphs. Dark morph: upperparts brown, streaked and mottled black and white on upperparts; reddish-brown tail with several black bars; chin, throat and chest white with bold black mesial line and streaks; lower abdomen, thighs and undertail rufous banded with white. Light morph: upperparts greyish-brown; underparts whitish with darkish eye-stripe and moustachial. Iris—yellow to brown; bill—blackish; cere—greenish-yellow; feet—yellow.

Voice: Prolonged shrill screams.

Range: Indian subcontinent, Burma, China and SE Asia.

Distribution and status: Race *nipalensis* is a rare resident in S Xizang and W Yunnan up to 4000 m. Race *fokiensis* is a rare resident in SE China, Taiwan and Hainan up to 2000 m. Race *orientalis* breeds in NE China and winters to Taiwan.

Habits: Frequents forest and open woodlands. Hunting from perches or from the air.

FALCONS Family: Falconidae

A medium-sized, worldwide family of fast-flying predatory birds with long, pointed-sickle-shaped wings and long narrow tails. Falcons are the 'jet fighters' among avian raptors, stooping on prey with superior speed. The powerful beak is hooked at the tip and has two additional small lateral 'hook teeth' on the upper bill. Thirteen species of falcon occur in China of which 11 are resident.

505. Collared Falconet *Microhierax caerulescens* Hongtui xiaosun   PLATE 55

Description: Tiny (15 cm) black and white falcon with rufous throat, thighs, vent and undertail diagnostic. Black cap separated from black of back by white nape. Black band extends from eye through ear coverts. Undertail barred black and white. Immature has white throat and rufous forehead and supercilium. Iris—brown; bill—grey; feet—grey.

Voice: Shrill screams and low chattering.

Range: E Himalayan foothills and SE Asia.

Distribution and status: Extremely rare in China. Race *burmanicus* is resident in W Yunnan (Yingjing). Nominate race resident in NE Indian subcontinent, sometimes recorded up to 2000 m, may occur in SE Xizang.

Habits: Forms small parties hunting from open trees. Dashes after insects like a wood swallow. Bobs head and flicks tail when watching for prey.

506. Pied Falconet *Microhierax melanoleucos* [White-legged Falconet]   PLATE 55
[Baitui] Xiaosun

Description: Tiny (15 cm) black and white falcon. Upperparts black with white spots on innermost secondaries. Underparts white; sides of face and ear coverts black, ringed with a white line or patch. Juvenile face suffused reddish. Iris—dark brown; bill—greyish; feet—grey.

Voice: A hard, high-pitched cry *shiew* and a fast repeated *kli-kli-kli-kli*.

Range: NE Indian subcontinent, S China, N Indochina.

Distribution and status: Globally Near-threatened (Collar *et al.* 1994). Uncommon in wooded lowlands of China up to 1500 m in W and S Yunnan, Guangxi, Guangdong, Jiangxi, Fujian, S Anhui and S Jiangsu.

Habits: Sits on exposed perches at edge of forest or in open country, including paddy fields. Makes sudden dashes to catch dragonflies and other insects, and sometimes boldly attacks small birds and other prey. Nests in tree-holes.

507. Lesser Kestrel *Falco naumanni* Huangzhua sun   PLATE 55

Description: Small (30 cm) brown falcon. Male has grey head, unmarked rufous upperparts with bluish-grey rump and tail. Underparts pale rufous with white chin and vent. Breast lightly spotted black. Tail with black subterminal band and white tip. Female is browner and barred and speckled above with dark streaks on underparts. Like respective sexes of Common Kestrel but smaller and more delicate with brighter colours on male and generally fewer black spots. Male has blue-grey upperwing coverts, and lacks dark stripe below eye of Common Kestrel. In flight has a more wedge-shaped tail. Has pale claws (c.f. black in Common Kestrel). Iris—brown; bill—grey with black tip and yellow cere; feet—yellow.

Voice: Calls like Common Kestrel but faster *kikikiki . . .*; also rasping *chay-chay-chay*. Young scream as in Common Kestrel.

Range: S Europe and N Africa to C Asia, Indian subcontinent, Burma, Laos and N China. Winters to south.

Distribution and status: Globally Vulnerable (Collar *et al.* 1994). Uncommon seasonal migrant. Breeds in N and W Xinjiang, Nei Mongol and Hebei. Migrant through Shangdong, S Sichuan and Henan and winters in Yunnan.

Habits: Colonial nester on cliffs. Hovers less than Common Kestrel. Mostly eats insects. Wing beat very fast. Migrates in large flocks.

508. Common Kestrel *Falco tinnunculus* [Eurasian/European/   PLATE 55
Rock/Old World/The Kestrel] Hong sun

Description: Small (33 cm) brown falcon. Male has grey crown and nape, bluish-grey unbarred tail and rufous upperparts lightly barred black; underparts buff streaked black. Larger female: all-brown upperparts; less rufous and more boldly barred than in male. Immature: like female but with heavier streaking. Distinguished from Lesser Kestrel by rounded tail, larger size, moustachial stripe, spotting on back of male, more streaking on underparts and paler cheeks on male. Iris—brown; bill—grey, tipped black with yellow cere; feet—yellow.

Voice: Piercing cries *yak yak yak yak yak*.

Range: Africa, Palearctic, Indian subcontinent and China; wintering S to Philippines and SE Asia.

Distribution and status: Fairly common resident and seasonal migrant. Nominate race breeds in NE and NW China; *interstinctus* is resident over most of the rest of the country except dry deserts. Northern populations winter in S China, Hainan and Taiwan.

Habits: Superb aerial grace—circling lazily or hovering motionless when hunting. Dives on prey, often taken on the ground. Perches on poles and in dead trees. Prefers open country.

509. Red-footed Falcon *Falco vespertinus* [Red-legged Falcon]   PLATE 55
Hongjiao sun

Description: Small (30 cm) grey falcon with rufous vent. Similar to Amur Falcon but underwing coverts and axillaries dark grey not white. Female very different from Amur Falcon with brownish upperparts, rufous-orange crown and underparts with sparse small black streaks. Ocular area blackish and distinctive white chin, patch under eye and collar. Wings and tail grey and barred below. Underwing coverts orange-buff. Juvenile with boldly streaked whitish underparts; evenly black barred underwing; and black line under eye like Eurasian Hobby. Iris—brown; bill—grey with orange cere; feet—orange-red.

Voice: A high-pitched *ki-ki-ki*; also shrill screaming *keewi-keewi*.

Range: E Europe and W Siberia.

Distribution and status: Rare in China. Breeds in Ulungu valley in NW Xinjiang.

Habits: As Amur Falcon. Colonial. Hovers. Sometimes hunts insects in flocks before dusk like a pratincole.

510. Amur Falcon *Falco amurensis* [Eastern Red-footed Falcon]   PLATE 55
Amu'er sun

Description: Small (31 cm) grey falcon with rufous thighs, belly and vent. Similar to Red-footed Falcon but has diagnostic white underwing coverts in flight. Female: forehead white; crown grey with black streaks; back and tail grey, barred black; throat white with blackish line below eye. Underparts: creamy white, broadly streaked black on breast and barred black on belly; underwing white with black spots and bars. Immature: as female but rufous-brown markings on underparts in place of black. Iris—brown; bill—grey with red cere; feet—red.

Voice: Shrill cries like a Kestrel.

Range: Breeds Siberia to N Korea and C and NE China, also once in NE India. On passage through Indian subcontinent and Burma, and occasionally in Arabia; winters Africa.

Distribution and status: Fairly common in breeding range. Rare passage migrant over eastern and southern China.

Habits: Hawks after insects before dusk, sometimes hunts in flocks like a pratincole. Migrates in large flocks of up to several hundred, often mixed with Lesser Kestrels. Rests on telephone wires.

511. Merlin *Falco columbarius* Huibei sun   PLATE 55

Description: Small (30 cm) compact falcon without moustachial streak. Male has lavender-grey crown and upperparts with faint black streaks; lavender-grey tail with black subterminal bar and white tip; cinnamon underparts with fine black streaks and rufous nape. Has white supercilium. Female and immature have greyish-brown upperparts with grey rump, white supercilium and throat and whitish underparts with heavy dark brown streaks on breast and belly and dusky and white bars on tail. Silhouette in flight is like miniature Peregrine Fulcon. Race *pallidus* is paler than other races. Iris—brown; bill—grey with yellow cere; feet—yellow.

Voice: Alarm is rapid, accelerating series of shrill piercing notes. Begging call of young is *yeee-yeee*.

Range: Holarctic; winters S as far as N India.

Distribution and status: Uncommon seasonal migrant. Race *lymani* breeds in Tianshan and winters in Kashi region in W Xinjiang. Birds of races *insignis* and *pacificus* pass through N and E China. Race *insignis* winters in China S of the Changjiang and in E Qinghai; *pacificus* winters in SE China. Race *pallidus* winters in S Xizang.

Habits: Lives on moors and open grassland. Dashes low over ground in pursuit of small birds.

512. Eurasian Hobby *Falco subbuteo* [European/Northern/   PLATE 55
The Hobby] Yan sun

Description: Small (30 cm) long-winged black and white falcon with rufous thighs and vent. Upperparts dark grey; chest creamy-white streaked with black. Female larger than male, browner and with more streaking on thighs and undertail coverts. Readily distinguished from Oriental Hobby by whitish breast. Iris—brown; bill—grey with yellow cere; feet—yellow.

Voice: Repeated shrill *kick*.

Range: Africa, Palearctic, Himalayas, China and Burma; migrating S in winter.

Distribution and status: Locally not uncommon resident and seasonal migrant. Nominate race breeds over northern China and Xizang, wintering in S Xizang; *streichi* breeds or is summer visitor in China S of 32 °N; sometimes wintering in Guangdong and Taiwan.

Habits: Catches insects and birds on the wing in fast flight over open and wooded areas up to 2000 m.

513. Oriental Hobby *Falco severus* Meng sun   PLATE 55

Description: Small (25 cm) long-winged, rufous and black falcon. Head and upperparts dark grey with a bluish hue; underparts rich chestnut; chin buffy. Immature has black streaks on rufous chest. Iris—brown; bill—grey with yellow cere; legs and feet—yellow.

Voice: Similar *kekekeke* cry to Lesser Kestrel.

Range: NE Indian subcontinent, SE Asia to Indonesia, New Guinea and Solomon Is.

Distribution and status: Nominate race is rare migrant in W and S Yunnan, Guangxi and Hainan in lowland forest.

Habits: Flies very fast over forest, chasing insects, looking rather like a large swift. Rests on trees rather than rocks.

514. Saker Falcon *Falco cherrug* Lie sun PLATE 55

Description: Large (50 cm) heavy-chested pale falcon with whitish nape and pale brown crown. Head has little contrast and poorly-defined black line under eye with white supercilium. Upperparts rich brown and lightly barred with contrasting darker brown wing ends. Tail has narrow white tips. Below whitish with narrow dark wing tips and fine black streaks on greater underwing coverts. Paler and blunter-winged than Peregrine Falcon. Juvenile is darker brown above and heavily streaked black below. Distinguished from Peregrine Falcon by white undertail coverts. Some northern Peregrine Falcons can look very like Saker Falcon. Altai Falcon *F. c. altaicus* is darker and more slaty than race *milvipes* with rufous bands on wing coverts and more heavily streaked underparts. Iris—brown; bill—grey with pale yellow cere; feet—pale yellow.
Voice: Like Peregrine Falcon but harsher.
Range: C Europe, N Africa, N India, C Asia to Mongolia and China.
Distribution and status: Uncommon seasonal migrant. Race *milvipes* breeds in Altai and Kashi regions of Xinjiang, Xizang, Qinghai, N Sichuan, Gansu and Nei Mongol as far as Hulun Nur. Passage birds are recorded in Liaoning and Hebei. Winters in S Xizang and C China. Population in decline due to serious poaching for falconry trade. Race *altaicus* is a very rare seasonal migrant. Breeds in Tianshan in NW Xinjiang. Winters in Kashi area of W Xinjiang and around Qinghai Lake (Koko Nor) and C Nei Mongol.
Habits: Typical large falcon of high mountains and plateaux.
Note: Race *altaicus* sometimes treated as a separate species or even as a race of Gyrfalcon.

515. Gyrfalcon *Falco rusticolus* Mao sun PLATE 55

Description: Very large (56 cm) or brownish-grey falcon with black streaking on underparts and black streaks and barring on upperparts. Has black wing tips. Juvenile is greyish-brown above with white tips to feathers and white spots; tail barred with white; head whitish. Separated from Peregrine and Saker Falcons by paler colour; broader rounder wings; no pure brown or rufous tones on upperparts; head pattern less distinct. Iris—yellow; bill—grey with yellow cere; feet—yellow.
Voice: Alarm is scolding, nasal *GEHe-GEHe-GEHe* . . . more drawn-out than Peregrine Falcon.
Range: Arctic Europe, Asia and N America.
Distribution and status: Very rare winter visitor. Race *obsoletus* recorded in Heilongjiang (Heilong River and Hulan).
Habits: Typical large falcon of northern tundra and moorlands.

516. Peregrine Falcon *Falcon peregrinus* [Peregrine] You sun PLATE 55

Description: Large (45 cm) heavily-built, dark falcon. Adult: crown and cheeks blackish or with black stripe; upperparts dark grey, spotted and barred with black; underparts white with black streaks on chest and fine black bars across the belly, thighs and undertail. Female notably larger than the male 'tiercel'. Immature browner with streaked belly.

Races differ in darkness. Race *peregrinator* shows more of a hood than a moustachial stripe with reduced white on cheeks and finer barring underneath. Iris—black; bill—grey, cere yellow; legs and feet—yellow.

Voice: In breeding season a shrill *kek-kek-kek-kek*.

Range: Cosmopolitan.

Distribution and status: Uncommon resident and seasonal migrant. Race *calidus* migrates across NE and E China, wintering in S China, Hainan and Taiwan; *japonensis* winters in SE China and *peregrinator* is resident over most of China south of Changjiang River.

Habits: Lives in pairs. Shows great speed in flight making breathtaking, spiralling stoops onto its prey from high in sky. One of the fastest birds in the world. Sometimes performs acrobatics. Nests on cliff ledges.

517. Barbary Falcon *Falco pelegrinoides* Beibei sun PLATE 55

Description: Largish (42 cm) grey-backed falcon with whitish underparts and narrow black line below eye. Similar to Peregrine Falcon but more contrast between black 'hand' of wing and grey coverts and back, paler grey rump and uppertail coverts, paler underparts and diagnostic rufous nape patch. Some rufous also on crown and rear eyebrow. Juvenile is browner with more black streaking on underparts and rufous wash on pale areas of nape. Iris—brown; bill—grey with yellow cere; feet—yellow.

Voice: Harsh *keck-keck-keck-keck*, like Peregrine Falcon.

Range: N Africa, Middle East to W China.

Distribution and status: Race *babylonicus* breeds in Tianshan and Qinghai; winters in Kashi area of W Xinjiang.

Habits: As Peregrine Falcon.

Note: Formerly treated as a race of Peregrine Falcon (see Amandon and Bull 1988).

GREBES—Family: Podicipedidae

A worldwide family of small to medium-sized duck-like waterbirds. Grebes have pointed bills, short wings, very short tails, erect necks, lobed rather than webbed feet and long silky feathers. Grebes are excellent divers, able to stay under water for several minutes at a time. They feed on fish and water insects and make nests on rafts of floating vegetation. Five species occur in China.

518. Little Grebe *Tachybaptus ruficollis* [Dabchick/ PLATE 56
Red-throated Grebe] Xiao piti

Range: Africa, Eurasia, India, China, Japan, SE Asia, Philippines, Indonesia to N New Guinea.

Distribution and status: Resident and parially migratory over whole country including Taiwan and Hainan. Race *capensis* is resident in NW China, *philippensis* on Taiwan and *poggei* over the rest of China. Occasional up to 2000 m.

Description: Small (27 cm) dark, squat grebe. Breeding: throat and foreneck reddish; crown and back of neck dark greyish-brown; upperparts brown; underparts greyish; conspicuous yellow rictal patch. Non-breeding: upperparts greyish-brown; underparts white. Iris—yellow; bill—black; feet—bluish-grey with pale tip.

Voice: Repeated high-pitched chittering *ke-ke-ke-ke* commonly given during courtship chases.

Range: Africa, Eurasia, India, China, Japan, SE Asia, Philippines, Indonesia to N New Guinea.

Distribution and status: Resident and parially migratory over whole country including Taiwan and Hainan. Race *capensis* is resident in NW China, *philippensis* on Taiwan and *poggei* over the rest of China. Occasional up to 2000 m.

Habits: Frequents lakes, swamps and flooded rice fields, where there is clear water and plenty of water plants. Generally singly or in small dispersed groups. In breeding season birds chase each other, running over the water and calling.

519. Red-necked Grebe *Podiceps grisegena* Chijing piti   PLATE 56

Description: Smaller (45 cm), less elongated and more rounded than Great Crested Grebe. Bill is also shorter, stouter and less dagger-like. In all plumages base of bill is diagnostic yellow. Has slight crest. Summer: combination of black cap, chestnut neck and whitish-grey cheeks diagnostic. Winter: Distinguished from Great Crested Grebe by greyer cheeks and foreneck and bill shape and colour. Iris—brown; bill—blackish with yellow base; feet—olive-black.

Voice: Silent outside breeding season. Very vocal when nesting; wailing howls *uooh, uooh, uooh* ending in harsh squeals; also harsh *cherk*.

Range: Holarctic: Scandanavia to Siberia wintering to Iran and N Africa; N America and NE Asia wintering to China and Japan and S USA.

Distribution and status: Breeds in wetlands of NE China; migrates through NE China and recorded wintering in Hebei, Fujian and Guangdong.

Habits: Often jumps clear of water when diving.

520. Great Crested Grebe *Podiceps cristatus* Fengtou piti   PLATE 56

Description: Large (50 cm) elegant grebe with slender neck and pronounced dark crest. Underparts whitish; upperparts uniform greyish-brown. In breeding season, adults have chestnut nape and mane-like ear tufts. Distinguished from Red-necked Grebe by white on side of face extending over eye and longer bill. Iris—reddish; bill—yellow with reddish base to lower mandible and blackish culmen; feet—blackish.

Voice: Adults have deep resonant call. Young beg with piping *ping-ping*.

Range: Palearctic, Africa, Indian subcontinent, Australia and New Zealand.

Distribution and status: Nominate race is locally common and widespread on larger lakes in N and Xizang Plateau. Partially migratory wintering to south of country.

Habits: In breeding season pairs perform elaborate courtship dance, facing each other, rising tall and nodding heads together, sometimes carrying vegetation in bill.

521. Horned Grebe *Podiceps auritus* [Slavonian Grebe]   PLATE 56
Jiao piti

Description: Medium-sized (33 cm) compact grebe with slight crest. Breeding: unmistakable eye-stripe and crest of long orange-yellow plumes contrast with black of head and extend beyond nape. Foreneck and flanks dark chestnut; upperparts mostly black.

Winter: more white on face than Black-necked Grebe; bill less upturned. Head larger and flatter than Black-necked Grebe. In flight distinguished from Black-necked Grebe by white wing coverts. White tip to bill distinguishes it from all Grebes except much smaller Little Grebe. Iris—red with white eye-ring; bill—black with whitish tip; feet—blackish-blue or grey.

Voice: Silent outside breeding season. Duet trilling similar to Little Grebe but more nasal; also hoarse and more guttural notes.

Range: Breeds on fresh water across northern temperate zone, dispersing in winter to about 30 °N, including coastal waters.

Distribution and status: Very rare. Breeds in W Tianshan and recorded on passage in NE China; wintering in SE China and lower Changjiang. One record Hong Kong.

Habits: Forms small flocks in winter.

522. **Black-necked Grebe** *Podiceps nigricollis* Heijing piti   PLATE 56

Description: Medium-sized (30 cm) grebe. Breeding adult is distinctive with fluffy yellow ear tufts and black foreneck. Ear tufts extend beyond ear coverts. Bill more upturned than Horned Grebe. Winter: distinguished from Horned Grebe by all-dark bill and dark cap extending below eye. White of chin extends as crescent behind eye. In flight lacks white wing coverts. Juvenile like winter adult but browner with dusky band across breast, and white eye-ring. Iris—red; bill—black; feet—blackish-grey.

Voice: When breeding gives plaintive, flute-like *poo-eeet* and shrill trill.

Range: Disjunct distribution in western N America, Eurasia to W Mongolia, Africa, C America and NE China. Winter birds disperse south to about 30 °N.

Distribution and status: Nominate race is rare breeder and winter visitor. Breeds in W Tianshan, Nei Mongol and NE China; recorded on passage across much of the country and winters along S and SE coasts and on rivers in SW China. May breed on Erhai Lake in N Yunnan. Vagrant Hong Kong.

Habits: Gregarious breeder on fresh waters. Wintering flocks found on lakes and along coasts.

TROPICBIRDS—Family: Phaethontidae

A small family comprising three species of elegant white seabirds characterised by a wedge-shaped tail with two elongated central streamers. These birds range far out to sea and are excellent divers, feeding largely on squids so are often active at night. They swim with cocked tail. Three species are recorded in Chinese waters.

523. **Red-billed Tropicbird** *Phaethon aethereus*   PLATE 57
Duanwei meng

Description: Largish (46 cm excluding tail) Adult: combination of red bill, barred upperparts and long white tail streamers diagnostic. Outer primaries black. Juvenile/immature differs from White-tailed and Red-tailed Tropicbirds in finer, denser barring of upperparts and diagnostic broad eye-stripe extending across neck as continuous nuchal collar. Iris—dark; bill—red; feet—yellowish with black webs.

Voice: Loud screams at nest and when circling ships.

Range: Tropical and subtropical Pacific, Atlantic and NW Indian Oceans.
Distribution and status: Breeds on Xisha (Paracel) Islands (Delacour and Jabouille 1930) in S China Sea. Strays reach SE China and Hainan.
Habits: Flies high over ocean with graceful flight. Feeds by hovering and then plunging on half-closed wings in gannet-like manner.

524. Red-tailed Tropicbird *Phaethon rubricauda* PLATE 57
Hongwei meng

Description: Largish (46 cm, excluding tail streamers) white or pinkish tropicbird. Adult pink in fresh plumage but bleaches quickly to white. Distinguished from White-tailed Tropicbird by red bill, less black in plumage and red tail streamers; from Red-billed Tropicbird by dark streamers and white outer primaries with only the shafts black. Immature usually has blackish bill and black barred upperparts. Iris—dark; bill—red; feet—blue with black webs.
Voice: Ratchet-like *pirr-igh* call in flight, and loud screams at nest.
Range: Tropical and subtropical Indian and Pacific oceans.
Distribution and status: Rarely recorded in Pacific Ocean off Taiwan.
Habits: Keeps well to sea. Flight similar to White-tailed Tropicbird.

525. White-tailed Tropicbird *Phaethon lepturus* [Yellow-billed PLATE 57
Tropicbird] Baiwei meng

Description: Smallish (37 cm, excluding tail streamers) white seabird with trailing, long, white tail streamers. Adult: mainly white with black eyebrow, black wing tips and black bar on upperwing. Immature: lacks streamers and has coarse black barring on upperparts but more black on primaries than immature Red-tailed Tropicbird. Smaller than other two tropicbirds. Iris—dark; bill—orange or yellow; feet—greyish with black webs.
Voice: Rattling *tetetete* and *tik* calls in flight and loud screams at nest.
Range: Tropical and subtropical Atlantic, Indian and Pacific oceans.
Distribution and status: Strays recorded off Taiwan.
Habits: Flies high over sea with fast wing beat circling, or twisting and turning sharply to plunge onto food in the sea.

BOOBIES—Family: Sulidae

This is a small, worldwide family of oceanic diving birds characterised by their large size; long, thin pointed wings; cigar-shaped bodies; and sharp powerful bills. They wander far out to sea in flocks and make the most spectacular vertical plunge-dives onto fish shoals. Often fly high with alternate periods of flapping and gliding. Three species have been recorded in Chinese waters.

526. Masked Booby *Sula dactylatra* [Blue-faced/White booby] PLATE 57
Lanlian jianniao

Description: Large (86 cm) black and white booby. Adult distinctive with white front to, and coverts on upperwing; white back; black mask on white head. Juvenile is like

Brown Booby but has white collar, paler brown upperparts and barred underwing. Iris—yellow; bill—yellow; feet—yellow to grey.

Voice: Silent at sea.

Range: Breeds on oceanic islands in tropical zone; ranges through most tropical seas.

Distribution and status: Race *personata* breeds on Senkaku Is (Diaoyu Dao) off NE Taiwan and ranges into China Sea.

Habits: As other boobies.

527. Red-footed Booby *Sula sula* Hongjiao jianniao PLATE 57

Description: Large (48 cm) black and white or ashy-brown booby with diagnostic red feet and white tail. Light, dark and intermediate morphs occur. Light morph: plumage mostly white with black primaries and secondaries. Dark morph: head, back and chest ashy-brown but tail white. All morphs have diagnostic red feet and pink base to bill. Immature: ashy-brown all over. Iris—brown; bill—greyish with pink base; naked skin at base of bill blue; naked skin under bill black; feet—bright red (diagnostic) but juveniles have yellowish-grey feet in all morphs.

Voice: Silent at sea. Calls only when nesting.

Range: Tropical oceans.

Distribution and status: Breeds on Xisha Is (Paracels) and locally common in S China Sea. Sometimes reaches SE coast in winter and recorded off Hong Kong and SE Taiwan.

Habits: As other boobies.

528. Brown Booby *Sula leucogaster* [White-bellied Booby] PLATE 57
He jianniao

Description: Large (48 cm) dark brown and white booby with dark head and tail. Adult: dark sooty-brown with white belly; Immature: light ashy-brown instead of white. Naked skin of face is red-yellow in female, bluish in male. Iris—grey; bill—yellow in adult, grey in juvenile; feet—yellowish-green.

Voice: Silent at sea. Calls only when nesting with growls, quacks and hisses.

Range: Tropical and subtropical oceans.

Distribution and status: Breeds on Xisha Is (Paracels) and Lanyu Islet off Taiwan. Locally common in S China Sea and rarely recorded along coasts from Shanghai to Hainan.

Habits: As other boobies but a more inshore bird especially in winter.

ANHINGAS—Family: Anhingidae

A small family of cormorant-like birds with slender form, which chase fish under water. One species lives in Asia and formerly occurred in China.

529. Darter *Anhinga melanogaster* [Snakebird>Oriental Darter] PLATE 58
Heifu sheti

Description: Unmistakable, large (84 cm) cormorant-like waterbird with very long, slender neck and small narrow head. Head and neck brown with white chin stripe extending down sides of neck. Rest of plumage blackish with white plume-like scapular feathers edged black. Iris—brown; bill—yellowish-brown with black culmen ridge; feet—grey.

Voice: Rattling and clicking calls. Screams during courtship.

Range: India, SE Asia, Philippines, Sulawesi and Sundas.

Distribution and status: Globally Near-threatened (Collar *et al.* 1994). Only one record from S China (1931) near Lungtian in Yunnan. Formerly probably resident in tropical zone.

Habits: Lives in large stretches of clean fresh water in lakes and big rivers.

CORMORANTS—Family: Phalacrocoracidae

A medium-sized, worldwide family of large, fish-eating birds with long, sharp terminally hooked bills. Cormorants chase their prey by swimming for long periods underwater. This is facilitated by the absence of waterproofing oils on the feathers causing them to quickly become waterlogged and have low buoyancy. They spend long periods after fishing with their wings spread, drying in the sun. Three breeding species and two visitors occur in China.

530. Little Cormorant *Phalacrocorax niger* [Javan Cormorant]   PLATE 58
Heijing luci

Description: Smallish (56 cm) black cormorant. Breeding: blackish-green with a few tiny white plumes on sides of head, above eye and on sides of neck. Non-breeding: lacks plumes but has whitish chin and sometimes throat. Immature: whiter on chest and browner on upperparts. Iris—blue-green; bill—brown with black tip and purplish base; feet—black.

Voice: Long drawn-rat calls *keh-eh-eh-eh-eh-eh* at breeding site.

Range: India, SW China, SE Asia and Greater Sundas.

Distribution and status: Extremely rare breeding resident and summer visitor to extreme S and SW Yunnan.

Habits: Inhabits lakes, flooded marshes and river banks. Generally in small flocks, swimming with only the head exposed and diving repeatedly for fish. Nests in colonies in trees over water or swamps.

Note: Some authors subsume this species within Pygmy Cormorant *P. pygmeus* of Eurasia.

531. Great Cormorant *Phalacrocorax carbo* [Common/Big Black/   PLATE 58
Large Black Cormorant] Putong luci

Description: Large (90 cm) glossy blackish cormorant with heavy bill and whitish cheeks and throat. In breeding season, displays white silky plumes on neck and head and white patches on flanks. Immature: dark brown with dirty whitish underparts. Iris—blue; bill—black; bare gular skin yellow; feet—black.

Voice: Guttural groans when breeding otherwise generally silent.

Range: E coast of N America, Europe, S Russia, S Siberia, NW and southern Africa, Middle East, C Asia, Indian subcontinent, China, SE Asia, Australia and New Zealand. Some seasonal migrations.

Distribution and status: Breeds in suitable areas throughout China. Large colony on Qinghai Lake. Migrates through C China to winter in southern provinces, Hainan and Taiwan. Common at breeding sites otherwise rare. A large wintering group is found in Hong Kong (Mai Po) with some birds staying all year.

Habits: Breeds on gravel islets of lakes or on islets along coast. Chases fish underwater. Swims half-submerged like other cormorants and frequently stands on rock or branch perch drying outstretched wings. Flies in V-formation or lines. Caught and trained by Chinese fishermen who fish with birds.

532. Japanese Cormorant *Phalacrocorax capillatus* [Temminck's PLATE 58
Comorant] Anlübei luci

Description: Large (81 cm) black cormorant. Similar to Great Cormorant but with greenish gloss on wings and back. Breeding adult has glossy green head and neck with sparse white filoplumes on sides of head. Has larger white face patch than Great Cormorant and white patch on thighs. Winter birds blackish-brown with whitish chin and throat. Bare skin at base of bill yellow. Juvenile has paler breast. Iris—blue; bill—yellow; feet—grey-black.

Voice: Generally silent except when breeding.

Range: Breeds Korea, Japan, Kurile and Sakhalin Is; ranges south in winter through coastal seas to SE China.

Distribution and status: Rare and irregular winter visitor to coasts of Taiwan and Fujian. Straggler to S Yunnan. Summer visitor to Liaoning, Hebei and Shandong.

Habits: Frequents steep, rocky cliffs. Almost totally marine. Used by fisherman for fishing.

533. Red-faced Cormorant *Phalacrocorax urile* Honglian luci PLATE 58

Description: Medium-sized (76 cm) glossy black cormorant with red face. Has purplish and green sheen. Breeding adult has two tufts on head; a few wispy white filoplumes on sides of head and white patch on thighs; juvenile is brown with red face. Very similar to Pelagic Cormorant, but red of face extends onto forehead above bill and is less extensive on malar area. Juvenile face redder than juvenile Pelagic. Breeding crest thicker and more bluish than Pelagic. Iris—blue; bill—yellow; feet—grey.

Voice: Silent at sea.

Range: Breeds in E Siberia, Kuril Is, Aleutians and Japan.

Distribution and status: Very rare. Stragglers recorded in Bohai Sea and off Taiwan.

Habits: Typical pelagic cormorant.

534. Pelagic Cormorant *Phalacrocorax pelagicus* Hai luci PLATE 58

Description: Medium-sized (70 cm) glossy black cormorant with red face. Very similar to Red-faced but breeding crests thinner and more floppy; red of face does not extend to forehead and is more extensive in malar area; juvenile and non-breeding adult have face pinkish-grey; smaller size. Iris—blue; bill—yellow; feet—grey.

Voice: Silent at sea.

Range: Breeds Alaska to Siberia and Japan, winters to California, S Japan and China.

Distribution and status: Uncommon. Nominate race is recorded on passage in NE

China and winters along coast from Liaodong Gulf in Bohai Sea, along east coast of China to Guangdong. Stragglers reach Taiwan.

Habits: Typical pelagic cormorant.

HERONS AND BITTERNS—Family: Ardeidae

A large, worldwide family of long-legged wading birds. Herons have long necks and long, straight, spear-like bills used for striking at fish, small vertebrates and invertebrate prey. Herons can be separated in flight from spoonbills and storks because they hold the neck bent and not straight. Several species exhibit long, fine, erectile plumes during the breeding season. Nests are generally large twig platforms made in trees. Most of the 21 species occurring in China are fairly distinctive, but care must be taken to separate the white egrets.

535. Little Egret *Egretta garzetta* Bai lu PLATE 59

Description: Medium-sized (60 cm) white heron. Distinguished from Cattle Egret by larger size, slimmer build, black bill, black legs with yellow toes, and in breeding season pure white colour and long tapering feathers on nape and floppy plumes on back and chest. Iris—yellow; facial skin greenish-yellow but pinkish in breeding season; bill—always black; legs and feet—black with yellow toes.

Voice: Silent apart from croaking calls at breeding colonies.

Range: Africa, Eurasia and Australasia.

Distribution and status: Nominate race is a common resident and migrant in S China, Taiwan and Hainan. Vagrants reach as far N as Beijing. Some migrate S in winter to tropical zone.

Habits: Frequents paddy fields, river banks, sand and mudbars and small coastal streams. Feeds in scattered flocks, often mixed with other species. Sometimes dashes after prey across coastal shallows. Birds fly in V-formation when returning to night roosts. Nests in colonies with other waterbirds.

536. Chinese Egret *Egretta eulophotes* [Swinhoe's Egret] PLATE 59
Huangzui bailu

Description: Medium-sized (68 cm) white egret with greenish legs and black bill with yellow base to lower mandible. In winter, distinguished from Little Egret by larger size and leg colour and from pale form of Pacific Reef Egret by longer legs and more dusky bill. Birds coming into breeding plumage have yellow bill and black legs. Have blue facial skin when breeding. Iris—yellow-brown; bill—black with yellow base to lower mandile; feet—yellow-green to blue-green.

Voice: Generally silent. Low croaks when disturbed.

Range: Globally Endangered (Collar *et al.* 1994). Breeds on islands off W coast of N Korea and off E China. Apparently winters mainly in Philippines, rarely to Borneo and Malay Peninsula.

Distribution and status: Formerly widespread but now a rare bird breeding on islets off the coasts of Jiangsu and Shandong and on a reservoir in Henan. Formerly recorded as breeder in Hainan and ceased breeding in Hong Kong in 1980s. Decline was due to collecting for feather plumes at end of last century but the species has never fully recovered. Vagrants recorded along E coast as far N as Hebei and Liaoning. Recorded on passage in Xisha archipelago (Paracels).

Habits: Like Little Egret, actively chasing prey through shallow water.

537. Pacific Reef Egret *Egretta sacra* [Eastern Reef Egret/   PLATE 59
Reef Heron] Yan lu

Description: Largish (58 cm) white or charcoal-grey heron. Dimorphic: the commoner grey form is distinguished by uniform grey plumage with a short crest and whitish chin often invisible in field. White form is distinguished from Cattle Egret by larger size and narrow head and neck; from other egrets by relatively short greenish legs, pale bill and different habits. Iris—yellow; bill—pale yellow; feet—green.

Voice: A hoarse grunted croak when feeding and harsher *arrk* when alarmed.

Range: Coasts of E Asia, W Pacific and Indonesia to New Guinea, Australia and New Zealand.

Distribution and status: Occasionally breeds in Hainan, Hong Kong, Taiwan, Penghu Is (Pescadores) and islets in S China Sea. Migrants reach the coast of Fujian, Zhejiang and Guangdong.

Habits: Almost always encountered on the shoreline, resting on rocks or cliff-sides or hunting at water's edge. Nests under large boulders on rock stacks of small islets.

538. Pied Heron *Egretta picata* Baijing heilu   PLATE 59

Description: Unmistakable smallish (50 cm) dark grey heron with white head and neck; black cap and floppy plumes; yellow bill and legs. Juvenile has entirely white head and no crest. Iris—yellow; bill—yellow; feet—yellow.

Voice: Not reported in region.

Range: Sulawesi to New Guinea and N Australia.

Distribution and status: Vagrant recorded S Taiwan.

Habits: Frequents rocky sea shores.

Note: Some authors place in genus *Ardea*.

539. Grey Heron *Ardea cinerea* [The Heron] Cang lu   PLATE 60

Description: Large (92 cm) white, grey and black heron. Adult: black eye-stripe and crest; black flight feathers, bend of wing and two chest bars; head, neck, chest and back white with some black streaks down the throat; otherwise grey. Young birds are greyer on head and neck and lack black on head. Iris—yellow; bill—greenish-yellow; feet—blackish.

Voice: Deep guttural *kroak* and goose-like honk.

Range: Africa, Eurasia, Korea and Japan to Philippines and Sundas.

Distribution and status: Locally common resident throughout China, Hainan and Taiwan where there is suitable habitat. Wintering birds from N migrate to S and C China.

Habits: Solitary hunter in shallow water. Wintering parties sometimes from large groups. Ponderous wing beat in flight. Roosts in trees.

540. White-bellied Heron *Ardea insignis* Baifu lu   PLATE 60

Description: Very large (127 cm) grey heron with white throat, belly, vent, axillaries, inner thighs and long plumes on lower neck. Much larger than Grey Heron. A few long plumes from crown are grey and white. Iris—yellow; bill—grey; feet—grey.
Voice: Harsh bark.
Range: E Himalayan foothills to NE Indian subcontinent and N Burma.
Distribution and status: Globally Endangered (Collar *et al.* 1994). Resident along rivers and tropical and subtropical forests in marshy areas at lower altitudes in areas of NE India claimed by China. Sight record (perhaps dubious) in SE Xizang in 1938.
Habits: Typical heron of marshes and ponds.
Note: Formerly called *A. imperialis*.

541. Purple Heron *Ardea purpurea* Cao lu   PLATE 60

Description: Large (80 cm) grey, chestnut and black heron. Black cap with droopy crest, and black stripe down side of rufous neck diagnostic. Back and wing coverts grey; flight feathers black; rest of plumage reddish-brown. Iris—yellow; bill—brown; feet—reddish-brown.
Voice: Harsh croaks.
Range: Africa, Eurasia to Philippines, Sulawesi and Sundas.
Distribution and status: Locally common resident throughout lowlands of E, C and S China, Hainan and Taiwan. Less common than Grey Heron.
Habits: Frequents paddy fields, reed beds, lakes and streams. Solitary birds creep through shallow weedy water with head cocked low and sideways to strike at fish and other food. Flies with a slow heavy wing beat. Nests in large colonies.

542. Great Egret *Casmerodius albus* [Great White/Large Egret/   PLATE 59
Great White Heron] Da bailu

Description: Large (95 cm) white heron. Much larger than other white egrets, with heavier bill and characteristic kink in neck. In breeding season the bare facial skin is blue/green; bill—black; bare thighs red and feet black. In non-breeding plumage the bare facial skin is yellowish; bill yellow, usually with dark tip; feet and legs black. Iris—yellow.
Voice: A low croaked *kraa-a* given in alarm.
Range: Cosmopolitan.
Distribution and status: Locally common at breeding areas otherwise uncommon. Nominate race breeds in Heilongjiang and NW Xinjiang, migrates through N China to winter in S Xizang. Race *modesta* breeds in Hebei to Jilin and in Fujian and SE Yunnan, wintering in S China, Hainan and Taiwan.
Habits: Generally singly or in small parties in wet or flooded areas. The birds stand rather upright, stabbing down on prey from above. Flies with a graceful and powerful slow stroke.

543. Intermediate Egret *Mesophoyx intermedia* [Plumed/Lesser/   PLATE 59
Yellow-billed/Smaller Egret] Zhong bailu

Description: Large (69 cm) white heron, intermediate in size between Little and Great Egrets and distinguished by rather short bill and S shaped neck. In breeding plumage it

has long fluffy plumes on back and chest; bill and thighs briefly pink and facial skin grey. Iris—yellow; bill—yellow often tipped brown; legs and feet—black.
Voice: Fairly silent but a deep rasping *kroa-kr* on take-off when disturbed.
Range: Africa, India, E Asia to Australasia.
Distribution and status: Fairly common in lowland wet areas of southern China. Nominate race is resident in Changjiang valley, through SE China and on Taiwan and Hainan. Birds S Yunnan are placed in the dubious subspecies *palleuca*. Vagrants reach Huang He valley.
Habits: Frequents paddy fields, lake sides, swampy areas, mangroves and mudflats. Nests in colonies with other waterbirds.

544. Cattle Egret *Bubulcus ibis* [Puff-backed Heron] Niubei lu   PLATE 59

Description: Smallish (50 cm) white heron. Breeding: white with head, neck and chest washed orange; iris, bill, legs and lores briefly bright red otherwise yellow. Non-breeding: white except for an orange wash on forehead in some birds. Distinguished from other egrets by stockier shape with shorter neck and rounder head and thicker shorter bill. Iris—yellow; bill—yellow; feet—dull yellow to blackish.
Voice: Silent apart from croaks at nesting colonies.
Range: E North America, C and N South America, Iberia to Iran, Indian subcontinent to S China, S Japan, SE Asia to Philippines and Sundas and Moluccas.
Distribution and status: Fairly common throughout lowland areas in the southern half of China including Hainan and Taiwan. Summer visitors occasionally reach as far N as Beijing.
Habits: Associates with grazing cattle and buffalo, catching the flies those beasts attract or disturb as they walk through the grass. Small parties fly in formation low over water courses each evening to communal roosting sites. Nests in colonies; often above water.

545. Chinese Pond Heron *Ardeola bacchus* Chi lu   PLATE 60

Description: Smallish (47 cm) white-winged, streaky brown heron. Breeding: head and neck dark chestnut; breast maroon. Winter: streaky brown when standing or white with a dark brown back in flight. Iris—brown; bill—yellow (in winter); legs and feet—greenish-grey.
Voice: Normally silent; low deep croaks in disputes.
Range: Bangladesh to China and SE Asia. Winters to Malay Peninsula, Indochina and Greater Sundas. Wanders to Japan.
Distribution and status: A common bird of rice paddies in S and C China. Occasional in S Xizang and lowland areas of NE China. Straggler to Taiwan.
Habits: Lives in paddy fields or other flooded areas, hunting singly or in small dispersed flocks. Flies in twos and threes to communal roosts each evening with a slow, short-winged beat. Nests in colonies with other waterbirds.
Note: Similar Indian Pond Heron *A. grayii* could occur in lowland areas of SE Xizang.

546. Little Heron *Butorides striatus* [Striated/Green-backed/   PLATE 60
Mangrove Heron] Lü lu

Description: Small (43 cm) dark grey heron. Adult: crown and long floppy crest glossy greenish-black with a black line running from base of bill, under eye and across cheek.

Wings and tail slaty-blue, glossed green and edged buff. Abdomen pinkish-grey, chin white. Female slightly smaller than male. Young birds are streaky brown. Iris—yellow; bill—black; feet—greenish.

Voice: In alarm a loud explosive *kweuk*, also rattling *kee-kee-kee-kee*.

Range: Americas, Africa, Madagascar, Indian subcontinent, China, NE Asia, SE Asia, Sundas, Philippines, New Guinea, Australia.

Distribution and status: Race *amurensis* breeds in NE China and migrates to southern coastal regions in winter; *actophilus* is quite common throughout S and C China; *javanicus* is fairly common on Taiwan and Hainan.

Habits: A solitary, shy bird of ponds, streams and paddies, staying in or close to thick cover of reeds, bushes or mangroves. Nests in small colonies.

547. Black-crowned Night Heron *Nycticorax nycticorax*   PLATE 60
[Common Night-Heron] Ye lu

Description: Medium-sized (61 cm) large-headed, stocky, black and white heron. Adult: crown black; neck and chest white; two long plumes from nape white; back black; wings and tail grey. Female smaller than male. During breeding season legs and lores become red. Immature: streaked and spotted brown. Iris—yellow in immature, bright red in adult; bill—black; feet—dirty yellow.

Voice: Deep throaty croak *wok or kowak-kowak* uttered in flight and hoarse croaks when disturbed.

Range: Americas, Africa, Europe to Japan, Indian subcontinent, SE Asia, Greater Sundas.

Distribution and status: Locally common in lowlands of E, C, and S China. Migrates S in winter to S China coast and Hainan.

Habits: Rests by day in tree colonies. At dusk the birds disperse to feed, giving deep croak calls. Feeds in paddy fields, pastures and along waterways. Nests in noisy colonies in trees over water.

548. White-eared Night Heron *Gorsachius magnificus*   PLATE 61
Hainan yan

Description: Medium-sized (58 cm) night heron with dark brown upperparts, crown, markings on side of head, crest and line on sides of neck. Has long, lanceolate buffy feathers scaled with dark edges over breast; sides of upper neck are orange-rufous. Some white spots on wing coverts. Wings grey. Adult male has bold white eye-stripe, white neck, black sides of breast and rufous 'shoulder' patch on wing. Iris—yellow; bill—yellowish with dark tip; feet—greenish-yellow.

Voice: No information.

Range: S China(one record E Tonkin).

Distribution and status: Listed as globally Critical (Collar *et al.* 1994). Formerly resident in Anhui, Zhejiang, Fujian, Guangxi and Hainan. Two records on Hainan in 1960s and found at three localities in Guangxi in 1990–92. One turned up in a bird market in Guangdong in 1998.

Habits: Lives in dense herb undergrowth in marshy patches around small streams in forest. Flies up into canopy when flushed.

549. Japanese Night Heron *Goraschius goisagi* [Japanese Bittern]　　PLATE 61
Litou yan

Description: Smallish (49 cm) squat, brown heron. Similar to Malayan Night Heron but differs in having smaller bill and crest; nape slaty-brown to chestnut, not black; wing tips not white. Has a characteristic black and white patch on the 'shoulder' of wing. Upperparts dark brown with paler vermiculations; underparts buff with central line of dark brown streaks. Grey flight feathers contrast with brown coverts in flight. Iris—yellow; bare skin of face—yellow; bill—horn; feet—dark green.
Voice: Deep, reverberating owl-like hooting both during breeding and on migration. Croaks while feeding.
Range: Breeds Japan. Winters to Philippines and Sulawesi.
Distribution and status: A rare passage visitor to coastal areas in Shanghai, Taiwan and Guangdong.
Habits: Favours wooded areas but feeds in open grassy areas in early morning and evening.

550. Malayan Night Heron *Gorsachius melanolophus* [Tiger Bittern/　　PLATE 61
Malay Night Heron] Heiguan yan

Description: Smallish (49 cm) stoutly-built, dark reddish-brown and black heron with diagnostic stubby bill with downcurved culmen. Adult: crown and short crest black: upperparts chestnut-brown finely speckled black; underparts rufous-buff streaked black and white; chin white with central row of black streaks. In flight black flight feathers and white tips to wings distinguish it from Cinnamon Bittern. Immature: dark brown upperparts spotted with white and barred buff; underparts whitish, spotted and barred brown. Distinguished from immature Black-crowned Night Heron by more stubby bill. Iris—yellow; naked area round eye—olive; bill—olive; feet—olive.
Voice: The call is a series of deep *oo* notes at about 1.5 second intervals at dawn and dusk from high canopy; also hoarse croaks and a rasping *arh, arh, arh*.
Range: India, S China, SE Asia and Philippines. Winters S to Greater Sundas.
Distribution and status: A rare lowland resident or summer visitor in SW Yunnan, Guangxi and Hainan.
Habits: This is a shy, nocturnal bird hiding in dense vegetation on or close to ground by day and feeding at night in open areas. Flies up into nearby trees when disturbed.

551. Little Bittern *Ixobrychus minutus* Xiao weiyan　　PLATE 61

Description: Small (35 cm) yellowish or black and white bittern. Adult male is fluffy white with black cap and black wings with large white panel. Has red bill. Female is yellowish-brown with brown streaks on upperparts and slight streaking on underparts; wing is brown with buff panel. Juvenile is mottled and streaked like a miniature Great Bittern. Iris—yellow to orange; bill—red in male, yellow in female; feet—yellow.
Voice: Mating song is grunting *gook* repeated every 2–3 seconds. Also gives loud, nasal *kekekeke* call when agitated.
Range: Eurasia, Africa, Madagascar, Australia, New Zealand.
Distribution and status: Very rare. Migrates through Tianshan area and winters in Kashi region of W Xinjiang.
Habits: Flies with jerky, quick jay-like strokes.

552. Yellow Bittern *Ixobrychus sinensis* [Little Yellow/Chinese   PLATE 61
Little Bittern] Huang weiyan

Description: Small (32 cm) buff and black bittern. Adult: black cap, light fulvous-brown upperparts and buff underparts. Black flight feathers contrast strongly with buff wing coverts. Immature is as adult but browner and heavily streaked all over, with black wings and tail. Iris—yellow; bare skin around eye—greenish-yellow; bill—greenish-brown; feet—greenish-yellow.

Voice: Generally silent. Slight screeching and soft, staccato *kakak kakak* in flight.

Range: India, E Asia to Philippines, Micronesia and Sumatra. In winter to Indonesia and New Guinea.

Distribution and status: Common wetland bird breeding from NE through C to SW China, Taiwan and Hainan. Winters in tropical zone.

Habits: Favours reed thickets along rivers and canals in swampy areas, also rice paddy.

553. Von Schrenck's Bittern *Ixobrychus eurhythmus*   PLATE 61
[Schrenck's Bittern] Zibei weiyan

Description: Small (33 cm) dark brown bittern. Male: crown black; upperparts purplish-chestnut, underparts streaky buff with a line of dark streaks down the throat and chest. Female and immature browner with white, black and brown flecks above and streaks below. In flight the grey underwing is characteristic. Iris—yellow; bill—greenish-yellow; feet—green.

Voice: Low squawks in flight.

Range: Breeds SE Siberia, E China, Korea and Japan. Winters S to SE Asia, Philippines and Indonesia.

Distribution and status: Globally Near-threatened (Collar *et al.* 1994). Not uncommon, breeding from Heilongjiang through E and C China to Yunnan and Guangdong. Migrates through Hainan and Taiwan.

Habits: A solitary, secretive bird of reed beds, rice paddies and marshes.

554. Cinnamon Bittern *Ixobrychus cinnamomeus* [Chestnut Bittern]   PLATE 61
Li weiyan

Description: Small (41 cm) cinnamon-orange bittern. Adult male: upperparts chestnut; underparts buffy-orange with a central stripe of black streaks down throat and breast and black streaks on flanks; whitish streaks on sides of neck. Female: duller and browner. Immature; streaky below and barred and spotted above. Iris—yellow; naked area at base of bill—orange; bill—yellow; feet—green.

Voice: A croak when disturbed into flight and a low courtship call *kokokokoko* or *geg-geg*.

Range: India, China, SE Asia, Sulawesi and Sundas.

Distribution and status: Common lowland resident of fresh water swamp and rice paddy from Liaoning through C to E China to SW China, Hainan and Taiwan. Winters in tropical zone.

Habits: A shy solitary bird of rice or grass by day, more active at night. When disturbed it jumps up and flies low with a slow, powerful rhythm. Nests in reeds or long grass.

555. Black Bittern *Dupetor flavicollis* [Mangrove Bittern] Hei yan PLATE 61

Description: Medium-sized (54 cm) blackish bittern. Adult male: general colour slaty-grey (looks black in the field) with yellow side of neck and black and yellow-streaked throat. Female: browner and more whitish below. Immature: black crown and buffy rufous feather tips to back and wings give brown scaling pattern. Long dagger-like bill separates this species from others of similar coloration. Iris—red or brown; bill—yellow-brown; feet—black-brown variable.

Voice: Loud, hoarse croak in flight and a deep booming sound during the breeding season.

Range: India, S China, SE Asia, Philippines and Indonesia to Australasia.

Distribution and status: Nominate race is uncommon summer breeder in lower and middle Changjiang valley, coastal areas of SE and S China, Xijiang valley and Hainan. Race *major* is rare in Taiwan.

Habits: A shy bird favouring forest and dense tangled swamps by day, flying to other feeding areas at night. Nests in bushes above water in dense swamps.

556. Great Bittern *Botaurus stellaris* [Common/The Bittern] PLATE 61
Da mayan

Description: Large (75 cm) golden-brown and black bittern with black cap and white chin and throat bordered by conspicuous black malar stripe. Sides of head golden but rest of plumage barred and mottled with black. In flight brown barred flight feathers contrast with gold of coverts and back. Iris—yellow; bill—yellow; feet—greenish-yellow.

Voice: The famous booming call is heard only in breeding season. Wintering birds are silent.

Range: Africa, Eurasia. Winter visitor in SE Asia and Philippines.

Distribution and status: Breeds in Tianshan Mts, Hulun Lake and NE provinces. Migrates S to winter in Changjiang valley, SE coastal provinces, Taiwan and S Yunnan. Not uncommon in breeding areas but otherwise rare.

Habits: Secretive bird of tall reeds. Sometimes freezes when spotted, holding bill vertically. Occasionally flushed and seen flying low over reeds.

FLAMINGOS—Family: Phoenicopteridae

Small family in Americas, Africa and Eurasia of extraordinarily specialised waterbirds with reddish colour, long necks and bill designed for filtering plankton from soda and saline lakes. Only one species is recorded in China as a vagrant.

557. Greater Flamingo *Phoenicopterus ruber* Da huolieniao

Description: Unmistakable, large, very tall (130 cm) pinkish waterbird with black-tipped pink, rather shoe-shaped, bill, very long neck, long red legs and reddish wings. Immature is pale brown with grey bill. Iris—whitish; bill—red with black tip; feet—red.

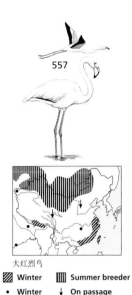

557

大红烈鸟

▨ Winter	▥ Summer breeder
• Winter	↓ On passage

Voice: Short nasal grunts and goose-like honks.
Range: C and S America, Africa, S Europe, C Asia and W India.
Distribution and status: Vagrant. Race *roseus* has been recently recorded as flocks and individuals in NW China and one record on Dongting Lake. Possible irruptions from breeding population in Afghanistan or C Kazakhstan.
Habits: Lives in flocks. Flies with neck outstretched. Stands in saline lakes, sifting inverted bill from side to side, filtering for food.

IBISES—Family: Threskiornithidae

A small worldwide, tropical family; similar and closely related to the storks but generally slightly smaller and with bills modified for probing in water or mud, rather than stabbing and smashing of prey. Ibises detect their food by touch rather than sight. The feet are partly webbed. The flight of most species consists of slow, flapping bouts alternated with short glides. Six species are found in China.

558. Glossy Ibis *Plegadis falcinellus* Cai huan   PLATE 62

Description: Smallish (60 cm), glossy, blackish-chestnut ibis, looking like a large dark curlew. Upperparts have a green and purple gloss. Iris—brown; bill—blackish; feet—greenish-brown.
Voice: Nasal grunts and bleating and cooing sounds at nest.
Range: Cosmopolitan.
Distribution and status: Breeding not confirmed in China. Occasionally seen around lakes of lower Changjiang valley and in SE China, Guangdong, Hong Kong and Hainan.
Habits: Small flocks inhabit marshes, paddy fields, and flooded grasslands. In the evening flies in line or formation to communal roosts. Nests in colonies with egrets and herons.

559. Black-headed Ibis *Threskiornis melanocephalus* [Oriental/   PLATE 62
Indian Black-necked Ibis] Heitou baihuan

Description: Large (76 cm) unmistakable white ibis with black head and long, decurved beak and bushy 'tail' of elongated grey tertiary plumes. Iris—red-brown; bill and feet—black.
Voice: Generally silent but makes curious grunts in the breeding season.
Range: India, S and E China, Japan, SE Asia and Greater Sundas.
Distribution and status: Globally Near-threatened (Collar *et al.* 1994). Possibly breeds in Heilongjiang but breeding not recorded in China. Winter visitors are rare along E and S coasts, occasionally inland as far as Sichuan, Yunnan and SE Xizang.

Habits: Frequents reedy swamps, and flooded grassy areas. Generally in small flocks, stalking actively in search of food or flying in formation. Nests in colonies with storks and other waterbirds.

560. **White-shouldered Ibis** *Pseudibis davisoni* Hei huan   PLATE 62

Description: Medium-sized (75 cm) black ibis with bare head, white patch on shoulder and red legs and pale blue nape patch distinctive. General plumage is dark brown with glossy black wings and tail. Lacks chestnut on underparts. Iris—dark; bill—black; feet—red.

Voice: Harsh *kyee-ahh.*

Range: Formerly SW China and SE Asia; now restricted to Indochina and Borneo.

Distribution and status: Globally Endangered (Collar *et al.* 1994). Status in China uncertain. Recorded from SW Yunnan in last century.

Habits: As Glossy Ibis but favouring forested swamp forests and streams.

Note: We disagree with Inskipp *et al.* (1996) in treating this form as conspecific with Black Ibis *P. papillosa.*

561. **Crested Ibis** *Nipponia nippon* [Japanese Crested Ibis]   PLATE 62
Zhu huan

Description: Unmistakable, large (55 cm) pinkish ibis with vermilion-red face; long, red-tipped decurved bill; long nuchal crest white or grey (breeding) and crimson legs. Young birds are grey and some adults retain this plumage. Also greyer in summer with longer crest. In flight underside of flight feathers red. Iris—yellow; bill—black with red tip; feet—crimson.

Voice: Harsh grunts.

Range: Formerly resident across E China to Korea and Japan; now extinct in wild except for one population in C China.

Distribution and status: Globally Critical (Collar *et al.* 1994). Breeds in S Qinling Mts (Yang county) in S Shaanxi. Only about 70 birds in the wild (was as low as 20). A captive population of about the same size has been built-up in Beijing Zoo and efforts may be made to reintroduce the species to Japan.

Habits: Colonial nester in large oak trees. Feeds in nearby farmland among cultivated crops and natural swampy areas.

562. **Eurasian Spoonbill** *Platalea leucorodia* Bai pilu   PLATE 62

Description: Large (84 cm) white spoonbill with long grey spatulate bill. Naked skin on head yellow with black line through lores to eye. Distinguished from wintering Black-faced Spoonbill by larger size, less black on face with white feathers extending further down base of bill and paler bill colour. Iris—red or yellow; bill—grey with yellowish 'spoon'; feet—blackish.

Voice: Silent when not breeding.

Range: Eurasia and Africa.

Distribution and status: Generally uncommon. Summers and possibly breeds from Tianshan Mts of NW Xinjiang to NE provinces. Migrates S in winter through C China to Yunnan, SE coastal provinces and Taiwan and Penghu islet. Flocks of over a 1000 winter on Poyang Lake (Jiangxi).

Habits: Frequents muddy pools, lakes or mudbars where it wades slowly, scything its bill from side to side in the water, sifting out food. Generally singly or in small parties; partially nocturnal.

563. Black-faced Spoonbill *Platalea minor* [Lesser Spoonbill]   PLATE 62
Heilian pilu

Description: Large (76 cm), white spoonbill with long, blackish-grey spatulate bill. Similar to winter Eurasian Spoonbill but bill entirely grey and facial skin black and less extensive. Iris—brown; bill—dark grey; legs and feet—black.

Voice: Silent outside breeding season.

Range: Breeds in islands off Korea and off Shandong coast. Winters Taiwan, S China, N Vietnam and formerly Philippines.

Distribution and status: Globally Vulnerable (Collar *et al.* 1994). Increasingly rare. On passage and may breed in E and NE China. Spring record from E Nei Mongol. Winters S to Jiangxi, Guizhou, Fujian, Guangdong and Hainan. Most of the world population, of about 600 birds, winters in Taiwan and Hong Kong.

Habits: As Eurasian Spoonbill.

PELICANS—Family: Pelecanidae

A small family comprising eight species of unmistakable huge waterbirds with very large bills and large distensible pouches stretching the full length of the bill. Pelicans inhabit lakes, large rivers and lagoons and are generally found in gregarious parties that cooperate in scoop-net fishing by side-sweeping of their beaks. They also fish by plunge-diving in flight. The flight, with recurved neck, appears laboured but is powerful and some species undertake long migrations. Pelicans make large stick nests in trees or on reed tussocks. Three species occur in China.

564. Great White Pelican *Pelecanus onocrotalus* [Eastern White/   PLATE 58
European White/Rosy Pelican] Bai tihu

Description: Very large (160 cm) white pelican. Plumage pinkish-white except for brownish-black primaries and secondaries. There is a small crest on back of head and a tuft of yellow feathers on the breast. Immature are brown. Iris—red; bill—lead blue; naked pouch—yellow; naked facial skin—pink; feet—pink.

Voice: Generally silent but can make throaty grunts.

Range: Breeds Africa, SC Eurasia, S Asia.

Distribution and status: Rarely winters and possibly breeds on lakes of Tianshan in NW Xinjiang and upper Huanghe (Yellow River) and Qinghai Lake. Stragglers recorded in Henan and Fujian.

Habits: Typical of family. Frequents lakes and large rivers.

565. Dalmatian Pelican *Pelecanus crispus* Juanyu tihu   PLATE 58

Description: Very large (175 cm) pelican with greyish-white plumage, pale yellow eyes, and orange or yellow gular pouch. Underwing is white with only tips of flight feathers black (Great White Pelican has more black in wing). Nape has curly crest. Feathers across forehead do not protrude forward as in Great White Pelican but form crescentic line. Iris—pale yellow with pink orbital skin; bill—grey above, pink below; feet—greyish.

Voice: Husky hisses in breeding season.

Range: Localised from SE Europe to China.

Distribution and status: Globally Vulnerable (Collar *et al.* 1994). Rare and local in northern China migrating to winter in the south. A few birds regularly winter in Hong Kong.

Habits: Gregarious and fishes in groups.

566. Spot-billed Pelican *Pelecanus philippensis* [Grey/Philippine   PLATE 58
Pelican] Banzui tihu

Description: Very large (140 cm) grey pelican. Grey plumage and blue-spotted pink bill distinctive. Wings darkish grey. No black in plumage. Bill pouch is purple clouded with black. Iris—pale brown; naked skin round eye—pinkish; bill—pink; feet—brown.

Voice: Husky, hissing calls only during breeding period.

Range: Breeds in SW India, Sri Lanka, Burma and E China; doubtfully SE Asia, Philippines. Migrant southward.

Distribution and status: Globally Vulnerable (Collar *et al.* 1994). Status in China uncertain. Formerly regarded as rare resident along E China and S China Sea coasts from Jiangsu to Guangxi, S Yunnan, and Hainan. Migrants occasionally recorded in Shandong. However, many of these records are doubtful due to confusion with Dalmatian Pelican. Possibly now extinct in China.

Habits: Lives in large flocks. Inhabits sheltered coastal bays, estuaries, lakes and large rivers.

STORKS—Family: Ciconiidae

A small, worldwide family of very large birds with long powerful beaks. They have long legs, broad wings and short tails and feed mostly on fish or small animals, which they catch while stalking quietly about in open wet areas. Some species feed on carrion. Storks are strong fliers and several species migrate over large distances. Fly with neck extended. They are experts at soaring on thermals and often circle high in the sky, gaining height for easy travel or searching for likely feeding places. There are four resident species in China and one visitor.

567. Painted Stork *Mycteria leucocephala* Baitou caiguan   PLATE 62

Description: Large (100 cm) whitish stork with black band across breast, black and white wings, black tail and yellow decurved bill. Bare skin of head reddish. In breeding season the back plumage is tinged pink. In flight black wings have broad white band on upper greater coverts and wing lining and narrow white bands on other upper coverts. Immature is brown with black wings and white rump and vent. Iris—brown; bill—orange; feet—pink.

Voice: Croaking of young and bill clapping.

Range: Indian subcontinent to SW China and Indochina.

Distribution and status: Globally Near-threatened (Collar *et al.* 1994). Regarded as common in the 1930s but rare by Cheng (1987). Now probably extinct in China. Recorded as summer visitor to the lower Changjiang valley and S China across to S Yunnan.

Habits: Colonial breeder in trees above water. Feeds along edge of ponds, lakes and rivers.

568. Black Stork *Ciconia nigra* Hei guan   PLATE 62

Description: Unmistakable, large (100 cm) black stork with white lower breast, belly and undertail; red bill and red legs. Has greenish and purplish gloss on black parts. In flight, underwing is black except for white axillaries and inner secondary coverts. Bare skin round eye red. Young have brown upperparts, white below. Iris—brown; bill—red; feet—red.

Voice: Melodious, guttural notes when breeding.

Range: Europe to N China; wintering to Indian subcontinent and Africa.

Distribution and status: Rare and declining seasonal migrant. Breeds across northern parts of China and winters S of Changjiang and on Taiwan. Regular wintering bird at Mai Po in Hong Kong during the 1960s, now very rarely recorded.

Habits: Uses marshy areas and ponds and edges of lakes, rivers and estuaries. Rather shy. Sometimes forms small flocks in winter.

569. White Stork *Ciconia ciconia* Bai guan   PLATE 62

Description: Very large (10 cm) white stork with black flight feathers. As Oriental Stork but bill is red instead of black. Iris—brown; bill—red; feet—red.

Voice: Bill clapping.

Range: Europe, N Africa to C Asia; wintering in Africa and India.

Distribution and status: Rare. Race *asiatica* is summer visitor and possible breeder in Tianshan and Kashi regions of W Xinjiang. Vagrant to NE Nei Mongol.

Habits: Nests in trees or on poles and chimney stacks. Forms flocks in winter. Feeds in wetland sites and often seen spiralling in thermals.

Note: Formerly included Oriental Stork (e.g. Cheng 1987).

570. Oriental Stork *Ciconia boyciana* Dongfang baiguan   PLATE 62

Description: Large (105 cm) pure white stork with black wings and heavy straight black bill. Legs are red and bare orbital skin is pink. In flight black primaries and secondaries contrast sharply against otherwise white plumage. Distinguished from White Stork by black instead of reddish bill. Immature is dirty yellowish-white. Iris—whitish; bill—black; feet—red.

Voice: Clapping of bill.

Range: NE Asia and Japan.

Distribution and status: Globally Vulnerable (Collar *et al.* 1994). Breeds in NE China and inhabits open country and forest. Winters in lakes of lower Changjiang valley with occasional birds regularly wintering in S Shaanxi, SW China and Hong Kong. Summer visitor to Ordos Plateau in W Nei Mongol. Shows some response to artificial breeding 'trees' constructed in breeding grounds in Heilongjiang.

Habits: As White Stork.

571. Lesser Adjutant *Leptoptilos javanicus* [Lesser Adjutant Stork]   PLATE 62
Tu guan

Description: Huge (110 cm) black and white stork with massive bill. Wings, back and tail black; underparts and neck collar white; naked head and throat pink; naked neck yellow with some fine white downy feathers on crown. Iris—bluish-grey; bill—grey; feet—dark brown.

Voice: Silent apart from buzzing sound at nesting sites and audible wing-beats and bill clapping.

Range: India, S China, SE Asia and Greater Sundas.

Distribution and status: Formerly recorded from Hainan, Jiangxi, Yunnan and Sichuan. Not comfirmed to breed in China and now probably extinct or close to extinction in China.

Habits: Frequents grassy wetlands, mud banks and mangroves. Sometimes seen soaring with other storks and even with eagles in thermals.

Note: The rare Greater Adjutant *L. dubius* could stray into SE Xizang.

FRIGATEBIRDS—Family: Fregatidae

A small family comprising five species of large, tropical oceanic birds characterised by gliding flight and unique silhouette with bow-shaped, long, pointed wings and long, forked tail (often closed and pointed). These birds are superb gliders, soaring and spiralling effortlessly on thermals or circling and diving over fish shoals. They frequently harry other seabirds to piratise disgorged food. Identification is difficult because plumage is very variable and there are many immature stages. Three species occur in Chinese waters.

572. Great Frigatebird *Fregata minor* [Greater Frigatebird]   PLATE 57
Xiao junjianniao

Description: Large (95 cm) dark frigatebird. Male: plumage distinct all-black except for pale bar across upperwing coverts and crimson gular pouch. Female: distinct with greyish-white chin and throat; white upper breast; little or no white on base of underwing; pinkish-red orbital ring. Immature: dark brown above with whitish head, neck and underparts stained rusty and distinguished from Lesser Frigatebird only by larger size, posteriorly convex white abdomen and less white at base of underwing. Iris—brown; bill—slate-blue in male, pinkish in female; feet—reddish in adult or blue in juvenile.

Voice: Braying, clapping and rattling calls recorded from nest areas. Silent at sea.

Range: Tropical oceans.

Distribution and status: Breeds on islets off Hainan, Xisha Is (Paracels) and Nansha Is (Spratlys). Locally common in S China Sea. Occasionally along S China coast N as far as Jiangsu and Hebei. Rarely to Lanyu Islet off Taiwan.

Habits: Pelagic habits similar to other frigatebirds, but more regular visitor along to coasts.

573. Lesser Frigatebird *Fregata ariel* [Least Frigatebird] PLATE 57
Baiban junjianniao

Description: Large (76 cm) dark frigatebird. Male: all-blackish with white patches on flanks and under base of each wing and red pouch. Female: black with brownish head, white breast and concave abdominal patch with some white on base of underwing; pink or bluish-grey orbital ring; black chin. Immature: brownish-black upperparts but head, neck, breast and flanks whitish stained rufous. Distinguished from immature Great Frigatebird by smaller size, concave shape to white underparts and more white on base of underwing. Iris—brown: bill—grey; feet—red-black.

Voice: Silent at sea.

Range: Tropical oceans. Wanders N in summer to China and Japan.

Distribution and status: Rare summer visitor recorded off Guangdong to Fujian and rarely Taiwan. More common in S China Sea, Xisha and Nansha islands (Paracels and Spratlys).

Habits: Pelagic; spirals in thermals and sometimes flies fast and low over the water with heavy deliberate strokes. Birds sometimes roost or rest on bamboo fish platforms or trees on small islands.

574. Christmas Island Frigatebird *Fregata andrewsi* PLATE 57
[Christmas Island Frigatebird] Baifu junjianniao

Description: Large (95 cm) dark frigatebird. Male: glossy green-black with red gular pouch and diagnostic white abdomen. Female: breast and belly white with white 'spur' extending onto underwing and white collar; pink orbital ring. Juvenile: browner, head pale rusty-brown, broad darkish band across chest. Iris—dark brown; bill—black (male) or pinkish (female and juvenile); feet—purplish-grey with flesh-coloured soles.

Voice: Silent at sea.

Range: Breeds on Christmas Island in the Indian Ocean. Recorded N to Malay Peninsula and S China Sea.

Distribution and status: Globally Vulnerable (Collar *et al.* 1994). Accidental through S China Sea to islets off Guangdong coast.

Habits: Pelagic, soaring high over sea on thermals or spiralling over fish.

LOONS—Family: Gaviidae

A small family of diving seabirds within a single genus. They dive to great depths after fish and swim underwater propelled by their feet. They have short tails and look superficially like cormorants and live in both fresh water and sea water. Flight is swift and powerful with neck extended and head held low. Four species have been recorded in China, another may also occur.

575. Red-throated Loon *Gavia stellata* [AG: Diver]   PLATE 56
Honghou qianniao

Description: Smallest diver (61 cm). Summer adult has grey face, throat and sides of neck with diagnostic chestnut band down centre of throat and front of neck, and back of neck finely striped with white spots. Upperparts otherwise brownish without white marks; underparts white. Winter adult: white extends to chin, sides of neck and face. Upperparts blackish with white streaks. Small head, slender neck and upward tilt of bill, when swimming, distinguish from Black-throated Loon as well as by faster and higher wing beats. Iris—red; bill—greenish-black; feet—black.
Voice: Goose-like cackle *gwuk-gwuk-gwuk* in flight.
Range: Circumpolar. Breeds in northern latitudes, winters S to about 30 °N.
Distribution and status: Rare. Recorded on migration from Heilong River in NE China and along E coast through Beidaihe, Lushun to Guangdong, Hainan and N Taiwan.
Habits: Breeds on fresh water but winters, sometimes in flocks, on coastal waters.

576. Black-throated Loon *Gavia arctica* [Arctic Loon,   PLATE 56
AG: Diver] Heihou qianniao

Description: Largish diver (68 cm). Breeding: grey head with blackish-green throat patch and white spangles on black upperparts. Sides of neck and breast finely striped black and white. Separated, with great difficulty from Pacific Loon, by green rather than purple gloss on throat patch. Non-breeding: white extends from underparts up front and sides of neck to chin and lower half of face. Conspicuous whitish flank patch. Distinguished from Red-throated Loon by larger head, thicker neck, cleeper-based bill held more horizontally and lack of white marks on upperparts. First-winter birds have white scaling on upperparts. Iris—red; bill—blackish-grey; feet—black.
Voice: Repeated croaking snore and yelping gull-like *aah-ouw*.
Range: Breeds Northern Hemisphere from N Scotland to Siberia. Winters south to about 30 °N.
Distribution and status: Race *viridigularis* is rare migrant recorded in Liaodong Peninsula of Liaoning, probably off Beidaihe (Hebei) and once from Fuzhou. Confusion with Pacific Loon possible.
Habits: Breeds singly on fresh water but usually winters in loose flocks in coastal waters.
Note: Pacific Loon is sometimes treated as a race of this species (e.g. Harrison 1983).

577. Pacific Loon *Gavia pacifica* [AG: Diver]   PLATE 56
Taipingyang qianniao

Description: Largish (66 cm) diver. Slightly smaller than Black-thorated Loon and similar in all plumages. Distinguished in breeding plumage by purple rather than green gloss to blackish throat and more white on nape. Iris—red; bill—grey to black; feet—black.
Voice: Generally silent.

Range: Breeds E Siberia to Alaska and Canada, wintering off coast of Japan and western N America as far S as 23 °N.

Distribution and status: Rare. Recorded on migration through Liaodong Peninsula of Liaoning, perhaps Beidaihe and Shandong. Possible record Hong Kong.

Habits: As Black-throated Loon.

Note: Sometimes treated as a race of Black-throated Loon (e.g. Harrison 1983).

[**578. Common Loon** *Gavia immer* [Great Northern Diver]   PLATE 56
Baizui qianniao

Description: Large (76 cm) diver. Breeding: distinctive with black bill; blackish head and neck with white necklace. Non-breeders can be confused with Black-throated Loon but larger heavier bill, crown flatter. Crown and hindneck darker than upperparts. Flight is more goose-like with slow steady beat. Irregular separation of white and dark down sides of neck. Generally lacks white flank patch. Separated from yellow billed Loon by dark culmen to white bill. More horizontal angle of bill and head darker than upperparts. Iris—red; bill—whitish (winter) to black; feet—black.

Voice: Silent on sea.

Range: Breeds from Aleutian Is across N America as far E as Iceland. Winters S to about 30 °N.

Distribution and status: Not definitely recorded in China but may range into Chinese waters.

Habits: Breeds singly on fresh water but mostly winters in small groups in coastal waters.]

579. Yellow-billed Loon *Gavia adamsii* [White-billed Diver,   PLATE 56
AG: Diver] Huangzui qianniao

Description: Very large (83 cm) thick-necked diver. Breeding: diagnostic ivory bill with black head and white necklace. Non-breeding: distinguished from other loons by larger size, upward tilt of bill, pale culmen and head paler than upperparts. Lacks white flank patch. White primary shafts diagnostic. Iris—red; bill—ivory yellow; feet—black.

Voice: Silent at sea but yodelling tremulous screams on breeding areas.

Range: Breeds in high Arctic from Murmansk E to Siberia and Alaska to northern Canada. In winter moves S to about 50 °N but occasionally much further south.

Distribution and status: Very rare. Recorded on migration from Liaodong Peninsula of Liaoning and Fuzhou.

Habits: Breeds singly on fresh water but winters in coastal waters.

SHEARWATERS, ALBATROSSES AND STORM PETRELS—Family: Procellariidae

A moderately large family of oceanic gull-like birds with curiously structured tube-nose bills, hooked at the tip and with the nostrils opening in a double tube on top of the bill. Three subfamilies are recognised.

SHEARWATERS—Subfamily: Procellariinae

Shearwaters acquire their English name from their habit of flying just above the water surface or even touching the water. They dive for fish, squid, plankton and crustaceans and nest on cliff-ledges or in burrows on rocky islets. They are silent at sea. Shearwaters are predominantly temperate in distribution with nine species seen in Chinese waters.

580. Northern Fulmar *Fulmarus glacialis* Baoxue hu PLATE 63

Description: Large (48 cm) variable petrel. Has pale and dark morphs but dark morph much commoner in Pacific race *rogersi*. Light morph resembles gulls but distinguished by stubby bill, thick neck and flap and glide flight. Flight pattern similar to a shearwater so that dark morph can be confused for large shearwater. Distinguished from shearwaters by body proportions and tail shape. Iris—very dark; bill—yellow with bluish base; feet—pink.

Voice: Noisy cackling at breeding sites. Generally silent at sea but guttural cackling given by feeding groups.

Range: Northern Hemisphere seas between 34 °N to Arctic. Race *rogersi* breeds on islands off E Siberia. Winter visitor to Japan.

Distribution and status: Accidental. Stragglers recorded along NE coasts.

Habits: Follows ships.

581. Tahiti Petrel *Pterodroma rostrata* [Peale's Fulmar/Beck's Petrel] PLATE 63
Gouzui yuanweihu

Description: Smallish (39 cm) black and white gadfly petrel. White belly and entirely dark head and breast distinctive. Tail wedge-shaped and blackish. Iris—dark brown; bill—black; feet—pinkish with black webs.

Voice: Silent at sea.

Range: Tropical and subtropical western Pacific, probably dispersing NW toward Taiwan.

Distribution and status: Collected NE Taiwan in May 1937.

Habits: Does not normally follow ships so rarely recognised. Swoops and soars close to sea surface.

582. Bonin Petrel *Pterodroma hypoleuca* Dian'e yuanweihu PLATE 63

Description: Small (30 cm) black and white petrel. Plumage pattern distinctive. Upperside dark but underparts mostly white with black trailing edge to wing and bold black diagonal bar from carpal inward across coverts. Tail blackish grey. No other petrels with this wing pattern range into China's seas. Flight fast and swooping. Iris—dark brown; bill—black; feet—pink with black toes.

Voice: Silent at sea.

Range: Breeds on Sakhalin, islands off S Japan and Hawaii ranging over western N Pacific.

Distribution and status: Nominate race accidental on Taiwan and off Fujian coasts.

Habits: Colonial breeder; does not usually follow ships.

583. Bulwer's Petrel *Bulweria bulwerii* Chunhe hu PLATE 63

Description: Small (28 cm) sooty-brown petrel with paler brown underparts. Pale bar on upperwing coverts usually visible. Distinguished from Swinhoe's Storm-petrel by larger size and long, wedge-shaped tail (looks long and pointed in flight). Iris—brown; bill—black; feet—pinkish with black webs.

Voice: Repeated low barking *chuff* at breeding colonies.

Range: Breeds on islands in Atlantic and Pacific Oceans. Range includes all of S China Sea.

Distribution and status: Very rare. Breeds on islands off Fujian and recorded off Taiwan and in S China Sea off Guangdong and Hong Kong.

Habits: Flies more strongly than storm-petrels with fluttery erratic swooping and high wheeling loops.

584. Streaked Shearwater *Calonectris leucomelas* [White-faced/ PLATE 63
White-fronted/Streak-headed Shearwater] Bai'e hu

Description: Large (48 cm) shearwater, dark brown above, white face and underparts with dark streaks on head and breast. Distinguished from the pale form of Wedge-tailed Shearwater by white face and bill colour. Iris—brown; bill—horn; feet—pinkish.

Voice: Silent at sea.

Range: Breeds on small islands in NW Pacific and winters S to the Equator.

Distribution and status: Breeds on Qingdao Is off Shandong. Not uncommon in S China Sea, recorded offshore from Shandong to Fujian, Hong Kong, Penghu Is, Taiwan and islands of S China Sea.

Habits: Similar to Wedge-tailed Shearwater.

585. Wedge-tailed Shearwater *Puffinus pacificus* Yiwei hu PLATE 63

Description: Medium-sized (43 cm) long-winged, wedge-tailed shearwater. Occurs in light and dark colour morphs. Dark morph: dark chocolate all over; pale morph; brown above and whitish below with underwings showing dark borders and undertail coverts. Iris—brown; bill—dark grey; feet—flesh-coloured.

Voice: Silent at sea.

Range: Breeds on islands in tropical Indian and Pacific Oceans.

Distribution and status: Stray birds can occur anywhere in SE Asian waters. Recorded off Penghu Is, Taiwan and through S China Sea.

Habits: Flies low over the sea, banking from side to side with occasional swoops and frequently gliding so low over water that the wing tips touch surface at bottom of beat.

586. Flesh-footed Shearwater *Puffinus carneipes* [Pale-footed PLATE 63
Shearwater] Rouzu hu

Description: Largish (43 cm) heavily-built, chocolate-brown shearwater with long wings and short rounded tail. Often confused with Wedge-tailed Shearwater but distinguished by thicker, dark-tipped, pale bill and whitish bases to underside of primaries. Iris—brown; bill—straw with brown tip; feet—yellow to pink.

Voice: Silent at sea.

Range: Breeds on islands off Australia and New Zealand.

Distribution and status: Indian Ocean populations disperse N in winter and occasional birds reach S China Sea.

Habits: Flies low over sea with deliberate beat and long glides; more graceful than Wedge-tailed Shearwater.

587. Sooty Shearwater *Puffinus griseus* Hui hu PLATE 63

Description: Large (44 cm) sooty-brown, slender shearwater. Variable amount of silver-white on underwing coverts appears as white flash during flap and glide flight. Separated from Short-tailed Shearwater by longer bill and more white in underwing coverts; from Wedge-tailed and Flesh-footed Shearwaters in white on underwing coverts, dark feet and faster more direct flight. Irish—dark brown; bill—dark grey; feet—blackish.

Voice: Occasional subdued cackling at sea.

Range: Breeds off Chile, Australia and New Zealand but ranges through equatorial oceans and throughout Atlantic and Pacific.

Distribution and status: Accidental; recorded off Fujian coast and Penghu Is and Taiwan. Hong Kong records regarded as doubtful.

Habits: Gregarious at sea; does not usually follow ships.

[588. Short-tailed Shearwater *Puffinus tenuirostris* PLATE 63
Duanwei hu

Description: Medium-sized (42 cm) sooty-brown shearwater. Separated from Sooty Shearwater by shorter bill and less silvery underwing coverts. Smaller, shorter-billed and with paler underwing than Wedge-tailed, Flesh-footed or Sooty Shearwaters. Feet darker and extend further beyond tail. Dark head may appear hooded at a distance. Flight rapid and rhythmic with stiff-winged glides between bursts of short fast beats. Iris—dark brown; bill—blackish-grey; feet—blackish-grey.

Voice: Silent at sea. Wailing, crooning and sobbing call at breeding colony.

Range: Breeds during southern summer on islands off S and SE Australia dispersing N through Pacific into Bering Sea. Spring migrant and summer visitor around Japan.

Distribution and status: Out of breeding season migrates to northern Pacific. Probably ranges into seas off NE coasts.

Habits: Gregarious at sea and sometimes follow ships.]

ALBATROSSES—Subfamily: Diomedeinae

A small subfamily of large aerial seabirds with very long straight wings. Three species range into Chinese waters.

589. Short-tailed Albatross *Diomedea albatrus* PLATE 57
Duanwei xintianweng

Description: Adult: large (length: 89 cm, wingspan: 216 cm) white-backed albatross with feet extending well beyond black tail in flight. Plumage sequence progresses from dusty-brown juvenile to pale-bellied immature with white upperwing patches and scaly back. Adult is the only white-bodied albatross in the Pacific. Adult has yellowish nape. Juvenile and immature plumages can be confused with smaller Black-footed Albatross but bill paler pinkish and feet bluish and lacks white at base of bill. Irish—brown; bill—pinkish with bluish tip; feet—bluish-grey.

Voice: Usually silent at sea.

Range: Ranges over N Pacific where now extremely rare.

Distribution and status: Very rare. Breeds on Diaoyu Dao and Chiwei Yu (Senkaku) islands to N of Taiwan and formerly on Penghu Is. Recorded on passage off Shandong and along E coast.

Habits: Colonial breeder, occasionally follows ships.

590. Black-footed Albatross *Diomedea nigripes* PLATE 57
Heijiao xintianweng

Description: Medium-sized (81 cm) mostly dusky-brown albatross except for narrow whitish areas around base of bill, over base of tail and undertail coverts. Old adults sometimes become bleached with whitish heads and breast. Juvenile as adult but lacks white over tail and on undertail coverts. Distinguished from juvenile Short-tailed Albatross by dark bill and feet. Iris—blackish-brown; bill—blackish-grey; feet—black.

Voice: Silent at sea.

Range: N Pacific; breeds Leeward, Marshall, Johnston and Torishima Islands.

Distribution and status: Very rare. Found in Taiwan Straits throughout the year. Ranges through S China Sea in spring and winter.

Habits: Colonial breeder. Often follows ships, feeding on garbage. Pelagic.

[591. Laysan Albatross *Diomedea immutabilis* PLATE 57
Heibei xintianweng

Description: Medium-sized (80 cm) black and white albatross. Distinctive with white from chin to vent but upperwings and back dark. Underwings mainly white with dark margins and blackish streaks on coverts. Has dark eye patch. In flight feet extend only marginally beyond tail. Juvenile as adult but bill greyer. Iris—dark brown; bill—yellow with dark tip; feet—pinkish-grey.

Voice: Usually silent at sea.

Range: N Pacific. Breeds on small islands of S Japan and Hawaii; ranges from 30 °N to 55 °N.

Distribution and status: Regular winter visitor to seas S of Japan throughout the year. Not recorded in China but probably ranges into Chinese waters.

Habits: Colonial breeder, follows ships for garbage.]

STORM-PETRELS—Subfamily: Hydrobatinae

A small subfamily of oceanic birds similar to shearwaters but smaller with a more fluttery flight and with the nostril tubes joined into a single aperture. These are the smallest oceanic birds and their weak butterfly-like flight and habit of hovering and treading the water with their webbed feet makes them easy to recognise among other seabirds.

Storm-petrels feed on small crustaceans or floating organic debris. They nest in rock crevices and burrows on rocky shores and islands. Only four species are recorded in Chinese waters but individual species are often difficult to distinguish.

[**592. Wilson's Storm-petrel** *Oceanites oceanicus*   PLATE 63
Yanhei chaweihaiyan

Description: Small (17 cm) dainty dark brown petrel with unforked tail and white rump and undertail coverts. Wings broad and short and lacking definite angles on either leading or trailing edges. Tail short so feet trail beyond tail in flight. Significant contrast between dark flight feathers and pale greater coverts above and below. Iris—brown; bill—brown; feet—black with yellow webs to toes.
Voice: Occasional querulous sparrow-like chattering in feeding flocks.
Range: Pelagic range virtually worldwide.
Distribution and status: Status uncertain. Recorded in S China Sea and migratory route close to Hong Kong (Lamont 1992). Some sightings of 'white-rumped' petrels in Hong Kong waters were probably this species.
Habits: Singly or in small parties. Flies low over sea with short glides interspersed with loose wing beats, tilting and rolling from side to side. When feeding, it hovers and paddles on water with feet dangling. Often follows ships.]

593. Leach's Storm-petrel *Oceanodroma leucorhoa*   PLATE 63
Baiyao chaweihaiyan

Description: Small (20 cm) dark brown petrel with pronounced bend in wing and long forked tail extending beyond feet in flight. Sides of rump and uppertail coverts white; vent dark brown. Pronounced pale diagonal bar across upperwing on greater coverts. Iris—brown; bill—brown; feet—brown.
Voice: Prolonged purring at breeding colonies but usually silent at sea.
Range: N Pacific and N Atlantic.
Distribution and status: Accidental. Stragglers have been recorded in Heilongjiang.
Habits: Fast, buoyant flight with steep swoops between short glides and fast wing beats. Sometimes follows ships.

594. Swinhoe's Storm-petrel *Oceanodroma monorhis*   PLATE 63
Hei chaweihaiyan

Description: Small (20 cm) dark petrel. Plumage all-dark brown with inconspicuous paler grey wing bar and slightly forked tail. Iris—dark; bill and feet—black.
Voice: Trilling chatter in flight over breeding colony but silent at sea.
Range: Breeds on islands off Japan, Korea and NE Taiwan and migrates W to the northern Indian Ocean in winter.
Distribution and status: Occasional visitor off Shandong, Fujian, Guangdong and Hong Kong coasts. Passes through S China Sea.
Habits: The flight is distinctive and tern-like with much bounding and swooping over the water, never pattering along the surface—sometimes follows ships.
Note: Perhaps conspecific with Leach's Storm-Petrel *O. leucorhoa.*

595. Matsudaira's Storm-petrel *Oceanodroma matsudairae* PLATE **63**
Riben chaweihaiyan

Description: Smallish (25 cm) dark, fork-tailed petrel. Similar to Swinhoe's Storm-petrel but larger and with whitish patch near wing tips produced by white shafts of primaries. Flight is slower. Iris—dark; bill and feet—black.

Range: Breeds in S Japan but winters to S China Sea, and Philippines N Indian Ocean.

Distribution and status: Must pass through Chinese waters. A probable record of four birds in Hong Kong waters on 21 August 1993.

Habits: The flight is heavier than Swinhoe's Storm-petrel.

Note: Perhaps conspecific with Black Storm-petrel *O. melania.*

PASSERIFORMES

PITTAS—Family: Pittidae

Pittas are a family of colourful, ground-living birds found from Africa to Australia. They are plump, short-tailed, long-legged birds that hop about on the forest floor or in low vegetation in search of invertebrates. They give simple plaintive calls or whistles, some species calling from high in the canopy. Pittas nest in hollow ball-like structures of vegetation, often close to the ground. They fly with a rapid wing beat when disturbed, keeping close to the forest floor and, in some species, revealing conspicuous white wing flashes. Some species are migratory. Almost all species are beautifully coloured with rich blue, gold, red or green patterns. Eight species occur in China.

596. Eared Pitta *Pitta phayrei* Shuangbian basedong PLATE **64**

Description: Medium-sized (25 cm) golden-brown pitta with diagnostic long 'ear' tufts formed by protruding pale lateral head stripes. Has bold black and rufous speckling on wing coverts. Male has black coronal band, eye band and moustachial stripe. In female these black bands are brown. Iris—brown; bill—blackish; feet—pinkish-brown.

Voice: Eerie whistle *wheeow-whit* like Blue Pitta but first note prolonged. Calls mostly at dusk. Short dog-like whine in alarm.

Range: SW China and SE Asia.

Distribution and status: Rare resident in Xishuangbanna, S Yunnan in evergreen and shady secondary forests above 800 cm.

Habits: Typical pitta of forest floor.

597. Blue-naped Pitta *Pitta nipalensis* Lanzhen basedong PLATE **64**

Description: Largish (28 cm) olive pitta with blue crown and nape. Underparts plain fulvous; tail brown tinged with green. Female as male but hindcrown fulvous instead of blue and hindneck green; throat more whitish. Very similar to Blue-rumped Pitta but distinguished, with difficulty, by blue rump of latter and by black line behind eye less pronounced than in Blue-rumped Pitta. Iris—brown; bill—brown; feet—pinkish-brown.

Voice: Magnificent double whistle given from ground and in trees. Also softchuckles.

Range: Nepal to SW China and SE Asia.

Distribution and status: Globally Near-threatened (Collar *et al.* 1994). Accidental in SE Yunnan (Hekou) and resident in SE Xizang up to 2000 m.

Habits: Skulks on forest floor, flicking leaves in search of food.

598. Blue-rumped Pitta *Pitta soror* Lanbei basedong PLATE 64

Description: Medium-sized (25 cm) olive and fulvous pitta with whitish face, olive-brown forehead and crown, dull blue nape and rump and cinnamon eyebrow. Female as male but duller olive with greener crown and nape. Similar to Rusty-naped and Blue-naped Pittas but distinguished from both by blue rump and from Blue-naped by black eye-stripe sometimes formed by a spot behind eye and patch on ear coverts. Race *douglasi* has lores and forehead light pinkish-brown. Iris—brown; bill—grey; feet—pinkish.

Voice: Sharp, breathless *tew* and longer, mellow descending *tiuu*.

Range: SE China, Hainan, SE Thailand and Indochina.

Distribution and status: Globally Near-threatened (Collar *et al.* 1994). Rare resident on Hainan (*douglasi*) and on Mt Yaoshan in Guangxi (*tonkinensis*).

Habits: Inhabits evergreen forest above 900 m.

599. Rusty-naped Pitta *Pitta oatesi* Litou basedong PLATE 64

Description: Medium-sized (26 cm) green and fulvous pitta with black eye stripe. Crown, nape and hindneck bright golden-rufous. Throat and sides of neck pinkish and rest of underparts tawny-rufous. Upperparts dull green with faint black streaks. Rump is bluish-green. Female is duller with duller blue on rump. Distinguished from Blue-rumped Pitta by golden head. Iris—dark brown; bill—black; feet—pinkish-brown.

Voice: Sharp, breathless *chow-whit* similar to Blue Pitta but first syllable truncated. Liquid, falling tone *poouw*. Alarm is explosive *tchick*.

Range: SW China, Burma, Indochina and Malay Peninsula.

Distribution and status: Rare in S Xishuangbanna and Pingbian areas of S Yunnan (*castaneiceps*) and in W Yunnan (*oatesi*) from 800–1370 m.

Habits: Typical pitta of primary and secondary forests. Often calls at night.

600. Blue Pitta *Pitta cyanea* Lan basedong PLATE 64

Description: Medium-sized (24 cm) deep blue pitta with broad orange base to sides of crown extending to nape. Has broad black eye band, buff cheeks and black moustachial stripe. Tail bright blue. Female is duller than male with white underparts but still unmistakable. The only Chinese pitta with cross-barred underparts. Iris—brown; bill—blackish; feet—pinkish-brown.

Voice: Liquid *pleoow-whit*. Raspy, squeaky *skyeew* in alarm.

Range: E India, Burma, Indochina.

Distribution and status: Globally Vulnerable (Collar *et al.* 1994). Nominate race is rare resident in S Xishuangbanna in S Yunnan. Found of 500–1500 m.

Habits: Inhabits evergreen forest, semi-deciduous forest and bamboo thickets. Typical pitta habits.

601. Hooded Pitta *Pitta sordida* [Black-headed Pitta]   PLATE 64
Lüxiong basedong

Description: Unmistakable medium-sized (18 cm) green pitta with black head. Upperparts green; wings blue with white wing patches; hood dark brown; chest and belly apple-green; vent bright red. Iris—brown; bill—black; feet—flesh-coloured.

Voice: Double whistle *pih-pih* repeated at short intervals.

Range: Indian subcontinent to SW China, SE Asia, Philippines, Sulawesi, Greater Sundas and New Guinea.

Distribution and status: Race *cucullata* is rare summer breeder in S and SE Yunnan and areas of Arunachal Pradesh claimed by China up to 2000 m.

Habits: Hops about on forest floor, turning over leaves and probing dead wood in search of invertebrates. Conspicuous in flight with large white patches on dark wings.

602. Fairy Pitta *Pitta nympha* Xian basedong   PLATE 64

Description: Medium-sized (20 cm) colourful pitta. As Blue-winged Pitta but with paler and greyer underparts and sky-blue wing and rump patches; also more contrasting head pattern. Iris—brown; bill and feet—blackish.

Voice: Clear double whistle, comprising two disyllabic notes: *kwah-he kwa-wu*, similar to Blue-winged Pitta but longer and slower.

Range: Breeds in Japan, Korea and E and SE China; winters to Borneo.

Distribution and status: Globally Vulnerable (Collar *et al.* 1994). Nominate race is summer breeder in Guangxi, Guangdong, Hong Kong, Fujian and Taiwan. Generally uncommon to 1000 m. Winters near coasts from Hebei and E coast. Recorded March in S Yunnan (Xishuangbanna).

Habits: As Blue-winged Pitta.

603. Blue-winged Pitta *Pitta moluccensis* [Moluccan Pitta]   PLATE 64
Lanchi basedong

Description: Medium-sized (18 cm) plump, colourful pitta with rufous chest. Head black with pale brown eyebrow; back green; wings bright blue with white wing patch; throat white; vent red. Darker than Fairy Pitta and with violet-blue rather than azure-blue wing patch. Has larger white wing patch and less contrast in head pattern. Iris—brown; bill—blackish; feet—pale brown.

Voice: Loud, clear, fluty whistle *tae-laew, tae-laew* lasting less than one second. Also has harsh *skyeew* alarm call.

Range: SW China, SE Asia, wintering to the Malay Peninsula, Sumatra and Borneo.

Distribution and status: Very rare in S Yunnan (Xishuangbanna).

Habits: Frequents lowland scrub and secondary forest. Hops along the ground like a thrush.

BROADBILLS—Family: Eurylaimidae

Broadbills are a small African and Asian family, with large heads, heavy broad bills, short legs and generally elongated tails. Most species are rather colourful. They are forest birds which from a perch catch flying insects in mid-air with a loud snap of the large bill. Some species also eat fruits. They nest in elaborate, hanging, purse-shaped nests often overhanging streams. Only two species occur in.

604. Silver-breasted Broadbill *Serilophus lunatus* [Gould's/   PLATE 64
Hodgson's Broadbill] Yinxiong siguanniao

Description: Small (15 cm) relatively gracile-billed, pinkish-grey broadbill with arched black eyebrow and blue wing flash. Scapulars, back and rump chestnut; tail black, narrowly edged and tipped white. Female has narrow white band across greyish chest. Iris—brown and green; bill—blue with yellow base; feet—yellowish-green.
Voice: Clear whistled *piu*.
Range: Nepal to SW China, SE Asia, Malay Peninsula and Sumatra.
Distribution and status: An occasional bird of subtropical hill forests at 800–700 m. Race *rubropygius* is resident in areas of Arunachal Pradesh claimed by China; *elisabethae* occurs in S Yunnan and SW Guangxi; *polionotus* is resident on Hainan.
Habits: Lives in small flocks in the subcanopy and understorey of open forests along streams and river banks. Mixes with other species flocks. Catches insects in the foliage and on the wing.

605. Long-tailed Broadbill *Psarisomus dalhousiae*   PLATE 64
Changwei kuozuiniao

Description: Unmistable elongate (25 cm) green broadbill with long blue graduated tail, yellow throat and face with black cap and nape. Wings black with a prominent blue patch. Has small blue spot on the top of the head and a yellow spot behind the eye. Immature is mostly green. Iris—green and grey; bill—green tipped with blue above, yellow beneath; feet—green.
Voice: Loud, high-pitched whistling calls of 5–8 notes on the same pitch.
Range: Himalayan foothills, S China, SE Asia, Sumatra and Borneo.
Distribution and status: Uncommon in primary and mature secondary subtropical hill forests, usually at 700–1500 m, but up to 2000 m in Himalayas. Nominate race is resident in SE Xizang, W and S Yunnan, SW Guangxi and SW Guizhou
Habits: Lives in flocks working through the middle canopy, sometimes mixing with other species.

FAIRY BLUEBIRDS AND LEAFBIRDS—Family: Irenidae

A small Oriental family of small to medium-sized, green-coloured, sweet-voiced birds. Leafbirds have short, thick legs and long, slightly curved bills. The plumage is long, thick and fluffy, especially on the rump. Formerly Leafbirds were placed in their own family with ioras, but now rearranged on basis of DNA studies and combined with *Irena*. Most species eat fruit and/or insects and make neat cup-shaped nests in the leafy terminal branches of trees and bushes. They do not migrate. Four species occur in China.

606. Asian Fairy Bluebird *Irena puella* [Fairy Bluebird/   PLATE 69
Blue-mantled/Blue-backed Fairy-bluebird] Hepingniao

Description: Medium-sized (25 cm) blue and black bird. Male: unmistakable with crown, nape, back, upperwing coverts, rump, uppertail coverts and vent rich shining blue; rest of plumage black. Female: all-over dull greenish cobalt-blue with brighter rump and vent. Iris—red; bill and feet—black.

Voice: Loud ringing drawn-out liquid whistles *whee-eet* on rising scale, sometimes preceded by several introductory notes on a descending scale. Often calls in flight.

Range: Indian subcontinent to SW China, SE Asia, Palawan and Greater Sundas.

Distribution and status: In areas of Arunachal Pradesh claimed by China and S Yunnan this is a common lowland resident in undisturbed forests up to 1100 m.

Habits: Seen singly or in small flocks. Keeps to the tops of tall trees and is most frequently seen when visiting fruiting figs where it mixes with other birds. Flies with undulating beat.

Note: Placed by some authors in its own family Irenidae but DNA evidence suggests it should be placed with leafbirds.

607. Blue-winged Leafbird *Chloropsis cochinchinensis*   PLATE 65
[Golden-hooded/Yellow-headed Leafbird] Lanchi yebei

Description: Smallish (17 cm) bright green leafbird with blue wings and black throat (male). Distinguished from other leafbirds by blue wings and sides of tail. Female lacks yellow eye-ring and has blue throat. Male has yellowish ring round black throat patch. Both sexes have purplish-blue malar stripe. Blue primaries distinguish from Golden-fronted Leafbird. Iris—dark brown; bill—black; feet—bluish-grey.

Voice: Clear, liquid, musical *chee, chee, cheeweet* or *chee, cheeweet*, a twittering call and a sweet song.

Range: Indian subcontinent to SW China, SE Asia, Malay Peninsula and Greater Sundas.

Distribution and status: Race *kinneari* is locally common resident in W and S Yunnan up to 1000 m.

Habits: Inhabits woodland, primary and tall secondary forests keeping to the tops of larger trees. Occurs singly, in pairs or sometimes in small parties and mixes readily with other species.

608. Golden-fronted Leafbird *Chloropsis aurifrons*   PLATE 65
Jin'e yebei

Description: Medium-sized (19 cm) bright green leafbird with yellowish forehead (male) and shiny blue 'shoulder' flash. Both sexes have black chin and throat. Female is slightly duller and in race *pridii* lacks the golden band surrounding the black throat. Immature lacks black of chin and throat and has green crown. Iris—dark brown; bill and feet—blackish.

Voice: Musical song of rising and falling liquid chirps with bulbul-like tone; also mimics other birds. Calls include many harsh whistles.

Range: Indian subcontinent to SW China, SE Asia and Sumatra.

Distribution and status: Race *pridii* is common in the hill forests of SE Xizang and SW Yunnan at 300–2300 m.

Habits: Actively searches insects in upper and middle canopy with systematic patrolling of branches. Often joins mixed-species flocks.

609. Orange-bellied Leafbird *Chloropsis hardwickii*   PLATE 65
Chenfu yebei

Description: Male: largish (20 cm) colourful leafbird with green upperparts and rich orange underparts with blue wings and tail and a black mask and bib with blue moustachial streak. The female is less conspicuous being mostly green with bluish moustachial streak and only a narrow ochre band in the centre of the abdomen. Iris—brown; bill—black; feet—grey.

Voice: Loud clear song and whistled call notes, often mimicking other birds calls.

Range: Himalayas, SE Asia and S China.

Distribution and status: The commonest and most widespread leafbird in the country, found throughout hill and montane forests of S China and Hainan. Nominate race is resident in SE Xizang and SW China; *melliana* in S and SE China; and *lazulina* on Hainan.

Habits: An active insectivore of all canopy layers.

SHRIKES—Family: Laniidae

A moderately large family found throughout the Old World and N America. Shrikes are medium-sized, powerfully built, predatory birds. They have large heads and powerful, deeply notched bills with a strong hooked 'tooth' at the tip. Shrikes perch on low bushes, telegraph wires or poles and pounce on their prey, usually large insects and small vertebrates; some species impale their prey on thorns. The nests are open, cup-shaped structures built in branch forks. In China there are 11 species. Most are migratory.

610. Tiger Shrike *Lanius tigrinus* [Thick-billed Shrike]   PLATE 66
Huwen bolao

Description: Medium-sized (19 cm) rufous-backed shrike. Noticeably thicker bill, shorter tail and larger eye than Brown Shrike. Male: crown and nape grey; back, wings and tail rich chestnut with fine black bars; broad black stripe through eye; underparts white, faintly barred brown on flanks. Female: similar but with white lores and eyebrow line. Immature is duller brown with indistinct barred black eye-stripe; pale eyebrow line; buffy underparts and stronger barring on belly and flanks than Brown Shrike. Iris—brown; bill—blue with black tip; feet—grey.

Voice: Harsh grating chatter, similar to Brown Shrike.

Range: Breeds E Asia, China and Japan; migrates S in winter to Malay Peninsula and Greater Sundas.

Distribution and status: Breeds from Jilin and Hebei through C and E China; migrating south. Fairly common up to 900 m.

Habits: Typical shrike behaviour of hawking for insects from a prominent perch in wooded areas, generally on forest edge. Less conspicuous than Brown Shrike, keeping more to forest.

611. Bull-headed Shrike *Lanius bucephalus* niutou bolao PLATE **66**

Description: Medium-sized (19 cm) brown shrike. Distinguished by brown crown and white-tipped tail. White patch at base of primaries conspicuous in flight. Male: black eye-stripe and white eyebrow, greyish-brown back. Whitish underparts lightly barred black (heavily in race *sicarius*) and washed rufous on flanks. Female: browner; distinguished from female Brown by rufous-brown ear coverts. Paler and less rufous in summer. Iris—dark brown; bill—grey with black tip; feet—plumbeous-grey.

Voice: Harsh grating calls, similar to Striated Grassbird; chattering *ju ju ju* or *gi gi gi* and mimicked calls of other birds.

Range: NE Asia, E China.

Distribution and status: Fairly common resident. Nominate race breeds in NE China from S Heilongjiang, Liaoning, Hebei and Shangdong migrating to winter in S and E China and Taiwan. Montane race *sicarius* is restricted to extreme S Gansu. Vagrant to Taiwan.

Habits: Favours secondary growth and cultivation.

612. Red-backed Shrike *Lanius collurio* Hongbei bolao PLATE **66**

Description: Smaller (19 cm) brown shrike with entire upperparts reddish-brown and uppertail coverts and tail rufous. Eye-stripe and side of head black; eyebrow white; underparts whitish washed pinkish on flanks in male or scalloped finely with black in female. Iris—brown; bill—grey; feet—black.

Voice: Harsh grating calls.

Range: C and E Asia, Russia; wintering to Indian subcontinent and Africa.

Distribution and status: Race *pallidifrons* is a passage migrant through NW China.

Habits: Favours bushes in steppe and desert country, open woodland and hedgerows.

613. Rufous-tailed Shrike *Lanius isabellinus* Zongwei bolao PLATE **66**

Description: Smaller (19 cm) greyish shrike with rufous tail. Male: entire upperparts light sandy-grey; black eye-stripe (but no loral band); white eyebrow; tail rufous with rufous-buff uppertail coverts; white ponel on wing. Female: duller than male with fine black scalloping on underparts. Iris—brown; bill—grey; feet—dark grey.

Voice: Song is rich and varied with chattering and mimicked sections. Call is harsh grating sound.

Range: Turkestan, Pakistan and W China; wintering to NE Africa and S Asia.

Distribution and status: Common shrike of NW China. Race *phoenicuroides* summer breeder in Altai and N Tianshan Mts; race *speculigerus* breeds in E Xinjiang and Ordos and Helanshan ranges of Nei Mongol; nominate race breeds in Tianshan and from Kashi, through Turpan to NW Gansu and Ningxia; race *tsaidamensii* is summer breeder in Qinghai.

Habits: As Red-backed Shrike.

Notes: Placed by some authors within Red-backed Shrike after Vaurie (1959), e.g. Cheng, (1987, 1994).

614. Brown Shrike *Lanius cristatus* [Red-tailed Shrike]   PLATE 66
Hongwei bolao

Description: Medium-sized (20 cm) plain brown shrike with white throat. Adult: forehead grey; supercilium white; broad black mask; crown and upperparts brown; underparts buff. Race *superciliosus* is greyer above with grey crown. Races *lucionensis* and *confusus* have whitish forehead. Immature: similar but back and sides barred with wavy dark brown lines. Black eyebrow distinguishes from immature Tiger Shrike. Iris—brown; bill—black; feet—blackish-grey.

Voice: Generally silent in winter. Harsh chattering *cheh-cheh-cheh* calls and song in breeding season.

Range: Breeds in E Asia; migrates S in winter to India, SE Asia, Philippines, Sundas, Sulawesi, Moluccas and New Guinea.

Distribution and status: Generally common up to 1500 m. Race *confusus* breeds in Heilongjiang, migrating through E China; *lucionensis* breeds in Jilin, Liaoning and N, C and E China, migrating S with some birds wintering in S China, Hainan and Taiwan. Nominate race is winter visitor and passage migrant over most of eastern China; *superciliosus* migrates across China to winter in Yunnan, S China and Hainan.

Habits: Frequents open cultivated and secondary habitats including gardens and plantations. Single birds perch on bushes, wires and small trees, chase flying insects or pounce on insects or small animals on the ground.

615. Burmese Shrike *Lanius collurioides* [Chestnut-backed Shrike]   PLATE 66
Libei bolao

Description: Medium-sized (20 cm) slim shrike with chestnut upperparts, grey crown, nape and mantle, black mask and no eyebrow. Male has black forehead, female forehead streaked with white. Wings and tail black with white patch on primaries visible in flight and white edge and tip to tail. Distinguished from Long-tailed shrike by shorter tail and white edges to tail. Iris—reddish-brown; bill—dark grey; feet—blackish.

Voice: Staccato rattle like Brown Shrike.

Range: NE India, Burma, Indochina and S China.

Distribution and status: Uncommon in secondary vegetation and cultivation in W and S Yunnan, S Guizhou, Guangxi and Guangdong.

Habits: Not shy.

616. Long-tailed Shrike *Lanius schach* [Schach/Rufous-headed/   PLATE 66
Black-capped/Black-headed/Rufous-backed Shrike] Zongbei bolao

Description: Largish (25 cm) long-tailed, brown, black and white shrike. Adult: forehead, mask, wings and tail black with a white wing spot; crown and nape grey or grey and black; back, rump and sides reddish-brown; chin, throat, breast and centre of belly white. Extent of black on head and back varies with race. Immature: duller with barred flanks and back and greyer head and nape. Melanistic, dark charcoal-grey form 'dusky shrike' is not rare in Hong Kong and Guangdong and occasional elsewhere within range. Iris—brown; bill and feet—black.

Voice: Harsh screeches *terrr* and a warbled song when it sometimes mimics other bird calls.

Range: Iran to China, India, SE Asia, Philippines, and Sundas to New Guinea.

Distribution and status: A common resident up to 1600 m. Race *tricolor* is resident in N, W and S Yunnan and S Xizang; nominate race in C, E, S and SE China; *formosae* on Taiwan and *hainanus* on Hainan.

Habits: Frequents open spaces, grassland, scrub, tea and other plantations and other open areas. Sits on a low perch and makes darting sallies after flying insects, or more commonly pounces on grasshoppers and beetles on the ground.

617. Grey-backed Shrike *Lanius tephronotus* Huibei bolao   PLATE 66

Description: Largish (25 cm) long-tailed shrike. Similar to Long-tailed Shrike but distinguished by dark grey upperparts with only a narrow rufous buff band across the rump and uppertail coverts. Bold white patch on primaries lacking or small. Iris—brown; bill—grey; feet—green.

Voice: Harsh grating calls and mimics other bird calls.

Range: Himalayas to S and W China; winters to SE Asia.

Distribution and status: Replaces Long-tailed Shrike in Xizang and C China with some overlap in the latter. Locally common up to 4500 m in the Himalayas, in scrub, open areas and cultivated lands.

Habits: Behaviour and screeching, strident calls are similar to Long-tailed Shrike. Quite tame.

618. Lesser Grey Shrike *Lanius minor* Hei'e bolao   PLATE 66

Description: Smallish (20 cm) grey shrike. Smaller than Great Grey or Chinese Grey Shrikes. Male has more extensive black on forehead and blacker secondaries. Flanks pinkish. Iris—brown; bill—grey; feet—black.

Voice: Harsh grating calls; chattering with some thrush-like whistles.

Range: S and E Europe, C Asia; wintering to Africa.

Distribution and status: Race *turanicus* is a rare summer breeder in forested steppes and scrub grasslands of extreme NW China in Junggar Basin and NW Alatau Mts.

Habits: Rather upright posture with tail held sharply down. Less undulating flight than most other shrikes.

619. Great Grey Shrike *Lanius excubitor* [Northern Shrike]   PLATE 66
Hui bolao

Description: Large (24 cm) grey, black and white shrike. Male: grey crown, nape, back and rump, bold black eye-stripe topped by white eyebrow; wings black with white bar; tail black with white edges; underparts whitish. Race *funereus* is paler grey above, has more vermiculations below and more white on wing; Female and immature: duller with buff scaly underparts. Iris—brown; bill—black; feet—blackish.

Voice: Shrill clear, ringing *schrreea* and drawn-out nasal *eeh*; also harsh *ga-ga-ga*.

Range: Northern Eurasia.

Distribution and status: Uncommon seasonal migrant in N China. Race *funereus* breeds in NW China. Races *mollis* and *sibiricus* winter in N and NE China. Races *homeyeri* and *leucopterus* winter in NW China.

Habits: Hunts from prominent tree perch or wire in open and wooded country. Sometimes hovers. Often impales prey on tree thorns.

619a. Southern Grey Shrike *Lanius meridionalis* Nan huibolao　　PLATE 66

Description: As Great Grey Shrike but with white or pinkish un-vermiculated underparts, white lores and broad white wing bar and paler bill. Iris—brown; bill—grey or horn; feet—blackish.

Voice: Similar to Great Grey Shrike.

Range: Africa, Middle East, Indian subcontinent, C Asia to Mongolia.

Distribution and status: Race *pallidirostris* breed in Helan Mts of Ningxia.

Habits: As Great Grey Shrike.

Note: Sometimes treated as conspecific with Great Grey Shrike (e.g. Cheng 1987).

620. Chinese Grey Shrike *Lanius sphenocercus* [Long-tailed　　PLATE 66 Grey Shrike] Xiewei bolao

Description: Very large (31 cm) grey shrike with black mask, white eyebrow and bold white wing bar on black wings. Larger than Great Grey Shrike. Central three tail feathers black with narrow white tips; outer tail feathers white. Race *giganteus* is darker than nominate and lacks white eyebrow. Iris—brown; bill—grey; feet—black.

Voice: Harsh *ga-ga-ga* calls similar to Great Grey Shrike.

Range: C Asia, SE Siberia, Korea, N and E China.

Distribution and status: Uncommon. Race *giganteus* breeds in Qaidam Basin of Qinghai, NE Xizang, N and W Sichuan and nominate race breeds in Nei Mongol and NE China, Shanxi, Shaanxi, Ningxia and Gansu. Recorded on passage through Liaoning and Qinghai to winter in Fujian and Guangdong. Found in steppe, scrub, semi-desert and edge of forests or trees along rivers.

Habits: Hovers and will take prey such as insects or small birds in the air. Hunts from prominent tree, bush or wire perch in open country, often close to farms or villages.

CROWS AND RELATIVES—Family: Corvidae

A very large family combining the old crow family with such diverse groups as woodswallows, orioles, cuckooshrikes, drongos, monarchs, fantails, ioras and woodshrikes.

CROWS, WOODSWALLOWS, CUCKOOSHRIKES, MINIVETS AND ORIOLES—Subfamily: Corvinae

CROWS AND JAYS—Tribe: Corvini

Crows and jays are a large tribe of generally large birds with powerful, straight bills and strong feet. They occur almost worldwide. They are intelligent wily birds and several species have learned to live with man as commensals. Most species have a lot of black in their plumage although some of the jays and magpies are colourful with bright blue, green, and brown feathers. They have harsh calls, make large, untidy stick nests and feed on a mixture of fruit and animal material. Some are scavengers. Thirty species occur in China.

621. Eurasian Jay *Garrulus glandarius* Song ya PLATE 67

Description: Small (35 cm) pinkish crow with diagnostic black and electric blue patterned wing panel and white rump. Moustachial stripe is black and wings are black with white patch. Wings in flight are broad and rounded. Flies with laboured irregular beat. Iris—pale brown; bill—grey; feet—fleshy-brown.

Voice: The call is a harsh *ksher* or plaintive mewing.

Range: Europe, NW Africa, Himalayas, Middle East to Japan, SE Asia.

Distribution and status: Widespread and quite common over much of N, C and E China. Many races occur in China—*brandtii* in Altai and central NE China; *bambergi* in W and E parts of NE China; *pekingensis* in Hebei area; *kansuensis* in Qinghai (Zedog) and W Gansu; *interstinctus* in SE Xizang; *leucotis* in S Yunnan; *sinensis* over most of C, E, S and SE China; and *taivanus* on Taiwan.

Habits: A noisy bird of deciduous woodlands and forest. Eurasion Jays feed on fruits, birds eggs and carrion and are major consumers of acorns. They are bold mobsters of raptors.

622. Siberian Jay *Perisoreus infaustus* Bei zaoya PLATE 67

Description: Small (28 cm) fluffy, grey and rufous magpie with relatively short tail and hint of short crest. Head dark brown; wings, rump and edges of tail rufous. Has buffy nasal tuft. Distinguished from Sichuan Jay by rufous in wing and tail. Race *opicus* is darker rufous in tone with dark head and extensive rufous in wing. Race *maritimus* is greyer but also dark with extensive rufous in wing. Iris—brown; bill—black; feet—black.

Voice: Rather quiet. Calls include buzzard-like mewing *geeak*, harsh *hearrr-hearrr* and hoarse nasal *skaaaak*. Song is jumble of whistles, creaks and trills with some mimicked calls.

Range: Scandinavia and boreal Palearctic to NE and NW China.

Distribution and status: Uncommon in northern cold conifer forests. Race *maritimus* occurs in Heilongjiang in the Lesser Hinggan Mts and Wusuli River area. Race *opicus* occurs in Altai Mts of N Xinjiang.

Habits: Inconspicuous inhabitant of forests—singly, in pairs or small parties.

623. Sichuan Jay *Perisoreus internigrans* [Szechwan Grey Jay] PLATE 67
Heitou zaoya

Description: Small (30 cm) grey fluffy magpie with rather short tail. Distinguished from Siberian Jay by all-grey plumage, stouter bill and lack of rufous in wings, rump and tail. Iris—brown; bill—yellowish-olive to horn; feet—black.

Voice: Alarm call is high-pitched *kyip, kyip* rising in inflection and sometimes repeated fast into longer series. Also gives plaintive mewing note like buzzard and Siberian Jay.

Range: C China.

Distribution and status: Globally Vulnerable (Collar *et al.* 1994). Uncommon. Endemic to SE Qinghai, W Gansu, N Sichuan and E Xizang.
Habits: Lives in subalpine conifer forests from 3050–4300 m. Little-known but similar to Siberian Jay.

624. Taiwan Blue Magpie *Urocissa caerulea* [Taiwan Magpie,   PLATE 67
Formosan Blue Magpie] Taiwan lanque

Description: Unmistakable elongated (69 cm) azure-blue magpie with black head and upper breast. Tail feathers tipped black and white, central two feathers very long. Iris—pale yellow; bill—scarlet; feet—red.
Voice: Alarm call is cackling chatter *kyak-kyak-kyak-kyak*, higher-pitched than Black-billed Magpie. Harsh ringing *ga-kang, ga-kang*. Also softer *kwee-eep* or *gar-suee* calls.
Range: Endemic to Taiwan.
Distribution and status: Absent from cleared western lowlands, but otherwise not uncommon from 300–1200 m in wooded hills.
Habits: A social breeder. Lives in small parties; similar habits to Red-billed Magpie of mainland. Descends to lower altitudes in winter.

625. Yellow-billed Blue Magpie *Urocissa flavirostris*   PLATE 67
[Golden-billed Magpie] Huangzui lanque

Description: Long (69 cm) colourful blue magpie with extremely long, drooping, graduated tail and black head with yellow bill. Yellow bill, yellow feet and black crown distinguish from Blue Magpie. Has white patch on otherwise black nape. Iris—brown; bill—yellow; feet—yellow.
Voice: Calls consist of loud harsh whistles and screeches.
Range: Himalayas, NE India, China, Burma and N Vietnam.
Distribution and status: Familiar and noisy bird of forests at 1800–3300 m in the Himalayas, SE Xizang and SW China.
Habits: Found in areas where Red-billed Magpie is absent but is more solitary in its habits and never as common as that species. Favours open forests and orchards. Sometimes in small parties.

626. Red-billed Blue Magpie *Urocissa erythrorhyncha*   PLATE 67
[Blue Magpie] Hongzui lanque

Description: Long-tailed (68 cm) bright blue magpie with white-crowned black head. Distinguished from yellow-billed Magpie by scarlet bill and red feet. Abdomen and vent are white and the outer tail feathers are graduated and black with white tips. Iris—red; bill—red; feet—red.
Voice: Gives harsh screeched contact calls plus a wide range of other calls and whistles.
Range: Himalayas, NE India, China, Burma and Indochina.
Distribution and status: A common and widespread species of forest edge, scrub and even villages. Nominate race is resident in C, SW, S and SE China and Hainan; *alticola* occurs in NW and W Yunnan; and *brevivexilla* from S Gansu and S Ningxia to Shanxi, Hebei, SE Nei Mongol and W Liaoning.

Habits: Noisy bird which lives in small flocks. Feeds on fruits, small birds and eggs, insects and carrion and often feeds on the ground. Aggressively mobs raptors.

627. White-winged Magpie *Urocissa whiteheadi* PLATE 67
Baichi lanque

Description: Large (46 cm) distinctly patterned black and white magpie with graduated tail and orange bill. Heavy build. Juvenile is greyish-brown instead of black, with bill and iris browner. Race *xanthomelana* is larger than nominate with greener iris, yellower wing bands and tail edge. Iris—yellow; bill—orange with greenish or brownish base; feet—black.

Voice: Loud and noisy with a variety of shrieking Eurasian Jay-like calls. Repeated hoarse rising *shurreek*; low purring *churrree*; repeated soft, liquid, rippled *brrriii . . . brrriii . . .* and rising *errreep*.

Range: Tonkin, Annam, N Laos, Hainan and SW China.

Distribution and status: Globally Near-threatened (Collar *et al.* 1994). Uncommon up to 1400 m in evergreen and mixed forests. Nominate race is resident on Hainan. Mainland form *xanthomelana* is recorded only sporadically in China. Recorded in S Sichuan (Leibo), SW Guangxi and S Xishuangbanna in S Yunnan.

Habits: Lives in noisy parties or extended family groups. Birds glide from tree to tree with wings outstretched displaying bold black and white pattern. Occupies forest, forest edge and secondary growth and crosses agricultural areas with trees.

628. Common Green Magpie *Cissa chinensis* [Hunting Cissa/ PLATE 67
Chinese Green Magpie] Lan lüque

Description: Large (38 cm) bright green magpie with long tail, red bill, and chestnut wings. There is a black eye-line and the graduated green tail feathers are tipped black and white. Distinguished from Indochinese Green Magpie by longer tail, more yellow on crown and black-tipped tertials. Iris—red; bill—red; feet—red.

Voice: A series of shrill shrieking whistles *keep keep keep* followed by harsh *chuck*. Also harsh trumpeting chack notes, rapid chattering and mimicked calls of other species. Similar to calls of Indochinese Green Magpie but shriller and less plaintive.

Range: Himalayas, S China, SE Asia, Sumatra and Borneo.

Distribution and status: Nominate race is uncommon resident of subtropical forests at 400–1800 m in SE Xizang, S Yunnan and Guangxi.

Habits: A shy bird of dense forest, more often heard than seen. Lives in small noisy families in primary and disturbed forests.

629. Indochinese Green Magpie *Cissa hypoleuca* [Eastern/ PLATE 67
Yellow-breasted Magpie, AG: Green Magpie] Duanwei lüque

Description: Medium-sized (32 cm) green jay with black eye-stripe, red bill and chestnut wings. Wing tip shows pale tertials. Distinguished from Common Green Magpie by shorter tail, yellow on breast and lack of black barring on tertials. Race *jinji* has longer tail tipped buff. Race *katsumatae* has central tail yellower and tipped grey. Iris—brown; bill—bright red; feet—red.

Voice: Variable, loud, plaintive, penetrating shriek *peu-peu-peu* followed by harsh *chuk*; also high-pitched scolding chatter.

Range: SW and S China and Indochina.

Distribution and status: Globally Near-threatened (Collar *et al.* 1994). Uncommon in evergreen forest up to 1000 m. Race *katsumatae* in Hainan and race *jinji* in Guangxi (Yaoshan) and S Sichuan (Mabian).

Habits: Travels in pairs or small groups, often calling noisily but surprisingly hard to see despite colourful plumage. Hunts for insects in the lower storeys of the forest.

Note: Formerly treated as race of Short-tailed Magpie *C. thalassina* (e.g. Cheng 1987) as per Vaurie (1962).

630. Azure-winged Magpie *Cyanopica cyanus* Hui xique   PLATE 67

Description: Small (35 cm), slender grey crow with black hood, azure-bluish wings and long blue tail. Iris—brown; bill—black; feet—black.

Voice: The call is a harsh rolling *zhruee* or clear *kwee*.

Range: NE Asia, China, Japan and Iberian Peninsula.

Distribution and status: Common and widespread in E and NE China. Introduced into Hong Kong but declining and possibly now extinct there. Nominate race winters in extreme northern NE China; *pallescens* is resident in Lesser Hinggan and *stegmanni* is found in the Greater Hinggan Mts and Changbai Mts of NE China; *interposita* is resident in eastern N China; *swinhoei* in lower Changjiang valley to S Gansu; *kansuensis* in S Qinghai to NW Gansu.

Habits: Lives in noisy groups in open pine and broadleaf woodlands, parkland and even towns. Flight characterised by quick wing beats interspersed by long silent glides. Feeds on fruits, insects and carrion in trees, on ground and even off tree trunks.

631. Rufous Treepie *Dendrocitta vagabunda* Zongfu shuque   PLATE 67

Description: Long (44 cm) brown treepie. Face is black grading into grey crown, nape and breast. Back and rump are rufous-brown and underparts are rufous-buff. This is the most southerly of China's five treepies. Readily distinguished from Grey Treepie and Collared Treepie by grey tail and whitish wing patch. Iris—red; bill—grey; feet—blackish.

Voice: Call is loud, flute-like, ringing *ko-ki-la* and other harsh, metallic and mewing notes.

Range: Indian subcontinent, Burma and N Indochina.

Distribution and status: Common in scrub and forest up to 2000 m in SW China and claimed areas of Arunachal Pradesh.

Habits: Keeps to crowns of small trees and hunts insects, nests and small vertebrates like a malcoha.

632. Grey Treepie *Dendrocitta formosae* Hui shuque   PLATE 67

Description: Large (38 cm) brownish-grey magpie with grey nape and very long graduated tail. Underparts grey with rufous vent; mantle brown; tail is black or black with central feathers grey; rump and lower back pale grey or white and wings black with a white patch at the base of the primaries. Races vary in details—*sapiens, sinica* and *insulae* have white rump and all black tail; nominate has central tail feathers basally grey and *himalayensis* has grey rump and longer tail with central feathers grey except for black tip. Race *insulae* is small and dull with smaller white patch on primaries. Iris—reddish-brown; bill—black with grey base; feet—dark grey.

Voice: Metallic, clucking, squawked *klok-kli-klok-kli-kli*. Also harsh and musical notes and alarm chatter.

Range: Himalayas, E and NE India, Burma, N Thailand, N Indochina and C, S and SE China.

Distribution and status: Race *himalayensis* is recorded in SE Xizang and Yunnan; *sapiens* in Mt Emei and Qionglai Mts in Sichuan; *insulae* on Hainan; nominate in Taiwan; and *sinica* from S and SE China. Rather common in open forests at moderate to high altitudes from 400–1200 m in SE China but up to 2400 m in Himalayas.

Habits: A shy but noisy bird that sits on low perches waiting for prey which it catches on ground, in foliage or by working steadily and jerkily through crowns of middle and upper canopy. Sometimes travels in noisy groups and mixed-species flocks.

633. Collared Treepie *Dendrocitta frontalis* [Black-browed/ Black-faced/White-naped Treepie] Hei'e shuque   PLATE 67

Description: Smallish (38 cm) crow with very long black tail and rufous mantle, back, lower belly and tail coverts. Black face contrasts with clear grey nape and upper breast. Distinguished from Rufous Treepie by black forecrown, ear coverts, throat and tail and rufous rump. Distinguished from Grey Treepie by rufous rump and lack of white wing spot or pale subterminal band on tail. Iris—reddish-brown; bill—black; feet—black.

Voice: Wide range of typical treepie range of discordant and musical metallic notes.

Range: E Himalayas, N Burma and NE Indian hills.

Distribution and status: Only marginally distributed in China. Recorded in SE Xizang (Medog) and in Gaoligong Mts along Myanmar–Yunnan border.

Habits: As other treepies. Rather rare from 300–2100 m. Less shy than Rufous Treepie.

634

634. Racket-tailed Treepie *Crypsirina temia* [Black Treepie] Panwei shuque

Description: Medium-sized (35 cm including 18 cm-long tail), blackish magpie with a very long, terminally flared tail. Entire plumage glossy dark grey with a bronze-green sheen. Bill is heavy and slightly hooked. Blue eye is prominent feature. Iris—blue; bill—black; feet—black.

Voice: Harsh, unmusical, whining, metallic calls of 2–3 syllables.

Range: SW China, SE Asia to Java

Distribution and status: Rare in China. Recorded only in S Yunnan lowlands, especially in Hong Ha (Red River) valley. Absent Xishuangbanna.

盘尾树鹊 ■ Resident

Habits: Travels singly or in pairs through secondary forest, bamboo, scrub and even gardens.

635. Ratchet-tailed Treepie *Temnurus temnurus* Tawei shuque   PLATE 67

Description: Unmistakable, small (30 cm) black crow with unique-patterned saw-like, ratchet tail. Head is black and body dark grey. The head is large and the bill heavy and decurved. Iris—brown; bill—black; feet—black.

Voice: Makes an amazing range of extraordinary cranking and honking calls and whistles.

Range: S Tennasserim, SW Thailand, C Laos, Vietnam, Hainan.

Distribution and status: Common in forests of Indochina but occurs in China only on Hainan (*nigra*) where it now appears to be rare. This is a monotypic genus that in China is confined to Hainan.

Habits: Lives in pairs or small flocks that work steadily and rather sneakily through the canopy feeding on insects and some fruits. The flight is flappy and clumsy with occasional long glides.

636. Black-billed Magpie *Pica pica* Xi que   PLATE 67

Description: Unmistakable, small (45 cm) pied crow with long black tail. Black of wings and tail have bluish gloss. Iris—brown; bill—black; feet—black.

Voice: The call is a rolling harsh cackle.

Range: Eurasia, N Africa, W Canada and W California.

Distribution and status: It is a widespread and common bird in China where it is regarded as a good luck omen and generally not persecuted. Race *bactriana* occurs in N and W Xinjiang and NW Xizang; *bottanensis* in S, SE and E Xizang to W Sichuan and Qinghai; *leucoptera* in Hulun Nur area of NE Nei Mongol; and *sericea* over the rest of China, Taiwan and Hainan.

Habits: Highly adaptable, as much at home in the open farmland of N China as in the high-rise skyscrapers of Hong Kong. Omnivorous scrounger feeding mostly on the ground. Lives in small parties. Nest is a domed and irregular pile of sticks, used year after year.

637. Mongolian Ground-jay *Podoces hendersoni* [Henderson's   PLATE 68
Ground-jay] Heiwei diya

Description: Smallish (30 cm) pale brown jay. Upperparts sandy-brown with vinous-shaded back and rump. Crown black with blue gloss. Wings glossy black with large white patch in primaries; tail bluish-black. Iris—dark brown; bill—black; feet—black.

Voice: Call similar to *clack-clack-clack* of wooden rattle, also harsh whistles.

Range: Mongolia and NW China.

Distribution and status: Quite common. Distributed from Badain Jaran desert of W Nei Mongol and N Gansu; N Xinjiang and NE Qinghai in desert foothills to N and S of Tianshan, Qaidam and Junggar basins of NE Qinghai from 2000–3000 m.

Habits: Lives in open rocky ground and scrub feeding on seeds and chasing invertebrates on the ground. Nests on the ground but perches in bushes.

638. Xinjiang Ground-jay *Podoces biddulphi* [White-tailed/   PLATE **68**
Biddulph's Ground Chough] Baiwei diya

Description: Small (29 cm) brown jay with decurved bill and short broad purplish-black crest. Cheeks and throat blackish. Lores, eye-ring and sides of head and neck buff; wing coverts black with purple gloss; flight feathers mostly white with blackish shafts and tips. Iris—dark brown; bill—black; feet—black.

Voice: Thrice repeated *chui-chui-chui* rising on last syllable and rapid descending series of low whistles.

Range: Endemic to Xinjiang.

Distribution and status: Globally Vulnerable (Collar *et al*. 1994). Uncommon. Distributed around and locally in the interior of the Taklimakan Desert as far E as Lop Nur foothills, associated with desert poplars from 900–1300 m.

Habits: Inhabits desert scrub and shrubby wastes. Runs quickly on ground but perches in bushes.

639. Hume's Groundpecker *Pseudopodoces humilis* [Tibetan/   PLATE **68**
Little Ground-jay AG: Groundpecker] Hebei nidiya

Description: Very small (19 cm) sandy-grey chough with whitish underparts and dusky loral stripe. Central tail feathers brown, outer feathers buff-white. Juvenile is more buff with buffy nuchal collar. Iris—dark brown; bill—black; feet—black.

Voice: Call is weak, prolonged *cheep*. Also short *chip* followed by rapidly whistled *cheep-cheep-cheep-cheep*.

Range: Tibetan Plateau and Kunlun Mts of W China.

Distribution and status: Locally common across the whole Qinghai–Xizang plateau at 4000–5500 m from SW Xinjiang as far E as Gansu and Ningxia through W Sichuan to NE Yunnan.

Habits: Lives in grassy plains and foothills with scattered bushes above treeline. Favours yak pastures. Often close to monasteries or settlements. Excavates nest holes. Flicks wings and tail energetically. Flight weak, low and fluttery. Behaves like wheatear.

640. Spotted Nutcracker *Nucifraga caryocatactes* [Eurasian Nutcracker] PLATE **68**
Xing ya

Description: Smallish (33 cm) unmistakable dark brown crow with heavy white spotting, white vent and corners of tail. Short tail gives curious stocky proportions, accentuated by stout, pointed bill. Several doubtful races are recognised in China. Northern races have smaller white spots confined to sides of head, breast and mantle. Race *hemispila* is paler brown; *owstoni* is sooty-brown. Iris—dark brown; bill—black; feet—black.

Voice: Dry, harsh *kraaaak*, sometimes repeated as rattle. Less screeched than Eurasian Jay. Has quiet piping song with whistles, clicks and whines interspersed with mimicked calls. Young birds give nasal bleat.

Range: N Palearctic, Japan and Taiwan; also Himalayas to SW and C China.

Distribution and status: Rather common in the subalpine conifer forest of China. Race *rothschildi* is resident in Tianshan of NW Xinjiang; *macrorhynchos* in the Hinggan and Changbai Mts of NE China; *interdicta* in conifer forest of Liaoning, Hebei, Shangdong and Henan; *macella* across C China to SW China and SE Xizang; *hemispila* in western S Xizang; *owstoni* on Taiwan.

Habits: Solitary or in pairs; occasionally small parties. Especially associated with pine forest and adapted to feeding on pine seeds. Also takes nuts and buries them for winter use. Movements jaunty and flight rhythmic and undulating.

641. Red-billed Chough *Pyrrhocorax pyrrhocorax*   PLATE 68
Hongzui shanya

Description: Smart, smallish (45 cm) black crow with slender, bright red decurved bill and red feet. Young birds as adult but bill is blacker. Distinguished from yellow-filled Chough by shorter, red rather than yellow bill. Iris—reddish; bill—red; feet—red.

Voice: The call is a harsh sharp *kee-ach*.

Range: Palearctic, Middle East, Baluchistan, Himalayas, C and E Asia.

Distribution and status: Race *brachypus* occurs across northern and E China; *himalayanus* is resident over the entire Qinghai–Xizang Plateau to Sichuan and NW Yunnan; *centralis* occurs in Taxakorgan area of extreme SW Xinjiang.

Habits: Very agile in flight, playing on thermals and gliding with short broad wings and prominent 'fingers' of spread primaries. It lives in small to large groups, often around buildings and farms at moderate to high altitudes over much of the Qinghai–Xizang plateau and Himalayas, or at lower altitudes in NE China.

642. Yellow-billed Chough *Pyrrhocorox graculus* [Alpine Chough]   PLATE 68
Huangzui shanya

Description: Small (38 cm) glossy black crow with short, decurved yellow bill and red legs. Similar to Red-billed Chough but bill shorter and not red. Tail more rounded in flight and tail longer at rest, projecting well beyond wing tips. Wings less rectangular in flight. Juvenile has grey legs and less yellow bill. Iris—dark brown; bill—whitish-yellow; feet—red.

Voice: Boisterous calls. Sweet rippling *preeeep* and descending whistled *sweeeoo*; more squeaky than Red-billed Chough. Also rolling *churrr* in alarm and quiet warbling and chittering notes when feeding.

Range: Spain, N Africa, Mediterranean, Middle East to C Asia, Himalayas and W and C China.

Distribution and status: Race *digitatus* is uncommon across the entire Qinghai–Xizang Plateau and Xinjiang as far E as N and W Sichuan.

Habits: Similar to Red-billed Chough but generally at higher altitudes. Gregarious. Flocks soar on thermals.

643. Eurasian Jackdaw *Corvus monedula* [Common Jackdaw]   PLATE 68
Hang ya

Description: Smallish (37 cm) black and grey crow with small, short bill. Distinguished from House Crow by much smaller size, petite bill and blue iris. Juvenile Daurian Jackdaw is darker with less contrast, has dark iris and fine silver streaks behind eye.

Juvenile Eurasian Jackdaw also has dark eyes and darker grey areas but lacks silver streaking on ear coverts. Iris—blue; bill—black; feet—black.

Voice: Typical call is abrupt, high *chjak*, repeated if excited. Also drawn-out *chaairurr* followed by *chak* and a slurred, high *kyow*.

Range: Europe, N Africa, Middle East to C Asia and W China.

Distribution and status: Common within restricted Chinese range. Nominate race breeds in Kashi, Tianshan, Altai and Turpan regions of W Xinjiang wintering in Zanda–Burang region of W Xizang.

Habits: Inhabits woodlands, moors, rocky areas and towns and villages. Forms small noisy gregarious flocks, often mixing with Rooks in fields.

644. **Daurian Jackdaw** *Corvus dauuricus* Dawuli hangya   PLATE 68

Description: Smallish (32 cm) pied crow with whitish pattern extending well down breast. Distinguished from Collared Crow by smaller size, daintier bill and more extensive white on breast. Juvenile has less colour contrast but distinguished from adult Eurasian Jackdaw by dark eye and from juvenile Eurasian Jackdaw by silver streaks on ear coverts. Iris—dark brown; bill—black; feet—black.

Voice: Flight call is *chak* as in Euraeian Jackdaw. Other calls also similar.

Range: E Russia and Siberia, eastern edge of Tibetan Plateau, C, NE and E China.

Distribution and status: Common, especially in N up to 2000 m. Breeds across N, C and SW China, wintering to S of range plus SE China and vagrant to Taiwan.

Habits: Less gregarious than Eurasian Jackdaw. Breeds in open nests and treeholes and crevices in cliffs or buildings. Often feeds among grazing animals.

Note: Treated by some authors as race of Eurasian Jackdaw (e.g. Cheng 1987).

645. **House Crow** *Corvus splendens* Jia ya   PLATE 68

Description: Large (43 cm) black crow. Similar to Carrion Crow but with grey collar and shorter bill. Iris—brown; bill—black; feet—black.

Voice: Very noisy. Typical call is flat, dry *kaaa-kaao*, weaker, softer and flatter than Rook. Many other calls in social interactions.

Range: Indian subcontinent, W and S Burma and SW China, and introduced via ship-assisted vagrancy in E Africa, Middle East and on various islands.

Distribution and status: Locally common (*insolens*) within very restricted Chinese range in Chumbi Valley of S Xizang and in extreme S Yunnan.

Habits: Social urban and village crow, raiding rubbish tips and feeding on agricultural land.

646. **Rook** *Corvus frugilegus* Tubi wuya   PLATE 68

Description: Largish (47 cm) black crow with diagnostic pale whitish-grey bare skin at base of bill. Juveniles have face fully feathered and can be confused with Carrion Crow but distinguished by more domed crown, more conical and sharper bill and looser 'baggy' plumage around 'thighs'. In flight has distinct wedge-shaped tail tip and also longer, narrower wings with more prominent 'fingers' and more protruding head. Iris—dark brown; bill—black; feet—black.

Voice: Drier and flatter *kaak* than call of Carrion Crow. Also high-pitched plaintive *kraa-a* and other calls. Song combination of crowing sounds with chuckles, gurgles and strange clicks accompanied by back and forth head movements.

Range: Europe to Middle East and E Asia.

Distribution and status: Formerly common but now much declined. Nominate race occurs in W Xinjiang. Race *pastinator* breeds over most of NE, E and C China wintering to S of breeding range plus SE coastal provinces, Taiwan and Hainan.

Habits: Very social both in feeding parties and nesting rookeries. Flocks mix with jackdaws. Feeds on fields and short grassland. Often follows domestic animals.

647. Carrion Crow *Corvus corone* Xiaozui wuya   PLATE 68

Description: Large (50 cm) black crow distinguished from Rook by black-feathered base to bill and from Large-billed Crow by low-browed profile and more slender, though still powerful, bill. Iris—brown; bill—black; feet—black.

Voice: The call is a harsh croaking *kraa*.

Range: Eurasia, NE Africa, Japan.

Distribution and status: Race *orientalis* breeds across C and N China (some populations migrate S to winter in S and SE China. Race *sharpii* recorded on passage in NW China.

Habits: Gathers in large roosting flocks but does not nest in colonies like rooks. It feeds in short grassland and agricultural fields, mostly taking invertebrates but is fond of carrion and is frequently seen on roads feeding on roadkills. Does not generally populate urban habitats as does Large-billed Crow.

648. Large-billed Crow *Corvus macrorhynchos* [Thick-billed Crow]   PLATE 68
Dazui wuya

Description: Large (50 cm) glossy black crow with a very heavy bill. Smaller and much squarer tail than Common Raven. From Carrion Crow by heavier bill and more rounded tail and more domed crown. Iris—brown; bill—black; feet—black.

Voice: Harsh, throaty *kaw* and higher-pitched *awa, awa, awa*; also low gurgling.

Range: Iran to China, SE Asia, Philippines, Sulawesi, Malay Peninsula and Sundas.

Distribution and status: Common resident over most of China except NW. Race *mandschuricus* occurs in NE China; *colonorum* over eastern and southern China, Hainan and Taiwan; *tibetosinensis* in SW and E Xizang and the eastern slopes of the Qinghai–Xizang Plateau from E Qinghai, W Sichuan and W Yunnan; *intermedius* occurs in S Xizang.

Habits: Pair-living crow frequent around villages.

Note: Formerly included within Jungle Crow *C. levaillantii*.

649. Jungle Crow *Corvus levaillantii* Conglin ya   PLATE 68

Description: Large (47 cm) all black crow with heavy bill. Very similar to Large-billed Crow but smaller with less massive bill and slightly squarer tail. Also has higher-pitched voice. Smaller and much squarer tail than Common Raven. From Carrion Crow by heavier bill and more rounded tail and more domed crown. Iris—dark brown; bill—black; feet—black.

Voice: Dry nasal *quank-quank-quank*, harsher and deeper than House Crow but higher-pitched than large-billed Crow or Comman Raven. Also short croaks and some musical notes.

Range: Indian subcontinent to E Himalayas, Burma and N Thailand.

Distribution and status: Jungle Crow (race *culminatus*) overlaps with Large-billed Crow in foothills of E Himalayas in SE Xizang and perhaps extreme SW Yunnan.

Habits: Inhabits wooded areas at low to moderate altitudes up to 1850 m. Lives in pairs or small parties. Replaces House Crow in rural areas.

650. Collared Crow *Corvus torquatus* Baijing ya

650

白颈鸦 ☐ Resident

Description: Large (54 cm) glossy black and white crow with heavy bill. Contrasting white nape and breast band distinctive from all other crows in region except much smaller Daurian Jackdaw which has more extensive white on underparts. Iris—dark brown; bill—black; feet—black.

Voice: Calls less rolling and more hoarse than Carrion Crow. Usual call is loud, often repeated *kaaarr*. Also several creaking and clicking sounds. Calls generally higher-pitched than Large-billed Crow.

Range: C, S and SE China to N Vietnam.

Distribution and status: Common especially in south of range. Resident over most of E, C and SE China and Hainan.

Habits: Inhabits plains, cultivated fields, river beds, towns and villages. Replaces Carrion Crow in E China. Sometimes seen in mixed flocks with Large-billed Crow.

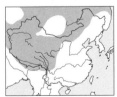

651

651. Common Raven
Corvus corax
渡鸦
☐ Resident

651. Common Raven *Corvus corax* [Northern Raven]
Du ya

Description: Very large (66 cm) all-black crow with heavy bill. Separated from other crows, especially Large-billed Crow by shaggy throat, less domed crown, long 'fingers' to outstretched wing, wedge-shaped tail and deep croaking calls. Race *tibetanus* is larger and glossier than other races. Iris—dark brown; bill—black; feet—black.

Voice: Distinctive deep, hollowcroaking *honk pruk-pruk-pruk-pruk* unlike any other corvid. Many other calls including guttural rattle and dry, grating *kraa*.

Range: N America, Palearctic, NW Indian subcontinent, Himalayas.

Distribution and status: Common and widespread in rugged terrain in open mountains and plateaux of

N and W China. Race *kamtschaticus* is resident across extreme N China and *tibetanus* is resident over the entire Qinghai–Xizang Plateau and W Xinjiang as far E as Sichuan, Gansu and W Nei Mongol.

Habits: Lives in pairs or small parties. Occasionally forms large flocks. Flight is powerful with soaring in thermals and sometimes tumbling and rolling upside down in mid-air. Sometimes attacks and kills prey species.

WOODSWALLOWS—Tribe: Artamini

A small, mostly Australasian, tribe (formerly treated as a family) of medium-sized insectivorous birds with short tails, long triangular wings and powerful beaks. Woodswallows catch insects on the wing in circular gliding flights, reminiscent of true swallows although they are unrelated. Woodswallows tend to be gregarious, congregating on bare high perches where they huddle close together. The nest is a simple cup in a tree fork. Only one species occurs in China.

652. Ashy Woodswallow *Artamus fuscus* [AG: Swallow-Shrike] PLATE 119
Hui yanbei

Description: Medium-sized (18 cm) greyish, swallow-like bird. Heavy bill bluish-grey; head, chin, throat and back grey; wings black; tail black with narrow white tip; rump white; rest of underparts buff. Distinguished from true swallows in flight by the broad triangular wings, square tail and much heavier bill. At rest, wings protrude well beyond tail. Iris—brown; bill—bluish-grey; feet—grey.

Voice: Unmusical chattering notes *tee-tee, chew-chew-chew*.

Range: Indian subcontinent to S China and SE Asia.

Distribution and status: A locally common bird of open spaces up to 1500 m in W, S and SE Yunnan, S Guangxi and S Guangdong and Hainan.

Habits: Perches on bare tree or other perch and makes circular hawking flights catching insects on the wing, sometimes over water. The flight is swallow-like with effortless glides. Birds sit close together and preen each other and wag their tails together. Boldly mobs hawks and crows.

ORIOLES, CUCKOOSHRIKES AND MINIVETS—Tribe: Oriolini

Orioles are small robust birds with often colourful plumage and powerful straight bills. They feed on fruits and insects. Nests are suspended from tree forks, being woven cups of roots and fibres twisted around supporting twigs. They have clear loud melodious songs. The flight is leisurely and undulating. Six species occur in China. Cuckooshrikes are neither closely related to the cuckoos nor the shrikes. Some species resemble cuckoos in shape and plumage while all are somewhat shrike-like in having strong, hooked bills for catching insects. They have soft fluffy feathers and short legs. Most species are noisy, conspicuous, flocking birds of the forest canopy. Most are dull-coloured, black, white and grey, but minivets are colourful with mostly bright red and yellow plumage.

All eat insects, some also take fruit. They do not visit the ground. Cuckooshrikes make cup-shaped nests in the canopy. There are 10 species in China.

653. Eurasian Golden Oriole *Oriolus oriolus* Jin huangli   PLATE 69

Description: Medium-sized (24 cm) yellow and black oriole with all yellow head. Adult male has fine black lores and black wings and base of tail but is mostly bright yellow. Female is duller and greener. Juvenile is greenish with fine streaks on underparts. Iris—red; bill—red; feet—grey.

Voice: Loud, cheery, fluty *oh wheela whee* song. Call is harsh, nasal, jay-like *kwa-kwaaek*.

Range: Breeds from S Europe to Indian subcontinent, N Mongolia and Siberia; some winter to Africa.

Distribution and status: Uncommon to locally common in extreme W China. Nominate race breeds in Tianshan, Tekes and Turpan areas of NW Xinjiang; *kundoo* breeds from Kashi S to Karakorum Mts in W Xinjiang. Recorded on passage to Indian subcontinent in SW Xizang.

Habits: Shy oriole of woods, poplar plantations and open country with scattered trees. Highly vocal in breeding season. Undulating flight. Keeps to upper canopy.

654. Black-naped Oriole *Oriolus chinesis* Heizhen huangli   PLATE 69

Description: Medium-sized (26 cm) yellow and black oriole with black stripe through eye and nape; largely black flight feathers. Male otherwise bright yellow. Distinguished from Slender-billed Oriole by thicker bill and broader black head band on nape. Female duller with olive-yellow back. Immature olive back; underparts whitish with black streaks. Iris—red; bill—pink; feet—blackish.

Voice: Clear, liquid, fluty whistle *lwee, wee, wee-leeow* and variations. Also a very harsh scolding call note and steady, plaintive quiet whistle.

Range: India, China, SE Asia, Sundas, Philippines and Sulawesi. Northern populations move S in winter.

Distribution and status: Race *diffusus* occurs over the eastern half of China including Hainan and Taiwan. Locally common up to 1600 m.

Habits: Inhabits open forests, plantations, parkland, villages and mangroves. Lives in pairs or family parties. Keeps to the trees but will come quite low down in search of insects. Conspicuous, slow, powerful wing beat in undulating flight.

655. Slender-billed Oriole *Oriolus tenuirostris* Xizui huangli   PLATE 69

Description: Medium-sized (25 cm) yellow oriole with black eye-stripe extending to nape. Very similar to Black-naped but has more slender bill and narrower black eye-stripe especially on nape. The back is more olive. Female: like male but more greenish with some dark streaking on underparts. Juvenile has heavier streaking and less distinct eye-stripe. Iris—red; bill—pinkish-orange; feet—grey.

Voice: Diagnostic, high-pitched, woodpecker-like *kich* plus typical fluty oriole calls.

Range: Himalayas to Burma, S China and Indochina.

Distribution and status: Breeds in Yunnan in montane forests from 2500–4300 m. Winters lower in open evergreen broadleaf forest. Rare due to trapping for cagebird trade.

Habits: Lives in pine forest, open woods, plantations and open country with scattered trees.

Note: Formerly treated as race of Black-naped Oriole (e.g. Cheng 1987, 1994).

656. Black-hooded Oriole *Oriolus xanthornus* [Asian/Indian/ PLATE 69
Oriental Black-headed Oriole] Heitou huangli

Description: Medium-sized (23 cm), unmistakable, black-hooded yellow oriole with yellow underparts. Wings and tail are black. Sexes alike. Immature similar to adult but with yellow forehead, whitish bill, whitish eye-ring and blackish stripes on dirty white throat. Iris—red; bill—red; feet—black.

Voice: Similar to Black-naped, liquid whistled call *wye-you* or *yiu-hu-a-yu* with middle notes stressed. Gives harsh nasal *kwaak* call. Male in display gives soliloquy of mellow fluty notes interspersed with harsh notes, while bent low with tail fully spread. Also mimics other bird calls.

Range: India, S China, SE Asia, Sumatra and Borneo.

Distribution and status: Limited distribution in China. Nominate race is resident in SE Xizang and W Yunnan. Not uncommon in forest and wooded country up to 1000 m.

Habits: As Black-naped Oriole but prefers forest edge, open forests, cultivated areas and secondary forest.

657. Maroon Oriole *Oriolus traillii* Zhu li PLATE 69

Description: Medium-sized (26 cm) black and maroon oriole with pale eye. Male is maroon with black head, upper breast and wings. Female has dark grey mantle and back, maroon tail coverts and tail but white belly and lower breast, heavily streaked black. Distinguished from female Silver Oriole by darker grey mantle and broader streaking on less white underparts. Juvenile is paler than female with whiter throat. Race *ardens* has maroon feathers basally crimson rather than white of other races. Iris—yellow; bill—bluish-grey; feet—grey.

Voice: Rich, fluty *pi-lo-i-lo* song. Nasal squawking call-note *kee-ah* is less harsh than Black-naped Oriole; also a rather catlike *mew* and woodpecker-like rattling 'laugh'.

Range: Himalayas, SW China, Taiwan, Hainan and Indochina.

Distribution and status: Uncommon in deciduous, mixed and evergreen forests in hills and montane regions from 600–4000 m. Nominate race is found in SE Xizang and W Yunnan; *nigellicauda* in SE Yunnan and Hainan; and *ardens* in Taiwan.

Habits: Moves S, or to lower altitudes and to more deciduous forests in winter. Usually singly or in pairs. Keeps to canopy and sometimes joins mixed-species flocks.

658. Silver Oriole *Oriolus mellianus* [Mell's Maroon Oriole] PLATE 69
Quese li

Description: Medium-sized (28 cm) black and silvery white oriole with magenta tail. Male is quite distinctive; female is like female Maroon Oriole but black head contrasts with grey back, also underparts whiter with narrower streaks. Iris—yellow; bill—grey; feet—grey.

Voice: Similar to Maroon Oriole.

Range: C China, occasionally migrating as far as SW Thailand.

Distribution and status: Globally Vulnerable (Collar *et al.* 1994). Rare. Breeds in S Sichuan, Guangxi and N Guangdong, dispersing S in winter.

Habits: As Maroon Oriole.

Note: Some authors consider this a race or form of Maroon Oriole *O. traillii.*

659. Large Cuckooshrike *Coracina macei* Da juanbei   PLATE 70

Description: Large (28 cm) grey shrike-like bird with black face mask and chin. Male: upperparts and breast grey, flight feathers black with whitish edges; tail black with dark grey central rectrices and brownish-grey tips; belly whitish; lores and eye-ring black and throat dark grey. Female paler, with grey barring on lower breast and flanks. Immature similar to female but browner and with bolder barring on underparts and rump. Iris— reddish; bill—black; feet—black.

Voice: Loud piercing whistle *pee-eeo-pee-eeo, tweer* or *twee-eet.*

Range: India, Himalayas, S China and SE Asia.

Distribution and status: Locally common in low-lying areas up to 1500 m. Race *siamensis* is resident in W and S Yunnan; *rexpineti* in SE China and Taiwan; and *larvivora* on Hainan.

Habits: Generally singly or in pairs. Keeps to the tops of the tallest trees often at the edge of forest clearings.

660. Black-winged Cuckooshrike *Coracina melaschistos*   PLATE 70
Anhui juanbei

Description: Medium-sized (23 cm) grey and black cuckooshrike. Male is uniform grey with glossy black wings, white undertail coverts and white tips to three outermost black tail feathers. Female is as male but paler with white barring on underparts, white streaking on ear coverts and broken white eye-ring. Female usually has small white patch on underwing. Iris—reddish-brown; bill—black; feet—lead blue.

Voice: Song is descending 3–4 whistled notes in slow measured sequence *wii wii jeeow jeeow.*

Range: Himalayas, China, Taiwan, Hainan and SE Asia.

Distribution and status: Uncommon to locally common in lowlands and mountains up to 2000 m. Nominate race resident in SE Xizang to NW Yunnan; *avensis* in SW China; *intermedia* in C, SE and S China with some northern birds wintering in Yunnan, S China and Taiwan; and *saturata* resident on Hainan.

Habits: Inhabits rather open woodlands and bamboo. Wintering birds move downslope from montane forests.

661. Rosy Minivet *Pericrocotus roseus* Fenhong shanjiaoniao   PLATE 70

Description: Smallish (20 cm) red or yellow patterned minivet with diagnostic white chin and throat and grey crown and mantle. Male: from other minivets by grey head and rosy breast. Female: from other minivets by rump and uppertail coverts only slightly paler than back and only faintly washed yellow; underparts very pale yellow. Iris—brown; bill—black; feet—black.

Voice: Squeaky, trill similar to Ashy Minivet.

Range: Himalayas to S China wintering to Indian subcontinent and parts of SE Asia.

Distribution and status: Fairly common in forest up to 1500 m in Yunnan, SW Sichuan (Xichang), Guangxi and SW Guangdong. Probably Hainan.
Habits: Forms large flocks in winter.
Note: Some authors include Swinhoe's Minivet *P. cantonensis* in this species (e.g. Cheng 1987). Hybrids seem to occur.

662. Swinhoe's Minivet *Pericrocotus cantonensis* [Brown-rumped   PLATE 70
Minivet] Xiao huishanjiaoniao

Description: Small (18 cm) black, grey and white minivet with distinctive white fore-crown. Distinguished from Ashy Minivet by pale buff rump and uppertail coverts; greyer nape and usually bolder white wing bar. Female as male but browner while wing patch sometimes absent. Iris—brown; bill—black; feet—black.
Voice: Trill like Ashy Minivet.
Range: Breeds C, S and E China; wintering to SE Asia.
Distribution and status: Globally Near-threatened (Collar *et al.* 1994). Locally common resident in C, E and SE China, passes through S and SE China on passage.
Habits: Forms largish flocks in winter. Lives in deciduous and evergreen forests up to 1500 m.
Note: Considered by some authors to be a race of Rosy Minivet (e.g. Cheng 1987). Hybrids seem to occur.

663. Ashy Minivet *Pericrocotus divaricatus* Hui shanjiaoniao   PLATE 70

Description: Smallish (20 cm) minivet with distinctive black, grey and white plumage. Distinguished from Swinhoe's Minivet by black lores. Distinguished from cuckooshrikes by white underparts and grey rump. Male: cap, eye-stripe and flight feathers black; otherwise grey above and white below. Female: paler and greyer. Iris—brown; bill and feet—black.
Voice: Metallic, jingling trill given in flight.
Range: NE Asia and E China. Winters S to SE Asia, Philippines and Greater Sundas.
Distribution and status: Nominate race breeds in Lesser Hinggan Mts of Heilongjiang and possibly on Taiwan. On passage through E and S China. Uncommon in deciduous woodlands and forest edge up to 900 m.
Habits: Hunts insects in tree canopy. Less showy in flight than bright coloured minivets. Forms flocks of up to 15 birds.

664. Grey-chinned Minivet *Pericrocotus solaris* [Mountain/   PLATE 70
Grey-throated/Yellow-throated Minivet] Huihou shanjiaoniao

Description: Small (17 cm) red or yellow minivet. Red male is distinguished from other minivets by dull dark grey throat and ear coverts. Yellow female is distinguished by lack of yellow on forehead, ear coverts and throat. Male of race *montpelieri* has dark olive mantle and olive-yellow rump with red tail coverts. Iris—dark brown; bill and feet—black.
Voice: Soft, slightly rasping *tsee-sip*.
Range: Himalayas, S China, SE Asia to Greater Sundas.
Distribution and status: Race *griseogularis* is a common resident in SE and S China

and Taiwan; *montpelieri* in N and NW Yunnan; and nominate in SE Xizang and S Yunnan. Generally in montane forests at 1200–2000 m.

Habits: As other minivets.

665. Long-tailed Minivet *Pericrocotus ethologus* Changwei shanjiaoniao **PLATE 70**

Description: Large (20 cm) black minivet with red or yellow markings and longish tail. Red male distinguished from Rosy and Grey-chinned Minivets by black throat and from Short-billed by shape of wing patch and by less saturated, pinkish-red of underparts. Female distinguished only with difficulty from Grey-chinned by indistinct dull yellowish suffusion above base of bill. Iris—brown; bill—black; feet—black.

Voice: Diagnostic, sweet double whistle *pi-ru*, the second note lower.

Range: Afghanistan to China and SE Asia.

Distribution and status: Common at 1000–2000 m. Nominate race breeds in C and SW China with one record from Hebei; *laetus* breeds in S and SE Xizang; and *yvettae* is confined to W Yunnan.

Habits: Lives in large twittering flocks that swirl over canopy to alight in tall trees of open and evergreen forests.

666. Short-billed Minivet *Pericrocotus brevirostris* Duanzui shanjiaoniao **PLATE 70**

Description: Medium-sized (19 cm) black minivet with red or yellow markings. Red male very bright like Scarlet Minivet but slimmer and more long-tailed with also a double slash of red on wing. Female: from Grey-chinned and Long-tailed Minivets by bright yellow on forecrown and from Scarlet by simpler wing patch. Iris—brown; bill—black; feet—black.

Voice: Loud, sweet, monosyllabic whistle. Rapid and complex song.

Range: Himalayas to S China, Burma and N Indochina.

Distribution and status: Uncommon in deciduous and secondary hill forests from 1000–2400 m. Nominate race breeds in SE Xizang and NW Yunnan; *affinis* breeds in S Sichuan and W Yunnan; and *anthoides* breeds in SE Yunnan, Guizhou, Guangxi and N Guangdong.

Habits: Typical pair-living minivet, generally less common than Long-tailed where both species are present.

667. Scarlet Minivet *Pericrocotus flammeus* [Indian/Flame Minivet] **PLATE 70**
Chihong shanjiaoniao

Description: Largish (19 cm) colourful minivet. Male is blue-black with red breast and belly, rump, edge of tail feathers and double wing patches. The female is greyer on the back, has yellow in place of red and yellow extends to the throat, chin, ear coverts and forehead. Stouter and less long-tailed than long-tailed Minivet and with more complex wing patches. Iris—brown; bill and feet—black.

Voice: Soft *kroo-oo-oo-tu-tup, tu-turr* or repeated *hurr*, also a higher-pitched *sigit, sigit, sigit.*

Range: India, S China, SE Asia, Philippines and Greater Sundas.

Distribution and status: Race *fohkiensis* is resident in S and SE China; *elegans* in Yunnan; and *fraterculus* on Hainan. A locally common bird in lowlands and hills to 1500 m.

Habits: Prefers primary forest where it flits between tree-tops of the finer-leaved trees in pairs or small groups.

668. Bar-winged Flycatcher-shrike *Hemipus picatus*　　　PLATE 70
[Pied Flycatcher-Shrike/Bar-winged Pygmy-Triller] Hebei quebei

Description: Small (15 cm) pied cuckooshrike with a broad white wing bar. Distinguished from Ashy Minivet by smaller size and whitish rump. Distinguished from Little Pied Flycatcher by lack of white eyebrow. Iris—brown; bill and feet—black.

Voice: Noisy, high-pitched *chir-rup, chir-rup.*

Range: India, SW China, SE Asia, Sumatra and Borneo.

Distribution and status: Race *capitalis* is a common bird of hill and montane forests at 500–2000 m in SE Xizang, W and S Yunnan, W Guangxi and S and C Guizhou. Race *pelvica* lives up to 1300 m in SE Xizang.

Habits: Lives in flocks, often mixed with other species, working through forest canopy, crouching to spot hiding or disturbed insects then swooping on them like a shrike.

FANTAILS—Tribe: Rhipidurini

Small tribe formerly included in flycatcher family, with fan-like tails, which they flick in a conspicuous manner. Three species occur in China.

669. Yellow-bellied Fantail *Rhipidura hypoxantha*　　　PLATE 72
Huangfu shanweiweng

Description: Small (12 cm) fantail with yellow forehead, supercilium and underparts. Broad face black mask in male is dark green in female. White tips to long fan-like tail distinguish from Black-faced Warbler. Iris—brown; bill—black; feet—black.

Voice: Sweet, high-pitched trilling notes or single high chirps.

Range: Himalayas to SW China and N Indochina.

Distribution and status: Uncommon in hills and montane forests of S and SE Xizang, S Sichuan and Yunnan at 800–3700 m.

Habits: Active, nervous bird constantly flicking open and cocking fan-tail.

670. White-throated Fantail *Rhipidura albicollis* [Spot-breasted/　　PLATE 72
White-spotted Fantail, AG: Fantail-Flycatcher] Baihou shanweiweng

Description: Medium-sized (19 cm) dark fantail. Almost totally dark grey (looks black in field) with white chin and throat, eyebrow and tips to tail. Dark grey underparts distinguishes from White-browed Fantail but some individuals are paler below. Iris—brown; bill and feet—black.

Voice: A song of high-pitched thin notes; three evenly-spaced *tut* notes followed by three or more descending notes; also sharp *cheet.*

Range: Himalayas, S China, SE Asia and Greater Sundas.

Distribution and status: Nominate race breeds in SE Xizang, SW China and Hainan in moist montane forests up to 3000 m.

Habits: As other fantails. Joins mixed bird flocks and often in bamboo thickets.

671. White-browed Fantail *Rhipidura aureola* Baimei shanweiweng   PLATE 72

Description: Medium-sized (17 cm) grey fantail with broad white supercilium and forehead. Tail is broadly edged white. Moustachial stripe white against black face and throat; underparts otherwise white. Fantail behaviour and colour pattern distinctive. Iris—brown; bill—brownish-black; feet—brownish-black.

Voice: Four ascending and 2–3 descending, spaced, melodious whistled notes.

Range: Pakistan, India, Sri Lanka, Bangladesh, SE Asia.

Distribution and status: Rare resident in W Yunnan. Lives in savanna habitat up to 1500 m.

Habits: Typical fantail behaviour.

DRONGOS—Tribe: Dicrurini

A small tribe, formerly regarded as a family, of blackish, insectivorous birds found from Africa to Asia, Australia and the Solomons. Most species have glossy black plumage, powerful beaks and long forked tails. They hunt larger insects in the air from a tree perch. Their calls are loud and sometimes melodious but usually harsh with discordant screeching. They are also excellent mimics of other birds. Drongos boldly attack hawks and cuckoos. Their nests are neatly-woven cups placed in low branch forks. Two species have extraordinary elongated outer tail feathers with terminal racket webs. Seven species occur in China.

672. Black Drongo *Dicrurus macrocercus* Hei juanwei   PLATE 71

Description: Medium-sized (30 cm) dully-glossed blue-black drongo. Beak relatively small, tail very long and very deeply forked and often held at a curious angle in wind. Immature has whitish barring on lower underparts. Taiwan race *harterti* has shorter tail. Iris—red; bill and feet—black.

Voice: Varied ringing calls, *hee-luu-luu, eluu-wee-weet* or *hoke-chok-wak-we-wak.*

Range: Iran to India, China, SE Asia, Java and Bali.

Distribution and status: A common breeding migrant and resident in low-lying open country, occasionally up to 1600 m. Race *albirictus* occurs in SE Xizang; *harterti* is resident on Taiwan. Migratory *cathoecus* breeds from S Jilin and S Heilongjiang through E and C China to Qinghai, SW China, Hainan and S China; migrates through SE China.

Habits: A bird of open spaces, often sitting on small trees or telegraph wires.

673. Ashy Drongo *Dicrurus leucophaeus* [Pale/Grey Drongo]   PLATE 71
Hui juanwei

Description: Medium-sized (28 cm) grey drongo with whitish facial area and long, deeply forked tail. Races vary in paleness: *leucogenis* is paler and *hopwoodi* darker than other races; *salangensis* has black lores; *hopwoodi* lacks pale facial patch. Iris—red; bill—grey-black; feet—black.

Voice: Clear loud song *huur-uur-cheluu* or *wee-peet, wee-peet*. Mewing and mimics the calls of other birds and is reported to sometimes call in the night.

Range: Afghanistan to China, SE Asia, Palawan and Greater Sundas.

Distribution and status: A common resident and seasonal migrant of open woodlands and forest edge in hills and mountains from 600–2500 m, but up to nearly 4000 m in Yunnan. Race *leucogenis* occurs from Jilin and S Heilongjiang through E China to SE China; *salangensis* in C and S China, wintering in Hainan; *hopwoodi* in SW China and S Xizang; and *innexus* is resident on Hainan.

Habits: Lives in pairs and sits on bare branches or vines in open clearings or gaps in the forest, hawking after passing insects, climbing after rising moths or diving after flying prey.

674. Crow-billed Drongo *Dicrurus annectans* Yazui juanwei   PLATE 71

Description: Medium-sized (29 cm) black drongo with heavy crow-like bill and forked tail with upturned outer feathers. Distinguished from Black Drongo by heavier bill and broader, less deeply forked tail and greenish rather than blue gloss. Immature has white spotting on breast. Iris—red-brown; bill and feet—black.

Voice: Typical drongo calls with loud musical whistles and harsh churs; also characteristic descending series of harp-like notes.

Range: Mostly summer breeding visitor in Himalayas and NE India; migrant to SE Asia and Greater Sundas.

Distribution and status: Breeds in SE Xizang. Rare winter visitor to W Yunnan and Hainan.

Habits: Prefers open woodland and coastal scrub and low mangrove forest. Typical drongo hunting behaviour.

675. Bronzed Drongo *Dicrurus aeneus* Gutongse juanwei   PLATE 71

Description: Small (23 cm) glossy black drongo with greenish sheen. Distinguished from Black Drongo by smaller size, more glossy plumage, only moderately forked tail, also by habits and habitat; from Crow-billed Drongo by shorter tail and from Drongo Cuckoo *Surniculus lugubis* by lack of white barring on undertail. Iris—brown; bill and feet—black.

Voice: Loud calls including clear notes and harsh discords.

Range: India, S China, SE Asia, to Sumatra and Borneo.

Distribution and status: Nominate race is common resident is SE Xizang, S Yunnan, E Guangxi and Hainan, in primary and secondary lowland forests up to 2000 m, *braunianus* on Taiwan.

Habits: Sits on prominent perches, sallying after insects in hawking flights through upper and middle canopy. Joins birdwaves. Mobs raptors and cuckoos boldly. Several birds may chase each other noisily. Favours gaps in canopy.

676. Lesser Racket-tailed Drongo *Dicrurus remifer* Xiao panwei   PLATE 71

Description: Medium-sized (26 cm without rackets) glossy black drongo with amazing elongated outer tail feathers with rackets at their ends. A tuft of short feathers forms a ridge above the bill. Smaller than Greater Racket-tailed Drongo and lacks the crest but most easily distinguished by square-ended tail. Generally separated from Greater Racket-tailed by altitude as the two species overlap only from 1000–1500 m. Worn birds may lack rackets. Iris—red; bill and feet—black.

Voice: Varied and melodious whistling notes *weet-weet-weet-weet-chewee-chewee* and occasional harsh screeches. Mimics the calls of other birds.

Range: India, S China, SE Asia to Sumatra and Java.

Distribution and status: Race *tectirostris* breeds in SE Xizang, W and S Yunnan, where locally common at 800–2000 m.

Habits: Inhabits dense rain forest, secondary forest and forest edge. Follows wildfires to catch fleeing grasshoppers. Boldly mobs raptors and crows.

677. Spangled Drongo *Dicrurus hottentottus* [Hair-crested Drongo]   PLATE 71
Faguan juanwei

Description: Largish (32 cm) velvety-black drongo with crest of long filaments from crown. Plumage spangled with light iridescent spots. Tail long and forked with blunt strongly upcurved outer feathers giving a lyre shape. Nominate race has heavier bill. Iris—red; bill and feet—black.

Voice: Melodious loud singing with occasional harsh screeches.

Range: India, China, SE Asia and Greater Sundas.

Distribution and status: Race *brevirostris* breeds across C and E China and Taiwan. Northern birds migrate S in winter. Nominate race breeds in SE Xizang and W Yunnan. A common bird of lowland and submontane forests, especially in drier areas.

Habits: Prefers rather open parts of the forest and sometimes gathers in noisy parties singing and chasing insects in the sky, especially at dawn and dusk. Hawks after insects from low perch, mixes with other species and also follows monkeys, catching insects disturbed by their movements.

678. Greater Racket-tailed Drongo *Dicrurus paradiseus*   PLATE 71
Da panwei

Description: Large (35 cm without rackets) glossy black drongo with amazing elongated outer tail feathers with terminal rackets. Rackets were webbed only on the outer side of the feather shafts and are twisted. Forked tail separates from Lesser Racket-tailed Drongo. Crest of elongated feathers on crown of adults not readily visible in the forest. Iris—red; bill and feet—black.

Voice: Glorious variety of lusty, resonant melodious warbles, whistles and bell-like notes with typical harsh drongo churrs. Often mimics other bird calls.

Range: India to China, SE Asia and Greater Sundas.

Distribution and status: Race *grandis* is resident in SE Xizang and W and S Yunnan; common in lowland forests up to 1400 m. Race *johni* is resident on Hainan.

Habits: Pair-living forest bird sometimes gathering in groups to display, giving noisy lusty song and hawking for insects from low exposed perches in the forest. Joins mixed-species flocks.

MONARCHS AND PARADISE FLYCATCHERS—Tribe: Monarchini

Small tribe formerly included in flycatcher family. Some species have greatly elongated tails. Only three species occur in China.

679. Black-naped Monarch *Hypothymis azurea*   PLATE 72
[Pacific/Small Monarch] Heizhen wangweng

Description: Medium-sized (16 cm) greyish-blue flycatcher. Male: head, chest, back and tail blue; greyer on wings; whitish belly; short crest; small patch above bill and narrow black throat band. Female: head blue-grey; greyer on chest; back, wing and tail brownish-grey; lacks black crest and throat bar of male. Iris—dark brown; orbital skin—bright blue; bill—bluish with black tip; feet—bluish.

Voice: Song is ringing *pwee-pwee-pwee-pwee*, contact call is harsh chirping *chee, chweet*.

Range: India to China, SE Asia, Philippines, Sundas and Sulawesi.

Distribution and status: Race *styani* breeds in SE Xizang and SW China to N Guangdong and Hainan. Some birds winter in breeding range. Race *oberholseri* is resident on Taiwan. A common resident up to 900 m, locally to 1500 m.

Habits: A lively, inquisitive bird of lowland and secondary forests. Readily attracted to an imitation of its contact call. Often joins mixed-species flocks. Usually in the lower parts of the forest, especially thickets near streams.

680. Japanese Paradise-flycatcher *Terpsiphone atrocaudata*   PLATE 72
[Black Paradise-Flycatcher] Zi shoudai

Description: Medium-sized (20 cm plus 20 cm-long tail of male) black-crested flycatcher. Male distinguished from Asian Paradise-flycatcher by black wings and tail and purplish back. Female similar to female Asian Paradise-flycatcher, but duller crown lacks metallic gloss. Iris—dark brown; bare skin round eye-blue; bill—blue; feet—bluish.

Voice: As Asian Paradise-flycatcher.

Range: Breeds Japan, Korea and Taiwan; winters in SE Asia.

Distribution and status: Nominate race breeds on Taiwan and recorded on passage through E China. Race *periophthalmica* breeds on Lanyu islet off SE Taiwan.

Habits: As Asian Paradise-flycatcher.

681. Asian Paradise-flycatcher *Terpsiphone paradisi* [Paradise/   PLATE 72
Asiatic Paradise Flycatcher] Shoudai

Description: Medium-sized (22 cm plus 25 cm-long tail of male) sexually dimorphic, glossy black-headed flycatcher with prominent crest. Male is notable with greatly elongated central pair of tail feathers up to 25 cm beyond rest of tail. Male has two colour morphs both quite different from Japanese Paradise Flycatcher: a) upperparts white streaked with black shafts and underparts pure white; wings black, or b) upperparts rufous, underparts greyish. Rufous males of race *saturatior* and *indochinensis* have only crown glossy black; *indochinensis* is more rufous and *saturatior* has an olive crest. White phase *saturatior* has heavier black streaking on upperparts. Most males of *saturatior* are white phase, whilst less than half are white in *incei* and almost none in *indochinensis*.

Female is rufous-brown with glossy black head but lacks elongated tail feathers. Iris—brown; orbital skin—blue; bill—blue with black tip; feet—blue.

Voice: Both ringing whistled song and very loud *chee-tew* contact call, similar to but stronger than calls of Black-naped Monarch.

Range: Turkestan, India, China, SE Asia and Sundas.

Distribution and status: Race *incei* breeds over most of N, C, S and SE China; *saturatior* is a winter visitor to Guangdong and S and W Yunnan; *indochinensis* breeds in S Yunnan. Generally fairly common in lowland forest, locally up to 1200 m.

Habits: The white male is conspicuous in flight. Generally hunts from a perch in the lower half of the canopy, often within mixed-species flocks.

IORAS—Subfamily: Aegithininae

Small subfamily of arboreal green passerines, formerly included within leafbird family. Only two species occur in China.

682. Common Iora *Aegithina tiphia* [Black-winged/Small Iora]   PLATE 65
Heichi quebei

Description: Small (14 cm) green and yellow iora with two conspicuous whitish wing bars. Upperparts olive-green; wing blackish but feathers edged white; eye-ring yellow; underparts yellow. Some males are more blackish on head and mantle but others resemble female. Iris—greyish-white; bill and feet—bluish-black.

Voice: Several calls including a musical monotonous trill or whistled *cheepow* or *cheepow cheepow* with the explosive *pow* dropping with whiplash ending.

Range: India, SW China, SE Asia and Greater Sundas.

Distribution and status: Race *philipi* is resident in S Yunnan and SW Guangxi. Locally common in lowlands up to 1000 m.

Habits: Inhabits gardens, mangroves, open woodlands and secondary forest. Generally singly or in pairs, hopping about in the leafy branches of small trees and keeping well concealed.

683. Great Iora *Aegithina lafresnayei* Da lüquebei   PLATE 65

Description: Large (17 cm) green iora without pale wing bars. Separated from leafbirds by yellow underparts and blue-grey lower mandible. Male upperparts vary from olive-green to blackish with green rump. Iris—brown; bill—bluish-grey; feet—black.

Voice: Song is rapid *chew chew chew....*

Range: SW China and SE Asia.

Distribution and status: Not uncommon in Xishuangbanna in S Yunnan.

Habits: Lives in canopy of small trees at forest edge, in secondary forests and scattered trees around villages.

PHILENTOMAS AND WOODSHRIKES—Subfamily: Malaconotinae

Small subfamily comprising woodshrikes which were formerly classed with cuckooshrikes, plus *Philentoma* flycatchers which were formerly classed in the flycatcher family with monarchs. Bill is stout and slightly hooked. Usually have dark face mask. Only one species is recorded in China.

684. Large Woodshrike *Tephrodornis gularis* [Brown-tailed/ PLATE 66
Hook-billed Woodshrike] Gouzui linbei

Description: Medium-sized (20 cm) grey-brown shrike-like bird with white rump and underparts. Upperparts grey-brown in male with grey crown and nape, brown in female; underparts white with greyish wash over breast. Has dark mask. Bill hooked at tip. Iris—yellow to brown; bill and feet—black.

Voice: Repetitive *wit wit wit . . .*, loud *chew-chew*, incessant noisy *kee-a, keea* or a harsh *chreek-chreek chee -ree* or *chee-ree-ree chee ree reeoo-reeoo.*

Range: India, China, SE Asia and Greater Sundas.

Distribution and status: Race *latouchei* is uncommon lowland resident in China S of 25 °N; *hainanus* is resident on Hainan.

Habits: Occurs in pairs or small noisy flocks, travelling from tree crown to tree crown. Flies after disturbed insects and often hunts from a perch. Also takes insects off water surfaces; favours forest edge and clearings.

Note: DNA studies indicate *Tephrodornis* is not orioline like Cuckooshrikes but related to Helmetshrikes. Related Common Woodshrike *T. pondicerianus* may occur in southern SE Xizang; smaller with white edges to tail and broad white supercilium.

WAXWINGS—Family: Bombycillidae

A small family of three small-sized birds with broad bills, soft plumage, crested heads and short, strong legs. Some species have waxy, red tips to secondaries which give the family its English name. Two species occur in China.

685. Bohemian Waxwing *Bombycilla garrulus* [Waxwing] PLATE 119
Taipingniao

Description: Largish (18 cm) pinkish-brown waxwing easily distinguished from Japanese Waxwing by yellow rather than crimson tip to tail. Undertail coverts chestnut and yellow strip on wing formed by tips of outer web of primaries and white bars formed by tips of tertials and tips of outer wing coverts. Adults have small red plastic-like projections on tips of secondaries. Iris—brown; bill—brown; feet—brown.

Voice: Characteristic silvery buzzing *sirr* flock call and song of simple call-like phrases.

Range: N Eurasia and NW America, wintering south.

Distribution and status: Uncommon. Winters in NE and NC China occasionally as far S as Changjiang also Kashi district of extreme W Xinjiang. Vegrants reach Taiwan, Fujian and Sichuan.

Habits: Flock-living. Feeds on rose hips, *Cotoneaster, Sorbus* and other berries. Eats insects in spring and summer. Birds may gorge till they can hardly fly.

686. Japanese Waxwing *Bombycilla japonica*   PLATE **119**
Xiao taipingniao

Description: Smallish (16 cm) waxwing distinguished by prominent crimson tip to tail. Also differs from Bohemian Waxwing in black through eye extending around crest to back of head and by having crimson vent. Lacks wax-like appendages on tips of secondaries but these feathers are tipped crimson. Lacks yellow wing bar. Iris—brown; bill—blackish; feet—brown.
Voice: High-pitched lisping flock calls.
Range: E Siberia and NE China, wintering to Japan and Riukyu Is.
Distribution and status: Irregular breeder in Lesser Hinggan Mts of Heilongjiang. Wintering flocks sometimes visit Hubei and Shandong. Rarely recorded in Fujian, Taiwan and C China.
Habits: Gregarious flocks gather in fruiting trees and bushes.

DIPPERS—Family Cinclidae

Dippers are a small family of only five species worldwide. They are compact thrush-like birds with short wings and tails that are often cocked. Colours are blackish-brown with or without white. All live along fast-flowing mountain streams and are specialise at swimming under water to find insect food. Dippers fly low over the water with fast wing beats and have shrill loud calls. Two species occur in China.

687. White-throated Dipper *Cinclus cinclus* Hewu

Description: Smallish (20 cm) dark brown dipper with diagnostic large white bib from chin and throat to upper breast. Lower back and rump greyish. In dark phase bib may be ashy-brown, occasionally streaked pale. Normal light phase has clean white bib. Race *leucogaster* has belly white. Juvenile is greyer with whiter underparts. Iris—reddish-brown; bill—blackish; feet—brown.

687

Voice: Squeaky, scratchy song. Gives harsh *zrets* call in flight.
Range: Palearctic; Himalayas, W China and NE Burma.
Distribution and status: Fairly common in suitable habitat from 2400−4250 m. Race *leucogaster* is found in Altai, Tianshan, Kashi and W Kunlun regions of W Xinjiang; also possibly extreme NE China; *cashmeriensis* is found in Himalayas and S Xizang as far E as NW Yunnan; race *przewalskii* is found along rivers draining from E of Qinghai−Xizang Plateau to Gansu and N Sichuan.
Habits: Lives along fast-flowing clear mountain streams in forest and open terrain. Shows seasonal movements in some areas. Bobs up and down. Has wing whirring display. Swims and dives in water to pop up again like a cork.

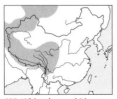

687. White-throated Dipper
Cinclus cinclus
河乌

⬛ Resident

688. **Brown Dipper** *Cinclus pallasii* He hewu

688

688. Brown Dipper
Cinclus pallasii
褐河乌
▨ Resident

Description: Largish (21 cm) dark brown dipper with no white or pale bib. Sometimes small white patch above eye is visible. Race *tenuirostris* is paler brown than other races. Iris—brown; bill—dark brown; feet—dark brown.

Voice: Shrill *dzchit, dzchit*. Slightly less shrill than White-throated Dipper. Full rich short song more musical than White-throated Dipper.

Range: S and E Asia, Himalayas, China, Taiwan and N Indochina.

Distribution and status: Common on fast flowing streams at 300–3500 m. Race *tenuirostris* is resident in W Tianshan, Himalayas and extreme S Xizang. Nominate race is resident through C, SW, S, E and NE China and Taiwan.

Habits: Lives in pairs. Shows some seasonal movements from high altitude breeding areas. Perches on boulders, bobbing and raising tail, jerked occasionally. Swims on surface then dips under surface like Little Grebe. Displays erect with whirring wings.

THRUSHES, SHORTWINGS, OLD WORLD FLYCATCHERS AND CHATS—Family: Muscicapidae

A large diverse family divided into several distinct subfamilies and tribes.

THRUSHES AND SHORTWINGS—Subfamily: Turdinae

A large, worldwide group including true thrushes, whistling thrushes, rock thrushes and shortwings. Food consists of insects, other invertebrates and berries. Most species feed at least partly on or close to the ground. Thrushes make solid, cup-shaped, fibrous nests often reinforced with mud and decorated with moss. Many species have melodious songs. There are 37 species in China.

689. **Rufous-tailed Rock Thrush** *Monticola saxatilis*   PLATE 73
Baibei jidong

Description: Smallish (19 cm) dimorphic rock thrush. Summer male: separated from Chestnut-bellied Rock Thrush by lack of black face mask, white feathers on back, brownish wings and blue central tail feathers on otherwise chestnut tail. In winter, male is heavily scalloped with black and white feather edges. Female: paler than female Blue Rock Thrush, pale-spotted above with rufous tail as in male. Young: similar to female but paler and more mottled. Iris—dark brown; bill—dark brown; feet—brown.

Voice: Calls include clear *diu*, a *chak* and soft, rattling, shrike-like *ks-chrrr*. Song is similar to Blue Rock Thrush but softer and more flowing.

Range: Europe, N Africa to Turkestan and Russian Transbaikalia and N China. Migrates for winter through NW India, Iran and Arabia to Sudan and S Tanzania.

Distribution and status: Fairly common in suitable habitat in NW Xinjiang, Qinghai, Ningxia, Nei Mongol and Hebei. Occasional migrants seen south of this.

Habits: Singly or in pairs, often perched on prominent rock or top of bare tree. Sometimes joins mixed flocks. In display, male flutters up with tail outstretched and floats down again with outstretched wings and tail.

[690. Blue-capped Rock Thrush *Monticola cinclorhynchus* PLATE 73
Lantou jidong

Description: Small (19 cm) dimorphic rock thrush. Male: smaller and slimmer than Rufous-bellied and distinguished by rufous rump, blacker wings and mantle and white wing patch. Very similar to White-throated male but lacks white throat patch and rufous chin. Female: plain olive-brown above; scaled white and dark brown below. Distinguished from female Blue Rock Thrush by olive-brown back and tail and white undertail coverts. Iris—dark brown; bill—black; feet—greyish-brown.

Voice: Harsh grating alarm call. Song is monotonously repeated varied and thrush-like but notes not very clear-cut.

Range: Himalayas and hills of NE India wintering to W Ghats and SW Burma.

Distribution and status: Not recorded in China but expected in SE Xizang and breeding known in parts of NE India claimed by China.

Habits: Secretive forest bird; sits very upright when alarmed. In display, flutters down from tree-top to lower perch with outstretched wings while singing. Generally solitary but joins mixed flocks in winter.

Notes: Some authors (e.g. Cheng 1987, 1994) include White-throated Rock Thrush as a race of this species.]

691. White-throated Rock Thrush *Monticola gularis* PLATE 73
Baihou jidong

Description: Small (19 cm) dimorphic rock thrush. Male: blue confined to crown, nape and a shoulder flash; sides of head black, underparts mostly chestnut-orange. Separated from other rock thrushes by white throat patch, and all but Blue-capped Rock Thrush by white wing patch. Female: separated from other female rock thrushes by bold black scaling on upperparts; and from Scaly Thrush by smaller size, white throat, pale lores and blackish ear coverts. Iris—brown; bill—blackish; feet—dull orange.

Voice: Harsh alarm call. Beautiful melancholy song in evenings.

Range: Breeds in NE Palearctic, wintering in S China and SE Asia. Accidental to Japan.

Distribution and status: Fairly common. Breeds in NE China plus Hebei and S Shanxi. Migrates S to winter in extreme S and SE China, also in Xishuangbanna, S Yunnan.

Habits: Rather quiet and tame. Often sits motionless for long periods. Uses mixed forests, conifers or grassy rocky terrain. Forms flocks in winter.

Notes: Treated as a race of Blue-capped Rock Thrush *M. cinclorhynchus* by some authors (e.g. Cheng 1987, 1994).

692. Chestnut-bellied Rock Thrush *Monticola rufiventris*   PLATE 73
Lixiong jidong

Description: Large (24 cm) dimorphic thrush. Breeding male has black face mask, blue upperparts, tail and throat and rest of underparts bright chestnut. Separated from red-bellied race of Blue Rock thrush by black face mask and glossy bright blue forehead. Coloration similar to some *Niltava* flycatchers but shape of head thrush-like and lacks shimmering blue on neck and shoulder. Female: brown, lightly scaled blackish above and heavily scaled dark brown and buff below. Distinguished from other female rock thrushes by broad whitish-buff crescent behind dark ear coverts also broader buff eye-ring. Juvenile with ochraceous spots and scaled brown. Iris—dark brown; bill—black; feet—blackish-brown.

Voice: Contact note *quock*, alarm is harsh jay-like rasping *chhrrs* interspersed with high shrill *tick*. Has pleasant warbling song uttered from tree-top *teetatewleedee twet tew* and variants.

Range: W Pakistan to S China and N Indochina.

Distribution and status: Fairly common in S and SE Xizang, Sichuan, W Hubei, Fujian, Yunnan, Guizhou, Guangxi and Guangdong at moderate altitudes. Breeds in forests from 1000–3000 m but winters in lowlands using forest and open rocky slopes.

Habits: Perches upright rocking tail slowly up and down. Sometimes faces along branch with cocked tail.

693. Blue Rock Thrush *Monticola solitarius* [Red-bellied   PLATE 73
Rock Thrush] Lan jidong

Description: Medium-sized (23 cm) slaty thrush. Male is dull bluish-grey with faint black and whitish scaling. Belly and undertail dark chestnut or, in race *pandoo*, blue. Distinguished from male Chestnut-bellied Rock Thrush by lack of black face mask and duller blue of upperparts. Female has bluish wash on grey upperparts with buffy underparts heavily scaled black. Immature are like female but scaled black and white on upperparts. Iris—brown; bill—black; feet—black.

Voice: Quiet croaks, harsh grating cries and a short sweet whistled song.

Range: Widespread resident and migrant in Eurasia, China, Philippines, SE Asia, Malay Peninsula, Sumatra and Borneo.

Distribution and status: Generally common, especially in E. Race *longirostris* breeds in SW Xizang; race *pandoo* in NW Xinjiang, S Xizang, Sichuan, S Gansu, Ningxia, S Shaanxi, Yunnan, Guizhou, and China S of Changjiang valley, vagrant on Taiwan and Hainan; *philippensis* breeds from NE China to Shandong, Hebei and Henan; passing through most of southern China and Taiwan on migration.

Habits: Uses prominent open perches such as rocks, houses, poles and dead trees from which to pounce on insects on the ground.

694. Blue Whistling Thrush *Myophonus caeruleus* PLATE 73
Zi xiaodong

Description: Large (32 cm) blackish thrush. Plumage all over blue-black with a few pale flecks on wing coverts. Wings and tail have a purplish iridescent wash. Feathers of head and neck have small reflective spangles at tips. Races vary in details. Nominate has black bill; *temminckii* and *eugenei* have yellow bill; *temminckii* has white tips to median coverts. Iris—brown; bill—yellow or black; feet—black.

Voice: A whistling song and imitations of other birds. In alarm it gives a high-pitched screech *eer-ee-ee* similar to a forktail.

Range: Turkestan to India and China, SE Asia, Malay Peninsula, Sumatra and Java.

Distribution and status: A common resident in hill forests at moderate altitudes up to 3650 m. Race *temminckii* is resident in S and SE Xizang; *eugenei* is resident in SW China; and nominate race is resident through eastern N China, C, E, S, and SE China.

Habits: Lives close to rivers and streams or among rocky outcrops in dense forest. Feeds on the ground, coming out into the open but fleeing for cover with alarm shrieks when disturbed.

695. Taiwan Whistling Thrush *Myophonus insularis* PLATE 73
[Formosan Whistling Thrush] Taiwan zixioadong

Description: Smallish (28 cm) whistling thrush. Entire plumage blackish-blue. Upperparts without shimmering spangles; breast has shining blue spangles. Iris—reddish-brown; bill—black; feet—black.

Voice: Screeched, far-carrying *zi* or *sui yi*.

Range: Endemic to Taiwan

Distribution and status: Frequent along dense forest streams at 600–1500 m.

Habits: As other whistling thrushes.

696. Orange-headed Thrush *Zoothera citrina* [White-throated
Thrush] Chengtou didong PLATE 73

Description: Medium-sized (22 cm) orange-headed ground thrush. Male: head, nape, and underparts rich orange-rufous; vent white; upperparts bluish-grey with a white bar on the upperwing (absent in *innotata*). Races *courtoisi, melli* and *aurimacula* have two dark vertical bars on cheeks. Female has upperparts olive-grey. Young like female but with streaks and scaling on back. Iris—brown; bill—blackish; feet—flesh-coloured.

Voice: One of the region's best song birds with a fine clear song. Also a loud screeching alarm whistle *teer-teer-teerrr*.

Range: Pakistan to S China, SE Asia and Greater Sundas. Some races are migratory.

Distribution and status: Uncommon resident and migrant up to 1500 m. Race *aurimacula* is resident on Hainan; *melli* breeds in S Guizhou, Guangxi and Guangdong; *innotata* is recorded in W and S Yunnan; and *courtoisi* breeds in Anhui (Huoshan).

Habits: A shy bird which prefers shady forest where it skulks in thick cover on the ground. It sings from tree perches.

697. Siberian Thrush *Zoothera sibirica* Baimei didong PLATE 73

Description: Medium-sized (23 cm) blackish (male) or brown (female) thrush with a conspicuous distinctive supercilium. Male slaty-black with white supercilium and white tips to feathers of tail and vent. Female olive-brown with buffy-white and rufous underparts and buffy white supercilium. Race *davisoni* is darker. Iris—brown; bill—black; feet—yellow.

Voice: In winter grounds gives only a quiet whistled contact note *chit* or *stit*. Song of two short phrases *chooeloot . . . chewee* followed by twittering *sirrr*.

Range: Breeds in N Asia; migrant in winter through SE Asia to Greater Sundas.

Distribution and status: Uncommon seasonal migrant. Race *sibirica* breeds in NE China. Nominate and *davisoni* pass through eastern provinces of China on migration.

Habits: An active bird of forest floor and canopy, sometimes in flocks.

698. Plain-backed Thrush *Zoothera mollissima* PLATE 73
Guangbei didong

Description: Largish (26 cm) thrush with uniform rufous-brown upperparts; outer tail feathers tipped white and with conspicuous pale eye-ring. White wing patch conspicuous in flight but not at rest. Breast has buff wash. Distinguished from Long-tailed Thrush by shorter tail; breast scaled rather than barred black and wing bars narrower and darker. Iris—brown; bill—blackish-brown with paler base to lower mandible; feet—flesh-coloured.

Voice: Rattling alarm call similar to Eurasian Blackbird. Single call note.

Range: Pakistan to SW China, N Burma and N Vietnam.

Distribution and status: Uncommon. Nominate race is resident in SW Sichuan, NW Yunnan and S Xizang; *griseiceps* is resident in C Sichuan to NW Yunnan. One record from N Guangdong.

Habits: Breeds in rocky areas with scattered bushes near the tree-line.

699. Long-tailed Thrush *Zoothera dixoni* Changwei didong PLATE 73

Description: Large (26 cm) long-tailed thrush with plain olive-brown upperparts and bold black scaling on whitish underparts. Buff wing patch conspicuous in flight. Has pale eye-ring and two buff wing bars. Distinguished from Scaly Thrush by lack of scaling on upperparts. Distinguished from Chinese Thrush and other *Turdus* thrushes by white tips to outer tail feathers, scaled rather than spotted markings on underparts and by characteristic *Zoothera* underwing pattern of two bold white bars. Distinguished from Plain-backed Thrush by more prominent wing bars; more olive colour; longer tail; buff rather than white wing patch and less crescent-like black barring on underparts. Iris—brown; bill—brown with yellow base to lower mandible; feet—flesh to dull yellow.

Voice: Song of flute-like notes.

Range: N India to SW China, winters to N and W Burma and N Indochina.

Distribution and status: Uncommon. Breeds in SE Xizang, S and W Yunnan and Sichuan. Recorded on migration in W Guangxi. Lives in evergreen forests from 1200–4000 m.

Habits: Often associates in mixed thrush flocks, usually feeding on ground.

700. Scaly Thrush *Zoothera dauma* [White's/White's Scaly/   PLATE 73
Golden/Small-billed/Tiger Thrush] Huban didong

Description: Large (28 cm) scaly brown thrush. Upperparts brown and underparts white, entirely laced with black and golden-buff scaly feather edges. Iris—brown; bill—dark brown; feet—pinkish.

Voice: Soft monotonous whistle and thin short *tzeet*. Nominate race differs—song is slow, broken *chirrup . . . chwee . . . chueu . . . weep . . . chirrol . . . chup . . .* etc.

Range: Widely distributed from Europe and India to China, SE Asia, Philippines, Sumatra, Java, Bali and Lombok.

Distribution and status: Fairly common resident and seasonal migrant up to 3000 m. Northern race *aurea* breeds in NE China and passes across all of China on passage, wintering in S and SE China and Taiwan; southern *socia* breeds in S and E Xizang to Sichuan, NW Yunnan, Guizhou and W Guangxi, wintering to S Yunnan (Xishuangbanna) and SE Xizang. Race *horsfieldi* is resident on Taiwan. Japanese *toratugumi* winters on Taiwan. Nominate in Himalayas may occur in SE Xizang.

Habits: Inhabits dense forest where it feeds on the ground.

701. Dark-sided Thrush *Zoothera marginata* [Lesser Brown/   PLATE 73
Lesser Long-billed Thrush] Changzui Shandong

Description: Medium-sized (25 cm) dark brown rather short-tailed thrush with long rather bulbous bill, Upperparts dark rufescent brown. Rufescent primaries contrasts with darker brown of coverts and mantle. Diffuse pale area on lores and sides of head. Dark crescent on ear coverts not as conspicuous as on other brown *Zoothera* thrushes. Outer tail feathers not tipped white. Iris—brown; bill—dark brown; feet—brown.

Voice: Utters short series of dull, hoarse *kuk* notes when flushed. Deepguttural *tchuck*. Song is thin whistle like Scaly Thrush but softer and shorter (0.5 sec) with downward inflection.

Range: C and E Himalayas, SE Asia.

Distribution and status: Rare. Recorded only from S Xishuangbanna in S Yunnan. May occur in SE Xizang. Inhabits evergreen forest of all elevations.

Habits: Very shy, keeping close to ground and digging in soft mud near streams.

Note: Larger, Long-billed Thrush *Z. monticola* has never been recorded in China but could occur in SE Xizang and SE Yunnan (Jinping). Distinguished from Lesser Long-billed Thrush by sides of head dark brown, and underparts dark spotted rather than scaly.

702. Grey-backed Thrush *Turdus hortulorum*   PLATE 74
Huibei dong

Description: Smallish (24 cm) grey thrush with rufous flanks. Male: entire upperparts rather uniform grey; throat grey or whitish; breast grey; centre of abdomen and under-tail coverts white with orange flanks and underwing. Female: upperparts browner; throat and breast white with black spots on side of breast and flanks. Distinguished from female Japanese Thrush by greyer upperparts and yellow bill and from female Black-breasted by whiter breast. Iris—brown; bill—yellow; feet—flesh-coloured.

Voice: Beautiful melodious song. Quiet chuckles and wheezy *chuck chuck* call in alarm.

Range: Breeds in E Siberia and NE China, winters to S China.

Distribution and status: Fairly common. Breeds in E Heilongjiang in NE China. Migrates over most of eastern China and winters S of Changjiang valley. Accidental on Hainan and Taiwan.

Habits: Hops among leaf litter in woods and gardens. Rather shy.

703. Black-breasted Thrush *Turdus dissimilis*   PLATE 74
Heixiong dong

Description: Smallish (23 cm) compact, dark thrush. Male: entire head, mantle and breast black. The back is dark grey and wings and tail are black. Lower breast and flanks are diagnostic bright chestnut and central belly and vent are white. Female is dark olive above with white chin and throat streaked black and white. Breast is olive-grey spotted with black. Grey breast distinguishes from female Grey-backed Thrush. Vent is white. Wings blackish and tail dark olive. Iris—brown; bill—yellow to orange; feet—yellow to orange.

Voice: Sweet mellow song. Calls include thin *see* and resounding series of *tup tup . . . tup* notes.

Range: NE India to SW China and N Indochina.

Distribution and status: Fairly common in mountains and hills in Yunnan, Guangxi and Guizhou in scrub, forest and woodlands.

Habits: Shy and solitary thrush. Feeds mostly on the ground.

704. Japanese Thrush *Turdus cardis* [Grey Thrush]   PLATE 74
Wuhui dong

Description: Small (21 cm) dimorphic thrush. Male: upperparts uniform blackish-grey with black head and upper breast; rest of underparts white with black spots on abdomen and flanks. Female: upperparts greyish-brown; underparts white with greyish bar across upper breast and rufous wash on sides of breast and flanks with black spots on breast and sides. Juvenile is browner with more rufous on underparts. Female distinguished from Black-breasted Thrush by grey rump and black spots extending to belly. Iris—brown; bill—yellow in male, blackish in female; feet—flesh-coloured.

Voice: Song is rich with long warbled phrases given from top of high tree.

Range: Breeds in Japan and E China wintering in S China and N Indochina.

Distribution and status: Uncommon. Breeds in S Henan, Hubei, Anhui (Yingshan) and Guizhou. Migrates S to winter in Hainan, Guangxi and Guangdong.

Habits: Inhabits deciduous forest and keeps to dense thickets and woods. Rather shy. Generally singly but forms small parties on passage.

705. White-collared Blackbird *Turdus albocinctus*   PLATE 74
Baijing dong

Description: Medium-sized (27 cm) blackbird with diagnostic complete white collar and upper breast. Female as male but duller and browner. Iris—brown; bill—yellow; feet—yellow.

Voice: Alarm note is loud, throaty *tuck, tuck, tuck, tuck*. Song is mellow but less varied than Eurasian Blackbird, often descending such as *tew-i, tew-u, tew-o*.

Range: Himalayas to W China, vagrant NW Burma

Distribution and status: Locally common in conifer and rhododendrons at edge of alpine zone in S and E Xizang and W Sichuan (Kangding); probably also in NW Yunnan. Altitudinal migrant. In summer lives along the tree-line feeding on open alpine meadows from 2700–4000 m. Winters at 1500–3000 m.

Habits: Generally singly or in pairs. Rather shy. Feeds on ground and in canopy.

706. Grey-winged Blackbird *Turdus boulboul* PLATE 74
Huichi dong

Description: Largish (28 cm) blackbird. Male: similar to Eurasian Blackbird but broad grey wing patch contrasts with black rest of plumage. Abdomen black with grey scaling. Bill; more orange than Eurasian Blackbird. Eye-ring yellow. Female uniform olive-brown with pale reddish-brown wing panel. Iris—brown; bill—orange; feet—dull brown.

Voice: Alarm call is chuckling *chook, chook, chook* like Eurasian Blackbird and angry *churr* near nest. Song generally one soft grace note followed by four clear notes on descending scale. Rich, fluty and mellow; reminiscent of Eurasian Blackbird.

Range: Himalayas, S China, N Indochina, Burma.

Distribution and status: Rare. Nominate race recorded wintering in S Yunnan, S Sichuan and Guizhou and endemic race *yaoschanensis* breeds on Yaoshan in Guizhou. Lives in dry scrub or evergreen montane forests from 640–3000 m, lower in winter.

Habits: Partially migratory.

707. Eurasian Blackbird *Turdus merula* [Common Blackbird] PLATE 74
Wu dong

Description: Largish (29 cm) dark uniform blackbird. Male is all black with orange-yellow bill, slight pale eye-ring and black feet. Female is blackish-brown above, dark brown below and has a dark greenish-yellow to black bill. Distinguished from Grey-winged Blackbird by uniformly dark wing. Iris—brown; bill—yellow in male, black in female; feet—brown.

Voice: Sweet song is less musical than European races but rattling alarm call is much the same. Has *dzeeb* flight call.

Range: Eurasia, N Africa, India to China; winters to Indochina.

Distribution and status: Common in woodlands, parks and gardens over most of China up to 4000 m. Race *maximus* is resident in S and SE Xizang; *sowerbyi* in C Sichuan; *intermedia* in NW China (Tianshan, Kashi, Lop Nur and Qaidam Basin); and *mandarinus* is resident through C, E, S, SW and SE China with some birds wintering on Hainan.

Habits: Feeds on the ground, quietly turning over leaves in search of invertebrates, probing for worms and, in winter, eating fruits and berries.

708. Island Thrush *Turdus poliocephalus* Dao dong PLATE 74

Description: Smallish (21 cm) rufous-bellied thrush with diagnostic white head and throat. Female has crown and upperparts dark olive-brown with white supercilium and whitish throat and nape mottled brown. Has pale bar on median wing coverts. Iris—brown; bill—yellow; feet—yellow.

Voice: Rattling alarm call. Song is loud, clear and melodious; starting slowly with alternating high and low notes, then speeding gradually to glorious lusty finale.

Range: Taiwan, Philippines, Borneo to New Guinea and Pacific islands.

Distribution and status: Race *niveiceps* is uncommon resident on Taiwan in temperate montane and pine forests from 1800–2500 m.

Habits: Keep to dense cover. Sings from tree perch.

709. Chestnut Thrush *Turdus rubrocanus* Huitou dong   PLATE 74

Description: Smallish (25 cm) chestnut and grey thrush. Colour pattern distinctive—head and neck grey, wings and tail black and most of body chestnut. Distinguished from Kessler's Thrush by grey rather than black head, lack of whitish margin between chestnut body and dark head and breast and by undertail coverts black with white tips rather than black with rufous tips. Has yellow eye-ring. Iris—brown; bill—yellow; feet—yellow.

Voice: Alarm chuckle like Eurasian Blackbird. Other calls include hard *chook-chook* and rapid, strident *sit-sit-sit*. Song is fine and reminiscent of Song Thrush but less sustained; given from top of tree.

Range: E Afghanistan, Himalayas to NE India, N Burma and Tibetan Plateau to C China.

Distribution and status: Race *gouldii* is common resident in SE and E Xizang, S Qinghai, Sichuan, NW Yunnan, S Gansu, S Ningxia, S Shaanxi, and Shennongjia area of W Hubei. Accidental in Shangdong. Nominate race could occur in S and SW Xizang. Inhabits subalpine deciduous and conifer forests from 2100–3700 m; lower in winter.

Habits: Generally singly or in pairs but forms small flocks in winter. Often feeds on ground. Rather shy.

710. Kessler's Thrush *Turdus kessleri* [White-backed Thrush]   PLATE 74
Zongbei heitoudong

Description: Large (28 cm) black and rufous thrush. Head, neck, throat, breast, wings and tail black; rest of plumage chestnut except buffy-white mantle extending to buffy-whitish breast band. Female paler than male with whitish streaky throat. Similar to Chestnut Thrush but distinguished by black not grey head, neck and throat. Iris—brown; bill—yellow; feet—brown.

Voice: Alarm note is harsh chuckle similar to White-collared Blackbird; call note is soft *dug dug*. Song composed of short phrases reminiscent of Mistle Thrush.

Range: Tibetan Plateau, NC China; stragglers reach Nepal and E Himalayas in winter.

Distribution and status: Rather rare resident in E Xizang, Gansu, Qinghai, Sichuan and NW Yunnan. Winters in S Xizang. Breeds in scrub above timber line from 3600–4500 m in rocky terrain. Descends to 2100 m in winter.

Habits: Forms flocks in winter and also feeds in fields. Flies low over ground, gliding between short bursts of wing beats. Eats juniper berries.

711. Grey-sided Thrush *Turdus feae* Hetou dong PLATE 75

Description: Medium-sized (23 cm) rich brown thrush with white belly and vent. Similar to respective sexes of Eyebrowed Thrush but breast and flanks grey not cinnamon. Similar to Pale Thrush but with short white eyebrow and without white tips to outer tail feathers. Iris—brown; bill—blackish-brown, yellow at gape and base of lower mandible; feet—brownish-yellow.

Voice: Call is slightly but distinctly thinner than Eyebrowed Thrush: *zeee* or *sieee*.

Range: Breeds in N China, winters to E India and E Asia.

Distribution and status: Rare. Breeds in Hebei, Shanxi and Beijing in mixed conifer/deciduous broadleaf forests, generally above 1000 m.

Habits: Forms flocks in winter often mixed with Eyebrowed Thrush.

712. Eyebrowed Thrush *Turdus obscurus* [White-browed/ PLATE 75
Dark/Grey-headed Thrush] Baimei dong

Description: Medium-sized (23 cm) brownish thrush with a conspicuous white eyestripe. Upperparts olive-brown with darker greyish head and white supercilium; breast brownish; belly white with rufous wash on the sides. Iris—brown; bill—yellow at base, black at tip; feet—yellowish to dark fleshy brown.

Voice: Thin *zip-zip* or drawn-out *tseep* contact call.

Range: Breeds in C and E Palearctic; migrant in winter to NE Indian subcontinent, SE Asia, Philippines, Sulawesi and Greater Sundas.

Distribution and status: Fairly common passage migrant, in open forest and secondary forests up to 2000 m, over all of China except Qinghai–Xizang plateau with some birds wintering in extreme S and SW China.

Habits: Moves through low bushes and trees in active noisy flocks. Quite tame and inquisitive.

713. Mistle Thrush *Turdus viscivorus* Hu dong PLATE 75

Description: Large (28 cm) brown thrush with buffy-white underparts heavily spotted black. Distinguished from Song Thrush by larger size, greyer-brown upperparts, white tips to outer tail feathers and white underwing. Wing coverts edged white. Female as male. Iris—brown; bill—black with yellow base; feet—pinkish-brown.

Voice: Call is dry, churring *zerrrrr*. Song is desolate and rich with sliding fluty notes. Similar to Eurasian Blackbird.

Range: Europe and N Africa to C Asia, C Siberia and W China.

Distribution and status: Rare. Race *bonapartei* breeds in SW Xinjiang and Tianshan in NW Xinjiang.

Habits: Shy and wary. Stands very upright. Feeds on cultivated fields, open areas and forest floor as well as canopy.

714. Pale Thrush *Turdus pallidus* Baifu dong PLATE 75

Description: Medium-sized (24 cm) brown thrush with white belly and vent. Male has greyish-brown head and throat, female has brown head and whitish throat with faint streaking. Wing lining grey or white. Similar to Brown-headed Thrush but breast and flanks brownish-grey not cinnamon and outer two tail feathers broadly tipped white.

Distinguished from Grey-sided Thrush by lack of pale eyebrow. Iris—brown; bill—grey above, yellow below; feet—pale brown.

Voice: Call similar to Brown-headed Thrush's *chuck-chuck*, also harsh bubbling alarm note. High-pitched *tzee* when flushed.

Range: Breeds in NE Asia; winters S to SE Asia.

Distribution and status: Breeds in NE China migrating through C China to winter S of Changjiang valley to Guangdong, Hainan and occasionally Yunnan and Taiwan.

Habits: Inhabits lowland forest and secondary growth, parks and gardens. Shy and keeps to undergrowth.

Note: Some authors include Eyebrowed and Brown-headed Thrushes within this species (e.g. Cheng 1987).

715. Brown-headed Thrush *Turdus chrysolaus* PLATE 75
Chixiong dong

Description: Medium-sized (24 cm) warm brown thrush with white abdomen and vent. Upperparts, wings and tail uniform brown. Male has greyish head and throat; female has brown head and whitish throat. Both sexes have cinnamon breast and flanks. Similar to Eyebrowed Thrush but lacks white eyebrow. Iris—brown; bill—horn, paler below; feet—yellowish-brown.

Voice: Call is a series of harsh *chuck-chuck* notes. Song is three-syllable *krrn-krrn-zee*.

Range: Breeds in S Japan, winters in Taiwan and E China, Hainan and Philippines.

Distribution and status: Fairly uncommon. Migrates through Hebei and Shandong. Wintering populations in Hong Kong, Hainan and Taiwan.

Habits: Favours mixed scrub, woodland and open land with clumps of trees. Feeds in open but stays close to cover.

716. Dark-throated Thrush *Turdus ruficollis* [Red-throated/ PLATE 75
Black-throated Thrush] Chijing dong

Description: Medium-sized (25 cm) thrush with plain grey-brown upperparts, clean white belly and vent and rufous wing lining. Two distinct races occur. Race *ruficollis* has rufous face, throat and upper breast, finely mottled white in winter; rufous edges to pale tail feathers. Race *atrogularis* has black face; throat and upper breast, finely streaked white in winter; lacks rufous edges to tail feathers. Female and juveniles have pale eyebrow and more streaked underparts. Iris—brown; bill—yellow with black tip; feet—brownish.

Voice: Flight note is thin *tseep*. Throaty alarm chuckle like Eurasian Blackbird but softer; throaty *which-which-which*.

Range: Breeds NC Asia (*ruficollis*) and NW Asia (*atrogularis*); winters S to Pakistan, Himalayas, N and W China and SE Asia.

Distribution and status: Quite common in evergreen forests above 1000 m up to 3000 m. Black-throated form *atrogularis* breeds in Altai, Tianshan, Kashi and W Kunlun regions of NW China. Birds of both red-throated form *ruficollis* and *atrogularis* pass through WC and NE China to winter in SE Xizang and W Yunnan.

Habits: Lives in loose flocks. Sometimes mixed with other thrushes. Makes long hops when on ground.

Note: These two races are sometimes treated as separate species (e.g. Stepanyan 1990 and Knystautas 1993).

717. Dusky Thrush *Turdus naumanni* Ban dong PLATE 75

Description: Medium-sized (25 cm) boldly patterned black and white thrush with pale rufous wing lining and broad rufous wing panel. Male (*eunomus*): black ear coverts and breast bar contrasting with white throat, supercilium and vent. Lower belly black scaled white. Female duller brown and buff but pattern as male, with smaller black spots on lower breast. Rarer nominate race has reddish tail and orange underparts and supercilium. Iris—brown; bill—blackish above, yellow below; feet—brown.

Voice: Soft, quite musical, squawking *chuck-chuck* or *kwa-kwa-kwa*. Also a starling-like *swic* note. In alarm a rapid *kveveg* note.

Range: Breeds in NE Asia migrating to Himalayas, China and Taiwan.

Distribution and status: Common on migration. Birds of nominate race and *eunomus* pass through China and Taiwan to winter S of 33°N.

Habits: Inhabits open grassy areas and fields. Forms large flocks in winter.

Note: Some authors treat the two races as separate species (e.g. Stepanyan 1990).

718. Fieldfare *Turdus pilaris* Tian dong PLATE 75

Description: Largish (26 cm) thrush with grey head and rump contrasting with chestnut-brown back. Underparts white, heavily streaked black on breast and flanks with variable rufous wash on flanks. Tail dark. Iris—brown; bill—yellow; feet—dark brown.

Voice: Call is loud, frothy *shak-shak* and thin nasal *geeh*. Song is squeaky chatter without pause while flying from tree to tree. Harsh rattle when mobbing crows.

Range: Breeds N Europe to Siberia; winters to S Europe, N Africa, Middle East, N India and W China.

Distribution and status: Rare. Race *subpilaris* breeds in Tianshan in NW China and recorded wintering in Kashi area of W Xinjiang and Qaidam Basin of Qinghai. Vagrant in Gansu and SE Nei Mongol.

Habits: Noisy flocking thrush of woodland and open fields. Favours subalpine birch forest. Strong flight.

719. Redwing *Turdus iliacus* Baimei gedong PLATE 75

Description: Smallish (20 cm) rich brown thrush with prominent pale eyebrow, streaked underparts and rusty-red flanks and underwing. Distinguished from red-tailed form of Dusky Thrush, Eyebrowed Thrush and Pale Thrush by streaked underparts and from female Grey Thrush by streaked rather than spotted underparts and by conspicuous pale eyebrow. Iris—brown; bill—black with yellow base; feet—greyish-brown.

Voice: Song is low squeaky chatter following short series of melancholy notes on descending scale. Call is *gak* or persistent *trett-trett-trett* . . . in alarm. Resting flocks give buzzing chorus. In winter a thin *tseee*.

Range: Breeds N Europe to Siberia.

Distribution and status: Rare. Occasional wintering birds reach Altai Mts in NW Xinjiang.

Habits: Forms flocks in winter often mixed with Fieldfares. Feeds in open fields.

720. Song Thrush *Turdus philomelos* Ou gedong PLATE 75

Description: Medium-sized (22 cm) olive-brown thrush with white underparts speckled with black spots on breast tinged buff and on abdomen and flanks. Has buffy tips to median and greater wing coverts. Combination of even brown upperparts, spotted underparts and buff underwing diagnostic. Iris—brown; bill—brown-horn; feet—pinkish-brown.

Voice: Call is sharp *zit* or persistent *xellxellxell* . . . in alarm. Song is powerful with fluted notes interspersed with sharp shrill notes. Many elements repeated about three times. Mimics other birds calls.

Range: Europe, Middle East to Lake Baikal.

Distribution and status: Only marginally reaches China. Breeds in extreme NW China.

Habits: Not as shy nor gregarious as other thrushes. Feeds in open grassy areas or forest floor as well as in canopy.

721. Chinese Thrush *Turdus mupinensis* [Eurasian/ PLATE 75
Mongolian Song Thrush] Baoxing gedong

Description: Medium-sized (23 cm) thrush with brown upperparts and bold black spots on buff underparts. Differs from Song Thrush in black patch behind ear coverts and bolder white wing bars. Iris—brown; bill—dirty yellow; feet—dull yellow.

Voice: Song is measured series of pleasant, usually 3–5 note phrases punctuated by 3–11-second intervals. Notes mostly even-pitched, sometimes rising, occasionally slurred.

Range: C China.

Distribution and status: Occasional in mixed and coniferous forests from lowlands to 3200 m from Hubei to S Gansu and south to NW Yunnan. Strays reach Shandong.

Habits: Generally in undergrowth. Singly or in small flocks. Rather shy.

722. Gould's Shortwing *Brachypteryx stellata* PLATE 76
Libei duanchidong

Description: Medium-sized (13 cm) shortwing. Distinguished by chestnut upperparts; underparts finely vermiculated grey and black with diagnostic triangular white spots on lower breast and belly. Flanks and vent tinged rufous. Has narrow grey supercilium. Race *fusca* is darker than nominate. Iris—dark brown; bill—black; feet—pinkish-brown.

Voice: Alarm note is sharp *tik-tik*. Song is series of fast, high-pitched notes with undulating quality.

Range: Nepal to SW China, N Burma and N Tonkin.

Distribution and status: Nominate race is uncommon in SE Xizang (Qamdo) and rare in Yunnan (Yanjing) in rhododendron, bamboo, juniper and subalpine forest from 2750–4200 m. Race *fusca* probably occurs in SE Yunnan (Jinping). Recently found at Juizhaigou, N Sichuan.

Habits: Keeps to undergrowth. Skulks in bamboo thickets, often near streams. Can be quite tame.

723. Rusty-bellied Shortwing *Brachypteryx hyperythra* PLATE 76
Xiufu duanchidong

Description: Small (13 cm) shortwing with bluish-grey upperparts and diagnostic deep ferruginous underparts. Has fine white supercilium (partially concealed) and black lores. Female: olive-brown above with pale ferruginous underparts and white centre of belly. Iris—brown; bill—black; feet—pinkish-brown.

Voice: Song faster and more musical than Lesser Shortwing—warble of slurred notes introduced by two spaced notes *tu-tiu*. Ends abruptly.

Range: Sikkim, NE India to SW China.

Distribution and status: Rare resident recorded in Gongshan area of NW Yunnan, also SE Xizang from 1100–3000 m.

Habits: Rather tame. Inhabits dense undergrowth and thickets.

724. Lesser Shortwing *Brachypteryx leucophrys* PLATE 76
[Mrs La Touche's Shortwing] Baihou duanchidong

Description: Small (13 cm) shortwing (looks like a babbler) with inconspicuous short, fine, white semi-concealed supercilium, buffy eye-ring and heavy bill. Shorter-tailed than White-browed Shortwing. Nominate male: upperparts slaty-blue with blue-grey breast band and flanks and white throat and centre of belly. Female: rufous-brown above with rufous-brown wash and scaling on breast and flanks; throat and belly white. Lacks rufous-orange forehead and lores of White-browed Shortwing. Race *carolinae* lacks 'blue' male. Male: rufous or olive-brown above, with white throat and centre of belly, buff wash on breast and flanks and white mottling on upper breast. Female similar to male but more rufous. Immature birds are streaked and spotted. Iris—brown; bill—dark brown; feet—pinkish-purple.

Voice: A hard *tock, tock* and a high, plaintive whistle. Song is fast, sweet, high warble preceded by two or three single emphatic notes and ending in a jingle.

Range: Himalayas, S China, SE Asia and Sundas.

Distribution and status: Race *nipalensis* is resident in SE Xizang and W Yunnan and on Mt Emei in Sichuan; and *carolinae* is resident in SE and S China. Common in moist montane forests from 1000–3200 m.

Habits: Shy, keeping to undergrowth and forest floor generally at lower altitudes than White-browed Shortwing.

725. White-browed Shortwing *Brachypteryx montana* PLATE 76
[Blue/Himalayan Blue/Indigo-blue Shortwing] Lan duanchidong

Description: Medium-sized (15 cm) dark bluish (male) or brown (female) shortwing. Races vary. Male of race *sinensis* is dark slate-blue above with a conspicuous white supercilium; pale grey below; black tail and wings with white patch at 'shoulder' of wing. Male *cruralis* has black lores and frontal band, lacks white at 'shoulder' and has dark blue underparts. Male *goodfellowi* is brown like female. Females of *sinensis* and *goodfellowi* are dull brown with pale brown breast, whitish centre of belly and rufous wings and tail. Females have smaller concealed white supercilium. Immature is mottled brown. Iris—brown; bill—black; feet—flesh with tinge of grey.

Voice: Song commences slowly with several single notes, quickens to a plaintive babble, then stops abruptly.

Range: Himalayas to S China, Philippines, SE Asia, Greater Sundas and Flores.

Distribution and status: Locally common at 1400–3000 m. Race *sinensis* occurs in mountains of S Shaanxi (Qinling) and SE China, including N Guangdong and Hong Hong; *cruralis* is resident in SE Xizang, NW Yunnan and Mt Emei in S Sichuan, wintering in S Yunnan; and *goodfellowi* occurs on Taiwan.

Habits: Shy, stays in dense thickets close to the ground, often near streams. Comes out into open clearings and even the bare rocky slopes of mountain tops. Variable in its habits according to availability of suitable foods.

Note: Taiwan form *goodfellowi* is probably a distinct species.

OLD WORLD FLYCATCHERS—Subfamily: Muscicapinae

A very large and varied Old World subfamily of smaller insectivorous birds. Flycatchers have rounded heads and small, broad-based pointed bills with a wide gape and fringe of stiff bristles which help them to snap up small insects. They have short slender legs and small feet. Males of most flycatchers are brightly coloured but most females are drab. They regularly join mixed bird flocks. The nests are neat, cup-shaped structures lined with hair and decorated with moss. Typical flycatchers have an upright posture and tend to hawk after insects from a perch. A total of 34 species occur in China, some of which are migratory.

726. Brown-chested Jungle Flycatcher *Rhinomyias brunneata* PLATE 72
[AG: Flycatcher] Baihou linweng

Description: Medium-sized (15 cm) nondescript brownish flycatcher with pale brown breast band and usually slight dark scaling on whitish neck; has pale lower mandible. Immature has scaly buff upperparts and black tip to lower mandible. Looks shortish winged and long-billed. Iris—brown; bill—blackish above, yellowish base below; feet—pink.

Voice: Harsh churrs.

Range: S China, migrating S in winter as far as Malay Peninsula and Nicobars.

Distribution and status: Nominate race is uncommon summer breeder in SE China.

Habits: Keeps to lower canopy of forest edge, bamboo thickets, secondary forest and plantations up to 1100 m.

727. Spotted Flycatcher *Muscicapa striata* Ban weng PLATE 79

Description: Medium-sized (15 cm) streaky grey flycatcher. Upperparts ashy-grey streaked with black on crown; underparts white streaked with grey on breast and sometimes flanks. Wings and tail brown with feathers edged pale. Distinguished from Grey-streaked Flycatcher by crown streaking and pale edges to tail feathers. Iris—brown; bill—blackish; feet—blackish.

Voice: Thin scratchy notes, *tseet . . . chup, chup*. Song consists of squeaky call notes.

Range: Europe and Asia to Lake Baikal and S to Pakistan. Winters in Africa.

Distribution and status: Rare resident in extreme NW Xinjiang in open woodlands and gardens.

Habits: Sits upright and hunts in aerial sallies from perch.

728. Grey-streaked Flycatcher *Muscicapa griseisticta*   PLATE 79
[Grey-Spotted/Spot-breasted Flycatcher] Huiwen weng

Description: Smallish (14 cm) brownish-grey flycatcher with white eye-ring and white underparts heavily streaked dark grey on breast and flanks. Has a narrow white band across forehead (barely visible in field) and a narrow white wing bar. Long wing almost reaches tip of tail. Lacks half collar of Dark-sided and smaller and much more heavily streaked on breast than Spotted Flycatcher. Iris—brown; bill—black; feet—black.

Voice: Call is described as loud melodious *chipee, tee-tee.*

Range: Breeds in NE Asia; migrates in winter to Borneo, Philippines and Sulawesi to New Guinea.

Distribution and status: Uncommon. Breeds in larch woods of extreme NE China but migrates through E, C and S China and Taiwan.

Habits: A shy bird near streams in dense forest, open forest, forest edge and even urban parks.

729. Dark-sided Flycatcher *Muscicapa sibirica* [Siberian/   PLATE 79
Sooty Flycatcher] Wu weng

Description: Smallish (13 cm) ashy-grey flycatcher with dark flanks. Upperparts dark grey; faint buff wing bar; underparts white with sooty-grey mottled flanks and mottled grey band across upper chest; conspicuous white eye-ring; white throat and usually white half collar; malar area streaked black. Long wings reach two thirds of way down tail. Races vary in degree of grey on underparts. Immature is spotted white on face and back. Iris—dark brown; bill—black; feet—black.

Voice: Lively, metallic tinkle, *chi-up, chi-up, chi-up;* not as harsh as Asian Brown Flycatcher. Song is complicated series of repetitive thin notes with melodious trills and whistles.

Range: Breeds in NE Asia and Himalayas; migrates in winter to S China SE Asia, Palawan and Greater Sundas.

Distribution and status: Nominate race breeds in NE China and winters in S and E China, Hainan and Taiwan; *rothschildi* breeds in Qinling Mts of S Shaanxi, SE Gansu, SE Qinghai, E Xizang and Sichuan and winters in S China; *cacabata* breeds in S Xizang. Fairly common in evergreen forests and woodlands up to 4000 m and in lowlands in winter.

Habits: Inhabits undergrowth and middle storeys of montane or submontane forests. Sits rather upright on a low bare branch, making dashes to catch passing insects.

730. Asian Brown Flycatcher *Muscicapa dauurica* [Grey-breasted/Brown
Flycatcher > Brown-streaked/Chocolate Flycatcher] Beihui weng   PLATE 79

Description: Smallish (13 cm) greyish-brown flycatcher. Upperparts grey-brown; underparts whitish with brownish-grey on sides of chest and flanks; eye-ring white; diffuse whitish areas in front of eye in winter. Race *cinereoalba* is greyer. Bill longer than

Dark-sided or Ferruginous Flycatchers and lacks half collar. Fresh birds have narrow white wing bar. Wing tips reach about halfway down tail. Iris—brown; bill—black with yellow base to lower mandible; feet—black.

Voice: Call is sharp, dry, trill *tit-tit-tit-tit*. Song is short trill interspersed with short whistled phrases.

Range: Breeds NE Asia and Himalayas and locally in SE Asia; migrant S in winter to India, SE Asia, Philippines, Sulawesi and Greater Sundas.

Distribution and status: Breeds in N and NE China, migrating through E and C China and Taiwan to winter in S China and Hainan. Race *latirostris* is common in woodlands and gardens at all altitudes but reaches lowlands in winter; *cinereoalba* may also visit China.

Habits: Catches insects from perch and shivers tail in a characteristic manner on returning to perch.

731. Brown-breasted Flycatcher *Muscicapa muttui* PLATE 79
Hexiong weng

Description: Smallish (14 cm) brownish flycatcher with buff-brown chest bar. Distinguished from similar-coloured flycatchers by combination of white lores and eye-ring; dark streak separating white malar stripe from white chin and throat, no wing bar, pale legs, yellow lower mandible, rufescent rump and buff vent. Edges of wing feathers are rufous. Iris—dark brown; bill—dark above, yellow below; feet—dull yellow.

Voice: Thin *sit* call and pleasant feeble song.

Range: Breeds NE India, SW, W and S China and W Tonkin; winters to SW India, Sri Lanka and recorded N and E Burma and NW Thailand.

Distribution and status: Rather rare resident in SE Gansu, Sichuan, Guizhou, Guangxi and Yunnan in subtropical forests. In hills in summer but lowlands in winter.

Habits: Quiet, solitary and partially crepuscular, hiding in dense thickets and bamboo by day.

732. Ferruginous Flycatcher *Muscicapa ferruginea* Zongwei heweng PLATE 79

Description: Smallish (13 cm) reddish-brown flycatcher with a buffy eye-ring and white throat patch. Head slaty; back brown; rump rufous; underparts white with brown chest bar, rufous flanks and undertail coverts. Usually has white half collar. Tertials and greater wing coverts edged rufous. Iris—brown; bill—black; feet—grey.

Voice: Soft low trill *si-si-si*. Generally silent in winter. Probable song is high-pitched, harsh *tsit-tittu-tittu*.

Range: Breeds in Himalayas, S China and NW Tonkin; migrates S in winter as far as Greater Sundas.

Distribution and status: Breeds in Taiwan, S Gansu, S Shaanxi, Sichuan, W Yunnan and SE Xizang. Migrates S in winter with some birds remaining on Taiwan and Hainan. Rare on mainland but quite common on Taiwan.

Habits: A shy bird of forest clearings and stream sides.

733. Yellow-rumped Flycatcher *Ficedula zanthopygia*   PLATE 79
[Tricoloured/Korean Flycatcher] Baimei [ji]weng

Description: Male; small (13 cm) yellow, white and black flycatcher. Has yellow rump, throat, breast and upper belly; lower belly and undertail coverts white; otherwise black except for white supercilium and wing flash. Female: dull brown above, paler below and has dull yellow rump. White supercilium and blacker back of male and yellow rump of female distinguish from respective sexes of Narcissus Flycatcher. Iris—brown; bill—black; feet—black.

Voice: Deep grating *tr-r-r-rt*, lower-pitched than Red breasted Flycatcher.

Range: Breeds in NE Asia; migrates S in winter to S China, SE Asia and Greater Sundas.

Distribution and status: Breeds in NE, C and E China north of 29°N. Migrates through S China. Fairly common up to 1000 m.

Habits: Frequents scrub and woods near water.

734. Narcissus Flycatcher *Ficedula narcissina*   PLATE 79
Huangmei [ji]weng

Description: Male: small (13 cm) black and yellow flycatcher. Nominate race has upperparts black with yellow rump, white wing patch and diagnostic yellow supercilium; underparts mainly orange-yellow. Race *elisae* has greenish back and yellow lores but no supercilium. Lower belly and undertail coverts yellow. Female: olive-grey upperparts with rufous tail; pale brown underparts with yellowish wash. Distinguished from female Yellow-rumped by lack of yellow rump. Female *elisae* is greenish on upperparts with yellowish underparts. Iris—dark brown; bill—bluish-black; feet—lead blue.

Voice: Generally silent in winter. Melodious song of repeated warbles and trisyllabic whistles such as *o-shin-tsuk-tsuk* and copied calls of other birds.

Range: Breeds in NE Asia; migrates in winter to S Thailand and Peninsular Malaysia, Philippines and Borneo.

Distribution and status: Nominate race breeds in Siberia and Japan and migrates through E and S China and Taiwan, through to Philippines; some birds winter on Hainan. Race *elisae* breeds in Hebei and Shaanxi and migrates to SE Asia. Generally uncommon.

Habits: Typical flycatcher, hawking insects from canopy perches and middle storey.

Note: Race *elisae* may prove to be a separate species.

735. Mugimaki Flycatcher *Ficedula mugimaki*   PLATE 79
Qu [ji]weng

Description: Male: smallish (13 cm) orange, black and white flycatcher. Upperparts blackish-grey with narrow white supercilium behind eye; bold white patch on wing and white edge to base of tail; throat, breast and sides of belly orange; centre of belly and undertail coverts white. Female: upperparts, including rump, brown; underparts as male but paler; lacks white on tail. Immature: plain brown above with buff underparts, wing bar and white belly. Iris—dark brown; bill—dark horn; feet—dark brown.

Voice: Soft *turrrr* calls.

Range: Breeds in N Asia; migrates S in winter to SE Asia, Philippines, Sulawesi and Greater Sundas.

Distribution and status: Breeds in NE China. Passage migrant through E and C China and Taiwan. Uncommon wintering bird in Guangxi, Guangdong and Hainan.

Habits: Frequents the canopy of forest edge, clearings and hill forest. Flicks and spreads tail.

736. Slaty-backed Flycatcher *Ficedula hodgsonii* PLATE 80
Xiuxiong lan[ji]weng

Description: Male: small (13 cm) slaty-blue flycatcher with orange breast. Upperparts lack iridescence. Outer tail feathers basally white. Orange-rufous breast grades to buffy belly. Distinguished from Hill Blue Flycatcher by duller back, white bases to tail, longer wings, shorter bill and lack of eyebrow or wing bar. Female distinguished from female Little Pied Flycatcher by lack of median pale bar on breast, from female Sapphire Flycatcher by duller underparts. Iris—brown; bill—black; feet—dark brown.

Voice: Hard *tchat* and rattled *terrht*. Song short, fast, flute-like and meandering, in whistled descending series.

Range: Nepal to W China and N Indochina.

Distribution and status: Uncommon resident in SE Xizang, E Qinghai, Yunnan, Sichuan, SE Gansu to Shanxi (Pangquangou) in dense moist forests at 2400–4300 m; descending in winter.

Habits: A quiet, arboreal flycatcher.

737. Rufous-gorgeted Flycatcher *Ficedula strophiata* PLATE 79
Chengxiong [ji]weng

Description: Smallish (14 cm) forest-living flycatcher with white flashes at base of black tail. Upperparts mostly uniform greyish-brown; wings olive; underparts grey. Adult male has narrow white forehead and small dark orange throat gorget (often inconspicuous). Female is similar to male but gorget is smaller and paler. Immature is streaky brown with black scaling on rufous flanks. Iris—brown; bill—black; feet—brown.

Voice: The usual call is a low *tik-tik* or high-pitched, repeated *pink*. Also gives low churring. Song is thin, metallic *tin-ti-ti*, with loud first note and last two notes soft.

Range: Breeds from Kashmir and Himalayas to S China and Vietnam; winters to SE Asia.

Distribution and status: Common altitudinal migrant in Xizang, C and SW China at 1000–3000 m. Some winter to S China.

Habits: This is a shy bird of the ground and lower bushes in closed forest.

738. Red-throated Flycatcher *Ficedula parva* [Red-breasted PLATE 79
Flycatcher] Hongtou[ji]weng

Description: Small (13 cm) brown flycatcher with conspicuous white lateral flashes on base of dark tail. Breeding male has red breast tinged with grey, but this is rarely seen in wintering grounds. Female and non-breeding male are dull grey-brown with whitish throat and narrow white eye-ring. Black tail and uppertail coverts distinguish from Asian Brown Flycatcher. Iris—dark brown; bill—black; feet—black.

Voice: Harsh, sharp *trrrt* in alarm; quiet *tic* and harsh *tzit*.

Range: Breeds in Palearctic; migrates in winter to China, Philippines, SE Asia and Borneo.

Distribution and status: Migrates through eastern half of China. Common wintering bird in Guangxi, Guangdong and Hainan.

Habits: Keeps to smaller trees at forest edge and along rivers. Dashes to cover when alarmed. Flicks dark tail to reveal white base patches and gives harsh clicking notes.

739. White-gorgeted Flycatcher *Ficedula monileger* Baihou [ji]weng   PLATE 79

Description: Small (13 cm) olive-brown flycatcher with diagnostic white bib on chin and throat bordered by black moustachial stripe and gorget. White eyebrow, rufescent wings and tail and buffy wash on breast and flanks. Nominate race has buff eyebrow. Iris—brown; bill—black; feet—greyish.

Voice: High-pitched whistling song; metallic *dik*; scolding rattle, short whistle.

Range: Himalayas, Burma and N Indochina.

Distribution and status: Rare. Race *leucops* is vagrant in SW Yunnan; nominate race probably occurs in SE Xizang in forested valleys from 1000–2000 m.

Habits: Keeps to undergrowth.

740. Snowy-browed Flycatcher *Ficedula hyperythra*   PLATE 80
[White-fronted/Snow-browed/Thicket/Dull Flycatcher] Zongxiong lan[ji]weng

Description: Male: small (12 cm) grey-blue and rufous flycatcher. Upperparts slaty-blue with prominent but short white eyebrows that almost meet on forehead; underparts orange-buff on throat, breast and flanks. Taiwan race has chestnut flanks. Female: brown above, buffy below with rusty-buff forehead, eyebrow and eye-ring. Immature: mottled brown. Distinguished from shortwings by smaller size and delicate tarsi. Iris—dark brown; bill—black; feet—fleshy.

Voice: Quiet song of three or four wheezy shrill notes *tsit-sit-si-sii*; single repeated *sip*.

Range: N India to S China, Philippines, SE Asia and Sundas.

Distribution and status: Nominate race is a not uncommon bird in montane forests of SE Xizang, Sichuan, Yunnan, Guangxi, Guangdong, Hainan and in S Yunnan (Xishuangbanna); *innexa* is endemic to Taiwan.

Habits: Unobtrusive. Spends much time on the ground where it hops like a robin.

741. Little Pied Flycatcher *Ficedula westermanni*   PLATE 80
Xiaoban [ji]weng

Description: Small (12 cm), pied or brown and white flycatcher. Male: upperparts black with broad long white eyebrow, white wing bar and edge to base of tail; underparts white. Female: greyish-brown upperparts, buffy wing bar and whitish underparts. Distinguished from Slaty-backed flycatcher by whiter centre of breast and belly which forms a pale median band. Immature is brown mottled tawny. Iris—brown; bill and feet—black.

Voice: Song is a high-pitched thin descending *swit, swit, swit* followed by a low rattle *churr-r-r-r*. Call is mellow *tweet* and low *churr*.

Range: India to S China, Philippines, SE Asia and Indonesia.

Distribution and status: Locally common in montane forest from 900–2600 m through SE Xizang, S Guizhou and Guangxi. Recently recorded Hong Kong; could have been escape or range extension.

Habits: Feeds at all levels in the canopy and sometimes joins mixed-species flocks.

742. Ultramarine Flycatcher *Ficedula superciliaris*   PLATE 80
Baimei lan[ji]weng

Description: Male: small (12 cm) blue flycatcher with white underparts. Iridescent crown and ultramarine back; sides of head, patch on side of breast and wing dark dull blue (looks black in poor light) diagnostic. Tail sometimes with small white patch at base; sometimes has narrow white supercilium. Female has same breast pattern but underparts buffy; upperparts greyish with brown wash on head. Lacks white on tail base. Sometimes has greyish or blue wash on uppertail coverts. Sometimes has white wing bar and white edges to tertials. Immature: brown with rusty spots and black scaling. Iris—dark brown; bill—dark grey; feet—grey.

Voice: Low rattling *trrrt*. Attracted to 'pishing'. Song is feeble, disjointed and high-pitched with short trills and chirps.

Range: Himalayas, N Burma, SW China, winters N Thailand.

Distribution and status: Race *aestigma* is uncommon from S Xizang to Sichuan in hill and montane forests up to 3000 m in summer. Winter visitor to S Yunnan.

Habits: Keeps to middle and upper storeys.

743. Slaty-blue Flycatcher *Ficedula tricolor*   PLATE 80
Hui lan[ji]weng

Description: Small (13 cm) slaty-blue flycatcher with whitish underparts. Tail black with white flashes at base. Sides of head and throat dark grey, extending onto side of breast. Underparts of race *cerviniventris* (includes *minuta*) are tinged rufous. Olive triangular patch on throat of male. Iris—brown; bill—black; feet—black.

Voice: Alarm call is *ee-tick*; also rapid *ee-tick-tick-tick-tick*.

Range: Himalayas, NE India, S China, Burma; winters to N Indochina.

Distribution and status: Nominate race resident in SW Xizang; *cerviniventris* in SE Xizang and *diversa* in C and SW China. Not uncommon in evergreen montane forests. Vagrant Beidaihe (Hebei).

Habits: Keeps mostly to undergrowth. Stays in conifer forests through winter. Droops wings and constantly flicks tail.

Note: Formerly listed as *F. leucomelanura* (e.g. Cheng 1987, 1994).

744. Sapphire Flycatcher *Ficedula sapphira* Yutou [ji]weng   PLATE 80

Description: Male: small (11 cm) flycatcher with brilliant ultramarine upperparts, whitish underparts with orange- rufous patch down centre of throat and breast. Sides of head and sides of breast brilliant blue. First-year male has entire breast orange-rufous and crown, nape and mantle olive-brown. Female: upperparts rufescent olive-brown; underparts distinctive with orange-rufous centre to throat and breast, buffy sides to throat, breast and flanks with white belly and undertail coverts. Rump is rufous. Iris—brown; bill—black; feet—bluish-grey.

Voice: Low *tit-tit-tit* rattle, deeper than most *Ficedula* spp. Song is high-pitched, metallic, insect-like *chiki-riki-chiki*.

Range: Nepal to SW China, N Burma, N Indochina.

Distribution and status: Race *tienchuanesis* resident from S Gansu through W Sichuan to S Qinling Mts in S Shaanxi. Nominate race resident in SE Xizang, S Sichuan and Yunnan. Race *laotiana* occurs in W Yunnan. Generally rare in hill forests from 900–2000 m.

Habits: Rather active and tame. Keeps to middle and upper storeys. Can be attracted by 'pishing'.

745. Blue-and-white Flycatcher *Cyanoptila cyanomelana* PLATE 80

[Japanese Blue-Flycatcher] Baifu[ji]weng

Description: Male: large (17 cm) blue, black and white flycatcher. Face, throat and upper breast diagnostically blackish; upperparts glossy cobalt-blue; lower breast, belly and undertail coverts white. Outer tail feathers basally white. Dark breast sharply demarcated from whtie abdomen. Race *cumatilis* is turquoise-blue and has black areas replaced by dark greenish-blue. Female: grey-brown upperparts with brown wings and tail, white centre of throat and belly. Distinguished from Asian Brown Flycatcher by larger size and lack of pale lores. Juvenile male has ashy-brown head, nape and breast band but blue wings, tail and uppertail coverts. Iris—brown; bill and feet—black.

Voice: Harsh *tchk, tchk* note. Generally silent in winter grounds.

Range: Breeds in NE Asia; migrates S in winter to China, Peninsular Malaysia, Philippines and Greater Sundas.

Distribution and status: Nominate race migrates through eastern half of China; some birds winter on Taiwan and Hainan. Race *cumatilis* breeds in NE China and passes through S and SW China to winter in SE Asia. Uncommon in tropical submontane forests up to 1200 m.

Habits: Frequents wooded areas in primary and secondary forests, feeding quite high in the canopy.

746. Verditer Flycatcher *Eumyias thalassina* PLATE 80

[Indian Verditer Flycatcher] Tonglan weng

Description: Largish (17 cm) uniform, greenish-blue flycatcher. Male has black lores; female duller with dusky lores. Both sexes have whitish scaling on undertail coverts. Immature grey-brown with greenish wash, scaled and spotted buff and blackish. Distinguished from male Pale Blue Flycatcher by shorter bill, greener colour and whitish scaling on blue-grey vent. Iris—brown; bill—black; feet—blackish.

Voice: High-pitched, hurried and sustained musical song with little variation in pitch or gradually descending scale; less husky than Pale Blue Flycatcher. Call is *tze-ju-jui*.

Range: India to S China, SE Asia, Sumatra and Borneo.

Distribution and status: Breeds in S Xizang, C, S and SW China. Some populations winter in SE China. Uncommon in pine and open forests up to 3000 m. Lower in winter.

Habits: Hawks after flying insects from exposed perch in canopy of open forest or at edge of forest clearings.

747. Large Niltava *Niltava grandis* [Great Niltava]   PLATE 81
Da xianweng

Description: Male: large (21 cm) dark flycatcher with blue upperparts with shining blue crown and stripe on side of neck, shoulder patch and rump; black underparts. Female: rufous olive-brown with blue-grey crown, pale blue shining neck patch and triangular buffy throat patch. Distinguished from Fujian and Rufous-bellied Niltava females by size and lack of white gorget. Immature: brown with white speckling on head, rusty spots on back and black scaling on underparts. Iris—dark brown; bill—black; feet—horn.

Voice: Clear rising whistle of three ascending notes introduced by a grace note *k'tu-tu-ti*; also scolding rattle and nasal *djuee*.

Range: Nepal to SW China, SE Asia and Sumatra.

Distribution and status: Nominate race breeds in SE Xizang above 2000 m; *griseiventris* breeds in SE Yunnan. Wintering birds descend to lowlands. Generally uncommon.

Habits: Solitary flycatcher keeping to mid-storey in submontane and montane forests. Occasionally flicks wings and tail.

748. Small Niltava *Niltava macgrigoriae* Xiao xianweng   PLATE 81

Description: Small (14 cm) dark flycatcher. Male is very dark blue flycatcher with black side of face and throat and white vent. Forecrown, side of neck and rump are shining blue. Distinguished from Large Niltava by small size, blue breast and white vent. Female is brown with rufous wings and tail, shiny blue patch on side of neck, buff throat and pale buff gorget. Distinguished from Large Niltava by small size, brown nape and pale gorget. Iris—brown; bill—black; feet—black.

Voice: Song is thin high-pitched *twee-twee-ee-twee* with second note highest. Scolding, churring, and descending *see-see* calls.

Range: Himalayas to NE India, S China and parts of SE Asia.

Distribution and status: Fairly common in evergreen forest from 900–2400 m in SE Xizang and S China.

Habits: Keeps to dark dense undergrowth of forest.

749. Fujian Niltava *Niltava davidi* Zongfu daxianweng   PLATE 81

Description: Medium-sized (18 cm) brightly coloured flycatcher. Male is deep blue above and rufous below. Side of face is black and forehead, small patch on neck, shoulder of wing and rump are bright iridescent blue. Distinguished with difficulty from Rufous-bellied Niltava by darker colours. Female is greyish-brown with rufous-brown tail and wings; white gorget on throat and small iridescent blue patch on side of neck. Distinguished from Rufous-bellied by whiter belly. Iris—brown; bill—black; feet—black.

Voice: Very high-pitched *ssssew* or *siiiii* repeated after shortish intervals. Alarm call is sharp metallic *tit tit tit . . . , trrt trrt trrt . . . trrt trrt tit tit . . .* etc. (C. Robson).

Range: S China; winters to Thialand and Indochina.

Distribution and status: This is a fairly common flycatcher of dense forest undergrowth in mountains of Sichuan, Guizhou, Yunnan, Fujian, Hainan and Guangxi.

Habits: Typical forest flycatcher.

750. Rufous-bellied Niltava *Niltava sundara*   PLATE 81
Zongfu xianweng

Description: Male: medium-sized (18 cm) large-headed flycatcher with blue upperparts, rufous underparts and black face mask. Crown, neck spots, shoulder patch and rump shining blue. Distinguished from Blue-throated Flycatcher by black throat, orange breast grading to buff vent. Distinguished with difficulty from Fujian Niltava by brighter plumage, more rufous vent and shining blue of forehead extending over crown. Female: brown with rufescent tail and rump. White gorget, light blue iridescent neck spot and buffy lores and eye-ring distinguishes from all female flycatchers except Fujian Niltava but vent more buffy and wings shorter. Iris—brown; bill—black; feet—grey.
Voice: Hard robin-like *tic*, thin *see* and soft *chacha*. Probable song is raspy *zi-i-i-f-chachuk*.
Range: Himalayas to W China and N Indochina.
Distribution and status: Race *denotata* breeds W Hubei, S Shaanxi, SE Gansu, Sichuan, Yunnan, Guizhou; nominate race breeds in Himalayas to SE Xizang and W Yunnan. Uncommon in open woodland and hill forests from 1500–3000 m.
Habits: Quiet and solitary. Sits low in bushes to drop on insects on ground. Bobs body forward and flicks tail every few seconds.

751. Vivid Niltava *Niltava vivida* Zongfu lanxianweng   PLATE 81

Description: Male: medium-sized (18 cm) blue and rufous flycatcher. Very similar to Rufous-belied Niltava but bright blue parts of plumage duller and rufous of breast extends onto throat as a triangle. Female lacks white gorget or blue neck patch and has grey crown and nape and buffy throat patch. Immature is like immature Rufous-bellied Niltava but more rufous. Iris—brown; bill—black; feet—black.
Voice: Simple slow song of mellow whistles interspersed with scratchy notes. Call is clear whistled *yiyou-yiyou*.
Range: NE India to SW China, Taiwan and SE Asia.
Distribution and status: Uncommon montane resident in evergreen and mixed forests from 2000–2700 m in summer and lower in winter. Race *oatesi* occurs in S and SE Xizang. Yunnan and S Sichuan and nominate race occurs on Taiwan. Vagrant Hong Kong.
Habits: Keeps to middle storey and canopy.

752. White-tailed Flycatcher *Cyornis concretus* [Short-tailed   PLATE 81
Flycatcher, AG: Blue Flycatcher] Baiwei lanxianweng

Description: Largish (19 cm) dark flycatcher with white patches in spread tail. Male: dark blue upperparts with black sides to head and black flight feathers. Underparts: dark blue breast grading to white vent. Distinguished from Hainan Blue Flycatcher by larger size and tail flashes; from White-tailed Robin by white belly and vent. Female: brown with broad white gorget, belly and undertail coverts. Distinguished from females of Fujian and Rufous-bellied Niltavas by size and lack of blue neck patch. Immature: brown with rusty spots on upperparts and black scaling on underparts. Iris—dark brown; bill—black; feet—dark grey.

Voice: Loud, sibilant whistles, second note higher and last note lowest '*where are you*' or *tuu tii*. Harsh *scree* in alarm.

Range: NE India, SW China, SE Asia, Sumatra and Borneo.

Distribution and status: Race *cyanea* is rare in S Yunnan (Xishuangbanna).

Habits: Solitary flycatcher of undergrowth in hill and submontane forests.

753. Hainan Blue Flycatcher *Cyornis hainanus*   PLATE 81
Hainan lanxianweng

Description: Small (15 cm) dark flycatcher. Male: dull blue, fading into white under-parts, brighter on forehead and 'shoulder'. Immature male has whitish throat. Female: brown above with rufous wash on rump, tail and secondaries; buffy lores and eye-ring. Underparts warm buff on breast grading to white belly and undertail. Iris—brown; bill—black; feet—pink.

Voice: Sweet melodious voice like Oriental Magpie Robin. Distinctive call of three rising notes followed by a falling note and last rising note '*hello mummy*'.

Range: S China, Hainan and SE Asia.

Distribution and status: Locally common resident on Hainan and summer breeder on mainland in S Yunnan, S Guizhou, E Guangxi, Guangdong and Hong Kong with some birds wintering. This is the commonest flycatcher in Xishuangbanna, S Yunnan.

Habits: Typical forest flycatcher of middle and upper storeys in lowland evergreen forest.

754. Pale-chinned Flycatcher *Cyornis poliogenys*   PLATE 81
Huijia xianweng

Description: Male: medium-sized (15 cm) brownish flycatcher with greyish head, rufous breast and flanks and white throat. Race *cachariensis* has pale orange throat and lacks grey cast to head and can be mistaken for female female Hill Blue Flycatcher. Female: brown with white throat and rufous breast band; similar to female Blue-throated Flycatcher but less rufous on wings and tail. Iris—brown; bill—black; feet—pink.

Voice: Repeated *tic* call. Song is high-pitched series of notes with rising and falling pitch and interspersed by harsh call notes.

Range: Nepal to NE India, SW China and N Burma.

Distribution and status: Race *laurentei* is an uncommon resident in W and SE Yunnan. Race *cachariensis* is fairly common in E Himalayan foothills of SE Xizang up to 1500 m.

Habits: Prefers rather open forest and hunts on or near ground.

755. Pale Blue Flycatcher *Cyornis unicolor*   PLATE 80
Chun lanxiangweng

Description: Largish (17 cm) pale blue (male) or brownish (female) flycatcher. Male: upperparts bright cobalt-blue with black lores; throat and breast pale blue; belly greyish-white and undertail coverts whitish. From Verditer Flycatcher by longer bill and lack of green in coloration. Has cinnamon eye-ring. Female: upperparts grey-brown, tail more rufous-brown; underparts greyish-brown; eye-ring and lores fulvous. Sometimes shows a narrow band of dull turquoise above base of bill. Immature: brown, mottled black and fulvous-buff. Iris—brown; bill—brown; feet—brown.

Voice: Loud sweet, thrush-like song running down scale then up again on last three notes, usually ending with harsh *chizz*; also occasional husky notes.

Range: Himalayas to S China, SE Asia and Greater Sundas.

Distribution and status: Nominate race breeds in SE Xizang, Yunnan and Guangxi; endemic *diaoluoensis* is found on Hainan. Generally uncommon up to 1400 m.

Habits: Keeps to primary forest where it stays in the canopy and is rather shy. Sometimes cocks tail.

756. Blue-throated Flycatcher *Cyornis rubeculoides* PLATE 81
[>Chinese Flycatcher] Lanhou xianweng

Description: Male: medium-sized (18 cm) blue flycatcher with black lores, white belly and orange-red upper breast. Nominate form is distinguished from Hill Blue Flycatcher by lack of black mask and by blue chin and throat and from Fujian and Rufous-bellied Niltavas by white belly. Female: greyish-brown above with orange-buff throat and buffy eye-ring; separated with difficulty from female Hill Blue Flycatcher by buffy lores and more rufescent tail Race *glaucicomans* has centre of throat orange-red. Iris—brown; bill—black; feet—pink.

Voice: Harsh *chek* calls. Song is sweet, high-pitched trill, *ciccy, ciccy, ciccy, ciccy, see*.

Range: India to SW China and SE Asia.

Distribution and status: Race *glaucicomans* is an uncommon summer breeder in C and SW China, nominate race breeds in SE Xizang up to 2000 m. Vagrants recorded Hong Kong and Beidaihe (Hebei).

Habits: Prefers open forests and hunts near to ground.

Note: Some authors separate race *glaucicomans* as a full species, Chinese Flycatcher, on basis of colour differences and differences in vocalisations (e.g. Viney *et al.* 1994).

757. Hill Blue Flycatcher *Cyornis banyumas* PLATE 81
Shan lanxianweng

Description: Medium-sized (15 cm) blue, orange and white (male) or brownish (female) flycatcher. Male: upperparts dark blue; forehead and short supercilium cobalt; lores, around eye, forecheeks and chinspot black; throat, breast and flanks orange; belly white. Orange chin and entire throat, and lack of shining rump distinguishes from all other orange-breasted blue flycatchers. Female: upperparts brown; eye-ring buff; underparts as male but paler. Distinguished from female Blue-throated by more rufous breast and rufous not buffy throat. Juvenile: brown, mottled with buffy-orange spots on upperparts. Iris—brown; bill—black; feet—brown.

Voice: Sweet melodious warbling song with quite complex phrases. In alarm a harsh *chek-chek*.

Range: Nepal to SW China, Palawan, SE Asia and Greater Sundas.

Distribution and status: Generally uncommon in S Sichuan, Yunnan and S Guizhou in deciduous and open forests up to 2400 m.

Habits: Sits quietly, hunting from low perches.

758. Pygmy Blue Flycatcher *Muscicapella hodgsoni*   PLATE 80
Zhu lanxianweng

Description: Very small (10 cm) narrow-billed flycatcher. Male: blue upperparts with shining blue crown and rump and black mask; orange underparts with paler vent. Female: brown upperparts with rufous rump and tail; underparts pale yellow with buff wash on breast. Iris—dark brown; bill—black; feet—blackish.
Voice: Distinctive high-pitched song *tzzit-che-che-che-cheee*. Call note is feeble *tzip* or low *churr*.
Range: Himalayas, SE Asia, Sumatra and Borneo.
Distribution and status: Nominate race is found in mountains of SE Xizang up to 3000 m and also in W Yunnan (Gaoligong) and SE Yunnan (Jinping).
Habits: Prefers understorey of forest, sometimes visiting ground but usually in middle storey. Has habit of flicking wings open.

759. Grey-headed Canary Flycatcher *Culicicapa ceylonensis*   PLATE 72
[AG: Flycatcher] Fangwei weng

Description: Small (13 cm) distinctive flycatcher with greyish head and slight crest; olive upperparts and yellow underparts.
Iris—brown; bill—black above, horn below; feet—yellowish-brown.
Voice: Song is clear sweet whistle *chic . . . chiree-chilee* with stress on first syllable of each pair and with upward inflection on last note; also rattly *churrru* call and soft *pit pit*.
Range: India to S China, SE Asia, Malay Peninsula and Sundas.
Distribution and status: Race *calochrysea* breeds in C, S and SW China and SE Xizang. Generally common in forests, most common in submontane forests at 1000–1600 m but recorded from the lowlands to 2000 m in the Himalayas. Vagrant Beidaihe (Hebei).
Habits: A noisy active bird, flitting from branch to branch, hunting keenly and chasing flying insects. It regularly flicks its tail open. Generally in lower or middle storeys. Often joins mixed flocks.

ROBINS, FORKTAILS, COCHOAS AND CHATS—Tribe: Saxicolini

A large tribe of thrush-like birds including robins, chats, forktails and other groups. The birds vary greatly in coloration but are mostly small to medium-sized, round-headed, with longish legs, sharp slender bills and broad wings. The tail varies from short to very long but in all species shows some tendency to be cocked periodically. There are 53 species recorded in China. Many are migratory.

760. Eurasian Robin *Erithacus rubecula* Ouya qu   PLATE 77

Description: Medium-sized (14 cm) plump, upright robin very familiar to European birdwatchers. Adult has red face and breast with grey at sides of face and breast; dirty white underparts and brownish above. Juvenile brown, with buff spots above, speckled and scalloped below with rufous-brown wash on breast.
Iris—dark brown; bill—black; feet—brown.

Voice: Song is strong rippling flow of clear plaintive notes with abrupt changes of pitch and tempo. Calls include sharp ticking *tic* or *tic-ic-ic* and thin metallic indrawn *seeek*.

Range: Temperate Europe with migrants from N moving S in winter on N Africa coast and in Middle East.

Distribution and status: Nominate race is occasional visitor to Altai, Tianshan and other parts of N Xinjiang. Recorded Golmud (Qinghai).

Habits: Inhabits mixed woodlands and secondary growth, gardens and in shade. Hops on ground, generally not shy. Sits on low perch to wait for insects and worms but sings from high perch. Nests in holes.

761. **Japanese Robin** *Erithacus akahige* Riben gequ   PLATE 77

Description: Small (15 cm) robin with brown upperparts, orange face and breast and greyish flanks. Male has narrow black gorget surrounding orange bib. Female similar but duller. Immature is scaly and brown. Iris—brown; bill—black; feet—pink.

Voice: The song is distinctive single high note followed by a sweet trill *peen-kararararara*.

Range: Sakhalin and Japan, winters to S China and locally Indochina.

Distribution and status: Scarce winter visitor to S China in forest and woodlands.

Habits: Cocks tail repeatedly.

762. **Ryukyu Robin** *Ericthacus komadori* Liuqiu gequ   PLATE 77

Description: Medium-sized (15 cm) unmistakable robin with rufous-brown upperparts, black face and breast and white underparts with blackish patch on flanks. Black lower breast decorated with fine concentric white necklaces. Female similar to male but duller and with white chin and throat. Iris—dark brown; bill—black; feet—pinkish.

Voice: Song is similar to Japanese Robin but musical although weaker in volume. Also high penetrating *tsee* location call and harsh *kwrick* alarm note given with tail flick and quivering wings.

Range: Endemic to Ryukyu Is chain of Japan.

Distribution and status: Nominate race is migrant and stragglers have reached Taiwan.

Habits: Typical of genus.

763. **Rufous-tailed Robin** *Luscinia sibilans* Hongwei gequ   PLATE 77

Description: Small (13 cm) rufous-tailed robin; graceful but rather nondescript. The upperparts are olive-brown with a rufous tail and the underparts are whitish scalloped olive on the breast. Distinguished from other female robins and flycatchers by rufous tail. Iris—brown; bill—black; feet—pinkish-brown.

Voice: Has a short sweet song.

Range: NE Asia wintering to S China.

Distribution and status: Uncommon migrant through most of eastern China in autumn and winter.

Habits: Rather terrestrial, keeps to floor or low vegetation in dense shady patches of forest and shivers tail energetically.

764. Common Nightingale *Luscinia megarhynchos*   PLATE 77
Xinjiang gequ

Description: Large (16.5 cm) warm brown robin with rufous tail and whitish underparts. Side of neck and flanks grey buff; vent yellowish-buff. Narrow pale eye-ring and indistinct short greyish supercilium. Rounded body with slim bill. Iris—dark brown; bill—brown; feet and legs—pale yellowish-pink.

Voice: Song is famous and admired—far-carrying clear whistles and rattling with much variety and fast tempo. Calls include creaking *errrk*, loud indrawn *hweet* and hard *chack*.

Range: S Europe and N Africa through Middle East to Turkestan, NW India and extreme W China.

Distribution and status: Race *hafizi* is a rare breeding bird of W Tianshan, C Turpan and N Fuhai in W Xinjiang in dense undergrowth of open deciduous forests.

Habits: Secretive in dense low scrub and bushes usually within 2 m of ground. Strong hops on ground with flicked wings and tail half erect and flicked from side to side. Song from dense cover often at night, hence English name.

765. Siberian Rubythroat *Luscinia calliope* Honghou gequ   PLATE 77

Description: Medium-sized (16 cm) plump brown robin with bold white supercilium and malar stripe. Upperparts: brown without rufous in tail; underparts: buff on flanks; buffy-white on belly. Female has brownish breast band. Black and white striped head pattern distinctive. Adult male has diagnostic red throat. Iris—brown; bill—dark brown; feet—pinkish-brown.

Voice: Loud, falling double whistle *ee-uk*, soft deep *ischuck* in alarm. Song is long scratching warble.

Range: Breeds in NE Asia. Migrates in winter to India, S China, SE Asia.

Distribution and status: Breeds in NE China and from NE Qinghai to S Gansu and Sichuan. Winters in S China, Taiwan and Hainan. Locally not uncommon.

Habits: A skulker of thickets in forest and secondary growth; generally near streams.

766. White-tailed Rubythroat *Luscinia pectoralis*   PLATE 77
Heixiong gequ

Description: Male: compact (14.5 cm) elegant grey robin with ruby throat and broad black breast band and white supercilium. Upperparts uniform grey. Tail has black central feathers but otherwise white base and tip. Underparts whitish tinged grey ventrally. Widespread race *tschebaiewi* has bold white malar streak in male less obvious in female. Female is browner with white throat and grey breast band. Races *ballioni* and *confusa* have no white malar streaks; a smaller red throat patch and *confusa* has darker grey upperparts. Iris—dark brown; bill—black; feet—brownish-black.

Voice: Loud, shrill, attractive song repeated for long periods all through the day. Sings with head thrown back, throat puffed out, wings held low and tail half cocked and fanned and flicked. Calls include harsh metallic notes, harsh *it-it* near nest and long drawn-out *siiii siiii* in alarm.

Range: Himalayas, Turkestan, NE India, Bangladesh, NW Burma and W, SW and C China.

Distribution and status: Race *tschebaiewi* is fairly common in S and E Xizang to E Qinghai, Gansu, W and NW Sichuan and NW Yunnan. Migrates S in winter. Race *confusa* is rarer in S and SW Xizang. Race *ballioni* is not uncommon in W Kashi and Tianshan region of W Xinjiang. Occupies scrub and bushes of subalpine forest to above tree-line in summer; descends in winter.

Habits: Sings from prominent low perch but otherwise shy and keeps close to or hops on ground.

767. Bluethroat *Luscinia svecica* Lanhou gequ   PLATE 77

Description: Male: colourful medium-sized (14 cm) robin with diagnostic chestnut, blue, black and white pattern on throat and whitish supercilium. Upperparts grey-brown; underparts white; diagnostic tail dark brown with side feathers basally rufous also visible in flight. Female lacks orange and blue on throat and has white throat with malar streaks of black spots connected to breast band of merging black spots. Distinguished from female rubythroats by tail pattern. Races vary in size of red throat spot (smallest in *abbotti*), deepness of blue (dark in *saturator*, pale in *svecica*) and presence or not of black band between blue and chestnut breast bands (*svecica*). Juvenile is warm brown with rusty yellow spots. Iris—dark brown; bill—dark brown; feet—pinkish-brown.

Voice: Rich song full of bell-like sounds increasing in tempo and including mimicked part of other species songs. Sometimes calls at night. Alarm call is *heet* like a wheatear. Contact call is harsh *truk*.

Range: Palearctic and Alaska migrating S in winter to India, China and SE Asia.

Distribution and status: Races *saturatior* and *kobdensis* breed in NW China and nominate breeds in NE China with wintering populations in SW China and SE China. Other races pass through China on migration—notably *przevalksii* which breeds in Siberia and perhaps Inner Mongolia and Qinghai and migrates through C China and *abbotti* which breeds in the W Himalayas and has been recorded in W Xizang. The species can be quite common during migration in all types of scrub from tundra, forest, swamps and edges of desert.

Habits: Shy and keeps to dense cover usually close to water. Feeds mostly on ground. Runs with hopping gait; halting to raise head and flash tail. Upright stance. Flight quick and direct to sweep into cover.

768. Rufous-headed Robin *Luscinia ruficeps* Zongtou gequ   PLATE 77

Description: Male: beautiful, colourful medium-sized (14.5 cm) robin with diagnostic chestnut crown and nape with white chin and throat bordered by black band. Upperparts brownish-grey; tail chestnut with blackish tip and central feathers like Bluethroat. Underparts whitish with grey band across breast and down flanks. Female similar to female Siberian Blue Robin but sides of head and neck dark brown and throat scaly. Iris—dark brown; bill—black; feet—pink.

Voice: Outstanding. Very powerful, loud, rich, throaty, melodious well-spaced phrases, preceded by short introductory note, *ti chulululu ti chewtchewtchewt ti titichewtchewtchewt ti cho-chutchutchut ti tiii-chuchutwilili*, etc. (C. Robson).

Range: C China, recorded as winter visitor or vagrant to Malay Peninsula.

Distribution and status: Probably more widespread than formerly believed in Qinling, N Minshan and perhaps Qionglai Mts. Generally rare but fairly common in parts of Jiuzhaigou reserve, N Sichuan, in dense scrubby subalpine forest from 2000–3000 m.

Habits: Sings in May; feeds on worms and plant material.

769. Blackthroat *Luscinia obscura* [Black-throated Blue Robin]   PLATE 77
Heihou gequ

Description: Male: small dark robin (14 cm) with yellowish-white belly and white flashes on base of tail. Crown, back, wings and rump slaty-blue; face, breast, uppertail coverts, centre and tip of tail black. Female dark olive-brown with pale buffish underparts, distinguished from female Siberian Blue Robin by unscaled underparts, buffy undertail coverts and rufous tinge to tail. Iris—dark grey; bill—black; feet—pinkish-grey.

Voice: Relatively simple phrases repeated after shortish intervals, *drreee-drreee drreee-drreee drreeedreee huti-huti huti-huti huti-huti*. Also combinations of more slurred, then higher notes, *duriiii'hutu, drrii'hitu, drrrii'huti-huti-huti* and some short trills and simple warbling. Song often interspersed with weak low single or double *tuc* call notes (C. Robson).

Range: NC China, vagrant to Indochina.

Distribution and status: Very rare in SE Gansu, Qinling Mts of S Shaanxi in subalpine conifer forest 3000–3400 m. On migration in Yunnan, once in extreme N Thailand.

Habits: Stays near ground in bamboo thickets. Flicks tail.

Note: Sometimes treated as race of *L. pectardens* (e.g. Meyer de Schauensee 1984).

770. Firethroat *Luscinia pectardens* Jinxiong gequ   PLATE 77

Description: Male: small chat (14.5 cm) with dark upperparts; dirty white belly, fiery orange-red breast and throat and whitish patch on side of neck. Upperparts slaty-brown with blackish-brown wings and tail and black sides of head and neck. Base of tail has white flashes. Female: brown; lacks tail flashes; underparts ochraceous with centre of abdomen white. Juvenile dark brown, spotted and with no white tip to tail. Iris—dark brown; bill—black; feet—brownish-pink.

Voice: Lengthy song is loud, sweet and varied with each note repeated several times, full of phrases mimicked from other species and interspersed with harsh notes. Alarm note is throaty *tok*.

Range: C and SW China; winters in NE India and NE Myanmar.

Distribution and status: Rare. Breeds in W Sichuan (Emei, Baoxing, Kanding, Maowen, Wenchuan), Yunnan (Lijiang) in forest from 3000–3500 m. Winters and may breed in SE Xizang.

Habits: Skulks in dense scrub and bamboo. Feeds on insects on forest floor. Flicks tail.

771. Indian Blue Robin *Luscinia brunnea* Lifu gequ   PLATE 77

Description: Male: small dark robin (15 cm) with slaty-blue upperparts; white supercilium and chestnut throat, breast and flanks. Lores and cheeks black. Centre of belly and undertail coverts white. Female: olive-brown above with whitish underparts washed

ochraceous on breast and flanks. Juvenile is dark brown with buff spots. Iris—brown; bill—black in summer, brown above and pinkish below in winter; feet—pinkish-brown.
Voice: Song is composed of 3–4 deliberate deep notes rising in volume followed by explosive trill of 4–5 quick sweet notes. Alarm is guttural *tuk-tuck*, high-pitched *tsee* and churring.
Range: Himalayas to C China and C Myanmar, migrates S in winter to SW India, Sri Lanka and Bangladesh.
Distribution and status: Uncommon resident in SE Gansu, N and W Sichuan, Mt Taibai of Shaanxi, NW Yunnan and SE Xizang in montane oak forests from 1600–3200 m. Descends in winter to evergreen forests.
Habits: Keeps to dense bamboo and rhododendron thickets. Flicks wings and tail.

772. Siberian Blue Robin *Luscinia cyane* [Siberian Bluechat] PLATE 77
Lan gequ

Description: Medium-sized (14 cm) blue and white or brown robin. Male unmistakable with slaty-blue upperparts and broad black band through eye and down side of neck to side of breast; underparts white. Female olive-brown upperparts with throat and breast brown scaled buff; rump and uppertail coverts suffused with blue. Immature birds and some females have some blue on tail and rump. Iris—brown; bill—black; feet—pinkish-white.
Voice: In winter a hard, low *tak*, also loud *se-ic*.
Range: Breeds in NE Asia; migrant in winter to India, S China, SE Asia and Greater Sundas.
Distribution and status: Nominate race breeds in Heilongjiang. On passage through C China to winter in SW and S China. Race *bochaiensis* on passage through E and SE China to winter in S China. Seasonally common in forest up to 1800 m.
Habits: Stays on or near the ground in dense forest.

773. Orange-flanked Bush Robin *Tarsiger cyanurus* PLATE 76
[Siberian/Red-flanked Bush-Robin/Blue-tailed Robin, AG: Bluetail]
Honglei lanweiqu

Description: Smallish (15 cm) white-throated robin with diagnostic orange flanks contrasting with white belly and vent. Male upperparts blue with white supercilium; immature and female brown with blue tail. Female is distinguished from female Siberian Blue Robin by white mesial stripe on brown throat rather than entire throat white, also orange rather than buff flanks. Race *rufilatus* has bright ultramarine on rump, lesser coverts and supercilium; greyer throat. Iris—brown; bill—black; feet—grey.
Voice: Call is quiet single or double croaking *chuck* Song is soft and weak *churr-chee* or *dirrh-tu-du-dirrrh*.
Range: Breeds in NE Asia and Himalayas. Migrates in winter to S China and SE Asia.
Distribution and status: Nominate race breeds in Heilongjiang; on passage through E China to winter S of Changjiang valley and on Taiwan and Hainan. Race *rufilatus* breeds from E Qinghai to S Gansu, S Shaanxi, Sichuan and E Xizang to winter in S Yunnan and SE Xizang.
Habits: Keeps low in undergrowth of damp montane forest and secondary growth.

774. Golden Bush Robin *Tarsiger chrysaeus* Jinse linqu   PLATE 76

Description: Small (14 cm) elegant robin. Male crown and upper back olive-brown. Yellow supercilium; black band from lores through eyes and cheeks. Scapulars, sides of back and rump bright orange; wings olive-brown. Tail orange; central rectrices and terminal band black. Underparts entirely orange. Female: above olive with ill-defined yellowish supercilium and buff eye-ring; underparts ochre-yellow. Iris—brown; bill—dark brown with yellow below; feet—pale flesh.

Voice: Alarm note is croak, *trrr*, also scolding *chirik chirik*. Wispy, high-pitched song is *tse, tse, tse, tse, tse, chur-r-r* or *tze-du-tee-tse chur-r-r*.

Range: Himalayas, NE India, Burma and SW China, wintering to NE Burma, N Thailand and N Vietnam.

Distribution and status: Uncommon. Nominate race breeds in S and E Xizang, W Sichuan, S Qinghai, S Gansu, S Shaanxi and NW Yunnan. Recorded in March in Xishuangbanna, S Yunnan. Found in summer in conifer forests, and rhododendron scrub near treeline from 3000–4000 m; descending in winter to lowland scrub.

Habits: Altitudinal migrant, skulker in winter.

775. White-browed Bush Robin *Tarsiger indicus*   PLATE 76
Baimei linqu

Description: Small dark robin (14 cm) with bold white supercilium. Male: slaty-blue upperparts, black side of head and orange-rufous underparts. Whitish centre of belly and undertail coverts. Female olive-brown above; white supercilium; brown cheeks; pale eye-ring and dull rufous-ochre underparts, paler on belly and buffy undertail coverts. Distinguished from Indian Blue Robin and Snowy-browed Flycatcher by larger and broader supercilium. Male of race *indicus* sometimes breeds in brown, female-like plumage. Male of race *formosanus* has yellowish-olive crown and olive-brown underparts. Iris—dark brown; bill—blackish; feet—greyish-brown.

Voice: Distinctive churring titter. Call is sweet *tiut-tiut* answered by sharper note. Song is rapidly repeated sharp note on ascending and descending scale or double-phrased bubbling *shri-de-de-dew . . . shri-de-de-dew*.

Range: Nepal to SW China, Taiwan, NE Burma and NW Vietnam.

Distribution and status: Race *indicus* is resident in SE Xizang; *yunnanensis* in NW and W Sichuan and NW Yunnan; and *formosanus* in Taiwan. Uncommon in mixed and conifer forests 2400–4300 m.

Habits: On or close to ground in dense undergrowth, fairly tame. Male displays with drooped quivering wings and flicking tail.

776. Rufous-breasted Bush Robin *Tarsiger hyperythrus*   PLATE 76
[Rufous-bellied Bush Robin] Zongfu linqu

Description: Small (14 cm) dark blue and orange robin. Male: upperparts dull blue with bright ultramarine forehead, eyebrow, 'shoulder' and uppertail coverts, side of head black; underparts orange-rufous with white centre of belly and undertail coverts. Female: upperparts rufescent olive-brown with slaty-blue rump and uppertail coverts and blackish blue-edged tail; underparts olive-brown with rufous tinge on flanks and vent; brown down centre of breast; and undertail coverts white. Iris—brown; bill—black; feet—grey.

Voice: Call is a *duk-duk-duk-squeak*; song is lisping warble, *zeew . . . zee . . . zwee . . . zwee.*

Range: E Himalayas to NE India and N Burma.

Distribution and status: Rare Breeds in SE Xizang and extreme W Yunnan.

Habits: Similar to Orange-flanked Bush Robin. Uses lower forest storey or sits in open on perch. Quite tame.

777. Collared Bush Robin *Tarsiger johnstoniae* PLATE 76
Taiwan linqu

Description: Small (12 cm) robin. Male: long white eyebrow on otherwise sooty black head and broad orange-red gorget which divides to form orange-red hind collar and orange-red bar across scapulars. Back, wings and tail sooty. Abdomen pale grey and vent white. Female duller, upperparts olive-grey, chin grey; underparts yellowish-buff. Eyebrow fainter than male. Immature: streaky brown with long buff eyebrow. Iris—dark; bill—horn; feet—dark brown.

Voice: Gives continuous *pi-pi-pi . . .* call.

Range: Endemic to Taiwan.

Distribution and status: Fairly common in main mountains at 2000–2800 m.

Habits: Frequents forest undergrowth and forest edge.

778. Oriental Magpie Robin *Copsychus saularis* PLATE 82
[<Magpie Robin] Que qu

Description: Medium-sized (20 cm) black and white robin. Male: head, chest and back shiny blue-black; wings and central tail feathers black; outer tail feathers and stripe across wing coverts white; belly and vent white. Female as male but dull grey instead of black. Immature similar to female but mottled. Iris—brown; bill and feet—black.

Voice: Plaintive *swee swee* call and harsh *chrrr.* Varied lusty singing including imitations of other birds but lacking the rich tone of White-rumped Shama.

Range: India, S China, Philippines, SE Asia and Greater Sundas.

Distribution and status: Common in lowlands to 1700 m. Race *prosthopellus* is resident in most of China S of 33 °N. Race *erimelas* occurs in SE Xizang and W Yunnan. Rare in some areas as a result of trapping for cagebird trade.

Habits: A familiar bird of gardens, villages, secondary forest, open forest and mangroves. Conspicuous in flight and perches conspicuously to sing or display. Feeds mostly on the ground where it constantly lowers and fans its tail before jerking it shut and upright again.

779. White-rumped Shama *Copsychus malabaricus* PLATE 82
Baiyao quequ

Description: Largish (27 cm) long-tailed, black, white and rufous robin. Male: head, neck and back black with blue gloss; wings and central tail feathers dull black; rump and outer tail feathers white; belly orange-rufous. Female: similar but black replaced by grey. Iris—dark brown; bill—black; feet—pale flesh.

Voice: Call is *chur-chi-churr* and harsh scolding note. Rich, complex melodious song including imitations of other birds.

Range: India to SW China, SE Asia and Greater Sundas.

Distribution and status: Uncommon up to 1500 m in tropical forests. Resident in SE Xizang (*indicus*), SW and S Yunnan (*interpositus*) and on Hainan (*minor*). Becoming rare as a result of trapping for cagebird trade.

Habits: Shy, keeps to thickets in denser forest. Sings lustily in morning and evening from a low perch with wings drooped and tail held high. Hops on the ground or makes short flights through undergrowth, flicking its long tail on landing.

780. Alashan Redstart *Phoenicurus alaschanicus* PLATE 78
Helanshan hongweiqu

Description: Medium-sized (16 cm) redstart with rufous breast. Male: crown, nape, sides of head to upper back blue-grey. Lower back and tail, except brown central tail feathers, rufous-orange. Chin, throat and breast rufous-orange; abdomen paler orange almost white. Wing brown with white patch. Very similar to Rufous-backed Redstart but with crown, sides of head and nape bluish-grey. Female: browner and duller above; underparts grey instead of rufous; wings brown with buff patch. Iris—brown; bill—black; feet—blackish.

Voice: No information. Probably similar to Rufous-backed Redstart.

Range: NC and W China endemic.

Distribution and status: Rare breeding bird in montane conifer forests of Qinghai (Xining, Tianjin and Qaidam Basin), Ningxia (Helan Mts) and E Gansu. Winters in S Shaanxi, and along Hebei–Shanxi border and accidentally Beijing.

Habits: Prefers dense bushes and loose rocky slopes in mountains.

Notes: Probably should be regarded as a subspecies of Rufous-backed Redstart.

781. Rufous-backed Redstart *Phoenicurus erythronota* PLATE 78
[Eversmann's Redstart] Hongbei hongweiqu

Description: Male: medium-sized (15 cm) colourful redstart with rufous throat, breast, back and uppertail coverts. Crown and nape grey with pale supercilium contrasting with black lores, band through eye, cheeks and 'shoulders'. Wings blackish with long white streak. Tail rufous with central two feathers brown. Belly and undertail coverts white. Female rich brown with tail as in male: buff eye-ring, throat, wing streak and edges to tertials; undertail white. Iris—brown; bill—black; feet—greyish-black.

Voice: Loud lively song; croaking *gre-er* in alarm and slurred call note, also louder whistled call *few-eet*.

Range: C Asia and S Siberia. Winters from Turkestan to S Iraq, Baluchistan and W Himalayas.

Distribution and status: Rare; breeding in W Xinjiang in Kashi, Aksu, Tianshan, Yutian, Urumqi and Altai Mts in subalpine conifer forests. Winters in plains.

Habits: Males highly vocal in spring. Tail is flitted up and down, not shivered sideways. Winters in arid scrub and woodlands at moderate altitudes.

782. Blue-capped Redstart *Phoenicurus coeruleocephalus* PLATE 78
[Blue-headed Redstart] Lantou hongweiqu

Description: Male: medium-sized (15 cm) beautiful, black and white redstart with blue-grey crown and nape diagnostic. White wing bar, edges of tertials and lower breast, belly and undertail coverts. Upperparts tinged brown in winter. Female brown with buffy eye-ring, white belly and undertail coverts, rufous uppertail coverts and brown tail narrowly edged rufous. Iris—brown; bill—black; feet—blackish.

Voice: Eurasian Robin-like *tik-tik* call note; piping *tit, tit, tit* in alarm near nest. Song is loud fast and high-ringing reminiscent of Rock Bunting.

Range: C Asia from Afghanistan to Altai Mts and Himalayas.

Distribution and status: Rare resident in W Xinjiang (Kashi, Tianshan, W Altai and Urumqi) from 2400–4300 m. Winters 1200–3000 m.

Habits: Breeds in rock crevices and lives in scrub on rocky hillsides in montane conifer forests. Winters in pine forests, scrub and olive groves. Feeds in canopy and on ground. Shakes rather than shivers tail.

783. Black Redstart *Phoenicurus ochruros* PLATE 78
Zhe hongweiqu

Description: Medium-sized (15 cm) dark redstart. Male (*rufiventris*) typically has black head, throat, upper breast, back, wings and central tail feathers; grey crown and nape; rufous lower breast, belly, undertail coverts, rump and outer tail feathers. Western race *phoenicuroides* is paler with black plumage replaced by dark grey and wings brownish-grey. Race *xerophilus* is intermediate between these two. Female: like female Daurian but lacks white wing patch and buffy eye-ring less conspicuous. Underparts sometimes washed rufous, and lores buffy (c.f. Rufous-tailed Flycatcher). Iris—brown; bill—black; feet—blackish.

Voice: Call note *tseep* frequently precedes alarm *tucc-tuee* or *tititicc*. Song is loud power-ful trill of 6–7 notes followed by curious crackling ending. Described as pouring lead shot into a bottle; generally given at night or at dawn from prominent perch.

Range: Palearctic; winters to NE Africa and SE China.

Distribution and status: Generally common and widespread breeder and winter vis-itor. Race *phoenicuroides* across W Xinjiang and W Xizang; *rufiventris* breeds E and W Xizang, Qinghai and Gansu to Shanxi, Sichuan and NW Yunnan. Winter birds recorded in Hebei, Shandong, Hong Kong and Hainan. Dubious race *xerophilus* resident in Kunlun Mts, Qilian Mts and Donghe Nanshan of S Xinjiang.

Habits: Found in open terrain at all altitudes. Highly territorial. Hunts from perches. Bobs head and shivers tail. Familiar around houses, gardens and farmland. Hops briskly and stands up erect.

784. Common Redstart *Phoenicurus phoenicurus* PLATE 78
Ouya hongweiqu

Description: Smallish (15 cm) colourful redstart. Male: white forehead and eyebrow separates grey crown, nape and mantle from black lores, face and throat. Wing brown without white patch. Breast, rump and outer tail feathers rufous. Central tail feathers dark brown. Belly and undertail coverts buff. Combination of white vent with no white wing patch diagnostic. Female is brown with rufous rump and outer tail feathers; buff lores and eye-ring; buff chin and buff belly and under tail coverts. Iris—brown; bill—black; feet—blackish.

Voice: Song is clear melancholy verse with *hiiit* grace note and rolling *tuee-tuee-tuee-tuee*. Alarm call is diagnostic *hueet-tic-tic*.

Range: N Europe to N Africa and east to Lake Baikal, Transcaspia and Altai Mts. Winters to Arabia, Africa, Middle East and Baluchistan.

Distribution and status: Marginally extends into China; breeding in extreme W Xinjiang in Tarbagatai Mts, perhaps also Altai Mts.

Habits: Similar to Black Redstart but generally in more wooded habitats. Shivers tail in typical redstart manner. Flies in short, quick dashes with tail flash. Undulates on longer flights. Trails tail.

785. Hodgson's Redstart *Phoenicurus hodgsoni*   PLATE 78
Heihou hongweiqu

Description: Medium-sized (15 cm) colourful redstart. Male: similar to Daurian but white eyebrow; grey of nape extends onto mantle and white wing patch a narrower streak. Separated from *phoenicuroides* race of Black Redstart by white forecrown and wing patch. Female: similar to female Daurian but has whitish rather than buff eye-ring, greyer breast and lacks white wing patch. Rather darker above than female Black Redstart. Iris—brown; bill—black; feet—blackish.

Voice: Call is clicking *prit*; alarm note is rattling *trrr, tschrrr*. The song is short and tinny with little modulation.

Range: Himalayas, Tibetan Plateau to C China. Winters to NE India and N Burma.

Distribution and status: Breeds from 2700–4300 m in S and SE Xizang, E Qinghai, Gansu, S Shaanxi, W Sichuan, NW Yunnan. Winters to Hubei, Hunan, E Sichuan and E Yunnan. Fairly common.

Habits: Prefers grass and scrub in open forests, often near streams where behaviour reminiscent of Plumbeous Redstart. In canopy, hunts like a flycatcher.

786. White-throated Redstart *Phoenicurus schisticeps*   PLATE 78
Baihou hongweiqu

Description: Medium-sized (15 cm) colourful redstart with diagnostic white throat patch and rufous on outer tail feathers only on basal half. Male: Crown and nape dark slate-blue with brighter blue forehead and eyebrow; mantle greyish-black; tail mostly black; lower back rufous; centre of belly and vent buffy-white; extensive white streak on wings with white edges to tertials. Female: crown and back tinged with brown in winter; buffy eye-ring; tail white, throat patch and long white wing streak as in male. Spotted juvenile easily distinguishable by white throat patch. Iris—brown; bill—black; feet—black.

Voice: Alarm note is long drawn-out *zieh* followed by rattling note. Song undescribed.

Range: C China, Tibetan Plateau. Some winter in NE India and N Burma.

Distribution and status: Breeds from 2400–4300 m in Qinling Mts of S Shaanxi, S Gansu, E and SE Qinghai, Sichuan to N W Yunnan and SE Xizang.

Habits: Singly or in pairs occupying thick scrub in subalpine conifer forests in summer descending villages and lowlands in winter. Rather flighty and wild. Forms small flocks during migration.

787. Daurian Redstart *Phoenicurus auroreus* PLATE 78
Bei hongweiqu

Description: Medium-sized (15 cm) colourful redstart with prominent broad white wing patch. Male: lores, side of head, throat, mantle and wings brownish-black apart from white wing patch. Crown and nape grey with silvery edges. Rest of plumage rufous-chestnut with darker rufous-black central tail feathers. Female brown with prominent white wing patch, buffy eye-ring and tail like male but duller. Vent sometimes rufous. Iris—brown; bill—black; feet—black.

Voice: Call is series of soft whistles followed by soft *tac-tac* notes; also short sharp whistled *peep* or *hit, wheet*. Song is cheerful series of whistled notes.

Range: Resident NE Asia and China migrating to Japan, S China, Himalayas, Burma and N Indochina.

Distribution and status: Nominate race breeds in NE China (Hebei) and recorded in Shangdong and Jiangxi Mts. Winters in S and SE China, Taiwan and Hainan. Dubious race *leucopterus* breeds in E Qinghai, Gansu, Ningxia, Qinling Mts of Shaanxi, N and W Sichuan, N Yunnan and SE Xizang. Winters in S Yunnan. Generally common.

Habits: Occupies subalpine forest, scrub and clearings in summer but deciduous bushy lowlands and cultivated lands in winter. Sits on prominent perch and shivers tail.

788. White-winged Redstart *Phoenicurus erythrogaster* PLATE 78
[Güldenstädt's Redstart] Hongfu hongweiqu

Description: Large (18 cm) boldly coloured redstart. Male: like Daurian but larger with greyish-white crown and nape and chestnut tail feathers. White wing patch very large. Black areas fringed ashy in winter. Female: like female Common Redstart but larger and with much less contrast between brown central feathers and rufous-brown feathers of tail. No wing patch in Common Redstart. Spotted young have prominent white wing patch. Iris—brown; bill—black; feet—black.

Voice: Calls include weak *lik* and harder *tek*. Short and clear whistled song *tit-tit-titer* followed by wheezy burst of short notes, delivered from prominent perch or during display flight.

Range: Caucasus, C Asia, Turkestan, Himalayas, NW and C China and Tibetan Plateau.

Distribution and status: Mountains of W and NW China—Xizang and Qinghai to S Gansu and Qinling Mts of S Shaanxi in open rocky alpine country from 3000–5500 m. Winters to Hebei, Shanxi, S Sichuan and N Yunnan.

Habits: Hardy, high altitude redstart. Shy and solitary. In display, male soars from prominent perch, quivering wings displaying striking white wing patch. Sometimes scavenges insects from dead animals. In winter females descend lower altitudes but males remain at quite high altitudes, sometimes feeding in snow.

789. Blue-fronted Redstart *Phoenicurus frontalis* PLATE 78
Lan'e [changjiao]diqu

Description: Medium-sized (16 cm) colourful redstart. Both sexes have distinctive tail pattern with T-shaped black (brown in female) pattern formed by terminal portion of central feathers and tips of other feathers contrasting with bright rufous rest of tail. Male:

head, breast, nape and mantle dark blue with cobalt forehead and short eyebrow; wings blackish-brown with feathers edged brown and buff but no white patch; abdomen, vent, back and uppertail coverts rufous-orange. Female: brown with buff eye-ring. Distinguished from similar female redstarts by dark-tipped tail. Iris—brown; bill—black; feet—black.

Voice: Single clicking *tic*. Alarm is soft, frequently repeated *ee-tit, ti-tit* from perch or in flight. Song is series of sweet warbling and harsh grating notes, similar to but less wheezy than Black Redstart.

Range: C China, Tibetan Plateau, Himalayas; wintering to SW Burma and N Indochina.

Distribution and status: Fairly common. Breeds in S Xizang, E and S Qinghai, S Gansu, Ningxia, Qinling Mts of S Shaanxi, Sichuan, Guizhou and Yunnan at high altitudes in alpine zone. Moves to lower altitudes within breeding range in winter with some southerly movement.

Habits: Generally solitary but forms small groups during migration. Pounces on insects from perch. Flicks tail up and down rather than shivers it. Rather tame.

790. White-capped Water Redstart *Chaimarrornis leucocephalus*   PLATE 78
[AG: Redstart/Chat/River Chat] Baiding xiqu

Description: Unmistakable, large (19 cm) black and chestnut redstart with white crown and nape. The rump, base of tail and belly are chestnut. Sexes alike. Immature is duller and brownish with the crown scaled black. Iris—brown; bill—black; feet—black.

Voice: The call is shrill and loud, a rather plaintive, rising *tseeit tseeit*. Song is weak undulating whistle.

Range: C Asia, Himalayas, China, wintering to India and Indochina.

Distribution and status: Very common along mountain streams and rivers over most of China and in the Himalayas. Breeds in headwaters up to high altitudes of over 4000 m but moves downriver in winter.

Habits: Habitually sits on prominent rocks in or close to water, bobbing head on alighting and flicking black-tipped tail continuously. Courtship involves curious head-waving display.

791. Plumbeous Water Redstart *Rhyacornis fuliginosus*   PLATE 78
Hongwei shuiqu

Description: Small (14 cm) sexually dimorphic chat living along streams. Male: rump, vent and tail rufous-chestnut; rest of plumage dark slaty-blue. Distinguished from most other redstarts by lack of dark central tail feathers. Female: grey above with pale eye-ring; white underparts scaled with grey edges with white vent and rump and white bases to outer tail feathers. Rest of tail black. Wings black with narrow white tips to coverts and tertials. Distinguished from Little Forktail by lack of notch in tail and lack of white crown and wing bar. Both sexes constantly flirt conspicuous tails. Juvenile grey spotted with white above. Daurian race *affinis* has rufous uppertail coverts in male, and female has less white in tail and scaling on underparts confined to centre of belly. Iris—dark brown; bill—black; feet—brown.

Voice: Sharp whistled *ziet, ziet* call note; threatening *kree* in territorial threat. Song is a short, rapid, metallic jingle *streee-treee-tree-treeeh* given from rock perch or in flight.
Range: Pakistan and Himalayas to China, Hainan, Taiwan and N Indochina.
Distribution and status: Common altitudinal migrant along fast-flowing streams and clear rivers from 1000–4300 m. Race *fuliginosus* in S Xizang, Hainan and all of southern China as far N as Qinghai, Gansu, Shaanxi, Shanxi, Henan and Shandong. Race *affinis* on Taiwan from 600–2000 m.
Habits: Solitary or in pairs. Almost always beside boulder-strewn streams and rivers, or perched on boulder in water. Flits from rock to rock flirting tail. Displays consist of fluttering hover with outspread tail, spiralling back down to perch. Highly territorial but often with dippers, water redstarts or forktails.

792. White–bellied Redstart *Hodgsonius phaenicuroides* PLATE 78
[Hodgson's Shortwing] Baifu duanchiqu

Description: Large (18 cm) long-tailed, redstart-like bird with rufous base to outer tail feathers. Short wings barely reach base of tail. Male: head, breast and upperparts slaty-blue; abdomen white and undertail coverts black tipped white. Tail long and wedge-shaped. Wings greyish-black with two small but conspicuous white spots on primary coverts. Brown phase males also occur. Female: olive-brown with buff eye-ring and paler underparts. Iris—brown; bill—black; feet—black.
Voice: Low *chuck*. Alarm is robin-like *tsiep tsiep tk tk* or *tck-tck sie*. Song is loud, melancholy whistled phrase of three notes with middle note more prolonged and higher-pitched and last note a half tone lower *he did so*. Sings in summer at dawn, dusk and on moonlit nights.
Range: Himalayas, Burma, N Indochina to C China; winters in E Burma and NW Thailand.
Distribution and status: Fairly common altitudinal migrant, resident in suitable habitat in E Qinghai, S and W Gansu, Ningxia, Qinling Mts of S Shaanxi, Hubei, Hebei, Sichuan, W Guizhou, W Yunnan and SE Xizang. In summer lives above or close to treeline 2200–4300 m, but as low as 1300 m in winter.
Habits: Keeps to dense scrub on or near ground and hard to flush or see except when calling from perch with tail erect and fanned. Rather noisy.

793. White–tailed Robin *Myiomela leucura* Baiwei landiqu PLATE 76

Description: Male is a large (18 cm), very dark blue robin usually appearing black with white flashes in base of tail and cobalt-blue forecrown. Throat and breast are dark blue with white spots on side of neck and breast usually concealed Female is brown with whitish transverse band on base of throat. Tail has same white flashes as male. Immature is like female but with rufous streaks. Iris—brown; bill—black; feet—black.
Voice: Call is weak sweet song of 7–8 whistled notes. Calls include thin whistles and low *tuc*.
Range: India, SE Asia and S China.
Distribution and status: Nominate race is resident in SE Xizang, C and SW China, N Guangdong and Hainan. Probably also SE China. Race *montium* is resident on Taiwan. Breeds in montane forests above 1000 m but migrates to lowlands in winter. Recorded once Hong Kong, possibly as an escape.

Habits: This is a shy bird of very dark thickets in evergreen forests.
Note: Formerly classed in genus *Cinclidium* (e.g. Cheng 1987).

794. **Blue-fronted Robin** *Cinclidium frontale* PLATE 76
[Blue-fronted Long-tailed Robin] Lan'e [changjiao]diqu

Description: Large (19 cm) dark robin with no white in long graduated tail. Similar to respective sexes of White-tailed Robin but lacks white patches on neck and tail. Shining blue of forehead, eyebrow and 'shoulder' duller than White-tailed Robin or *Niltava* flycatchers. Also distinguished from niltavas by lack of blue on rump or side of neck. Female has whitish full eye-ring and whitish centre of throat and belly. Iris—dark brown; bill—black; feet—black.

Voice: Song of short, melodic clear phrases e.g. *tweee-ke-tui*; harsh buzzing alarm call.

Range: E Himalayas to C China and N Indochina.

Distribution and status: Rare. Recorded in Sichuan (*orientale*) at Shimien on the Tatuho river and Dafending Panda reserve, probably occurs in S Yunnan. Must also occur in SE Xizang (nominate).

Habits: Occupies bamboo in broadleaf evergreen forest. Secretive habits and dark colour may explain why so rarely seen.

Note: Meyer de Shauensee (1984) ascribed Sichuan record to nominate race *contra* Cheng (1987, 1994).

795. **Grandala** *Grandala coelicolor* Lan dachiqu PLATE 82

Description: Medium-sized (21 cm) thrush-like bird. Male: unmistakable being entirely bright purple-blue with silky sheen, except black lores, wings and tail. Tail slightly forked. Female: greyish-brown above with buff streaking on head to mantle. Underparts greyish-brown, streaked buff on throat and breast. Wings with basal inner remiges white, conspicuous in flight. Wing coverts with small white tips. Rump and uppertail coverts tinged blue. Iris—brown; bill—black; feet—black.

Voice: Rather quiet. Call note is *tji-u* and song is soft clear *tji-u tji-u ti-tu tji-u* audible only at close range, uttered from rock perch.

Range: E Himalayas, E Xizang plateau.

Distribution and status: Uncommon or locally common resident in mountains of S and SE Xizang, NW Yunnan, E Qinghai, W Gansu and W Sichuan. Altitudinal migrant with summer range 3400–5400 m but 2000–4300 m in winter.

Habits: Found on alpine meadows and bare rocky screes above the scrub zone. Prefers rain-swept ridges and heights. Perches on rocks like a chat. Forms small to large flocks sometimes of single sex. Flight and posture similar to a rock thrush but large wheeling flocks reminiscent of Common Starlings. In winter flocks settle in trees

796. **Little Forktail** *Enicurus scouleri* Xiao yanwei PLATE 82

Description: Small (13 cm) black and white forktail without long tail. Similar in colour to White-crowned but with short notched tail. Distinguished easily from female Plumbeous Redstart by white crown, white band in wing extending across lower back and forked tail. Iris—brown; bill—black; feet—pinkish-white.

Voice: Short high-pitched whistle, quieter than other forktails.

Range: Turkestan and Pakistan to Himalayas, NE India, S and C China, Taiwan, W and N Burma and N Indochina.

Distribution and status: Fairly common on mountain streams from 1200–3400 m in S Xizang, Yunnan, Sichuan, S Gansu, S Shaanxi and China S of Changjiang River plus Taiwan. At lower altitudes in Taiwan and in winter. Taiwan race generally ascribed to race *fortis* but not recognised by Cheng (1987, 1994).

Habits: Very active. Lives on small rocky streams in forests especially around waterfalls. Wags skimpy tail slowly up and down or flicks it open in rhythmic motion similar to Plumbeous Water Redstart. Behaviour also more like Plumbeous Water Redstart than other forktails. Nests behind waterfalls.

797. **Black-backed Forktail** *Enicurus immaculatus*   PLATE 82
Heibei yanwei

Description: Smallish (22 cm) forktail with black back. Separated from White-crowned Forktail by smaller size and white breast and from Slaty-backed Forktail by darker back. Juvenile has slaty, or brownish back and grey scaling on breast similar to juvenile Slaty-backed. Iris—brown; bill—black; feet—pinkish.

Voice: Short whistle *aut-see*; second note higher than the first. Less piercing than Slaty-backed Forktail. Also short song.

Range: Himalayas to N Burma and N Thailand.

Distribution and status: Not listed by Cheng (1994) for China but certainly occurs in areas of NE India claimed by China and recently recorded in Tengchong, W Yunnan.

Habits: Typical of genus. Singly or in pairs along stony streams, incessantly flicking long tail.

798. **Slaty-backed forktail** *Enicurus schistaceus*   PLATE 82
Huibei yanwei

Description: Medium-sized (23 cm) pied forktail. Distinguished from other forktails by grey crown and back. Juvenile has slaty dark brown crown and back and some scaling on breast. Iris—brown; bill—black; feet—pink.

Voice: Sharp high-pitched metallic *teenk*

Range: Himalayas to S China and Indochina.

Distribution and status: Common along mountain forest streams from 400–1800 m in SE Xizang, Sichuan, Yunnan, Guizhou, Guangxi, Guangdong, Hunan and Fujian.

Habits: As other forktails. Patrols rocky streams in wooded areas.

799. **White-crowned Forktail** *Enicurus leschenaulti*   PLATE 82
Baiding heibei yanwei

Description: Medium-sized (25 cm) black and white forktail. Forehead and forecrown white (feathers sometimes raised into small crest); rest of head, nape and chest black; belly, lower back and rump white; wings and tail black except for white tips to feathers of very long, forked and graduated tail; two outermost tail feathers entirely white. Iris—brown; bill—black; feet—pinkish.

Voice: A loud, thin, shrill double whistle, extremely sharp to the human ear, *tsee-eet*.

Range: N India, S China, SE Asia and Greater Sundas.

Distribution and status: A common bird along clear mountain streams up to 1400 m. Race *sinensis* is resident from Hunan to Shaanxi, S Gansu and all China S of the Changjiang River plus Hainan. Race *indicus* is resident in SE Xizang and SW Yunnan. **Habits:** Active restless bird of fast-flowing rocky streams and rivers. Settles on rocks or walks along the water's edge, pecking for food and constantly spreading the long forked tail. Flies with undulating flight close to the ground, calling as it goes.

800. Spotted Forktail *Enicurus maculatus*   PLATE 82
Banbei yanwei

Description: Large (27 cm) pied forktail. Generally similar to White-crowned but distinguished from all other forktails by white spots on back. Race *maculatus* has white tips to breast feathers. Iris—brown; bill—black; feet—pinkish-white.
Voice: Shrill rasping *kree* or *tseek* very similar to whistling thrush, given during flight. Also sharp creaky *cheek-chik-chick-chick-chik* given at rest or in flight.
Range: Himalayas to S and C China, Burma and NW and S Vietnam.
Distribution and status: Fairly common on mountain streams. More common than White-crowned at higher altitudes. Race *maculatus* in extreme S Xizang, *guttatus* from S Xizang to Sichuan and NW Yunnan and *bacatus* in mountains of S Yunnan, Guangdong and Fujian from 1200–3650 m.
Habits: More montane than other forktails; found along small rocky streams. Generally in pairs.

801. Purple Cochoa *Cochoa purpurea* Zi kuanzuidong   PLATE 82
Description: Male: large (28 cm) brownish-purple thrush-like bird with lavender-blue crown, edges to greater coverts and flight feathers. Wings tipped black. Tail lavender with black tip. Female: as male but reddish-brown above and pale brown below. Immature: crown black, spotted white; upperparts brown; throat buff and belly barred black and rufous. Iris—dark brown; bill—black; feet—black.
Voice: Low chuckling. Song is flute-like *peeeeee* or *peeee-you-peeee*.
Range: Himalayas to S China, Burma and Indochina.
Distribution and status: Resident in mountains of Sichuan as far N as Wenchuan (Wolong), Yunnan and SE Xizang from 900–2800 m.
Habits: Arboreal and lethargic. Generally in the tallest trees searching for fruits but sometimes feeds on ground. Its serrated beak is used to tear flesh off fruits. Also takes insects.

802. Green Cochoa *Cochoa viridis* Lü kuanzuidong   PLATE 82
Description: Large (28 cm) black and iridescent green thrush-like bird. Head greenish-blue with black face mask; wing black with blue wing coverts and wing patch; tail blue with black tip; rest of plumage glossy green. Female has some green feathering in wing patch. Iris—dark brown; bill—black; feet—pink.
Voice: Pure monotone whistle of about 2 seconds; also harsh note.
Range: Himalayas to SW China, Burma and Indochina.
Distribution and status: A rare resident in Fujian, S Yunnan and SE Xizang from 300–1600 m.
Habits: Arboreal and lethargic; forages for fruit and insects in canopy.

803. Hodgson's Bushchat *Saxicola insignis* PLATE 83
[White-throated Bushchat] Baihou shiji

Description: Male: largish (14.5 cm) chat with red breast, whitish vent and black and white upperparts. Similar to Common Stonechat but chin, throat and side of neck white forming a broken nuchal collar, and remiges basally white. Female: similar to Common Stonechat but greyer back and basally remiges white. Iris—brown; bill—black; feet—black.

Voice: Metallic *tek, tek* in alarm.

Range: Breeds in Kazakhstan to W Mongolia migrating in winter to the terai zone of N India, Nepal and Sikkim.

Distribution and status: Rare records of migrating birds in Qinghai and Alashan region of Nei Mongol. Mostly in alpine and subalpine meadows with bushes.

Habits: Solitary. Typical bushchat habits but feeds mostly on ground. Perches on top of bushes and pounces on insects.

804. Common Stonechat *Saxicola torquata* PLATE 83
[>Siberian Stonechat] Heihou shiji

Description: Medium-sized (14 cm) black, white and rufous chat. Male has black head and flight feathers, dark brown back, bold white patches on neck and wing, whitish rump and rufous breast. Female is duller without black; buffy underparts and white patch only on wing. Race *presvalskii* has buff throat and cinnamon underparts. Distinguished from female Pied Bushchat by paler colour and white wing patch. Iris—dark brown; bill—black; feet—blackish.

Voice: Scolding *tsack-tsack* like striking of two stones together.

Range: Breeds in Palearctic, Japan, Himalayas and northern SE Asia; migrant in winter to Africa, S China, India and SE Asia.

Distribution and status: Race *stejnegeri* breeds in NE China, wintering in China S of the Changjiang River and on Hainan; *presvalskii* breeds from S Xinjiang through Qinghai, Gansu, Shaanxi, Sichuan to S Xizang and SW China; northern birds migrate S in winter; *maura* breeds in N and W Xinjiang.

Habits: Prefers open habitat such as farmland, gardens and secondary scrub. Uses prominent low perches from which to pounce on prey on ground.

Note: Some authors separate N Asian forms as Siberian Stonechat *S. maura*.

805. Pied Bushchat *Saxicola caprata* [AG: Chat] PLATE 83
Baiban heishiji

Description: Small (13.5 cm) black and white chat. Male is all over sooty-black except for conspicuous white wing streak and white rump. Female is streaky brown with a pale brown rump. Immature birds are brown and spotted. Iris—dark brown; bill—black; feet—black.

Voice: A scolding alarm *chuh* and a pretty little whistling song *chip-chepee-cheweechu*.

Range: Iran to SW China, SE Asia, Philippines, Sulawesi, Sundas and New Guinea.

Distribution and status: Race *burmanica* is resident in SW China and SE Xizang. A common bird from lowlands to as high as 3300 m.

Habits: A bird of dry open grassy country. Perches prominently on the top of a bush, rock, post or wire and flutters down onto small insect prey. When singing or excited the male cocks its tail.

806. Jerdon's Bushchat *Saxicola jerdoni* Heibai linji   PLATE 83

Description: Male: medium-sized (15 cm) chat with entirely glossy black upperparts and white underparts. Female: upperparts brown with rufous-brown rump. Throat white; rest of underparts pale rufous, darker on breast and flanks. Has much shorter and less distinct pale supercilium than female Grey Bushchat and longer, more rounded tail. Iris—brown; bill—black; feet—black.

Voice: Short plaintive whistle, higher-pitched than most other chats. Low *chit-churr* or *churr* in alarm.

Range: Bengal, Bangladesh, N and E Burma, N Thailand, N Laos, NW Vietnam and SW China.

Distribution and status: Rare resident in extreme SW Yunnan (S Xishuangbanna) in lowland grasslands in flood plains.

Habits: Shy, singly or in pairs. Perches among grass stems diving to catch insect prey. Sometimes flicks tail open.

807. Grey Bushchat *Saxicola ferrea* Hui linji   PLATE 83

Description: Medium-sized (15 cm) greyish chat. Male: distinctive with mottled grey upperparts, bold white supercilium and black face mask contrasting with white chin and throat. Underparts whitish with smoky-grey band across breast and on flanks. Wing and tail black with flight feathers and outer tail feathers edged grey and inner wing coverts white (visible in flight). Back feathers are fringed brown; much greyer when worn. Female: brown version of male pattern with chestnut-rufous rump. Juvenile like female but scaly brown underparts Iris—dark brown; bill—grey; feet—black.

Voice: Call note a rising *prrei*; in alarm a soft *churr* followed by plaintive piping *hew*. Song is short feeble trill ending with a rolling whistle.

Range: Himalayas, southern half of China and N Indochina migrating to subtropical lowland zone in winter.

Distribution and status: Rather common. Race *ferrea* in SE Xizang and W Yunnan; dubious race *haringtoni* in the rest of China S of 34 °N, wintering in Taiwan.

Habits: Uses open scrub and cultivated lands, perching on same favourite perches for long periods. Flits tail. Catches insects on ground and on the wing.

808. Northern Wheatear *Oenanthe oenanthe* Sui ji   PLATE 83

Description: Small (15 cm) sandy-brown chat with dark wings and white rump. Summer male: white forehead and supercilium, black face mask. Wintering male has dark eye-stripe and white supercilium, buffy-brown crown and back, blackish wings and centre and tip of tail and rufous breast. The rump and sides of tail are white. Female similar but duller. Iris—brown; bill—black; feet—black.

Voice: Call is sharp whistled *heet* and in alarm a hard *chak*. Song of short crackly phrases

at fast tempo with whistled *heet* interspersed, given from rock perch or in flight; often at night.
Range: Breeds in Palearctic; migrant to Africa.
Distribution and status: Fairly common in deserts, high plateaux and rocky pastures. Nominate race breeds in Altai, Tianshan and Kashi areas of W Xinjiang, also Hulun Nur area of NE Nei Mongol, Ningxia, Ordos Plateau and Shanxi. Recorded as straggler in S Jiangsu and Hebei (Beidaihe).
Habits: A bird of open country. Mostly territorial. Stands tall with alert self-confident jizz. Bobs body and flits wings. Runs and hops. Flight quick and low with flurry before landing.

809. Variable Wheatear *Oenanthe picata*　　　　　　　　　　PLATE 83
[Eastern Pied Wheatear] Dongfang banji

Description: Largish (14.5 cm) black and white chat in at least three colour phases. Race *picata* has entire upperparts black, except white rump and base to outer tail feathers, and entire underparts white except black chin, throat and upper breast. Race *opistholeuca* has black on underparts extending to include entire abdomen. Race *capistrata* is very difficult to separate from male Pied Wheatear but whiter breast and crown. Female very variable. Generally similar to male forms but black replaced by sooty or greyish-black. Iris—dark brown; bill—black; feet—black.
Voice: A sweet rather scratchy song with twittering and mimicked part of other species' songs, given from perch or in song flight.
Range: Iran to Turkestan and Himalayas to Kashmir, wintering from S Iran and E Arabia to NW India.
Distribution and status: Occasional in extreme W China in the Kashi region of SW Xinjiang
Habits: Display song with hovering wings and drooping spread tail. Perches and bobs, diving on insect prey. Sometimes takes insects on the wing like a flycatcher. Very aggressive territorially even against other species such as chats and robins.

810. Pied Wheatear *Oenanthe pleschanka* Baiding ji　　　　　PLATE 83

Description: Medium-sized (14.5 cm) long-tailed wheatear often perched in bush. Male: all black upperparts except white rump, crown and nape; base of outer tail feathers greyish-white; all-white underparts except black chin and throat. Separated from male Variable Wheatear of race *capistrata* by greyer crown and buff wash on breast. Female: brownish above with buff supercilium and white base to outer tail feathers. Chin and throat dark scaled with white tips; breast reddish flanks buff and vent white. Rare colour phase *vittata* occurs in which both sexes have white throat. Iris—brown; bill—black; feet—black.
Voice: Call includes hard and dry *tritt tack*. Short musical song with twitterings and mimicked phrases, given from rock perch or in song flight. More vocal than Northern Wheatear.
Range: Romania and S Russia to Transbaikalia and N China. Winters S to Iran, Arabia and E Africa.

Distribution and status: Fairly common in suitable arid habitat in W Xinjiang, Qinghai, Gansu, Ningxia, Nei Mongol, Shaanxi, Shanxi, Henan, Hebei and Liaoning.
Habits: Lives on stony, barren land with bushes, also wasteland, farms, and towns. Perches upright; wags tail up and down. Hunts insects from perch. Male in song flight circles very high then dives suddenly to ground.
Note: Treated by some authors as a race of Black-eared Wheatear *Oenanthe hispanica* (e.g. Cheng 1987, 1994).

811. Desert Wheatear *Oenanthe deserti* Mo ji   PLATE 83

Description: Smallish (14–15.5 cm) sandy-buff wheatear with black tail and blackish wings. Male has black sides of face, neck and throat. Female has blackish sides of head but white chin and throat. Blacker wings than female Northern Wheatear. Southern race *oreophila* is larger than northern *atrogularis*. Almost entirely black tail distinguished from all other wheatears in flight. Iris—brown; bill—black; feet—black.
Voice: Harsh alarm *chrt-tt-tt*, shrill whistled call. Song is mournful descending trill *teee-ti-ti-ti* repeated by male monotonously.
Range: Arabia, Middle East to Mongolia, W Himalayas, W, N and C China and Tibetan Plateau, Winters Arabia to NE Africa and NW India.
Distribution and status: Fairly common in desert areas. Race *atrogularis* in NW Xinjiang, W Xizang, Ningxia and N Shaanxi, Gansu. Race *oreophila* in SW Xinjiang, Qinghai and Xizang.
Habits: Found in stonier desert and wasteland than Isabelline Wheatear. Often perches on low vegetation. Fairly shy. Male gives brief fluttering display flight near nest. Hops on ground. Flies off to dip behind rock out of sight.

812. Isabelline Wheatear *Oenanthe isabellina* Sha ji   PLATE 83

Description: Large (16 cm) longish-billed, plain, pinkish, sandy-brown wheatear without black face mask. Paler-winged than most other wheatears. Tail blacker than autumn Northern Wheatear. Sexes alike but male has blacker lores. Supercilium and eye-ring whitish. Distinguished from female Desert Wheatear by plumper body, larger head, longer legs, less black in wing coverts and whiter rump and base of tail. Juvenile is spotted pale above and has dusky edges to breast feathers. Iris—dark brown; bill—black; feet—black.
Voice: High, piped *cheep*. Chattering song of up to 15 seconds is longer than other wheatears with mimicked phrases and series of clear *wee-wee-wee-wee-wee*. Low soft, sweet song also heard in winter.
Range: SE Europe through Middle East to NW Himalayas, SE Russia, N China and Mongolia. Winters to NW India and C Africa.
Distribution and status: Fairly common up to 3000 m in steppe and desert areas of Xinjiang, Qinghai, Gansu, N Shaanxi and Nei Mongol.
Habits: Solitary or in pairs on sandy desert with bushes. Stands slightly more upright than Northern Wheatear. Runs swiftly over ground and stops to bob. Male display is to leap in air and hover with outspread tail then flutter back down to land.

STARLINGS AND MYNAS—Family: Sturnidae

STARLINGS—Tribe: Sturnini

A large Old World tribe, formerly treated as a family, of robust birds with powerful, sharp, straight beaks and long legs. They are mostly gregarious and most species feed on the ground where they stride about in a characteristic jaunty manner. They feed on fruits and invertebrates. Most species nest in tree holes. They are noisy, garrulous birds with harsh calls and a capacity to imitate other birds calls. Nineteen species occur in China.

813. Chestnut-tailed Starling *Sturnus malabaricus*   PLATE 84
[Grey-headed Starling] Huitou liangniao

Description: Medium-sized (20 cm) pale grey starling with silky pearly mane. Lacks any white shoulder stripe. Distinguished from juvenile White-shouldered and Red-billed Startings by chestnut outer tail feathers, darker rump and some rufous on flanks. Iris—white; bill—olive-green with yellow tip and cobalt-blue at base; feet—brownish-yellow.
Voice: Chattering in feeding flocks. High-pitched disyllabic call notes, single, tremulous whistle and short sweet song.
Range: India, S China, Burma, Indochina.
Distribution and status: Race *nemoricolus* is uncommon in open forest, farmland and gardens of low hills in S Sichuan, SE Xizang, SW Guizhou, Yunnan, SW Guangxi and Hong Kong; also Taiwan.
Habits: Forms flocks; often found feeding in flowering *Erythrina* and *Bombax* trees.

814. Brahminy Starling *Sturnus pagodarum*   PLATE 84
Heiguan liangniao

Description: Medium-sized (21 cm) starling with black crown and long crest and cinnamon underparts. Tail and primaries blackish. Upperparts otherwise pale greyish-brown. Iris—white; bill—slate-blue base with yellow tip; feet—brownish-yellow.
Voice: Gurgling cry followed by bubbling yodels.
Range: India.
Distribution and status: Accidental in SW Yunnan. Occasional in SE Xizang.
Habits: In India it is a common bird of towns, villages and railway lines. Uses communal roosts.

815. Red-billed Starling *Sturnus sericeus* [Silky/Ashy Starling]   PLATE 84
Siguang liangniao

Description: Largish (24 cm) grey, black and white starling with red bill. Wings and tail glossy black with white patch in primaries conspicuous in flight. Head is whitish and silky; upperparts otherwise grey. Iris—black; bill—red with black tip; feet—dull orange.
Voice: Not recorded.
Range: China, Vietnam, Philippines.
Distribution and status: Resident over most of S and SE China, Taiwan and Hainan, dispersing to N Vietnam and Philippines in winter. Not uncommon in farmland and orchards up to 800 m.
Habits: Forms large flocks on migration.

816. **Purple-backed Starling** *Sturnus sturninus*   PLATE 84
[Daurian Starling/Daurian Starlet] Bei liangniao

Description: Smallish (18 cm) dark-backed starling. Adult male: back glossy, iridescent purple; wings iridescent green-black with pronounced white wing bars; head and breast grey with black patch on nape; belly white. Distinguished from Chestnut-cheeked Starling by black nape patch and lack of chestnut on sides of neck. Female: ashy-grey above; brown nape spot; black wings and tail. Immature: pale brown; underparts mottled brown. Iris—brown; bill—blackish; feet—green.

Voice: Typical harsh starling whistling and whickering notes.

Range: Breeds from Transbaikalia to NE China, migrates in winter to SE Asia and Greater Sundas.

Distribution and status: Breeds in NE and N China; wintering through SE China to S and SW China and Hainan. Generally uncommon up to moderate altitudes.

Habits: Feeds on the ground in open coastal areas.

817. **Chestnut-cheeked Starling** *Sturnus philippensis*   PLATE 84
[Red-cheeked/Violet-backed Starling] Zibei liangniao

Description: Smallish (17 cm) dark-backed starling. Male has pale grey or buffy head, whitish underparts and glossy dark violet back, black wings and tail with white shoulder bar. Distinguished from Purple-backed Starling by chestnut ear coverts and side of neck. The female is greyish-brown above and whitish below with black wings and tail. Iris—brown; bill—black; feet—dark green.

Voice: Loud squealing calls and shrieks.

Range: Breeds in Japan's Kuril Is; migrates in winter to Philippines and Borneo.

Distribution and status: Uncommon passage migrant on E China coast. Winters on Taiwan and Lanyu islet.

Habits: Lives in small flocks, preferring open country. Feeds in trees.

818. **White-shouldered Starling** *Sturnus sinensis*   PLATE 84
[Chinese/Grey-backed Starling] Huibei liangniao

Description: Smallish (19 cm) grey starling. Male: distinguished from other starlings by wholly white upperwing coverts and scapulars. General plumage grey; crown and belly whitish flight feathers black and white tips to outer tail feathers. Female has less white on wing coverts. Immature birds are browner. Iris—bluish-white; bill—grey; feet—grey.

Voice: Harsh squawks and squeals.

Range: Breeds S China and N Vietnam; migrates in winter to SE Asia, Philippines and Borneo.

Distribution and status: Breeds in S and SE China and Taiwan. Partially migratory with wintering populations in Taiwan and Hainan.

Habits: Lives in noisy flocks that congregate to feed in figs and other fruiting and flowering trees in open country and gardens.

819. **Rosy Starling** *Sturnus roseus* Fenhong liangniao   PLATE 85

Description: Medium-sized (22 cm) distinctive pink and black starling. Breeding male: unmistakable glossy black with pink back, breast and flanks. Female: similarly patterned but duller. Juvenile: buff above with brown wings and tail; pale underparts; yellow bill. Iris—black; bill—pinkish; feet—pinkish.

Voice: *Ki-ki-ki* call in flight and harsh *shrr*. Rattling *chik-ik-ik-ik* call when feeding in flocks.

Range: E Europe to C and W Asia, wintering to India; vagrant Thailand.

Distribution and status: Common resident in open areas of NW China, migrates to Gansu and W Xizang. Vagrants have reached Shanghai and Hong Kong.

Habits: Lives in large flocks in dry, open areas. Follows domestic animals to catch disturbed insects.

820. Common Starling *Sturnus vulgaris* Zichi liangniao   PLATE 85

Description: Medium-sized (21 cm) blackish, glossy, purplish-green starling with varying degrees of white spotting. Fresh feathers of body are lanceolate in shape with rusty edges giving a scalloped and spotted pattern which mostly disappears with wear. Iris—dark brown; bill—yellow; feet—reddish.

Voice: The calls are harsh screeches and whistles.

Range: Eurasia.

Distribution and status: Common in western China in cultivated areas, around towns and edges of desert. Race *poltaratskyi* breeds in the N Junggar Basin and *porphyronotus* in Tianshan and Kashi region of W Xinjiang. Migrants of *poltaratskyi* are recorded on passage across W China with occasional accidental records along E and S Coasts.

Habits: Lives in small to large flocks and feeds on the ground in open areas. Congregates in winter into large flocks which migrate to southern limits of range.

821. White-cheeked Starling *Sturnus cineraceus*   PLATE 84
Hui liangniao

Description: Medium-sized (24 cm) brownish-grey starling with white streaking on sides of black head. Vent, tips of outer tail feathers and narrow wing bar on secondaries white. Female paler and duller than male. Iris—reddish; bill—yellow with black tip; feet—dull orange.

Voice: Monotonous creaking *chir-chir-chay-cheet-cheet*.

Range: Siberia, China, Hainan, Taiwan, Japan, N Vietnam and N Burma, Philippines.

Distribution and status: Breeds in N and NE China, migrating to southern China in winter. A common bird of open country and farmland with scattered trees.

Habits: Lives in flocks and feeds on farmland, replacing Common Starling in Far East.

822. Asian Pied Starling *Sturnus contra*   PLATE 84
[Pied/Asiatic Pied Starling] Ban liangniao

Description: Medium-sized (24 cm), black and white starling. White crown, sides of head, wing bar, rump and abdomen; throat, breast and upperparts otherwise black (brown in immature). Iris—grey; bare orbital skin—orange; bill—yellow with red base; feet—yellow.

Voice: Noisy, discordant, jaunty cries.

Range: India, SW China, SE Asia, Sumatra, Java and Bali.

Distribution and status: Locally common in cultivated lowlands. Race *superciliaris* is resident in SE Xizang and NW Yunnan; *floweri* resident in SW and S Yunnan.

Habits: Lives in small parties, inhabiting open land. Feeds mostly on the ground, probing for earthworms and other small invertebrakes. Congregates in communal roosts at night.

823. Black-collared Starling *Sturnus nigricollis*   PLATE 84
Heiling liangniao

Description: Large (28 cm) black and white starling. Head white with black collar and throat. Back and wings black with white edgings. Tail black with white tip. Bare skin round eye and legs yellow. Female like male but browner. Juvenile lacks black collar. Distinguished from Asian Pied Starling by larger size, white throat and crown, yellow orbital skin, lack of clear white wing bar and black bill. Iris—yellow; bill—black; feet—pale grey.

Voice: Call is harsh screeches and whistles.

Range: S China and SE Asia.

Distribution and status: A common bird of farmland in southern China generally feeding on the ground in small flocks in paddy fields, grazing areas or open ground.

Habits: Sometimes feeds among herds of buffalo or cattle.

824. Vinous-breasted Starling *Sturnus burmannicus*   PLATE 84
[Jerdon's Starling] Hongzui liangniao

Description: Largish (25 cm) greyish starling with whitish head, yellow bill and blackish eye-stripe. Breast and belly vinous; wings dark grey with white patch at base of primaries conspicuous in flight. Iris—whitish-yellow; bill—red with black base; feet—brownish-yellow.

Voice: Utters chattering note when feeding in parties.

Range: Burma and Indochina.

Distribution and status: Rare in extreme SW Yunnan.

Habits: Prefers open, dry country, cultivation and gardens. Feeds in flocks and joins communal roosts at night.

825. Common Myna *Acridotheres tristis* [Indian Myna]   PLATE 85
Jia bage

Description: Jaunty (24 cm) brownish myna with dark head. Distinguished from other mynas by lack of crest, and bare yellow orbital skin. In flight the white wing flash is conspicuous. Immature birds are duller. Iris—reddish; bill—yellow; feet—yellow.

Voice: Liquid gurgles, sharp screeches and a musical whistle, including mimicked notes.

Range: Afghanistan to SW China and SE Asia and Malay Peninsula.

Distribution and status: Nominate race is resident in SW Sichuan, W and S Yunnan, SE Xizang and Hainan. Common in agricultural areas and villages.

Habits: Generally in flocks on the ground. Prefers towns, fields and gardens.

[**826. Bank Myna** *Acridotheres ginginianus* Huibei anbage   PLATE **85**

Description: Medium-sized (25 cm) myna with red orbital skin, black head and yellow bill. Greyer than Common Myna. Iris—yellow; bill—yellow; feet—yellow.
Voice: Tuneless song with gurgles and whistles. Less harsh than Common Myna.
Range: India.
Distribution and status: Recorded only as feral escapes especially in Hong Kong and Guangdong area. Probably occurs in SE Xizang.
Habits: Lives in towns.]

827. White-vented Myna *Acridotheres cinereus*   PLATE **85**
[Javan/Great Myna AG: Mynah] Lin bage

Description: Medium-sized (26 cm) black starling. Plumage dark grey almost black except for conspicuous white patch on primaries especially in flight, white vent and white tip of tail. Has a slight crest. Distinguished from Crested Myna by broader white tip to tail, all-yellow bill and white vent. Immature browner. Iris—orange; bill—orange; feet—yellow.
Voice: Garrulous, harsh, creaky notes; whistles and rattles. Sometimes mimics other birds.
Range: NE India, E and SE Asia, Sulawesi, Sumatra (introduced), Java and Bali.
Distribution and status: Race *grandis* is uncommon resident in W and S Yunnan and SW Guangxi.
Habits: Lives in small to large flocks, feeding largely on the ground in open grassy areas and paddy fields. Frequently settles on or around domestic animals, catching insects disturbed by their movements.
Note: Sometimes treated as a race of Jungle Myna *A. fuscans* but more likely to be conspecific with forms *javanicus* and *grandis*. Treated as separate species by King *et al.* (1975) but as conspecific by Inskipp *et al.* (1996).

828. Collared Myna *Acridotheres albocinctus* Bailing bage   PLATE **85**

Description: Medium-sized (26 cm) black myna with slight crest and diagnostic broad white nuchal collar. Collar is buff in winter. Undertail coverts broadly edged white. Juvenile: paler and browner. Iris—bluish or yellow; bill—yellow; feet—bright yellow.
Voice: Not recorded.
Range: NE India, SW China, Burma.
Distribution and status: Local migrant in moist open spaces up to 1200 m in NW Yunnan.
Habits: Favours marshes and cattle fields.

829. Crested Myna *Acridotheres cristatellus* [Chinese Jungle Myna]   PLATE **85**
Bage

Description: Large (26 cm) black myna with prominent crest. Distinguished from White-vented Myna by longer crest, red or pink base to bill, narrow white tail tip and undertail coverts barred black and white. Iris—orange; bill—pale yellow with red base; feet—dull yellow.
Voice: Like Common Myna. Captive birds can learn to 'talk'.

Range: China and Indochina. Introduced into Philippines and Borneo.

Distribution and status: Nominate race is resident in mid-Changjiang catchment from E Sichuan and S Shaanxi to S China; *brevipennis* on Hainan; and *formosanus* on Taiwan. Common in farmland and villages.

Habits: Lives in small flocks, generally seen strutting on ground in open fields or in towns and gardens.

830. Golden-crested Myna *Ampeliceps coronatus*　　　　　　　PLATE 85
Jinguan shubage

Description: Smallish (23 cm) glossy black myna with diagnostic yellow wing patch, chin and pinkish-yellow bare facial skin. Adult male has bright golden-crested crown. Female similar but yellow area smaller. Iris—brown; bill—pink with bluish base; feet—orange-yellow.

Voice: High-pitched, metallic whistle and bell-like note.

Range: NE India, Burma, S China, Indochina.

Distribution and status: Irregular vagrant to Xishuangbanna in SW Yunnan and recorded off coast of Guangdong.

Habits: Lives in small parties in canopy of forest.

831. Hill Myna *Gracula religiosa* [Talking Myna/Common　　　　PLATE 85
Grackle/Grackle > Eastern Hill Myna] Liaoge

Description: Large (29 cm) glossy black starling with conspicuous white wing patches; and diagnostic orange wattles and lappets on sides of head. Iris—dark brown; bill—orange; feet—yellow.

Voice: Loud, clear, piercing *tiong* and an enormous range of clear whistles and calls mimicking other birds.

Range: India to China, SE Asia, Palawan and Greater Sundas.

Distribution and status: Race *intermedia* is resident in tropical lowlands of SE Xizang, S China and Hainan. Locally common but mostly in reduced numbers by trapping for cagebird trade.

Habits: Keeps to tall trees and lives in pairs, sometimes gathering in flocks.

NUTHATCHES AND WALLCREEPERS—Family: Sittidae

A small family; mostly in Old World. They are small, non-migratory insectivorous forest birds found in Europe, Asia and North America. They specialise in foraging on tree trunks and branches, clinging to bark with one foot held above the body. Twelve species occur in China which is the centre of their distribution.

NUTHATCHES—Subfamily: Sittinae

832. Eurasian Nuthatch *Sitta europaea*　　　　　　　　　PLATE 86
[Wood/Common Nuthatch] Putong shi

Description: Medium-sized (13 cm) neatly-coloured nuthatch with bluish-grey upperparts, black eye-stripe, white throat, buffish belly and rich chestnut flanks. Races vary in details: *asiatica* has white underparts, *amurensis* has narrow white eyebrow and pale buff underparts, *sinensis* has entire underparts pinkish-buff. Iris—dark brown; bill—black with pinkish base to lower mandible; feet—dark grey.

Voice: Gives loud sharp *seet, seet* calls, scolding *twet-twet, twet* and musical whistled song.

Range: Palearctic.

Distribution and status: Rather common in deciduous woodlands over much of the country. Race *scorsa* is resident in NW China; *asiatica* in Greater Hinggan Mts of NE China; *amurensis* in the rest of NE China and *sinensis* through E, C, S and SE China and Taiwan.

Habits: Acorns and nuts are drilled into crevices and holes in tree trunks. Flight is jerky and undulating. Occasionally feeds on ground. Lives in pairs or small flocks.

833. Chestnut-vented Nuthatch *Sitta nagaensis*　　　　PLATE 86
Litunshi

Description: Medium-sized (13 cm) grey nuthatch similar to Eurasian Nuthatch but with underparts pale buff, washed grey on throat, ear coverts and breast contrasting strongly with deep brick-red flanks. Deep rufous undertail coverts have a line of bold white scallops down each side. Iris—dark brown; bill—black with grey base to lower mandible; feet—variable greyish-brown.

Voice: Simple *sit* or *sit-sit*, wren-like trills, mews and fast, stony, rattling song *chichichichichi* . . .

Range: NE India, Burma, N Thailand, S Vietnam, S Laos, SW and SE China.

Distribution and status: Race *montium* is fairly common in SE Xizang, W Sichuan, SW Guizhou and Yunnan in mixed forests from 1400–2600 m. There is a small isolated population in Wuyishan, Fujian province.

Habits: Typical of genus.

Note: Treated by Cheng (1987, 1994) as race of Eurasian Nuthatch and synonymous with race *nebulosa* of Meyer de Shauensee (1984)

834. Chestnut-bellied Nuthatch *Sitta castanea* Lifu shi　　　　PLATE 86

Description: Small (13 cm) grey and rufous nuthatch with contrasting white cheek patch and richly coloured underparts. Male is distinct with brick-red underparts. Black eye-stripe broadens widely at rear. Female can be confused with darker-bellied races of Eurasian Nuthatch but white cheek patch larger and more prominent. Tail has small white subapical patches. Undertail coverts vary with race being black with orange scalloping in *tonkinesis* and white with orange scalloping in *cinnamoventris*. Iris—brown; bill—black with pale base to lower mandible; feet—horn to blackish.

Voice: Short telephone-like trills *prrt prrt*. Clear squeaks, metallic *chit* notes, sparrow-like *cheep cheep*. Song is clear repeated whistle and melodious trills.

Range: Himalayas, India, Burma, Indochina and SW China.

Distribution and status: Rare in extreme S Yunnan in woodland and pine forest from 1000–2200 m. Race *cinnamoventris* west of the Lancang (Mekong) and *tonkinensis* between the Lancang and Red rivers.

Habits: Usually found in pairs or small mixed parties.

835. White-tailed Nuthatch *Sitta himalayensis* PLATE 86
Baiwei shi

Description: Small (12 cm) grey and cinnamon nuthatch with diagnostic white base to central tail feathers and uniform rufous undertail coverts. Similar to some races of Eurasian, Chestnut-vented and Chestnut-bellied Nuthatches, the latter two of which also have white subterminal spots on outer tail feathers. If white tail base not visible, the unscalloped uniform rufous undertail coverts are diagnostic. Iris—brown; bill—blackish with pale base to lower mandible; feet—greenish-brown.

Voice: Single or double sharp *chak* notes, rapid rattling *chip-chip-chip-chip* ... about 10 notes per second. Song is slow and fast variants of a whistled crescendo.

Range: Himalayas, Burma, N Laos, NW Vietnam and SW China.

Distribution and status: Nominate race is uncommon in forests of S Xizang and extreme W Yunnan.

Habits: Typical of genus.

836. Chinese Nuthatch *Sitta villosa* [Snowy-browed Nuthatch] PLATE 86
Heitou shi

Description: Small (11 cm) nuthatch with white supercilium and thin black eye-stripe. Crown black in male, grey in fresh female plumage. Upperparts otherwise lavender-grey. Throat and side of face whitish becoming greyish-buff or cinnamon on rest of underparts. Similar to Yunnan Nuthatch but eye-stripe narrower and not flared at rear, underparts much richer. Race *bangsi* is richer orange-cinnamon on underparts than nominate race. Iris—brown; bill—blackish with paler base to lower mandible; feet—grey.

Voice: Scolding harsh *schraa*, melodic piping series *wip wip wip*; short nasal *quir quir* notes. Song is series of pure upward-inflected whistles.

Range: Endemic to N and NE China extending marginally into Korea, Ussuriland and Sakhalin I.

Distribution and status: Race *bangsi* is uncommon in C Gansu and adjacent Qinghai and Sichuan. Nominate race extends from S Gansu to Jilin and Hebei in pine forests in hills.

Habits: Typical of genus.

837. Yunnan Nuthatch *Sitta yunnanensis* Dian shi PLATE 86

Description: Small (12 cm) grey, black and buff nuthatch with diagnostic broad black eye-stripe broadening at rear topped by narrow white supercilium. Sides of face and throat white; rest of underparts pinkish-buff. Juvenile: reduced and less bold eye-stripe and supercilium. Iris—brown; bill—blackish; feet—grey.

Voice: Rather vocal. Various nasal *nit* and *toik* calls, high-pitched *tit*, scolding *schri-schri-schri* ... and high nasal *ziew-ziew-ziew* ...

Range: Endemic to SW China.

Distribution and status: Rare to locally common in S and SW Sichuan, W Guizhou, Yunnan and SE Xizang in pine forests from 2000–3350 m in summer and as low as 1200 m in winter.

Habits: Typical of genus.

838. White-cheeked Nuthatch *Sitta leucopsis* Bailian shi   PLATE 86

Description: Medium-sized (13 cm) nuthatch with diagnostic prominent buffy cheek patch enclosing eye. Upperparts lavender-grey with black crown and half collar. Underparts rich cinnamon. Nominate race may occasionally occur in China in W Himalayas and has paler underparts with cheek patch almost white. Iris—brown; bill—black with grey base to lower mandible; feet—greenish-brown.

Voice: Contact call is repeated, bleated, nasal *kner-kner*. When excited, gives long series of single similar notes. Song is rapid repetition of wailing squeaks or clear loud *ti-tui ti-tui ti-tui* . . .

Range: Himalayas and W China.

Distribution and status: Race *przewalskii* uncommon in subalpine forests along the eastern edge of the Qinghai–Xizang plateau in NE Qinghai, SW Gansu, N and W Sichuan, N Yunnan and E Xizang. Summers at 2000 m to the tree-line but winters as low as 1000 m.

Habits: Found in pairs or small parties, sometimes mixing with other species.

839. Velvet-fronted Nuthatch *Sitta frontalis* Rong'e shi   PLATE 86

Description: A small (12 cm) colourful nuthatch with red bill. Forehead velvety-black; back of head, back and tail violet with a bright blue flash on the primaries. Male has a black supercilium behind eye. Underparts pinkish with a whitish chin. Juvenile duller with blackish bill. Iris—yellow with reddish orbital skin; bill—red with black tip; feet—reddish-brown.

Voice: A shrill insistent *chip-chip* or sharp chittering call. In flight a *seep-seep-seep* call is given. Song is fast rattle of *sit* notes.

Range: India, S China, SE Asia to Philippines and Greater Sundas.

Distribution and status: Nominate race is resident in SE Xizang, S and W Yunnan, Guangxi and Guizhou. Increasingly common in S Guangdong (Dinghushan) and Hong Kong (Tai Po Kau) where it may have become established from escapes.

Habits: Travels in pairs or family parties working over the trunks and branches of forest trees, often from top to bottom.

Notes: Records of *S.f. chiengfengensis* from Hainan are referred to Yellow-billed Nuthatch *S. solangiae*.

840. Yellow-billed Nuthatch *Sitta solangiae* [Lilac Nuthatch]   PLATE 86
Danzi shi

Description: A colourful medium-sized (13 cm) nuthatch with yellow bill. Plumage almost identical to Velvet-fronted Nuthatch but distinguished by colour of bill. Iris—yellow with reddish orbital skin; bill—yellow with black tip; feet—reddish-brown.

Voice: A shrill insistent *chit-chit* or sharp chittering call *sit-it-it-it-it*.

Range: Vietnam and Hainan.

Distribution and status: Race *chienfengensis* is not uncommon in montane forests of Hainan.

Habits: As Velvet-fronted Nuthatch but more montane.

Note: Treated by Cheng (1987) as a race of *S. frontalis* but in Vietnam the two species are sympatric. Harrap and Quinn (1996) suggest this form may be conspecific with Sulphur-billed Nuthatch *S. oenochlamys* of Philippines.

841. Giant Nuthatch *Sitta magna* Jü shi   PLATE 86

Description: Very large (20 cm) grey, black and white nuthatch with chestnut vent. Appears comparatively long-tailed. Patterned like Chestnut-vented Nuthatch but much larger, black eyestripe much broader up sides of head with crown stripe noticeably paler grey than back. Male has black streaking on crown. In flight from below, white base of flight feathers contrast with black carpals. Iris—dark brown; bill—black with paler base to lower mandible; feet—greenish-brown.

Voice: Black-billed Magpie-like harsh *get-it-up*, single clear piping *kip*.

Range: SW China, E Burma and NW Thailand.

Distribution and status: A local and occasional bird in pine forests at 1200–1800 m in extreme S Sichuan, S and W Yunnan and extreme SW Guizhou.

Habits: Undulating flight with whirring broad wings. Tail sometimes half-cocked. Generally similar to other nuthatches but more deliberate and less restless.

842. Beautiful Nuthatch *Sitta formosa* Li shi   PLATE 86

Description: Unmistakable large (16 cm) nuthatch with blackish upperparts marked with brilliant blue. Underparts cinnamon-orange. In flight from below white bases to primaries contrast against blackish underwing coverts. Iris—reddish-brown; bill—black with pale base to lower mandible; feet—greenish-brown.

Voice: Typical nuthatch call but sweeter and less harsh than Eurasian Nuthatch.

Range: E Himalayas, NE India, Burma, Laos, N Vietnam and SW China.

Distribution and status: Very rare and status in China uncertain. Collected in Ailaoshan in S Yunnan in April and reported by Meyer de Schauensee (1984) to have been found in NW Yunnan.

Habits: Typical of the genus.

WALLCREEPERS—Subfamily: Trichodominae

843. Wallcreeper *Tichodroma muraria* [Red-winged Wallcreeper]   PLATE 65
Hongchi xuanbique

Description: Smallish (16 cm) elegant grey bird with short tail, long bill and striking crimson wing patch. Breeding male has black face and throat, reduced in female. Non-breeding adults have whitish throat and brown wash on crown and cheeks. Flight feathers are black with prominent white tips to outer tail feathers and two rows of white spots on primaries give striking effect in flight. Iris—dark brown; bill—black; feet—brownish-black.

Voice: Thin piping notes and whistles, less harsh than nuthatch. Song is variable series of repeated high whistles *ti-tiu-tree*, increasing in speed.

Range: Spain and S Europe to C Asia, N India, China and S Mongolia.

Distribution and status: Race *nepalensis* is uncommon and irregular in extreme W

China, Qinghai–Xizang plateau, Himalayas, C and N China. Wintering birds visit much of S and E China.

Habits: Creeps about on cliffs and rocks, flicking wings to display red patches. Winters at lower altitudes and even feeds on buildings.

TREECREEPERS AND WRENS—Family: Certhiidae

TREECREEPERS—Subfamily: Certhiinae

A small subfamily of seven species mostly in Europe and Asia but with one American and one African species. Formerly treated as a family but Sibley and Monroe (1990) have placed this group as a subfamily Certhiinae within a much enlarged Certhiidae family that includes several other New World subfamilies.

844. Eurasian Treecreeper *Certhia familiaris*   PLATE 15
[Northern/Common Treecreeper] Xuanmuque

Description: Smallish (13 cm) mottled brown treecreeper. Underparts white or buff with only a slight rufous wash on flanks and rufous uppertail coverts. Whitish breast and flanks and paler supercilium distinguish it from Rusty-flanked Treecreeper and smaller size and pale throat distinguish from Brown-throated Treecreeper. Plain brown tail distinguishes from Bar-tailed Treecreeper. Races vary only in minor details. Iris—brown; bill—brown above, pale below; feet—brownish.

Voice: Quiet contact *zit*, loud piercing rolling *zrreeht*. Song is wren-like in tone with introductory scraping notes ending in thin trill.

Range: Eurasia, Himalayas to N China, Siberia and Japan.

Distribution and status: Several races are all fairly common within their respective distributions in temperate broadleaf and conifer forests at all altitudes. Race *tianshanica* in NW China, *khamensis* in C China, S Xizang and SW China, *bianchii* in Qinghai, Gansu and S Shaanxi, *orientalis* in NE China and *daurica* in extreme NE China.

Habits: Typical of genus. Regularly joins mixed bird flocks.

845. Bar-tailed Treecreeper *Certhia himalayana*   PLATE 15
[Himalayan Treecreeper] Gaoshan xuanmuque

Description: Medium-sized (14 cm) dark greyish mottled treecreeper. Easily distinguished from all other treecreepers by lack of rufous on rump or underparts; greyer colour and prominent banding on tail. Throat white, belly and breast smoky-buff. Bill is longer and more curved than other treecreepers. Iris—brown; bill—brown with pale lower mandible; feet—brownish.

Voice: Thin descending *tsiu* sometimes in series, thin *tsee* or rising *tseet*. Song is lilting trill.

Range: C Asia to N Afghanistan, Himalayas, Burma and SW China.

Distribution and status: Race *yunnanensis* is uncommon in mixed deciduous and conifer forests of S Gansu, S Shaanxi, N and W Sichuan, SW Guizhou, N and W Yunnan and SE Xizang. Lives at high altitudes at 2000–3700 m.

Habits: Sometimes joins mixed flocks.

846. Rusty-flanked Treecreeper *Certhia nipalensis*   PLATE 15
[Nepal/Stoliczka's Treecreeper] Xiuhongfu xuanmuque

Description: Medium-sized (14 cm) mottled dark brown treecreeper. Underparts buff with rufous flanks and tail coverts. Rufous flanks distinguish from Eurasian Treecreeper and smaller size and pale throat distinguish from Brown-throated Treecreeper. Plain brown tail distinguishes from Bar-tailed Treecreeper. Iris—brown; bill—dark brown above, paler below; feet—brownish.

Voice: Call is thin *sit*. Song is high-pitched accelerating trill preceded by three silvery notes *si-si-sit'st't't't't*.

Range: Himalayas to W Yunnan and N Burma.

Distribution and status: Common in coniferous and mixed forests at 1800–3500 m in Himalayas, SE Xizang and Gaoligongshan on the Yunnan–Myanmar border west of Salween (Nujiang) River.

Habits: Typical treecreeper behaviour.

847. Brown-throated Treecreeper *Certhia discolor*   PLATE 15
[Sikkim Treecreeper] Hehou xuanmuque

Description: Medium-sized (14 cm) mottled dark rufous brown treecreeper with dull underparts and brownish wash on throat. Juveniles and non-breeding adults may have grey throat but lack of white throat distinguishes from all other treecreepers. Lack of rufous flanks distinguishes from Rusty-flanked; unbarred brown tail distinguishes from Bar-tailed Treecreeper. Iris—brown; bill—dark brown with paler lower mandible; feet—brownish.

Voice: Explosive *tchip*, higher, thinner *tsit* or *seep* and rattling *chi,r,r,it*. Song is monotonous rattle or trill.

Range: Himalayas, SW China and Indochina.

Distribution and status: Nominate race is rare in forests of S Yadong region of Xizang and *shanensis* is rare in forests of Gaoligongshan in W Yunnan.

Habits: Typical of genus.

WRENS—Subfamily: Troglodytinae

Large American subfamily of 60 species but only one found in Old World. Wrens are very small birds that live in dense undergrowth and conifer forests. Many are altitudinal migrants. They have loud, beautiful songs and build globular nests with a side entrance.

848. Winter Wren *Troglodytes troglodytes* Jiaoliao   PLATE 105

Description: Tiny (10 cm) brown, barred and spotted babbler-like bird with cocked tail and fine bill. Dark russet-brown plumage with narrow black bars and an indistinct buffy eyebrow diagnostic. Races vary in ground colours. Palest race *tianshanicus* in NW China and darkest race *nipalensis* in Himalayas. Iris—brown; bill—brown; feet—brown.

Voice: Harsh scolding *chur*, hard *tic-tic-tic* and powerful melodious song of high clear notes and trills.

Range: Holarctic S to NW Africa, N India, NE Burma, Himalayas, China and Japan.

Distribution and status: Breeds in conifer forest and moors of China in NE, NW, N, C and SW China, Taiwan and eastern edge of Qinghai–Xizang Plateau. Seven races recognised: *tianshanicus* in NW China; *nipalensis* in C Xizang; *szetschuanus* in SE and E Xizang, Sichuan, S Qinghai, S Gansu, S Shaanxi and W Hubei; *talifuensis* in Yuman; *idius* in E Qinghai, N Gansu, E Nei Mongol, Hebei, Hunan and Shaanxi; *dauricus* in NE China; and *taivanus* on Taiwan. Northern birds migrate to winter in E and S coastal provinces.

Habits: Cocky behaviour. Constantly flicks erect tail. Creeps under cover to pop out and scold observer then flick away. Flight low and whirring for only short distances. Roosts in communal huddles in crevices in winter.

TITS—Family: Paridae

True tits are small perching birds. They are agile, active and acrobatic with sharp little bills with which they will hammer furiously to open cracks to get at insects or seeds. They are aggressive toward other birds. Tits nest in tree holes, except penduline tits which make hanging nests. The tits are well represented in N America and Eurasia and there are 23 species in China.

849. White-crowned Penduline Tit *Remiz coronatus*   PLATE 88
Panque

Description: Male: small (11 cm) delicate, pale tit with black forehead and face mask, sometimes extending onto rear crown but separated from chestnut mantle by white collar. Female: duller with grey on crown and collar. Juvenile: more uniform buffy-brown with slight dark mask. Distinguished from Chinese Penduline Tit by paler underparts and broad black forehead and whitish collar in adults. Race *stoliczkae* is slightly paler than nominate race. Iris—reddish-brown; bill—dark brown to grey; feet—dark grey.

Voice: Thin *pseee* call; *swee-swee* phrase; plaintive *tsi* notes and in flight *ti-ti-ti-ti-ti* call.

Range: C Asia to NW China and SE Russia.

Distribution and status: Race *stoliczkae* is uncommon in riverine forest and willow scrub of Xinjiang and Ningxia provinces up to 2400 m. Race *coronatus* occurs in West Ili valley of Junggar Basin, NW Xinjiang.

Habits: Forms flocks in winter. Generally more tree-living than other penduline tits.

Note: Formerly sometimes treated as a subspecies of Eurasian penduline Tit *R. pendulinus* (e.g. Cheng 1987, 1994).

850. Chinese Penduline Tit *Remiz consobrinus*   PLATE 88
Zhonghua panque

Description: Male: small (11 cm) delicate tit with grey crown, black face mask, rufous back and notched tail. Female and juvenile patterned like male but duller with faint dark mask. Iris—dark brown; bill—greyish-black; feet—bluish-grey.

Voice: High-pitched, soft, thin penetrating whistle *tsee*; fuller *piu* and fast series of *siu* notes. Song is finch-like with *si-si-tiu* phrases on '*tea-cher*' theme.

Range: Extreme E Russia and NE China, migrating to Japan, Korea and E China.

Distribution and status: Considered uncommon in N China but increasingly common in winter in E China, Japan and as far S as Hong Kong.

Habits: Forms flocks in winter and much attached to reed bed habitat.

Note: Sometimes treated as a subspecies of Eurasian Penduline Tit *R. pendulinus* (e.g. Cheng 1987, 1994).

851. Fire-capped Tit *Cephalopyrus flammiceps* PLATE 87
Huoguanque

Description: Very small (10 cm) tit, looks like flowerpecker. Male: combination of rufous forecrown and centre of throat, yellow sides of throat and breast, olive upperparts with yellow wing bar diagnostic. Female: dull yellowish-olive with buff underparts, yellow wing bar and pale line through eye. Immature has white underparts. Race *olivaceus* is greener than nominate. Iris—brown; bill—black; feet—grey.

Voice: High-pitched *tsit, tsit* and soft *whitoo-whitoo*. Song comprised of thin, high-pitched phrases rather like Coal Tit.

Range: Himalayas, SW and C China, rare visitor to N Thailand.

Distribution and status: Nominate race occurs in extreme SW Xizang; race *olivaceus* is uncommon resident in Yunnan, Sichuan, E Xizang, Guizhou and S Gansu; in hill and montane forest and forest edge up to 3000 m.

Habits: Lives in flocks, feeding in tree-tops.

852. Marsh Tit *Parus palustris* [>Black-bibbed Tit] PLATE 88
Zhaoze shanque

Description: Small (11.5 cm) tit with black cap and chin, brownish or olive upperparts and whitish underparts with buffy flanks. Lacks wing bars or nuchal patch. Difficult to separate from Willow Tit but generally lacks pale wing panel and has glossier black cap. Race *hypermelaena* has olive-green tinge to upperparts and may show short ragged crest; *dejeani* is similar but has less glossy crown; *brevirostris* has greyer upperparts, paler underparts and pale wing panel; while *hellmeyeri* has browner upperparts. Iris—dark brown; bill—blackish; feet—dark grey.

Voice: Explosive *pitchou* separates from Willow Tit. Repeated whistle *chiu-chiu-chiu*, churring and typical tit *tseet* calls. Song is repeated monosyllabic or disyllabic motifs.

Range: Disjunct distribution in temperate Europe and E Asia

Distribution and status: Rather common in NE China (*brevirostris*), E China (*hellmayri*), C China (*hypermelaena*) and SW China (*dejeani*).

Habits: Generally singly or in pairs; sometimes joining mixed flocks. Prefers deciduous woodlands of oak and other trees; also thickets, hedgerows, riverine woods and orchards.

Note: Black-bibbed Tit *P. p. hypermelaena* (including form *dejeani*) is sometimes treated as a separate species (Harrap and Quinn 1996) but is certainly part of the *palustris* superspecies.

853. Willow Tit *Parus montanus* [>Songar Tit]   PLATE **88**
Hetou shanque

Description: Small (11.5 cm) tit with brownish black cap and chin, brownish-grey upperparts and whitish underparts with buffy flanks. Lacks wing bars or nuchal patch. Difficult to separate from Marsh Tit but usually has pale wing panel and has matt, and more extensive, black cap. The head is proportionally larger. Races vary slightly: *songarus* has ochre-brown back and cinnamon flanks; *stoetzneri* is similar but with brown cap; *affinis* also has brown cap and pinkish wash on underparts; *weigoldicus* has blackish-brown cap and pinkish underparts; and *baicalensis* has grey upperparts and only a slight buff wash on flanks; Iris—brown; bill—blackish; feet—dark bluish-grey.

Voice: Nasal *dzee* and *tchay* notes, introduced by thin *si-si* phrases and often preceded by loud sharp *tzit* or *tzit-tzit* in contrast to explosive *pitchou* of Marsh Tit. Song varies across distribution, generally of long notes of almost constant pitch e.g. *duu-duu-duu-duu* and *s'pee-s'pee-s'pee-s'pee*.

Range: Europe and N Asia to Japan

Distribution and status: Rather common in conifer forest at moderate altitudes of NE China and Altai Mts of NW China (*baicalensis*), Tekes Valley of Tianshan (*songarus*), NC China (*affinis*), SC and SW China (*weigoldicus*) and N China (*stoetzneri*).

Habits: As Marsh Tit but preferring damp forests.

Note: Songar Tit *P. m. songarus* (including form *stoetzneri*) is sometimes treated as a separate species (Harrap and Quinn 1996) but is certainly part of the *montanus* superspecies.

854. White-browed Tit *Parus superciliosus* Baimei shanque   PLATE **88**

Description: Small (13 cm) tit with prominent white eyebrow. Crown and bib black; forehead white, extending as long white eyebrow line. Sides of head, flanks and abdomen cinnamon; vent buff. Upperparts dark grey washed olive. Separated from Rusty-breasted Tit by lack of white cheeks and occurs in less forested habitat. Iris—brown; bill—black; feet—blackish.

Voice: Noisy clear ringing whistles, insect-like rattles and trills. Song variable and complex.

Range: Endemic to WC China.

Distribution and status: Not uncommon in Nanshan and E Qinghai to S Gansu and Songpan area of N Sichuan and south to Yirong range in Xizang. Also recorded in Lhasa area of S Xizang.

Habits: Small flocks sometimes mix with tit-warblers foraging in dwarf juniper and rhododendron scrub of alpine zone.

855. Rusty-breasted Tit *Parus davidi* [Red-bellied/   PLATE **88**
Pére David's Tit] Hongfu shanque

Description: Small (13 cm) distinctive tit with dull black head and bib, fluffy white cheek patches, rufous nuchal collar and chestnut-orange underparts. Back, wings and tail olive-grey; flight feathers edged pale. Iris—dark brown; bill—black; feet—dark grey.

Voice: Sibilant *psit*; chattering *chit'it'it* and simple 'chick-a-dee' call *t'sip't-zee*.

Range: Endemic to C China.

Distribution and status: An occasional bird in mountains of S Shaanxi, S Gansu and N and W Sichuan from 2400–3400 m. May occur in W Hubei.

Habits: Lives in small flocks working in active and agile manner through canopy of broadleaf, birch, mixed and conifer forests.

[856. Siberian Tit *Parus cinctus* Xiboliya shanque   PLATE 88

Description: Small (13 cm) tit with diagnostic ashy-brown crown, nape and mantle. Otherwise like a pale Willow Tit. Iris—brown; bill—blackish; feet—dark grey.

Voice: Variable with *chik-chik-chik* call, repeated *psiup* notes and insect-like vibrant phrases as a song.

Range: Alaska, N Scandinavia, N Russia to E Siberia and as far S as Altai Mts and N Mongolia.

Distribution and status: Reported by David and Oustalet (1877) and by Shaw (1936) as resident in NE China in Nei Mongol and Hebei respectively but there are no recent records and status in China remains uncertain. Possibly a rare winter visitor or possibly the earlier records were misidentified Willow Tits.

Habits: Can semi-hibernate at night, roosts in tree holes and has special barb adaptations for fluffing-out plumage for long periods to keep warm.]

857. Rufous-naped Tit *Parus rufonuchalis* [Dark Grey/Black/   PLATE 88
Simla Black/Black Crested Tit] Zongzhen shanque

Description: Dark, robust, (13 cm) crested tit with black bib reaching upper belly. Lower belly grey with rufous undertail coverts. Flight feathers black. White cheek patch prominent below eye. Narrow white nuchal patch. Distinguished from Rufous-vented Tit by more extensive black bib and blackish flight feathers. Juvenile duller. Iris—brown; bill—black; feet—dark grey.

Voice: Vocal with plaintive *cheep*, more complex phrases, deeper *chut chut* call, squeaky *trip'ip'ip*. Songs include territorial trill and whistling song.

Range: C Asia, W Himalayas, NW China.

Distribution and status: Uncommon in conifer forest and juniper scrub at 2900–3500 m in W Tianshan and Kashi, W Xinjiang; also western S Xizang.

Habits: Lives at high altitudes; uses stone 'anvils' for cracking nuts.

Note: Formerly treated as a race of Rufous-vented Tit (e.g. Cheng 1987).

858. Rufous-vented Tit *Parus rubidiventris* [Sikkim Black/   PLATE 88
Rufous-bellied Crested/Tit, AG:Crested Tit] Heiguan shanque

Description: Small (12 cm) crested tit with black crest and bib, white cheeks, grey upperparts without wing bar and grey underparts with rufous vent. Distinguished from Rufous-naped Tit by less extensive black bib and grey flight feathers. Juvenile duller with shorter crest. Iris—brown; bill—black; feet—bluish-grey.

Voice: Thin high-pitched *seet*, sharp *chit*, querulous scolding *chit'it'it'it* and more complex phrases. Stony 'rattle song' *chip,chip,chip,chip* . . . and pure notes and slurred whistles and trills.

Range: Himalayas and WC China.

Distribution and status: Race *beavani* is uncommon in conifer forests from 2500 m to tree-line in the eastern Himalayas and the eastern connecting ranges of the Qinghai–Xizang Plateau in S Xizang, NW Yunnan, W and N Sichuan, S Gansu, S Shaanxi (Qinling Mts), and Qinghai.

Habits: Found in pairs or small parties often in mixed flocks.

859. Coal Tit *Parus ater* Mei shanque   PLATE 87

Description: Small (11 cm) tit with black crown, sides of neck, throat and upper breast. There are two white wing bars and a large white patch on the nape which distinguishes it from Willow and Marsh Tits. Back is grey or olive-grey and belly is white with or without buff. Most races have a pointed black crest. Distinguished from Great and Green-backed Tits by lack of black breast stripe. Crest is very small in *ater* and *insularis*, short in *rufipectus*, medium in *pekinensis*, long in *aemodius* and *kuatunensis* and very long in *ptilosus*. Underparts are whitish in *ater* and *ptilosus*, buffish in *pekinensis, insularis* and *kuatunensis* and pinkish-buff in *aemodius* and *rufipectus*. Race *rufipectus* has cinnamon undertail coverts. Iris—brown; bill—black with grey cutting edges; feet—slaty-grey.

Voice: Foraging *pseet* call; alarm *tsee see see see see*; song like weak Great Tit.

Range: Europe, N Africa and Mediterranean E to China, Siberia and Japan.

Distribution and status: Common in conifer forests of NE China (*rufipectus*), S Xizang and C China (*aemodius*), NE China (*ater*), eastern N China (*pekinensis*), Wuyishan and other SE China mountains (*kuatunensis*) and Taiwan (*ptilosus*). Japanese race *insularis* sometimes winters along coast of NE China.

Habits: Hardy tit. Caches food to use in winter. Feeds on underside of snow-covered branches.

860. Yellow-bellied Tit *Parus venustulus* Huangfu shanque   PLATE 87

Description: Small (10 cm) short-tailed tit with yellow underparts, two rows of white spots on wings and very short bill. Male has black head and bib with white cheek patch and nuchal spot; blue-grey upperparts and silvery rump. Female has greyer head with white throat separated from cheek patch by grey malar stripe and slight pale superciliary spot. Juvenile: like female but duller and more olive on upperparts. Smaller and lacks black abdominal stripe of Great and Green-backed Tits. Iris—brown; bill—blackish; feet—bluish-grey.

Voice: High-pitched, nasal *si-si-si-si*. Song is repeated mono or disyllabic motif like Coal Tit but more powerful.

Range: Endemic to SE China.

Distribution and status: Locally common in mixed deciduous forests of S, SE, C and E China as far north as Beijing; up to 3000 m in summer, lower in winter.

Habits: Lives in flocks in wooded areas. Subject to periodic irruptions.

861. Grey-crested Tit *Parus dichrous* [Brown Crested Tit]   PLATE 87
Heguan shanque

Description: Small (12 cm) plain-coloured tit with prominent crest but lacking any black or yellow in plumage and with contrasting buff-white half collar. Upperparts generally dull grey; underparts vary with race from buff to cinnamon. Race *dichrous* has whitish submoustachial stripe, greyish-brown throat contrasting with cinnamon-buff breast and dark grey upperparts; *wellsi* is paler buff below with no contrast between throat and breast; *dichroides* is similar to *wellsi* with crown and crest a paler grey than rest of upperparts. Iris—reddish-brown; bill—blackish; feet—bluish-grey.

Voice: Varied calls include rapid *ti-ti-ti-ti*; thin *sip-pi-pi*, plaintive *pee-di* and alarm *cheea, cheea*. Song includes trills and elements of other calls.

Range: Himalayas and WC China.

Distribution and status: Locally common in conifer forests from 2480–4000 m in E Himalayas of SE Xizang (*dichrous*); N and W Sichuan and N and W Yunnan (*wellsi*) and from Qinling Mts of S Shaanxi, S Gansu, S and E Qinghai and extreme N Sichuan (*dichroides*).

Habits: Shy and quiet in pairs or small parties.

862. Great Tit *Parus major* [Grey/Cinereous/Japanese Tit] PLATE 87
Da shanque

Description: Large (14 cm) plump black, grey and white tit. Head and throat glossy black except for large contrasting white patch on side of face and nape patch. Has single bold white wing bar; black band down centre of breast, broader in male and reduced to bib in juvenile. Six races vary slightly but *kapustini* in extreme N China has yellowish underparts and greenish back. The latter race can be confused with Green-backed Tit but there is no overlap in range and the latter has two white wing bars.

Voice: Highly vocal. Contact call is variant of cheery *pink tche-che-che*. Song of noisy whistles *chee-weet* or *chee-chee-choo*.

Range: Palearctic, India, China, Japan, SE Asia to Greater Sundas.

Distribution and status: Six races in three groups are generally common in open woodlands and gardens: *major* group—*kapustini* in extreme NE and NW China; *minor* group—*minor* in C, E, N and NE China, *tibetanus* on Xizang Plateau, *subtibetanus* in SW China and SE Xizang, *comixtus* in S and SE China and Taiwan; *cinereus* group—*hainanus* on Hainan.

Habits: Frequents mangroves, gardens and open forests. Active, versatile little bird which may be active in tree-tops or down at ground level. In pairs or small parties.

863. Turkestan Tit *Parus bokharensis* Xiyu shanque PLATE 87

Description: Large (15 cm) grey tit. Similar to and difficult to separate from grey races of Great Tit but with longer and graduated tail and purer grey upperparts. However, in China this species only overlaps with yellow-bellied race of Great Tit (*kapustini*) so confusion is unlikely. Iris—dark brown; bill—black; feet—slate-grey.

Voice: Similar to Great Tit but thinner and more plaintive. Song phrases on 'tea-cher' motif e.g. *pid-du, pid-du, pid-du*.

Range: C Asia to NW China and SW Mongolia.

Distribution and status: Uncommon or locally common in willow thickets, orchards and woods of extreme NW China around Junggar basin and Manas River onto slopes of Altai and Tianshan Mts to over 2000 m.

Habits: Similar to Great Tit.

864. Green-backed Tit *Parus monticolus* Lübei shanque PLATE 87

Description: Largish tit (13 cm) similar to yellow-bellied races of Great Tit but separated by green mantle and two white wing bars. Distribution in China only overlaps with white-bellied races of Great Tit. Race *yunnanensis* has brighter green upperparts than nominate form. Iris—brown; bill—black; feet—slaty-grey.

Voice: Similar to Great Tit but louder, shriller and clearer.

Range: Pakistan and Himalayas to S China, C Laos, Vietnam and Burma.

Distribution and status: Common in montane forests and forest edge in S Xizang (*monticolus*), C and SW China (*yunnanensis*) and Taiwan (*insperatus*) from 1100–4000 m.

Habits: Similar to Great Tit. Forms flocks in winter.

865. Yellow-cheeked Tit *Parus spilonotus* [Black-spotted Yellow Tit]   PLATE 87
Huangjia shanque

Description: Large (14 cm) tit with pronounced crest, showy black and yellow head pattern. Rest of plumage black, grey and white in race *rex* but with yellow wash on mantle and underparts in nominate race. Female: more greenish-yellow with two yellow wing bars. Immature has less black on underparts. Iris—brown; bill—dark grey or black; feet—bluish-grey.

Voice: Similar to Great Tit. Harsh churring; sharp *si-si-si*; *tee cher, tsee tsee-chi chi chi*; lisping *witch-a-witch-a-witch-a*. Song is repeated, ringing, three-note motif *chee-chee-piu*.

Range: E Himalayas to S China and Indochina.

Distribution and status: Common in open forests up to 2400 m in S Xizang and extreme W Yunnan (*spilinotus*) and S China provinces (*rex*).

Habits: Similar to Great Tit.

Note: Treated by Cheng (1987) as race of Black-lored Tit *P. xanthogenys*.

866. Yellow Tit *Parus holsti* [Taiwan Yellow Tit]   PLATE 87
Taiwan huangshanque

Description: Small (13 cm) tit with long crest and lemon-yellow underparts. Male: back and wing coverts black. Rear crest white. Wings light blue, tail blue with white outer edges. Yellow loral patch and black ventral spot. Female: duller with olive-green back. Lacks black ventral spot. Juvenile paler with whitish underparts. Iris—dark brown; bill—black; feet—grey.

Voice: Lively resonant repeated *chichishui-chichishui*; *zijide, zijide* or *jijiang, jijiang, jijiang*; scolding alarm call.

Range: Endemic to Taiwan.

Distribution and status: Rare to locally common in lower mountains of main range at 1000–2300 m in temperate broadleaf and conifer forests.

Habits: Lives in canopy in flocks.

867. Azure Tit *Parus cyanus* Huilan shanque   PLATE 88

Description: Small (13 cm) short-billed, fluffy whitish tit with longish tail. Markings similar to European Blue Tit, with which it can hybridise, but underparts and head white and body markings grey and lavender-blue. Wing bar, broad tips to secondaries and edge of tail white. Juvenile may have slight yellow wash on underparts. Distinguished from similar Yellow-breasted Tit by lack of contrasting yellow on breast. Iris—dark brown; bill—dark bluish-grey; feet—dark grey.

Voice: Contact call is slurred *tsirrup*, nasal *tsee-tsee-dze-dze*, alarm is scolding *chr-r-r-r-rit*. Song is varied and involves trills and '*tea-cher*' phrases.

Range: E Europe, Russia, Siberia, N Kazakhstan, Mongolia and N China.

Distribution and status: Race *tianschanicus* is locally common in woods in Heilongjiang, E Nei Mongol, N and W Xinjiang including Altai and Tianshan ranges.

Habits: Active noisy flocks keep to shelter of low trees and bushes in willow thickets, scrubby vegetation and orchards.

868. Yellow-breasted Tit *Parus flavipectus*   PLATE 88
Huangxiong shanque

Description: As Azure Tit (13 cm), with which some races hybridise, but with distinctive yellow breast patch. Iris—dark brown; bill—black; feet—dark grey.

Voice: Like Blue Tit. High-pitched buzzing and scolding notes. Complex song with high pitched trills.

Range: C Asia and N China.

Distribution and status: Race *berezowskii* is a disjunct, uncommon and poorly-known population in the Qilian Mts of Qinghai.

Habits: Singly, in pairs or small flocks, sometimes mixing with other species.

Note: Treated by Cheng (1987) as a race of Azure Tit. Harrap and Quinn (1996) suggest this isolated race might be best treated as a full species.

869. Varied Tit *Parus varius* Zase shanque   PLATE 88

Description: Small (12 cm) distinctive tit with forehead, lores and cheek patches pale buff to rufous; bib and crown dull black with pale medial line on back of head; rufous nuchal collar and grey upperparts. Underparts are chestnut-rufous with buffy ventral line. Taiwan race is smaller and darker without median line on head or pale ventral line. Juvenile duller and paler. Iris—brown; bill—black; feet—grey.

Voice: Thin sharp *pit*, high-pitched *spit-spit-see-see* and scolding *ch-chi-chi*. Has variant of '*chick-a-dee*' call pattern. Song is rich and varied with monotonal pure whistled *peee* a characteristic component.

Range: Resident in NE China, Taiwan, Korea, Japan and Kuriles.

Distribution and status: Race *varius* is uncommon in E Liaoning and SW Jilin. Race *castaneoventris* is rather local in Taiwan from moderate altitudes to peaks of main mountain range. Recorded in Babaoshan and Mangdang, Guangdong Province.

Habits: Shy and secretive. Lives in pairs and occasionally small flocks. Feeds in canopy and stores nuts.

870. Yellow-browed Tit *Sylviparus modestus*   PLATE 87
Huangmei linque

Description: Small (10 cm) aberrant tit which looks like a leaf-warbler or flower-pecker. Plumage nondescript olive with short crest, narrow yellow eye-ring and short pale yellow eyebrow which may be concealed. Legs rather stocky. Distinguished from Fire-capped Tit by crest and lack of contrasting pale rump. Iris—dark brown; bill—horn with greyish base; feet—blue-grey.

Voice: High-pitched trills *si-si-si-si-si*; more mellow whistled units *piu-piu-piu* . . .

Range: Himalayas, Indochina and S China.

Distribution and status: Rather common in mixed conifer, evergreen and deciduous forests up to 2000 m in S Xizang, W Yunnan, Sichuan, Guizhou and Wuyi Mts of SE China.

Habits: Acrobatic and tit-like in movements. Erects crest and reveals pale eyebrow when alarmed or excited.

871. Sultan Tit *Melanochlora sultanea* Mianque   PLATE **88**

Description: Unmistakable large (20 cm) yellow and black tit with spectacular long fluffy yellow crest. Female similar to male, but throat and breast are dark olive-yellow and upperparts washed olive. Race *flavocristata* has shorter crest than nominate race and tail only rarely tipped white. Race *serosa* has less brightly yellow crest. Iris—brown; bill—black; feet—grey.

Voice: Repeated, loud, squeaky whistle *tcheery-tcheery-tcheery* and shrill chattering alarm calls. Song is series of about five clear whistled *chiu* notes, repeated at intervals.

Range: Himalayas, S China, SE Asia and Malay Peninsula.

Distribution and status: Generally uncommon in lowland evergreen forest. Race *sultanea* in Xishuangbanna, SW Yunnan; *flavocristata* on Hainan, and *serosa* in S and SE China.

Habits: Lives in canopy of primary and secondary forest in mixed flocks actively chasing large insects.

LONG-TAILED TITS—Family: Aegithalidae

Long-tailed tits are small agile perching birds with small, sharp, conical beaks and longish to very long tails. They are active foragers of insects and seeds and usually live in small flocks. They make nests in suspended pouch-like nests. There are four species in China.

872. Long-tailed Tit *Aegithalos caudatus*   PLATE **88**
Yinhou [changwei] shanque

Description: Beautiful fluffy little tit (16 cm) with tiny black bill and very long white-edged black tail. Different races vary in colour pattern. Birds in NE China (nominate) are almost completely white-bodied but juveniles have black sides to head. Birds in Yangtze valley (*glaucogularis*) have a broad black brow, brown and black wing pattern and pinkish tinge on the underparts. Juvenile has paler underparts and rufous breast. Birds in E China (*vinaceus*) are similar to *glaucogularis* but paler. Iris—dark brown; bill—black; feet—dark brown.

Voice: The call is a single short *ssrit* or in alarm a thin metallic trill *seehwiwiwiwi*. Birds also give dry churring calls and a high-pitched *seeh-seeh-seeh* especially as a contact call in flight.

Range: Many races occur across Europe and temperate Asia.

Distribution and status: Common in open forest and forest edges over NE (nominate), SW to C and N China (*vinaceus*) and C to E China (*glaucogularis*).

Habits: Lives in small active flocks feeding on insects and some seeds in canopy and low bushes. Roost huddled in a row.

873. **Black-throated Tit** *Aegithalos concinnus*   PLATE 88
Hongtou [changwei] shanque

Description: Small (10 cm) active, elegant tit. Different races vary. Crown and nape rufous, broad eye-stripe black, chin and throat white with round black bib. Underparts white with varying amounts of chestnut. Races *talifuensis* and *concinnus* have white lower breast and belly with rich chestnut breast band and flanks; the former being slightly darker. Race *iredalei* has deeper buff underparts with a wash of cinnamon across breast and on flanks. Mantle and wings grey and tail blackish with white edges. Juvenile has paler crown, white throat and narrow black gorget. Iris—yellow; bill—black; feet—orange.

Voice: Similar to Long-tailed Tit. Thin contact *psip, psip*; low churring notes *chrr, trrt, trrt*; sibilant *si-si-si-si-li-u* and high twittering song.

Range: Himalayas, Burma, Indochina, S and C China.

Distribution and status: Common in open pine and broadleaf forest at 1400–3200 m. Races *iredalei* in S Xizang, *talifuensis* in SW China and *concinnus* across C, S, SE China and Taiwan. Grey crowned race *pulchellus* may occur in Xishuangbanna W of the Mekong (Lancang) in SW Yunnan.

Habits: Active in large flocks often mixed with other species.

874. **Rufous-fronted Tit** *Aegithalos iouschistos*   PLATE 88
[Black-headed Tit] Heitou [changwei] shanque

Description: Small (11 cm) delicate tit. Sides of head black; coronal stripe, moustachial stripe, ear coverts and side of neck cinnamon-buff; Back, wings and tail uniform grey. Underparts cinnamon-rufous with silvery bib lightly streaked black and edged by inverted black V. Juvenile is paler and lacks contrasting bib. Iris—yellow; bill—black; feet—brown.

Voice: Similar to Long-tailed Tit. Calls include repetitive *see-see-see-see* and *trrup*. Alarm is shrill *zeet, zeet* or churred *trr-trr-trr*.

Range: E Himalayas and Burma.

Distribution and status: Nominate race is a common bird of broadleaf and montane conifer and hill forests up to 3600 m in the E Himalayas of SE Xizang.

Habits: Feeds in flocks in small trees and undergrowth.

875. **Black-browed Tit** *Aegithalos bonvaloti*   PLATE 88
Heimei [changwei] shanque

Description: Small (11 cm), delicate tit; similar to Rufous-fronted Tit but paler with white forehead and edge to bib and white lower breast and belly. Race *obscuratus* is similar to nominate but duller, darker and browner. Iris—yellow; bill—black; feet—brown.

Voice: Similar to Long-tailed Tit.

Range: SE Tibetan Plateau, W and N Burma, C and SW China.

Distribution and status: Common resident in SE Xizang and SW China (*bonvaloti*) and NC Sichuan (*obscuratus*).

Habits: As Rufous-fronted Tit.

Note: Formerly included as race of Rufous-fronted Tit (e.g. Cheng 1987) but separated as *per* Wunderlich (1991) and Inskipp *et al.* (1996).

876. Sooty Tit *Aegithalos fuliginosus* [Silver-faced/White-necklaced Tit] PLATE 88
Yinlian [changwei] shanque

Description: Small (12 cm) tit with grey throat and white upper breast forming contrasting necklace. Sides of crown and face silvery-grey; nape buff-brown; crown and upperparts brown. Tail is brown laterally edged white. Abdomen has greyish-brown gorget band and rufous flanks. Underparts otherwise white. Juvenile paler with white forehead and coronal band. Iris—yellow; bill—black; feet—pinkish to blackish.
Voice: Typical of genus. Thin *sit*; high silvery *si-si-si, si-si*; a rolling *sirrrup* and hard rattling *chrrr* calls.
Range: Endemic to WC China.
Distribution and status: Resident in SW Hubei (Shennongjia), S Shaanxi (Qinling Mts), S Gansu (Baishuijiang) and south to Minshan, mid-Yalong and Dafengding, Mabian areas of Sichuan. An uncommon bird from 1000–2600 m.
Habits: Flocks live in deciduous broadleaf forests and prickly oak forests.

SWALLOWS—Family: Hirundinidae

Familiar, worldwide family of graceful birds with slender bodies and long, pointed wings. Swallows are gregarious and catch insects in mid-air, hawking to and fro along waterways or circling high in sky. The superficially resemble swifts but the flight is not so fast; swallows glide with wings half closed, unlike swifts which glide with sickle wings fully extended. The sexes look alike. Unlike swifts, swallows frequently perch in trees, on telegraph wires, television aerials, poles or houses and will settle on the ground to drink at pools, collect mud for nests and occasionally to catch ants and other insects. Swallows nest in cup-shaped nests of mud built under house roofs or on cliff underhangs; some species burrow in banks. Swallows are famous for their migratory abilities. There are 11 species of swallows recorded in China. One other species may occur or may already be extinct.

RIVER MARTINS—Subfamily: Pseudochelidoninae

[877. White-eyed River Martin *Pseudochelidon sirintarae* PLATE 89
Baiyan heyan

Description: Largish (15 cm) martin with diagnostic white spectacles around eyes, broad yellow bill and rounded tail with streamers. Broad triangular wings similar to woodswallow. Rump white. Iris—whitish-yellow; bill—yellow; feet—pink.
Voice: Not recorded.
Range: Winters in C Thailand; breeding area unknown.
Distribution and status: Not recorded in China but birds migrate S to Thailand and it has been speculated the species must breed somewhere in S China. Not recorded for several years in Thailand and now possibly extinct.
Habits: Roosts in reed beds but probably lives along shingle banks of large rivers.]

SWALLOWS AND MARTINS—Subfamily: Hirundininae

878. Sand/Pale Martins *Riparia riparia/diluta*   PLATE 89
[Common/Gorgeted Sand-Martin/Bank Swallow] Ya shayan

Description: Small (12 cm) brown swallow with white underparts and a diagnostic brown breast band. Immature has buff throat. Iris—brown; bill and feet—black.
Voice: Shrill twitters.
Range: Cosmopolitan (except Australia). Eurasian birds winter S to SE Asia and Philippines.
Distribution and status: Locally common along sand-banked rivers at all altitudes. Race *diluta* breeds in NW China; *tibetana* on the Qinghai–Xizang plateau; *ijimae* in NE China; and *fokiensis* in C and E China. All races migrate S in winter with populations of *ijimae* and *fokiensis* wintering in S China.
Habits: Lives over marshes and rivers making sweeping flights over water or perches on prominent dead branches.
Note: Races *diluta, tibetana* and *fokiensis* are now treated as a separate species Pale Martin *R. diluta.*

879. Plain Martin *Riparia paludicola* Hehou shayan   PLATE 89

Description: Small (12 cm) dull grey-brown swallow with slightly forked tail. Distinguished from crag martins by lack of white spots at end of tail and from Sand Martin by pale greyish-brown throat and breast, without dark breast band. Iris—brown; bill—black; feet—black.
Voice: Call is weak rasped *tschree.*
Range: Africa, S Asia, SE Asia, Philippines, SW China and Taiwan.
Distribution and status: Locally common up to 1000 m. Race *chinensis* occurs in tropical S Yunnan and Taiwan.
Habits: Colonial, living on marshes and along rivers especially with sand banks. Nests in holes in banks.

880. Eurasian Crag Martin *Hirundo rupestris*   PLATE 89
[Mountain Crag Martin] Yanyan

Description: Smallish (15 cm) dusky-brown martin with square-ended tail with white spots near tip. Similar to Dusky Crag Martin but paler with contrast from below in flight between dark underwing, undertail and tail and paler crown flight feathers, throat and breast. Iris—brown; bill—black; feet—brownish-flesh.
Voice: Call is rather weak *tshree.*
Range: S Palearctic wintering to S Asia and N Africa.
Distribution and status: Nominate race is rather uncommon across a wide area of W, N, C, and SW China from 1800–4600 m. Some birds from N of range winter in SW China.
Habits: Inhabits cliffs in mountainous regions and dry river gorges. Occasionally on buildings.

881. Dusky Crag Martin *Hirundo concolor* PLATE 89
Chunse yanyan

Description: Small (12.5 cm) all-blackish martin with white spots near tip of rather square tail. Tail and wing broader than swifts. Distinguished from Eurasian Crag Martin by being darker and abdomen as dark as undertail coverts. Iris—dark brown; bill—blackish-brown; feet—brown.

Voice: Soft low *chit-chit* uttered at rest and in flight.

Range: S and SE Asia.

Distribution and status: Uncommon resident in S Yunnan and SE Xizang at 1000–2000 m.

Habits: Inhabits mountains with cliffs. Similar to Eurasian Crag Martin but more southerly distribution. Non-migratory.

882. Barn Swallow *Hirundo rustica* [Common/House/ PLATE 89
Rustic/The Swallow] Jiayan

Description: Medium-sized (20 cm including elongated tail feathers) glossy blue and white swallow. Upperparts steely-blue; breast reddish-edged with a blue band, belly white; tail very elongated and with white spots near tips of feathers. Distinguished from Pacific Swallow by cleanerwhite belly, more elongated tail and blue chest bar. Immature has duller plumage, lacks tail streamers and is more difficult to distinguish from Pacific Swallow. Iris—brown; bill and feet—black.

Voice: High-pitched *twit* and twittering calls.

Range: Nearly worldwide. Breeds in northern latitudes and migrates S in winter through Africa, Asia, SE Asia, Philippines and Indonesia to New Guinea and Australia.

Distribution and status: Nominate race breeds in NW China; *tytleri* and *mandschurica* breed in NE China; and *gutturalis* breeds over the rest of the country. Most birds migrate S in winter but resident populations remain in S Yunnan, Hainan and Taiwan.

Habits: Glides and circles high in the sky, or low over land or water to catch small insects. Alights on dead branches, poles and telegraph wires to perch. Feeds independently but with large numbers feeding at same site. Sometimes congregate in large roosting flocks, even in cities.

883. Pacific Swallow *Hirundo tahitica* [Eastern House/ PLATE 89
Small House Swallow/Least Swallow] Yangban yan

Description and status: Smallish (14 cm) blue, red and buff swallow. Upperparts steely-blue; forehead chestnut. Distinguished from Barn Swallow by dirty white underparts, less elongated tail without streamers, lack of dark blue chest bar and slightly smaller and less smart appearance. Iris—brown; bill—black; feet—black.

Voice: Pleasant twittering calls and high-pitched *tweet* in alarm.

Range: S India, SE Asia, Philippines, Malay Peninsula and Sundas to New Guinea and Tahiti.

Distribution and status: A common resident of open spaces especially over water on Taiwan, Lanyu and Huoshao islets and islands of S China Sea.

Habits: Generally found in small, loose parties, working independently in circles or gliding low over water. The nest is a cup of mud pellets stuck under a roof or bridge or rock overhang. It has an open entrance in the rim.

884. Red–rumped Swallow *Hirundo daurica* [Lesser Striated/   PLATE 89
Ceylon Swallow] Jinyao yan

Description: Large (18 cm) swallow with pale chestnut rump contrasting with steely dark blue upperparts; underparts white, finely streaked with black. The tail is long and deeply forked. Cannot be reliably separated in field from Striated Swallow, but Striated only has limited distribution in China. Iris—brown; bill and feet—black.
Voice: Shrill calls in flight.
Range: Breeds in Eurasia and parts of India; migrating S in winter to Africa, S India and SE Asia.
Distribution and status: Very common over most of China. Nominate race breeds in NE China; *japonica* breeds over all eastern China with resident populations in Guangdong and Fujian; *nipalensis* breeds in S Xizang and W Yunnan; *gephrya* breeds in eastern parts of Qinghai–Xizang plateau to Gansu, Ningxia, Sichuan and N Yunnan; recorded on migration in SE China.
Habits: Like Barn Swallow.
Note: Some authors place the Striated Swallow in this species but see Vaurie (1951).

885. Striated Swallow *Hirundo striolata* [Greater Striated Swallow]   PLATE 89
Banyao yan

Description: Large (20 cm) swallow with streaky chest and red rump. Upperparts steely-blue; underparts dirty white with black streaks; tail deeply forked. Cannot be reliably separated in field from Red-rumped Swallow. Iris—brown; bill—black; feet—dark brown.
Voice: Usually silent but sometimes gives loud *chew-chew* or vibrating *schwirrr*.
Range: NE India, SE Asia, Philippines, Java, Bali and Lesser Sundas.
Distribution and status: Nominate race breeds in W and S Yunnan; race *stanfordi* is resident on Taiwan, Generally common up to 1500 m.
Habits: Similar to other swallows but keeps more to lowlands near cultivated areas. Lives in pairs or small flocks and flies with slower beat and more soaring than other swallows. The nest is a cup made of mud pellets with a tunnel entrance, plastered on the undersurface of a ceiling or overhang.

886. Northern House Martin *Delichon urbica*   PLATE 89
[Common House-Martin] Maojiaoyan

Description: Small (13 cm) steel-blue martin with forked tail. Whitish underparts and white rump. Distinguished with difficulty from Asian House Martin by pure white rather than dusky white breast, more extensive white on rump and more forked tail. Iris—dark brown; bill—black; feet—pink with white feathering as far as the toes.
Voice: Call is dry rolling scratchy *prreet*, less rasped than Sand Martin. Alarm call is high *seerr* or *jeet*. Song is soft twittering, on same pitch as call note.
Range: Africa, Eurasia, NE Pakistan and NW India and N China; winters in SE Asia, India and Nepal.

Distribution and status: Fairly common. Nominate race breeds in extreme NW and W China, wintering to India; *lagopoda* breeds in NE China and winters in E, SE and S China. All Hong Kong records now attributed to Asian House Martin.

Habits: Gregarious; nests on cliffs. Mixes to feed with other swallows and swifts.

887. Asian House Martin *Delichon dasypus* PLATE 89
[Asiatic House-Swallow] Yanfu maojiaoyan

Description: Small (13 cm) plump black swallow with white rump, shallow forked tail and greyish underparts. Upperparts steely-blue; rump white; breast dusky-white. Distinguished from Northern House Martin by black wing linings. Race *nigrimentalis* has white underparts. Iris—brown; bill—black; feet—pink with white feathering as far as the toes.

Voice: Excited whickering calls, similar to Northern House Martin.

Range: Breeds Himalayas to Japan and winters S to SE Asia, Philippines (rarely) and Greater Sundas.

Distribution and status: Locally very common. Race *cashmiriensis* breeds on Qinghai–Xizang plateau and EC China, migrating S in winter; race *nigrimentalis* is resident on Taiwan and in S and SE China; nominate race is recorded on passage on E coast.

Habits: Single birds or small flocks mix with other swallows or swiftlets. More aerial than other swallows and mainly seen in soaring flight.

888. Nepal House Martin *Delichon nipalensis* PLATE 89
Heihou maojiaoyan

Description: Small (12 cm) steely blue-black and white martin. Like Northern and Asian House Martins but smaller with narrower white rump band, narrow squarer-cut tail, white nuchal collar, dull black chin and throat and glossy black undertail coverts. Iris—brown; bill—black; feet—brown with white feathering as far as the toes.

Voice: Rather silent. Occasional short, high-pitched *chi-i* in flight.

Range: Himalayas, W Burma, Laos and NW Tonkin.

Distribution and status: Uncommon in China in extreme SW Yunnan and SE Xizang. Lives from 2000–4000 m in summer, down to 350 m in winter.

Habits: Inhabits river valleys and mountain cliffs. Gregarious. Flight is smooth-flowing with swoops, glides and tight turns.

CRESTS—Family: Regulidae

Tiny warbler-like passerines with bright crown bands. Mostly live singly in canopy of conifer forest. Two species occur in China.

889. Goldcrest *Regulus regulus* Daijü PLATE 99

Description: Tiny (9 cm) brightly coloured, greenish warbler-like bird with black and white patterned wing and diagnostic golden-yellow or orange (male) median coronal stripe, edged by black lateral stripes. Upperparts uniform olive-green to yellow-green; underparts greyish or buffish-white with yellow-green flanks. Pale area around dark eye gives bland, beady-eyed look. Unlikely to be confused with any other Chinese warbler. Races vary in details. Race *coatsi* is paler than other races; *japonensis* is darker with greyer nape and broader white wing bars; *himalayensis* has whiter underparts; *sikkimensis* is darker and greener than *himalayensis*, while *yunnanensis* is even darker and greener above with buffy underparts and grey flanks; *tristis* almost lacks black lateral crown stripes with duller underparts. Juvenile lacks coronal stripes and could be confused for some *Phylloscopus* warblers but lacks eye-stripe or supercilium and characterised by large head, grey ocular region and beady-eyed appearance. Iris—dark brown; bill—black; feet— brownish.

Voice: Thin, high-pitched *sree sree sree*. Alarm is emphatic *tseet*. Song is high-pitched with repeated phrases ending in flourish.

Range: Palearctic from Europe to Siberia and Japan including C Asia, Himalayas and China.

Distribution and status: Common in most temperate and subalpine conifer forests. Race *coatsi* winters in Nanshan and perhaps Altai Mts; *japonensis* is resident or summer breeder in NE China wintering to E China and Taiwan; *sikkimensis* is resident in E Himalayas to S Xizang and W China; *yunnanensis* in S Gansu and S Shaanxi S through Sichuan to Yunnan; and *tristis* in Tianshan of NW Xinjiang.

Habits: Usually alone in lower canopy of conifer forest. Joins bird waves.

890. Flamecrest *Regulus goodfellowi* Huoguan daijü — PLATE 99

Description: Tiny (9 cm) warbler-like bird with blackish crown and bright orange (yellow in female) median coronal stripe, posteriorly yellow; black eye patch surrounded by white ring which extends into white eyebrow and lores. Has black moustachial stripe and white throat. Nuchal collar and breast grey; back olive; abdomen, rump and vent yellow. Wings and tail are black with white wing bars and yellow edges to primaries. Iris—brown; bill—black; feet—dark grey.

Voice: Regular call is shrill *see-see*.

Range: Endemic to Taiwan.

Distribution and status: Frequent bird of coniferous and montane forests of the main mountain range at 2000–3000 m.

Habits: Regularly mixes with tits and other species in mixed flocks. Descends in winter.

BULBULS—Family: Pycnonotidae

A large African and Asian family of short-necked, short-winged birds with longish tails and rather slender bills. They have soft fluffy plumage and several have an erectile crest. Plumage of the sexes is similar and most bulbuls are rather dull in colour with, at most, yellow, orange, black or white patterns. Bulbuls are primarily frugivorous although they also eat insects. They are confident birds with lively, and in some species, very musical songs. They tend to be arboreal and make rather untidy cup-shaped nests in trees. Generally not migratory, though the Black Bulbul is partially migratory. There are 27 species in China.

891. Crested Finchbill *Spizixos canifrons*   PLATE 91
Fengtou quezuibei

Description: Distinctive, large (22 cm) olive-green bulbul with thick, ivory-coloured, finch-like bill and prominent crest. Underparts are greenish-yellow. Distinguished from Collared Finchbill by grey forehead and cheeks, lack of white throat collar, and longer crest. Has broad black terminal band to tail. Iris—brown; bill—ivory; feet—pink.

Voice: Strident but melodious *purr-purr-prruit-prruit-prruit* in alarm. Chatters and long dry bubbling trill.

Range: NE India, SE Tibet, SW China, Burma and N Indochina.

Distribution and status: Race *ingrami* is fairly common in open country, secondary forest and farmland of SW Sichuan and Yunnan. Nominate race occurs in SE Xizang.

Habits: Birds live singly or in small flocks in open woodland, clearings, scrub and gardens up to 3000 m. Sometimes perch on telephone wires.

892. Collared Finchbill *Spizixos semitorques*   PLATE 91
Lü quezuibei

Description: Large (23 cm) greenish bulbul with heavy ivory-coloured bill and short crest. Similar to Crested Finchbill but with shorter crest; head and throat blackish with grey nape. White throat diagnostic; also whitish around base of bill and white streaks on cheeks. Tail green with black tip. Iris—brown; bill—pale yellow; feet—pinkish.

Voice: Cheerful fluty calls. Hurried, loud whistles *ji de shi shei, ji de shi shei, shi shei.*

Range: S China, Taiwan and N Indochina.

Distribution and status: Common in hills from 400–1400 m in S and SE China (nominate) and Taiwan (*cinereicapillus*).

Habits: Generally in secondary growth and scrub. Perches on telegraph wires or bamboo in small groups. Catches insects in flight.

893. Striated Bulbul *Pycnonotus striatus* Zongwen lübei   PLATE 90

Description: Medium-sized (20 cm) crested, olive-green bulbul with underparts heavily streaked pale yellow. Upperparts olive with fine white streaking. Has yellow eye-ring. The only other streaked green bulbul is Stripe-throated, which lacks any streaks on back and abdomen. Iris—red-brown; bill—black; feet—greyish-brown.

Voice: Sharp, mellow whistles *tyiwut*; trisyllabic *whee-too-wheet* call and loud *pyik . . . pyik*. Song is group warbling and chattering.

Range: Himalayas, NE India, N Burma, N Indochina to SW China.

Distribution and status: Race *paulus* is rare in SW Yunnan and race *arctus* occurs in extreme SE Xizang and Yunnan.

Habits: Lives in active, noisy flocks of 6–15 birds. Inhabits evergreen forest in mountains.

Note: Cheng (1987, 1994) places all Chinese specimens in nominate race.

894. Black-headed Bulbul *Pycnonotus atriceps*   PLATE 90
Heitou bei

Description: Medium-sized (17 cm) yellowish bulbul with diagnostic glossy black head, black throat and blue eye. Upperparts yellowish-olive; wings blackish; tail blackish with conspicuous yellow tips; underparts greenish-yellow. A rare colour form is grey with a white edge to tail. Distinguished from Black-crested Bulbul by lack of crest and yellow tip to tail. Iris—pale blue; bill—black; feet—dark brown.

Voice: Hesitant series of short whistles; ringing metallic *chewp*.

Range: NE India, SE Asia, Palawan and Greater Sundas.

Distribution and status: Nominate race is an uncommon resident of lowland areas up to 900 m in S Yunnan (Xishuangbanna).

Habits: Frequents forest edge, secondary forest and coastal scrub. Singly or in small flocks, often mixing with other species.

895. Black-crested Bulbul *Pycnonotus melanicterus* PLATE 90
[Black-headed/Black-crested Yellow Bulbul] Heiguan huangbei

Description: Medium-sized (18 cm) yellowish bulbul with black head and crest. Upperparts brownish-olive; underparts yellow. Race *flaviventris* has more yellow on belly. Iris—reddish; bill and feet—black.

Voice: Noisy, cheery, slurred whistles *hee-tee-hee-tee-weet* with last note falling.

Range: India, S China, SE Asia and Greater Sundas.

Distribution and status: Common in lowland and hill forests up to 1200 m. Race *flaviventris* occurs in S Yunnan (Xishuangbanna) and Guangxi. Nominate race occurs in SW Yunnan to west of Lancang (Mekong) River and also in SE Xizang.

Habits: Rather shy, favouring densely-leaved taller trees of the forest edge and secondary forests. Occasionally chases flying insects but generally actively searches for fruits. Erects crest when excited.

896. Red-whiskered Bulbul *Pycnonotus jocosus* PLATE 90
Hong'er bei

Description: Medium-sized (20 cm) bulbul with long narrow forward-pointing black crest and diagnostic red ear patch on black and white patterned head. Rest of upperparts are brownish and underparts buff with red vent. Tip of tail is edged white. Race *monticola* has complete black breast band. Immature lacks red ear patch and has pink vent. Iris—brown; bill and feet—black.

Voice: Loud incessant chattering and a sweet short whistled 2–3 note song *wit-t-waet*; also musical *prroop*.

Range: India, S China and SE Asia. Introduced to Australia and other regions.

Distribution and status: Very common in gardens, parks and secondary forest and scrub. Nominate race occurs in S China; *monticola* from SE Xizang to S Yunnan and dubious race *hainanensis* in S Guangdong and Naozhou.

Habits: Lives in noisy active flocks. Perky birds sit on prominent perches, often the very highest point of small trees, to sing and chatter. Favours open wooded areas, forest edge and secondary growth and villages.

897. Brown-breasted Bulbul *Pycnonotus xanthorrhous* PLATE 90
[Anderson's Bulbul] Huangtun bei

Description: Medium-sized (20 cm) greyish-brown bulbul with black crown and nape. Distinguished from Sooty-headed Bulbul by brown ear coverts, grey-brown breast band and lack of white tip to tail. Distinguished from Light-vented Bulbul by brown ear coverts, lack of yellow in wing and deeper yellow undertail coverts. In race *andersoni* the brown breast band is virtually absent. Iris—brown; bill—black; feet—black.

Voice: Harsh *brzzp*.

Range: S China, Burma and N Indochina.

Distribution and status: Fairly common from 800–4300 m. Nominate race recorded from W Sichuan, W and S Yunnan and SE Xizang. Race *andersoni* from C, E and S China.

Habits: Typical flocking bulbul of secondary bramble and bracken scrub in hills.

898. Light-vented Bulbul *Pycnonotus sinensis* [Chinese Bulbul]   PLATE 90
Baitou bei

Description: Medium-sized (19 cm) olive bulbul distinguished by a broad white stripe immediately behind the eye extending right round the nape, slightly crested black crown, black moustachial and white vent. Juvenile has olive head and grey breast bar. Iris—brown; bill—blackish; feet—black.

Voice: Typical chattering chirps and simple phrased, non-musical calls.

Range: S China, Taiwan, N Vietnam and Nansei Is.

Distribution and status: This is a common gregarious bird of forest edge, scrub, mangroves and gardens. One of the commonest birds in Hong Kong. Race *hainanus* is resident in S Guangxi, SW Guangdong and Hainan; *formosae* resident on Taiwan; and nominate through C, E, S and SE China. Northern populations migrate S in winter. Now regularly reported in Hebei (Beidaihe) and Shandong.

Habits: Active flocks crowd fruit trees. Sometimes flycatches from perch.

899. Styan's Bulbul *Pycnonotus taivanus* Taiwan bei   PLATE 90

Description: Medium-sized (19 cm) bulbul with black cap and moustachial on otherwise white head. There is an orange spot at base of lower bill. Upperparts olive, wings and tail brown edged olive-yellow. Underparts whitish with grey breast and brownish flanks. Iris—dark brown; bill—black; feet—black.

Voice: Similar to light-vented Bulbul. Loud variable *qiao-keli, qiao-keli*.

Range: Endemic to Taiwan.

Distribution and status: Globally Near-threatened (Collar *et al.* 1994). Occasional bird of lowlands up to 600 m in SE parts of island (Huangdong and Hengchun Peninsula). Occurs in lowland habitats where light-vented Bulbul is absent. There is a small overlap zone near Fanliao.

Habits: Typical forest bulbul.

[900. Himalayan Bulbul *Pycnonotus leucogenys* Baijia bei   PLATE 90

Description: Medium-sized (20 cm) olive-brown bulbul with long forward-curving brown crest. Face, chin and throat black with white cheek patch. Underparts whitish with pale yellow undertail coverts. Tail black with white tip. Distinctive and not readily confused with any other bulbul. Iris—dark brown; bill—black; feet—black.

Voice: Calls similar to Red-vented Bulbul; chattering; also angry *pit-pit*. Song is three or four phrases such as *tea-for-two* and *take-me-with-you* in endless combinations.

Range: NW India and Himalayas.

Distribution and status: Not recorded in China but nominate race occurs in E Himalayas in NE of claimed areas of Arunachal Pradesh, west of Zangpo (Brahmaputra). Common in foothills from 300–1800 m in drier valleys.

Habits: Typical, jaunty bulbul habits.]

901. Red-vented Bulbul *Pycnonotus cafer* PLATE 90
Heihou hongtunbei

Description: Medium-sized (20 cm) brownish bulbul with crested black head and crimson undertail coverts. Uppertail coverts whitish and brown ear coverts, black throat and dark breast. Iris—dark brown; bill—black; feet—dark horn to black.

Voice: Cheery call notes sometimes expressed as *be-care-ful* with emphasis on last note. Alarm is sharp, loud *peep*. Agonistic chattering and *peep-a-peep-a-lo*. Has sweet low subsong.

Range: India, Burma to SW China. Introduced Fiji.

Distribution and status: Race *stanfordi* is locally common in extreme W Yunnan to West of Nujiang (Salween) River.

Habits: Typical, flock-living, noisy bulbul.

902. Sooty-headed Bulbul *Pycnonotus aurigaster* [Black-capped/ PLATE 90
Golden-vented/Red-vented Bulbul] Baihou hongtunbei

Description: Medium-sized (20 cm) black-capped bulbul with whitish rump and red vent; chin and top of head black; collar, rump, chest and belly white; wings black; tail brown. Juveniles have yellowish vent. Distinguished from Red-whiskered Bulbul by shorter crest and lack of red on cheeks. Iris—red; bill and feet—black.

Voice: Melodious fluty calls and loud rasping notes *chook, chook*.

Range: S China, SE Asia and Java. Introduced to Sumatra and Sulawesi.

Distribution and status: Fairly common lowland species up to 500 m in SE China and Hong Kong (*chrysorrhoides*); SW Guangdong and S Guangxi (*resurrectus*) and SW China (*latouchei*).

Habits: Lives in noisy, active flocks, often mixing with other bulbuls. Prefers open wooded or bushy habitats, forest edge, secondary growth, parks and gardens.

903. Stripe-throated Bulbul *Pycnonotus finlaysoni* PLATE 90
Wenhou bei

Description: Medium-sized (19 cm) greenish bulbul with diagnostic broad yellow stripes on crown, cheeks, chin and throat. Undertail coverts bright yellow. Unmistakable for any other bulbul. Iris—brown; bill—black; feet—pinkish-brown.

Voice: Musical squawking *whic-ic, whic-ic*. Characteristic loud melodious song *ding-da-ding-ding-da-ding-ding* rising in crescendo to fourth note then dying away. Also bubbling alarm call.

Range: SW China, and SE Asia.

Distribution and status: Rare. Race *eous* recorded in S Yunnan.

Habits: Typical bulbul of secondary forest and forest edge in lowland mixed evergreen and deciduous forests.

904. Flavescent Bulbul *Pycnonotus flavescens* PLATE 90
[Blyth's/Pale-faced Bulbul] Huang lübei

Description: Medium-sized (20 cm) olive-green bulbul with pale yellow vent, white lores, grey face and throat and greyish streaky breast. There is a short full crest. Immature has duller lores. Iris—brown; bill—black; feet—grey.
Voice: A short song of five notes and harsh buzzing *tcherrp* notes.
Range: NE India, SW China, SE Asia and Borneo.
Distribution and status: Race *vividus* is a local montane resident in W and SW Yunnan at 600–2800 m.
Habits: Typical flock-living bulbul of open forest, forest edge and secondary growth of mid-elevations.

905. White-throated Bulbul *Alophoixus flaveolus* [Round-tailed PLATE 91
Green/Ashy-fronted Bearded Bulbul] Huangfu guanbei

Description: Largish (22 cm) crested, brown bulbul with white puffy throat and yellow underparts. Similar to Puff-throated Bulbul but brighter with browner upperparts and brighter lemon-yellow abdomen. Iris—brown; bill—black; feet—pink.
Voice: Noisy. Loud, harsh nasal croaks and loud clear *teek, da-te-ek, da-te-ek*, also sharp whip-like notes and sweet whistles.
Range: Himalayas to NE Burma.
Distribution and status: Occasional in limited area of China up to 1800 m. Nominate race is found west of Mekong (Lancang) River in Xishuangbanna, S Yunnan and SE Xizang (Medog).
Habits: As Puff-throated Bulbul. Perches with well fanned tail.

906. Puff-throated Bulbul *Alophoixus pallidus* PLATE 91
Baihou guanbei

Description: Large (23 cm) noisy bulbul with long, rather straggly crest of pointed feathers, olive upperparts, grey side of head, yellow underparts and white, puffy bearded throat. Distinguished with some difficulty from White-throated Bulbul by having duller underparts with paler yellow on belly. Iris—brown; bill—black; feet—brown.
Voice: Flocks make constant discordant cries and occasional weak song.
Range: SW China, Hainan, Burma and Indochina.
Distribution and status: Nominate race is a common resident in lowland evergreen and open forests of SW China (*henrici*) and Hainan (*pallidus*).
Habits: Lives in small active parties. It is aggressive in mobbing birds of prey. Often joins in mixed flocks and is generally active in lower layers of the forest.

907. Grey-eyed Bulbul *Iole propinqua* PLATE 91
Huiyan duanjiaobei

Description: Medium-sized (19 cm) plain, olive bulbul with diagnostic white or pale grey eyes. Has slight crest, indistinct supercilium and cinnamon undertail coverts. Upperparts olive and underparts yellowish-buff. Iris—grey or white; bill—pinkish-grey; feet—pinkish.

Voice: Constantly repeated, distinctive, nasal mewing *cheer-y* cry, with second note lower.
Range: S China, SE Asia.
Distribution and status: Common up to 1200 m. Nominate race in S Yunnan west of Yuan Jiang (Red River); *aquilornis* in SE Yunnan and Guangxi.
Habits: Frequents secondary forest and scrub in evergreen tropical zone.

908. Brown-eared Bulbul *Ixos amaurotis* [Chestnut-eared/   PLATE 91
Eurasian Brown-eared Bulbul] Li'er duanjiaobei

Description: Large (28 cm) grey bulbul with slight spiky crest and chestnut ear coverts and side of neck. Crown and nape grey; wings and tail brownish-grey. Throat and breast grey with pale streaks. Abdomen whitish with grey-spotted flanks and black and white barred vent. Taiwan race *nagamichii* has browner underparts with dark rufous breast. Iris—brown; bill—dark grey; feet—blackish.
Voice: Rather noisy *peet, peet, pii yieyo.*
Range: Japan, Taiwan and Philippines.
Distribution and status: Abundant. Smaller brown race *nagamachii* is resident in S Taiwan and Lanyu islet. Larger greyer nominate race and *hensoni* are occasional passage migrants through NE China; wintering in Jiangsu, Zhejiang and Taiwan in hills and lowlands in flocks.
Habits: A canopy dweller of forest, deciduous woodland, plantations and gardens.

909. Ashy Bulbul *Hemixos flavala* [Brown-eared Bulbul]   PLATE 91
Hui duanjiaobei

Description: Medium-sized (20 cm) bulbul with slight crest, deep sepia-brown or black (*bourdellei*) crown and pinkish brown ear coverts, greyish upperparts and white throat. Has yellowish panel on dark brown folded wing caused by pale edges to greater coverts. Iris—brownish-red; bill—dark brown; feet—dark brown.
Voice: Loud ringing call of 4–5 notes rising then falling in pitch; also harsh *trrk note.*
Range: Himalayas, SW China, SE Asia and Greater Sundas.
Distribution and status: Common in hills of S Yunnan (*bourdellei*) and W Yunnan and SE Xizang (nominate), up to 1000 m.
Habits: Typical forest bulbul living in small parties and keeping to middle and lower storeys of open submontane forest and scrub. Puffs out throat feathers like *Alophoixos* bulbuls.
Note: Some authors include races of Chestnut Bulbul within this species (e.g. Cheng 1987).

910. Chestnut Bulbul *Hemixos castanonotus*   PLATE 91
Libei duanjiaobei

Description: Largish (21 cm) smart bulbul with chestnut-brown upperparts and slightly crested black crown, white throat and whitish abdomen. Breast and flanks are pale grey. Wings and tail are greyish-brown with greenish-yellow edges on wing coverts and tail feathers. The white throat is sometimes puffed out like that of an *Alophoixos* bulbul but this species is rather distinctive. Race *canipennis* is more rufous and lacks greenish-yellow edges to feathers of wing and tail. Iris—brown; bill—dark brown; feet—dark brown.

Voice: Loud scolding notes and sharp ringing call '*tickety boo*'.

Range: S China, Hainan and Tonkin.

Distribution and status: Race *canipennis* is common in lowland forests of S and SE China; nominate race is resident on Hainan and hybridises with *canipennis* in S Guangxi. Becoming commoner in Hong Kong as forests mature.

Habits: Lives in small active parties. Keeps to rather dense thickets.

Note: Treated as races of Ashy Bulbul by some authors (e.g. Cheng 1987) but subsequently Cheng (1994) placed only race *canipennis* in Ashy Bulbul, retaining nominate as distinct species.

911. Mountain Bulbul *Hypsipetes mcclellandii* PLATE 91
Lüchi duanjiaobei

Description: Large (24 cm) noisy olive-coloured bulbul with a spiky short crest, a rufous nape and upper breast and streaky whitish throat. Crown is dark brown with whitish streaking. Back, wings and tail are greenish. Abdomen and vent whitish. Iris—brown; bill—blackish; feet—pink.

Voice: The song is a sibilant monotonous three-note call or a rising three-note call; also has various mewing calls.

Range: Himalayas to S China, Burma, Indochina and SE Asia.

Distribution and status: A common flock or pair-living bird of montane forests and scrub at 1000–2700 m. Nominate race resident in SE Xizang; *similis* in Yunnan and Hainan; and *holtii* over most of S and SE China.

Habits: Birds feed on small fruits and insects and sometimes form large flocks. Aggressively mobs raptors and cuckoos.

912. Black Bulbul *Hypsipetes leucocephalus* PLATE 91
Hei [duanjiao]bei

Description: Medium-sized (20 cm) black bulbul with slightly forked tail and bright red bill, feet and eye. Some races have white head and western races have greyish foreparts. Distinguished from Red-billed Starling by darker breast and back. Immature is greyish. Has slight flattish crest. Iris—brown; bill—red; feet—red.

Voice: Calls very variable include loud squawks, twitters and strident whistles. A common call is a nasal cat-like mewing.

Range: India, S China, Taiwan, Hainan, Burma and Indochina.

Distribution and status: Race *psaroides* is resident in SE Xizang; *ambiens* in NW Yunnan; *sinensis* in NW Yunnan S of *ambiens*; *stresemanni* in N Yunnan; *concolor* in W and S Yunnan; *leucothorax* in C China; *perniger* in S Guangxi and Hainan; *nigerrimus* on Hainan; and the nominate race over the rest of S and SE China. This is a common bird of evergreen hill forest.

Habits: Feeds on fruits and insects and shows some seasonal movements. In winter large flocks of several hundred birds can be found in southern China.

Cisticolas and Prinias—Family: Cisticolidae

Large family of African and Asian passerines formerly included in larger family of Old World Warblers. There are 10 species in China.

913. Zitting Cisticola *Cisticola juncidis* [Fan-tailed/   PLATE 92
Straw-headed/Common/Streaked Cisticola, AG: Fantail-Warbler]
Zong shanweiying

Description: Small (10 cm) streaked brown warbler with buffy-rufous rump and distinctive white-tipped tail. Distinguished from non-breeding Bright-headed Cisticola by white supercilium noticeably paler than sides of neck and nape. Iris—brown; bill—brown; feet—pink to reddish.
Voice: Series of clicking *zit* notes given in undulating display flight.
Range: Africa and S Europe to India, China, Japan, Philippines, SE Asia, Sundas, Sulawesi and N Australia.
Distribution and status: Race *tinnabulans* breeds in C and E China, wintering to S and SE China. Common up to 1200 m.
Habits: Lives in open grassland, paddy fields, and sugar cane beds in generally wetter areas than Bright-headed Cisticola. In courtship flight the male hovers and circles high over his mate calling. In non-breeding season this is a shy inconspicuous bird.

914. Bright-headed Cisticola *Cisticola exilis* [Bright-capped/   PLATE 92
Golden-capped/Golden-headed Cisticola, AG: Fantail-Warbler]
Jintou shanweiying

Description: Small (11 cm) streaky brown warbler with a bright golden crown and brown rump in breeding male. Female and non-breeding male have crown heavily streaked black but distinguished from Zitting Cisticola by buffish supercilium the same colour as side of neck and nape. Underparts buff, whitish on throat; tail dark brown tipped buff. Iris—brown; bill—black above, pink below; feet—light brown.
Voice: Breeding male gives a scratching *buzz* followed by a loud liquid *plook* from perch or in flight, also harsh high-pitched scolding.
Range: India, China, Philippines, SE Asia, Sundas, Sulawesi and Moluccas to New Guinea and Australia.
Distribution and status: Race *curtoisi* is resident in S and SE China; *volitans* in Yunnan; and *tytleri* in Yunnan west of the Nujiang (Salween) River and SE Xizang. Locally common in suitable habitat up to 1500 m.
Habits: Inhabits tall grassland, reeds and rice fields. A secretive bird, sometimes seen perched on a tall grass stem or bush. Fluttering flight.

915. White-browed Chinese Warbler *Rhopophilus pekinensis*   PLATE 92
Shan mei

Description: Large (17 cm) long-tailed, brown streaky warbler with greyish eyebrow and blackish moustachial stripe. Resembles stocky prinia. Upperparts ashy-brown, heavily streaked blackish; outer tail feathers edged white. Chin, throat and breast white; rest of underparts white, boldly streaked chestnut on flanks and belly and sometimes washed cinnamon. Western race *albosuperciliaris* is much paler with white eyebrow; upperparts ashy-grey, streaked brown and underparts white with slight cinnamon streaking on flanks and belly; undertail buff. Race *leptorhynchus* is intermediate. Bill is slender and rather decurved. Iris—brown; bill—horn; feet—yellowish-brown.

Voice: Mellow *chee-anh* calls in duet. Song described as sweet, continuous '*dear, dear, dear*' opening on high note, falling away quickly before rising again at start of second syllable.

Range: N and W China.

Distribution and status: Generally uncommon in dry, stony, bush-covered hills and montane scrub. Race *pekinensis* from S Liaoning west to Huang He (Yellow River) valley as far as Helan Mts in Ningxia; *leptorhynchus* from Qinling Mts of S Shaanxi to S Gansu; race *albosuperciliaris* from Qinghai and W Nei Mongol to Kashi region of W Xinjiang.

Habits: Inhabits scrub and reeds. Flits from cover to cover and runs well on ground. Not shy. Forms flocks outside breeding season and sometimes associates with babblers.

Note: Sometimes treated as a babbler (e.g. Cheng 1987).

916. Striated Prinia *Prinia criniger* Shan jiaoying PLATE 92

Description: Largish (16.5 cm) dark brown streaked prinia with long graduated tail. Upperparts grey-brown with black and dark brown streaking. Underparts whitish with fulvous wash on flanks, breast and undertail coverts and distinctive black streaking on breast. Non-breeding birds browner with less black on breast and buff and black streaked crown. Non-breeding Brown Prinia is similar but lacks any black speckles on side of breast. Race *catharia* is browner and more streaked than nominate; *parvirostris* is darker with greyer underparts; *parumstriata* is greyer with brown speckles, streaked forehead and whiter underparts; *striata* is paler and greyer. Iris—pale brown; bill—black (brown in winter); feet—pinkish.

Voice: Song is monotonous wheezy, scraping 2–4-note series like a saw blade being sharpened on a grindstone. Call note is sharp *tchack, tchack*.

Range: Afghanistan to N India, Burma, S China and Taiwan.

Distribution and status: Common up to 3100 m. Nominate race is resident in SE Xizang; *catharia* in SW China; *parvirostris* in SE Yunnan; *parumstriata* in S and SE China; and *striata* on Taiwan.

Habits: Lives in tall grass and scrub, often on fallow agricultural fields. Males call from prominent perch. Weak flight.

917. Brown Prinia *Prinia polychroa* He shanjiaoying PLATE 92

Description: Largish (15 cm) long-tailed, dull rufous-brown prinia. Upperparts brown, slightly streaked on crown, mantle and coverts; tail very graduated with pale buff tips and dark subterminal band; underparts whitish with buff flanks and undertail coverts. More rufous, paler and less heavily streaked than Striated Prinia and lacks streaking on breast. Iris—reddish-brown; bill—brown above, pale below; feet—whitish.

Voice: Call is loud *twee-ee-ee-ee-eet*. Short song is wheezy *chirt-chirt-chirt-chirt* or *chook-chook-chook-chook*.

Range: SW China, SE Asia and Java.

Distribution and status: Race *bangsi* is only marginal in China. Uncommon resident up to 1500 m in SE Yunnan.

Habits: Inhabits tall grasslands and low scrub. A shy elusive bird keeping to thick cover. Lives in pairs or family parties.

Note: Some authors formerly included races of Striated Prinia *P. criniger* in this species (e.g. Cheng 1987).

918. Hill Prinia *Prinia atrogularis* [Black-throated/White-browed   **PLATE 92**
Prinia, AG: Wren-Warbler] Heihou shanjiaoying

Description: Largish (16 cm) long-tailed brown warbler with diagnostic black-streaked breast. Upperparts brown, flanks buffy-rufous and belly buffy-white. Grey cheeks, white supercilium and very long tail are distinctive. Race *superciliaris* has olive-green upperparts and lacks submoustachial stripe. Black of underparts less extensive. Iris—pale brown; bill—dark above, pale below; feet—pinkish.

Voice: Loud piercing *cho-ee, cho-ee, cho-ee* like Common Tailorbird but slower.

Range: Himalayas, S China, SE Asia, Malay Peninsula and Sumatra.

Distribution and status: Common in hills and mountains at 600–2500 m. Nominate race is resident in S and SE Xizang; *superciliaris* in W Yunnan, Guangxi, Guangdong and Fujian.

Habits: Noisy and active in grass and low vegetation of submontane and montane forests including dwarf moss forest and scrub.

919. Rufescent Prinia *Prinia rufescens* Anmian jiaoying   **PLATE 92**

Description: Smallish (11.5 cm) rufescent-brown prinia with not so long tail and whitish lores and supercilium. Upperparts rufous-brown with greyish head in breeding season. Underparts white with buff wash on lower belly, flanks and undertail coverts. Distinguished from Plain Prinia by shorter tail and more rufescent upperparts. Distinguished from non-breeding Grey-breasted Prinia by more prominent supercilium extending beyond eye, browner bill and more rufescent upperparts. Iris—brown; bill—horn-brown; feet—pinkish.

Voice: Call is feeble, high-pitched *seep, seep, seep*; churring trill and excited repeated chip. Song is repeated series of squeaky notes, *chewp, chewp, chewp*.

Range: NE Indian subcontinent and E India to S China, Burma and SE Asia.

Distribution and status: Nominate race is common in lowland secondary forest and scrub in SE Xizang and W and S Yunnan.

Habits: Secretive in low thickets with jerky, active movements. Forms small flocks in autumn and winter. Responds to 'pishing'.

920. Grey-breasted Prinia *Prinia hodgsonii*   **PLATE 92**
Huixiong jiaoying

Description: Smallish (12 cm) greyish-brown prinia with longish, graduated tail. Breeding adult: greyish upperparts with brown wing panel formed by rufous fringes to flight feathers; underparts white with distinctive grey breast band. Non-breeding adult and juvenile difficult to separate from non-breeding Rufescent Prinia but with shorter pale supercilium (indistinct behind eye); smaller darker bill and the tips of tail white rather than buff. Tail is much shorter than Plain Prinia. Non-breeding race *rufula* has more rufous cap and creamy-rufous underparts. Iris—orange-yellow; bill—black (brown in winter); feet—pinkish.

Voice: Song is loud, squeaky *chiwee-chiwee-chiwi-chip-chip-chip* rising in pitch and volume to a sudden halt. Call is thin *chew-chew-chew* or incessant tinkling *zee-zee-zee*.

Range: Indian subcontinent to SW China and SE Asia.

Distribution and status: Common in secondary forest undergrowth, scrub and grass up to 1800 m. Race *confusa* is resident in S Sichuan and W and S Yunnan; *rufula* in SE Xizang and NW Yunnan.

Habits: Forms flocks in winter. Shy and skulking. Similar to habits of Rufescent Prinia but in drier habitats.

921. Yellow-bellied Prinia *Prinia flaviventris* PLATE 92
Huangfu jiaoying

Description: Largish (13 cm) long-tailed, olive-green warbler with white throat and chest and diagnostic yellow lower breast and belly. Head grey with sometimes faint, short whitish supercilium; upperparts olive-green; thighs buff or rufous. Some variation due to moult. Tail shorter in breeding season and male mantle blacker (charcoal in female); pinkish-grey in winter. Race *sonitans* is browner above and paler below. Iris—pale brown; bill—black to brown above; pale below; feet—orange.

Voice: Weak, harsh *schink-schink-schink* and soft mewing like young cat, *twee twee*. Song is hurried bubbling *tidli-idli-lia* with stress on last falling syllable, preceded by opening *chirp*.

Range: Pakistan to S China, SE Asia and Greater Sundas.

Distribution and status: Common up to 900 m. Race *delacouri* is resident in SW and S Yunnan; *sonitans* in S and SE China, Hainan and Taiwan.

Habits: Inhabits reedy swamps, tall grasslands and scrub. A fairly shy bird keeping out sight in long grass or reeds except when perched on a tall stem singing. Makes clicking sound when flicking its wings.

922. Plain Prinia *Prinia inornata* [Tawny/Plain-coloured/ PLATE 92
Prinia, AG: Wren-Warbler] Hetou jiaoying

Description: Largish (15 cm) long-tailed brownish warbler with a pale supercilium. Upperparts dull greyish-brown; underparts yellowish-buff to rufescent. Back paler and more uniform than Brown Prinia. Taiwan race *flavirostris* is paler with yellow bill. Iris—light brown; bill—blackish; feet—pink.

Voice: Song is monotonous, insect-like reeling in long bursts up to one minute at a rate of 3–4 notes per second. Call is rapidly repeated *chip* or *chi-up* note.

Range: India, China, SE Asia and Java.

Distribution and status: Common resident to about 1500 m; race *extensicauda* in C, SW, S and SE China and Hainan; and *flavirostris* on Taiwan.

Habits: Inhabits areas of long grass, reed beds, marshes, maize and paddy fields. A cocky active bird often in small parties and regularly calling from trees, grass stems or in flight. Less common than Yellow-bellied Prinia in Hong Kong.

Note: Sometimes treated as race of Tawny-flanked Prinia *P. subflava* (e.g. Cheng 1987, 1994).

WHITE-EYES—Family: Zosteropidae

A large family occurring in Africa, Asia and Australasia. White-eyes derive their name from the ring of silvery feathers around the eye in most species. White-eyes are generally small, tit-like birds with greenish-olive or yellowish plumage, small, slender, slightly curved bills, short wings and small strong feet. They are extremely agile, restless birds, often forming mixed flocks which work through the tree-tops in search of small fruits and insects; they visit flowers to feed on nectar like sunbirds. The calls are shrill twitters and chirps and the nests are neat cup-shaped structures woven into a branch fork. Three species occur in China.

923. Chestnut-flanked White-eye *Zosterops erythropleurus* PLATE 115
[Red-flanked] Hongxie xiuyanniao

Description: Medium-sized (12 cm) white-eye. Separated from Japanese and Oriental white-eyes by greyer upperparts and chestnut flanks (sometimes absent), also paler lower mandible and more restricted yellow throat patch and lack of yellow on forecrown. Iris—reddish-brown; bill—olive; feet—grey.
Voice: Twittering *dze-dze* calls typical of genus.
Range: E Asia, E and S China and Indochina.
Distribution and status: Breeds in NE China wintering S over C, S and E China. Locally common in primary and secondary forests, generally above 1000 m.
Habits: Sometimes mixes with flocks of Japanese White-eyes.

924. Oriental White-eye *Zosterops palpebrosus* [Indian/Small PLATE 115
White-eye] Huifu xiuyanniao

Description: Small (11 cm) olive-green white-eye. Similar to Japanese White-eye but distinguished by a narrow lemon-yellow bar down centre of belly, black lores and ocular area and narrower white eye-ring. Iris—yellow-brown; bill—black; feet—olive-grey.
Voice: Soft high twittering *dzi-da-da* or repeated metallic *dza dza*. Flocks twitter continuously.
Range: Indian subcontinent to S China, SE Asia and Sundas.
Distribution and status: Race *siamensis* is a common resident of the lowlands and hills up to 1400 m in SE Xizang and SW China from S Sichuan to SW Guangxi. Race *joannae* recorded in W China but status uncertain. Performs seasonal movements.
Habits: Frequents primary and secondary vegetation. Forms large flocks which join freely with other birds such as minivets, travelling through the tops of the highest trees.

925. Japanese White-eye *Zosterops japonicus* PLATE 115
Anlü xiuyanniao

Description: Attractive small (10 cm) flocking bird with bright green-olive upperparts, conspicuous white eye-ring and yellow throat and vent. Breast and flanks are grey with white abdomen. Lacks chestnut flanks of Chestnut-flanked White-eye and ventral yellow band of Oriental White-eye. Iris—pale brown; bill—grey; feet—greyish.

Voice: Birds constantly emit soft *tzee* note and quiet trills.

Range: Japan, China, Burma and N Vietnam.

Distribution and status: Race *simplex* is resident or summer breeder over E, C, SW, S and SE China and Taiwan, with northern populations migrating S in winter. Race *hainana* is resident on Hainan and *batanis* resident on Lanyu and Huoshao islets off SE Taiwan. Common in woodland, forest edge, parks and towns. Frequently captured as cagebird, resulting in some escapes.

Habits: Active and noisy bird of tree crowns, feeding on small insects, tiny fruits and nectar.

WARBLERS, GRASSBIRDS, LAUGHINGTHRUSHES AND BABBLERS— Family: Sylviidae

A huge, diverse family containing most of the former warbler family plus laughingthrushes and babblers. Divided into several subfamilies and tribes.

OLD WORLD WARBLERS—Subfamily: Acrocephalinae

These warblers are a very large Old World subfamily of small, very active insectivorous birds with narrow, pointed bills; formerly treated as a family including sylvian warblers and African warblers and grassbirds. Most are drab in colour and difficult to identify in the field. They have generally clear, pretty songs and make neat cup-shaped or domed nests. Tailorbirds make elaborate nests of leaves stitched together with plugs of spiders web. Most species are migratory. China is one of the world's main distribution centres of warblers with no less than 84 species. The subfamily can be divided into five groups.

1. Ground Warblers—almost tail-less ground skulkers, of which only three occur in China (*Tesia*).
2. Bush Warblers—small, drab brown warblers with shortish tails and usually marked eye-stripes, no wing bars or crown bars. They are shy skulkers (*Urosphena, Cettia* and *Bradypterus*).
3. Grass Warblers—drab brownish birds inhabiting scrub, swamps and grassland. Generally longish-tailed and good singers. Specialised to live among reeds and tall grasses (*Locustella, Acrocephalus, Hippolias* and *Megalurus*).
4. Tailorbirds and Tit Warblers—colourful small warblers with cocky tails and reddish-coloured heads and loud calls (*Orthotomus* and *Leptopoecile*).
5. Leaf Warblers—small, canopy-feeding birds, including several winter migrants (*Seicercus, Tickellia, Abroscopus* and *Phylloscopus*).

1. Ground Warblers

926. Chestnut-headed Tesia *Tesia castaneocoronata*   PLATE 94
[AG: Ground warbler] Litou diying

Description: Small (10 cm) rather upright, colourful, wren-like warbler with short tail and diagnostic chestnut head and nape. Upperparts green; underparts yellow. There is a white spot above and behind the eye. Juvenile is olive-brown above, orange-chestnut below. Race *abadiei* has darker green upperparts and greenish nape; race *repleyi* is paler than nominate. Iris—brown; bill—brown with pale base to lower mandible; feet—olive-brown.

Voice: Loud shrill four-note song *sip, sit-it-up*. Common call is single, ventriloqual, piercing *tzeeet*. Also makes chattering *chiruk, chiruck* notes interspersed with soft *wee*.

Range: Himalayas, N Burma, N Indochina to SW and C China.

Distribution and status: Locally common. Nominate race in Himalayas, SE Xizang, S Sichuan and Guizhou; *abadiei* occurs in NW Tonkin but must be expected in SE Yunnan (Jinping); race *ripleyi* is confined to Yunnan.

Habits: Keeps to dense humid forest undergrowth near streams. Curious sideways movement when shuffling along branches or fallen logs. Altitudinal migrant, generally from 2000–4000 m in summer but below 2000 m in winter.

927. Slaty-bellied Tesia *Tesia olivea* [AG: Ground Warbler]   PLATE 94
Jinguan diying

Description: Small (9.5 cm) rather upright, grey, wren-like warbler with very short tail. Underparts slaty-grey; upperparts olive-green with crown scaled yellow to bright golden-yellow in breeding male. There is a black eye-stripe. Distinguished from Grey-bellied Tesia by darker underparts, lack of pale supercilium and yellow on crown. Juvenile has olive-brown upperparts with grey underparts. Distinguished from juvenile Grey-bellied Tesia by darker grey underparts and lack of contrasting supercilium. Iris—dark brown; bill—dark above, orange with yellow tip below; feet—brownish.

Voice: Short, shrill, piped alarm note. Call is rattling wren-like *tchiriok*.

Range: E Himalayas, Burma, N Thailand, N Indochina to S Sichuan.

Distribution and status: Locally common in SE Xizang, S Sichuan and W Yunnan. Up to 2700 m in summer but descends to 1000 m in winter.

Habits: Secretive skulker in dense undergrowth near streams. Moves like Chestnut-headed Tesia. Has some strange behaviour. Throws ground debris in frenzied foraging and jumps back and forth in the air. Moves sideways along branches and claps wings over head.

928. Grey-bellied Tesia *Tesia cyaniventer* Huifu diying   PLATE 94

Description: Small (9.5 cm) grey, rather upright, wren-like warbler. Very similar to Slaty-bellied Tesia but underparts paler grey; lacks yellow on crown and has distinct pale supercilium above black eye-stripe. Iris—dark brown; bill—dark above, yellow below with dark tip; feet—olive-brown.

Voice: Call is rapid rattling, wren-like, *tchirik* similar to Slaty-bellied. Song, preceded by soft twitter, is three trilling notes in quick succession ending in lower note *tsitsitsitjutjutju*.

Range: Himalayas to S China, N Burma and N Indochina.

Distribution and status: Locally common in SE Xizang, W and SE Yunnan and Guangxi. Up to 2550 m in summer but below 1800 m in winter.

Habits: Like other tesias, although occasionally as high as 6 m off ground. Inquisitive and restless.

2. BUSH WARBLERS

929. Asian Stubtail *Urosphena squameiceps* Lintou shuying PLATE 94

Description: Small (10 cm) very short-tailed, bush warbler with prominent dark eye-stripe and pale supercilium. Upperparts uniform brown; underparts whitish with buffy flanks and vent. Crown scaly. Looks dumpy with broad wings and slender, spiky bill. Distinguished from other bush warblers by short tail. Iris—brown; bill—dark above, pale below; feet—pinkish.

Voice: High-pitched, insect-like song *see-see-see-see-see-see-see-see-see-see* becoming louder at end. Also makes low *chip-chip-chip* call.

Range: Breeds in NE Asia, wintering in SE Asia.

Distribution and status: Quite common. Breeds in NE China (SE Heilongjiang) but passes through C and E China to winter in SE and S China and Taiwan.

Habits: Solitary or in pairs. Skulks on or close to ground in dense underbrush of conifer and deciduous forests below 1300 m in breeding areas and in more open scrubby habitats up to 2100 m in wintering areas.

930. Pale-footed Bush Warbler *Cettia pallidipes* PLATE 93
Danjiao shuying

Description: Smallish (12.5 cm) olive-brown warbler without rufous in upperparts, with buff eyebrow and white underparts. Flanks and vent are buffish. Characterised by fleshy pale feet and nearly square tail. Difficult to identify in the field but distinguished from Dusky Warbler by lack of rufous tinge on eyebrow and flanks. Distinguished from Asian Stubtail by longer tail and lack of dark edges to crown feathers. Iris—brown; bill—brown; feet—pink.

Voice: Voice is a series of sharp chirps.

Range: Himalayas, Burma, S China and Indochina.

Distribution and status: Nominate race is resident in W Yunnan, *laurentei* resident in SE Yunnan and vagrant to Guangdong (Zhuhai) and Hong Kong. Locally common in lowlands up to 1500 m.

Habits: Shy skulker in undergrowth of forest and secondary scrub, often feeding on ground.

931. Manchurian Bush Warbler *Cettia canturians* PLATE 94
Manzhou shuying

Description: Large (17 cm) plain rufous bush warbler with prominent buff eyebrow and dark brown eye-stripe. Lacks wing bar or coronal stripes. Female smaller than male. Distinguished from Thick-billed Warbler by pale supercilium, smaller size, slender bill, rufescent crown and less buffy underparts. Distinguished with difficulty from Japanese Bush Warbler by more rufous colour; underparts with dark buff suffusion on flanks and undertail coverts. Iris—brown; bill—brown above, pale below; feet—pink.

Voice: Rich musical chuckle starting with a low trill and ending with a terminal flourish *tu-u-u-teedle-ee-tee.*

Range: Breeds in E Asia, winters to Assam, S China, Taiwan, Philippines and SE Asia.

Distribution and status: Quite common. Breeds in S Gansu, S Shaanxi (Qinling Mts), Sichuan, Henan, S Shanxi, Hubei, Anhui, Jiangsu and Zhejiang. Winters S of Changjiang valley in S and SE China and Hainan.

Habits: Usually cocks slightly notched tail. Inhabits secondary scrub up to 1500 m.

Note: Treated as a separate species by Deignan (1963) and Sibley and Monroe (1990) but many authors treat this as a race of Japanese Bush Warbler (e.g. Cheng 1987 and Inskipp *et al.* 1996). There are minor differences in both plumage and song.

932. Japanese Bush Warbler *Cettia diphone* PLATE 94
[Singing Bush Warbler] Riben shuying

Description: Medium-sized (15 cm) uniform olive-brown bush warbler with prominent creamy buff eyebrow and blackish eye-stripe. Underparts creamy-white with diffuse buffish breast band and olive-brown flanks and undertail coverts. Less rufescent than Manchurian Bush Warbler with whiter underparts. Race *riukiuensis* is greyer and smaller than *borealis*. Juvenile has creamy-yellow underparts. Iris—brown; bill—brown above, pinkish below; feet—pink.

Voice: Pulsating whistle followed by three notes *hot-ket-kyot.* Descending series of double whistles *pe-chew, pe-chew, pe-chew.* Call note is dry ticking sound. Song of *borealis* is rendered as short, loud, melancholy whistles, *koo-goo-oo-oo-ook . . . tulee-tulee.*

Range: Breeds Japan and NE China, wintering in E China and Taiwan.

Distribution and status: Quite common. Race *borealis* breeds in Lesser Hinggan and Changbai Mts and Kaidao areas of NE China; migrates through Shandong and eastern provinces to winter in Taiwan. Race *cantans* also winters in Taiwan and race *riukiuensis* has been recorded in Jiangsu.

Habits: Inhabits dense thickets of scrub bamboo and grass up to 3000 m. Generally solitary and secretive.

933. Brownish-flanked Bush Warbler *Cettia fortipes* PLATE 93
[Strong-footed Bush Warbler] Qiangjiao shuying

Description: Smallish (12 cm) drab brown bush warbler with long buff supercilium. Underparts whitish with brown-buff wash, especially on sides of breast, flanks and undertail coverts. Juvenile is more yellow. Very similar to Yellow-bellied Bush Warbler but darker brown above with browner, less yellow underparts, less white belly and less grey throat. Call is also distinct. Iris—brown; bill—dark brown above, basally pale below; feet—brownish-flesh.

Voice: Song is sustained rising *weee* followed by an explosive *chiwiyou*. Also gives persistant *tack tack* call.

Range: Himalayas to S China, SE Asia and Greater Sundas.

Distribution and status: Fairly common resident. Nominate race occurs in S Xizang; *davidiana* over C, S, SE and SW China; and *robustipes* on Taiwan.

Habits: Skulks in thick scrub, easily heard but hard to flush into view. Generally solitary.

Note: We have retained the form *robustipes* within this species e.g. Cheng (1987), Meyer de Shauensee (1984) and Inskipp *et al.* (1996). Sibley and Monroe (1990) and Watson *et al.* (1986) elevated *robustipes* to species status and included the races of Yellowish-bellied Bush Warbler within it. This treatment was followed by Cheng (1994) and Viney *et al.* (1994). While *robustipes* may merit species rank as Taiwan Bush Warbler, it appears mistaken to include it within *acanthizoides* (see Sibley and Monroe (1993) and Inskipp *et al.* 1996)).

934. Chestnut-crowned Bush Warbler *Cettia major*   PLATE 94
Da shuying

Description: Smallish (13 cm) rather colourful bush warbler with rufous crown and long white upward-slanted supercilium (starting only marginally in front of eye, with rufous lores). Upperparts dull olive-brown; ear coverts streaked olive. Underparts whitish with buff wash on sides of breast and flanks. Juvenile has dull brown crown and more buff on breast. Japanese Bush Warbler is warmer brown above with paler rufous cap and buffy lores. Grey-sided Bush Warbler is smaller and more gracile with slender bill, less white on underparts and lacks rufous lores. Iris—brown; bill—dark above, pale below; feet—pink.

Voice: Song is introductory note followed by 3–4-noted explosive warble. Call is sharp, bunting-like *peep*.

Range: Himalayas to SW China; vagrant to Thailand.

Distribution and status: Uncommon. Nominate race breeds in SE Xizang, W and NW Yunnan and S Sichuan. Winters to S of range.

Habits: Skulks in scrub and forest undergrowth, generally at 1500–2200 m.

935. Aberrant Bush Warbler *Cettia flavolivacea*   PLATE 94
Yise shuying

Description: Medium-sized (13.5 cm) drab, olive-coloured bush warbler with dirty yellow underparts and pale yellow supercilium and narrow eye-ring. Unlikely to be confused with other adult bush warblers but some juvenile forms are yellow such as Brownish-flanked and Yellowish-bellied. Distinguished from Brownish-flanked by yellow rather than buff supercilium and from Yellowish-bellied by lack of grey throat and upper breast; also less rufous upperparts and lack of wing panel. Iris—pale brown; bill—dark with pink base to lower mandible; feet—yellow.

Voice: Song is short sweet warble followed by long inflected whistle, *dir dir-tee teee-weee*. Call is soft, shivering *brrt-brrt* similar to Yellowish-bellied, also sudden *chick*; grating alarm call.

Range: Himalayas to C and SW China, Burma and N Indochina.

Distribution and status: Uncommon resident. Nominate race occurs in SE Xizang; *dulcivox* from W Yunnan to Sichuan and *intricatus* breeds in SE Shanxi, S Shaanxi (Mt Taibai) and NW Sichuan. The latter race migrates S in winter. Vagrant in Shandong.

Habits: Lives in tall grass, scrub, bamboo and thickets in forest from 1200–4900 m down to 700 m in winter.

936. Yellowish–bellied Bush Warbler *Cettia acanthizoides* PLATE 93
[Verreaux's/Hume's Bush Warbler] Huangfu shuying

Description: Small (11 cm) plain brown bush warbler. Upperparts rather uniform brown but cap sometimes slightly rufous and rump sometimes more olive. Rufous edges of flight feathers give contrasting wing panel. White or buff supercilium is long behind eye. Underparts: grey throat and upper breast; yellowish wash on sides; buffy-white flanks, undertail coverts and centre of belly. Similar to large Brownish-flanked Bush Warbler but paler and with more yellow on belly; greyer throat and upper breast and whiter lower abdomen. Smaller than Aberrant Bush Warbler with browner upperparts and greyer throat. Iris—brown; bill—dark above, pink below; feet—pinkish-brown.

Voice: Strange song of 3–4 thin, drawn-out, human-like whistles, each lasting about two seconds and climbing in scale, then followed by several fast repeated, up and down, *chee-chew* notes. Call notes include vibrant *brrrr* and sharp *tik tik tik*.

Range: Himalayas to SE China and E Burma.

Distribution and status: Common resident. Nominate race (includes *concolor*) distributed over C, SW and E China and Taiwan, and *brunnescens* in S and SE Xizang. Vagrant to Hong Kong.

Habits: Skulker of dense scrub and forest undergrowth and bamboo thickets in hills from 1500–4000 m in summer and to 1000 m in winter.

Note: Some authors have (erroneously) placed this species under the name *C. robustipes* (e.g. Cheng 1994 and Viney *et al.* 1994) but see Sibley and Monroe (1993) and Inskipp *et al.* (1996).

937. Grey-sided Bush Warbler *Cettia brunnifrons* PLATE 94
Zongding shuying

Description: Small (11 cm) quite colourful bush warbler with pale rufous crown and bold creamy supercilium. Underparts greyish-white with grey wash on sides of breast and buff wash on flanks and undertail coverts. Similar colour and pattern to Chestnut-crowned Bush Warbler but smaller and more gracile with finer bill, more grey on underparts and creamy rather than rufous lores. Iris—brown; bill—dark above, pale below; feet—pinkish-grey.

Voice: Song is high-pitched, sweet, short phrase continuously uttered *dzit-su-ze-sizu*, followed immediately by nasal *bzeeuu-bzeeuu*. Call note is metallic, ping-like *tzip*. Alarm is rattle of fast call notes.

Range: Himalayas to W China and Burma.

Distribution and status: Nominate race breeds in S and western SE Xizang; *umbraticus* breeds in eastern SE Xizang, Sichuan and W and NW Yunnan. Migrates S in winter.

Habits: Lives in dense thickets, bamboo, rhododendron or bracken scrub in subalpine conifer zone from 2600–4300 m. Winters in low hills. Skulking and shy. Rather vocal.

938. Cetti's Bush Warbler *Cettia cetti* [Cetti's Warbler] PLATE 94
Kuanwei shuying

Description: Medium-sized (14 cm) drab, brown, robust bush warbler with short greyish-white supercilium, whitish half ring below eye and buff vent and undertail coverts. Rest of underparts white with grey wash on sides of breast. Has long, graduated, plain brown tail. Lower back slightly more rufous than mantle. Female is smaller than male. Distinguished from reed warblers by short, rounded wings, rounder tail and domed head. Separated from Japanese Bush Warbler by lack of rufous crown. Iris—brown; bill—dark above, pink below; feet—pink.

Voice: Diagnostic sudden loud outburst of song. Starts with pinging *chip* followed by rich warbled phrase *CHUti-CHUti-CHUti* or variants, ending in warbles and rattles before starting again. Call note is sharp, explosive *chik* or alarm trill.

Range: Europe, N Africa, Middle East to C Asia.

Distribution and status: Rare resident. Race *albiventris* is resident in Kashi, Tianshan and Mt Bogda regions of NW Xinjiang.

Habits: Lives in dense vegetation of tall herbs and reeds in marshy areas and scrubby hillsides. Mostly in lowlands but sometimes up to 2450 m. Skulking habits typical of genus.

939. Spotted Bush Warbler *Bradypterus thoracicus* PLATE 93
[AG: Scrub Warbler] Banxiong duanchiying

Description: Medium-sized (13.5 cm) drab brown warbler with short, broad wings and faint white supercilium. Upperparts brown with rufous tinge on crown; underparts whitish with blackish spots on throat, grey breast band and brownish flanks. Undertail coverts brown with white tips forming broad chevrons. Black spotting on throat, bold in spring forming complete gorget but sparse in winter and almost absent in race *przevalskii*. Bill shorter and straighter than Long-billed Bush Warbler, also has less extensive and less diffuse supercilium. Race *davidi* is darker and *przevalskii* is paler than nominate. Iris—dark brown; bill—black; feet—pinkish to brown.

Voice: Calls include drawn-out, harsh *tzee-eenk*, also chacking and explosive *pwit*. Song varies with race—cicada-like, dry buzzing *dzzzzzzzr, dzzzzzzzr, dzrrr* or rhythmic *trick-i-di, tric-i-di* etc.

Range: C Asia and Himalayas to S Siberia and N and C China, wintering to N India, Burma and Thailand.

Distribution and status: Locally common. Nominate race breeds in SE Xizang; *przevalskii* in SW China through Qinling Mts of S Shaanxi; and *davidi* in NE China. Some populations winter in SE China and SW Yunnan.

Habits: Breeds in juniper and rhododendron scrub above treeline up to 4300 m but descends to foothills and plains in winter. Very secretive and elusive.

940. **Long-billed Bush Warbler** *Bradypterus major*   PLATE 94
[Large-billed Bush Warbler, AG: Scrub Warbler] Juzui duanchiying

Description: Medium-sized (13 cm) drab, olive-brown warbler with short broad wings, shortish tail and long diffuse white supercilium. Similar to Spotted Bush Warbler but distinguished by longer bill with more decurved culmen, distinctive white lores and lack of grey wash on breast. Spotted juvenile generally more rufous than more olive juvenile of Long-billed. Race *innae* is paler than nominate with almost no throat spots. Iris—brown; bill—blackish; feet—pinkish-brown.

Voice: Call is quiet *tic* or grating *trrr* in alarm. Song is monotonous, metallic *pikha pikha pikha* ... about three notes per second and sometimes delivered without pause for several minutes.

Range: W Himalayas, N India and W China.

Distribution and status: Globally Vulnerable (Collar *et al.* 1994). Rare resident at 2400–3600 m, to 1200 m in winter. Nominate race occurs in W Kunlun Mts of Xinjiang and W Xizang. Race *innae* is resident in E Kunlun and E Xinjiang.

Habits: Secretive skulker, difficult to flush. Lives in rank grass and low scrub on hillsides of valleys.

941. **Chinese Bush Warbler** *Bradypterus tacsanowskius*   PLATE 94
[AG: Scrub Warbler] Zhonghua duanchiying

Description: Medium-sized (14 cm) brown warbler with pale underparts, pale eyebrow, white lores and longish, graduated tail. Upperparts uniform brown; underparts variable from white to yellow with buff-brown on sides of breast and flanks with or without brown spots on throat and upper breast. Undertail coverts light brown with broad pale tips giving broad chevron pattern. Distinguished from summer Spotted Bush Warbler by paler olive-brown upperparts, lack of grey on breast, fewer spots on throat and spots brown not black. Also paler lower bill. From winter Spotted by less contrasting chevron pattern on undertail and, where present, yellow wash on underparts. Distinguished from Russet Bush Warbler by paler upperparts, browner sides to neck and breast and thinner, paler bill. Distinguished from Brown Bush Warbler by paler undertail coverts, yellow tinge on underparts and less brown on sides of breast. Iris—brown; bill—pale; feet—pinkish.

Voice: Call is *chirr, chirr* similar to Lanceolated Warbler. Song is cricket-like rasping, *dzzzeep-dzzeep-dzeep*, lower-pitched than Spotted Bush Warbler.

Range: Breeds from S and E Siberia to NE and C China, wintering to SE Asia and NE India.

Distribution and status: Locally common. Breeds from 2800–3600 m in NE China S to Guangxi, Yunnan, Sichuan, E Qinghai and SW Gansu. Winters in S Yunnan and Probably SE Xizang.

Habits: Secretive skulker. Favours dense scrub in larch forest glades in summer and grass and reed beds in winter.

942. **Brown Bush Warbler** *Bradypterus luteoventris*   PLATE 93
[AG: Scrub Warbler] Zonghe duanchiying

Description: Medium-sized (14 cm) drab brown warbler with short broad wings and poorly-defined buffy supercilium. Underparts: white chin, throat and upper breast; sides of face, sides of breast, belly and undertail coverts rich buffy-brown. Whitish tips to undertail coverts give scaly appearance. Summer birds may have dusky streaks on throat. Juvenile has buff throat. Bill slim and slightly hooked and forehead rounded. Bill more slender than Russet Bush Warbler. Chinese Bush Warbler has more yellowish underparts. Race *ticehursti* is less rufous than nominate. Iris—brown; bill—dark above, pink below; feet—pink.

Voice: Calls include rasping, reeling *tic-tic-tic-tic-tic-tic*.

Range: Himalayas to S China and SE Asia.

Distribution and status: Fairly common resident. Nominate race is widely distributed from SE Xizang across to E and SE China. Race *ticehursti* is recorded in SW Yunnan (Cangyuan).

Habits: Lives in secondary scrub, grass and bracken on bare hills and in gaps in pine forests from 1200–3300 m. Skulking and with rather horizontal carriage.

943. Russet Bush Warbler *Bradypterus seebohmi* [Mountain   PLATE 93
Bush-Warbler > Javan/Timor Bush-Warbler, AG: Scrub-Warbler]
Gaoshan duanchiying

Description: Medium-sized (13.5 cm) dark brown warbler with a longish, broad, graduated tail. Upperparts olive-brown with a rufous tinge; tail more olive; chin and throat white streaked with black; rest of underparts white, washed grey on sides of neck and olive-brown on sides of breast and belly. Whitish tips of undertail coverts give scaly appearance. Iris—brown; bill—black above, pinkish below; feet—pinkish.

Voice: Song a mechanical, endlessly repeated, rasping *zee-ut, zee-ut*, in long series. Call is excited chacking and explosive *rink-tink-tink*.

Range: E Himalayas, NE India, S China, SE Asia, Taiwan, Philippines, Java and Timor.

Distribution and status: Race *melanorhynchus* is resident on mountains of SE China and Taiwan; also S Sichuan and NE Yunnan. Not uncommon in suitable habitat but apparently often overlooked. Recent records of this species from Bhutan suggest this species probably also occurs in SE Xizang.

Habits: Skulks in dense scrub at forest edge and on open, scrubby hillsides up to 2800 m.

3. GRASS WARBLERS

944. Lanceolated Warbler *Locustella lanceolata* [Streaked Warbler,   PLATE 95
AG: Grasshopper-Warbler] Maoban huangying

Description: Smallish (12.5 cm) brown-streaked warbler. Upperparts olive-brown streaked blackish; underparts white washed ochraceous and streaked black on chest and flanks; supercilium buff; tail lacks white tip. Distinguished from Grasshopper Warbler by smaller size, bolder streaking on upperparts and on breast and blacker crown. Iris—dark brown; bill—brown above, yellowish below; feet—pinkish.

Voice: Song is a prolonged, rapid, high-pitched trill; sharper and slower than Grasshopper Warbler. Call note is *churr-churr* and low *chk*.

Range: Breeds in Siberia, E Palearctic; migrates S in winter to Philippines, SE Asia, Greater Sundas and N Moluccas.

Distribution and status: Uncommon seasonal migrant. Breeds in NE China; recorded on passage across eastern China and also NW China.

Habits: Frequents wet rice fields, swampy scrub, fallow fields and bracken near water.

945. Grasshopper Warbler *Locustella naevia* [Common   PLATE 94 Grasshopper Warbler] Haiban huangying

Description: Medium-sized (13 cm) olive-brown warbler, heavily streaked blackish. Underparts grey-buff with creamy throat and warmer wash across upper breast, black spotting in summer across upper breast and sometimes as streaks on flanks. Rather indistinct pale supercilium. Very similar to Lanceolated Warbler but belly greyer, buffy undertail coverts marked with black arrow-shaped streaks; tail darker; slightly larger size; upperparts less heavily streaked and different song. Iris—brown; bill—dark with pink base to lower mandible; feet—pinkish.

Voice: Call is hard *sit*. Song from low perch is fast, dry, endless reeling like alarm clock with muffled clapper.

Range: Breeds W Europe to NW Mongolia; winters Spain, N Africa and India.

Distribution and status: Rare. Race *straminea* breeds in Tekes valley of W Tianshan in NW Xinjiang. Migrates S in winter.

Habits: Skulks in dense ground vegetation; rarely takes flight.

946. Rusty-rumped Warbler *Locustella certhiola* [Pallas's/   PLATE 95 Grey-naped Warbler, AG: Grasshopper Warbler] Xiao huangying

Description: Medium-sized (15 cm) streaked brown warbler with a buff eye-stripe and white tips to rufous tail. Upperparts brown streaked grey and black; wings and tail reddish-brown, the latter with blackish subterminal bar; underparts whitish with buff chest and flanks, in juveniles washed yellow and with triangular black spots on chest. Race *centralasiae* is palest and *rubescens* the darkest. Nominate has heavier black streaking on mantle and secondaries. Iris—brown; bill—brown above; yellowish below; feet—pinkish.

Voice: Prolonged harsh trill *chir-chirrrr*, also thin alarm note *tik tik tik*.

Range: Breeds in N and C Asia; migrates S in winter to China, SE Asia, Palawan, Sulawesi and Greater Sundas.

Distribution and status: Uncommon summer breeder and passage migrant. Race *centralasiae* breeds in W and N Xinjiang, Qinghai, N Gansu and W Nei Mongol, wintering in S China. Race *minor* breeds in NE China, recorded on passage through eastern provinces. Race *rubescens* also recorded on passage through eastern China, and nominate race accidental in Hebei.

Habits: Inhabits reed beds, swamps, paddy field and grassy thickets and bracken near water, also forest edge. Skulks in dense vegetation and when flushed flies only a few metres before diving into cover again.

947. Middendorff's Warbler *Locustella ochotensis* PLATE 95
[AG: Grasshopper-Warbler] Bei huangying

Description: Largish (16 cm) olive-brown warbler with buffy-brown flanks and whitish belly. Immature are streaked on breast and flanks. Distinguished from Rusty-rumped Warbler by unstreaked upperparts and from Pleske's Warbler by browner upperparts, paler underparts, darker eye-stripe and shorter bill. Iris—brown; bill—dark above, pale below; feet—pinkish.

Voice: Shrill, grinding call, *viche ... viche ... viche.*

Range: Breeds in NE Asia; in winter S to S China, Philippines, Sulawesi and Borneo.

Distribution and status: Nominate race recorded as rare passage migrant on E coast, Taiwan and Guangdong.

Habits: Prefers patches of grass or reeds.

948. Pleske's Warbler *Locustella pleskei* [Styan's Grasshopper Warbler] PLATE 95
Shishi huangying

Description: Largish (16 cm) plain, greyish-brown warbler with short buffy supercilium and white underparts with grey wash on sides of breast and flanks. Outer tail feathers tipped whitish. Shows thin silver edges to wing coverts. Crown and mantle slightly speckled dark. First-winter birds have yellow tinge to throat. Similar to Middendorf's Warbler but greyer; lacks rufous rump; less distinct supercilium. Iris—brown; bill—dark above, pink below; feet—pink.

Voice: Song said to differ from Middendorf's Warbler but no details available.

Range: Breeds in SE Siberia, Japan and S Korea; migrating through SE China. Winters in Hong Kong.

Distribution and status: Scarce on passage or wintering in SE coastal provinces. Could breed in NE China and should be searched for.

Habits: Lives in open thickets on exposed headlands and hillsides. Winters in reed beds, scrub, and mangroves. Skulking in summer but less shy in winter.

949. Savi's Warbler *Locustella luscinioides* Qu huangying PLATE 95

Description: Medium-sized (14 cm) plain olive-grey warbler. Upperparts concolorous, plain with indistinct supercilium and white half ring under eye. Underparts whitish with pale, pinkish-buff on upper breast, sides, flanks and undertail coverts. Side of breast has some diffuse brownish streaks. Undertail coverts slightly chevroned by white tips. Iris—brown; bill—dark horn above, pink below; feet—pale flesh-coloured.

Voice: Call is *ching ching* like a Great Tit, also soft *puitt.* Song starts gently then builds up to hard buzzing *surrrrrrrr ... like a mole cricket.

Range: Europe and Ukraine, wintering to W Africa; C Asia to Mongolia, wintering to E Africa.

Distribution and status: A few stray records of race *fusca* from W Xinjiang, but could well breed in Kashi and Tianshan areas and should be searched for.

Habits: Lives in thickets near fresh or brackish water. Skulking, although not shy. Shuffles and runs on ground in jerky fashion like a mouse with tail raised in the air.

950. Gray's Warbler *Locustella fasciolata* [Gray's Grasshopper Warbler]   PLATE 95
Cangmei haungying

Description: Largish (15 cm) plain warbler with olive-brown upperparts; white supercilium; dark eye-stripe and dusky cheeks. Underparts white with grey or rufous buff band across breast and on flanks, with faint whitish feather edges. Undertail coverts buff. Juvenile has yellowish underparts with streaked throat. Bill is larger than other *Locustella* species and could be confused with Great Reed Warbler apart form graduated tail, smaller head and greyer colour. Iris—brown; bill—black above, pink below; feet—pinkish-brown.

Voice: Elaborate, long song of rising and falling phrases. Calls include trilling *cherr-cherr . . . cher* and loud, snarling *tschrrok tschrrok*.

Range: Breeds in NE Asia and Japan; migrating through E China to Philippines, Sulawesi and New Guinea.

Distribution and status: Uncommon. Nominate race breeds in NE Nei Mongol and N Heilongjiang in Greater and Lesser Hinggan Mts. On passage across eastern provinces and Taiwan. Darker race *amnicola* should also be looked for.

Habits: Occurs in lowland and coastal woods, thickets, hill grasslands and scrub. Creeps, runs and hops through undergrowth. Carriage horizontal but stands tall like a chat on the ground.

951. Japanese Swamp Warbler *Locustella pryeri*   PLATE 100
[Streak-backed Swamp Warbler/Marsh Grassbird, AG: Marsh Warbler]
Banbei daweiying

Description: Small (12 cm) reed-living warbler with rufous-brown upperparts, heavily streaked with black. Has long, broad, graduated tail and diffuse whitish eyebrow. Underparts whitish with pale copper flanks and sides of breast and buff undertail coverts. Separated from Lanceolated Warbler by lack of streaks on breast. Separated from Zitting Cisticola by larger size, pale crown and longer tail. Iris—brown; bill—shiny black above, pink below; feet—pink.

Voice: Song is low-pitched *djuk-djuk-djuk* in display flight or from reed perch. Call is *chuck*.

Range: Breeds Japan and NE China; winters C China; vagrant Korea.

Distribution and status: Globally Vulnerable (Collar *et al.* 1994). Rare seasonal migrant. Race *sinensis* breeds in Liaoning (Chaoyang) and Hebei coast. Recorded on migration in Hubei; winters along mid-Changjiang valley in Hubei and Jiangxi.

Habits: Lives in reedbeds. Shy and retiring.

Note: Formerly treated within genus *Megalurus* (e.g. Cheng, 1987, 1994).

952. Sedge Warbler *Acrocephalus schoenobaenus*   PLATE 95
Shuipu weiying

Description: Small (12.5 cm) olive-brown reed warbler with broad white eyebrow topped by black band; olive crown streaked black; brown upperparts, heavily streaked blackish; chestnut lower back, rump and uppertail coverts and white underparts with rufous-buff wash on sides of breast and flanks. Winter birds more rufous and with fine black spots in band across breast. head profile very flat and bill short and sharp. Should not be confused within Chinese region. Iris—brown; bill—dark above, pale below; feet—pinkish.

Voice: Loud medley of harsh, sweet phrases, faster than Great Reed Warbler, interspersed with mimicked phrases. Song may last several minutes without pause. Calls include churring, harsh *tue* and scolding rattle.

Range: Breeds Europe, W and C Asia; migrates to Africa.

Distribution and status: Rare. Breeds in Tianshan in W Xinjiang.

Habits: Inhabits boggy areas with tall grasses, reeds and bushes. Keeps low. Flicks tail when calling. Male has circling song flight.

953. Streaked Reed Warbler *Acrocephalus sorghophilus* PLATE 95
[Speckled Reed Warbler] Xiwen weiying

Description: Medium-sized (13 cm) reed warbler. Upperparts ochraceous-brown with faint speckling on crown and mantle. Underparts buff with whitish throat. Cheeks yellowish; eyebrow yellowish-buff outlined above with broad black line. Paler above and more streaked than Black-browed Reed Warbler and with more robust and longer bill. Iris—brown; bill—black above, yellowish below; feet—pink.

Voice: Not recorded.

Range: Breeding endemic to NE China. Breeds in NE China and winters in Philippines.

Distribution and status: Globally Vulnerable (Collar *et al.* 1994). Presumed to breed in Liaoning and NE Hebei. Migrates through Hebei, Jiangsu, Hubei and Fujian to Luzon.

Habits: Breeding and summer feeding areas are presumed to be reed beds but migrants found in fields of millet.

954. Black-browed Reed Warbler *Acrocephalus bistrigiceps* PLATE 95
[Von Schrenck's Reed Warbler] Hemei weiying

Description: Medium-sized (13 cm) brown warbler with buffy white supercilium bordered above and below with distinctive black stripes. Underparts whitish. Iris—brown; bill—dark above, pale below; feet—pinkish.

Voice: Harsh *chur* in alarm. Call is sharp *tuc* or thin *zit*. Song is sweet and varied with many repeated phrases; less scratchy than Eurasian Reed Warbler.

Range: Breeds in NE Asia; migrates in winter to India, S China, and SE Asia.

Distribution and status: Breeds in NE China, Hebei, Henan, S Shaanxi and lower reaches of Changjiang valley. On passage over S and SE China with some birds wintering in Guangdong and Hong Kong. Accidental on Taiwan.

Habits: Typical reed warbler living among tall reeds and grasses close to water.

955. Paddyfield Warbler *Acrocephalus agricola* PLATE 96
Daotian weiying

Description: Smallish (14 cm) plain rufous-brown reed warbler with rather short white supercilium topped by indistinct short black line. Back, rump and uppertail coverts rufous. Underparts white with rufous-buff wash on flanks, undertail coverts and usually across breast. Line through eye and ear coverts brown. Iris—brown; bill—black above, pink below; feet—pinkish.

Voice: Song is long and fluent, typical of genus; lacking harsh notes of some other reed warblers. Call is sharp *chik-chik*, slurred *zack-zack* or harsh churring.

Range: Breeds C Asia to W China; winters Iran, India and Africa.

Distribution and status: Uncommon. Race *capistrata* breeds in W and S Xinjiang and Qaidam basin of Qinghai (Cheng 1987, 1994, uses name *brevipennis* for this race). Accidental to Hong Kong.

Habits: Feeds low down in vegetation close to lakes and rivers. Has distinctive habit of constant tail-flicking, tail-cocking and raising crown feathers.

Note: Some authors include Manchurian Reed Warbler and Blunt-winged Reed Warbler within this species (e.g. Cheng 1987). Alström *et al.* (1991) and Inskipp *et al.* (1996) include Manchurian Reed Warbler within this species but not Blunt-winged.

956. **Manchurian Reed Warbler** *Acrocepahlus tangorum* PLATE 96
Manzhou weiying

Description: Medium-sized (14 cm) plain, greyish-brown reed warbler with dark eye-stripe, broad white supercilium and large, long bill. In fresh winter plumage much more rufous and with rufous wash on breast, flanks and undertail coverts. Very similar to Paddyfield Warbler with longer bill and bolder black brow above supercilium. Similar to Blunt-winged Warbler but has longer bill, longer tail and less contrast between black second brow and centre of crown. Iris—brown; bill—dark above, pink below; feet—orange-brown.

Voice: Sharp *chi chi* and copied phrases of other birds.

Range: Breeds NE China; winters locally in SE Burma, SW Thailand and S Laos.

Distribution and status: Uncommon. Breeds Hulun Nur area of Nei Mongol and Songhua valley of Heilongjiang, also Liaoning. On passage through Liaoning and presumably on E coast. One record Hong Kong.

Habits: As Paddyfield Warbler.

Note: Variously treated as a race of Paddyfield Warbler (e.g. Alström *et al.* 1991, Cheng, 1994, Inskipp *et al.* 1996) or a race of Black-browed Reed Warbler (e.g. Howard and Moore 1980) or as a full species (e.g. Viney *et al.* 1994).

957. **Blunt-winged Warbler** *Acrocephalus concinens* PLATE 96
[Swinhoe's Reed Warbler] Dunchi [daotian] weiying

Description: Medium-sized (14 cm) drab, rufous-brown, unstreaked reed warbler with short round wings and short white supercilium, hardly extending beyond eye. Upperparts dark olive-brown with rufous rump and uppertail coverts. Has dark brown eye-stripe but no dark brow above supercilium. Underparts white with buff wash on sides, flanks and undertail coverts. Distinguished from Paddyfield and Manchurian Reed Warblers by shorter eyebrow and lack of second dark brow. Iris—brown; bill—dark above, pale below; feet—pinkish with blue soles.

Voice: Strident but scratchy song. Call is vibrant *thrrak* or *tschak*.

Range: Breeds C Asia and E China; winters from India to SW China and SE Asia.

Distribution and status: Uncommon. Nominate race breeds in N and C China passing through SW and SE China on migration. Once recorded in Hong Kong in April.

Habits: Lives in reed beds; also tall grasslands in hills.

958. Eurasian Reed Warbler *Acrocephalus scirpaceus* PLATE 96
[European Reed Warbler] Lu weiying

Description: Medium-sized (13 cm) drab, uniform, unstreaked reed warbler. Has indistinct white eyebrow; short dark eye-line; darkish ear coverts; long bill. Upperparts olive-brown with warmer tone to rump and uppertail coverts. Underparts white with rufous-buff sides of breast and flanks. Has peaked head jizz. Very similar to Blyth's Reed Warbler, but has darker upper mandible; broader and better developed pale lores; darker and warmer colour. Separated from Manchurian Reed and Paddyfield Warblers by lack of dark brow and from Blunt-winged by less rufous, more olive tone, dull rump and longer wings. Iris—pale brown; bill—dark above, pink below; feet—yellowish-brown.

Voice: Repetitive droning song of low-pitched, even notes, delivered in erratic bursts of 2–4 notes and sometimes mimics other bird calls. Call is quiet *churr*.

Range: Breeds Europe, Asia Minor, C Asia; winters to Africa.

Distribution and status: Accidental. Race *fuscus* is vagrant recorded in Jiangsu, but known to reach Tianshan. Recently found in W Xinjiang.

Habits: Lives in tall reeds. Sidles up reed stems with head up and tail down. Inquisitive but nervous. Flight low with spread tail.

Note: Race *fuscus* may need upgrading to species status.

959. Blyth's Reed Warbler *Acrocephalus dumetorum* PLATE 96
Bushi weiying

Description: Medium-sized (14.5 cm) drab, greyish-brown, unstreaked reed warbler with short rounded wings. Has thin dark eye-stripe and short but broad white supercilium with well-developed pale lores and no dark superior brow line. Bill is long and has dark culmen and tip. Colour colder and greyer than Eurasian Reed or Paddyfield Warblers and rump concolorous with rest of upperparts. Underparts white with buff wash on side of neck, upper breast and flanks. Iris—olive; bill—pinkish with dark tip; feet—brownish.

Voice: Long, slow song; rich and mimetic with croaks, chirping and chattering but recurrent *tack tack* and *see-see-hue* phrases. Calls include penetrating *thik*, hard *chak* and scraping *cherr*.

Range: Breeds Europe to NW Asia; winters NE Africa, Himalayas, India and Burma.

Distribution and status: Vagrant. Trapped three times since 1986 at Mai Po in Hong Kong. May breed in NW China.

Habits: Lives in damp thickets and marshy scrub and grassland. Very vocal.

960. Great Reed Warbler *Acrocephalus arundinaceus* PLATE 96
Da weiying

Description: Large (20 cm) bulky, unstreaked reed warbler with large, thick bill with dark tip. Upperparts warm brown with rufous rump and uppertail coverts. Underparts white with warm buff wash on sides of breast, flanks and undertail coverts. Head has peaked jizz; well-developed white or buff (fresh) supercilium; no dark superior brow. Similar to Clamorous Reed Warbler but darker with less rufous tail and uppertail coverts and whiter underparts. These species do not overlap within China. Larger than Oriental Reed Warbler without streaking on throat and with longer wings. Iris—brown; bill—dark with pale base to lower mandible; feet—greyish-brown.

Voice: Loud, strident, discordant song; rather throaty and guttural, interspersed with squealing cries and low croaks. Calls include hard *tack* and croaking *churr*.

Range: Africa, Eurasia, India to W China.

Distribution and status: Locally common up to 2000 m. Race *zarudnyi* breeds in the Kashi and Tianshan region of W Xinjiang.

Habits: As Oriental Reed Warbler. Lives in reed beds and brush near water. Crashes heavily through reeds; lumbers up stems. Thrush-like on ground. Flies with tail fanned.

Note: Some authors include Oriental Reed Warbler within this species (e.g. Cheng 1987; Sibley and Monroe 1993).

961. Oriental Reed Warbler *Acrocephalus orientalis* [Eastern Reed Warbler, AG: Great Reed Warbler] Dongfang daweiying   PLATE 96

Description: Largish (19 cm) brown warbler with conspicuous buff supercilium. Distinguishable in field from Clamorous Reed Warbler by blunter, shorter, stouter bill; shorter tail with pale tips; more richly coloured underparts and dark streaking on breast; or in the hand by the outer (ninth) primary longer than sixth. Gape is pinkish not yellow. Distinguished from allopatric Great Reed Warbler by smaller size and shorter primary projection and more streaking on side of breast. Iris—brown; bill—brown above, pinkish below; feet—grey.

Voice: In winter quarters it only utters a single, harsh, grating *chack* at intervals.

Range: Breeds in E Asia; migrating S in winter to India and SE Asia, Philippines, Indonesia and occasionally as far as New Guinea and Australia.

Distribution and status: Breeds from N and E Xinjiang and south through C, E and SE China. On passage over southern provinces and Taiwan.

Habits: Favours reed beds, rice fields, marshes and secondary scrub in lowlands.

962. Clamorous Reed Warbler *Acrocephalus stentoreus*   PLATE 96
Southern Great Reed-Warbler>Heinroth's/Large-billed Reed-Warbler] Zao weiying

Description: Largish (19 cm) brown warbler with elongated tail and whitish supercilium. Upperparts uniform olive-brown; underparts whitish with fawn flanks and undertail coverts. Distinguished from Great Reed Warbler by more rufous rump, tail and uppertail coverts and less richly coloured underparts. Distinguished from Oriental Reed Warbler by larger, more slender, stouter and sharper bill; lack of dark streaking on throat and larger tail without pale tips. Iris—brown; bill—greyish-brown above, pale base to lower mandible; feet—greyish-brown.

Voice: Harsh alarm note *chack* or *churr*, sweet song, interspersed with higher notes and squeaky cackles with characteristic *ro-do-peck-kiss* phrase. More melodious and less raucous than Great Reed Warbler. Generally calls at night.

Range: Resident Egypt and Middle East; breeds C Asia and Himalayas, wintering to India; summer breeder S China and resident SE Asia, N Philippines, Sundas and Sulawesi to Australia.

Distribution and status: Race *amyae* is summer breeder in SE Xizang, Sichuan and Guizhou.

Habits: Inhabits swampy reed beds and paddy fields near reed beds, also mangroves. Clings sideways to reed stems when perched and puffs out throat feathers when singing. Generally singly or in pairs in reeds or other vegetation close to the ground.

963. Thick-billed Warbler *Acrocephalus aedon* Houzui Luying PLATE 96

Description: Large (20 cm) olive-brown or rufous, unstreaked reed warbler with short but stout bill. Distinguished from other large reed warblers by bland expression caused by absence of dark eye-stripe and virtually no pale supercilium. Has long, graduated tail. Race *stegmanni* is much more rufous than nominate race. Iris—brown; bill—dark above, pale below; feet—greyish-brown.

Voice: Loud, full song commences with ticking *tschok tschok* then bursts into melodious chatter of whistled phrases and mimicked calls, Call is insistent *chack chack* and harsh chatter.

Range: Breeds N Palearctic; winters India to S China and SE Asia.

Distribution and status: Uncommon but widespread. Nominate race recorded breeding in Bugt and Zalantun in NE Nei Mongol. More common race *stegmanni* (includes *rufescens*) breeds widely in NE China and C Nei Mongol. Both races migrate across E China but probably generally overlooked.

Habits: Lives in forest, woods and secondary scrub. Rather skulking, often in dark thickets.

964. Booted Warbler *Hippolais caligata* Xue liying PLATE 93

Description: Small (11 cm) brown warbler. Size and shape of *Phylloscopus* leaf warbler but with colour pattern like a reed warbler. Bill is very small. Whitish supercilium is long and broad but diffuse, extending well beyond eye. Upperparts concolorous grey-brown; underparts creamy-white with buff wash on flanks and undertail coverts. Has white eye-ring. Tail is square-cut with white outer feathers. Differs from very similar Sykes's Warbler by smaller size, browner upperparts, more buffy underparts and smaller bill. Similar to Olivaceous Warbler but browner, with smaller bill, longer supercilium and more square-cut tail. Iris—brown; bill—dark above, pink below; feet—pinkish-grey.

Voice: Sweet, liquid bubbly song consisting of several subsongs each introduced by soft *clik* note.

Range: Breeds Russia and C Asia, wintering in India.

Distribution and status: Rare resident up to 2000 m in Junggar Basin and Manas River areas of W Xinjiang.

Habits: Inhabits dry scrub and shrubby habitats. Tail held still when singing, not pumped up and down like Olivaceous warbler. Skulking and elusive. Looks clumsy in movements.

Note: Sykes's Warbler *H. rama* is often included within this species (e.g. Cheng 1987, Inskipp *et al.* 1996). For reasons of exclusion see that species.

965. Sykes's Warbler *Hippolais rama* Saishi xueliying   NOT ILLUSTRATED

Description: Small (12 cm) brown warbler. Size and shape of *Phylloscopus* leaf warbler but with colour pattern like a reed warbler. Very similar to Booted Warbler but larger, has less brown upperparts and whiter underparts and larger bill. Iris—brown; bill—dark above, pink below; feet—pinkish.

Voice: Song is louder, slower and more monotonous than that of Booted Warbler with more whistles and churrs.

Range: Breeds Russia and C Asia, wintering in India.

Distribution and status: Rare resident up to 2000 m in Altai Mts, Shache, Kashi, Qinghe and Turpan regions of W Xinjiang. Accidental Hong Kong.

Habits: As Booted Warbler.

Note: Often included within Booted Warbler (e.g. Cheng 1987, Inskipp *et al.* 1996). Despite apparent hybrid zone in area of overlap (Cramp 1992) we consider this valid species on the basis of morphology, sympatoy and song as per Glutz and Bauer von Blotzh (1991) and Sibley and Monroe (1993).

966. Olivaceous Warbler *Hippolais pallida* Caolü liying   PLATE 93

Description: Smallish (13 cm) uniform brown warbler with short whitish supercilium, rather flat crown and whitish underparts. Tail squarish with short undertail coverts giving long-tailed appearance. Upperparts concolorous rich brown. Larger-billed and shorter eyebrow than Booted or Sykes's Warblers. Similar to Eurasian Reed Warbler but tail square not rounded or graduated and outer feathers edged and tipped whitish. Rump less rufous; smaller bill. Iris—brown; bill—dark above, pale below; feet—bluish-grey to greyish-brown.

Voice: Song has rasping tone with much rise and fall of pitch but little mimicry. Call is short *tec*, sparrow-like chattering and alarm of *tick-tick-tick*.

Range: Spain, S Europe, Iran and N Africa.

Distribution and status: Rare. Race *elaeica* recorded breeding in extreme W Xinjiang.

Habits: Has diagnostic habit of constantly dipping and waving its tail in a downward movement.

4. TAILORBIRDS AND TIT WARBLERS

967. Mountain Tailorbird *Orthotomus cuculatus* [Golden-headed   PLATE 99
Tailorbird] Jintou fengyeying

Description: Small (12 cm) rufous-capped, yellow-bellied forest warbler with a pronounced yellow supercilium. Upperparts olive-green; chin, throat and upper chest greyish-white; lower breast and belly bright yellow. Iris—brown; bill—black above, pale below; feet—pink.

Voice: A variable sweet tinkling song of 2−3 repeated notes followed by a trill, *pee-pee-cherrrree* quite different from other tailorbirds. Call is buzzing *kiz-kiz-kiz*.

Range: N India to S China, Philippines, SE Asia, Malay Peninsula and Indonesia.

Distribution and status: Race *coronatus* is resident in SE Xizang, W and S Yunnan, SW Guangxi (Yaoshan) and N Guangdong (Babaoshan) to N Fujian (Wuyi Mts). Not uncommon on higher mountains of 1000–2500 m.

Habits: Inhabits montane forest, open montane scrub and bamboo thickets. A gregarious bird, often found in small parties but generally skulking in thick cover and difficult to see. Easily recognised by its song. Does not make a leaf-purse nest.

968. Common Tailorbird *Orthotomus sutorius* [Long-tailed Tailorbird] Changwei fengyeying
PLATE 99

Description: Small (12 cm) rufous-crowned, white-bellied warbler with a long, often cocked tail. Forehead and forecrown rufous; lores and side of head whitish; hindcrown and nape greyish; back, wings and tail olive-green; underparts white with grey flanks. In breeding plumage the central tail feathers of the male become further elongated due to moult. Race *inexpectatus* lacks streaking on cheeks and ear coverts; *longicauda* has darker upperparts. Iris—pale buff; bill—black above, pinkish below; feet—pinkish-grey.

Voice: Very loud, repetitive, strident call *te-chee-te-chee-te-chee* or single *twee*.

Range: India to China, SE Asia and Java.

Distribution and status: Common resident up to 1500 m. Race *inexpectatus* occurs in SE Xizang and Yunnan; *longicauda* in S and SE China and Hainan.

Habits: Frequents light forest, secondary forest and gardens. A lively bird, always on the move or cockily giving its penetrating call. Keeps to the understorey and generally stays in thick cover.

969. Dark-necked Tailorbird *Orthotomus atrogularis* [Black-necked/Dark-cheeked Tailorbird] Heihou fengyeying
PLATE 99

Description: Small (11.5 cm) rufous-crowned, white-bellied warbler with a long, often upturned, tail, yellow vent and diagnostic blackish throat (lacking in immatures). Upperparts olive-green; sides of head grey. Female is duller with less red on head and less black on throat. Distinguished from Common Tailorbird by rufous hindcrown, lack of white supercilium, greener back and yellow undertail coverts and thighs. Iris—brown; bill—black above, pinkish below; feet—pinkish-grey.

Voice: Sweet blubbering clear *kri-ri-ri* unlike other tailorbirds.

Range: N India to SW China, Philippines, SE Asia, Sumatra and Borneo.

Distribution and status: Locally common resident in S Yunnan (Xishuangbanna) up to 1200 m.

Habits: Frequents light forest, secondary forest, river banks and gardens.

970. White-browed Tit Warbler *Leptopoecile sophiae* [Stoliczka's Tit Warbler] Huacai queying
PLATE 96

Description: Small (10 cm) fluffy, purplish tit warbler with rufous crown and white eyebrow. Male has violet breast and rump, blue tail and black face mask. Female paler with upperparts yellowish-green and little blue in rump; underparts whitish. Distinguished from Crested Tit Warbler by white supercilium, lack of crest, rufous crown and white edges to outer tail feathers. Race *obscura* is darker than nominate with all purple underparts and blue not violet rump; *major* is paler with bluish-pink belly, extending to breast; *stoliczhae* is the palest race with buff underparts extending to base of throat. Iris—red; bill—black; feet—grey-brown.

Voice: Song is sweet, loud chirping cry. Call note is high-pitched metallic *tzret*.

Range: C Asia, Himalayas, W China.

Distribution and status: Uncommon resident. Nominate race in Kashi, Tianshan and Hami regions of Xinjiang, N Gansu, NE Qilian Mts in Qinghai and S Qinghai; *major* in Tarim basin of Xinjiang and Qaidam basin of Qinghai; *stoliczkae* in mountains of Tarim basin, Altun and around Qaidam basin, also Shiquan river in W Xinjiang; and *obscura* in E Xizang to Gansu, Sichuan and E Qinghai.

Habits: Lives in dwarf scrub above tree-line, to 4600 m in summer, descending to 2000 m in winter. Lives in flocks when not breeding. Flight is weak. Often visits ground.

971. Crested Tit Warbler *Leptopoecile elegans*   PLATE 96
Fengtou queying

Description: Breeding male is a small (10 cm) fluffy purple and maroon warbler with lilac-grey crown, white forehead and crest and all-blue tail. Female has white throat and upper breast, grading lilac ventrally; grey ear coverts; pinkish nape and mantle separated by black line from grey crown, and whitish crest. Distinguished from White-browed Tit Warbler by prominent crest, no white in tail and grey crown. Iris—red; bill—black; feet—black.

Voice: Soft peeping call and shrill wren-like chatters.

Range: Tibet and C China.

Distribution and status: Globally Near-threatened (Collar *et al.* 1994). Uncommon resident in Qinghai, Gansu, N and W Sichuan, E and SE Xizang.

Habits: In summer inhabits fir forest and scrub above tree-line to 4300 m. Descends to 2800–3900 m in winter in subalpine forest zone. Forms small flocks and mixes with other species.

5. LEAF WARBLERS

972. Common Chiffchaff *Phylloscopus collybita*   PLATE 93
[Eurasian Chiffchaff] Jiza Liuying

Description: Smallish (11 cm) greenish-brown warbler with black eye-stripe, buffy supercilium but no coronal stripes and generally no prominent wing bar. Has primrose-yellow at bend of wing. Race *tristis* is greyer and whiter than other races. Separated from Mountain Chiffchaff with difficulty by olive wash on rump and uppertail coverts and edges in flight and tail feathers, also buff not white eye-ring. Underparts creamy with warm buff to flanks, sometimes shows faint wing bar of pale tips to greater coverts causing confusion with worn Greenish Warbler but lacks yellow, arched and flared supercilium of Greenish and has smaller bill. Distinguished from Dusky Warbler by longer wings, notched tail and black bill and black bill and legs. Lacks Dusky's tail-flicking behaviour and hard *chett* call. Distinguished from Booted Warbler by smaller size, darker bill and feet and lack of white tips to outer tail feathers. Iris—brown; bill—black; feet—black.

Voice: Song is more plaintive and faster than nominate *P. c. collybita: chi-vit, chi-vit-chi-vit* or *weechoo, weechoo, cheweechoo* run together as a series of tinkling musical notes. Call is distinctive thin *peep* or *heep* like Eurasian Bullfinch.

Range: Eurasia, wintering from Mediterranean, N Africa to India.

Distribution and status: Race *tristis* is uncommon in Altai and Tianshan Mts of NW Xinjiang. Vagrants recorded N Guangdong and Hong Kong.

Habits: Typical small leaf warbler.

Note: Mountain Chiffchaff *P. sindianus* is placed by some authors within this species but we follow Baker (1997) in treating this as a separate species. Look for Willow Warbler *P. tochilus*, not recorded in China, but which could also occur in Altai Mts.

973. Mountain Chiffchaff *Phylloscopus sindianus* PLATE 93
Dongfang jizalivying

Description: Smallish (11 cm) grey-brown, compact warbler with black eye-stripe and white supercilium lacking coronal stripes or wing bars. Shows bold border to ear coverts; whitish eye-ring and notched tail. Very similar to eastern race *tristis* of Eurasian Chiffchaff but lacks obvious olive-green in plumage especially on rump, uppertail coverts and edges of flight feathers and tail. Face pattern also stronger with darker eye-stripe and broader, longer white supercilium. Rictal bristles longer and stiffer than Eurasian Chiffchaff and the head is slightly smaller with daintier bill. Whitish underparts washed buff on flanks and breast. Difficult to separate from Eurasian Chiffchaff in worn plumage unless singing. Iris—brown; bill—blackish with brown base to lower mandible; feet—blackish.

Voice: Plaintive, slightly dissyllabic *huit* or *hweet*, higher in pitch than Eurasian Chiffchaff and rising more quickly; also loud dissyllabic *tiss-yip*.

Range: Caucasus, NE Turkey to NW Iran, Pamirs, NW Himalayas and W China.

Distribution and status: Nominate race is uncommon resident in Tianshan, Kunlun Mts to Astyn Tagh in Xinjiang and in extreme W Xizang.

Habits: Lives in willows, poplars and rhododendron and other scrub from 2500–4400 m, descending in winter.

Note: Sometimes classed as a subspecies of Eurasian Chiffchaff e.g. Sibley and Monroe (1993), but we follow Cheng (1994), Cramp (1992) and Inskipp *et al.* (1996) in considering the two specifically distinct

[974. Wood Warbler *Phylloscopus sibilatrix* Lin liuying PLATE 98

Description: Distinctive, largish (12.5 cm) long-winged leaf warbler with greenish upperparts and lemon supercilium, chin and breast. Wing panel yellowish-green. Tertials edged pale yellow. Has narrow dark eye-stripe and yellow eye-ring. Tail square or slightly notched. Yellow of breast sharply demarcated from silky white belly. First-winter upperparts slightly darker than adult and throat less yellow. Iris—brown; bill—dark above, yellow-flesh below; feet—pale yellow.

Voice: Call a liquid *tiuh* and soft *wit-wit-wit*. Song is series of ringing *zip* notes accelerating into metallic trill.

Range: Europe, Urals, Caucasus, migrating in winter to equatorial Africa.
Distribution and status: Accidental. Recorded in September in S Yangbaijian, Xizang. Record now considered doubtful. More likely to be encountered in Altai Mts.
Habits: A bird of mature forests, usually high in canopy, typically in crouched position.]

975. Dusky Warbler *Phylloscopus fuscatus* He liuying　　PLATE 93

Description: Medium-sized (11 cm) drab, brown leaf warbler. Rather compact and rounded in shape with short rounded wings and rounded, slightly notched tail. Underparts creamy with fulvous wash on breast and flanks. Upperparts greyish-brown with olive-green fringe to flight feathers. Small fine bill and long thin legs. Nominate race has rufous-buff wash on supercilium, no buff on cheeks and darker brown upperparts. Distinguished with difficulty from Radde's Warbler by fine, weak and darker bill; thinner legs; narrower and shorter supercilium (rufous toward rear in nominate race); dark brown margin above supercilium on lores, and supercilium well-defined between eye and bill; rump has no olive-greenish wash. Iris—brown; bill—dark above, yellowish below; feet—brownish.

Voice: Song is loud monotonous series of clear whistles concluding in a trill. Similar to, but slower than Radde's Warbler. Call is sharp *chett . . . chett* like striking of stones.

Range: Breeds N Asia in Siberia, N Mongolia, N and E China, migrating S in winter to S China, SE Asia and N Indian subcontinent.

Distribution and status: Nominate race breeds in NE and NC China, wintering in S China, Hainan and Taiwan. Race *weigoldi* breeds in S Qinghai, E Xizang and NW Sichuan, wintering in Yunnan and SE Xizang. Both forms are common, especially during migration.

Habits: Skulks in low dense vegetation in damp scrub along streams, around swamps and in forest up to 4000 m. Cocks tail and flicks tail and wings.

Note: Some northern birds, of race *weigoldi*, are larger and sometimes ascribed to a separate race *robustus*. Smoky Warbler is sometimes treated as conspecific with Dusky (e.g. Cheng 1987).

976. Smoky Warbler *Phylloscopus fuliginventer* Yan liuying　　PLATE 98

Description: Medium-sized (11 cm) drab brown leaf warbler. Very similar to Dusky Warbler but darker sooty-brown above with yellowish-green wash on underparts and yellowish supercilium. Race *tibetanus* is darker above and less yellow below with greyish supercilium. Worn specimens become greyish on underparts. Underparts less bright yellow than Tickell's Leaf Warbler and supercilium shorter. Buff-throated Warbler is not as dark and has buffish not yellow-green underparts. Sulphur-bellied Warbler differs in having pale grey-brown upperparts, an olive tinge to rump and orange tinge to front of supercilium. Iris—brown; bill—brown above, yellowish below; feet—brownish-black.

Voice: Song is repetitive single note *tsli-tsli-tsli . . .* or dissyllabic *tslui-tslui . . .* Call is sharp *tzik* or a soft *stup*. Also gives *chek* call, slightly sharper than *chett* call of Dusky Warbler.

Range: Himalayas, migrating in winter to plains of N India.

Distribution and status: Races *fuliginventer* (Himalayas) and *tibetanus* (SE Xizang from Tsari to Nujiang River) breed in rocky alpine pasture and scrub above treeline from 3500–4500 m. Generally scarce.

Habits: Skulks in cover, horizontal carriage, much tail-cocking and wing and tail-flicking. Sometimes feeds in open especially along streams.

Note: Sometimes treated as conspecific with Dusky Warbler (e.g. Cheng 1987).

977. Tickell's Leaf Warbler *Phylloscopus affinis* PLATE 98
Huangfu liuying

Description: Medium-sized (10.5 cm) pale and bright, compact-looking leaf warbler with longish wings and rounded but slightly notched tail. Upperparts olive-green with long, bold yellow supercilium, sometimes whitish towards rear; dusky-yellow ear coverts and no wing bar. Tail and flight feathers brown with olive fringes to outer webs. Underparts yellow with buff wash on sides of breast and olive wash on flanks and vent. Narrow white tip and inner border on three outer tail feathers. Worn plumage greyer and less yellow. Distinguished with difficulty from Buff-throated Warbler by longer bill with no dark tip to lower mandible, more prominent supercilium, shorter tail, more yellow on ear coverts, paler yellower belly. Smaller and more olive than Sulphur-bellied Warbler with brighter supercilium, lacks streaked throat of Yellow-streaked Warbler. Iris—brown; bill—brown above, yellowish below; feet—dark.

Voice: Song is rapid series of soft notes preceded by grace note, *chip chi-chi-chi-chi-chi-chi*. Call note is sharp *chep* and in alarm rapid repeated *tak-tak*.

Range: Breeds from N Pakistan through Himalayas to C China. Winters to India, Bangladesh, N Burma and SW China.

Distribution and status: Locally common in alpine scrub and rocky valleys of S Xizang, Qinghai, Gansu, S Shaanxi, Sichuan and N Yunnan at 2700–5000 m. Winter migrant in SE Xizang, W Yunnan and Guizhou in scrub and bamboo.

Habits: Skulker of low vegetation with quick, jerky movements. Sometimes forms small flocks in winter when also uses canopy more.

Note: Similar Tytler's Leaf Warbler *P. tytleri* has white supercilium, and could be found in disputed areas of Kashmir currently controlled by China.

978. Buff-throated Warbler *Phylloscopus subaffinis* PLATE 93
[Buff-bellied Leaf Warbler] Zongfu liuying

Description: Medium-sized (10.5 cm) olive-green leaf warbler with dull yellow supercilium and no wing bar. Narrow white tip and edges to three outer tail feathers difficult to see in the field. Very similar to Tickell's Leaf Warbler but more dusky on ear coverts, slightly shorter bill with dark tip to lower mandible. Supercilium less prominent especially in front of eye and lacks narrow dark line above. Wings shorter than Tickell's. Greener than Sulphur-bellied Warbler with paler, less orange supercilium. Lacks streaky throat of Yellow-streaked Warbler. Upperparts greener and underparts less green than Smoky Warbler. Iris—brown; bill—dark horn with yellowish cutting edge above and with yellow base to lower mandible; feet—dark.

Voice: Song is similar to Tickell's Leaf Warbler but slower, softer and weaker and lacking introductory grace note, *tuee-tuee-tuee-tuee*. Call note is soft rasping, cricket-like *chrrup* or *chrrip*.

Range: C and E China, wintering in subtropics of S China, N Burma and N Indochina.

Distribution and status: Not common breeder over C S and E China, wintering to S coast and SW China.

Habits: Altitudinal migrant in montane forest and scrub up to 3600 m in summer, wintering in hills and lowlands. Skulks in dense undergrowth; in pairs in summer and forming small flocks in winter. Droops and quivers wings when agitated.

979. Sulphur-bellied Warbler *Phylloscopus griseolus* PLATE 98
Hui liuying

Description: Medium-sized (11 cm) cold brown leaf warbler with sulphur-yellow underparts and greyish-brown wash across upper breast and on sides and flanks. No white in tail, no wing bar and no crown stripes. Has whitish chin, long pale supercilium, and darkish line through eye. Distinguished from Buff-throated Warbler by colder, less olive colour and fore-supercilium orange-yellow becoming yellow to rear. Brighter and paler than Smoky Warbler. From Yellow-streaked Warbler by colder colour and lack of yellow streaks on throat. Iris—brown; bill—pinkish with dark tip; feet—brown.

Voice: Song of about one second, consists of 4–5 fast notes on same pitch, *tsi-tsi-tsi-tsi-tsi*. Sometimes preceded by clicking grace note. Call is soft, distinctive *quip* which sounds like a drip.

Range: Breeds in mountains of S Asia and W China; winters to India.

Distribution and status: Rare summer breeder. Recorded in Kashi and Tianshan, Hejing and Kunlum Mts of Xinjiang and Qilian Mts of Qinghai.

Habits: Shuffles sideways along branches like a treecreeper. Flits among scrub like an accentor. Frequents boulder-strewn hillsides.

980. Yellow-streaked Warbler *Phylloscopus armandii* PLATE 98
[Buff-browed/Milne Edwards Warbler] Zongmei liuying

Description: Medium-sized (12 cm) plump, brown, plain-coloured leaf warbler with slightly forked tail and short, spiky bill. Upperparts olive-brown with olive edges to flight feathers, wing coverts and tail. Has long white eyebrow, buffy in front of eye. Sides of face mottled dark and dark lores and eye-stripe contrast with cream eye-ring. Underparts dirty yellowish-white with olive wash on sides and flanks. Diagnostic yellow streaks on throat often run faintly through breast to belly. Undertail coverts buffish-yellow. From Radde's Warbler by lack of breast band, sharper bill, whiter underparts and behaviour. From Dusky Warbler by brighter, paler plumage; from Buff-throated Warbler by larger size and paler bill and legs; from Sulphur-bellied Warbler by buff fore-supercilium, not orange-yellow and from Tickell's Leaf Warbler by lack of green hue to upperparts. Iris—brown; bill—brown above, paler below; feet—yellowish-brown.

Voice: Call is distinctive, sharp *zic*, like some buntings. Song is similar to Radde's Warbler but weaker.

Range: Breeds N and C China, N Burma; winters S China, S Burma and N Indochina. Accidental Hong Kong.

Distribution and status: Generally uncommon migrant. Nominate race breeds from Liaoning through N China and C China to N Sichuan, E Qinghai and E Xizang; vagrant to Shandong. Race *perplexus* has more southerly distribution in SE Xizang, NW and N Yunnan, SE Sichuan, Ningxia and Hubei; wintering to S and W Yunnan and Guizhou.

Habits: Frequents willow and poplar groves in subalpine spruce forest on mountain slopes. Often feeds on ground in low scrub.

981. Radde's Warbler *Phylloscopus schwarzi* Juzui Liuying PLATE 98

Description: Medium-sized (12.5 cm) brownish-olive, undecorated leaf warbler with largish, slightly forked tail and thick, tit-like bill. Fore-supercilium buff becoming creamy behind eye; eye-line dark brown with diffuse dark speckling on side of face and ear coverts. Underparts dirty white with buff wash across breast and on flanks and cinnamon undertail coverts. Has rather hunched posture. Larger and more robust than Dusky Warbler with longer, broader supercilium and more olive colour. Lacks streaking on throat of Yellow-streaked Warbler. Iris—brown; bill—brown above, pale below; feet—yellowish-brown.

Voice: Call is stuttered *check . . . check*. Song consists of a short build-up of melodious, low notes ending in a trill, *tyeee-tyeee-tyee-tyee-ee-ee*.

Range: Breeds NE Asia; winters S China, Burma and Indochina.

Distribution and status: Fairly common seasonal migrant. Breeds in Greater and Lesser Hinggan Mts of NE China and passes across E and C China on migration. Rarely winters in Guangdong and Hong Kong.

Habits: Skulking and often feeds on ground where it looks clumsy and heavy. Flicks tail and wings nervously.

982. Buff-barred Warbler *Phylloscopus pulcher* PLATE 98
Chengbanchi liuying

Description: Small (12 cm) leaf warbler with brownish-olive back and faint pale crown stripe. Diagnostic two rufous buff wing bars. White inner web of outer tail feathers. Rump is pale yellow and underparts dirty yellow. Eyebrow not pronounced. Iris—brown; bill—black with yellow base to lower bill; feet—pink.

Voice: The call is a fine *zip* followed by a fast shrill trill.

Range: Himalayas, Burma, S Tibet, C China; winters to N Thailand.

Distribution and status: Breeds in north of range and in high mountains but winters S and at lower altitudes. This is one of the commonest birds in conifer and rhododendron forests in Himalayas, Qinghai–Xizang plateau and C China from 2000–4000 m.

Habits: Lively warbler of forest canopy sometimes joining mixed flocks.

983. Ashy-throated Warbler *Phylloscopus maculipennis* PLATE 97
Huihou liuying

Description: Small (9 cm) green-backed leaf warbler with two yellowish wing bars; pale yellow rump; greyish-white face, throat and upper breast; yellow lower breast to undertail coverts and long, broad yellowish-white supercilium. Lateral crown stripes and eye-stripe dark greyish green with grey median crown stripe. Bill is small and delicate. Smaller than Buff-barred Warbler and with yellow underparts. Distinguished from Pallas's

and Chinese Leaf Warblers by pale grey throat and grey rather than yellow median crown stripe. Iris—brown; bill—black with flesh-coloured base; feet—pinkish.
Voice: Song is sweet whistle on monotone *wee-ty wee-ty wee-ty*. Call is sharp, repeated *zit* similar to Pallas's Leaf Warbler.
Range: Kashmir to SW China, Burma and Indochina.
Distribution and status: Uncommon seasonal and attitudinal migrant. Nominate race breeds in S and SE Xizang, Sichuan and NW Yunnan. Winters to S of breeding distribution and at lower altitudes.
Habits: Inhabits mixed oak and conifer forests with rhododendron undergrowth from 2100–3400 m.

984. Pallas's Leaf Warbler *Phylloscopus proregulus* PLATE 97
Huangyao liuying

Description: Small (9 cm) green-backed leaf warbler with lemon rump; two pale wing bars; greyish-white underparts with pale yellow wash on vent and undertail coverts; bold yellow eyebrow and median crown stripe. In fresh plumage, lores are orange. Bill is small and delicate. Upperparts brighter green and underparts more yellow than Lemon-rumped Warbler. Distinguished from Buff-barred and Ashy-throated Warblers by yellow coronal stripe. For separation from Chinese Leaf Warbler see under that species. Iris—brown; bill—black with yellowish base; feet—pink.
Voice: Powerful, resonant song of clear diverse notes repeated 4–5 times, *choo-choo-chee-chee-chee* etc, interspersed with trills and rattles. Calls include soft nasal *dju-ee* or *swe-eet* and soft *weesp*, less piercing than Yellow-browed Warbler.
Range: Breeds N Asia; winters in India, S China and N Indochina.
Distribution and status: Common seasonal migrant. Nominate race breeds in NE China; migrates across E China to winter in low-lying areas S of Changjiang River including Hainan.
Habits: Lives in subalpine forests in summer up to tree-line at 4200 m. Winters in lowland woodland and scrub.
Note: Some authors include Lemon-rumped Warbler within this species (e.g. Cheng 1987).

985. Lemon-rumped Warbler *Phylloscopus chloronotus* PLATE 98
[Pale-rumped Leaf Warbler] Yanhuangyao liuying

Description: Medium-sized (10 cm) greenish leaf warbler with long white supercilium and coronal stripe, pale rump, two yellowish wing bars and white tips to tertials. Sometimes has pale spot on ear coverts. Differs from Pallas's Leaf Warbler due to greyer greenish-olive upperparts, less pronounced yellow markings on head and face and paler yellow supercilium in front of eye. Underparts greyer, less white. Also slightly larger with different wing formula. For separation from Chinese Leaf Warbler see under that species. Iris—brown; bill—dark; feet—brown.
Voice: Song is drawn-out, thin rattle followed by rapid series of hammering notes on same pitch, *tsirrrrrrrrrr-tsi-tsi-tsi-tsi-tsi-tsi-tsi*, repeated every few seconds. Very different from Pallas's Leaf Warbler.
Range: Himalayas to C China; winters northern SE Asia.

Distribution and status: Common seasonal migrant. Breeds in Qinghai, Gansu, Sichuan, E and S Xizang and NW Yunnan. Winters in Yunnan.

Habits: Breeds in upper fir forest with spruce and junipers.

Note: Sometimes treated as a race of Pallas's Leaf Warbler (e.g. Cheng 1987, Viney *et al.* 1994). However, differences in song, breeding habitat and wing formula indicate this is a valid species.

986. Gansu Leaf Warbler *Phylloscopus kansuensis* PLATE 98

Gansu huangyaoliuying

Description: Medium-sized (10 cm) greenish leaf warbler with pale rump, hint of second wing bar, bold white supercilium, pale coronal stripe and whitish edges to tertials. Not distinguishable in the field from Lemon-rumped Warbler, except by voice. Iris—dark brown; bill—dark above, pale below; feet—pinkish-brown.

Voice: Song is faltering, thin, high-pitched, slightly harsh *tsrip*, followed by a series of slightly accelerating *tsip* notes and ends with a clear 1–2 second trill. Similar to Emei Leaf Warbler but very different from Pallas's Leaf Warbler.

Range: Breeds NW China, winters in SW China.

Distribution and status: On basis of current knowledge, should be added to list of globally threatened species. Rare and poorly known. Breeds in Gansu and SE Qinghai in the Xining to Lanzhou area; probably winters in Yunnan. Must be often overlooked.

Habits: As other leaf warblers. Breeds in deciduous forest with some spruce and junipers.

Note: Formerly included within Pallas's Leaf Warbler *P. proregulus* (e.g. Cheng 1987).

987. Chinese Leaf Warbler *Phylloscopus sichuanensis* PLATE 98

Sichuan liuying

Description: Medium-sized (19 cm) greenish leaf warbler with pale rump, long white eyebrow, pale coronal stripe, two white wing bars (second very slight) and pale edges and tips to tertials. Very similar to Lemon-rumped Warbler but differs in being larger, more elongated and with slightly larger and less rounded head; sides of crown paler and coronal stripe less distinctive and sometimes only present as pale spot on back of head; paler centres to greater wing coverts and paler lower bill. Lacks pale spot on ear coverts. Iris—brown; bill—dark above, pale below; feet—brown.

Voice: Song is monotonous, dry *tsiridi-tsiridi-tsiridi-tsiridi-tsiridi* . . . lasting more than one minute and similar to Striated Prinia. Distinct from other leaf warblers and delivered from tree-top. Call is irregular series of loud, clear, scolding whistles such as *tueet-tueet-tueet tueet tueet tueet*. Also soft *trr* on nest.

Range: C and E China; winters NW Thailand, N Laos, C Burma.

Distribution and status: Status uncertain. Appears to be widespread, breeding from E Qinghai and Sichuan to NE China. Many records attributed to Pallas's Leaf and Lemon-rumped Warblers may in fact be this species. Wintering area of the species is not known but one caught in Hong Kong.

Habits: Typical leaf warbler. Lives in low deciduous secondary forest, rarely exceeding 2600 m.

988. Yellow-browed Warbler *Phylloscopus inornatus* PLATE 97
[Inornate Warbler] Huangmei liuying

Description: Medium-sized (11 cm) bright olive-green warbler with usually two whitish wing bars visible, a clear white or cream supercilium and no visible crown stripes. Underparts vary from white to yellowish-green. Distinguished from Arctic Warbler by brighter upperparts, bolder wing bars and white tips to tertials. Distinguished from allopatric Hume's Warbler by brighter, greener upperparts and from Pallas's and Chinese Leaf Warblers by lack of pale median crown stripe and from Greenish Warbler by smaller size and dark lower bill. Iris—brown; bill—upper mandible dark, lower mandible with yellow base; feet—pinkish-brown.

Voice: Noisy. Frequently uttered call is loud *swe-eeet* with rising inflection. Song is weak series of call notes, falling and fading; also dysyllabic *tsioo-eee* dropping then rising in pitch on second note.

Range: Breeds in N Asia and NE China; migrant S in winter to India, SE Asia and Malay Peninsula.

Distribution and status: Nominate race breeds in NE China migrating across most of the country to winter in S Xizang and SW, S and SE China plus Hainan and Taiwan. Generally common in forest and wooded areas.

Habits: Forms active flocks often mixing with other small insect feeders, working through foliage in middle and upper canopy.

Note: Some authors include Hume's Warbler within this species.

989. Hume's Warbler *Phylloscopus humei* [AG: Leaf Warbler] PLATE 97
Xiumei liuying

Description: Medium-sized (10 cm) leaf warbler with olive-grey upperparts; two wing bars; no pale rump; no white in tail; long pale supercilium; dark eye-line; dull grey median crown stripe. Very similar to Yellow-browed Warbler but duller and greyer, with upper wing bar indistinct; edges of tertials less white and wing coverts paler. Iris—brown; bill—black with pale base to lower mandible; feet—brown.

Voice: Call note is short sweet *wesoo*, sparrow-like *chirp* or rising *pwis*. Song is lively repetition of call note *wesoo* followed by falling, nasal wheeze *zweeeeee*. Race *mandellii* is similar but less wheezy.

Range: C Asia and NW and C China, wintering to India, S China and SE Asia.

Distribution and status: Fairly common seasonal migrant. Nominate race breeds in NW China (Junggar Basin, Turpan, Kashi and Tianshan), wintering in S Xizang. Race *mandellii* breeds in C China from NW Yunnan to Sichuan, Qinghai, Gansu, Ningxia, S Shaanxi and SE Shanxi; wintering to SE Xizang. Vagrant to Hong Kong.

Habits: Inhabits larch and pine forests from 300–4000 m. Shy. Often joins mixed flocks. Active arboreal warbler of canopy.

Note: Treated by some authors as race of Yellow-browed Warbler (e.g. Cheng 1987).

990. Arctic Warbler *Phylloscopus borealis* [AG: Willow Warbler] PLATE 97
Jibei liuying

Description: Largish (12 cm) greyish-olive leaf warbler with conspicuous long yellowish-white supercilium. Upperparts dark olive with a prominent white wing bar and second indistinct bar across tips of median coverts; underparts whitish with brownish-olive flanks; lores and eye-stripe blackish. Distinguished from Yellow-browed Warbler by larger size, slightly stouter, upturned bill, shorter-looking tail and bolder head pattern. Distinguished from Pale-legged Leaf Warbler by brighter and greener colour and paler crown. Differs from Large-billed Leaf Warbler in pale base to lower mandible. Iris—dark brown; bill—dark brown above, yellow below; feet—brown.

Voice: Rattling series of *chweet* notes with last note on higher pitch and characteristic harsh, husky *dzit* occasionally given by wintering birds. Song is shivering trill of up to 15 notes becoming louder and faster after slow tentative start.

Range: Breeds N Europe, N Asia and Alaska; migrant S in winter to S China, SE Asia, Philippines and Indonesia.

Distribution and status: Race *hylebata* breeds in N and E Heilongjiang, migrating along E coast and some birds wintering in SE China. Nominate race and *xanthodryas* breed to N of China and migrate across eastern half of China and Taiwan. Fairly common in primary and secondary forests up to 2500 m.

Habits: Frequents open wooded areas, mangroves, secondary forest and forest edge. Joins in mixed-species flocks, working through the foliage of trees, searching for food.

991. Greenish Warbler *Phylloscopus trochiloides* PLATE 98
Anlü liuying

Description: Medium-sized (10 cm) leaf warbler with greenish back; usually only one yellowish-white wing bar; no white in tail; long yellowish-white supercilium and little contrast between greyish median crown stripe and green sides of crown. Has dark line through eye and dusky streaks on ear coverts. Underparts greyish-white with olive wash on flanks. Whitish eye-ring. Distinguished from Eurasian Chiffchaff by bold wing bar and broader eye-stripe. Smaller than Large-billed Leaf and Arctic Warblers with finer bill; smaller head and shorter primary projection. Arctic and Two-barred Warblers usually have second wing bars. Iris—brown; bill—horn above, pinkish below; feet—brown.

Voice: Call is loud, shrill *tiss-yip*, similar to call of White Wagtail. Also gives *pseeeoo* call. Song of tit-like, lively phrase is preceded by call note and ends with fast rattle.

Range: Breeds N Asia and Himalayas; winters to India, Hainan and SE Asia.

Distribution and status: Common seasonal migrant. Race *viridianus* breeds in NW China, wintering to India; nominate race breeds in C China to NW Yunnan, wintering in SE Xizang and S Yunnan; and *obscuratus* breeds in Qinghai, E and S Xizang and winters in Yunnan.

Habits: Lives in scrub and woods at high altitudes in summer but winters in lowland forest, scrub and farmland.

Note: Two-barred Warbler is placed by some authors (e.g. Sibley and Monroe 1990) as separate species or treated as race of this species (e.g. Cheng 1987, Baker 1997). There appears to be virtually no geographic overlap between these forms and we feel Two-barred is best treated as an allospecies within the superspecies *P. trochiloides*, pending more information.

992. Two-barred Warbler *Phylloscopus plumbeitarsus* PLATE 98
Shuangban lüliuying

Description: Medium-sized (12 cm) dark green leaf warbler with prominent long white supercilium, no coronal stripe, dark legs, double wing bar, white underparts and green rump. Distinguished from Greenish Warbler by broader greater coverts wing bar and presence of yellowish-white lesser coverts wing bar and darker greener upperparts with whiter underparts. Sometimes has a yellow tinge to side of head and neck. Smaller and plumper than Arctic Warbler. Distinguished from Yellow-browed Warbler by longer bill with pink lower mandible and lack of pale tertial tips. Iris—brown; bill—dark above, pink below; feet—bluish-grey.

Voice: Lond distinctive, dry, flat-sounding, sparrow-like, trisyllabic *chi-wi-ri*. Song similar to Greenish Warbler.

Range: Breeds NE Asia and NE China, wintering to Thailand and Indochina.

Distribution and status: Common. Breeds in NE China; migrates through most of China and winters on Hainan.

Habits: Breeds in deciduous–conifer mixed forest and birch and aspen groves up to 4000 m. Winters in secondary scrub and bamboo up to 1000 m.

Note: Placed by some authors (e.g. Sibley and Monroe 1990) as separate species or treated as race of Greenish Warbler (e.g. Cheng 1987, Baker 1997). There appears to be virtually no geographic overlap between these forms and we feel this form is best treated as an allospecies within the superspecies *P. trochiloides*, pending more information.

993. Pale-legged Leaf Warbler *Phylloscopus tenellipes* PLATE 97
Huijiao liuying

Description: Medium-sized (11 cm) dull but distinctive leaf warbler with brownish-olive upperparts; two buff wing bars; long white supercilium (buffy in front of eye) and olive eye-stripe. Bill fairly large and legs pale pink. Rump and uppertail coverts distinctive olive-brown. Underparts white with buffy-grey wash on flanks. Browner than Arctic Warbler, with smaller and paler bill than Large-billed Leaf Warbler. Iris—brown; bill—dark above, pinkish below; feet—pale pink.

Voice: Diagnostic short, very high-pitched, metallic *tink*. Song is cricket-like, dry chirps ending abruptly.

Range: Breeds NE Asia and Japan; winters E and S China and SE Asia.

Distribution and status: Uncommon seasonal migrant. Breeds in Changbai Mts, NE China; migrates through eastern China; occasional birds winter along S coast and Hainan.

Habits: Lives in hills up to 1800 m with dense undergrowth. Uses mangroves and scrub in winter. Keeps to lower storeys and flits about in a light, active manner, flicking tail downwards in characteristic fashion.

994. Large-billed Leaf Warbler *Phylloscopus magnirostris* PLATE 98
[AG: Willow Warbler] Wuzui liuying

Description: Largish (12.5 cm) leaf warbler with greenish-olive upperparts; no white in tail; one or usually two yellowish wing bars; dark eye-line and mottled ear coverts.

Underparts white with greyish wash on flanks and often a yellowish tinge. Supercilium is long and yellow at front, white at rear. Bill is large and dark with slight hooked tip. Distinguished from Arctic Warbler by dark bill, greener upperparts and green wing panel. Separated from Greenish Warbler by larger size; bigger bill; longer supercilium and more yellow on cheeks and underparts. Best distinguished by voice. Distinguished from Pale-legged by darker legs; larger bill; more yellow on supercilium, cheeks and underparts. Iris—brown; bill—dark with pink base to lower mandible; feet—greenish-grey or pink.

Voice: Call is double note *pe-pe* with second note on much higher pitch. Sometimes gives ascending *yaw-wee-wee*. Song is distinctive clear, loud whistle of five notes *tee-ti-tii-tu-tu* on descend scale with last notes prolonged.

Range: Breeds Himalayas, S and W China and NE Burma; winters to India.

Distribution and status: Uncommon seasonal migrant. Breeds in Himalayas of S and SE Xizang, NW Yunnan, Sichuan, E Qinghai and Gansu. Passage birds recorded in W Yunnan, Hubei and SE Yunnan.

Habits: Lives in open grassy glades and gaps in forest from 2000–4000 m; wintering to lower elevations. Feeds more on branches than in foliage. Flight flitting and light.

995. Eastern Crowned Warbler *Phylloscopus coronatus*   PLATE 99
[Temminck's Crowned Warbler, AG: Crowned Leaf-Warbler]
Mian liuying

Description: Largish (12 cm) yellowish-olive leaf warbler with whitish supercilium and median crown stripe. Upperparts greenish-olive with yellow edges to flight feathers including single yellowish-white wing bar; underparts whitish with contrasting lemon-yellow vent; lores and eye-stripe blackish. Distinguished from Blyth's Leaf Warbler by single wing bar, larger bill and yellower crown stripe and supercilium. Iris—dark brown; bill—brown above, pale below; feet—grey.

Voice: Soft call note is soft *phit phit*. Variable piercing song is *pichi pichu seu sweu* with last note highest.

Range: Breeds in NE Asia; migrant S in winter to China, SE Asia, Sumatra and Java.

Distribution and status: Nominate race breeds in Changbai Mts of Jilin, Hebei and in C China in Sichuan. Recorded on passage over eastern and southern provinces; accidental on Taiwan.

Habits: Frequents mangroves, wooded areas and forest edge from sea-level to highest peaks. Joins mixed-species flocks and is generally seen in the canopy of larger trees.

Note: Similar Western Crowned Leaf Warbler *P. occipitalis* of Himalayas could occur in S Xizang. Distinguished by larger darker bill, paler head stripes and whiter supercilium.

996. Emei Leaf Warbler *Phylloscopus emeiensis*   PLATE 98
Emei liuying

Description: Medium-sized (10 cm) greenish leaf warbler with yellowish supercilium, grey coronal stripe, greenish rump, two yellowish wing bars and dark tertials. Underparts whitish with yellow wash on side of head and flanks. Very similar to Blyth's Leaf Warbler; head pattern less distinct with dark lateral crown stripes paler and greener, also central stripe less distinct; ear coverts marginally darker; outer tail feathers only fractionally white, easily distinguished by song. Iris—brown; bill—dark above, fleshy below; feet—pinkish-brown.

Voice: Song is clear, slightly quivering, straight trill of 3–4 seconds, similar to Arctic Warbler but very different from tit-like song of Blyth's Leaf Warbler. Call is soft *tu-du-du; tu-du* or *tu-du-du-di* somewhat like Greenish Leaf Warbler.

Range: C China endemic.

Distribution and status: Rare, restricted-range species breeding on Mt Emei in S Sichuan and adjacent montane forests as far east as Fanjingshan in Guizhou. Winters to SE Burma and probably other areas not yet identified.

Habits: Breeds in subtropical broadleaf forests up to 1900 m. Feeds in canopy and shrub layers. Flicks wings fast, unlike Blyth's which flicks wings alternately and slower. Does not feed on undersides of branches as does Blyth's.

997. Blyth's Leaf Warbler *Phylloscopus reguloides* PLATE 99
[AG: Crowned Willow Warbler] Guanwen liuying

Description: Medium-sized (10.5 cm) brightly coloured leaf warbler with green upperparts; two yellow wing bars; bright yellow supercilium and median crown stripe and yellow wash on white underparts, especially side of face, flanks and undertail coverts. White border to inner web of two outer tail feathers. Distinguished from Sulphur-breasted Warbler by paler lateral crown stripes, bolder double wing bars and less yellow on underparts. From White-tailed Leaf Warbler by larger size and less yellow underparts also by alternate flicking of one wing. Races brighter green above and yellower below from west to east. Iris—brown; bill—dark above, pink below; feet—greenish to yellow.

Voice: Song is tit-like *chi chi pit-chew pit-chew* then breaking into wren-like trilling. Call is repeated two-note, loud *pit-cha* or three-note *pit-chew-a*.

Range: Breeds N Pakistan, Himalayas, W and S China, Burma and Indochina.

Distribution and status: Fairly common seasonal migrant and resident. Nominate race breeds in S and SE Xizang to N Yunnan and SW Sichuan; *claudiae* in Sichuan, S Gansu (Baishuijiang), S Shaanxi (Qinling Mts) and SE Shanxi; *fokiensis* in SE China; race *goodsoni* resident on Hainan. Nominate and *claudiae* winter in S Yunnan. Races *claudiae* and *fokiensis* migrate through S and SE China; vagrant Beidaihe (Hebei).

Habits: Race *fokiensis* displays yellow flanks when wings characteristically flicked up alternately. Sometimes hangs upside down to feed under branches.

Note: Race *goodsoni* has been sometimes treated as a race of Sulphur-breasted Warbler *P. ricketti* (e.g. Cheng 1987).

998. White-tailed Leaf Warbler *Phylloscopus davisoni* PLATE 99
Baibanwei liuying

Description: Medium-sized (10.5 cm) leaf warbler with bright green upperparts; two yellowish wing bars; white underparts with yellow wash; indistinct yellow median crown stripe and bold yellow supercilium and dark greenish eye-stripe. Three outer tail feathers have white on inner edges, extending on outer feather. Very similar to Blyth's and Emei Leaf Warblers. Best distinguished by voice and behaviour but more white in outer tail feathers. Median coronal stripe of Emei Leaf Warbler is less distinct and less yellow. Blyth's Leaf Warbler is less green above and whiter below. Race *disturbans* has less yellow on underparts and less white on tail feathers; *ogilveigranti* is darker with underparts dirty white with yellow streaks. Iris—brown; bill—dark above, pink below; feet—pinkish-brown.

Voice: Tit-like song, typically single, high-pitched *pitsu* note followed by trisyllabic *tit-sui-titsui-titsui* or dissyllabic *titsu-titsu-titsu*. Call note is similar single *pitsiu* or *pitsitsui*.

Range: S China, Burma and Indochina.

Distribution and status: Uncommon seasonal migrant. Nominate race breeds in W, NW and S Yunnan; *disturbans* in Sichuan, Guizhou, SE Yunnan, Guangdong, Hong Kong and probably Hainan; *ogilviegranti* breeds in mountains of SE China, on Wuyishan in NW Fujian and probably Babaoshan in N Guangdong. Migrants pass through S China.

Habits: Flicks wings fast and together *contra* Blyth's slower and alternate flicking.

999. Hainan Leaf Warbler *Phylloscopus hainanus* PLATE 98
Hainan liuying

Description: Smallish (10.5 cm) very yellow leaf warbler with green upperparts; yellowish-green lower back; yellow underparts and yellow eyebrow and median coronal stripe. Two wing bars usually visible. Lower mandible pink. Extensive white on penultimate and outer tail feathers and pale lateral crown stripes separate from similar White-tailed Leaf and Sulphur-breasted Leaf Warblers. Iris—brown; bill—brown above, pink below; feet—orange-brown.

Voice: Song is comprised of high, short, varied phrases similar to White-tailed Leaf Warbler. Call note similar to parts of song such as *pitsitsui, pitsitsui* or *pitsi-pitsu*.

Range: Hainan endemic.

Distribution and status: Globally Vulnerable (Collar *et al.* 1994). Locally common resident in W Hainan.

Habits: Inhabits forest above 600 m, mostly in scrub and secondary growth. Active in middle and upper canopy. Sometimes mixes in flocks.

1000. Yellow-vented Warbler *Phylloscopus cantator* PLATE 97
Huangxiong liuying

Description: Medium-sized (11 cm) distinctive leaf warbler with yellow median crown stripe and eyebrow contrasting with blackish lateral crown bands. Has prominent yellow wing bar with faint second bar. Throat, upper breast and undertail coverts yellow, contrasting with white lower breast and belly in white-eye pattern. Iris—brown; bill—dark above, pale below; feet—pinkish.

Voice: Song of several notes on same pitch ending with two slurred notes. Call is double note with accent on second note.

Range: E Himalayas and N Laos; wintering to Bangladesh, Burma, NW Thailand.

Distribution and status: Seen seven times by group of experienced Hong Kong birdwatchers and P. Round in March 1990 in Xishuangbanna (S Yunnan). This appears to be the only record for China. In evergreen forest up to 1700 m in winter, and up to 2500 m in summer.

Habits: Forms flocks in winter. Forages in bamboo and lower storey bushes in forest.

Note: The name *cantator* has been used in Chinese literature, but for races now treated in Sulphur-breasted Warbler or, in the case of *P. c. goodsoni*, a race of Blyth's Leaf Warbler (e.g. Cheng 1987).

1001. Sulphur-breasted Warbler *Phylloscopus ricketti*   PLATE 97
[Slater's Leaf Warbler] Heimei liuying

Description: Medium-sized (10.5 cm) colourful leaf warbler with bright green upperparts and bright yellow underparts and eyebrow. Two yellow wing bars usually visible. Eye-stripe and lateral coronal stripes blackish-green and median coronal stripe yellowish with grey streaks on nape. Similar to Yellow-vented Warbler but entire underparts yellow. Lateral crown stripes much darker than Hainan Leaf Warbler. Lacks yellow eyering of Golden Spectacled Warbler. Iris—brown; bill—dark above, yellowish below; feet—yellowish-pink.

Voice: Contact note denotes as *pitch-you, pitch-you.*

Range: Breeds C and SE China; winters to Indochina.

Distribution and status: Uncommon seasonal migrant and resident. Nominate race breeds in S Gansu, Sichuan, Guizhou, Hubei, Hunan, Guangxi, Guangdong, Fujian and Hong Kong.

Habits: Mixes with other warblers. Inhabits mixed forest on hills up to 1500 m.

Note: Sometimes treated as a race of Yellow-vented Warbler *P. cantator* (e.g. Cheng 1987). Form *goodsoni* listed for Hainan should be referred to Blyth's Leaf Warbler.

1002. Golden-spectacled Warbler *Seicercus burkii*   PLATE 99
Jinkuang wengying

Description: Small (13 cm) yellowish warbler with broad greenish-grey coronal stripe bordered on either side by black browline with yellow underparts. Inner web of outer tail feathers white. Yellow eye-ring separates from White-spectacled and Grey-cheeked Warblers. Some races have single yellow wing bar. Iris—brown; bill—black, pale below; feet—yellowish.

Voice: Call is loud *chip chiwoo.*

Range: Himalayas, Assam, Burma, N Thailand, C and S China and N Indochina.

Distribution and status: Nominate race breeds in S and SE Xizang; *distinctus* breeds in S Shaanxi, Sichuan, Guizhou, Yunnan and extreme SE Xizang, and sometimes winters to Guangdong; race *tephrocephalus* in NW Yunnan, winters to S Yunnan; and *valentini* in C and SE China, some birds winter to Yunnan. Breeds in montane forest and woodlands from 1800–3600 m. Regular visitor to Hong Kong.

Habits: Keeps mostly to lower storeys.

Note: Some forms warrant species status, pending analysis of songs.

1003. Grey-hooded Warbler *Seicercus xanthoschistos*   PLATE 100
Huitou wengying

Description: Smallish (11 cm) brightly coloured warbler with diagnostic grey crown and mantle; green wings, rump and tail; yellow underparts and white eye-ring. Has indistinct pale median coronal stripe between darker grey lateral stripes. Supercilium white. Yellow edges to primaries give yellow panel on closed wing. Distinguished from other *Seicercus* warblers by grey mantle and white supercilium and from all similar-coloured leaf warblers by lack of wing bars and white eye-ring. Iris—brown; bill—dark above, pale below; feet—pinkish-brown.

Voice: Call is distinctive, high, repeated *psit-psit*; also plaintive *tyee-tyee*.

Range: Himalayas to SW China and Burma.

Distribution and status: Locally common resident and altitudinal migrant. Nominate race in S Xizang; *flavogularis* may also occur in SE Xizang in Mishmi Hills area.

Habits: Lives in evergreen and mixed forests on hills and mountains from 1000–2700 m, to foothills in winter.

1004. White-spectacled Warbler *Seicercus affinis*   PLATE 100
Baikuang wengying

Description: Smallish (11 cm) colourful warbler with grey head and bold white eye-ring. Upperparts green; underparts yellow. Distinguished from Golden-spectacled Warbler by white rather than yellow eye-ring, grey sides of head and sometimes a yellow wing bar. Distinguished from Grey-cheeked Warbler by yellow chin and throat and yellow and black lores. From Grey-hooded Warbler by lack of white supercilium and greenish mantle. Race *intermedius* has prominent yellow tips to greater coverts to form wing bar; indistinct in nominate race. Iris—brown; bill—dark above; yellow below; feet—yellow.

Voice: Call is sharp *che-weet*.

Range: Nepal to China, Burma and Indochina.

Distribution and status: Uncommon seasonal migrant from foothills to 2300 m. Nominate race recorded breeding in SE Yunnan; also occurs in SE Xizang. Race *intermedius* breeds in Wuyishan (NW Fujian) and probably Babaoshan (N Guangdong); winters in lowlands of Fujian and Guangdong.

Habits: Inhabits bamboo thickets in moist forests in mountains. Winters to foothills and joins mixed flocks.

Note: Race *intermedius* is sometimes classed as a race of Golden-spectacled Warbler.

1005. Grey-cheeked Warbler *Seicercus poliogenys*   PLATE 100
Huilian wengying

Description: Small (10 cm) colourful warbler with grey head, green upperparts, yellow underparts and white eye-ring. Distinguished from Grey-hooded Warbler by green mantle and lack of white eyebrow. Distinguished from White-spectacled Warbler by darker grey crown and sides of head; less distinct dark lateral crown stripes and whitish chin and upper breast. Similar coloured Grey-headed Flycatcher has crest, lacks white eye-ring and has different jizz. Iris—brown; bill—dark above, pale below; feet—yellowish-brown.

Voice: Call is distinctive, tit-like note *chee-chee* and wren-like *tsik*.

Range: Sikkim to SW China and Burma.

Distribution and status: Rare altitudinal migrant in China. Breeds in NW Yunnan wintering to W Yunnan; resident in SE Xizang.

Habits: Lives in dense bamboo thickets in moist forests from 600–3000 m. Descends to foothills in winter. Joins mixed flocks. Keeps to lower storeys.

1006. Chestnut-crowned Warbler *Seicercus castaniceps* PLATE 99
[Chestnut/Chestnut-headed Warbler] Litou wengying

Description: Very small (9 cm) olive warbler with rufous-brown cap, black lateral crown stripe and black eye-stripe, white eye-ring, grey cheeks, yellow wing bars, yellow rump and flanks; grey breast and yellowish-grey belly. Race *sinensis* has greener back and yellower underparts than nominate; *laurentei* is similar but less yellow below with white centre of belly. Iris—brown; bill—black above, pale below; feet—greyish-horn.

Voice: Song is high-pitched, metallic and glissading; also double call note *chi-chi* and wren-like *tsik*.

Range: Himalayas to S China, SE Asia, Malay Peninsula and Sumatra.

Distribution and status: Uncommon resident. Nominate race found in S and SE Xizang and W Yunnan; *laurentei* in SE Yunnan and SW Guangxi (Yaoshan); and *sinensis* in C and S China. On mountains up to 2500 m. Probably more common than paucity of records suggests.

Habits: Actively searches the canopy of small trees in montane forest. Forms mixed flocks with other species.

1007. Broad-billed Warbler *Tickellia hodgsoni* PLATE 100
Kuanzui wengying

Description: Small (10 cm) colourful warbler with long legs and rufous crown. Upperparts green; face, throat and breast plain grey and belly, thighs and undertail coverts yellow. Flight feathers and tail blackish. Has faint white eyebrow and eye-ring. Superficially resembles Chestnut-crowned Warbler but lacks black lateral crown stripes, yellow rump and wing bars. Colour pattern similar to Mountain Tailorbird but has shorter bill, does not cock tail and has altogether different jizz. Iris—brown; bill—dark above, pale below; feet—yellowish.

Voice: Song is reported as single, long drawn-out, shrill whistle followed by two notes after pause of about 10 seconds, the second note on lower pitch.

Range: Nepal to SW China, N Indochina and W Borneo.

Distribution and status: Very rare in China. Race *tonkinensis* is resident in Jinping area of SE Yunnan at 2000–2500 m, but lower in winter; nominate race occurs in SE Xizang from 1100–2700 m.

Habits: Shy and skulking in undergrowth of dense thickets of moist montane forests.

1008. Rufous-faced Warbler *Abroscopus albogularis* PLATE 99
[White-throated/Fulvous-faced Flycatcher Warbler]
Zonglian wengying

Description: Smallish (10 cm) brightly coloured and distinctive warbler with chestnut head and black lateral coronal stripes. Upperparts green with yellow rump. Underparts white with black speckling on chin and throat and yellow wash on upper breast. Distinguished from Chestnut-crowned Warbler by chestnut side of head; less prominent white eye-ring and lack of wing bars. Race *flavifacies* has richer red face and darker upperparts. Iris—brown; bill—dark above, pale below; feet—pinkish-brown.

Voice: A shrill twitter.

Range: Nepal to S China, Taiwan, Burma and N Indochina.

Distribution and status: Fairly common resident. Nominate race occurs in Xishuangbanna and probably Jinping areas of S Yunnan; *fulvifacies* is widespread in C, S and SE China, Hainan and Taiwan.

Habits: Inhabits evergreen forest and bamboo thickets.

1009. Black-faced Warbler *Abroscopus schisticeps*   PLATE 100
Heilian wengying

Description: Distinctive smallish (10 cm) colourful warbler with bold yellow eyebrow, chin, throat and undertail coverts; black face mask; grey crown and nape; green upperparts; no wing bar; white belly. Worn plumage in summer extends yellow to upper breast in race *flavimentalis*. Race *ripponi* has greyish breast band. Pattern unlike any other Chinese warbler but see also Yellow-throated Fulvetta and Yellow-bellied Fantail. Iris—brown; bill—dark above, flesh below; feet—greenish.

Voice: Alarm call is high-pitched *tz-tz-tz-tz-tz-tz*.

Range: Nepal to SW China, Burma and NW Tonkin.

Distribution and status: Rare resident from 600–2800 m. Race *flavimentalis* in S Xizang (Nyalam) and NW Yunnan and probably present in SE Xizang; *ripponi* in W Yunnan and Sichuan (Kanding).

Habits: Lives in small parties in mossy thickets in montane evergreen forests.

1010. Yellow-bellied Warbler *Abroscopus superciliaris*   PLATE 100
[Bamboo/White-throated Warbler] Huangfu wengying

Description: Small (11 cm) yellow-bellied warbler with conspicuous white supercilium. Forecrown grey; back of head and back greenish-olive; chin, throat and upper breast white; rest of underparts yellow. Iris—brown; bill—black with whitish base; feet—pink.

Voice: Loud, piercing song, comprised of 3–4 separate stanzas each higher-pitched than the last. Call is short, loud chatter.

Range: E Himalayas, S China, SE Asia and Greater Sundas.

Distribution and status: Nominate race is locally common resident in hills up to about 1500 m in W and S Yunnan; *drasticus* in SE Xizang.

Habits: Frequents secondary forests particularly in bamboo areas. Generally in small flocks in low bushes and bamboo thickets.

GRASSBIRDS—Subfamily: Megalurinae

A small taxon of brownish, striped warblers with long tails, specialised for life among tall grasses. Formerly included within the Old World warblers. Males give display flights. There are only two species in China.

1011. Striated Grassbird *Megalurus palustris* [AG: Warbler/Canegrass-Warbler/Marsh Warbler] Zhaoze daweiying   PLATE 100

Description: Unmistakably large (male: 26 cm, female 23 cm) brown warbler with boldly black-streaked back, buffish supercilium and very elongated pointed tail. Upperparts bright reddish-brown with black streaks on back and wing coverts; underparts

whitish with narrow blackish streaks on chest and flanks and a rufous wash on flanks and undertail coverts. Iris—brown; bill—black above, pinkish below; feet—pink.
Voice: Harsh, musical, bubbly song uttered from perch and in flight and a sharp clicking call, *chak chak*. Also whistle ending in explosive *wheeechoo*.
Range: India, China, Philippines, SE Asia, Java and Bali.
Distribution and status: Race *toklao* is an uncommon resident up to about 1800 m in SE Xizang, Yunnan, S Guizhou and Guangxi.
Habits: Inhabits open grassy fields, bamboo clumps and secondary scrub. Lives partially on the ground where it runs under thick cover.

1012. Rufous-rumped Grassbird *Graminicola bengalensis* PLATE 95
[Large Grassbird] Dacaoying

Description: Large (17 cm) distinctive streaked grass warbler with long, blackish, graduated tail; dark tertials; blackish crown, nape and mantle streaked with brown and white and whitish supercilium. Face and underparts whitish with rufous-brown sides of breast and flanks; undertail coverts buffy with dark streaks. Outer tail feathers broadly tipped white. Fresh winter plumage is much browner due to fluffy brown edges to feathers of upperparts. Superficially like Striated Grassbird but smaller with shorter, more rounded tail. Iris—reddish-brown; bill—black above, pinkish below; feet—pinkish-brown.
Voice: Song said to resemble Rusty-rumped Warbler.
Range: Nepal to NE India, S China, Burma and Indochina.
Distribution and status: Rare resident. Race *sinica* occurs in Guangxi, N Guangdong and Hong Kong, *striata* on Hainan.
Habits: Skulks in tall reed beds in swampy low-lying areas; also tall grasslands on low hills. When flushed, dives quickly back to cover.

LAUGHINGTHRUSHES—Subfamily: Garrulacinae

A small subfamily of medium-sized, longish-tailed passerines formerly included within the babbler family. Many species live in flocks, particularly outside the breeding season. Several species have beautiful songs; others give flock chorus calls including chattering, squealing and 'laughing', particularly when alarmed. They often feed among the leaf litter on the forest floor, turning over leaves like a thrush. Flight is weak and in a follow-the-leader fashion. None are migratory. China is the world distribution centre for this group and there are 38 species in the region.

1013. Masked Laughingthrush *Garrulax perspicillatus* PLATE 101
Heilian zaomei

Description: Largish (30 cm) greyish-brown laughingthrush with diagnostic black forehead and mask. Upperparts dull brown; outer tail feathers broadly tipped dark brown; underparts greyish shading to whitish belly and cinnamon undertail coverts. Iris—brown; bill—blackish with paler tip; feet—reddish-brown.
Voice: Loud piercing contact and alarm calls; chattering group calls.
Range: Resident in E, C and S China and N Vietnam.

Distribution and status: A common bird in suitable lowland habitat of S and E China from S Shaanxi southward and C Sichuan and E Yunnan eastward excluding Hainan.
Habits: Lives in small parties in thick scrub, bamboo thickets, reeds and including fields and town parks. Feeds mostly on the ground. Noisy calls.

1014. White-throated Laughingthrush *Garrulax albogularis* PLATE 101
Baihou zaomei

Description: Medium-sized (28 cm) dull brown laughingthrush with diagnostic white throat and upper breast. Forehead narrowly rufous in *laebus*, broadly rufous in *albogularis* and entire crown and nape rufous in *ruficeps*. Upperparts otherwise dull ashy-brown. Outer four pairs of tail feathers tipped white. Underparts with greyish-brown band across breast; belly rufous. Undertail coverts of Taiwan race white. Iris—greyish or brown (Taiwan); bill—dark horn; feet—greyish.
Voice: Wheezy call in flocks; gentle *teer, teer* contact call and buzzing *tzzzzzzzzzzzzz* in alarm. Also squeals and laughing calls when excited.
Range: C and SW China, S Tibet, Himalayas, Tonkin and Taiwan.
Distribution and status: Fairly common in evergreen forests and middle elevations in S Xizang and Yunnan (nominate) and in mountains of S Qinghai, Sichuan (*laetus*) at 1200–4600 m. Race *ruficeps* is fairly common in primitive broadleaf and juniper forests of Taiwan from 850–1800 m.
Habits: A noisy bird living in small to large flocks in middle storey of forest.

1015. White-crested Laughingthrush *Garrulax leucolophus* PLATE 101
Baiguan zaomei

Description: Unmistakable largish (30 cm) laughingthrush with black forehead, lores and eye-stripe contrasting with bushy-crested white head and upper breast. Three races differ in body colour: nominate has greyish nape, chestnut nuchal collar and breast band and olive-brown back, wings and belly with blackish tail; *patkaicus* has white nape, more chestnut on back and less black tail; and *diardi* has white underparts with chestnut on flanks; nape and upper back grey; wings and lower back chestnut; tail olive-brown. Iris—reddish-brown; bill—blackish; feet—dark grey.
Voice: Clear whistles and low chattering lead into group chorus of loud cackling 'laughter'.
Range: Himalayas, SW China, SE Asia and W Sumatra.
Distribution and status: A fairly common bird in suitable habitat from sea-level to 1500 m in SE Xizang (nominate) and W Yunnan (*patkaicus*) and S Yunnan (*diardi*).
Habits: Lives in noisy flocks but keeps to denser thickets and is quite shy. Feeds mostly on or near forest floor, turning over leaves in search of insects. Prefers secondary forest.

1016. Lesser Necklaced Laughingthrush *Garrulax monileger* PLATE 101
Xiaoheiling zaomei

Description: Medium-sized (28 cm) rufous-brown laughingthrush with white underparts and bold black gorget and line behind eye. Similar to Greater Necklaced Laughingthrush with which it often associates but distinguished primarily by black line on lores and brown primary coverts. Several races vary in minor details. Iris—yellow; bill—dark grey; feet—greyish.

Voice: Similar to and mimicking Greater Necklaced Laughingthrush, strange piping calls.

Range: Himalayas to S China, Hainan, Burma and Indochina to Malay Peninsula.

Distribution and status: Common in montane forests at 350–1400 m, in W Yunnan (*monileger*), C and S Yunnan (*schauenseei*), Guangxi and SE Yunnan (*tonkinensis*), SE China (*melli*) and Hainan (*schmacheri*).

Habits: A gregarious and noisy bird generally feeding on forest floor turning over leaves. Sometimes mixed with other laughingthrushes including Greater Necklaced.

1017. Greater Necklaced Laughingthrush *Garrulax pectoralis* PLATE 101
Heiling zaomei

Description: Largish (30 cm) rufous-brown laughingthrush with complex black and white patterning on head and breast. Similar to Lesser Necklaced Laughingthrush but distinguished primarily by pale lores, also dark primary coverts which contrast with rest of wing. Five races in Yunnan and Hainan vary slightly but *picticollis* over CS and E China is the most distinct form with whiter throat and lores, and black band around throat replaced by broad grey band. Iris—chestnut; bill—black above, grey below; feet—bluish-grey

Voice: Soft squeaking group contact calls and loud orchestra of mournful descending 'laughter' mixed with short whistles.

Range: E Himalayas, NE India, E to C and E China and S to W Thailand, N Laos and N Vietnam.

Distribution and status: A fairly common bird in south faces of E Himalayas, through forested hills of Yunnan, C and SE China and Hainan from 200–1600 m.

Habits: Lives in noisy flocks; feeds mostly on ground; mixes with other laughingthrushes including similar Lesser Necklaced; performs dancing displays with birds hopping about bowing and spreading wings while calling. Flies in long glides.

1018. Striated Laughingthrush *Garrulax striatus* PLATE 101
Tiaowen zaomei

Description: Largish (30 cm) unmistable laughingthrush, with long bushy crest and dark brown plumage streaked finely by white shaft lines over head and back and more coarsely on breast and flanks. Bill is short and thick. Race *cranbrooki* has broad black supercilium, no streaks on crown and less distinct streaks on rest of plumage. Distinguished from Streaked Laughingthrush by larger size and white streaks. Iris—red-brown; bill—dark horn; feet—grey.

Voice: Loud discordant cackles, shrill contact cries and loud musical whistles *O-will you-will you-wit* and variants often repeated many times. Two noted whistle *teo-wo*.

Range: Himalayas E of Himachal Pradesh to NE India, NW Burma and extreme NW and W Yunnan.

Distribution and status: Apparently rare in Gaoligongshan range of W Yunnan (*cranbrooki*). Probably not uncommon in SE Xizang (*vibex*) and in evergreen zone of Himalayas (*striatus* in west and *sikkimensis* in central) at 750–2750 m.

Habits: Lives in pairs or small parties, sometimes mixing with other laughingthrushes. Keeps to dense thickets near streams in broadleaf and mixed forests. Feeds in fruit trees. Erects crest when calling.

1019. White-necked Laughingthrush *Garrulax strepitans*   PLATE 101
Baijing zaomei

Description: Medium-sized (29 cm) dark laughingthrush with dark brown face, cheeks, throat and breast contrasting with white patch on side of neck which grades into grey collar. Crown is rufescent brown and rear ear coverts dark rufous. Distinguished from Dark-throated and Grey Laughingthrushes by brown crown. Iris—brown; bill—black; feet—dark maroon.

Voice: Group calls of dry *chuh* notes followed by shrill maniacal 'laughter' with long trills.

Range: SW China, Laos, E Burma, N Thailand.

Distribution and status: A rare bird only marginally ranging into China in SW of Xishuangbanna prefecture of S Yunnan at 1000–1800 m.

Habits: Keeps to undergrowth and moist ravines of montane evergreen forest.

1020. Grey Laughingthrush *Garrulax maesi*   PLATE 101
Hexiong zaomei

Description: Medium-sized (27 cm) dark laughingthrush. Like Black-throated but ear coverts pale grey, bordered above and behind by white. Distinguished from White-necked by greyer colour. Hainan race *castanotis* has bright rufous ear coverts with almost no white behind, and dark brown throat and upper breast. Iris—brown; bill—black; feet—dark brown.

Voice: Loud calls.

Range: SC and S China to Hainan, Tonkin and N and C Laos.

Distribution and status: Globally Near-threatened (Collar *et al.* 1994). Uncommon in mountains of S Sichuan from SE Xizang to N Guizhou (*grahami*); on mountains of SE Guizhou, Guangxi and N Guangdong (*maesi*) and in SW Hainan (*castanotis*).

Habits: Keeps to undergrowth in evergreen montane forests.

Note: Treated by some authors as a race of *G. strepitans*.

1021. Rufous-necked Laughingthrush *Garrulax ruficollis*   PLATE 101
Lijing zaomei

Description: Smallish (23 cm) laughingthrush with black mask and throat and diagnostic rufous-chestnut patch on side of neck. Crown grey; tail blackish; lower belly and undertail coverts rufous. Rest of plumage olive-grey with pale edges to primaries. Iris—brown; bill—black; feet—brown.

Voice: Three-noted mellow whistle *weeoo-wihoo-wich*; sharp musical contact notes and bursts of raucous cackling.

Range: Himalayas, N and W Burma and W Yunnan.

Distribution and status: Uncommon in lowlands of SE Xizang and rare in extreme W Yunnan up to 800 m.

Habits: Noisy groups feed among floor litter in scrub, bamboo and secondary forests. Flicks tail and moves jerkily through low vegetation.

1022. Black-throated Laughingthrush *Garrulax chinensis* PLATE 101
Heihou zaomei

Description: Smallish (23 cm) dark grey laughingthrush with black sides to head and throat and olive-grey belly and undertail coverts. Black puffy forehead bordered above with white. Cheeks of mainland forms white but Hainan race has hindneck and side of neck rufous-brown. Edges of primaries pale. Distinguished from White-necked Laughingthrush by white on brow. Iris—red; bill—black; feet—yellow or grey.
Voice: Beautiful clear-noted thrush-like song and loud cackling chorus.
Range: S China and Indochina.
Distribution and status: Common in lowland forests up to 1200 m in SW Yunnan (*lochmius*); SE Yunnan to Guangdong (*chinensis*) and Hainan (*monachus*).
Habits: Small groups keep to dense scrub in bamboo thickets and semi-evergreen forest.

1023. Yellow-throated Laughingthrush *Garrulax galbanus* PLATE 101
Huanghou zaomei

Description: Smallish (23 cm) laughingthrush with bluish-grey crown; black mask and bright yellow throat diagnostic. Upperparts brown; tail terminally black with white border. Abdomen and undertail coverts buff grading to white. Simao race is paler. Iris—reddish-brown; bill—black; feet—grey.
Voice: Feeble chirping calls.
Range: Assam to Chin hills of Burma. Two isolated population in S Yunnan and SE China.
Distribution and status: Globally Near-threatened (Collar *et al.* 1994). Yunnan race (*simaoensis*) known only from the type-specimen. Rare in hills of Wuyuan district of Jiangxi province (*courtoisi*).
Habits: Keeps to dense scrub, feeds in floor litter.

[1024. Rufous-vented Laughingthrush *Garrulax gularis* PLATE 101
Ligang zaomei

Description: Smallish (23 cm) brown laughingthrush with black mask and yellow underparts. Crown and nape grey, sides of breast grey, lower abdomen, undertail coverts and edge of tail rufous. Distinguished from Yellow-throated by colour of vent and lack of white edge to tail. Iris—reddish-brown; bill—black; feet—yellowish-orange.
Voice: Loud, sweet whistles, chatters and group cackling 'laughter'.
Range: E Bhutan, Assam to N Burma and N and C Laos.
Distribution and status: Not yet recorded in China but could occur in SE Xizang up to 1220 m. In adjacent NE India it is locally common in evergreen forest and scrub.
Habits: Lives in large flocks but is a shy skulker and difficult to observe. Feeds on ground.
Note: Formerly classed as a race of *Garrulax delesserti*.]

1025. Plain Laughingthrush *Garrulax davidi* PLATE 101
[David's Laughingthrush] Shan zaomei

Description: Medium-sized (29 cm) greyish laughingthrush. Nominate: upperparts uniform greyish-brown; underparts paler. Has paler eyebrow and blackish chin. Race *concolor* is greyer and less brown overall. Iris—brown; bill—decurved bright yellow, tipped greenish; feet—pale brown.

Voice: Song is loud, quickly repeated series of short notes, introduced by weak whining notes and followed by lower, weaker notes, *wiau wa-WIKWIKWIK woitwoitwoitwoit.* Also gives series of *wiau* notes on their own as contact or alarm call by both sexes (C. Robson).

Range: Endemic to N and C China.

Distribution and status: An occasional bird of mountains at 1600–3300 m. From W Heilongjiang to Hubei westward to E Qinghai (*davidi*); on Qilianshan on Gansu–Qinghai border (*experrectus*); and S Qilian and Anyemaqen ranges of SE Qinghai, Min and Qionglai ranges of Sichuan (*concolor*).

Habits: As other laughingthrushes, preferring thickets and scrub.

1026. Snowy-cheeked Laughingthrush *Garrulax sukatschewi* PLATE 102
[Black-fronted/Sukatshev's Laughingthrush] Hei'e shanzaomei

Description: Medium-sized (28 cm) vinous grey-brown laughingthrush with conspicuous white cheeks and ear coverts underlined by blackish-brown stripe joining sooty-brown lores. Outer tail feathers mixed with grey and tipped white. Tertials tipped white. Uppertail coverts rufous; vent warm buff. Iris—brown; bill—yellow; feet—yellow.

Voice: Variable melodious calls. Birds shake head, flick tail and quiver feathers when calling.

Range: Endemic to NC China.

Distribution and status: Globally Vulnerable (Collar *et al.* 1994). Recorded only in extreme S Gansu (Bailongjiang and Baishuijiang region) and adjacent N Sichuan. A rare bird in forested hills from 2000–3500 m.

Habits: Small parties generally feed on ground in conifer forest and scrub.

1027. Moustached Laughingthrush *Garrulax cineraceus* PLATE 102
Huichi zaomei

Description: Smallish (22 cm) boldly patterned laughingthrush. Crown, nape, stripe behind eye and moustachial stripe and streaks on side of neck black. Both nominate race and *sternuus* have whitish to pale buff lores and cheeks; nominate is much more rufescent on body and has blackish to greyish-brown crown and nape. Primary coverts black, primaries edged grey. Tertials, secondaries and tail feathers terminally black with white crescent tips. Distinguished from White-browed Laughingthrush by tail and wing pattern. Iris—cream; bill—horn; feet—dull yellow.

Voice: Various low muscial calls; thrush-like alarm call and a loud song *diu-diuuid.*

Range: NE India and N Burma to E, C and SE China.

Distribution and status: Not uncommon in montane forests at 200–2570 m but mostly below 1800 m. Race *cinereiceps* extends from Hubei and Hunan to S Gansu and SE to Guangdong, Zhejiang and Fujian. Race *sternuus* found in SE Xizang, Sichuan, Yunnan and N Guangxi.

Habits: Lives in pairs or small parties in secondary scrub and bamboo thickets, sometimes close to villages.

[**1028. Rufous-chinned Laughingthrush** *Garrulax rufogularis*   PLATE 102
Zongke zaomei

Description: Smallish (22 cm) rufous-brown laughingthrush scaled on upperparts with black crescent feather tips. Forehead and moustachial area black contrasting with pale lores and orange-rufous chin. Primaries edged pale grey; tail feathers subterminally black tipped rufous. Distinguished from Barred Laughingthrush by rufous chin and tail tips and black forehead. Immature has white chin and less black scaling. Iris—straw; bill—grey; feet—dark grey.

Voice: Chuckles and chattering calls; loud squealing alarm call.

Range: Himalayas, NE India, N Burma and W Tonkin.

Distribution and status: Locally common in oak and rhododendron forests at 610–2200 m. Never recorded in China but race *rufiberbis* can be expected in Medog area (SE Xizang) and race *rufogularis* in Chumbi valley (Yadong) of S Xizang.

Habits: Shy skulker of forest undergrowth and dense thickets, in pairs or small parties.]

1029. Barred Laughingthrush *Garrulax lunulatus*   PLATE 102
[Bar-backed Laughingthrush] Banbei zaomei

Description: Smallish (23 cm) warm brown laughingthrush with conspicuous white eye patches and bold black and buff scaled pattern on upperparts (except crown) and flanks. Primaries and outer tail feathers edged grey. Tail has white tip and black subterminal bar. Distinguished from White-browed Laughingthrush by black barring on upperparts. Iris—dark grey; bill—greenish-yellow; feet—flesh-coloured.

Voice: Song, *wu-chi wi-wuoou*, repeated after short intervals (C. Robson).

Range: Endemic to C China.

Distribution and status: Globally Near-threatened (Collar *et al.* 1994). Found from Shennongjia of Hubei through Qinling Mts of S Shaanxi to Baishuijiang region of S Gansu and Min and Qionglai Mts of C Sichuan. An occasional bird at 1200–3660 m.

Habits: Lives in groups in bamboo understorey of broadleaf and coniferous forests.

1030. White-speckled Laughingthrush *Garrulax bieti*   PLATE 102
[Biet's Laughingthrush] Baidian mei

Description: Medium-sized (25 cm) laughingthrush. Similar to Barred Laughing thrush but back feathers have small black subapical mark tipped by white dot. Base colour of throat, upper breast and flanks noticeably darker. Has white speckling on sides of neck and sides of mantle. White margins on underparts more spot-like. Distinguished from Giant and Spotted Laughingthrushes by lack of black on head, smaller size and white spots on underparts. Iris—straw; bill—yellowish; feet—pinkish.

Voice: Not recorded.

Range: Endemic to SW China.

Distribution and status: Globally Vulnerable (Collar *et al.* 1994). Uncommon in Lijiang area of NW Yunnan and adjacent SW Sichuan.
Habits: Poorly-known. Inhabits bamboo thickets in conifer and secondary forests from 3050–3650 m.
Note: Treated by some authors as a race of Barred Laughingthrush.

1031. Giant Laughingthrush *Garrulax maximus*   PLATE 102
Da zaomei

Description: Large (34 cm) boldly spotted laughingthrush with long tail. Crown, nape and moustachial stripe dark greyish-brown, sides of head and chin otherwise chestnut. Chestnut back feathers spotted by black subapical patch tipped by a white spot. Wings and tail patterned like Spotted Laughingthrush. Distinguished from that species by larger size, longer tail and rufous throat. Iris—yellow; bill—horn; feet—pink.
Voice: Shrill loud calls like Large Hawk Cuckoo and jerky rattling chorus.
Range: C China to SE Xizang.
Distribution and status: Locally common on mountains from 2135–4115 m in extreme S Gansu, W Sichuan, NE Yunnan and SE Xizang.
Habits: Generally keeps to higher altitudes than Spotted Laughingthrush.
Note: Formerly placed within Spotted Laughingthrush.

1032. Spotted Laughingthrush *Garrulax ocellatus*   PLATE 102
[White-spotted Laughingthrush] Yanwen zaomei

Description: Large (31 cm) laughingthrush with black, crown, nape and throat and boldly spotted upperparts and sides of breast. Lores, subocular region and chin pale buff contrasting with black of head. Upperparts brown, each feather with black crescent-shaped subapical spot and white tip. White tips form conspicuous bars on wing and fringe to tail. S Xizang race has chestnut ear coverts. Distinguished from Giant Laughingthrush by shorter tail and black throat. Iris—yellow; bill—horn; feet—pink.
Voice: Beautiful, clear, piercing, whistled song and harsh grating chorus and alarm squawks.
Range: Himalayas, NE Burma, S Xizang and C China.
Distribution and status: From Shennongjia in Hubei and extreme S Gansu through C Sichuan mountains to NE Yunnan (*artemesiae*); Yarlung Zangbo valley of S Xizang (*ocellatus*) and W and NW Yunnan (*maculipectus*). Not uncommon in forested mountains from 1100–3100 m.
Habits: Pairs or small parties forage amongst leaf litter. Sometimes mixes with other laughingthrushes.
Notes: Some authors include White-speckled Laughingthrush and occasionally Giant Laughingthrush within this species.

1033. Grey-sided Laughingthrush *Garrulax caerulatus*   PLATE 102
Huixei zaomei

Description: Medium-sized (24 cm) rufous-brown laughingthrush with white underparts. Ear coverts greyish-white; crown rufous; lores and fine line behind eye black. Bare skin round eye bluish. Underparts white with grey flanks. Iris—rufous; bill—dark horn; feet—grey.

Voice: Chitters, varied sweet calls and discordant cries; loud *oh dear dear*.

Range: E Himalayas to SE Xizang, N Burma and SW China.

Distribution and status: Occasional in SE Xizang (*caerulatus*), rare in Salween–Mekong divide of Yunnan (*latifrons*) at 1500–2700 m.

Habits: An active bird keeping to forest undergrowth, bamboo thickets and hill scrub.

Notes: Some authors include Rusty Laughingthrush within this species.

1034. Rusty Laughingthrush *Garrulax poecilorhynchus*　　　　PLATE 102
[Rufous/Scaly-headed Laughingthrush] Zong zaomei

Description: Largish (28 cm) rufous-brown laughingthrush with conspicuous bare blue skin around eye. Head, breast, back, wings and tail olive chestnut-brown with faint black scaling on crown. Abdomen and edges to primaries grey; vent white. Iris—brown; bill—yellowish with blue base; feet—bluish-grey.

Voice: Loud melodious varied whistled song *hoo guo hoo hoo hoo*. Sometimes mimics other birds calls.

Range: C China to SE China and Taiwan.

Distribution and status: Rare on the mainland. Races *ricinus* in NW Yunnan and *berthemyi* from S Sichuan to Shanghai and as far S as N Guangdong. Race *poecilorhynchus* is not uncommon in Taiwan in lower mountains at 600–2100 m.

Habits: Lives in small flocks in undergrowth and bamboo understorey of primitive broadleaf hill and mountain forest. Shy and avoids open spaces.

Notes: Placed by some authors within Grey-sided Laughingthrush.

1035. Spot-breasted Laughingthrush *Garrulax merulinus*　　　　PLATE 102
Banxiong zaomei

Description: Medium-sized (24 cm) olive-brown laughingthrush with blackish streaks/spots on throat and upper breast and slight pale buff line behind eye. Underparts more buff than Hwamei and dark streaks more pronounced. Birds in SE Yunnan have paler underparts and bolder throat streaks. Iris—brown; bill—black; feet—brown.

Voice: Very melodious clear-noted song, also a coughing chuckle.

Range: Himalayas, NE India, N and W Burma, SW China and N Indochina.

Distribution and status: Globally Near-threatened (Collar *et al.* 1994). Marginal in China, recorded from Daweishan in SE Yunnan (*obscurus*) and W of the Salween (Nujiang) in extreme W Yunnan (*merulinus*). Expected in SE Xizang. Rare in hills at 900–1850 m.

Habits: Feeds on ground in forest edge and scrub; usually in pairs.

1036. Hwamei *Garrulax canorus* [Melodious Laughingthrush/　　PLATE 102
Chinese Thrush] Huamei

Description: Smallish (22 cm) rufous-brown laughingthrush with diagnostic white eye-ring prolonged behind eye as a narrow line. Crown and nape finely streaked blackish. Taiwan race *taewanus* lacks the white eye marking, is more grey in colour and is more heavily streaked. Hainan race *owstoni* has white eye markings but has paler underparts and more olive upperparts than *canorus*. Iris—yellow; bill—yellowish; feet—yellowish.

Voice: Melodious and lively, clear whistled, song much favoured by bird fanciers.

Range: C and SE China, Taiwan, Hainan and N Indochina.

Distribution and status: A common bird of scrub and secondary forest in most of C, S and SE China up to 1800 m.

Habits: Rather shy skulker feeding in the leaf litter. Lives in pairs or small parties.

Notes: Mainland race *canorus* has become established on Taiwan and some interbreeding with *taewanus* occurs.

1037. White-browed Laughingthrush *Garrulax sannio*   PLATE 102
Baijia zaomie

Description: Medium-sized (25 cm) greyish-brown laughingthrush with rufous undertail coverts and diagnostic buffy-white face pattern of eyebrow and cheeks divided by dark stripe behind eye. Races vary slightly. Birds of SE Xizang and SW China (*comis*) have whiter face markings and birds of C China (*oblectans*) are more olive than the nominate race of SE China and Hainan. Iris—brown; bill—brown; feet—grey-brown.

Voice: Harsh ringing and buzzing calls, discordant cackling.

Range: NE India, N and E Burma, C and S China including Hainan and N Indochina.

Distribution and status: All races are rather common at medium elevations up to 2600 m.

Habits: Less shy than most laughingthrushes. Keeps to secondary scrub, bamboo thickets and the edge of forest clearings.

1038. Streaked Laughingthrush *Garrulax lineatus*   PLATE 102
Xiwen zaomei

Description: Small (21 cm) greyish laughingthrush, heavily streaked brown and whitish. Upperparts ashy-olive streaked with white shafts on back; wings and tail rufous; tail tipped greyish. Underparts grey with feathers of breast and flanks with whitish shafts and rufous edges. The race in SE Xizang is darker than the nominate race. Iris—brown; bill—brown; feet—pale brown.

Voice: Constant murmuring and crying with clear calls *pity-pity-we are*. Alarm call is plaintive *sweet pea pea*.

Range: Afghanistan, Himalayas to SE Tibet.

Distribution and status: Common in scrub on open hillsides from 1700–3300 m, descending lower in winter. Resident in S Xizang (*lineatus*) and SE Xizang (*imbricatus*).

Habits: Lives in pairs or small groups rarely ascending higher than 2 m; keeps to dense cover and feeds on ground.

1039. Blue-winged Laughingthrush *Garrulax squamatus*   PLATE 103
Lanchi zaomei

Description: Medium-sized (26 cm) dark brown laughingthrush with black eyebrow and inner wing coverts and pale bluish-grey edges to primaries. Feathers of crown, nape, breast and flanks scalloped with black edges. Dimorphic: some birds with rufous olive-brown upperparts and bronze tail with chestnut tip; others have grey crown and blackish tail. Distinguished from Scaly Laughingthrush by black eyebrow and pale eye. Iris—white or blue; bill—black; feet—brown.

Voice: Cries and quiet song is a repeated *kri taboo* or single soft note.

Range: Himalayas, N Burma, SW China and N Indochina.

Distribution and status: Rare in W and S Yunnan in moist hills at 1000–2500 m.

Habits: Keeps to moist forest near streams. Very shy and secretive. Less noisy than other laughingthrushes.

1040. Scaly Laughingthrush *Garrulax subunicolor*   PLATE 103
[Plain-coloured Laughingthrush] Chunse zaomei

Description: Medium-sized (24 cm) dull brown laughingthrush with grey cap and body plumage scaled by black edges. Similar to Blue-winged Laughingthrush but has reddish eye, lacks black eyebrow, has dull brown inner wing coverts and more olive-yellowish rather than blue-grey edges to primaries. Tail feathers tipped white. Iris—reddish; bill—black; feet—brown.

Voice: Clear four-noted whistle, sharp alarm note and squeaky chatters.

Range: E Himalayas, NE Burma, S Tibet, SW China and NW Tonkin.

Distribution and status: Common resident in S Xizang (*subunicolor*). Occasional in W Yunnan (*griseatus*) and rare in S Yunnan (*fooksi*). Seasonal migrant at 1830–3400 m; generally at higher elevations than Blue-winged Laughingthrush.

Habits: Lives in small groups keeping close to ground in montane forest and open scrub.

1041. Elliot's Laughingthrush *Garrulax elliotii*   PLATE 103
Chengchi zaomei

Description: Medium-sized (26 cm) laughingthrush. General colour greyish-brown with faint darker and whitish markings on the mantle and breast. Face darker. Vent and lower belly cinnamon. Yellowish basal edges to outer primaries and bluish-grey edges on distal portion give pattern of yellowish and bluish-grey patches on closed wing. Tail feathers grey with white tips and yellowish outer web. Race *preswalskii* has paler crown, browner underparts and reddish instead of yellowish wing patch and edge of tail. Iris—pale cream; bill—brown; feet—brown.

Voice: Far-carrying double note and group chattering.

Range: C China to SE Tibet and NE India.

Distribution and status: Nominate race found through Daba, Qingling and Min mts S through W Sichuan, SE Xizang and NW Yunnan. Race *preswalskii* is distributed from Qilian range of N Gansu S through E Qinghai. Common in undergrowth of all forest types from 1200–4800 m.

Habits: Forages in small flocks in undergrowth and bamboo understorey of open and secondary forests and thickets.

1042. Variegated Laughingthrush *Garrulax variegatus*   PLATE 103
Zase zaomei

Description: Medium-sized (26 cm) laughingthrush with conspicuous black and white face pattern and complex-coloured wing pattern. Body generally grey-brown, ventrally chestnut. Tail basally black, terminally grey with narrow white edge. Iris— yellow; bill—black; feet—yellow

Voice: Loud musical whistles *weet-a-weer* or *weet-a-woo-weer* answered by other group members. Alarm call is muttering and squealing notes.

Range: E Afghanistan, W Pakistan, W Himalayas, SW Tibet.

Distribution and status: Only marginally ranging into China in extreme SW parts of Xizang. Rare at 2500–3300 m.

Habits: In pairs or flocks in undergrowth of open oak and mixed forests in ravines.

1043. Brown-cheeked Laughingthrush *Garrulax henrici* PLATE 103
[Black-cheeked Laughingthrush] Huifu zaomei

Description: Medium-sized (26 cm) greyish-brown laughingthrush with contrasting brown head sides, whitish malar stripe and slight whitish eyebrow. Wings and base of tail are edged bluish-grey. Primary coverts form black patch. Underparts grey ventrally dull chestnut. Tail has narrow white tip. Iris—brown; bill—orange; feet—yellow.

Voice: Fluty call *whoh-hee* and noisy chattering.

Range: S and SE Tibet and adjacent NE India.

Distribution and status: Locally common in S Xizang from Xigaze E to Brahmaputra (*henrici*) and SE Xizang (*gucenensis*) at 2800–4600 m, descending to 2000 m in winter.

Habits: Lives in pairs or small parties in forest and scrubby valleys. Keeps out of sight. Sometimes associates with Black-faced Laughingthrush.

1044. Black-faced Laughingthrush *Garrulax affinis* PLATE 103
Heiding zaomei

Description: Medium-sized (26 cm) dark laughingthrush with broad white moustachial patch and white patch on neck contrasting against blackish head. Body colour varies slightly between races but generally dull olive-brown. Wing and tail feathers edged yellowish. Iris—brown; bill—black; feet—brown.

Voice: A repetitive monotonous call of 3–4 mournful notes *to-wee-you*; long rolling alarm call *whirr whirrer* and harsh scolding notes.

Range: E Himalayas, Assam, to C China, N Burma and W Tonkin.

Distribution and status: Many races. Resident in SW Xizang (*affinis*), S Xizang (*bethelae*), SE Xizang and W Yunnan (*oustaleti*), S Yunnan (*saturatus*), NE Yunnan and SW Sichuan (*muliensis*) and S Gansu to C Sichuan (*blythii*). Generally uncommon from 1500–4500 m. This is the highest-living laughingthrush, although it descends as low as 550 m in winter.

Habits: Lives in mixed woods, rhododendron and juniper thickets, keeping to undergrowth.

Notes: Some authors place the White-whiskered Laughingthrush of Taiwan within this species.

1045. White-whiskered Laughingthrush *Garrulax morrisonianus* PLATE 103
[Mountain Morrison Laughingthrush] Yushan zaomei

Description: Medium-sized (26 cm) brown laughingthrush with prominent long white eyebrow and moustachial line. Crown grey, scaled white; sides of face and chin brown; throat and nape brown, scaled with grey edges. Abdomen grey, vent chestnut. Wings and tail slaty blue-grey with basally yellowish-brown edges to outer feathers and yellowish-brown edges to primaries. Iris—back brown; bill—yellow; feet—pinkish-brown.

Voice: Melodious loud whistles like repeated bell ring *di,di,di . . .*

Range: Endemic to Taiwan.

Distribution and status: Distributed along central mountain range. Quite common from 1800–3500 m to treeline.

Habits: Typical laughingthrush of forest undergrowth and thickets.

Notes: Sometimes placed within Black-faced Laughingthrush.

1046. Chestnut-crowned Laughingthrush *Garrulax* PLATE 103
erythrocephalus Hongtou zaomei

Description: Largish (28 cm) dull brown laughingthrush with chestnut crown and nape, grey ear coverts and side of neck, chestnut inner wing coverts and olive-yellowish edges to wing and tail feathers. Lores and chin blackish; throat brown; breast scaly. Races vary in colour of ear coverts and degree of streaking on head. Iris—brown; bill—black; feet—brown.

Voice: Several variable loud calls; a whistled *too-rit-a-reill* answered *wroo-wroo* and variants; hissing whistle, constant twitters and chuckles; churring alarm call.

Range: Himalayas to SE Tibet, N, W and E Burma, N Indochina and W and S Yunnan to Malay Peninsula.

Distribution and status: Rare in S Xizang (*nigrimentum*); common in SE Xizang (*imprudens*); uncommon in W Yunnan (*woodi*); rare in S Yunnan W of Mekong (*melanostigma*); and S Yunnan E of Mekong (*connectens*). Ranges from 1200–3350 m.

Habits: Lives in small parties, flies in glides from one patch of thick cover to another. Typical skulking laughingthrush of scrub, forest edge and bamboo thickets.

1047. Red-winged Laughingthrush *Garrulax formosus* PLATE 103
Lise zaomei

Description: Large (28 cm) laughingthrush with crimson wings and tail. Similar to Red-tailed Laughingthrush but distinguished by grey crown streaked black, and brown mantle, back and breast. Iris—brown; bill—black; feet—blackish.

Voice: Song consists of loud, quite thin, plaintive whistled phrases, repeated at 2.5–8-second intervals, *chu-weewu* or slightly rising *chiu-wee*. Duets include *chiu-wee—u-weeoo* (rising slightly in middle) and *u-weeoo—wueeoo* (quickly delivered). Also a louder *wu-eeoo*. Calls include a soft, subdued *wiiii*.

Range: C China to W Tonkin.

Distribution and status: A rare bird of C and W Sichuan, N Yunnan and Guangxi, in mountains at 900–3000 m.

Habits: Lives in shy groups on or near forest floor in dense evergreen forest, secondary growth and bamboo.

1048. Red-tailed Laughingthrush *Garrulax milnei* PLATE 103
Chiwei zaomei

Description: Medium-sized (25 cm) laughingthrush with crimson wings and tail. Similar to Red-winged Laughingthrush but distinguished by rufous cap and nape and grey and olive scaly plumage of back and breast. Pale grey ear coverts are more or less plain. Races vary slightly in colour of back and ear coverts. Iris—dark brown; bill—blackish; feet—blackish.

Voice: Loud strident call and group chatters.

Range: S China, N Burma to N Indochina.

Distribution and status: Globally Near-threatened (Collar *et al.* 1994). Local and generally uncommon in mountains of Yunnan (*sharpei*); Guangxi (Yaoshan) and Guizhou (*sinianus*); and Wuyi mountains of SE China (*milnei*) at 1000−2400 m.

Habits: Makes noisy dancing displays, jerking tail and flicking crimson wings. Lives in groups in dense undergrowth and bamboo thickets of evergreen forest.

1049. Red-faced Liocichla *Liocichla phoenicea* PLATE 103
Hongchi soumei

Description: Unmistakable medium-sized (23 cm) liocichla with crimson sides to face and primaries. Easily distinguished from Red-winged and Red-tailed Laughingthrushes by red sides to head. Rest of plumage generally greyish-brown. Squared tail is black, tipped orange. Iris—brown; bill—dark horn; feet—brown.

Voice: Loud musical 2−4 note call *chi-chweew*, or *tu-reew-ri* etc; low churring notes.

Range: E Himalayas to SW China, N, W and E Burma and N Indochina.

Distribution and status: Rare in NW Yunnan (*bakeri*), W Yunnan (*ripponi*) and SE Yunnan (*wellsi*) from 900−2200 m with seasonal shifts in altitude.

Habits: A shy bird of dense undergrowth in evergreen montane forest, forest edge and secondary growth.

1050. Emeishan Liocichla *Liocichla omeiensis* PLATE 103
Huixiong soumei

Description: Smallish (17 cm) liocichla. Upperparts grey-olive, underparts and side of face grey. Forehead, eyebrow and side of neck olive-yellow. Wing has conspicuous orange bar and yellow edges to black primaries and tertials. Squared tail is olive with black barring and red tip. Outer tail feathers edged yellow with reddish tip. Vent blackish with orange tips. Female lacks red edges to tail and wing feathers. Iris—brown; bill—brown; feet—brown.

Voice: Loud clear calls.

Range: Endemic to C China.

Distribution and status: Globally Vulnerable (Collar *et al.* 1994). Confined to mountains of S Sichuan and NE Yunnan, where it is an occasional bird in forest from 1000−2400 m. May be threatened by trapping for bird trade where surprisingly large numbers are recorded.

Habits: As other liocichlas.

Notes: Considered conspecific with Steere's Liocichla by some authors.

1051. Steere's Liocichla *Liocichla steerii*   PLATE 103
Huangzhi soumei

Description: Medium-sized (18 cm) liocichla. Crown and nape grey finely streaked with whitish; crescent-shaped yellow patch in front of eye; eyebrow black underlined posteriorly with yellow. Back olive-brown; rump grey; squared tail olive with dark slate-grey subterminal band and narrow white tip; wings with chestnut secondaries, tipped dark slate-grey; primaries black, edged yellow. Underparts grey with yellowish-olive lower breast and black vent scaled with bright yellow feather tips. Iris—dark brown; bill—blackish; feet—olive-brown.

Voice: Resounding call *ji, jurr* and husky murmuring *ga-ga-ga.*

Range: Endemic to Taiwan.

Distribution and status: Frequent in lower mountains and hills at 900–2500 m.

Habits: Keeps to lower storeys of broadleaf forest and orchards. Remains in thick cover but often seen by roadside.

BABBLERS AND SYLVIA WARBLERS—Subfamily: Sylviinae

A large diverse subfamily containing two main tribes—babblers and Sylvia warblers.

BABBLERS—Tribe: Timaliini

This is a large, poorly-defined tribe, formerly treated as a family with laughingthrushes. Babblers are generally gregarious and noisy and most have rather harsh, chattering calls. Many species tend to be active on, or close to, the ground. They have short wings and are not strong fliers. Mostly sedentary. They make cup-shaped nests in trees and bushes. DNA studies show that most babblers are related to the warblers.

Several groups are recognised and it is convenient to break down the tribe into five distinct groups.

1. Jungle babblers—inconspicuous, rather quiet babblers living on or close to the ground in thickets (*Pellorneum*).
2. Scimitar and wren babblers—mostly feed on or close to the ground in dense forest. The wren babblers have characteristic tails, varying from short to almost none; the scimitar babblers are distinct with strong decurved bills (*Pomatorhinus, Xiphirhynchus, Rimator, Napothera, Pnoepyga, Spelaeornis, Sphenocichla*).
3. Tree babblers and tit babblers—small agile birds found more in bushes, grass and bamboo and only rarely come to the ground. They have small tit-like beaks, short wings, long strong legs and long, soft, fluffy feathers (*Stachyris, Macronous*).
4. Song babblers—small to larger, often colourful birds, often with loud songs. They are mostly arboreal though will come to the ground. They hop with a characteristic jerky motion and make short, fluttery flights, calling and flicking to and fro (*Timalia, Chrysomma, Babax, Leothrix, Cutia, Pteruthius, Gampsorhynchus, Actinodura, Minla, Alcippe, Heterophasia, Yuhina, Myzornis*).

5. Parrotbills—specialised babblers with heavy, parrot-like bills. They live in flocks and are closely associated with bamboos (*Panurus, Conostoma, Paradoxornis*). Formerly treated by some authors as a separate family Paradoxornithidae (e.g. Meyer de Shauensee 1984).

1. JUNGLE BABBLERS

1052. Buff-breasted Babbler *Pellorneum tickelli*   PLATE **104**
[Tickell's Babbler AG: Jungle-Babbler] Zongxiong yamei

Description: Small (14.5 cm) rufescent brown babbler. Crown and upperparts olive-brown with conspicuous pale shafts. There is an inconspicuous buff supercilium. Underparts whitish with buffish breast and tawny flanks. Iris—light brown to red; bill—brown above, pink below; feet—pinkish.

Voice: Diagnostic rapid *pit-you* with emphasis on lower second note. Wheezy descending trill (often in duet) and metallic churring.

Range: Himalayas, Assam to SW China, SE Asia.

Distribution and status: Race *fulvum* is resident in S and SW Yunnan and *assamensis* is probably in SE Xizang. A common submontane resident found in hills from 500–1130 m.

Habits: Generally keeps to the undergrowth of forest and forest edge but rarely works up vine-covered trees to the canopy, hunting insects.

Note: May be placed in genus *Trichastoma* (e.g. Cheng 1987, 1994).

1053. Spot-throated Babbler *Pellorneum albiventre*   PLATE **104**
[White-bellied Jungle Babbler/Plain Brown Babbler] Baifu youmei

Description: Small (14 cm) warm brown babbler with greyish side of head, buffy underparts centrally white. Throat white with diagnostic brown chevron spots sometimes not very clear. Distinguished from Buff-breasted Babbler and fulvettas by short, more rounded, tail. Iris—reddish-brown; bill—blackish; feet—flesh-brownish.

Voice: Low chuckling, clear whistle and rich melodious warbling song with varied but repeated phrases. Alarm is low rippling note.

Range: E Himalayas, Burma, Indochina to SW China.

Distribution and status: Nominate race occurs in SE Xizang; *cinnamomeum* is a rare resident in SW Yunnan from 1000–2000 m.

Habits: Skulks on or close to ground in dense grass, bamboo, or secondary scrub.

1054. Puff-throated Babbler *Pellorneum ruficeps* [Spotted Babbler]   PLATE **104**
Zongtou youmei

Description: Smallish babbler (17 cm) with dark rufous cap, pale supercilium and streaked underparts. Upperparts generally olive-brown; underparts pale buff with white throat and heavy brown streaks on breast and flanks. Iris—reddish-brown; bill—brown; feet—pinkish.

Voice: Song is continuous undulating whistle *sweety-swee-sweeow* ... Call is diagnostic *pre-tee-sweet* rising on second note and falling on last. Alarm churrs.

Range: India to SW China and SE Asia.

Distribution and status: Three races in S Yunnan: *shanense* W of Lancang (Mekong); *oreum* between Lancang and Hong Ha (Red River); and *vividum* E of Hong Ha. Quite common in scrub and forest undergrowth up to 1250 m, possibly higher.

Habits: Skulks on or close to ground. Puffs out white throat feathers.

2. SCIMITAR AND WREN BABBLERS

1055. **Large Scimitar Babbler** *Pomatorhinus hypoleucos*   PLATE 104
Changzui gouzuimei

Description: Large (27 cm) brown scimitar-billed babbler with dark bill. Has rusty patch behind brown ear coverts. Race *tickelli* has long white eye-stripe behind eye. Chin, throat and centre of breast white. Sides and flanks ashy-brown with white streaks; undertail coverts rufous. Race *hainanus* is of more olive tone. Iris—brown; bill—greyish-brown; feet—greenish-grey.

Voice: Loud clear triple hoot *hu-hu-pek*, sometimes more complex when two birds call antiphonally.

Range: NE India to S China, Hainan and SE Asia.

Distribution and status: Rather uncommon resident. Race *tickelli* in S Yunnan, Guangxi and Guangdong; *hainanus* on Hainan.

Habits: Lives in bamboo thickets and undergrowth in evergreen and mixed forests up to 1200 m.

1056. **Spot-breasted Scimitar Babbler** *Pomatorhinus erythrocnemis*   PLATE 104
Banxiong gouzuimei

Description: Largish (24 cm) scimitar babbler without pale supercilium and with rufous cheeks. Very similar to Rusty-cheeked but with heavy black spots or streaks on breast. Many races differ in details (see table).

Race	Breast	Flanks	Crown/nape	Mantle
dedekeni	spotted black	rufous	brown with streaks	olive-brown
cowensae	spotted black	cinnamon	brown with streaks	rufous-streaked
decarlei	spotted black	rufous	brown with streaks	olive-brown
gravivox	broadly streaked black	grey	brown with streaks	olive-brown
abbreviatus	broadly streaked black	grey	brown with streaks	maroon
swinhoei	broadly streaked black	grey	brown with streaks	chestnut-brown
erythrocnemis	streaked black	grey	dark grey streaked	chestnut-brown
sowerbyi	spotted black	rufous	brown with streaks	olive-brown
odicus	unspotted	grey	brown with streaks	olive-brown

Iris—yellow to chestnut; bill—grey to brown; feet—fleshy-brown.

Voice: Duet, male with loud *queue pee* and immediate female response *quip*.

Range: NE India, N and W Burma, N Indochina, E, C and S China.

Distribution and status: Fairly common resident in scrub and thickets and forest edge. Race *decarlei* in SE Xizang, NW Yunnan and S Sichuan; *dedekeni* from E Xizang to W Sichuan; *odicus* in Yunnan and Guizhou; *abbreviatus* in SE China; *swinhoei* in E China; *erythrocnemis* in Taiwan; *cowensae* in C China; and *gravivox* from S Gansu, NE Sichuan, S Shaanxi, Shanxi and W Henan.

Habits: Typical scrub-living scimitar babbler.

Note: Previously included within Rusty-cheeked Scimitar Babbler.

[1057. Rusty-cheeked Scimitar Babbler *Pomatorhinus*   PLATE **104**
erythrogenys Xiulian gouzuimei

Description: Largish (25 cm) scimitar babbler with brown bill; rufous-brown crown and nape streaked dark olive-brown; back, wings and tail plain rufous-brown; cheeks, flanks and undertail coverts bright orange-rufous; underparts otherwise whitish with grey spotting and streaking on breast. Iris—reddish-brown; bill—brown; feet—brown.

Voice: Loud and distinctive duet, deep *callow-creee, callow-creee* with *cree* about four tones higher and given by female in answer to males *callow*. Alarm chatters.

Range: Himalayas, E Burma and NW Thailand.

Distribution and status: Not recorded in China but race *haringtoni* may occur in the Torsa–Chumbi valley between Sikkim and Bhutan.

Habits: Skulks close to ground in long grass or dense scrub but sometimes calls from tree crown.]

[1058. White-browed Scimitar Babbler *Pomatorhinus*   NOT ILLUSTRATED
schisticeps Baimei gouzuimei

Description: Medium-sized (22 cm) brown scimitar babbler with grey crown and yellow bill. Has long white supercilium and broad black eye-band. Underparts white with brown sides, flanks and undertail coverts. Distinguished from Streak-breasted Scimitar Babbler by larger size and lack of streaks on breast. Iris—yellow; bill—yellow with black culmen; feet—plumbeous.

Voice: Three to six hoots uttered at varying speeds; a harsh call.

Range: Himalayas, Burma and Indochina.

Distribution and status: Not recorded in China but race *salimalii* occurs in Mishmi hills of NE India and probably ranges into adjacent areas of SE Xizang. Nominate race occurs in Himalayan foothills up to 2000 m west of Dibang River.

Habits: Shy skulker, sometimes in small parties in dense undergrowth hopping on ground and occasionally ascending trees.]

1059. Streak-breasted Scimitar Babbler *Pomatorhinus ruficollis*   PLATE **104**
Zongjing gouzuimei

Description: Smallish (19 cm) brown scimitar babbler with chestnut nuchal collar, long white supercilium, black lores, white throat and streaked white breast. Many races vary in details, see table:

Race	Breast	Belly	Mantle
godwini	white, streaked grey-brown	grey-brown	brown
eidos	white, streaked olive-brown	olive-brown, streaked white	brown, with chestnut wash
similis	white, streaked brown	olive-brown, washed rufous	brown
albipectus	mostly white	olive-brown	brown
reconditus	white, streaked brown	olive-brown	brown
styani	rufous-brown, streaked white	olive-brown	brown, with chestnut wash
hunanensis	brown, streaked white	olive-brown	brown
stridulus	chestnut-brown, streaked white	rich brown	chestnut-brown
musicus	white, with dark chestnut central splodges	rich chestnut-brown	dark brown
nigrostellatus	dark chestnut streaked white	rich chestnut-brown	dark brown

Iris—brown; bill—black above yellow below (race *reconditus* pink below); feet—brownish-plumbeous.

Voice: Song is mellow two or three-noted hoot with emphasis on first note and last note on lower pitch. Female sometimes answers with squeaky response.

Range: Himalayas, N Indochina, N and W Burma, C to S China, Taiwan and Hainan.

Distribution and status: Quite common in mixed, evergreen or scrubby secondary forests with bamboo from 80–3400 m. Many geographical races recognised: *musicus* on Taiwan; *nigrostellatus* on Hainan; *stridulus* on Wuyi Mts of SE China; *hunanensis* on mountains of C and S China; *styani* in S Gansu to Zhejiang and S to N Sichuan and N Guizhou; *eidos* endemic to Emeishan region of Sichuan; *similis* in SW Sichuan and NW and W Yunnan; *albipectus* in S Yunnan between Lancang and Red rivers; and *godwini* in SE Xizang. Some gradation between forms is noted.

Habits: Typical of genus.

Note: Separation of this species from *P. schisticeps* as per Deignan (1964) is not accepted by all authors.

1060. Red-billed Scimitar Babbler *Pomatorhinus ochraceiceps* PLATE 104
Zongtou gouzuimei

Description: Medium-sized (23 cm) brown scimitar babbler with buff underparts and white throat and supercilium and diagnostic scarlet bill. Distinguished from Coral-billed Scimitar Babbler by lack of black above white supercilium; white breast (warm buff in *stenorhynchus*) and more orange and more slender bill. Iris—pale brown; bill—orange-red; feet—brown.

Voice: Deep liquid, single or double-note *tu-lip* and hollow double-hoot *hoop-hoop*. In alarm gives scolding metallic rattle, harsher than White-browed Scimitar Babbler.

Range: NE India to SW China and SE Asia.

Distribution and status: Uncommon. Nominate race is resident in Xishuangbanna (S Yunnan) and *stenorhynchus* occurs in W Yunnan (Luxi, Yongde) and SE Xizang.
Habits: Inhabits bamboo from 1220–2400 m. Lives in pairs or small parties. Joins mixed bird flocks.

1061. Coral-billed Scimitar Babbler *Pomatorhinus ferruginosus* PLATE 104
Hongzui gouzuimei

Description: Medium-sized (23 cm) brown scimitar babbler with white supercilium and stout red bill. Distinguished from Red-billed Scimitar Babbler by stouter, less orange bill; black band above white supercilium; blacker side of face. Nominate race has blackish crown. Iris—straw; bill—red; feet—brown.
Voice: Harsh churring rattle similar to White-browed Scimitar Babbler; also mellow, rising dissyllabic whistle.
Range: Himalayas.
Distribution and status: Rare. Nominate race is resident in SE Xizang and *orientalis* is resident in SE Yunnan and extreme W Yunnan.
Habits: Inhabits hill evergreen forests from 900–2000 m.

1062. Slender-billed Scimitar Babbler *Xiphirhynchus superciliaris* PLATE 104
Jianzuimei

Description: Smallish (20 cm) dark brown scimitar babbler with diagnostic extremely long, slender, decurved blackish bill. Head slaty with narrow white eyebrow. Upperparts dark rufescent brown; underparts rusty with buff tinge and whitish throat. Iris—grey to red; bill—black; feet—slaty.
Voice: Three-noted rippling whistle and single mellow high-pitched hoot. Song is soft hoot of 7–8 rapid notes on monotone. Alarm is harsh chittering.
Range: E Himalayas to W Yunnan, N and W Burma and W Tonkin.
Distribution and status: Globally Near-threatened (Collar *et al.* 1994). Rare above 1000 m. Race *foresti* is resident in W Yunnan. Nominate and *intextus* races probably occur in SE Xizang. Race *rothschildi* probably occurs in SE Yunnan (Jinping).
Habits: Noisy, shy, restless babbler in pairs or small parties. Lives in evergreen montane forest on or near ground in steep rocky terrain, almost exclusively in bamboo.

1063. Long-billed Wren Babbler *Rimator malacoptilus* PLATE 105
[White-throated/Sumatran Wren-Babbler] Changzui liaomei

Description: Small (13 cm) brown, short-tailed, fluffy, buff-streaked brown babbler with long decurved bill. Distinguished from Eye-browed Wren Babbler by long bill and lack of pronounced supercilium. Underparts dark brown with buffy white throat and whitish-buff streaks on breast. Iris—brown; bill—black; feet—pink.
Voice: In Burma reported as beautiful whistled call; also soft chatters.
Range: Himalayas and Sumatra.
Distribution and status: Globally Near-threatened (Collar *et al.* 1994). Nominate race is uncommon but probably often overlooked in SE Xizang Himalayas in montane forests from 900–2500 m. This race has recently been found in Gaoligong Mts of W

Yunnan (c. Robson pers. comm). Race *pasquieri* should also be looked for in Jinping area of SE Yunnan as known from the Vietnamese side of the border in Hoang Lian Son range.

Habits: Lives on or close to the ground hopping among dense ground vegetation.

1064. Limestone Wren Babbler *Napothera crispifrons*   PLATE 105
Huiyan jiaomei

Description: Largish (19 cm) dark greyish-brown wren babbler with dark scaling on upperparts; white throat with black streaks and whitish streaking on olive-grey belly. Usually shows grey supercilium behind eye. Larger and longer-tailed than Streaked Wren Babbler. Iris—brown; bill—brown; feet—pinkish.

Voice: Continuous loud, undulating, rich whistles *tuu-wii-chuu, tuu-wii-chuu* . . .

Range: SE Asia and SW China.

Distribution and status: Local and uncommon. Recorded in Menglun limestone forests of Xishuangbanna (S Yunnan).

Habits: Inhabits limestone areas in forest. Shy; moves jerkily on ground. Hides in rock crevices. Cocks tail.

1065. Streaked Wren Babbler *Napothera brevicaudata*   PLATE 105
Duanwei jiaomei

Description: Small (15 cm) brown wren babbler with dark scaling on crown, nape and mantle. Underparts rufous-brown with faint streaking and black and white streaking on throat. A few fine white spots visible on tips of greater coverts and tertials. Distinguished from Limestone Wren Babbler by smaller size, white spots on wing, more contrasting scaling on upperparts, less bold streaking on throat and shorter tail. Iris—brown; bill— brown; feet—pinkish.

Voice: Varied shrill whistles such as *pew-ii* or *pewii-uu*; emphasis on second note. Also falling, mournful, *piu*. Alarm is harsh chattering with soft piping note.

Range: NE India to SW China and SE Asia.

Distribution and status: Local and uncommon. Race *venningi* in extreme W Yunnan and *stevensi* in SE Yunnan and Guangxi (Longjin).

Habits: Lives in evergreen forest undergrowth in rocky terrain, particularly limestone areas. Secretive, skulking among boulders. Runs quickly and mouse-like.

1066. Eyebrowed Wren Babbler *Napothera epilepidota*   PLATE 105
Wenxiong jiaomei

Description: Very small (11 cm) short-tailed, brown babbler with a pronounced white supercilium. General colour dark brown but upperparts scaled dark; chin and throat buffy-white; centre of belly white; underparts streaked buff. Small whitish spots on tips of wing coverts and tertials. Iris—brown; bill—brown; feet—pale brown.

Voice: Loud prolonged one-second, almost flat *peeeow* and churrs.

Range: E Bhutan, NE India to SW China, SE Asia, and Greater Sundas.

Distribution and status: Race *laotiana* is resident in Guangxi (Yaoshan) and S

Yunnan; *hainanus* is resident on Hainan; and *guttaticollis* occurs in SE Xizang. Not uncommon but generally overlooked in hills and mountains up to 2000 m.
Habits: A shy inconspicuous bird of dense undergrowth.

1067. Scaly-breasted Wren Babbler *Pnoepyga albiventer*   PLATE 105
Linxiong jiaomei

Description: Small (10 cm) almost tail-less wren-like babbler in two colour morphs. Pale morph: upperparts olive-brown with slight scaly pattern and buff spot on tip of each feather; underparts white with dark centres to feathers of breast and scaled with darker edges, scaly olive-brown flanks. Fulvous morph: upperparts olive-brown with buff spot on tip of each feather; underparts as pale morph but buff instead of white. Sexes alike. Distinguished from Pygmy Wren Babbler by larger size and buff speckling on crown and neck. Iris—brown; bill—horn; feet—pinkish-brown.
Voice: Call is single or double *sik* or *seek . . . sik*. Alarm is piercing whistle and scolding *tsik, tsik*.
Range: Himalayas to SW China, W and N Burma and W Tonkin.
Distribution and status: Uncommon. Nominate race is resident in S and SE Xizang to NW Yunnan and Sichuan. May occur elsewhere in Yunnan (Jinping).
Habits: Secretive skulker. Flicks wings as it creeps about. Inhabits montane forest at 1500–3660 m, down to 1100 m in winter, along mossy and fern-lined stream banks.

1068. Pygmy Wren Babbler *Pnoepyga pusilla*   PLATE 105
[Lesser Scaly-breasted/Brown Wren-Babbler] Xiaolin jiaomei

Description: Very small (9 cm) almost tail-less, boldly scalloped babbler occurring in pale and fulvous colour morphs. Very similar to respective morphs of Scaly-breasted Wren Babbler, but distinguished by smaller size and distinctive song. Spotting on upperparts is less extensive and confined to lower back and wing coverts with no spots on crown. Iris—dark brown; bill—black; feet—pink.
Voice: Loud, piercing whistle of 2–3 well-spaced notes of decreasing pitch. High-pitched squeak followed by short chatter.
Range: Nepal to S China, SE Asia, Malay Peninsula, Sumatra, Java, Flores and Timor.
Distribution and status: Nominate race is locally common in SE Xizang, C, SW, S and SE China in mountain forest from 520–2800 m. Race *formosana* is resident on Taiwan.
Habits: Scuttles about the forest floor in thickets in a mouse-like manner. Shy and secretive except when calling.
Note: According to C. Robson (pers. comm.) Taiwan form *formosana* should be regarded as a separate species.

[1069. Rufous-throated Wren Babbler *Spelaeornis caudatus*   PLATE 105
[Short-tailed Wren Babbler] Duanwei liaomei

Description: Very small (9 cm) olive-brown wren babbler with short tail. Upperparts dark brown with black scaling. Underparts distinctive with rufous chin and throat; belly scaled with blackish and whitish spots. Iris—brown; bill—blackish; feet—brown.

Voice: High-pitched *tzit*; higher than Scaly-breasted Wren Babbler. Alarm note is quiet *birrh birrh birrh*.

Range: E Himalayas.

Distribution and status: Globally Vulnerable (Collar *et al.* 1994). Not recorded in China but range extends through Bhutan into area of NE India claimed by China as part of SE Xizang. Eastern limits of distribution not known.

Habits: Shy skulker of forest undergrowth in montane evergreen forest at about 1750–2440 m.]

[1070. Rusty-throated Wren Babbler *Spelaeornis badeigularis* NOT ILLUSTRATED [Mishmi Wren-Babbler] Xiuhou liaomei

Description: Small (9 cm) olive-brown wren babbler with short tail, white chin, dark chestnut throat and rest of underparts scaled blackish and whitish. Ear coverts olive-brown. Distinguished from Rufous-throated Wren Babbler by dark chestnut restricted to lower throat only. Iris—brown; bill—black; feet—brown.

Voice: Unknown.

Range: Mishmi Hills of Arunachal Pradesh (NE India).

Distribution and status: Globally Vulnerable (Collar *et al.* 1994). Rare and known only from the type-specimen taken in this small range of hills. Not recorded in China but presumably occurs on Chinese side of the hills. Recorded from disputed territory.

Habits: Shy and skulking in wet broadleaved evergreen forest.]

1071. Bar-winged Wren Babbler *Spelaeornis troglodytoides* PLATE 105 [Spotted Long-tailed Wren Babbler] Banchi liaomei

Description: Small (13 cm) longish-tailed wren babbler. Upperparts umber-brown spotted black and white; tail and wings narrowly barred black. Underparts rufous with distinctive white throat. Breast usually streaked faintly white. Race *souliei* has browner less rufous cheeks; *rocki* is larger and has deeper orange cheeks, marginally greyer back and less barring on wings. Iris—red-brown; bill—blackish above, pinkish below; feet—brown.

Voice: Low song of 4–5 notes. Alarm call is faint *churr*. Call is subdued *cheep*.

Range: E Himalayas to SW China and NE Burma.

Distribution and status: Rare and local. Nominate race is resident in C and SW Sichuan; *halsueti* in Baishuijiang of S Gansu and Qinling Mts of S Shaanxi; *rocki* confined to a small area of NW Yunnan E of the Lancang (Mekong) River; and *souliei* in SE Xizang and W and NW Yunnan to the W of the Lancang.

Habits: Lives in undergrowth of montane forest from 1500–3600 m.

1072. Spotted Wren Babbler *Spelaeornis formosus* PLATE 105 Lixing Liaomei

Description: Small (10 cm) short-tailed wren babbler. Upperparts diagnostic dark brown with small white spots and wings and tail barred rufous and black. Underparts peppery buffy-brown with black vermiculations and small white spots. Iris—dark brown; bill—horn-brown; feet—horn-brown.

Voice: Squeaky *sik . . . sik* similar to Scaly-breasted Wren Babbler. Sharper than Pygmy Wren Babbler.

Range: E Himalayas to SW, S and SE China, W and N Burma, N Indochina.

Distribution and status: Globally Near-threatened (Collar *et al.* 1994). Uncommon and local at 1100–2150 m. Recorded from SE Yunnan and Wuyishan in N Fujian but should occur in SE Xizang and may be overlooked elsewhere.

Habits: Secretive skulker in montane evergreen forest undergrowth.

1073. Long-tailed Wren Babbler *Spelaeornis chocolatinus* PLATE 105

Changwei liaomei

Description: Small (11 cm) slim-looking, dark brown wren babbler with relatively long tail, white throat and greyish side of head. Upperparts olive-brown scaled with black. Races vary in underpart pattern. Race *kinneari* has buffy-white throat, olive-brown breast with buffy and black barring and speckling and greyish belly speckled white. Race *reptatus* has whitish throat spotted light brown or tawny-buff spotted with reddish-brown and grading gradually to olive-brown on rest of underparts with grey centre of belly. Iris—reddish-brown; bill—black; feet—pinkish.

Voice: Loud explosive whistle *wheeuh*. Song is warbling of 2–3 notes repeated fast with accent on last note; also sharp quick *ticki-ticki-ticki-ticki* or rattling wren-like song.

Range: NE India, Burma, SW China and W Tonkin.

Distribution and status: Rare. Race *reptatus* is resident in W Yunnan and C and SW Sichuan; *kinneari* ranges into SE Yunnan (Jinping) from adjacent Vietnam.

Habits: Inhabits montane forest, forest edge and secondary growth from 1650–3000 m.

1074. Wedge-billed Wren Babbler *Sphenocichla humei* PLATE 105

Xietou liaomei

Description: Large (17 cm) chocolate-brown wren babbler with longish tail and pale grey supercilium behind eye, curving down behind ear coverts. Head scaled and streaked white. Plumage finely barred, especially on wings and tail. Distinguished by diagnostic broad, sharp wedge-shaped bill and large size. Iris—dark brown; bill—blackish; feet—brown.

Voice: Alarm call is dry rattling, *hrrt hrrt hrrt . . .; hrr'it*, etc. (C. Robson).

Range: E Himalayas, NE India and NE Burma.

Distribution and status: Globally Near-threatened (Collar *et al.* 1994). Very rare, recorded at 2010 m in China. Probably occurs in SE Xizang and recorded in extreme W Yunnan.

Habits: Lives in parties of 10–15 birds on ground, but also feeding on bark and in moss on tree stems.

3. Tree Babblers and Tit Babblers

1075. Rufous-fronted Babbler *Stachyris rufifrons*　　PLATE 106
[>Buff-chested/Yellow-throated Babbler AG: Tree Babbler]
Huanghou suimei

Description: Small (12 cm) brown tree babbler with rufous crown. Distinguished with difficulty from Rufous-capped Babbler by white chin; white throat with black streaks; more buffish and less yellowish underparts. Iris—reddish-brown; bill—grey; feet—greenish-yellow.

Voice: Mellow, musical four-noted whistle, *whi-whi-whi-whi* and chattering. Song is 6–7 note monotone with pause after first note and rest rapid; similar to Rufous-capped Babbler.

Range: NE Indian subcontinent, SE Asia, Sumatra, Borneo.

Distribution and status: Race *planicola* is rare in extreme NW Yunnan (Gongshan); race *ambigua* probably occurs in Himalayan foothills of SE Xizang.

Habits: Inhabits forest edge, secondary growth scrub and grass up to 1200 m. Generally in restless small parties creeping through undergrowth and bamboos.

Note: Considered by some authors to involve two species, with Buff-chested Babbler *S. ambigua* being elevated to specific status (e.g. Sibley and Monroe 1993).

1076. Rufous-capped Babbler *Stachyris ruficeps*　　PLATE 106
Hongtou suimei

Description: Small (12.5 cm) brown tree babbler with rufous crown. Upperparts dull greyish-olive; lores dull yellow; throat, breast and sides of head tinged yellow; underparts buffy olive-brown; throat streaked black. Distinguished from Rufous-fronted Babbler by yellower, less buff underparts. Race *praecognita* has less grey upperparts; *goodsoni* has yellow throat with dark streaks; *davidi* has yellow underparts. Iris—red; bill—blackish above, paler below; feet—brownish-green.

Voice: Song like Golden Babbler but without pause after first note, *pi-pi-pi-pi-pi-pi*. Low chittering and soft four-noted whistle *whi-whi-whi-whi* like iora.

Range: E Himalayas to C and S China, Taiwan, N Burma and Indochina.

Distribution and status: Common resident. Race *bhamoensis* resident in W Yunnan; nominate in SE Xizang; *davidi* in C, S and SE China; *goodsoni* on Hainan; and *praecognita* on Taiwan.

Habits: Inhabits forest, scrub and bamboo brakes.

1077. Golden Babbler *Stachyris chrysaea* [Golden-headed Babbler]　　PLATE 106
Jintou suimei

Description: Small (11.5 cm) yellowish-olive babbler with black lores and yellow throat distinctive. Crown golden-yellow, streaked black. Underparts light yellow. Iris—reddish; bill—black; feet—yellow.

Voice: Surprisingly loud, low-pitched whistle of 4–8 *toot* notes with first note or two emphasised by pause; also flock chatters.

Range: Nepal to SW China, SE Asia, Malay Peninsula and Sumatra.

Distribution and status: Nominate race is resident in SE Xizang and Yunnan W of Nujiang (Salween) River. Race *aurata* is resident in tropical S Yunnan (Xishuangbanna). Common in hill and montane forest at 950–2130 m.

Habits: Lives in small parties, often mixed with other species, in foliage of low bushes in primary, secondary and pine forests.

1078. Grey-throated Babbler *Stachyris nigriceps* [Black-throated/   PLATE 106
Grey Babbler, AG: Tree-Babbler] Heitou suimei

Description: Small (13.5 cm) olive-brown babbler with white-streaked blackish crown and nape, long black lateral crown stripe underlined by white line, grey over white malar stripe and dark grey chin and throat. Eye-ring white; underparts olive-buff. Iris—light brown; bill—blackish; feet—dull yellow.

Voice: Rattling *prrreee-prrreee* calls.

Range: Nepal to SW China, SE Asia, Malay Peninsula, Sumatra and Borneo.

Distribution and status: Nominate race occurs in SE Xizang; *coltarti* is resident in W Yunnan; and *yunnanensis* in SE Yunnan and SW Guangxi. A common bird of forest undergrowth in hills and mountains at 1000–1500 m.

Habits: Lives in small parties skulking close to the ground in moist undergrowth of hill and montane forests.

1079. Spot-necked Babbler *Stachyris striolata* [Spotted Babbler,   PLATE 106
AG: Tree-Babbler] Banjing suimei

Description: Smallish (16 cm) olive-brown babbler with chestnut crown and nape, white throat and black moustachial stripe. Supercilium, forehead and side of neck black, boldly spotted white. Underparts rufous-chestnut. Iris—red; bill—black; feet—greenish-black.

Voice: Simple song and chattering alarm.

Range: S China, SE Asia, Sumatra.

Distribution and status: Race *tonkinensis* is resident in S Yunnan and SW Guangxi (Yaoshan). Race *swinhoei* is found on Hainan. Locally common in hill and mountain forests and bamboo breaks to 950 m.

Habits: Flock-living skulker of forest floor and undergrowth in dense montane forests.

1080. Striped Tit Babbler *Macronous gularis* [Yellow-breasted/   PLATE 106
Striated/Stripe-throated Tit-Babbler] Wenxiongmei

Description: Small (13 cm) rufescent brown babbler with streaked underparts. Crown, wings and tail dark rufous; underparts pale greenish-yellow, with conspicuous dark streaks, on throat and breast. Iris—creamy-white; bill—grey; feet—olive.

Voice: A monotonous *chunk chunk chunk* in phrases of 3–10 or more notes repeated *ad nauseam* throughout the day.

Range: S and E India and Himalayas to SW China, SE Asia and Greater Sundas.

Distribution and status: Race *lutescens* is resident in S and SE Yunnan; *sulphureus* is found in Yunnan W of Nujiang (Saleween) River; and *rubricapilla* probably occurs in SE Xizang. A common bird of lowland areas in suitable habitat up to 1150 m.

Habits: Occurs in pairs or small flocks in dense secondary growth, forest edge and bamboo thickets, especially near streams. The birds spend most of their time within a few metres of the ground but sometimes climb higher in vine-laden trees.

4. SONG BABBLERS

1081. Chestnut-capped Babbler *Timalia pileata* PLATE 106
[Red-capped Babbler] Hongdingmei

Description: Medium-sized (17 cm) reddish-chestnut crowned babbler with short white supercilium. Upperparts warm brown; ear coverts white to grey; lores black; breast white with black shaft streaks; belly grey washed fulvous-brown on sides and vent. Iris—chestnut; bill—black; feet—greyish-yellow.

Voice: Wide variety of clear, loud, whistled notes, metallic trills, slurred notes, warbles, rising and falling in pitch; and loud, sharp dissyllabic whistles rising in pitch.

Range: Nepal to S China, SE Asia and Java.

Distribution and status: Race *smithi* is resident in S China. Common in lowlands at 340–880 m.

Habits: Skulks in thick undergrowth, thick grass and dense shrub layer in more open scrub areas. Often in small flocks keeping close to the ground and calling from thick cover.

1082. Yellow-eyed Babbler *Chrysomma sinense* PLATE 106
Jinyan meique

Description: Largish (19 cm) rufous-brown babbler with long graduated tail, distinctive sturdy black bill and orange eye-ring. Lores, chin, throat and upper breast clean white, becoming cinnamon towards vent. Iris—yellow; bill—black; feet—orange.

Voice: Loud warbling song with short strident phrases; plaintive piping series *pui pui pui* . . . and explosive chattering.

Range: Pakistan to S China and SE Asia.

Distribution and status: Locally common resident up to 1500 m in Yunnan, SW Guizhou, Guangxi (Yaoshan) and Guangdong (Xijiang valley).

Habits: Lives in small parties outside breeding season, skulking in scrub and tall grass. Often feeds on ground. Calls from tall stem before diving back into cover.

1083. Rufous-tailed Babbler *Chrysomma poecilotis* PLATE 106
Baoxing meique

Description: Medium-sized (15 cm) rufous-brown babbler with longish graduated chestnut-rufous tail. Upperparts rufous-brown with greyish eyebrow stripe becoming dark posteriorly and black and white moustachial streaks. Throat white, centre of breast buff; flanks and vent cinnamon. Wing and tail chestnut. Iris—brown; bill—brown; feet—pale brown.

Voice: Not recorded.

Range: Endemic to Sichuan and Yunnan mountains.

Distribution and status: Recorded from NE Sichuan in an arc W of Sichuan Basin and S to Likiang mountains of N Yunnan. Frequent at 1500–3810 m in grass and scrub.
Habits: Similar to Yellow-eyed Babbler.
Note: Sometimes placed in separate genus *Moupinia*.

1084. Chinese Babax *Babax lanceolatus* Maowen caomei　　PLATE 107

Description: Largish (26 cm) streaky babbler. Looks like a heavily-streaked greyish-brown laughingthrush with rather long tail, slightly down-curved bill and characteristic dark moustachial stripe. Iris—yellow; bill—black; feet—pink.
Voice: Call is loud creaky wailing note *ou-phee-ou-phee* repeated several times.
Range: NE India, W and N Burma and China.
Distribution and status: Nominate race is resident in C and SW China, *latouchei* in SE China and *bonvaloti* in N and W Sichuan, E Xizang and NW Yunnan. Generally common resident. Population in Hong Kong probably feral.
Habits: A noisy bird of scrub, thickets and undergrowth of open montane and hill forests. Lives in small groups and feeds mostly on the ground. Rather skulking in habits but perches in prominent place to call.

1085. Giant Babax *Babax waddelli* Dacaomei　　PLATE 107

Description: Large (31 cm) streaky, grey-brown babbler with pale eye. Similar to Chinese Babax but larger and greyer with blacker tail and bill longer and more decurved. Base colour of underparts whiter. Iris—grey-white; bill—blackish-horn; feet—blackish-horn.
Voice: Song is series of quavering whistled notes similar to thrush. Harsh grating call.
Range: S Tibet, doubtfully N Sikkim.
Distribution and status: Globally Near-threatened (Collar *et al.* 1994). Rare to locally common from 2700–4570 m. Nominate race resident in S and SE Xizang. Race *jomo* only recorded in Chumbi valley area of S Xizang.
Habits: Small parties of 5–6 birds skulk in dense undergrowth turning over leaves in search of food in arid scrub.

1086. Tibetan Babax *Babax koslowi* Zongcaomei　　PLATE 107

Description: Medium-sized (28 cm) cinnamon-coloured babax. Upperparts cinnamon-brown with pale scaling; lores blackish; sides of head greyish; mantle feathers edged grey; wings and tail cinnamon-brown. Primaries edged grey. Throat grey, breast pale cinnamon with grey scaling; underwing and undertail coverts pale cinnamon. Iris—yellow; bill—blackish; feet—blackish-brown.
Voice: Not recorded.
Range: Endemic to E Tibetan valley region.
Distribution and status: Globally Near-threatened (Collar *et al.* 1994). Recorded from S Qinghai, NW Sichuan and SE Xizang at 3350–4500 m.
Habits: As other babax's, a skulker of scrubland, rocky areas and abandoned agricultural fields.

1087. Silver-eared Mesia *Leiothrix argentauris*　　　　　PLATE 106
[Silver-eared Leothrix] Yin'er xiangsiniao

Description: Medium-sized (17.5 cm) colourful babbler. Black crown, silvery-white cheeks and orange forehead distinctive. Tail back and wing coverts olive; throat and breast reddish-orange; wings red and yellow; tail coverts red. Iris—red; bill—orange; feet—yellow.

Voice: Hollow rattled chatter and cheery whistled song *chi-uwi, chi-uwi, chi-uwi* or *chi-uwi-chiu*.

Range: Himalayas, S China, SE Asia and Sumatra.

Distribution and status: Race *vernayi* occurs in SE Xizang and Yunnan W of Nujiang (Salween) River, *ricketti* in S Yunnan; nominate race in Yunnan W of Red River and; *rubrogularis* in SE Yunnan, S Guizhou and Guangxi. Common at 350–2000 m.

Habits: A restless bird of dense thickets in lower and middle storeys of montane forests.

1088. Red-billed Leiothrix *Leiothrix lutea*　　　　　PLATE 106
Hongzui xiangsiniao

Description: Pretty and colourful, smallish (15.5 cm) babbler with conspicuous red bill. Upperparts olive-green with yellow patch around eye, underparts orange-yellow. Tail is blackish and slightly forked. Wing is blackish with red and yellow edges forming conspicuous bars at rest. Iris—brown; bill—red; feet—pink.

Voice: Has a fine but rather monotonous song.

Range: Himalayas, Assam, W and N Burma, S China and N Tonkin.

Distribution and status: Nominate race is resident in C and SE China; *kwantungensis* in S China; *yunnanensis* in W Yunnan; and *calipyga* in S and SE Xizang.

Habits: Lives in noisy chattering flocks in undergrowth of secondary forests. It is a favoured cagebird because of its active singing, beauty and 'loving' habits. Resting birds sit pressed close together and groom each other.

1089. Cutia *Cutia nipalensis* Banxie jimei　　　　　PLATE 107

Description: Medium-sized (19 cm) unmistakable, boldly patterned babbler. Forehead, crown, nape and edges of flight feathers bluish-grey; mantle, back, rump, and long upper-tail coverts rufous-chestnut; tail, rest of wings and broad eye-stripe black; underparts white with black barring on flanks. Female paler with large black streaks on olive-brown mantle and back; eye-stripe dark brown. Iris—red brown; bill—blackish; feet—yellow to orange.

Voice: Long loud *cheeeet* rising in inflection; double *chirp*, also loud monotonous *chipchip-chip-chipchip* repeated many times.

Range: Himalayas to SW China and SE Asia.

Distribution and status: Uncommon in montane evergreen forest from 1800–2600 m. Nominate race is resident in S and SE Xizang, W Sichuan and NW Yunnan; *melanchima* is recorded in hills of Xishuangbanna (S Yunnan)

Habits: Moves along epiphyte-laden branches searching for food. Usually in small parties or in mixed-species flocks.

1090. Black-headed Shrike Babbler *Pteruthius rufiventer* PLATE 107
[Rufous-billed Shrike Babbler] Zongfu beimei

Description: Large (21 cm) shrike babbler with chestnut upperparts; shiny black head, wings and tail; grey chin, throat and upper breast distinctive, and yellow patch on side of breast. Lower breast and vent vinous-brown. Tail and secondaries narrowly tipped chestnut. Female as male but sides of head grey; black crown marked with grey; rest of upperparts bright olive-green except chestnut rump, uppertail coverts and tips of secondaries; tail greenish above, blackish below with chestnut tip. Iris—grey; bill—black above, paler below; feet—brownish.

Voice: Generally silent. Curious *whirr-i-oh* call from flock containing one of these birds may be its call.

Range: Nepal to SW China, W and N Burma and W Tonkin.

Distribution and status: Globally Near-threatened (Collar *et al.* 1994). Rare resident from 1500–2500 m in W and NW Yunnan in montane evergreen forest.

Habits: Lives in small parties; often in 'bird waves' with tits and other babblers. Rather lethargic.

1091. White-browed Shrike Babbler *Pteruthius flaviscapis* PLATE 107
[Red-winged/Black-crowned/Greater Shrike-Babbler] Hongchi beimei

Description: Medium-sized (17 cm) babbler. Male: head black with white supercilium; mantle and back grey; tail black; wings black with white tips to primaries and gold and orange tertials; underparts grey. Female: duller with buff underparts; head greyish; coloured secondaries less bright. Iris—greyish-blue; bill—bluish-black above, grey below; feet—whitish-pink.

Voice: Loud and piercing monotone *too-too-too*, *klip klip* or *chip chip chap chip chap*.

Range: NE Pakistan to China, SE Asia and Greater Sundas.

Distribution and status: Resident on Hainan (*lingshuiensis*); SE and C China (*ricketti*) and SE Xizang (*validirostris*); occasional in montane forest at 350–2440 m.

Habits: Lives in pairs or in mixed-species flocks, moving through the lower and upper canopy catching insects. Shuffles sideways along small twigs searching keenly for food.

1092. Green Shrike Babbler *Pteruthius xanthochlorus* PLATE 107
Danlü beimei

Description: Small (12 cm) olive-green shrike babbler. Looks like a leaf warbler but chunky; sluggish in movements and with stout black bill. Grey head with white eye-ring, greyish throat and breast, yellow belly, vent and wing linings diagnostic. Primary coverts grey, pale wing bar. Nominate race has slaty crown in male, ashy-grey in female, no eye-ring. Race *pallidus* is paler; *obscurus* has grey crown in male. Iris—grey-brown; bill—blue-grey with black tip; feet—grey.

Voice: Usually silent. Quick repeated *whit*. Song is rapid, monotonous repetition of single note.

Range: NE Pakistan to SE China, W and N Burma.

Distribution and status: Uncommon resident in subalpine, mixed and conifer forests from 760–5600 m. Nominate race recorded in SE Xizang; *pallidus* in Yunnan, Sichuan, S

Gansu (Baishuijiang) and S Shaanxi (Qinling); and *obscurus* in Wuyishan NW Fujian. Descends in winter.

Habits: Joins mixed flocks of tits, babblers and leaf warblers. Looks like sluggish leaf warbler.

1093. Black-eared Shrike Babbler *Pteruthius melanotis* PLATE 107
Lihou beimei

Description: Male: small (11.5 cm) colourful shrike babbler with two bold white wing bars. Distinguished from Chestnut-fronted Shrike Babbler by black crescent-shaped line behind ear coverts and yellow forecrown. Has pale chestnut throat and upper breast. Female: as male but has yellow throat and dull buffish instead of white wing bars. Iris—red-brown; bill—dark grey above, paler below; feet—pale brown.

Voice: Usually silent. Call note is pleasant *too-weet, too-weet.*

Range: Nepal to SW China and SE Asia.

Distribution and status: Nominate race is uncommon resident in montane evergreen forest from 1200–2400 m in NW, W, S and SE Yunnan and SE Xizang.

Habits: Canopy dweller, mixing with tits, leaf warblers and fantails. Sluggish and rather upright.

1094. Chestnut-fronted Shrike Babbler *Pteruthius aenobarbus* PLATE 107
Li'e beimei

Description: Small (11.5 cm) brightly coloured babbler. Male has chestnut forehead, chin and throat; upperparts olive-green with bold double white bars on black upperwing coverts; white eye-ring and greyish-white supercilium; underparts yellow. Female has whitish underparts, chestnut only on forehead. Race *yaoshanensis* has chestnut forecrown and breast.

Voice: Thin high-pitched, piercing call *too-weet-weet-weet.*

Range: Assam to S China, SE Asia and Java.

Distribution and status: Race *yaoshanensis* in SW Guangxi and *intermedius* in S Yunnan. An uncommon bird of higher mountains from 500–2500 m.

Habits: Lives in the top of low trees in montane forest sometimes mixing within flocks with other species.

1095. White-hooded Babbler *Gampsorhynchus rufulus* PLATE 107
Baitou beimei

Description: Large (24 cm) unmistakable babbler with all-white head contrasting with rufous-brown upperparts. Tail long and graduated with narrow white tips. Underparts whitish with buff wash on belly. Race *torquatus* has black patch on side of breast and cinnamon wash on underparts. Iris—yellow; bill—plumbeous, paler on lower mandible; feet—pinkish.

Voice: Distinctive, loud, harsh chatter, *chr-r-r-r-uk.*

Range: E Nepal to SW China and SE Asia.

Distribution and status: Uncommon resident in lowlands. Nominate race in extreme W Yunnan; *torquatus* in S and SE Yunnan.

Habits: Inhabits bamboo.

1096. Rusty-fronted Barwing *Actinodura egertoni*
Xiu'e banchimei   PLATE 109

Description: Medium-sized (22.5 cm) rufous-brown babbler with long tail and fine black barring on wings and tail. Distinguished from Spectacled Barwing by lack of white eye-ring and by chestnut forecrown with grey streaks. Breast rufescent. Distinguished from Streak-throated Barwing by unstreaked underparts and chestnut forehead. Race *lewisi* differs from nominate in having pronounced dark edges on very grey head and being generally darker and greyer; *ripponi* is greyer-brown on upperparts than nominate. Iris—grey-brown; bill—brown above, pale below; feet—pinkish-brown.

Voice: Constant feeble cheeping. Song is loud, sharp three-note whistle with first note accentuated and last note lower, *ti-ti-ta*.

Range: Nepal to SW China and W and N Burma.

Distribution and status: Uncommon. Nominate race is expected in SE Xizang to W of Dibang branch of Brahmaputra; *lewisi* occurs to E side of Dibang in Mishmi Hills of NE India and can be expected on China side of Mishmi Hills; and *ripponi* occurs in Yunnan to west of Nujiang (Salween) River.

Habits: Occurs in dense thickets in montane evergreen forest. Lives in small noisy parties, sometimes mixed with laughingthrushes.

1097. Spectacled Barwing *Actinodura ramsayi*   PLATE 109
Baikuang banchimei

Description: Medium-sized (24 cm) rufous-brown babbler with slight crest and bold white eye-ring. Wings and tail are finely barred black and there is a large rufous patch in wings at base of flight feathers. Underparts dull cinnamon, throat streaked blackish. Distinguished from other Chinese barwings by white eye-ring. Iris—brown; bill—grey; feet—grey.

Voice: Call is loud but mournful *tu-tui-tui-tui-tuuui* rising then falling in pitch.

Range: E Burma, N Indochina and S Yunnan.

Distribution and status: Fairly common resident in scrubby forest at moderate altitude above 450 m in S Yunnan.

Habits: Active and noisy. Erects crest when calling from tops of small bushes.

1098. Hoary-throated Barwing *Actinodura nipalensis*   PLATE 109
Wentou banchimei

Description: Medium-sized (21 cm) dark brown babbler with fine black barring on wings and longish tail. Distinguished from other barwings by crested head streaked buff. Sides of head grey, narrow eye-ring whitish, black moustachial streak. Tail with black terminal band. Underparts pale brownish-grey becoming rufescent on belly. Iris—brown; bill—dark brown; feet—pinkish-brown.

Voice: Whistled *tui whee-er*. In alarm a loud rapid *je-je* ... repeated several times.

Range: Nepal to W Arunachal Pradesh and S Tibet.

Distribution and status: Race *vinctura* is rare resident in S Xizang (Bomi) from 2300–2800 m.

Habits: Lives in small flocks in oak and rhododendron forests. Sometimes joins mixed-species flocks.

1099. Streak-throated Barwing *Actinodura waldeni*   PLATE 109
Wenxiong banchimei

Description: Medium-sized (21 cm) brown barwing. Very similar to Hoary Barwing but crest feathers pale-edged giving scaly appearance. Underparts grey streaked rufous. Race *saturiator* has rufous-brown underparts streaked buff. Iris—brownish-grey; bill—dark brown; feet—brown.

Voice: Call note is soft *chup, chup*; also a mewing note and *churr*. Song of nominate race is a loud, strident, wavering phrase ending with tremolo.

Range: Assam to W China and Burma.

Distribution and status: Race *daflaensis* is very rare resident in SE Xizang and *saturiator* occurs in NW and W Yunnan from 1525–2745 m.

Habits: As Hoary Barwing. Creeps among mossy trunks. Appears relatively tame.

1100. Streaked Barwing *Actinodura souliei*   PLATE 109
Huitou banchimei

Description: Large (22 cm) floppy-crested barwing with streaking on body feathers. Lores and forecheeks black. Crest and ear-coverts pale grey. Sides of head dark chestnut, throat reddish-chestnut. Feathers of mantle, back, rump, abdomen and vent are black, lanceolate and edged cinnamon. Wing and tail chestnut with fine black barring. Outer tail feathers broadly tipped white. Iris—brown; bill—brown; feet—pinkish.

Voice: Soft contact calls or harsh loud churring alarm notes.

Range: SW China, W Tonkin.

Distribution and status: Globally Near-threatened (Collar *et al.* 1994). Uncommon. Nominate race is resident in S Sichuan (Mt Emei) and NW Yunnan; race *griseinucha* is found in SE Yunnan.

Habits: Noisy bird of deciduous forest undergrowth at 1100–3300 m.

1101. Taiwan Barwing *Actinodura morrisoniana*   PLATE 109
[Formosan Barwing] Taiwan banchimei

Description: Large (18 cm) brown babbler with floppy crest and sides of head dark chestnut, mantle and rump grey, throat reddish-chestnut. Centre of back rufous-brown and breast olive-brown with pale streaks. Abdomen and vent are rufous-brown. Wing feathers are barred with black and tail is tipped white. Iris—brown; bill—black; feet—pinkish.

Voice: Voice is soft *jiao jiao* or in alarm a hurried, husky *jia jia jia*.

Range: Endemic to Taiwan.

Distribution and status: Common resident of deciduous forest undergrowth at 1200–3000 m in the central mountain range of Taiwan.

Habits: Active and noisy. Lives in small flocks which shuffle agilely about among branches catching small insects.

1102. Blue-winged Minla *Minla cyanouroptera*　PLATE 106
Lanchi ximei

Description: Unmistakable long-tailed (15 cm) arboreal babbler with blue wings, tail and crown. Mantle, flanks and rump buffish, throat and belly whitish, cheeks greyish. White eyebrow and eye-ring. Tail is white with black edge from below and rather long, slender and square-cut. Iris—brown; bill—black; feet—pink.

Voice: Call is loud, long, two-note whistle *see-saw* or *pi-piu* repeated endlessly, rising in pitch at end of call; also loud *swit*.

Range: Himalayas, NE India, SE Asia and S China.

Distribution and status: Race *wingatei* is a common resident at 1000–2800 m in forests of S and SW China in Yunnan, Sichuan, Guizhou, Guangxi, Hunan and Hainan.

Habits: Lives in small active flocks working through upper and lower canopies.

1103. Chestnut-tailed Minla *Minla strigula*　PLATE 106
Banhou ximei

Description: Small (17.5 cm) active tit-like babbler with erectile rufous crown, black and white or yellow scaly pattern on throat, yellowish underparts and olive upperparts. Primaries are edged orange-yellow to give colourful panel and tail is centrally rufous, tipped black, but laterally black-tipped and edged yellow. Iris—brown; bill—grey; feet—grey.

Voice: Slurred whistle *chu-u-wee, chu-u-wee* with second note falling, otherwise on rising pitch; also metallic *chew*.

Range: Himalayas, NE India, SE Asia and S China.

Distribution and status: Nominate race is found in S Xizang and race *yunnanensis* (including *castanicauda*) in SE Xizang, Yunnan and W Sichuan from 2100–3600 m.

Habits: This is a common inquisitive bird keeping to lower trees and bushes of montane broadleaf and conifer forests. Lives in flocks and joins 'bird waves'.

1104. Red-tailed Minla *Minla ignotincta* Huowei ximei　PLATE 106

Description: Male: unmistakable small (14 cm) arboreal babbler with broad white supercilium contrasting with black crown, nape and broad eye-stripe and red edges to tail and primaries. Back is olive-grey, wings otherwise black with white edges, tail centrally black and underparts white with creamy tinge. Female and juvenile have paler-edged wing feathers and edge of tail pink. Iris—grey; bill—grey; feet—grey.

Voice: Has loud plaintive 3–4 note call, repeated *chik* and various high-pitched titters. Song is loud ringing *twiyi twiyuyi*..

Range: Nepal to S China and northern SE Asia.

Distribution and status: Nominate race is resident in SE Xizang and W and NW Yunnan; *jerdoni* is resident in C and S China from 1800–3400 m.

Habits: This is a common flock-living bird often mixing with 'bird waves' working through montane broadleaf forest.

1105. Golden-breasted Fulvetta *Alcippe chrysotis*　PLATE 108
Jinxiong quemei

Description: Smallish (11 cm) colourful fulvetta. Pattern diagnostic: yellow underparts with dark throat; blackish head with grey ear coverts, and white median crown stripe extending to upper mantle. Upperparts olive-grey. Wings and tail blackish with yellow edges to flight and tail feathers. Tertails tipped white. Race *forresti* has yellow-orange underparts. Iris—hazel; bill—greyish-blue; feet—pinkish.

Voice: Continuous low buzzing twitter. Thin descending series of five high-pitched notes.

Range: Nepal to SW China, NE Burma, W Tonkin and C Annam.

Distribution and status: Uncommon resident. Race *swinhoii* occurs in S Gansu (Baishuijiang), S Shaanxi (Qinling Mts), Sichuan, NW Guangxi, Guizhou, N Guangdong (Babaoshan) and NE Yunnan; *amoena* is resident in SE Yunnan; *forresti* in W and NW Yunnan; and the nominate race in SE Xizang.

Habits: Typical flocking fulvetta of scrub and evergreen forest from 950–2600 m.

1106. Gold-fronted Fulvetta *Alcippe variegaticeps* PLATE 108
Jin'e quemei

Description: Smallish (11 cm) colourful fulvetta with black moustachial line. Forecrown golden. Midcrown streaked black and white merging with hindcrown and nape streaked buff. Conspicuous white eye patch extends to hindcrown. Throat white, rest of underparts washed grey. Wing black with white shoulder and double yellow line formed by edges of primaries. Tail grey with yellow edges. Iris—dark brown; bill—brown; feet—orange.

Voice: Noisy chattering flock alarm calls; single note call (R. Williams).

Range: Endemic to C and S China.

Distribution and status: Globally Vulnerable (Collar *et al.* 1994). Recorded only from several localities in S and SC Sichuan and Yaoshan in Guangxi where it is locally common from 700–1900 m.

Habits: Lives in pairs in forest undergrowth.

1107. Yellow-throated Fulvetta *Alcippe cinerea* [Dusky Green PLATE 108
Fulvetta AG: Tit-Babbler] Huanghou quemei

Description: Small (10 cm) fulvetta with distinctive head pattern: yellow throat and supercilium, broad black eye-line and broad black stripe on side of crown. Upperparts olive-grey; crown scaled black; underparts yellow, flanks grey. Iris—brown; bill—blackish above, pale below; feet—dull yellow.

Voice: Low *chip* or *chip-chip* call and soft twittering. Song is descending, high-pitched trill.

Range: E Himalayas to SW China, NE Burma and N Laos.

Distribution and status: Globally Near-threatened (Collar *et al.* 1994). Uncommon; restricted-range in China in SE Xizang and NE Yunnan.

Habits: Inhabits evergreen forest undergrowth.

1108. Rufous-winged Fulvetta *Alcippe castaneceps* PLATE 108
Litou quemei

Description: Medium-sized (11.5 cm) brown fulvetta. Head and wing pattern distinctive: supercilium and streaks on ear coverts white; eye-stripe behind eye and narrow moustachial stripe black; crown rufous-streaked with pale feather shafts; wing with rufous primary edges and black wing coverts. Underparts white with buffy flanks. Iris—brown; bill—horn-brown; feet—olive-brown.

Voice: Three-noted crescendo *tu-twee-twe*; high-pitched wheezy descending trill *tsi-tsi-tsi-tsi-tsi-tsirr*, quiet *chut* or *chip* contact note. Song is rich, undulating, descending warble *ti-du-di-du-di-du-di*.

Range: Nepal to SW China and SE Asia.

Distribution and status: Uncommon resident in S Xizang and W Yunnan (nominate) also SE Yunnan (*exul*).

Habits: Typical fulvetta travelling in noisy excited flocks at 1800–3000 m in undergrowth of evergreen forest. Sometimes scales into canopy.

1109. White-browed Fulvetta *Alcippe vinipectus*   PLATE 108
Baimei quemei

Description: Medium-sized (12 cm) brown boldly marked fulvetta. Pattern of head and breast diagnostic: broad white supercilium bordered black above; crown and nape greyish-brown; side of head blackish; throat and upper breast whitish with black or rufous streaks. Underparts otherwise greyish-buff. Wing has pale panel formed by silvery-grey edges to primaries. Races vary in some details: *chumbiensis* has brown rather than black line above eyebrow; *perstriata* has heavier black streaking on throat; and the nominate lacks streaking on throat. Iris—whitish; bill—pale horn; feet—greyish.

Voice: Soft, high-pitched and incessant *chip, chip. Churr* in alarm. Song is faint *chit-it-it-or-key* given with head held forward and tail flicked.

Range: Himalayas and Tibet to C and SW China, N Burma and W Tonkin.

Distribution and status: Locally common resident. Nominate race in western S Xizang; *chumbiensis* in Chumbi valley of S Xizang; *perstriata* in W and NW Yunnan; and *bieti* in N and NE Yunnan and Sichuan as far N as Wenchuan (Wolong).

Habits: Lives in active flocks in prickly oak scrub and undergrowth of subalpine forests from 2000–3700 m.

1110. Chinese Fulvetta *Alcippe striaticollis*   PLATE 108
Gaoshan quemei

Description: Medium-sized (12 cm) grey fulvetta with white eye and whitish throat boldly streaked with brown. Upperparts grey-brown slightly streaked dark on crown and mantle; underparts pale grey; lores blackish; cheeks pale brown. Wings rufous-brown; Primaries edged with white to form pale wing patch. Iris—whitish; bill—horny brown; feet—brown.

Voice: Clear, ventriloqual *tsway ahh-tsway ahh* of variable loudness.

Range: Endemic to C China and SE Xinjiang

Distribution and status: A common bird of prickly oak scrub and thickets on mountains from 2200–4300 m. Distributed from S Gansu through Sichuan to NW Yunnan and SE Xinjiang.

Habits: Lives in small flocks.

1111. Spectacled Fulvetta *Alcippe ruficapilla*
Zongtou quemei

PLATE **108**

Description: Medium-sized (11.5 cm) brown fulvetta with rufous crown. Crown bordered by black line which reaches nape. Pale supercilium indistinct. Lores dusky contrasting with white eye-ring. Throat whitish with faint streaks; rest of underparts vinaceous with whitish centre of belly. Upperparts greyish-brown grading to rufescent on rump. Wing coverts edged rufous; primaries edged pale grey giving pale wing panel. Tail brown. Race *sordidior* has slightly paler and browner crown and blacker lateral crown stripes; race *danisi* has chocolate-brown crown. Iris—brown; bill—horn above. pale below; feet—pinkish.

Voice: Not recorded.

Range: SE Tibet to SW and C China and N Indochina.

Distribution and status: Uncommon to locally common resident. Nominate in S Gansu, S Shaanxi (Qinling Mts) and Sichuan; *sordidior* in W, C and N Yunnan, SW Sichuan and W Guizhou; and birds identified as *danisi* without comparative data in SE Yunnan and SW Guizhou.

Habits: Inhabits evergreen oak forest from 1250–2500 m.

1112. Streak-throated Fulvetta *Alcippe cinereiceps*
Hetou quemei

PLATE **108**

Description: Medium-sized (12 cm) brown fulvetta with pinkish-grey throat streaked dusky. Breast centrally white with pinkish-brown to chestnut sides. Primaries edged white, then black, then rufous to give striped effect. Distinguished from Spectacled Fulvetta by greyish sides of head; lack of supercilium and eye-ring; throat and breast tinged grey, and black and white streaks on wings. Races vary in colour of crown: vinous-brown bordered laterally by grey brown line in *guttaticollis*, without brown lateral lines in *fucata* and *berliozi*; chocolate-brown without lateral lines in *manipurensis*; sooty-brown with black lateral lines in *tonkinensis*; brown with grey lateral lines in *formosana* or light ashy-brown with no lateral lines in *fessa* and nominate. Iris—yellow to pink; bill—black (male), brown (female); feet—greyish-brown.

Voice: Rattling song of 3–4 notes. Call is tit-like *cheep*.

Range: NE India, S China, Taiwan, W and N Burma and W Tonkin.

Distribution and status: Common and widespread resident. Nominate race occurs in Sichuan, W Guizhou and NE Yunnan; *manipurensis* in W Yunnan; *fessa* in Gansu, S Shaanxi (Qinling Mts) and Ningxia (Mt Liupan); *fucata* in NE Guizhou and W Hubei; *berliozi* in S Hunan (Mt Qingdong); *guttaticollis* in N Guangdong and NW Fujian (Wuyishan); and *formosana* on Taiwan.

Habits: Inhabits undergrowth of evergreen forests from 1500–3400 m, locally down to 1100 m in S and thickets of mixed and conifer forests and bamboo brakes.

Note: Formerly included Ludlow's Fulvetta (e.g. Cheng 1987).

1113. Ludlow's Fulvetta *Alcippe ludlowi*
[Himalayan Brown-headed Tit Babbler] Lude quemei

PLATE **108**

Description: Largish (12 cm) brown fulvetta with chocolate-brown head. Similar to Streak-thorated Fulvetta but throat white with dark streaks rather than grey with dark streaks. Similar to White-browed Fulvetta but lacks white supercilium and brown line above it. Sexes alike. Iris—brown; bill—dark brown with pinkish base; feet—pinkish-brown.

Voice: Rattling alarm call.

Range: Bhutan, Arunachal Pradesh.

Distribution and status: Rather common through SE Xizang.

Habits: Lives in small flocks in bamboo thickets and rhododendron forest from 2100–3355 m.

Note: Formerly included as a race of Streak-throated Fulvetta (e.g. Cheng 1987).

1114. Rufous-throated Fulvetta *Alcippe rufogularis*   PLATE **108**
Zonghou quemei

Description: Largish (13 cm) rich brown fulvetta with boldly patterned head and diagnostic rufous-chestnut gorget across lower throat. Crown and nape rufescent brown bordered by black line contrasting with long white supercilium, brown ear coverts and white chin and upper thorat. Has white eye-ring. Centre of belly white with olive flanks. Iris—light brown; bill—brown; feet—light yellow.

Voice: Sweet *chi-chu-one-two-three* with last three notes ascending scale. Musical *chip chur*.

Range: Bhutan to SW China and SE Asia.

Distribution and status: Globally Near-threatened (Collar *et al.* 1994). Uncommon and limited distribution in China. Race *stevensi* is uncommon in Xishuangbanna (S Yunnan). Nominate race probably occurs in SE Xizang W of Zangpo (Dibang) River and *collaris* may occur E of Zangpo (Dibang).

Habits: Inhabits shrub layer of evergreen forest up to 900 m. Shy and secretive.

1115. Dusky Fulvetta *Alcippe brunnea* [Gould's Fulvetta   PLATE **108**
AG: Tit-Babbler] Heding quemei

Description: Largish (13 cm) brown fulvetta with rufous-brown cap. Similar to Rufous-throated Fulvetta but lacks rufous gorget and with tawny forehead. Underparts buffy. Separated from Rufous-winged by plain brown wings. Separated from Rusty-capped Fulvetta mainly by lack of white supercilium. Iris—light brown or yellowish-red; bill—dark brown; feet—pink.

Voice: Not recorded.

Range: S and C China and Taiwan.

Distribution and status: Common resident. Nominate race on Taiwan; *weigoldi* in Sichuan; *olivacea* in S Shaanxi (Qinling Mts), W Hubei, E Sichuan (Wujiang River), N Guizhou and NE Yunnan; *superciliaris* occurs widely in SE and S China; and *arguta* is confined to Hainan.

Habits: Lives in scrub layer of evergreen and deciduous forests at 400–1830 m.

Note: Some authors include races of Rusty-capped Fulvetta. *A. dubia* in this species but range overlaps suggest this is a valid species.

1116. Rusty-capped Fulvetta *Alcippe dubia* PLATE 108
[Rufous-headed Fulvetta] Hexie quemei

Description: Large (14.5 cm) brown fulvetta with rufous forehead and olive-brown upperparts. Has black stripe on side of crown above prominent white supercilium and unstreaked buffy underparts. Distinguished from Dusky Fulvetta by cheeks and ear coverts streaked black and white; also larger size. Iris—brown; bill—dark brown; feet—pinkish.

Voice: Chattering alarm note. Song is *chee-chee-chee-chee-chee-hpwit*.

Range: E Himalayas to Burma, N Indochina and SW China.

Distribution and status: Uncommon resident. Race *intermedia* occurs in Yunnan W of Nujiang (Salween); *genestieri* found in rest of Yunnan, S Sichuan, Guizhou, W Hunan and SW Guangxi.

Habits: Inhabits forest understorey.

Note: Some authors include this species within Dusky Fulvetta (e.g. Meyer de Schauensee 1984), but their sympatric range suggest this is a valid species.

1117. Brown-cheeked Fulvetta *Alcippe poioicephala* PLATE 108
Huilian quemei

Description: Large (16 cm) brown fulvetta with grey crown and nape; blackish lateral crown stripes and buffy underparts. Separated from Grey-cheeked and Nepal Fulvettas by warm brown ear coverts, lack of white eye-ring and larger size. Iris—light brown; bill—blue-grey; feel—light grey.

Voice: Song is distinctive, sweet, slurred whistling, *joey joey dii-wiu*, the penultimate note highest. Also various squeaks, bubbly buzzing and churring.

Range: India to SW China and SE Asia.

Distribution and status: Locally common in restricted range. Race *haringtoniae* in SW Yunnan and *alearis* in Xishuangbanna (S Yunnan) and SE Yunnan.

Habits: Mixes with other babblers working through low canopy in lowland tropical forests up to 950 m.

1118. Grey-cheeked Fulvetta *Alcippe morrisonia* PLATE 108
Huikuang quemei

Description: Largish (14 cm) noisy inquisitive flock-living fulvetta. Upperparts brown, head grey and underparts buff. Has conspicuous white eye-ring. Dark lateral crown stripes vary from prominent to barely visible. Distinguished from Brown-cheeked Fulvetta by greyish ear coverts and white eye-ring. Iris—red; bill—grey; feet—pinkish.

Voice: Song is a sweet whistling, *ji-ju ji-ju*, usually followed by undulating drawn-out squeaking note. Call is agitated churring when disturbed. Readily attracted by 'pishing'.

Range: China, Taiwan, Hainan, NE and E Burma and N Indochina.

Distribution and status: Common resident at moderate altitudes. Race *yunnanesis* in SE Xizang and NW Yunnan, *fraterculus* in SW Yunnan, *schaefferi* in SE Yunnan, *rufescentior* on Hainan, *morrisoniana* on Taiwan, *hueti* in Guangdong to Anhui and *davidi* in W Hubei and Sichuan.

Habits: Often mixed with other species in 'bird waves'. These birds are aggressive in mobbing small owls and other raptors.

1119. Nepal Fulvetta *Alcippe nipalensis* PLATE 108
Baikuang quemei

Description: Largish (13.5 cm) brown fulvetta with grey crown, broad white eye-ring and black supercilium. Separated from Grey-cheeked Fulvetta (where ranges overlap) by brownish-tinged crown and nape, more pronounced lateral crown stripes, more rufescent upperparts, more pronounced white eye-ring, whitish centre to throat and abdomen and less distinctly buff remainder of underparts. Iris—greyish-brown; bill—horn; feet—plumbeous-brown.

Voice: Constant twittering; metallic *chit*; shrill whinnying note, *dzi-dzi-dzi-dzi-dzi* and *p-p-p-p-jet*.

Range: Nepal to NE India and W and N Burma.

Distribution and status: Locally common in restricted range. Race *commoda* is found in SE Xizang.

Habits: Lives in restless flocks in hill and montane forests up to 2200 m. Mixes with other species in flocks.

1120. Rufous-backed Sibia *Heterophasia annectens* PLATE 109
Libei qimei

Description: Small (19 cm) sibia with black head, white throat and breast and rufous-chestnut back and uppertail coverts. Long graduated tail is black with white tips. Wings black with white edges to tertails and white edges to flight feathers. Flanks and undertail coverts buffy and nape and upper mantle black with white streaks. Iris—light brown; bill—dark with yellow base to lower mandible; feet—yellow.

Voice: Harsh alarm chatter and song of 3–4 whistled notes *chip, chu chu ii*, the last two notes dropping in pitch; sometimes introduced by a grace note.

Range: E Nepal to SW China and SE Asia.

Distribution and status: Uncommon to locally common from 600–1525 m, to 2300 m in NE India and W Yunnan (nominate) and in Xishuangbanna of Yunnan (*mixta*). May also occur in SE Xizang.

Habits: Active bird of canopy in evergreen submontane forest and adjacent scrub.

1121. Rufous Sibia *Heterophasia capistrata* Hetou qimei PLATE 109

Description: Large (23.5 cm) elegant, rufous-brown sibia with black, slightly crested head. Tail has subterminal black band and feathers are basally rufous-cinnamon. Wings appear mostly grey, with blackish secondaries and primary coverts, tip is grey. Iris—reddish-brown; bill—black; feet—pinkish-brown.

Voice: Song is flute-like *tee-dee-dee-dee-dee-o-lu* with first five notes on monotone, sixth note lowest and last note in between. Alarm is harsh *chrai-chrai-chrai-chrai-chrai*. Call is rapid *chi-chi*.

Range: Himalayas to NE India.

Distribution and status: Rare resident in Chumbi valley, S Xizang and probably SE Xizang (*bayleyi*) and Nyalam and Qomolangma region of S Xizang (*nigriceps*).
Habits: Occupies mixed forest zone from 2200–2600 m. Lives in pairs or small noisy parties. Strictly arboreal and very lively, searching mossy branches for food. Often joins mixed-species flocks.

1122. Grey Sibia *Heterophasia gracilis* Hui qimei PLATE 109

Description: Medium-sized (23.5 cm) greyish sibia with dark grey crown and head sides, blacker face and whitish throat and breast. Upperparts vinous-grey. Tail has black subterminal band and edge, and paler grey tip; wings blackish with mostly pale grey tertials and narrow pale edges to primaries. Iris—red; bill—black; feet—brown.
Voice: Melancholy song is shrill series of flute-like descending whistled notes; similar to Black-headed Sibia. Alarm is harsh *trrrit trrrit*. Contact call is squeaky *witwit-witarit* or soft sibilant *ti-ew*.
Range: NE India to SW China and W and N Burma.
Distribution and status: Very local resident in Yunnan W of Nujiang (Salween).
Habits: Inhabits montane evergreen forests from 900–2300 m. Active and restless; often in mixed-species flocks.

1123. Black-headed Sibia *Heterophasia melanoleuca* PLATE 109
Heiding qimei

Description: Long-tailed (24 cm) grey sibia with black head, tail and wings and brownish-tinged mantle. Crown is glossy. Tail is tipped grey on central feathers and white on outer feathers. Throat and central underparts are white but flanks are smoky-grey. Iris—brown; bill—black; feet—grey.
Voice: Call is five-note song with three notes on same pitch and last two lower.
Range: Burma, W and C China, N Thailand and Indochina.
Distribution and status: Race *desgodinsi* is common in montane forests of SC and S China above 1200 m.
Habits: Similar to a malkoha or squirrel, creeping about among mossy epiphytes in rather skulking and jerky manner.

1124. White-eared Sibia *Heterophasia auricularis* PLATE 109
Bai'er qimei

Description: Unmistakable, medium-sized (23 cm) arboreal babbler with black crown and unique white lores, eye-ring and broad white eye-stripe, prolonged backwards and upward-ending in long spreading filamentous plumes. Throat, breast and upper back grey; rest of underparts pinkish-cinnamon, lower back and rump rufous. Tail black with central feathers tipped whitish. Iris—brown; bill—black; feet—pink.
Voice: Call is resonant repeated *fei fei fei..* rising at the end or rattling *de de de de*.
Range: Endemic to Taiwan.
Distribution and status: Common in oak forest of Taiwan from 1200–3000 m, locally down to 200 m in winter.
Habits: Sometimes gathers in small parties to feed in fruiting and flowering trees. Active and not shy.

1125. Beautiful Sibia *Heterophasia pulchella* Lise qimei PLATE 109

Description: Medium-sized (23.5 cm) bluish-grey sibia with broad black eye-stripe. Upperparts and underparts bluish-grey contrasting with brown tertials and basal two-thirds of central tail feathers. Tail has black subterminal band. Distinguished from Grey Sibia by blue grey underparts and crown. Iris—red or brown; bill—black; feet—brown.
Voice: Usually silent but sometimes very vocal with great variation. One call sounds like jingling of a bunch of keys. Song is *ti-ti-titi-tu-ti*, descending toward end. Shriller and faster than Grey Sibia. About six notes with drop in pitch after each pair and after first note. Harsh *churr* in alarm.
Range: NE India to SW China and NE Burma.
Distribution and status: Uncommon resident in SE Xizang and NW Yunnan.
Habits: Lives in mossy forests from 1650–2745 m, down to 1050 m in winter.

1126. Long-tailed Sibia *Heterophasia picaoides* PLATE 109
Changwei qimei

Description: Large (33.5 cm) grey and white arboreal babbler with very long pointed tail. Plumage dull grey with darker crown, whitish vent and white wing patch conspicuous in flight. Tail feathers are tipped paler grey. Iris—brown; bill—black; feet—black.
Voice: A noisy bird constantly uttering shrill twittering calls *tsip-tsip-tsip-tsip*, interspersed with churring.
Range: Himalayas, S China, SE Asia, Malay Peninsula and Sumatra.
Distribution and status: Nominate race is resident in SE Xizang and NW Yunnan; *cana* occurs in S and SW Yunnan. Common in mountain forests between 600 and c. 2500 m.
Habits: Lives in small flocks, keeping to the tops of taller trees. Flies powerfully, calling.

1127. Striated Yuhina *Yuhina castaniceps* Li'er fengmei PLATE 110

Description: Medium-sized (13 cm) yuhina with greyish upperparts, whitish underparts and characteristics chestnut cheeks extending as nuchal collar. Has short crest and upperparts finely streaked with white feather shafts. Tail is dark brownish-grey with white edge. Iris—brown; bill—reddish-brown with dark tip; feet—pink.
Voice: Continuous *ser-weet ser-weet*.
Range: NE Indian subcontinent, China and SE Asia.
Distribution and status: Common from 400–2200 m in Himalayas and through S and E China. Race *torqueola* is resident in C, S and SE China; *plumeiceps* occurs in NW and W Yunnan and SE Xizang.
Habits: This is an active bird generally found in noisy flocks working for insects through the lower forest canopy.

1128. White-naped Yuhina *Yuhina bakeri* PLATE 110
Baixiang fengmei

Description: Medium-sized (13 cm) yuhina with thick crest; white chin and white nuchal patch. General plumage olive-brown with rufescent wash to vent and rufous-chestnut remainder of crown and nape. Iris—brown; bill—brown; feet—pinkish-brown.

Voice: Shrill *chip* and a soft chatter. Ringing *zee zee* and high-pitched alarm call.
Range: E Himalayas and N Burma.
Distribution and status: Locally common but very restricted resident. Recorded in Yunnan to W of Nujiang (Salween) and known from SE Xizang foothills to Mishmi hills.
Habits: Inhabits secondary and primary evergreen oak forests from 450–2400 m. Lives in parties.

1129. Whiskered Yuhina *Yuhina flavicollis* [Yellow-naped Yuhina] PLATE 110
Huangjing fengmei

Description: Medium-sized (13 cm) thick-crested yuhina with white eye-ring and buffy-rufous collar. Rear of head grey separated from white throat by black moustachial line. Upperparts uniform brown; sides and flanks tawny-brown with diagnostic white streaks on sides. Iris—brown; bill—dark brown above, light brown below; feet—yellowish-brown.
Voice: Thin squeaky *swii-swii-swii* and metallic ringing note. Constant twittering in flocks. Song rendered as *twe-tyurwi-tyawi-tyawa*.
Range: Himalayas to W China, W, N and E Burma and N Indochina.
Distribution and status: Uncommon montane resident. Nominate race is found in S and SE Xizang; *rouxi* is found in SW, S and SE Yunnan.
Habits: Inhabits evergreen forests from 1500–2285 m. Lives in noisy flocks.

1130. Stripe-throated Yuhina *Yuhina gularis* PLATE 110
Wenhou fengmei

Description: Largish (15 cm) dull brown yuhina with prominent crest, black streaking on pinkish-buff throat and orange-rufous streak in black wing. Rest of underparts are dull rufous-buff. Mt Emei race is paler with rufous-brown crest. Iris—brown; bill—dark above reddish below; feet—orange.
Voice: The call is a distinct, descending, nasal mewing *queee*. Flocks make constant buzzing calls.
Range: Himalayas, Assam to SW China. Common at 1100–3050 m, locally down to 850 m in winter in montane broadleaf forest.
Distribution and status: Nominate race is resident in S and SE Xizang, W and S Yunnan; *omeiensis* is found from NW Yunnan to SW Sichuan (Mt Emei)
Habits: Flocks mix with other species in 'bird waves' and work busily through the crowns of flowering trees.

1131. White-collared Yuhina *Yuhina diademata* PLATE 110
Bailing fengmei

Description: Large (17.5 cm) ashy-brown yuhina with rather floppy crest and large white nuchal patch linking with broad white eye-ring and white rear supercilium. Chin, nares and lores black. Flight feathers black with whitish edges. Lower belly white. Iris—reddish; bill—blackish; feet—pink.
Voice: Weak cheeping note like a white-eye.
Range: W China, NE Burma and NW Tonkin.

Distribution and status: Very common montane resident in S Gansu, S Shaanxi (Qinling Mts), Sichuan, W Hubei, Guizhou and Yunnan.

Habits: Lives in pairs or small noisy flocks in scrub from 1100–3600 m, down to 800 m in winter.

1132. Rufous-vented Yuhina *Yuhina occipitalis* PLATE 110
Zonggang fengmei

Description: Medium-sized (13 cm) brown yuhina with prominent crest, frontally grey posteriorly orange-rufous. Has greyish-olive upper mantle and black moustachial streak. Underparts are pinkish-buff with rufous undertail coverts. Has white eye-ring. Iris—brown; bill—pinkish; feet—orange-pink.

Voice: Gives a short buzzing call note. Alarm is *z-e-e . . . zit*. Song is high-pithced *zee-zu-drrrrr, tsip-che-e-e-e-e*.

Range: Nepal to N Burma and SW China.

Distribution and status: Nominate race is resident in S and SE Xizang; *obscurior* is resident in Yunnan and W Sichuan. Common resident in montane mossy forest from 1800–3700 m, down to 1350 m in winter.

Habits: Flocks generally mixing with other species and moving about actively in 'bird waves'.

1133. Taiwan Yuhina *Yuhina brunneiceps* [Formosan Yuhina] PLATE 110
Hetou fengmei

Description: Medium-sized (13 cm) yuhina with crested crown chestnut edged black and white sides of head. Black moustachial extends into black line around ear coverts to back of eye. Throat is white, finely streaked black. Rest of underparts whitish with grey wash on breast and chestnut mottling on flanks. Back wings and tail olive-grey. Iris—red; bill—black; feet—dully yellow.

Voice: Mellow, sweet call *too, mee, jeeoo*.

Range: Endemic to Taiwan.

Distribution and status: A common bird of temperate forests at 1000–2800 m.

Habits: Social, active. Keeps to lower forest and often joins other species in mixed flocks. Not shy.

1134. Black-chinned Yuhina *Yuhina nigrimenta* PLATE 110
Hei'e fengmei

Description: Small (11 cm) greyish yuhina with short crest, grey head, olive-grey upperparts and whitish underparts. Has black forehead, lores and diagnostic black upper chin. Iris—brown; bill—black with red below; feet—orange.

Voice: Emits a constant, squeaky, buzzing chatter. Calls include high *de-de-de-de*. Song is soft *whee-to-whee-de-der-n-whee-yer*.

Range: Himalayas, NE India, N Burma, S China and Indochina.

Distribution and status: Common montane resident. Race *intermedia* from SE Xizang to Sichuan, Hubei and SW China; *pallida* in SE China.

Habits: Active and gregarious, mostly in the canopy in montane forest, forest clearings and secondary scrub from 530–2300 m in summer but descending as low as 300 m in winter. Sometimes mixes with other species in large flocks.

1135. White-bellied Yuhina *Yuhina zantholeuca*   PLATE 110
[Erpornis] Baifu fengmei

Description: Small (13 cm) olive-green babbler with greyish-white underparts, yellow undertail coverts and prominent crest. Distinguished from similarly-coloured warblers by its crest. Iris—brown; bill—horn; feet—horn.

Voice: Metallic *chit*, nasal *na-na* and descending high-pitched trill song *si-i-i-i-i-i*.

Range: Himalayas, S China, SE Asia, Sumatra and Borneo.

Distribution and status: Race *griseiloris* is resident in SE and S China and Taiwan; nominate SE Xizang and Yunnan; and *tyrranula* on Hainan. Common in forests at 250–1600 m but to 2000 m and perhaps higher in Taiwan. Altitudinal migrant.

Habits: Lives in flocks and feeds in mid-to upper storeys often mixing with warblers and other species.

Note: Formerly sometimes placed in genus *Alcippe*.

1136. Fire-tailed Myzornis *Myzornis pyrrhoura*   PLATE 110
Huowei lümei

Description: Unmistakable small (12.5 cm) bright green babbler with red fringes to outer tail feathers and orange-red wing panel. Has blackish centres to crown feathers. Primaries, lores and orbital area black. Breast has reddish wash; undertail cinnamon. Female is slightly duller than male and lacks reddish wash on breast. Iris—red or brown; bill—black; feet—yellowish-brown.

Voice: Normally silent. High-pitched *tsi-tsit* contact call. A *trrrr-trrr-trrr* preceded by high-pitched squeak. Repeated *tzip* in alarm.

Range: Nepal to SW China and NE Burma.

Distribution and status: Rare resident recorded from SE Xizang, W and NW Yunnan.

Habits: Solitary or in small parties. Inhabits montane forests from 3000–3660 m, but to 2000 m in winter. Joins flocks of warblers and sunbirds. Probes rhododendrons for nectar.

5. PARROTBILLS

1137. Bearded Parrotbill *Panurus biarmicus* Wenxuque   PLATE 111

Description: Small (17 cm) slender, cinnamon parrotbill with grey head, fine bill and in male, diagnostic, vertical tapering black moustachial stripe. Body is cinnamon-buff with very long tail and black and white pattern on wings. Female lacks black on head but juvenile has black lores. Iris—pale brown; bill—orange; feet—black.

Voice: Call is a twanging lively *pching* or twittering song.

Range: Palearctic.

Distribution and status: Locally common in suitable reedy habitat in northern China.

Habits: Lives in active flocks clambering and hopping about in reed beds and flying with weak rapid wing beats, sometimes flocks high into sky before diving back into reeds.

1138. Great Parrotbill *Conostoma oemodium* [Giant Parrotbill]   PLATE 111
Hongzui yaque

Description: Very large (28 cm) brown parrotbill with strong, rather conical, yellow bill and greyish-white forehead distinctive. Lores dark brown. Underparts: pale grey-brown. Iris—yellow; bill—yellow; feet—greenish-yellow.
Voice: Clear musical *wheou, wheou* . . .; grating croak or churring.
Range: Himalayas to SW China and NE Burma.
Distribution and status: Uncommon resident from 2000–3300 m, but to 1400 m in winter, in S Gansu (Baishuijiang), S Shaanxi (Qinling Mts), Sichuan, NW Yunnan and S Xizang.
Habits: Inhabits bamboo and rhododendron thickets in subalpine forest.

1139. Three-toed Parrotbill *Paradoxornis paradoxus*   PLATE 111
Sanzhi yaque

Description: Largish (23 cm) olive-grey parrotbill with floppy crest and conspicuous white eye-ring. Chin, lores and broad eyebrow dark brown. Edge of primaries whitish forming pale patch on closed wing. Iris—whitish; bill—orange-yellow; feet—brown.
Voice: Song is quite high-pitched, plaintive *tuwi-tui* or *tuii-tew*, repeated after intervals (sometimes single note). Also weaker *tidu-tui-tui*. Subsong incorporates low chuntering with high *tuwii, tuwii-tu* and *tuuu* notes. Alarm calls are typically harsh low *chah* and *chao* notes.
Range: Endemic to C China.
Distribution and status: Found in Mt Taibai and Qinling Mts of S Shaanxi also in the Min and Qionglai Mts of Sichuan and Gansu (Baishujiang). Uncommon from 1500–3660 m.
Habits: Small flocks live in bamboo thickets in broadleaf and conifer forests.

1140. Brown Parrotbill *Paradoxornis unicolor*   PLATE 111
[Himalayan Brown Parrotbill, AG:Crowtit] He Yaque

Description: Large (20 cm) brown parrotbill with chunky yellow bill and long black lateral crown stripe. Underparts grey. Distinguished from Three-toed Parrotbill by grey cheeks and more uniform wing; also lack of eye-ring. Iris—grey; bill—yellow; feet—greenish-grey.
Voice: *Chirrup* call and alarm churrs.
Range: Nepal to SW China and NE Burma.
Distribution and status: Uncommon resident from 1850–3600 m (lower in winter). Found in SE Xizang, Sichuan and W and NW Yunnan.
Habits: Lives in bamboo thickets in small noisy parties, sometimes mixed with other parrotbills.

1141. Grey-headed Parrotbill *Paradoxornis gularis*   PLATE 111
Huitou yaque

Description: Large (18 cm) brown parrotbill with diagnostic grey head and orange bill. Has long black lateral crown stripe and black centre of throat. Underparts otherwise white. Iris—reddish-brown; bill—orange; feet—grey.

Voice: Single *jieu* notes or rapidly delivered phrase of about four ringing notes, *chiu-chiu-chiu-chiu*, with harsh chattering.

Range: Sikkim to S China and SE Asia.

Distribution and status: Common resident, nominate race can be expected in SE Xizang; *fokiensis* in China S of Changjiang plus Sichuan; and *hainanus* on Hainan.

Habits: Lives in forest canopy and undergrowth, bamboo and scrub from 450–1850 m. Forms flocks.

[1142. Black-breasted Parrotbill *Paradoxornis flavirostris*   PLATE 111
[Gould's/Black-throated Parrotbill AG:Crowtit] Banxiong yaque

Description: Large (19 cm) brown parrotbill with chunky yellow bill and diagnostic black breast band and black chin and patch behind ear coverts. Side of face and throat white with black scaling. Underparts pinkish-buff. Iris—brown; bill—yellow; feet—grey.

Voice: Striking whistle rendered as *phew, phew, phew, phuit* rising in tone and volume. Bleating mews and warble of three notes.

Range: Nepal to NE India and W Burma.

Distribution and status: Globally Vulnerable (Collar *et al.* 1994). Rare up to 1800 m in adjacent NE India, may occur in SE Xizang.

Habits: Found in scrub, tall grasses and bamboo. Lives in small parties. Rather shy.

Note: Some authors include Spot-breasted Parrotbill within this species (e.g. Cheng 1987) but differences in voice and sympatric ranges suggest these are valid species.

1143. Spot-breasted Parrotbill *Paradoxornis guttaticollis*   PLATE 111
Dianxiong yaque

Description: Distinctive large (18 cm) parrotbill with diagnostic dark inverted chevrons on breast. Crown and nape are rufous and there is a conspicuous black patch on rear ear coverts. Upperparts otherwise dull rufescent brown and underparts buffy. Iris—brown; bill—orange-yellow; feet—bluish-grey.

Voice: Rapid, loud, mellow whistle of 8–10 *whit* notes on the same pitch, also group chittering and sibilant *chut-chut-chut.*

Range: NE India, Burma, SW China and N Indochina.

Distribution and status: Nominate race is common resident at moderate to high altitudes in C, SW, and SE China; dubious race *gonshanensis* occurs in W Yunnan.

Habits: Inhabits scrub, secondary growth and tall grass.

1144. Spectacled Parrotbill *Paradoxornis conspicillatus*   PLATE 112
Baikuang yaque

Description: Small (14 cm) parrotbill with rufous-chestnut crown and nape and conspicuous white eye-ring. Upperparts olive-brown; underparts pinkish-brown with faint streaks on throat. Hubei race *rocki* is paler with larger bill. Iris—brown; bill—yellow; feet—yellowish.

Voice: Twangy, high-pitched *triiih-triiih-triiih-triiih* … and shorter *triit* notes (C. Robson).

Range: Endemic to C China.

Distribution and status: Nominate race is found from Qinling Mts of Shaanxi through E Sichuan and S Gansu to Qinghai Lake area. Race *rocki* is found in W Hubei. Not common in mountain areas from 1360–2900 m, to 1000 m in winter.

Habits: Small active flocks keep low in bamboo layer of montane forests.

1145. Vinous-throated Parrotbill *Paradoxornis webbianus*   PLATE 112
Zongtou yaque

Description: Tiny (12 cm) pinkish-brown parrotbill with small tit-like bill. Crown and wings are rufous-chestnut. Slight streaking on throat. Iris is brown and eye-ring inconspicuous. Wings edged rufous in some races. Iris—brown; bill—grey or brown with paler tip; feet—greyish-pink.

Voice: Song is high-pitched *tw'i-tu tititi* and *tw'i-tu tiutiutiutiu*, etc, repeated after short intervals and interspersed with short *twit* notes. Sometimes just *tiutiutiutiu*; call is tiny continuous chirpy rattling (C. Robson).

Range: China, Korea, Taiwan and N Tonkin.

Distribution and status: Common resident in scrub, thickets and forest edge at moderate altitudes. Seven races recognised: *mantschuricus* in NE China, *fulvicauda* in Hebei, Beijing and Henan, nominate in E China in Shanghai area, *bulomachus* on Taiwan, *ganluoensis* in mid-Sichuan, *stresemanni* in Guizhou and E Yunnan and *suffusus* over most of C, E, S and SE China.

Habits: Lives in active flocks, generally in undergrowth and low bushes. Easily attracted to soft 'pishing' calls.

Note: Some authors include races of Brown-winged Parrotbill and/or Ashy-throated Parrotbill within this species (e.g. Cheng, 1987 Cheng 1994).

1146. Brown-winged Parrotbill *Paradoxornis brunneus*   PLATE 112
Hechi yaque

Description: Small (12.5 cm) brown parrotbill with small bill. Distinguished from Vinous-thorated by darker and more chestnut crown to upper mantle and head sides, brown wings, much more vinous throat and upper breast with darker chestnut streaks and largely brownish-yellow bill. Iris—brown; bill—brownish-yellow; feet—pink.

Voice: Continuous twittering.

Range: NE Burma and SW China.

Distribution and status: Globally Near-threatened (Collar *et al.* 1994). Fairly common resident from 1830–2800 m. Nominate race in W Yunnan; *ricketti* in NW Yunnan (Jinsha and Lijiang valleys) and SW Sichuan (Muli, Xichang); and *styani* in NW Yunnan (Dali).

Habits: Forms large flocks of 30–50 birds. Lives in bamboo thickets and tall grasses and scrub.

Note: Sometimes treated as a race of Vinous-throated Parrotbill (e.g. Cheng 1987).

1147. Ashy-throated Parrotbill *Paradoxornis alphonsianus*   PLATE 112
Huihou yaque

Description: Small (12.5 cm) greyish-brown parrotbill with small pink bill. Differs from Vinous-throated Parrotbill in brownish-grey side of head and neck. Throat and breast indistinctly streaked grey. Iris—brown; bill—pink; feet—pink.
Voice: Soft twitters.
Range: SW China and NW Tonkin.
Distribution and status: Locally common resident on mountains from 320–1800 m and perhaps higher locally. Nominate race in N and W Sichuan and *yunnanensis* in S and SE Yunnan.
Habits: As Vinous-throated Parrotbill.
Note: Sometimes treated as race of Vinous-throated Parrotbill (e.g. Cheng 1987).

1148. Grey-hooded Parrotbill *Paradoxornis zappeyi*   PLATE 112
Anse yaque

Description: Small (13 cm) brown parrotbill with crested grey head and conspicuous white eye-ring. Grey crown has slight bushy crest. Upperparts rufous-brown with darker tertials and central tail feathers. Throat and breast pale grey; abdomen pinkish-brown. Iris—brown; bill—yellow; feet—greyish.
Voice: Call recorded as *shh . . . shh . . . shh.*
Range: Endemic to C China.
Distribution and status: Globally Vulnerable (Collar *et al.* 1994). Confined to Wushan and Mt Emei and Dafending in SW Sichuan and extreme W Guizhou, where locally common from 2350–3200 m.
Habits: Lives in small flocks in mountain bamboo understorey.

1149. Rusty-throated Parrotbill *Paradoxornis przewalskii*   PLATE 112
Huiguan yaque

Description: Small (13 cm) parrotbill with grey crown and nape and contrasting reddish-brown forehead, lores and eyebrow becoming posteriorly blackish. Upperparts greyish-olive; face, throat and upper breast cinnamon; rest of underparts pale brown with greyish flanks. Wings olive with rufous patch; tail olive-grey with bright edges. Iris—brown; bill—yellow; feet—flesh-coloured.
Voice: Contact call is short rattles, interspersed with thin high notes, *trr-trr-trr-trr . . . tsit tsit tsit . . . trr-trr-trr . . . tsit tsit tsit-it . . .*, etc (C. Robson).
Range: Endemic to NC China.
Distribution and status: Globally Vulnerable (Collar *et al.* 1994). Confined to SE Qinghai through S Gansu to Songpan area of NW Sichuan. A poorly-known and uncommon bird at 2440–3050 m.
Habits: Small active flocks wash through bamboos and grasses in open larch forests and montane scrub.

1150. Fulvous Parrotbill *Paradoxornis fulvifrons*   PLATE 112
Huang'e yaque

Description: Small (12 cm) rufous-brown parrotbill. Head with long dark greyish lateral crown stripe and boldly patterned wing with rufous panel contrasting with white-edged primaries. Tail long, mustard brown with rufous feather edges. Variable amount of white on side of neck; extending over side of face in race *albifacies*. Distinguished from Black-throated and Golden Parrotbills by lack of dark throat. Iris— red-brown; bill—horny-pink; feet—brown to plumbeous.

Voice: Continuous twittering and faint mouse-like *cheep* notes.

Range: Nepal to SW China and NE Burma.

Distribution and status: Uncommon resident. Race *chayulensis* occurs in S and SE Xizang; *albifacies* in W and NW Yunnan and SW Sichuan; and *cyanophrys* in S Gansu, S Shaanxi (Qinling Mts) and Sichuan.

Habits: Occupies bamboo thickets in mixed woodland and spruce or juniper forests from 1700–3500 m. Lives in large parties of 20–30 birds.

1151. Black-throated Parrotbill *Paradoxornis nipalensis* PLATE 112
Cheng'e yaque

Description: Pretty little (11.5 cm) parrotbill with rufous crown, grey cheeks and whitish underparts. Has distinctive black throat and upper breast and broad black eyebrow. Malar area is white, flanks are cinnamon. Back is yellowish-brown and tail is rufous-brown but wings are black with white edges and conspicuous rufous panel. Race *crocotius* has rufous ear coverts, greyish lower throat and whitish breast. Distinguished from Black-throated and Black-browed Tits by thick bill. Iris—brown; bill—pinkish-grey; feet—pink.

Voice: Calls are plaintive bleats and churring notes.

Range: Himalayas, NE Indian subcontinent, Burma, N Indochina and SW China.

Distribution and status: Uncommon in broadleaf montane forest from 1800–2745 m. Race *poliotis* is resident in SE Xizang and W and NW Yunnan; *crocotius* is resident in SE Xizang to W of Dibang River.

Habits: Lives in flocks in mid-storey, undergrowth and bamboo.

Notes: Some authors include races of Golden Parrotbill within this species (e.g. Cheng 1987).

1152. Golden Parrotbill *Paradoxornis verreauxi* PLATE 112
Jinse yaque

Description: Small (11.5 cm) ochraceous yellow parrotbill with black throat; orange crown, wing panel and edges to tail feathers. Race *morrisonianus* is greyer than nominate and lacks narow black line above short white supercilium. Race *craddocki* has orange-brown upperparts shaded olive-brown on nape and back. Similar to Black throated Parrotbill but more yellow and with white supercilium. Iris—dark brown; bill—grey above pinkish below; feet—pinkish.

Voice: High-pitched twitters; cheeps; churrs and purring chatter in alarm; squeaky trill as Black-throated Parrotbill.

Range: C and SE China, Taiwan, N Indochina and E Burma.

Distribution and status: Locally common in scrub and bamboo thickets from 1000–3050 m, locally to 330 m in winter. Nominate race in S Shaanxi (Qinling Mts), Hubei, Sichuan and NE Yunnan; *craddocki* in N Guangxi (Yaoshan) and probably SE

Yunnan (Jinping); *pallidus* in E Guangxi, S Hunan (Manshan), N Guangdong (Babaoshan) and N Fujian (Wuyishan); and *morrisonianus* on Taiwan.
Habits: Lives in small flocks in montane bamboo thickets in evergreen forest.
Note: Included by some authors within Black-throated Parrotbill (e.g. Cheng 1987, 1994).

1153. Short-tailed Parrotbill *Paradoxornis davidianus* PLATE 112
Duanwei yaque

Description: Tiny (10 cm) brown parrotbill with short rufous-edged tail and chestnut head. Race *thompsoni* has deeper colours; grey mantle and back and lacks white flecks on black chin and throat. Race *tonkinensis* like *thompsoni* but black throat grades gradually into grey of breast. Iris—brown; bill—pinkish; feet—pinkish.
Voice: Soft twitters.
Range: S and SE China, E Burma and N Indochina.
Distribution and status: Globally Vulnerable (Collar *et al.* 1994). Range poorly known. Nominate race resident in lowlands from 100–1830 m. Not common on Wuyishan (NW Fujian), also recorded S Hunan (Manshan) and will certainly be found on other hills in S China. Probably resident in SW Yunnan (*thompsoni*) and SE Yunnan (*tonkinensis*).
Habits: Lives in bamboo, often in flocks.

1154. Lesser Rufous-headed Parrotbill *Paradoxornis* PLATE 111
atrosuperciliaris [Black-browed Parrotbill] Heimei yaque

Description: Medium-sized (15 cm) brown parrotbill with creamy underparts and rufous head with distinctive short, black eyebrow. Iris—red-brown; bill—grey tipped white; feet—bluish-grey.
Voice: Distinctive, wheezy call note like twang of guitar. Loud chittering in alarm and mewing note.
Range: Sikkim to SW China, Burma and Indochina.
Distribution and status: Globally Near-threatened (Collar *et al.* 1994). Nominate race is rare resident up to 2200 m in W and NW Yunnan; also expected in SE Xizang.
Habits: Lives in small flocks in bamboo.

1155. Greater Rufous-headed Parrotbill *Paradoxornis ruficeps* PLATE 111
Hongtou yaque

Description: Largish (19 cm) brown parrotbill with rufous head and whitish underparts. Very similar to Black-browed Parrotbill but larger; lacks black eyebrow and has plumbeous-grey rather than pale bluish-pink skin on lores and around eye. Iris—reddish-brown; bill—orange to dark with grey tip and lower bill; feet—bluish-grey.
Voice: Characteristic, squirrel-like chittering, interspersed with slow double notes *tee-ur*. Constant *chir-chirrup* notes when feeding in groups and plaintive mewing call when separated. Audible bill snaps.
Range: NE Indian subcontinent to SW China, Burma and N Indochina.

Distribution and status: Globally Near-threatened (Collar *et al.* 1994). Nominate race is rare resident in SE Xizang and extreme W Yunnan from 900–1675 m.
Habits: Lives in small flocks in bamboo, sometimes scrub and tall grass. Sometimes mixes with other species. Often perches head down to feed like tit.

1156. Reed Parrotbill *Paradoxornis heudei* PLATE 111
Zhendan yaque

Description: Medium-sized (18 cm) parrotbill with heavy hooked yellow bill and prominent black lateral brow stripe. Forehead, crown and nape grey. Black brow stripe lined cinnamon above and edged white below. Mantle cinnamon, usually streaked with black; lower back cinnamon. There is a narrow white eye-ring. Central tail feathers sandy-brown otherwise black with white tips. Chin, throat and centre of abdomen whitish; flanks cinnamon. Wing: rich cinnamon on scapulars, paler on flight feathers with blackish tertials. Iris—reddish-brown; bill—greyish-yellow; feet—pinkish-yellow.
Voice: Not recorded.
Range: Endemic to E and NE China to SE Siberia (Ussuriland).
Distribution and status: Globally Near-threatened (Collar *et al.* 1994). Confined to reedbeds of lower Heilongjiang and Liaoning (*palivanovi*) and Changjiang valley and Jiangsu coastal reedbeds (*heudei*). Its habitat has largely been destroyed for agricultural purposes.
Habits: Lives in small active flocks in reedbeds.

SYLVIA WARBLERS—Tribe: Sylviini

Small, dull-coloured, insect-eating passerines with slender bills. Resemble true warblers in which family they were formerly classed. Feed on insects in summer months but also take berries in autumn and winter. There are five species in China

1157. Greater Whitethroat *Sylvia communis* [Common/ PLATE 100
Grey Whitethroat] Hui [baihou] linying

Description: Medium-sized (14 cm) colourful warbler with greyish-brown upperparts; rufous-brown wing panel created by rufous edges to greater coverts, secondaries and tertials; puffy white throat and white undertail coverts; underparts whitish with buff wash across breast, flanks and thighs. Outer tail feathers white. Sometimes a hint of whitish supercilium. Lesser Whitethroat has greyer and more uniform upperparts and lacks rufous wing panel. Iris—reddish-brown; bill—dark with yellow base below; feet—pinkish.
Voice: Call notes include churring; sharp *tac tack* and nasal *tcharr*. Song is staccato, scratchy warble of several phrases, typically starting with *che-che worra che-wi* and variants.
Range: Temperate Palearctic, winters to Africa.
Distribution and status: Rare seasonal migrant. Race *icterops* breeds in Kashi, Tianshan, mid-Turpan and Altai Mts regions of W Xinjiang. Race *rubicola* migrates across much of China.

Habits: Puffs out white throat in display and song. Skulking and keeps to tall scrub and bushes.

1158. Lesser Whitethroat *Sylvia curruca* Baihou linying PLATE 100

Description: Smallish (13.5 cm) warbler with grey head; brown upperparts; white throat and whitish underparts. Ear coverts dark blackish-grey and sides of breast and flanks washed buff. Outer tail feathers edged white. Similar to Desert Lesser Whitethroat but darker plumage, darker feet and larger bill. Darker and less rufous than Desert Lesser Whitethroat. Iris—brown; bill—black; feet—dark brown.

Voice: Song starts with delicate, musical warble, leading to penetrating rattle *chikka-chikka-chikka* . . . regularly repeated. Call is harsh *tic-titic* and buzzing *tz-tz-tz-tz-zz-zz-zz*.

Range: Temperate Palearctic wintering to tropics of Africa, Arabia and India.

Distribution and status: Uncommon breeder and passage migrant. Breeding recorded in Hulun Nur area of Nei Mongol and possibly Hebei near Beijing. Migrants recorded widely across China.

Habits: Inhabits open habitats in dense scrub. Rather skulking.

Note: Some authors include Desert Lesser Whitethroat within this species (e.g. Cheng 1987, Baker 1997, Inskipp *et al.* 1996).

1159. Desert Lesser Whitethroat *Sylvia minula* PLATE 100
Sha baihou linying

Description: Smallish (13 cm) plain warbler with uniform sandy-grey upperparts and white throat and underparts. Edge of tail white. Distinguished from Lesser Whitethroat by paler grey plumage, lack of blackish ear coverts and smaller bill. Distinguished from Desert Warbler by grey rather than rufous-brown plumage and grey uppertail coverts. Iris—brown; bill—black; feet—greyish-brown.

Voice: Pleasant variable warble, lacking rattle of Lesser Whitethroat. Call is harsh churring, buzzing or *tit-titic*.

Range: Kurghizstan to NW China; winters Pakistan and NW India.

Distribution and status: Locally common seasonal migrant. Nominate race breeds in Tianshan, W Xinjiang; *margelanica* breeds from E Xinjiang to Ningxia (Cheng 1987, Meyer de Schauensee 1984). Baker (1997) reverses the distributions of these two forms without explanation.

Habits: More active than other whitethroats. Regularly flicks tail.

Note: Often treated as race of Lesser Whitethroat (e.g. Cheng 1987). Distinct in China but apparently connected by wide zone of intermediates from Lower Volga to Mongolia (Roselaar in Cramp 1992).

1160. Desert Warbler *Sylvia nana* Mo [di] linying PLATE 100

Description: Smallish (11 cm) plain rufous-brown warbler with rufous tertials, rump and uppertail coverts and white underparts. Similar to Lesser Whitethroat and Desert Lesser Whitethroat but much paler and more rufous. Iris—yellowish-brown; bill—yellow with black culmen; feet—yellowish.

Voice: Song like Greater Whitethroat but more rambling, interspersed with scratchy phrases and lark-like trills. Call is dry purr *drrrrrrrrrrr* with emphasis on *d*, on descending scale and fading out; also vibrant *chee-chee-chee-chee*.

Range: NW Africa, SC Asia to NW China, winters to Arabia and Pakistan.

Distribution and status: Nominate race breeds in NW China from W Xinjiang to W Nei Mongol. Uncommon seasonal migrant.

Habits: Spends much time on ground, hops and flicks half cocked tail. Flies low over ground to skulk in bushes.

1161. Barred Warbler *Sylvia nisoria* Hengban linying PLATE 100

Description: Large (15.5 cm) robust grey warbler with distinctive barred plumage of white underparts with scaling of dark grey crescent feather tips. Has double white wing bar and distinctive yellow eye. Adult unlikely to be mistaken but unbarred juvenile could be confused with Garden Warbler *S. borin* or Orphean Warbler *S. hortensis*, neither of which have been recorded in China but are distributed quite close to NW Xinjiang. Iris—yellow; bill—dark with yellow base to lower mandible; feet—yellowish or brownish-grey.

Voice: Song of short, deep-toned phrases for 3–10 seconds, regularly repeated. Call is typical *chak chak* of genus; also harsh churring or softer two-noted *chad chad*.

Range: C Eurasia, wintering to E Africa.

Distribution and status: Race *merzbacheri* is rare seasonal migrant. Breeds in Kashi, Tianshan, Korla and Ruoqing areas of W Xinjiang. Vagrant Hebei.

Habits: Keeps to thickets near rivers and lakes up to 2300 m. Skulking behaviour.

LARKS—Family: Alaudidae

Larks are a moderate-sized family of worldwide distribution. They are short-legged, terrestrial birds of open spaces, looking superficially like pipits but have weaker flight, shorter tails and thicker bills and several have short, erectile crests. Larks sing while on the wing and several species hover in a fluttering manner giving their beautiful, melodious songs. They feed and nest on the ground. In China there are 14 species, although some are marginal or vagrants.

1162. Singing Bushlark *Mirafra cantillans* [Eastern Lark, PLATE 113
AG: Bushlark] Ge bailing

Description: Small (14 cm) rufous-brown lark with rufous crown, heavily mottled black. Underparts pale buff streaked with black on breast; outer tail feathers white. Superficially resembles a pipit but has a thicker bill and shorter tail and legs. Distinguished from skylarks by rufous on wings. Iris—dark brown; bill—brown above; yellowish below; feet—pinkish with very elongated hind claw.

Voice: Song is chat-like with short melodic whistled phrases; a *chirrup* in alarm.

Range: Africa, India, SE China, SE Asia.

Distribution and status: Race *williamsoni* is rare in Guangdong and Guangxi. One record Hong Kong.

Habits: Singly or in dispersed flocks; frequents open areas of short grass and paddy stubble. Usually walking on the ground or making a weak, fluttery, undulating flight. Sings on the ground or in the air during flight or hovering and while slowly descending vertically. Perches in bushes.

Note: Similar Rufous-winged Bushlark *Mirafra assamica* could occur in S Yunnan, S Guangxi and SE Xizang. Distinguished by larger bill and rufous secondaries.

1163. Bimaculated Lark *Melanocorypha bimaculata* PLATE 113
[Eastern Calandra Lark] Erban bailing

Description: Largish (16.5 cm) heavily built, short-tailed lark with stout bill. Supercilium and small white stripe below eye. Chin, throat and half neck collar white underlined by black gorget. Upperparts mottled rich brown; underparts white with rufous flanks and some streaking on side of breast. In flight has dull greyish underwing, lacks white trailing edge to wing and has narrow white tip to tail but no white edges. Iris—brown; bill—pinkish with dark culmen and tip; feet—orange.

Voice: Flight call is gruff, rolling note *trrelit*; also *dre-lit* call. Similar to Greater Short-toed Lark. Song consists of rich varied phrases with frequent use of drawn-out rolling call note given from ground or in circling flight.

Range: Asia Minor and SW Asia, wintering to Arabia, NE Africa, NW India and W China.

Distribution and status: Uncommon. Nominate race is resident in W Xinjiang (Kashi, N Junggar, Bogda Mts and S Tarim Basin). Vagrants recorded as far east as Japan.

Habits: Flight low and undulating.

1164. Tibetan Lark *Melanocorypha maxima* [Long-billed PLATE 113
Calandra Lark] Changzui bailing

Description: Large (21.5 cm) reddish, thick-billed lark. Similar to Bimaculated Lark but larger with much more white in tail and less prominent black spot on breast. Tertials and secondaries conspicuously tipped white. Outer tail feathers white. Juvenile washed yellow. Race *holdereri* has paler underparts than nominate and less heavily streaked above. Iris—brown; bill—whitish-yellow with black tip; feet—dark brown.

Voice: Rather feeble song of disconnected strophes interspersed with mimicked phrases of other birds calls such as *Tringa* spp. Utters loud musical whistles when disturbed.

Range: C Asia to W and C China.

Distribution and status: Common in breeding range. Nominate race resident over S Xizang to S Gansu and N and W Sichuan. Race *holdereri* is resident in Kunlun Mts of W Xizang, Qamdo area of NE Xizang, Qinghai and NW Sichuan.

Habits: Inhabits humpy tussock vegetation around high altitude lakes generally from 4000–4600 m. Flight looks leisurely but purposeful. In agonistic display male drops wings and raises tail displaying white V and sways from side to side.

1165. Mongolian Lark *Melanocorypha mongolica* PLATE 113
Menggu bailing

Description: Large (18 cm) rusty-brown lark with black breast bar and white underparts. Head pattern distinctive with pale cinnamon crown surrounded by chestnut ring above white eyebrow which extends to nape above chestnut nape collar. Wing has contrasting pattern of chestnut wing coverts above white secondaries and black primaries. Iris—brown; bill—pale horn; feet—orange.

Voice: Sweet song, favoured for cagebird trade.

Range: N and C China and Mongolia and S Siberia.

Distribution and status: Uncommon. Nominate race breeds in Hulun Nur and Linxi areas of E Nei Mongol, N Shaanxi and Hebei; *emancipata* breeds in SE Qinghai and W Gansu (Lanzhou).

Habits: Inhabits rocky hills, also dry and swampy short grassland. Commonly sold as a cagebird.

1166. White-winged Lark *Melanocorypha leucoptera*   PLATE 113
Baichi bailing

Description: Large (18.5 cm) long-winged lark with shortish, stout bill. Has conspicuous white panel in closed wing and diagnostic white underwing and broad white trailing edge in flight, contrasting with black secondaries and inner primaries. Adult has rufous forewing. Male has unstreaked rufous crown and ear coverts. Extensive white edges to tail. Distinguished from Snow Bunting by lack of white on wing coverts and different bill shape. Iris—brown; bill—greyish-horn with yellow base; feet—orange.

Voice: Flight call is repeated, metallic *wed* recalling skylark; also deep rolling *schirrl-schirrl-schirrl* call with steady rhythm.

Range: Breeds in S Russia to W Siberia, migrating to W China, N Iran and Turkestan.

Distribution and status: Rare in China. Winters in Tianshan Mts of NW Xinjiang.

Habits: Inhabits dry grassy steppes.

1167. Black Lark *Melanocorypha yeltoniensis*   PLATE 113
Hei bailing

Description: Large (20 cm) lark with very heavy bill. Male has unmistakable all-black plumage. Female similar to Bimaculated Lark but much more heavily streaked underparts, blackish wings at rest, black underwing in flight and slaty legs. Lacks white trailing edge of wing in flight and little white at edge of tail. Iris—dark brown; bill—yellowish; feet—slaty-grey.

Voice: Flight call resembles Eurasian Skylark. Wing beats slowed in song flight.

Range: Kazakhstan and S Russia.

Distribution and status: Accidental, recorded wintering in NW Xinjiang.

Habits: Rather nomadic, continually changing breeding area. Inhabits grassy steppes.

1168. Greater Short-toed Lark *Calandrella brachydactyla*   PLATE 113
[Short-toed Lark] Duanzhi bailing

Description: Medium-sized (14 cm) sandy lark with black streaking on upperparts and buffy-white underparts with fine dark streaking on upper breast. Has short white supercilium underlined black. Distinguished from Asian Short-toed Lark by indistinct black patch on side of neck; larger bill, less streaking on throat, and broader supercilium. Long tertials reach tip of primaries. Race *dukhunensis* has pale buff underparts and much darker brown upperparts. Iris—brown; bill—horn; feet—flesh-coloured.

Voice: Flight call is sparrow-like *tjirp* and skylark-like *drelt*. Song given as short repetitive bursts, during circling flight.

Range: S Palearctic to China and Japan.

Distribution and status: Rather common. Race *longipennis* is found in Xinjiang, while *dukhunensis* is found in Qinghai, S Xizang, NE Yunnan, Sichuan, S Gansu and E Nei Mongol.

Habits: Inhabits semi-desert, salt flats and dry steppes and, in winter, agricultural fields.

Note: Treated by some authors as a race of Red-capped Lark *C. cinerea* (e.g. Cheng 1987). Song flight has abrupt undulations coinciding with wing beats and song.

1169. Hume's Short-toed Lark *Calandrella acutirostris* PLATE 113
[Hume's Lark] Xizui bailing

Description: Medium-sized (14 cm) greyish-brown lark with small black patch on side of breast. Upperparts lightly streaked blackish; short buff-coloured eyebrow. Difficult to separate in field from Greater Short-toed Lark by greyer plumage and dark brown outer tail feather tipped white but much less so than *brachydactyla*. Supercilium finer; bill longer and more pointed. Iris—brown; bill—pink with tip; feet—pinkish.

Voice: As Greater Short-toed Lark but flight call distinct; a full rolling *tiyrr*.

Range: Turkestan, NC and W China and Xizang. Migrates in winter to Afghanistan and N India.

Distribution and status: Race *tibetana* breeds in N Sichuan, S Gansu and E Qinghai to NW Xinjiang and S Xizang. Nominate race breeds in Tianshan, Bogdo Olashan and W Kunlun Mts of Xinjiang, migrating through Xizang. Not uncommon in mountains and highlands from 3600–4900 m.

Habits: Lives on bare rocky mountain sides and grassy steppes. Behaviour as Greater Short-toed Lark.

1170. Asian Short-toed Lark *Calandrella cheleensis* PLATE 113
[Lesser Short-toed Lark] [Yazhou] duanzhi bailing

Description: Smallish (13 cm) mottled brown lark without crest. Similar to Greater Short-toed Lark but smaller without black neck patch, more stubby bill and streaking of breast more extensive. Stands rather upright. Heavily streaked upperparts and broad white edges to tail separate from most other small larks. Iris—dark brown; bill—horn-grey; feet—fleshy-brown.

Voice: Typical flight call a characteristic quiet purring *prrrt* or *prrr-rrr-rrr*. Song given in spiralling flight is varied, melodious, interspersed with mimicry.

Range: Turkey to Mongolia and China.

Distribution and status: Rather common. Race *seebohmi* in Tianshan and Kashi region of NW Xinjiang; *kukunoorensis* SE Xinjiang to Qinghai; *stegmanni* in N Gansu to Sogo Lake; *biecki* in E Qinghai, S Gansu, Ningxia and W Nei Mongol; *tangutica* in extreme S Qinghai and NE Xizang; *cheleensis* over NE and eastern N China.

Habits: Lives on arid steppes and pastures. Song flight lacks rising and falling undulations of Greater Short-toed Lark.

Note: Some authors treat Asian Short-toed Lark as race of Lesser Short-toed Lark *C. rufescens* (e.g. Cheng 1987).

1171. Crested Lark *Galerida cristata* Fengtou bailing PLATE **113**

Description: Largish (18 cm) streaky brown lark with long narrow crest. Upperparts sandy-brown with blackish streaks; tail coverts buff. Underparts pale buff, heavily streaked blackish on breast. Looks stocky with shortish tail and longish, curved bill. In flight wings are broad with rusty underwing; tail is dark brown with cinnamon sides. Juvenile is heavily spotted above. Distinguished from skylarks by bulkier appearance, spiky crest and longer, more curved bill, less rufous on ear coverts and lack of white trailing edge to wing. Race *magna* has whiter underparts. Iris—dark brown; bill—yellowish-pink with dark tip; feet—pinkish.

Voice: On rising into air, gives clear *du-ee* and fluty *ee* or *uu*. Song is sweet, plaintive phrases of 4–6 notes. Continually repeated and interspersed with trills. Slower, shorter and clearer than Eurasian Skylark.

Range: Europe to Middle East, Africa, C Asia, Mongolia, Korea and China.

Distribution and status: Common in summer in suitable habitat. Race *magna* resident in NW Xinjiang, Qinghai, Gansu, Ningxia (Helan Mts and along Huang He = Yellow River) and W Nei Mongol (Baotou). Race *leautungensis* occurs from Sichuan to Liaoning. Resident populations deplend in winter by some southward migration.

Habits: Inhabits dry steppes, semi-desert and cultivated fields. Sings from perch and high in air.

1172. Eurasian Skylark *Alauda arvensis* [Skylark/Northern Skylark] PLATE **113**
Yunque

Description: Medium-sized (18 cm) mottled greyish-brown lark with streaking, crown and erectile crest. Tail forked with white edges and trailing edge of wing is white in flight. Distinguished from pipits by shorter tail and legs, crest and less upright stance. Distinguished with difficulty from Japanese Skylark for which see that form. Distinguished with difficulty from Oriental Skylark by larger size, whiter trailing edge to wing and by call. Iris—dark brown; bill—horn; feet—flesh-coloured.

Voice: Song is given white fluttering high in sky; continuous stream of trilling and warbling. When flushed gives variable *chrriup*.

Range: Breeds across Europe to Transbaikalia, Korea, Japan and N China, wintering to N Africa, Iran, NW India.

Distribution and status: Very common in N China in winter. Race *dulcivox* breeds in NW Xinjiang; *intermedia* in the mountains of NE China and *kiborti* in the marshy plains of NE China. Races *pekinensis* and *lonnbergi* breed in Siberia but visit N and E and coastal S China in winter.

Habits: Famous for its melodious lively song delivered at great height in fluttering song fight, followed by spectacular dive to earth and cover. Inhabits grasslands, steppes, moorland and marshes. Normal flight erratic and undulating. Squats low when alarmed.

Note: Some authors include Japanese Skylark as a race of this species (e.g. Cheng 1987).

1173. Japanese Skylark *Alauda japonica* Riben yunque PLATE **113**

Description: Medium-sized (17 cm) lark with mottled brown upperparts and short crest. Has distinctive rufous 'shoulder' triangle of lesser wing coverts and rufous ear coverts surrounded by white supercilium meeting white half collar and white throat. Underparts white with blackish streaking on breast. Browner and warmer brown than Crested Lark and with shorter crest. Distinguished from Asian and Greater Short-toed Larks by rufous lesser wing coverts. Separated with difficulty from Eurasian Skylark by rufous 'shoulder'. Distinguished with difficulty from Oriental Skylark by larger size, whiter trailing edge to wing in flight and call. Iris—dark brown; bill—pale yellow with dark tip; feet—orange.

Voice: Song like Eurasian Skylark from high in the sky; also short *byur-rup* when flushed from cover.

Range: Breeds in Japan, Some birds winter S in China.

Distribution and status: Occasional on passage or wintering along S and E coasts of China, including Hong Kong. Distribution not well documented due to confusion with Eurasian Skylark.

Habits: As Eurasian Skylark.

Note: Treated by some authors as a race of Eurasian Skylark (e.g. Cheng 1987, 1994, Inskipp *et al.* 1996). Status requires further review.

1174. Oriental Skylark *Alauda gulgula* [Eastern/Small/   PLATE 113
Lesser Skylark] Xiao yunque

Description: Small (19 cm) mottled brown pipit-like bird with faint pale eyebrow and a slight crest. Distinguished from pipits by heavier bill, weaker flight and by posture and from bushlark by lack of rufous on wings and behaviour. Distinguished from Eurasian and Japanese Skylarks by smaller size, less white on trailing edge to wings in flight and call. Iris—brown; bill—horn; feet—flesh-coloured.

Voice: High-pitched sweet song on ground and in rising display flight. Call is dry buzz *drzz*.

Range: Breeds in E Palearctic and migrates S in winter.

Distribution and status: Very common in southern and coastal parts of China. Race *inopinata* is found through S and E Qinghai–Xizang plateau; *weigoldi* in C and E China; *coelivox* in SE China; *vernayi* in SW China; *sala* on Hainan and adjacent S Guangdong, and *wattersi* on Taiwan.

Habits: Prefers open land with short grass. Differs from bushlarks by never perching in trees.

1175. Horned Lark *Eremophila alpestris* [Shore Lark]   PLATE 113
Jiao bailing

Description: Medium-sized (16 cm) dark lark with distinctive head pattern. Male has bold black breast bar, black and white (or yellow) patterned face and diagnostic small black 'horns' extending from black bar across forecrown. Upperparts rather uniform dull brown; rest of underparts white with some brown streaking on flanks. Female and juvenile are duller (and lack 'horns') but head pattern recognisable. Underwing white in flight. Races show slight variation. Race *flava* has yellow on face; *brandti* and *przewalskii* have white forehead; *teleschowi* has no white on crown; and *albigula* has black pectoral band connected to black band below eye. Iris—brown; bill—grey, darker on upper bill; feet—blackish.

Voice: Flight-call is high-pitched, mournful *siit-di dit*. Song from perch or in flight is simple, high-pitched, sibilant or rippling *tu-a-li, tioli-ti* or similar phrases.

Range: Holarctic, N Africa, S Mexico, Colombia, China.

Distribution and status: Common in breeding range. Eight races recognised in China: *brandti* in N Xinjiang, Nei Mongol, E Qinghai, N Gansu, N Shaanxi and N Shanxi; *albigula* in Kashi and Tianshan region of W Xinjinang; *argalea* in Karakoram Mts of SW Xinjiang and in SW Xizang; *elwesi* in E Xizang, Qilian Mts of E Qinghai and NW Sichuan; *przewalskii* in Qaidam Basin of Qinghai; *teleschowi* in Kunlun and Altun Mts of S Xinjiang and *khamensis* in S and W Sichuan. Race *flava* breeds in Siberia but winters in drier regions of NE China. Some races rather dubious and hybrids occur.

Habits: Breeds at high altitude on desolate steppes and cold deserts. Moves to lower altitudes in winter, on short grass and lake shores.

FLOWERPECKERS AND SUNBIRDS—Family: Nectariniidae

Subfamily: Nectariniinae

Small, colourful birds feeding on insects, nectar and mistletoe berries. Formerly two families, now treated as two tribes.

FLOWERBIRDS—Tribe: Dicaeini

Flowerpeckers are a tropical family of very small active birds mostly found in the Oriental and Australian regions. Several species are brightly coloured with red and orange plumage. Bill shape is variable from sharp and pointed to thick. Flowerpeckers live in the tree-tops eating tiny insects and small fruits, but have a particular association with mistletoes *Loranthus*. They are the main distribution agents of the seeds of these plants and are most abundant where there are many mistletoes such as in gardens, mangroves and coastal scrub, but some species are more forest-loving than others. Beautiful purse nests are suspended from leafy twigs, made of leaves and grass fibres felted together with spider webs. There are six species in China. Females are often difficult to identify but flowerpeckers usually travel in pairs or small flocks so identification can usually be made from the males.

1176. Thick-billed Flowerpecker *Dicaeum agile*   PLATE 115

[Striped Flowerpecker] Houzui zhuohuaniao

Description: Small (9 cm) nondescript brownish flowerpecker. Crown olive-brown; cheeks grey; back olive; throat and underparts greyish-white; breast faintly streaked grey. The bill is noticeably thick. Underside of fail feathers are white tipped. Iris—orange; bill—grey; feet—dark slate-grey.

Voice: *Tchup tchup* call, similar to other flowerpeckers but not so hard.

Range: India, SE Asia, Sumatra and Java to Lesser Sundas.

Distribution and status: Overlooked but actually rather common in Menghai area of

Xishuangbanna in S Yunnan, also in lowlands of SE Xizang. Records questioned by Cheng (1994) but 17 Yunnan records in 1989 confirm species' presence in China. **Habits:** Similar to other flowerpeckers. A bird of lowland forest; wags tail from side to side when perched.

1177. Yellow-vented Flowerpecker *Dicaeum chrysorrheum*   PLATE 115
Huanggang zhuohuaniao

Description: Small (9 cm) white-bellied flowerpecker. Adult: olive-green upperparts; bright yellow or orange undertail coverts; rest of underparts white, heavily streaked with diagnostic bold black marks. Iris—red/orange; bill and feet—black. Juvenile is duller and paler with greyer underparts and paler streaks.
Voice: Repeated *zit-zit-zit* in flight and a repeated call *zip-a-zip-treee*.
Range: NE Indian subcontinent, SW China, SE Asia and Greater Sundas.
Distribution and status: Nominate race is locally common in S Yunnan (Xishuangbanna) in hills up to 800 m.
Habits: A bird of gardens and open forest. Typical busy forager of small fruits and insects, aggressively chasing other birds from its food trees.

1178. Yellow-bellied Flowerpecker *Dicaeum melanoxanthum*   PLATE 115
Huangfu zhuohuaniao

Description: Male: large (13 cm) flowerpecker with bright yellow lower belly and diagnostic white throat stripe contrasting with black head, sides of throat and upperparts. Outer tail feathers have white patch on inner web. Female: similar but duller. Iris—brown; bill—black; feet—black.
Voice: Harsh, agitated *zit-zit-zit-zit*.
Range: Nepal to SW China, Burma and Indochina.
Distribution and status: Uncommon in W and SW Sichuan, W and S Yunnan in open pine forest and forest gaps and edges in evergreen submontane forest from 1400–4000 m. Descends in winter.
Habits: Keeps to evergreen forest edge and gaps; feeds on mistletoes.

1179. Plain Flowerpecker *Dicaeum concolor*   PLATE 115
[Plain-colored Flowerpecker] Chunse zhuohuaniao

Description: Tiny (8 cm) nondescript flowerpecker. Olive-green above, pale greyish below with creamy central belly and fine white tufts at bend of wing. Distinguished from Thick-billed Flowerpecker by fine bill. Iris—brown; bill—black; feet—dark blue-grey.
Voice: Staccato, penetrating *tzik*; song is repeated *tzierrr*.
Range: India, S China, SE Asia and Greater Sundas.
Distribution and status: Race *olivaceum* is a common lowland resident in Hunan, E Sichuan, and China S of Changjiang River; *uchidai* is resident in S Taiwan; and *minullum* on Hainan.
Habits: Typical flowerpecker inhabiting hill forest, secondary growth and cultivated areas, and frequently visits mistletoes.

1180. Fire-breasted Flowerpecker *Dicaeum ignipectus*   PLATE 115
[Buff-breasted/Green-backed/Bronze-backed Flowerpecker]
Hongxiong zhuohuaniao

Description: Small (9 cm) dark flowerpecker. Male: upperparts dark glossy greenish-blue; underparts buffy with scarlet patch on breast and narrow black stripe down belly. Female: buffy-ochraceous below. Immature: like immature Plain Flowerpecker but separated altitudinally. Iris—brown; bill and feet—black.

Voice: Song is high-pitched, metallic chittering, *titty-titty-titty*; call is clicking *chip*.

Range: Himalayas, S China, SE Asia and Sumatra.

Distribution and status: A common resident of hill forests from 800–2200 m. Nominate race in C and S China and SE Xizang; *formosum* on Taiwan.

Habits: As other flowerpeckers, mostly seen visiting clumps of *Loranthus* mistletoe in tree crowns.

1181. Scarlet-backed Flowerpecker *Dicaeum cruentatum*   PLATE 115
Zhubei zhuohuaniao

Description: Small (9 cm) black and red flowerpecker. Male: scarlet crown, back and rump, black wings, sides of head and tail, grey flanks and white rest of underparts. Female is olive above with scarlet rump and uppertail coverts; tail black. Immature is plain grey with orange bill and dull orange tinge on rump. Iris—brown; bill—blackish-green; feet—blackish-green.

Voice: Typical call is hard, metallic *tip . . . tip . . . tip*; song is thin repeated *tissit . . . tissit . . .*

Range: India, S China, SE Asia, Sumatra and Borneo.

Distribution and status: Nominate race (= *erythronotum*) is uncommon in lowland forests of SE Xizang, S Yunnan, Guangxi, Guangdong and Fujian; *hainanum* is resident on Hainan.

Habits: Active and aggressive visitors of mistletoe clusters in secondary forest, gardens and plantations up to 1000 m.

SUNBIRDS AND SPIDERHUNTERS—Tribe: Nectariniini

Sunbirds and spiderhunters are an Old World tropical tribe, formerly treated as a family, of small, mostly very colourful, birds with long curved beaks. Their metallic plumage and ability to hover in front of flowers resemble those of American hummingbirds. Most of the family are nectar feeders but they also take some insects and pollen. The long-beaked spiderhunters have become partly insect feeders. All are active, restless birds incessantly on the move looking for food. Many tropical flowers are adapted to attract these birds as pollinating agents and typically have small trumpet flowers and red or orange colours. The nests of sunbirds are beautiful hanging structures made of fine grass-heads and other soft materials. Spiderhunter nests are sewn on to the underside of large leaves with spiders web threads piercing the leaf as plugs. Twelve species occur in China.

1182. Ruby-cheeked Sunbird *Anthreptes singalensis* PLATE 114
[Rubycheek] Zijia zhizui taiyangniao

Description: Small (10 cm) colourful sunbird. Male: crown and upperparts dark iridescent green; cheeks deep copper-red; belly yellow; throat and breast orange-brown. Female: upperparts greenish-olive; underparts like male but paler. Iris—red-brown; bill-black; feet—greenish-black.

Voice: A shrill chirp, *seet-seet*, also shrill rising trill ending in a brief double note, followed immediately by descending trill ending with two separated notes.

Range: Nepal to SW China, SE Asia and Greater Sundas.

Distribution and status: Race *koratensis* is resident in SE Xizang and W and S Yunnan. Uncommon in lowlands up to 1000 m.

Habits: Lives singly or in pairs, sometimes mixes with other species. Prefers forest edge, light undergrowth and coconut plantations where it feeds on pollen.

1183. Purple-naped Sunbird *Hypogramma hypogrammicum* PLATE 114
[Blue-naped Sunbird] Lanzhen huaminiao

Description: Large (15 cm) sunbird with diagnostic heavily-streaked yellow underparts. Male has metallic purple nape, rump and tail coverts. Iris—red or brown; bill—black; feet—brown or olive.

Voice: Strident single *schewp* call.

Range: SW China, SE Asia, Sumatra and Borneo.

Distribution and status: Race *lisettae* is locally common resident of lowland forest up to 1000 m in W Yunnan.

Habits: Prefers smaller trees and undergrowth of forest, swamp forest and secondary scrub. Fans and flicks tail.

1184. Olive-backed Sunbird *Nectarinia jugularis* PLATE 114
[Yellow-breasted/Yellow-bellied Sunbird > Black-throated Sunbird]
Huangfu huaminiao

Description: Small (10 cm) sunbird with greyish-white belly. Male has a black metallic purple chin and breast with crimson and grey breast band and bright orange plumes at wing 'shoulders' Upperparts olive-green. In eclipse, metallic purple reduced to a narrow stripe down centre of throat. Female lacks black and is olive-green above, yellow below; usually has pale yellow supercilium. Iris—dark brown; bill and feet—black.

Voice: Musical chirps *cheep, cheep, chee weet* and a short melody ending in a clear trill.

Range: Andaman and Nicobar islands, S China, SE Asia, Philippines and Indonesia to New Guinea and Australia.

Distribution and status: Race *rhizophorae* is uncommon in lowlands of S Yunnan and Guangxi but quite common on Hainan, especially in coastal scrub.

Habits: A noisy bird which flits from one flowering tree or bush to another in small parties. Males sometimes chase back and forth aggressively. Frequents gardens, coastal scrub, and mangroves.

1185. Purple Sunbird *Nectarinia asiatica* Zi huaminiao PLATE 114
Description: Small (11 cm) very dark sunbird. Male appears all-black in most lights but has green gloss and maroon band across breast. Pectoral tufts may show yellow and

orange. Female is olive above and has dull yellow underparts. Separated from female Olive-backed sunbird by paler underparts and narrower white tips to tail. Eclipse male is like female but with black streak down centre of throat and distinguished from male Olive-backed by iridescent blue wing coverts. Iris—brown; bill—black; feet—black.

Voice: Song is descending *swee-swee-swee-swit-zizi-zizi*, also twittering. Call is buzzing *tzit* or wheezy rising *swee*.

Range: Indian subcontinent to SW China and SE Asia.

Distribution and status: Rare resident in Xishuangbanna, SW and W Yunnan and SE Xizang in open country and gardens at low to moderate elevations.

Habits: Typical sunbird of scrub vegetation and forest edge.

1186. Mrs Gould's Sunbird *Aethopyga gouldiae* PLATE 114
Lanhou taiyangniao

Description: Male: largish (14 cm) scarlet, blue and yellow sunbird with elongated blue tail. Distinguished from Black-throated Sunbird by bright colours and scarlet breast; from Fire-tailed and Crimson Sunbirds by blue tail. Nominate race has only a little scarlet streaking on yellow breast. Female is olive above and greenish-yellow below, with ashy-olive chin and throat. Pale yellow rump separates from all other species except Black-throated Sunbird which has less distinct white tips to tail. Iris—brown; bill—black; feet—brown.

Voice: Quick, repeated *tzip*, rattling alarm and lisping *squeeeee* song rising in middle like see-saw.

Range: Himalayas and NE India to SW China and Indochina.

Distribution and status: Common montane bird in evergreen forest at 1200–4300 m in summer, lower in winter. Race *gouldiae* in Himalayas; *dabryii* in C and SW China.

Habits: Regular feeder on *Rhododendron* in spring and *Rubus* in summer.

1187. Green-tailed Sunbird *Aethopyga nipalensis* PLATE 114
Lühou taiyangniao

Description: Male: largish (14 cm) scarlet, green and yellow sunbird with elongated tail. Distinguished from Gould's and Fire-tailed Sunbirds by metallic green tail and grey belly. Female: olive above, dull greenish-yellow below grading to grey on throat and chin; lacks yellow rump; white tips to tail feathers. Distinguished from similar female sunbirds by graduated tail. Iris—brown; bill—black; feet—brown.

Voice: Call is loud *chit chit*. Song is *tchiss . . . tchiss-iss-iss-iss*.

Range: Himalayas and NE India to SW China, Burma and Indochina.

Distribution and status: Race *koelzi* is uncommon bird of mossy forests from 1800–3600 m in moist Himalayan valleys, SE Xizang, S and W Sichuan and W Yunnan.

Habits: Visits flowering bushes and chases other sunbirds away aggressively.

1188. Fork-tailed Sunbird *Aethopyga christinae* PLATE 114
Chawei taiyangniao

Description: Small (10 cm) dainty sunbird. Crown and nape metallic green, upperparts olive or blackish with yellow rump. Uppertail coverts and central tail feathers glossy metallic green. Central two tail feathers have slender elongated points. Outer tail feathers are black, tipped white. Side of head is black with iridescent green moustachial stripe and maroon throat patch. Underparts otherwise dirty olive-white. Female is very small with olive upperparts and pale greenish-yellow underparts. Nominate race has blacker wings. Iris—brown; bill—black; feet—black.

Voice: Noisy. Song is high-pitched trill; also gives bubbling twitters when feeding. Loud metallic *chiff-chiff-chiff* call.

Range: S China, Hainan, Vietnam.

Distribution and status: A common lowland bird in SE and S China (*latouchii*) and Hainan (nominate).

Habits: Lives in forest and wooded areas, even towns, visiting flowering bushes and trees.

1189. Black-throated Sunbird *Aethopyga saturata* PLATE 114
Heixiong taiyangniao

Description: Male: largish (14 cm) dark sunbird with elongated tail. In poor light looks blackish with pale rump and breast. In good light crown and tail are metallic blue and mantle is dull maroon, throat is black. Breast is greyish-olive with fine dark streaks. Race *saturata* has restricted yellow rump band; *petersi* has more yellow in underparts; and *assamensis* has more extensive black on breast. Female: very small with whitish yellow rump. Iris—brown; bill—black; feet—dark brown.

Voice: Rapid *ti-ti-ti-ti-ti-ti-ti* ... call.

Range: Himalayas and NE India to SW and S China, Burma and SE Asia.

Distribution and status: Race *saturata* found in Himalayas and SE Xizang, *assamensis* E of Medog through W Yunnan; and *petersi* in Xishuangbanna (S Yunnan) and Guangxi. Generally uncommon in hills and lower montane forest from 300–1800 m.

Habits: Frequents flowering bushes along streamsides.

1190. Crimson Sunbird *Aethopyga siparaja* [Yellow-backed/ PLATE 114
Scarlet-throated Sunbird] Huangyao taiyangniao

Description: Medium-sized (13 cm including long tail) bright red (male only) sunbird with dark grey belly. Female is dull dark olive-green without red wash on wings or tail. Races *tonkinesis* and *owstoni* lack the elongated central tail feathers. Iris—dark; bill—blackish; feet—bluish.

Voice: A soft *seeseep-seeseep*. Rapid tripping song of 3–6 clear notes, *tsip-it-tsip-it-sit*.

Range: Indian subcontinent to S China, SE Asia, Philippines, Sulawesi, Malay Peninsula and Greater Sundas.

Distribution and status: Race *viridicauda* is resident in W and S Yunnan; *tonkinensis* in SE Yunnan (Jinping); and *owstoni* is confined to Naozhou Is (S Guangdong). A common resident of lowlands, found up to 900 m.

Habits: Seen singly or in pairs visiting *Erythrina* bushes and similar flowering trees on estates and at the forest edge.

1191. Fire-tailed Sunbird *Aethopyga ignicauda* PLATE 114
Houwei taiyangniao

Description: Long (20 cm) colourful sunbird. Male: unmistakable, red with greatly elongated bright scarlet central tail feathers. Crown is metallic blue; lores and side of head black; throat and moustachial streak metallic purple. Underparts yellow with bright orange patch on breast. Female greyish-olive with yellow rump and is much smaller than male. Iris—brown; bill—black; feet—black.

Voice: The call is a quiet shrill note *shweet*. Song is monotonous *dzidzi-dzidzidzidzi*.

Range: Himalayas and NE India to Burma and S Tibetan Plateau.

Distribution and status: Not rare altitudinal migrant in mountains of SW China and S Xizang in clearings among subalpine conifer forest up to tree-line.

Habits: Feeds on the flowers of rhododendrons, brambles and flowering bushes.

1192. Little Spiderhunter *Arachnothera longirostra* PLATE 114
Changzui buzhuniao

Description: Smallish (15 cm) olive and yellow spiderhunter. Upperparts olive-green; underparts bright yellow. Whitish-grey throat diagnostic. Iris—brown; bill—black above, grey below; feet—bluish-indigo.

Voice: Sharp *weechoo* or *cheek-cheek-cheek* in flight or a simple high-pitched song, *tik-ti-ti-ti*, the first note higher and stressed, endlessly repeated, about three notes per second.

Range: Indian subcontinent to China, SE Asia, Philippines, Malay Peninsula and Greater Sundas.

Distribution and status: Nominate race is a fairly common resident in lowland and hill forests up to 2000 m in W Yunnan; *sordida* in SE Yunnan.

Habits: Secretive; keeping to dark thickets such as wild bananas and tall gingers. Frequently seen flying speedily across jungle trails giving its characteristic flight call. Also found in secondary forest, plantations and gardens.

1193. Streaked Spiderhunter *Arachnothera magna* PLATE 114
Wenbei buzhuniao

Description: Large (19 cm), heavily-streaked spiderhunter with bright orange legs. Feathers of upperparts are olive with black centres, giving bold streaked appearance. Underparts yellowish-white, streaked with black. Iris—brown; bill—black; feet—orange.

Voice: Sharp *cheet* call given in fast flight. Song is strident chatter.

Range: Himalayas and NE India, S China and SE Asia.

Distribution and status: Nominate race is a common resident in tropical zone of SE Xizang, W and S Yunnan, S Guizhou and SW Guangxi, up to moderate altitudes in evergreen forest.

Habits: Fiercely territorial and engages in agonistic chases. Feeds among wild bananas and gingers.

SPARROWS, FINCHES, PIPITS AND ALLIES—Family: Passeridae

A newly-created, diverse family of small passerines including several subfamilies formerly treated as full families.

SPARROWS—Subfamily: Passerinae

Small flock-living passerines with conical, seed-eating bills including true sparrows and snowfinches. There are 12 species in China.

1194. Saxual Sparrow *Passer ammodendri* Heiding maque   PLATE 116

Description: Medium-sized (15 cm) sparrow. Male distinct with black coronal stripe to nape; black eye-stripe and black chin; tawny-brown supercilium and side of nape; pale grey cheeks. Upperparts brown, heavily streaked black. Female drab but blackish streaks on mantle and pale tips to median and greater wing coverts distinctive. Male of race *nigricans* has blacker streaking on mantle and back. Race *stoliczkae* is richer cinnamon on back and sides of crown and nape. Iris—dark brown; bill—black in male, yellow with black tip in female; feet—pinkish-brown.

Voice: Melodic chirps and short whistle.

Range: C Asia to NW China and Mongolia.

Distribution and status: Locally common. Race *nigricans* in extreme NW Xinjiang (Junggar Basin to Manas River) and race *stoliczkae* in Tianshan, Kashi, Kunlun Mts to W Nei Mongol and Ningxia.

Habits: Lives in oases and river beds of deserts and arid foothills associated with saxual *Arthrophytum haloxylon*. Rather shy. Mixes with Spanish Sparrows in winter flocks.

1195. House Sparrow *Passer domesticus* Jia maque   PLATE 116

Description: Medium-sized (15 cm) sparrow. Male distinguished from Eurasian Tree Sparrow by grey crown and uppertail coverts, lack of black ear patch and by more extensive black on throat and upper breast. Female is plain with pale supercilium. Less rich in colour than female Russet Sparrow and wing bar less conspicuous, and from female Saxual Sparrow by lack of forked tail and paler breast. Has buff stripe down side of mantle, bordered by blackish streaks. Races *partini* and *bactrianus* have whiter cheeks and underparts and are smaller than nominate. Race *partini* has more black on breast. Iris—brown; bill—black (breeding male) or straw with dark tip; feet—pinkish-brown.

Voice: Monotonous chirps. When excited gives rolling *chur-r-rit-it-it-it* call or in alarm a shrill *chree*. Song is monotonous series of chirps.

Range: Palearctic and Oriental; introduced N and and S America, W, C and S Africa, New Zealand, Australia and many small island groups. Asian range includes SC Asia, S Russia, S Sibera, Mongolia, N China, Afghanistan, Indian subcontinent, Thailand (except S), N and S Laos; introduced Singapore.

Distribution and status: Locally common in towns and villages of arid regions, in oases and at edges of deserts of extreme W and NE China. Race *domesticus* is found in NE Nei Mongol; *bactrianus* in NW Xinjiang (including Altai and Tianshan region) E to Manas River and S to Kashgaria, recently recorded in W Qinghai; and *partini* is resident in SW Xizang in the Himalayas up to 4600 m. Races *partini* and *bactrianus* migrate in winter to N India.

Habits: Gregarious and communal rooster. Raids grain crops but also feeds on insects and some leaves. Generally commensal with human habitation.

1196. Spanish Sparrow *Passer hispaniolensis* PLATE 116
Heixiong maque

Description: Medium-sized (15.5 cm) chunky sparrow with heavy bill. Adult male has chestnut crown and nape, white cheeks and heavy black streaking on mantle and flanks. Chin and upper breast black. Female is plain and similar to female House Sparrow but distinguished by larger bill, longer supercilium, pale 'braces' on side of mantle and streaks on breast and flanks. Iris—dark brown; bill—black in male yellow with black tip in female; feet—pinkish-brown.

Voice: Song as House Sparrow but more rhythmic. Calls like House Sparrow but higher in pitch. *Chirrup* call is deeper.

Range: Cape Verde Is, S Europe, N Africa, Middle East, C Asia to W China.

Distribution and status: Race *transcaspius* is a locally common resident in Kashi and Tianshan and Kunlun region of NW Xinjiang at lower altitudes.

Habits: Inhabits open country and fields with trees. Where House Sparrow absent it lives in urban habitats.

1197. Russet Sparrow *Passer rutilans* [Cinnamon Sparrow] PLATE 116
Shan maque

Description: Medium-sized (14 cm) brightly coloured sparrow. Male has bright cinnamon or chestnut crown and upperparts, clean black streaking on mantle, black throat and dirty whitish cheeks. Female is duller with broad dark eye-stripe and long cream supercilium. Males of race *cinnamomeus* have yellow wash on side of head and underparts. Races *batangensis* and *intensior* are like *cinnamomeus* but paler yellow. Iris—brown; bill—grey (male), yellow with dark tip (female); feet—pinkish-brown.

Voice: Calls include *cheep*, rapid *chit-chit-chit* or repeated song, *cheep-chirrup-cheweep*.

Range: Himalayas, E Tibetan Plateau and C, S and E China.

Distribution and status: Common. Race *cinnamomeus* in E and SE Xizang to S Qinghai; *intensior* in SW China to SE Xizang and NW Sichuan; *batangensis* in W Batang area of S Sichuan and W Yunnan; and nominate race over most of C, S, and SE China and Taiwan. Clement *et al.* (1993) treat *batangensis* within *intensior*.

Habits: Lives in flocks in upland open forest, woodland or scrub near cultivation. In absence of House Sparrow it becomes a bird of towns and villages.

1198. Eurasian Tree Sparrow *Passer montanus* PLATE 116
[European/The Tree Sparrow] [Shu] maque

Description: Smallish (14 cm) plump, lively sparrow with brown crown and nape. Sexes alike. Adult has brownish upperparts and buffy-grey underparts. Has complete greyish-white collar around nape. Distinguished from House and Russet Sparrows by conspicuous black cheek spot and less extensive black on throat. Juvenile as adult but duller and paler with yellow base to bill. Iris—dark brown; bill—black; feet—pinkish-brown.

Voice: Call is chirpy hard *cheep cheep* or metallic *tzooit*, also *tet tet tet* given in flight. Song is repeated series of call note, interspersed with *tsveet* notes.

Range: Europe, Middle East, C Asia, E Asia, Himalayas and SE Asia.

Distribution and status: Common over the whole of China, Hainan and Taiwan up to moderate altitudes. Seven geographical races occur in China: *montanus* in NE China; *saturatus* in E, C and SE China and Hainan; *dilutus* in NW China; *tibetanus* on Xizang Plateau; *kansuensis* in Gansu and C Nei Mongol; *hepaticus* in SE Xizang; and *malaccensis* in tropical SW China and Hainan.

Habits: A bird of lightly wooded areas, villages and farmland, and pest of grain crops. Replaces the House Sparrow as the urban sparrow in the E of the country.

1199. Rock Sparrow *Petronia petronia* Shi que PLATE 116

Description: Medium-sized (15 cm) squat petronia with dark, lateral coronal stripes, pale supercilium and dark stripe behind eye. Sexes alike. Head pattern quite distinctive. In flight appears shorter-tailed and with broader wing base than House Sparrow. Race *brevirostris* is smaller and has less well-defined head pattern and short thick bill. Iris—dark brown; bill—grey with yellow base to lower mandible; feet—pinkish-brown.

Voice: Noisy with House Sparrow-like chirps and metallic *vi-veep*, sharp *cheeooee* and, in flight, a softer *sup* call. Song is repeated collection of call notes.

Range: S Palearctic to Middle East, C Asia to N China and Mongolia.

Distribution and status: Common. Race *intermedia* is resident in Tianshan and Kashi region of NW Xinjiang and *brevirostris* occurs in E Qinghai, Gansu, N Sichuan, and E to Beijing and Hulun Nur in E Nei Mongol.

Habits: Inhabits barren hills and rocky ravines up to 3000 m. Colonial and often associates with House Sparrow. Runs and hops on ground; strong flight.

1200. White-winged Snowfinch *Montifringilla nivalis* PLATE 116
[Snow Finch] Baibanchi xueque

Description: Largish (17 cm) plump and elongated finch with extensive white in wing and sides of notched tail. Adult has grey head and streaky brown upperparts with buffy belly. Throat is stippled black, particularly in breeding male. Juvenile similar to adult but head buffy-brown, and white area tinged sandy. Races *alpicola* and *kwenlunensis* have grey-brown head and upperparts paler than darker race *henrici* which has greyer belly and brownish flanks. Lives at much higher altitudes than similarly patterned Snow Bunting *Plectophenax nivalis*. Iris—brown; bill—black with yellow base to lower mandible (breeding) or yellow with black tip (non-breeding); feet—black.

Voice: Calls include sharp nasal *pschieu* in flock during flight, also *tsee* and softer *pruuk* or, in alarm, harsher *pchurrt*. Song is monotonously repeated *sitticher-sitticher* from perch or in circular song flight.

Range: Spain, Mediterranean region to Middle East, C Asia and W China.

Distribution and status: Three races occur, all fairly common: *alpicola* in Tianshan and Kashi region of NW Xinjiang, *kwenlunensis* in the Kunlun Mts of S Xinjiang; and *henrici* in Qinghai and C and E Xizang.

Habits: Lives at very high altitudes on rocky slopes between glaciers and melting snow. Forms large flocks in non-breeding season, mixing with other snowfinches and mountain finches. Fairly tame.

1201. Tibetan Snowfinch *Montifringilla adamsi* [Black-winged/   PLATE **116**
Adam's Snowfinch] Hechi xueque

Description: Large (17 cm) plump and elongated finch. Sexes alike. Very similar to White-winged Snowfinch but much browner on head and upperparts and with less visible white on wings in flight and at rest. Has blackish speckles on 'shoulder' of wing. Iris—brown; bill—black (breeding) or yellow with black tip; feet—black.
Voice: Sharp *pink pink* and softer mewing note. Flocks twitter. Song is monotonously repeated single note given from perch or in hovering song flight.
Range: Kashmir, Himalayas, Tibet Plateau and NW China.
Distribution and status: Common from 3500–5200 m. Nominate race occurs in W, S and E Xizang, S and E Qinghai, W Sichuan and probably NW Yunnan. Race *xerophila* occurs in Altun Mts, Qaidam Basin and Qilian Mts of E Xinjiang and N and E Qinghai.
Habits: Elaborate butterfly-like courtship flight. Feeds on ground, often close to villages on cultivated fields. Forms large flocks in winter.

1202. White-rumped Snowfinch *Pyrgilauda taczanowskii*   PLATE **116**
[Moudelli's Snowfinch] Baiyao xueque

Description: Largish (17 cm) grey finch with heavy mottling on mantle and black lores. Sexes alike. Adult is paler than other snowfinches with diagnostic large white rump. Juvenile is more sandy brown with off-white rump. Iris—brown; bill—horn or yellow with black tip; feet—black.
Voice: Call is sharp resounding *duid duid*. Song is short, loud *duid ai duid, duid*.
Range: Tibetan Plateau and W China.
Distribution and status: Fairly common from 3800–4900 m. Resident in Xizang, Qilian and Anyemaqen Mts of E Qinghai and SW Gansu and Min Mts of N and W Sichuan.
Habits: Inhabits high stony plateaux, cold deserts, steppes and edges of marshes. Has lark-like display flight and 'drumming' courtship display on ground. Lives in small flocks among pika colonies using pika holes for shelter and nesting. Forms larger flocks in winter. Bows and bobs tail on landing. Rather shy.

1203. Small Snowfinch *Pyrgilauda davidiana*   PLATE **116**
[Pere David's Snowfinch] Heihou xueque

Description: Medium-sized (15 cm) buff-brown finch with distinct black forehead, lores, chin and throat. Has white bases to primary coverts and whitish outer tail feathers. Juvenile is paler than adult and lacks black on face. Distinguished from juvenile Blandford and Rufous-necked Snowfinches by lack of supercilium or white on face. Race *potanini* is paler and less streaked than nominate. Iris—brown; bill—straw with black tip; feet—black.
Voice: No information.
Range: Russian Altai to Mongolia and N China.
Distribution and status: Uncommon at 1000–3000 m. Race *davidiana* occurs in Qilian Mts of E Qinghai, Gansu, Ningxia (Helan Mts), E Nei Mongol (Dalai Lake area) and around Lake Hulun Nur in NE Nei Mongol.

Habits: Inhabits stony mountains and semi-desert with sparse grass cover, usually close to water. Associated with pika colonies. Forms large flocks in winter and becomes quite tame, coming into farms and villages.

1204. Rufous–necked Snowfinch *Pyrgilauda ruficollis* PLATE 116
[Red-necked Snowfinch] Zongjing xueque

Description: Medium-sized (15 cm) brown finch with black lores and whitish sides of face. Sexes alike. Adult has distinctive head pattern with fine black moustachial stripe but white chin and throat. Nape and sides of neck richer chestnut than any other snowfinches. Has white tips to wing coverts. Juvenile is paler and duller but paler chestnut ear coverts already distinctive. Race *isabellina* is paler with greyish upperparts and buffy-yellow tinge. Iris—brown; bill—black (adult) or pinkish with dark tip (juvenile); feet—black.
Voice: Call is soft, repeated *duuid* or chattering alarm.
Range: Tibetan Plateau and NW China.
Distribution and status: Common with summer range of 3800–5000 m. Nominate race inhabits Xizang plateau, N Himalayas, SE Qinghai and W Sichuan. Vagrant once to Liaoning. Race *isabellina* ranges through E Kunlun, Altun to W Qilian Mts.
Habits: Lives among pika colonies like other snowfinches. Has elabarate buzzing courtship flight. Rather tame. Flight is weak and low. Mixes with other snowfinches in winter flocks.

1205. Plain-backed Snowfinch *Pyrgilauda blandfordi* PLATE 116
[Blandford's Snowfinch] Zongbei xueque

Description: Medium-sized (15 cm) brown snowfinch with distinctive black and white head pattern. Sexes alike. Adult has black lores, chin, bib, stripe down centre of forehead and diagnostic narrow 'horns' rising over eyes. Underparts whitish. Juvenile duller and paler and lacks black, but whiter face than Rufous-necked Snowfinch. Race *barbata* has greyer upperparts, without ginger hue; *ventorum* is even paler with yellow tinge on side of nape. Iris—brown; bill—black (adult) or straw (juvenile); feet—black.
Voice: Rather silent. Rapid contact twittering on ground and in air.
Range: Tibetan Plateau to W and NW China.
Distribution and status: Patchy distribution but locally common from 4200–5000 m. Nominate race occurs in Karakoram and W Kunlun Mts of Xinjiang, S Qinghai and S Xizang; *barbata* occurs in Qilian and Dakong Mts S to Lake Qinghai; and *ventorum* occurs in SE Xinjiang to W Qaidam Basin in Qinghai.
Habits: Inhabits dry, stony steppes with stunted grasses. Associated with pika colonies. Forms large winter flocks with other snowfinches. Has stiff-winged, hovering display flight. Quite tame. Runs mouse-like over ground.

WAGTAILS AND PIPITS—Subfamily: Motacillinae

A moderately large worldwide subfamily of slender, terrestrial birds which walk with a deliberate gait; formerly treated as a family. Many species 'wag' their tails, giving the wagtails their English name. They have slender bills and long thin legs. All species are insectivorous but also eat other small invertebrates and many are migratory. Most pipits superficially resemble larks but the longer legs and finer bills are diagnostic. Twenty species occur in China. Most are migratory.

1206. Forest Wagtail *Dendronanthus indicus* [Tree Wagtail]   PLATE 117
Shan jiling

Description: Medium-sized (17 cm) brown and pied forest wagtail. Upperparts grey-brown with white eyebrow; wings boldly patterned black and white; underparts white with two black bars across the chest, the lower sometimes incomplete. Iris—grey; bill—horn-brown with paler lower mandible; feet—pinkish.
Voice: A loud *chirrup* frequently uttered. Short *tsep* call in flight.
Range: Breeds in E Asia; migrates S in winter to India, SE China, SE Asia, Philippines and Greater Sundas.
Distribution and status: Breeds in NE, N, C and E China. Winters in S, SE and SW China, Hainan and SE Xizang up to 1200 m. Locally common.
Habits: Walks singly or in pairs on open parts of the forest floor. Gentle lateral swaying motion of tail, unlike the vertical tail-wagging of other wagtails. Rather tame; when disturbed flies in low undulating flight to alight again a few metres ahead. Also perches in trees.

1207. White Wagtail *Motacilla alba* [Pied/Common   PLATE 117
Pied Wagtail > Masked Wagtail] Bai jiling

Description: Medium-sized (20 cm) black, grey and white wagtail. General plumage grey above, white below. Wings and tail are marked black and white. In winter, hind-crown, nape and breast are marked black but less extensively than in breeding season. Extent of black varies considerably with race. Races *dukhunensis* and *ocularis* have black chin and throat and *baicalensis* has grey chin and throat, otherwise white. Race *ocularis* has black stripe through eye. Females similar to males but duller. Immatures are grey where adults are black. Iris—brown; bill and feet—black.
Voice: Clear, hard *chissick*.
Range: Africa, Europe and Asia. Birds breeding in E Asia winter S to SE Asia and Philippines.
Distribution and status: Race *personata* breeds in NW China and *baicalensis* in extreme N and NE China; *dukhunensis* is recorded on passage in NW China; while *ocularis* winters in S China, Hainan and Taiwan. Common at moderate altitudes up to 1500 m.
Habits: A bird of open spaces mainly near water, paddy fields, along stream edges and on roads. Flies with low dipping flight giving alarm call when disturbed.
Notes: Records of the larger White-browed Wagtail *M. maderaspatensis* from Yunnan in the last century must be regarded as doubtful. This species can, however, be expected in lowlands of SE Xizang. It is entirely black above with a broad white eyebrow.

1208. Black-backed Wagtail *Motacilla lugens*   PLATE 117
Heibei jiling

Description: Large (19 cm) black and white wagtail of White Wagtail complex with all-black back. Wings mostly white in flight. Wintering birds have back grey with black spots. Race *alboides* has black chin, nominate has black stripe through eye. Iris—brown; bill—black; feet—black.

Voice: Sharp disyllabic *chunchun, chunchun.*

Range: Breeds in Siberia, winters to E China. Race *alboides* breeds northern SE Asia, Himalayas; winters to S Indian subcontinent and SE Asia.

Distribution and status: Migrates through NE China provinces to winter in Fujian, Taiwan and Guangdong. Uncommon. Race *alboides* breeds through S and E Xizang, Yunnan, Sichuan and Guizhou; *leucopsis* breeds from NE China to N Qinghai, Sichuan, E, S and SE China and Taiwan, wintering in coastal provinces of S and SE China and Taiwan; and *lugens* is recorded on passage on E coast with some wintering on Taiwan.

Habits: As White Wagtail.

Note: Often treated as a race of White Wagtail.

1209. Japanese Wagtail *Motacilla grandis* Riben jiling   PLATE 117

Description: Large (20 cm) black and white wagtail. Above largely black; forehead, chin and eyebrow white. Underparts white. Wings black with white bars and feather edges. Tail black with white edges. Iris—dark brown; bill—black; feet—black.

Voice: Buzzing *bi* or *ji* flight call. Complex but distinctive song *tz tzui tztzui-tztzui pitz pitz tztzui pitz pitz-bitz bitzeen bitz bitzeen-bitz bitzeen tztzui tzigi chigi jijijiji.*

Range: Japan; one breeding record S Korea.

Distribution and status: Occasional winter visitors reach Taiwan and Hebei.

Habits: Prefers farmland, paddy fields and streams.

1210. Citrine Wagtail *Motacilla citreola* [Yellow-headed,   PLATE 117
Yellow headed Wagtail] Huangtou jiling

Description: Smallish (18 cm) wagtail with bright yellow head and underparts. Upperparts vary: race *citreola* has grey back and wings; *werae* is paler grey on back; and *calcarata* is black on back and wings. Two white wing bars. Female has grey crown and cheeks, but separable from Yellow Wagtail by grey back. Immature has dull white in place of yellow. Iris—dark brown; bill—black; feet—blackish.

Voice: Wheezy *tsweep* less harsh than Grey or Yellow Wagtails. Song from perch or in flight is repeated calls with warbled phrases.

Range: Breeds N Middle East, Russia, C Asia, NW Indian subcontinent, N China; winters to India and S China, SE Asia.

Distribution and status: Race *werae* breeds in NW China to N of Tarim Basin; *citreola* breeds in N and NE China, migrating in winter to S China coast; *calcarata* breeds across Tibetan Plateau and WC China, migrating to winter in SE Xizang and Yunnan.

Habits: Favours marshy meadows, tundra and willow thickets.

1211. Yellow Wagtail *Motacilla flava* [> Siberian/Siberian Yellow/   PLATE 117
Grey-headed/Dark-headed Wagtail] Huang jiling

Description: Medium-sized (18 cm) brownish or olive wagtail. Similar to Grey Wagtail but olive-green or olive-brown, not grey, on back and shorter tail, also no white wing bar or yellow on rump visible in flight. Races vary: male of commoner *simillima* has grey crown, white supercilium and throat; *taivana* has crown same olive colour as back, and yellow supercilium and throat; *tchutschensis* has dark blue-grey crown and nape, white supercilium and white throat; *macronyx* has grey head, no supercilium, white chin and yellow throat; *leucocephala* has white crown and sides of head; *plexa* has slaty-grey crown and nape; *melanogrisea* has olive-black crown, nape and sides of head. Non-breeding plumage is browner and duller than breeding but by March and April birds are assuming full coloration. Female and immature lack yellow vent. Immature has white belly. Iris—brown; bill—brown; feet—brown to black.

Voice: A thin musical *tsweep*, rising slightly at end, given in flock flight. Song is repetition of call interspersed with warbles.

Range: Breeds Europe to Siberia and Alaska; migrates S to India, China, SE Asia, Philippines and Indonesia to New Guinea and Australia.

Distribution and status: This is a common summer breeder, winter visitor and passage migrant to lowland areas. Races *plexa, angarensis, simillima, beema* and *tschuschensis* breed in E Siberia but are seen on passage through eastern provinces; *simillima* also passes through Taiwan. Race *macronyx* breeds in N and NE China, wintering in SE China and Hainan. Race *leucocephalus* breeds in NW China and winters in Kashi area. Race *melanogrisea* breeds in Tianshan and Tarbagatai Mts of W Xinjiang. Race *taivana* passes through E China and winters in SE China, Taiwan and Hainan.

Habits: Frequents rice fields and marsh edges and pastures. Often in very large flocks, feeding around cattle and buffalo.

1212. Grey Wagtail *Motacilla cinerea* Hui jiling   PLATE 117

Description: Medium-sized (19 cm) long-tailed, greyish wagtail with yellow-green rump and yellow underparts. Distinguished from Yellow Wagtail by grey mantle, white wing bar and yellowish rump in flight and longer tail. Underparts yellow in adult or whitish in young. Iris—brown; bill—brownish-black; feet—pinkish-grey.

Voice: Shrill *tzit-zee* or single hard *tzit* uttered in flight. Song is series of trilled phrases with shivering call notes in fluttery flight.

Range: Breeds Europe to Siberia and Alaska; migrates S to Africa, India, SE Asia, Philippines and Indonesia to New Guinea and Australia.

Distribution and status: Race *robusta* breeds in W Tianshan (possibly *cinerea*) in NW China and in NE to C China; also Taiwan. Winters in SW China, middle Changjiang, S and SE China, Hainan and Taiwan. Generally common at all altitudes.

Habits: Frequents rocky streams where it searches for food in damp gravel or sand; also on alpine meadows of highest mountains.

1213. Richard's Pipit *Anthus richardi* Tian liao   PLATE 118

Description: Large (18 cm) long-legged, brown, streaked pipit of open grassland. Upperparts streaked brown with pale buff eyebrow; underparts buff with dark streaks on upper breast. Iris—brown; bill—brown above, yellowish below; feet—yellow-brown with hind claw noticeably flesh-coloured.

Voice: Harsh, high-pitched, long *shree-ep* in flight and when flushed; also chirps. Song, given in spiralling flight, is ringing, monotonous *chee-chee-chee-chee-chia-chia-chia* with last three notes falling in pitch.

Range: C Asia, India, China, Mongolia and Siberia through SE Asia, Malay Peninsula and Sumatra.

Distribution and status: Common seasonal migrant up to 1500 m. Nominate race breeds in E Qinghai, Altai and Tarbagatai Mts of NW China, migrating S in winter; *centralasie* breeds from E Qinghai and N Gansu to Tianshan in W Xinjiang, migrating south in winter; and *sinensis* (including *ussuriensis*) breeds in N, NE, C, S and SE China, Hainan and Taiwan and is partially migratory.

Habits: Favours open coastal or montane grassy meadows, burnt grassland and dry paddy fields. Occurs singly or in small flocks. Stays on the ground where it stands very upright. Flies with undulating flight, calling with each dip.

Note: Formerly sometimes considered a single species with Common Pipit *A. novaeseelandiae*, together with Paddyfield Pipit, and the African and Australian forms.

1214. Paddyfield Pipit *Anthus rufulus* Daotian liao PLATE 118

Description: Large (16 cm) tall-standing pipit. Similar, but smaller than, migratory Richard's Pipit with shorter tail, shorter legs and hind claw, smaller bill and more horizontal carriage. Iris—brown; bill—brownish-pink; feet—pink.

Voice: Repeated *chew-ii, chew-ii* or *chip-chip-chip* in undulating song flight and weak, chattering *chup-chup* call.

Range: Indian subcontinent to Burma, SE Asia and SW China.

Distribution and status: Common in S Sichuan, and Yunnan wintering population also reaches Guangxi and Guangdong.

Habits: Found in rice fields and short grazed pastures.

Note: Often treated as a race of Richard's Pipit. Runs rapidly on ground wagging tail when feeding.

1215. Tawny Pipit *Anthus campestris* Pingyuan liao PLATE 118

Description: Large (18 cm) pipit very similar to Richard's but marginally smaller with shorter legs and more horizontal carriage. Has less pronounced streaking on sandy-grey upperparts and almost no streaking on pale buff underparts (except in immature). In hand, the hind claw is shorter and more curved than Richard's and the tarsus is shorter (<28 mm). Similar to Paddyfield Pipit but with longer tail. Iris—dark brown; bill—pinkish; feet—pale yellow.

Voice: Loud song is melancholy *cher-lee*; calls include clear loud *tchilip* or *tzeep* and hoarse, mellow *chep*.

Range: Europe to Asia Minor, Iran and N Africa, winters in N Africa, Arabia, Afghanistan, NW India.

Distribution and status: Breeds in NW and W Xinjiang, especially Tianshan. Migrates S in winter. Rare.

Habits: A bird of dry open ground and fields.

1216. Blyth's Pipit *Anthus godlewskii* Bulaishi yuanliao PLATE 118

Description: Large (18 cm) pipit very similar to Richard's and Paddyfield Pipits and immature Tawny Pipit. Smaller and more compact than Richard's Pipit with shorter tail, shorter legs and hindclaw, shorter and more pointed bill and shorter legs. Upperparts more heavily streaked and underparts often more uniformly buffish. Median wing coverts have broader pale tips giving distinct wing bar. Distinguished from Paddyfield Pipit by call, larger size, patterning of median coverts and more heavily streaked upperparts. In hand has shorter tarsus than Richard's or Paddyfield and paler lores and longer wing than immature Tawny, more curved hindclaw and shorter bill. Iris—dark brown; bill—flesh-coloured; feet—yellowish.

Voice: Mellow *chup* and diagnostic loud buzzy *spzeeu* with slight drop at end.

Range: Mongolia, Transbaikal Russia, Siberia and NE China; winters to India.

Distribution and status: Breeds on W side of Greater Hinggan Mts through Nei Mongol to Qinghai and Ningxia. Migrates S to SE Xizang, Sichuan and Guizhou. Rather rare up to 3400 m. Vagrant Hong Kong.

Habits: Favours open fields, lake shores and steppes.

1217. Tree Pipit *Anthus trivialis* Lin liao   PLATE 118

Description: Medium-sized (16 cm) buffish-brown or greyish-brown pipit. Strongly streaked black on head and mantle. Underparts buffy-white, heavily streaked on breast. In race *haringtoni* streaking extends to flanks. Browner than Olive-backed Pipit, without greenish-olive cast, and more heavily streaked on back. Has less striking face pattern. There is a small white triangle on inner web of second outermost tail feather. Bill is short and hindclaw short and heavily curved. Iris—brown; bill—brown above, pink below; feet—pinkish.

Voice: Harsher *teez* call than Olive-backed Pipit. Song is loud, far-carrying series ending in finch-like trill.

Range: Europe to Lake Baikal and W Himalayas. Winters in Africa, Mediterranean and Indian subcontinent.

Distribution and status: Race *haringtoni* breeds in NW Xinjiang, in W Tianshan; nominate breeds in Russia. Both winter south. Vagrant Guangxi. Generally rare.

Habits: Favours grassy and bushy habitat at forest edge.

1218. Olive-backed Pipit *Anthus hodgsoni* [Spotted Pipit/   PLATE 118
Olive/Indian/Oriental Tree-Pipit] Shu liao

Description: Medium-sized (15 cm) olive pipit with bold white supercilium. Distinguished from other pipits by less streaked upperparts, buffy throat and flanks and heavier black streaking on breast and flanks. Race *yunnanensis* is more lightly streaked on mantle and belly than nominate. Iris—brown; bill—pinkish below, horn above; feet—pink.

Voice: Thin, hoarse *tseez* call in flight and a single phrase *tsi.. tsi..* repeated at rest from tree perch or on ground. Song is faster and higher-pitched than Tree Pipit with hard wren-like trill.

Range: Breeds Himalayas and E Asia; migrates in winter to Indian subcontinent, SE Asia, Philippines and Borneo.

Distribution and status: Nominate race breeds in NE China and Himalayas; wintering in SE, C and S China, Taiwan and Hainan. Race *yunnanensis* breeds from S Shaanxi to Yunnan and S Xizang; wintering in S China, Hainan and Taiwan. Common up to 4000 m in open wooded areas.

Habits: Prefers more wooded habitat than other pipits and alights in trees when disturbed.

1219. Pechora Pipit *Anthus gustavi* [Siberian Pipit >   PLATE **118**
Menzbier's Pipit] Bei liao

Description: Medium-sized (15 cm) brown pipit. Similar to Olive-backed Pipit but white streaks on back form a double V-shape and distinguished by browner coloration. Black moustachial prominent. Race *menzbieri* has yellowish underparts. Distinguished from Red-throated Pipit by white bars on back and wing, whiter belly and lack of white edge to tail. Iris—brown; bill—horn above, pink below; feet—pink.
Voice: Hard *pwit* call.
Range: Breeds NE Asia and China; migrates in winter to SE Asia, Philippines, Sulawesi and Borneo.
Distribution and status: Rare. Race *menzbieri* breeds in NE Heilongjiang and passes through Shandong. Nominate race recorded on passage in Jiangsu.
Habits: Prefers open wet grassy areas and coastal forest. Sometimes alights in trees.

1220. Meadow Pipit *Anthus pratensis* Caodi liao   PLATE **118**

Description: Medium-sized (15 cm) olive-brown pipit with slender bill and fine black streaking on crown, broadly on back but unstreaked rump. Underparts buff streaked brown anteriorly. Tail brown with broad white wedge near tip of outer feathers and adjacent feathers with white tip. Breast less heavily streaked than Tree Pipit but flanks more streaked. Lacks white supercilium and bold wing bars of Rosy Pipit. Iris—brown; bill—horn; feet—pinkish.
Voice: Soft, squeaky *sip-sip-sip*.
Range: Breeds in W Palearctic. Winters to N Africa, Middle East and Turkestan.
Distribution and status: Rare winter visitor to W Tianshan, NW Xinjiang on grassland and stony semi-desert.
Habits: Has distinctive creeping walk. Lives in loose flocks.

1221. Red-throated Pipit *Anthus cervinus*   PLATE **118**
Honghou liao

Description: Medium-sized (15 cm) brown pipit. Distinguished from Olive-backed Pipit by browner upperparts and rump more heavily streaked and blotched with black, less bold black streaking of breast and pinkish throat coloration in summer. Distinguished from Pechora Pipit by pinkish-buff rather than white belly (in summer), lack of white bars on back and wing and by call. Iris—brown; bill—horn with yellow base; feet—flesh.
Voice: Thin, high-pitched *pseeoo* call in flight, more musical than other pipits.
Range: Breeds N Palearctic; migrates to Africa, N Indian subcontinent, SE Asia reaching Malay Peninsula, Philippines, Sulawesi and Borneo.
Distribution and status: A not uncommon visitor passing through N, E and C China to winter S of Changjiang valley, and Hainan and Taiwan.
Habits: Prefers cultivated wet areas including rice fields.

1222. Rosy Pipit *Anthus roseatus* Fenhongxiong liao PLATE **118**

Description: Medium-sized (15 cm) greyish, streaky pipit with prominent eyebrow. Breeding birds unmistakable with mostly unstreaked pink underparts and eyebrow. Non-breeding birds are identified by bold buffy-pink supercilium, grey back boldly streaked black, and breast and flanks more heavily spotted or streaked black. In hand, lemon-yellow axillaries diagnostic. Iris—brown; bill—grey; feet—pinkish.

Voice: Call is weak *seep-seep*. Song in display flight is *tit-tit-tit-tit-tit teedle teedle*.

Range: Himalayas, Tibetan Plateau and China; wintering to N Indian subcontinent plains.

Distribution and status: Breeds from W Xinjiang, around edges of Xizang Plateau, and E to Shanxi and Hebei and S to Sichuan and Hebei. Winters S to SE Xizang and Yunnan. Straggler on Hainan. Fairly common on alpine meadows and grassy plateaux from 2700–4400 m. Winters lower on rice fields.

Habits: Usually skulking near streams. More horizontal posture than most pipits.

1223. Water Pipit *Anthus spinoletta* Shui liao PLATE **118**

Description: Medium-sized (15 cm) streaky greyish-brown pipit with streaking on crown. Breeding: diagnostic orange-buff underparts, darker on breast, only faintly streaked on sides of breast and flanks. Winter: very dark grey-brown upperparts heavily streaked anteriorly; dull buff underparts heavily streaked anteriorly. Iris—brown; bill—blackish with pink below in winter; feet—black.

Voice: Sharp double note *tsu-pi* or *chu-i* repeated several times when flushed. Thinner and sharper than Rosy Pipit.

Range: Palearctic, NW India, China, Taiwan and N Indochina; winters to S of breeding range.

Distribution and status: Race *coutellii* breeds in NW Xinjiang, Qinghai and W Gansu but winters in S China and Taiwan. Quite common during migration.

Habits: Favours alpine pastures and grassy areas close to streams.

1224. Buff-bellied Pipit *Anthus rubescens* [American Pipit] PLATE **118**
Huangfu liao

Description: Smallish (15 cm) brown heavily streaked pipit. Similar to Olive-backed but upperparts browner and breast and flanks more heavily streaked, with blackish patch on side of neck. Edges of primaries and secondaries are white. Rare race *rubescens* is browner and less streaked. Iris—brown; bill—horn above, pinkish below; feet—dull yellow.

Voice: Flight call is sharp *jeet-eet*, less shrill than Water Pipit. Song is rapid series of *chee* or *cheedle* notes.

Range: Breeds: W Palearctic, NE Asia and N America, wintering southward.

Distribution and status: Race *japonicus* breeds in Siberia but winters over NE China to Yunnan and Changjiang valley. Quite common. N American race *rubescens* has been recorded as vagrant in Turkestan and may occasionally reach Xinjiang.

Habits: Uses moist grassy areas along streams and rice fields in winter.

1225. Upland Pipit *Anthus sylvanus* Shan liao   PLATE 118

Description: Large (17 cm) rich tawny-brown streaky pipit with white supercilium. Like Richard's and Paddyfield Pipits but browner, with more extensive streaking on underparts, shorter stouter bill, hindclaw shorter and different call note. Tail feathers narrow and pointed. Axillaries pale yellow. Iris—brown; bill—pinkish; feet—reddish.

Voice: High-pitched sparrow-like call *zip zip zip* given from ground and far carrying song *weeeee tch weeeee tch*, more like a bunting than pipit.

Range: Baluchistan, Himalayas to S China.

Distribution and status: Uncommon among grassy hills with bushes in Sichuan, Yunnan, and most of China S of Changjiang River.

Habits: Singly or in pairs. Flicks tail sharply rather than wagging it.

ACCENTORS—Subfamily: Prunellinae

A small Old World subfamily of 13 species of small birds with slender pointed bills formerly treated as a family. They feed on insects and fruits on the ground or in bushes and creep quietly among undergrowth or bare ground. Birds make quick wing flicks and tail jerks in display. Sexes are alike and most species form flocks in winter.

1226. Alpine Accentor *Prunella collaris* Ling yanliao   PLATE 119

Description: Large (17 cm) streaky brown accentor with contrasting black greater wing coverts tipped white forming two dotted wingbars. Head and central underparts ashy-brown, flanks streaked rich chestnut, undertail coverts black with white edging, throat white barred with black spots. Primaries are brown with rufous edges forming contrasting wing panel. Tail is dark brown with a white tip. Immature has underparts rufous-grey streaked with black. Iris—dark brown; bill—blackish with yellow at base of lower mandible; feet—reddish-brown.

Voice: Call is rolling *churrup* or *chu-chu-chu*; alarm is sharp *tchurrt* and song is clear and melodious with trills and some squeaky notes.

Range: Palearctic, Himalayas, N and W China and Taiwan.

Distribution and status: Common in scrub and bare areas of alpine meadows above tree-line through NE and NC China, Himalayas and Xizang plateau. Populations that breed in the NE winter in eastern provinces.

Habits: Generally singly or in pairs, rarely forms flocks. Often sits on prominent rock. Flight is quick and fluent; undulating before diving to cover. Birds are very tame.

1227. Altai Accentor *Prunella himalayana* [Himalayan/   PLATE 119
Rufous-streaked Accentor] Gaoyuan yanliao

Description: Medium-sized (16 cm) streaky brown accentor with underparts streaked rufous and white. Upperparts similar to Alpine. White throat edged black and spotted with brown at sides. Centre of belly creamy white. Iris—reddish; bill—blackish, lower bill with yellow base; feet—dull yellow to orange.

Voice: Call is silvery and finch-like *tee tee*; gives twittering contact call when feeding; song is sweet trilling warble.

Range: C Asia to NW Mongolia, Russia E of Lake Baikal, NW China and SW Tibetan Plateau. Winters in E Afghanistan, NW Indian subcontinent.

Distribution and status: Uncommon resident in NW China (Altai and Tianshan Mts of Xinjiang and S and W Xizang) at 3500–5500 m in rocky alpine meadows.

Habits: Lives in small to large flocks, sometimes mixed with other accentors and mountain finches.

1228. Robin Accentor *Prunella rubeculoides* [Robin Hedge   PLATE 119
Sparrow] Qu yanliao

Description: Medium-sized (16 cm) greyish accentor with cinnamon-chestnut breast. Head, throat, upperparts, wings and tail ashy-brown with faint black streaking on mantle; wing coverts narrowly edged white; wing feathers edged brown; narrow black collar between grey throat and rufous breast; underparts otherwise white. Iris—reddish-brown; bill—blackish; feet—dull reddish-brown.

Voice: Call is trill; alarm is sharp *zieh-zieh* and song a simple sweet chirping *si-ti-si-tsi, tsutsitsi* or *tzwe-e-you, tzwe-e-you.*

Range: Himalayas, S Tibetan Plateau and C China.

Distribution and status: Uncommon resident in N and E Qinghai, Gansu, W Sichuan and S Xizang in grassy meadows and rhododendron willow scrub from 3600–4900 m.

Habits: Typical of genus. Tame and confiding.

1229. Rufous-breasted Accentor *Prunella strophiata*   PLATE 119
Zongxiong yanliao

Description: Medium-sized (16 cm) streaky brown accentor with narrow white supraloral line and diagnostic cinnamon eyebrow behind eye. Underparts white streaked with black except for cinnamon breast band. Iris—pale brown; bill—black; feet—dull orange.

Voice: Call is high-pitched chattering *tirr-r-rit*; song is wren-like but less loud and interspersed with harsh notes.

Range: E Afghanistan, Himalayas, NE Burma, SE Tibetan Plateau and C China.

Distribution and status: Uncommon resident from 2400–4300 m in S and SE Xizang, Qinghai, Gansu, Qinling Mts of Shaanxi, W Sichuan and NW Yunnan. Descends in winer.

Habits: Favours upper forest and scrub above tree-line.

1230. Siberian Accentor *Prunella montanella*   PLATE 119
Zongmei shanyanliao

Description: Smallish (15 cm) mottled brown accentor with boldly patterned head. Top and sides of head blackish, otherwise ochre-yellow. Orange-buff supercilium and throat, separate it from Brown Accentor. Iris—yellow; bill—horn; feet—dull yellow.

Voice: Call a ringing *seereesee* or *si-si-si-si*; song is a squeaky warble.

Range: Breeds across Russia and Siberia, Korea to Japan. Accidental in Alaska and Europe.

Distribution and status: Winters in N and NE China. Rare in Qinghai and N Sichuan to Anhui and Shandong. Rarely south to Jiangsu.

Habits: Keeps to undergrowth of forest and scrub.

1231. Brown Accentor *Prunella fulvescens* He yanliao   PLATE 119

Description: Smallish (15 cm) dusky-streaked brown accentor with bold white eyebrow. Underparts white with pinkish wash on breast and flanks. Several geographic races vary in tone. Palest forms are found in Kunlun Mts. Iris—light brown; bill—blackish; feet—pale reddish-brown.

Voice: Song is a short low warble; alarm a tinny rattle; call note a bunting-like trill, *ziet, ziet, ziet*.

Range: C Asia, Afghanistan, Himalayas, Tibetan Plateau, NW and N China, S Siberia and Transbaikal Russia.

Distribution and status: Race *fulvescens* is found in NW China and W Xizang; *dresseri* ranges from Lop Lake, Xinjiang and Qinghai to S Gansu; *dahurica* is found in extreme NE as far E as Ergun River in Nei Mongol; *nanschanica* ranges through Ningxia, S Gansu, Sichuan to S and SE Xizang (sometimes considered to comprise three subspecies *nanschanica, khamensis* and *sushkini*). Generally rare or uncommon although *dahurica* can be quite common in suitable habitat.

Habits: Favours open alpine mountain slopes and screes with scrub to almost no vegetation.

1232. Black-throated Accentor *Prunella atrogularis*   PLATE 119
Heihou yanliao

Description: Smallish (15 cm) brown accentor with bold black and white patterned head. Crown brown or grey; sides of head and throat black (throat dirty white in first-winter); Bold eyebrow and fine moustachial streak white (tinged yellow in first-winter birds). Rest of upperparts brown with faint dusky streaking. Underparts pinkish on breast and flanks, ventrally whitish. First-winter birds can be mistaken for Siberian Accentor but dirty white throat diagnostic. Iris—pale brown; bill—black; feet—brownish-flesh.

Voice: Soft *trrt* and weak ringing trill *si-si-si-si*.

Range: Urals to Turkestan, NW India, NW China. Winters to Iran and NW India subcontinent and Himalayas.

Distribution and status: Race *huttoni* breeds in mountains of NW China up to 3000 m, migrating S to lower elevations in winter; *atrogularis* is winter resident in W Tianshan (W Xinjiang). Regarded as rare.

Habits: Lives in tangles of scrub in woodlands.

1233. Mongolian Accentor *Prunella koslowi*   PLATE 119
Helanshan yanliao

Description: Medium-sized (15 cm) brown accentor. Upperparts buffy-brown with faint dark streaking; throat grey; underparts buff. Tail and wings brown with buff edges. White tips to wing coverts give pale speckled lines. Iris—brown; bill—blackish; feet—pinkish.

Voice: Not described.

Range: NC China and Mongolia.

Distribution and status: Found on Helan Mts and near Zhongwei in Ningxia. Status uncertain.

Habits: An occasional bird of open scrub in arid mountains and semi-desert.

1234. Maroon-backed Accentor *Prunella immaculata*   PLATE 119

Libei yanliao

Description: Small (14 cm) unstreaked grey accentor with chestnut-cinnamon vent and maroon lower back and secondaries. Forehead frosted by whitish scalloping to feather edges. Iris—white; bill—horn; feet—dull orange.

Voice: Call note is feeble, high-pitched, metallic *zieh-dzit*. Song unrecorded.

Range: E Himalayas, N Burma, S Tibetan Plateau, N and C China.

Distribution and status: Breeds in SE Xizang, S Qinghai, S Gansu and N and W Sichuan, wintering in N and W Yunnan. Generally rare.

Habits: Lives in damp undergrowth of conifer forests at 2000–4000 m. Uses more open scrub in winter.

WEAVERS—Subfamily: Ploceinae

This is a medium-sized subfamily mostly distributed in Africa, with some Asian species, that includes the familiar finch-like weaverbirds. Weavers are small, plump, short-tailed birds with short, thick, seed-eating bills. They build covered ball-shaped nests which reach their most complex and elaborate construction among the weaverbirds. They are gregarious flock-forming birds and this, combined with their preferred diet of grass seeds, makes them serious pests to human agriculture, stealing rice and other cereals. They also eat insects. In China there are three species; two of which only doubtfully occur.

[1235. Black-breasted Weaver *Ploceus benghalensis*   PLATE 116

[Black-throated Weaver AG: Weaver Bird] Heihou zhibuniao

Description: Medium-sized (15 cm) chunky weaver with yellow crown and pale throat. Breeding male distinguished from Baya Weaver by breast band blackish-brown or brown with fulvous fringes, and from Streaked by lack of streaks on abdomen. Female and non-breeding male have brown crown but still have breast band black or fringed fulvous. Female distinguished from Baya by brown and yellow pattern on sides of head. Iris—brown; bill—bluish-plumbeous; feet—pinkish-brown.

Voice: Soft *chit-chit* by birds flying into colony. Song is soft chirping like a cricket.

Range: N and NE Indian subcontinent.

Distribution and status: Recorded by Cheng (1987, 1994) as uncommon in S Xishuangbanna in S Yunnan but not listed by Meyer de Shauensee who instead lists Streaked Weaver.

Habits: Inhabits reed swamps, grassy plains and rice fields. Lives in large, widely scattered, colonies in tall grass or reeds in groups of 2–3 nests. Nest has moderate length, vertical entrance tunnel and is non-pendant.

Note: As records are very close to known distribution of Streaked Weaver in N Laos, the specimens merit rechecking.]

[1236. Streaked Weaver *Ploceus manyar* [Striated/Manyar Weaver]   PLATE 116
Wenxiong zhibuniao

Description: Medium-sized (14 cm) golden-capped weaver. Breeding male: crown golden-yellow, rest of head, chin and throat black; underparts white with black streaks on chest; upperparts blackish-brown with fulvous edges to feathers. Non-breeding male and female: head brown with black streaks on crown and buff eyebrow and whitish patch on neck. Race *peguensis* is darker and more rufous above. Iris—brown; bill—blackish-grey to brown; feet—pale brown.

Voice: Constant chattering and whistling calls.

Range: Pakistan to SW China, SE Asia, Java and Bali.

Distribution and status: Race *peguensis* is recorded for NW Yunnan; race *williamsoni* recorded for extreme SW Yunnan (Meyer de Shauensee (1984), but records not accepted by Cheng 1987). Status uncertain but more likely to occur than Black-breasted Weaver.

Habits: Lives in large colonies around breeding trees or at other times of year in mobile flocks. Males are polygamous, each female making her own elaborated woven nest. Preferred habitats are grassy swamps and reed beds or paddy fields. Nest has short vertical tunnel.]

1237. Baya Weaver *Ploceus philippinus* [Common Weaver/Baya]   PLATE 116
Huangxiong zhibuniao

Description: Medium-sized (15 cm) golden-crowned weaver. Breeding male: crown and nape golden-yellow; side of face black; underparts buff; upperparts dark grey-brown with pale feather edges. Female: lacks yellow and black head markings; has tawny eyebrow stripe and tawny chest. Iris—brown; bill—blackish-grey to brown; feet—pale brown.

Voice: Constant raucous chattering calls, *chit-chit-chit*. Song is softer *chit-chit-chit* followed by high-pitched wheezy whistle *chee-ee-ee* given by males in chorus.

Range: India, China, SE Asia, Malay Peninsula, Sumatra, Java and Bali.

Distribution and status: Locally common in lowlands and hills up to 1000 m. Race *burmanicus* is resident in W and S Yunnan. Population in Hong Kong may be feral.

Habits: Lives in large social colonies centred on communal nest trees in open areas. Habits similar to Streaked Weaver.

MUNIAS AND JAVA SPARROW—Subfamily: Estrildinae

A large worldwide but largely tropical subfamily of small finch-like birds with stout conical seed-eating bills. They live in flocks and can be serious pests of cereal crops. There are five species in China.

1238. Red Avadavat *Amandava amandava* [Strawberry Finch/   PLATE 115
Strawberry Waxbill/Red Munia] [Hong] meihuaque

Description: Small (10 cm) red-rumped finch with white spots. Male: crimson with blackish wings and tail, regularly spotted on flanks, wings and rump with small white spots. Female: underparts grey-buff; mantle brown; rump red; wings and tail blackish; a few white spots on wings. Race *flavidiventris* is more orange with paler belly. Iris—brown; bill—red; feet—flesh.

Voice: Rather feeble thin *psheep* or *teei* notes and chirps. Song is feeble warble and twittering.

Range: Pakistan to SW China, SE Asia, Java, Bali and Lesser Sundas. Introduced Malay Peninsula, Sumatra, Borneo and Philippines.

Distribution and status: Uncommon up to 1500 m. Race *flavidiventris* occurs in tropical S Yunnan and *punicea* is resident on Hainan. Small feral populations have become established in Guangdong from escaped cagebirds.

Habits: Social, living in small flocks. Frequents scrub, grassland, cultivated areas, paddy fields and reed beds. Fast-flying restless flocks are conspicuous because of the crimson rump patch.

1239. White-rumped Munia *Lonchura striata* [Striated/   PLATE 115
White-backed/Sharp-tailed Munia, AG: Mannikin] Baiyao wenniao

Description: Medium-sized (11 cm) munia with dark brown upperparts and distinctive pointed black tail, white rump and buffy-white belly. The back is thinly streaked white and the underparts finely scaled and streaked darker buff. Immature is paler with buffy rump. Iris—brown; bill—grey; feet—grey.

Voice: Lively chirps and trilled *prrrit*.

Range: India, S China, SE Asia and Sumatra.

Distribution and status: Race *swinhoei* occurs over most of S China and Taiwan with *subsquamicollis* in tropical Yunnan and Taiwan. Locally common bird of forest edge, secondary scrub, farmland and gardens at lower altitudes up to 1600 m.

Habits: Lives in small noisy flocks; habits like other munias.

1240. Scaly-breasted Munia *Lonchura punctulata* [Nutmeg/   PLATE 115
Spotted/Spice Munia/Ricebird, AG: Mannikin] Ban wenniao

Description: Smallish (10 cm) warm brown munia. Sexes alike. Upperparts brown, streaked with white feather shafts; throat reddish-brown; underparts white scaled with dark brown on breast and flanks. Immature has underparts rich buff without scales. Race *subundulata* is darker with olive rump; *topela* has much fainter scaling on breast. Iris—reddish-brown; bill—bluish-grey; feet—grey-black.

Voice: Disyllabic chirps *ki-dee, ki-dee,* or in alarm *tret-tret*. Song is soft melody of flute-like whistles and lower slurred notes.

Range: India, S China, Philippines, SE Asia, Sundas and Sulawesi. Introduced into Australia and elsewhere.

Distribution and status: Locally common up to 2000 m. Race *subundulata* in SE Xizang; *yunnanensis* in Yunnan; *topela* in S and SE China, Hainan and Taiwan.

Habits: Frequents open grassy patches in cultivated lands, paddy fields, gardens and secondary scrub. Pairs or small flocks mix readily with other species of munia. Shows typical munia tail-wagging and active, flighty behaviour.

1241. Black-headed Munia *Lonchura malacca* [Chestnut/   PLATE 115
Tricoloured Munia AG: Mannikin] Lifu wenniao

Description: Medium-sized (11.5 cm) chestnut munia with black head, throat and vent. Sexes alike. Young birds are dirty brown all over. Race *formosana* has brownish brow and sides of face and greyish nape and crown. Iris—red; bill—blue-grey; feet—pale blue.
Voice: Shrill reedy *pwi-pwi*. Gives triple chirp in flight. Song is bill-snapping and quiet singing ending in long whistle.
Range: India, China to SE Asia, Sumatra, Borneo, Philippines and Sulawesi.
Distribution and status: Rather uncommon in tropical lowlands. Race *atricapilla* in SW and S China and Hainan; *formosana* on Taiwan.
Habits: Forms large flocks which do not mix with other species. Flocks move through paddy fields with a whirring of wings as they rise and settle.

[1242. Java Sparrow *Lonchura oryzivora* He que

1242

禾雀   ■ Resident

Description: Large (16 cm), colourful red-billed munia. Adult: head black with conspicuous white cheek patch; upperparts and breast grey; belly pink; undertail white; tail black. Immature has pinkish head with grey crown; breast pink. Iris—red; bill—deep pink; feet—red.
Voice: Low churring calls. Quiet *tup* and soft chattering song ending with whined phrase *ti tui*.
Range: Endemic to Java and Bali. Introduced widely from SE Asia to Australia.
Distribution and status: Listed as globally threatened (Collar *et al.* 1994) in endemic range. Introduced into E China and now established in Shanghai, Zhejiang, Fujian, Guangdong, Hong Kong, Guangxi and Taiwan.
Habits: A bird of towns, gardens and cultivated fields. Congregates in large flocks. Birds are highly social, rubbing next to each other at perches. In disputes over nest sites protagonists perform an elaborate body-weaving display. Used by Chinese fortune tellers.
Note: Often described with in its own genus *Padda*.]

FINCHES AND BUNTINGS—Family: Fringillidae

Finches and buntings are a large, almost worldwide, family of small, thick-billed seed-eating birds. They closely resemble the weavers but have longer, notched tails, slightly less massive bills and make open, cup-shaped, rather than covered, nests. They are flighty, flocking birds of open meadows and scrubland. Several northern breeding species migrate south into tropical Asia in winter. In China there are no less than 86 species including 19 of the beautiful rosefinch group for which China is the global centre of distribution.

FINCHES—Subfamily: Fringillinae

1243. Chaffinch *Fringilla coelebs* Cangtou yanque PLATE 121

Description: Prettily marked, medium-sized (16 cm) finch with bold white 'shoulder' patch and bold white wing bar. Breeding male has grey crown and nape, chestnut mantle and pinkish face and breast. Female and juvenile duller and greyer. Distinguished from Brambling by greenish rump and whiter shoulder bar. Iris—brown; bill—grey in male, horn in female; feet—pinkish-brown.

Voice: Call is distinctive metallic *chink* or loud *wheet*, also quieter *twit* notes. Song is musical rattle descending to rapid flourish.

Range: Europe, N Africa to W Asia.

Distribution and status: Occasional visitor. Nominate race recorded wintering in Tianshan of NW Xinjiang, Nei Mongol, Ningxia, Hebei and Liaoning.

Habits: Lives in pairs or flocks in deciduous and mixed woodlands, gardens and secondary scrub. Mixes with other finches. Often feeds on ground.

1244. Brambling *Fringilla montifringilla* Yan que PLATE 121

Description: Medium-sized (16 cm) boldly marked, robust finch with rufous breast and white rump. Adult breeding male has black head and nape, darkish back; white belly and black wings and forked tail, bold white 'shoulder' bar, rufous wing bar and white spot on base of primaries. Non-breeding male and breeding female similar but head pattern distinctively patterned brown, grey and blackish. Iris—brown; bill—yellow with black tip; feet—pinkish-brown.

Voice: Melodious song has several flute-like note followed by long *zweee* or descending rattle. Call is raspy loud monotonously repeated *zweee*. Also high-pitched notes and churrs. Flight call is *chuee*.

Range: N Palearctic.

Distribution and status: Uncommon. Winters in mixed deciduous forest and woods and clearings in conifer forest across eastern half of China, also Tianshan and W Qinghai in NW China and occasionally to S China.

Habits: Bounding, undulating flight. Lives in pairs or small flocks. Feeds on ground or trees like Chaffinch.

1245. Fire-fronted Serin *Serinus pusillus* [Red-fronted/ PLATE 120
Golden-fronted Serin] Jin'e sique

Description: Small (13 cm) mottled brown finch with blackish head and bright red frontal patch. Adults alike; plumage brighter in breeding season. Juvenile similar to adult but head paler with dull rufous forehead and cheeks and streaked dark on crown and nape. Tail forked. Bill short and conical. Iris—dark brown; bill—grey; legs—dark brown.

Voice: Song like European Goldfinch—melodious rippling trill with twittering. Call twittering or soft *duet* or trills in flight.

Range: Turkey through C Asia, Himalayas, Tibetan Plateau and NW China.
Distribution and status: Not common in China. Resident in extreme W and NW Xinjiang and NW Xizang.
Habits: Inhabits dwarf juniper zone above tree-line or rocky slopes with bushes from 2000–4600 m. Flocks fly with rapid beat interspersed by sudden bobbing undulations. Feeds mostly on ground.

1246. Grey-capped Greenfinch *Carduelis sinica* [Oriental/ PLATE 120
Chinese Greenfinch] Jinchi [que]

Description: Small (13 cm) yellow, grey and brown finch with broad yellow wing bar. Adult male has grey cap and nape, uniform brown back and yellow wing patch, basal outer tail feathers and vent. Female is duller and juvenile paler and more streaked. Distinguished from Black-headed Greenfinch by lack of dark head pattern and warmer brown plumage. Tail forked. Iris—dark brown; bill—pinkish; feet—pinkish-brown.
Voice: Song is similar to European Greenfinch but harsher with coarse *kirr* notes. Calls similar to European Greenfinch but has distinct twittering *dzi-dzi-i-dzi-i* flight call and nasal *dzweee* call.
Range: SE Siberia, Mongolia, Japan, E China, Vietnam.
Distribution and status: Common. Several resident races *chabovovi* in N Heilongjiang and Hulum Nur area of E Nei Mongol; *usuriensis* in SE Nei Mongol, S Heilongjiang, Liaoning and Hebei; nominate race over most of eastern and southern China as far W as E Qinghai, Sichuan, Yunnan and Guangzi; and *kawarahiba* breeds in Kamchatka (wintering in Japan), but occurs as vagrant to Taiwan.
Habits: Uses scrub, open fields, plantations, gardens and edges of forest up to 2400 m.
Note: European Greenfinch *Carduelis chloris* was added to the Chinese birdlist in 1999 based on a sighting in W Xinjiang. Distinguished from other Chinese greenfinches by larger size, greener coloration, green crown and heavier bill.

1247. Yellow-breasted Greenfinch *Carduelis spinoides* PLATE 120
[Himalayan Greenfinch] Gaoshan jinchique

Description: Small (14 cm) olive and yellow finch with distinctive striped head. Female as male but duller and more streaked. Separated from Black-headed Greenfinch by yellow rump and striped head. Juvenile is paler, more streaked and very similar to Grey-capped and Black-headed Greenfinches, but more yellow on underparts and side of neck. Iris—dark brown; bill—pink; feet—pink.
Voice: Song is like high-pitched European Greenfinch. Call note also similar to European Greenfinch with twittering followed by harsh *dzwee*. Another *sweee-tu-tu* call is like a sparrow.
Range: Himalayas, W and N Burma and SW China.
Distribution and status: Nominate race locally common in S Xizang, W Yunnan and SW Sichuan from 1600–4400 m.
Habits: Inhabits open conifer forest in pairs or small flocks. Altitudinal migrant. Feeds in trees. Has bat-like song flight.

Note: Some authors (e.g. Ali and Ripley 1983) include races of *C. ambigua* in this species hence erroneous map in Clement *et al.* (1993).

1248. Black-headed Greenfinch *Carduelis ambigua* [Tibetan/   PLATE 120
Yunnan Greenfinch] Heitou jinchi[que]

Description: Small (13 cm) yellowish greenfinch with blackish-green head. Similar to Yellow-breasted Greenfinch but head unstriped and rump and breast olive not yellow. Similar to Grey-capped Greenfinch but much greener without warm brown tone. Juvenile is paler than adults and streaky. Similar to juvenile of Yellow-breasted and Grey-capped Greenfinches but darker and greener. Iris—dark brown; bill—pink; feet—pink.
Voice: Song like European Greenfinch but shriller and drier. Call is thin, high-pitched twitter *tit-it-it-it-it*, usually in flight.
Range: SE Tibetan Plateau, SW China and N Indochina.
Distribution and status: Locally common resident from 1200–3100 m (lower in winter). Race *taylori* in SE Xizang; nominate race in S and W Sichuan, W Guizhou, W and SE Yunnan and extreme SW Xizang. Vagrant to Hong Kong.
Habits: Altitudinal migrant. Lives in pairs or small flocks in open conifer or deciduous forests and open areas with scattered trees. Sometimes feeds in fields.
Note: Treated by some authors (e.g. Ali and Ripley 1983) as race of *C. spinoides*.

1249. Eurasian Siskin *Carduelis spinus* [Siskin] Huang que   PLATE 120

Description: Very small (11.5 cm) finch with distinctive short bill and boldly banded black and yellow wings. Adult male has black cap and chin and bright yellow on side of head, rump and base of tail. Female is duller, more streaked and lacks black cap and chin. Juvenile as female but browner and wing bar more orange. Separated from all other small, similarly-coloured finches by sharp, pointed bill. Iris—dark brown; bill—pinkish; feet—blackish.
Voice: Song is jangling mixture of notes with metallic twitters, trills and wheezy notes, given from high perch or in bat-like display flight. Typical call is tinny *tsuu-ee* or dry *tet-tet*. Also twitters, chirps and sharp *tsooeet* in alarm.
Range: Discontinuous. Europe to Middle East and E Asia.
Distribution and status: Fairly common. Breeds in Greater and Lesser Hinggan Mts of NE China and rarely in Jiangsu province. On passage through E China, wintering on Taiwan and in Xizang and lower Changjiang valleys and S and E coastal regions.
Habits: Winters in large flocks with undulating flight. Feeds tit-like and acrobatic.

1250. Tibetan Siskin *Carduelis thibetana* Zang huangque   PLATE 120

Description: Small (12 cm) greenish-yellow finch like a canary. Breeding male is uniform olive-green with yellow supercilium, rump and belly. Female is duller green, more streaked on upperparts and flanks, with whitish vent. Juvenile as adult female but duller paler and more streaked. Iris—brown; bill—brown-horn to grey; feet—brownish-flesh.
Voice: Song is nasal buzzing *zeezle-eezle-eeze*, interspersed with trills. Call is soft dry chattering interspersed with occasional wheezy *twang*. Flocks in flight give twittering and wheezing commotion.
Range: E Himalayas to W China.

Distribution and status: Uncommon resident in S and SE Xizang, NW Yunnan, SW Sichuan at 2800–4000 m. Lower in winter.

Habits: Altitudinal migrant in subalpine forest zone in small to large flocks. Mostly feeds in trees.

1251. European Goldfinch *Carduelis carduelis* [Goldfinch/ Eurasian Goldfinch] Hong'e jinchi[que]   PLATE 120

Description: Small (14.5 cm) finch with fine bill, red forehead and bib and boldly marked black, white and yellow wing. Chinese races greyer than European taxa and lack black on head. Tail forked, black with narrow white tips. Juvenile is browner and streaked on crown, back and breast; lacks red on head but identified by broad yellow wing bar. Iris—dark brown; bill—pinkish-orange; feet—pinkish-brown.

Voice: Call is shrill *pee-uu* and liquid twittering contact notes. Song is mix of twittering, interspersed with repeated *tew-tew-tew* and *tewee-it* phrases. Popular cagebird.

Range: Europe and Middle East to C Asia and W China.

Distribution and status: Locally common. Race *caniceps* resident in extreme SW Xizang (Zanda–Burang); *paropansi* resident in Altai and Tianshan ranges of NW Xinjiang.

Habits: Inhabits conifer and mixed forests, in gaps, and at forest edge, also orchards up to 4250 m. Lives in pairs or small flocks. Feeds on seed heads of herbs.

1252. Hoary Redpoll *Carduelis hornemanni* [Arctic Redpoll] Jibei zhudingque   PLATE 120

Description: Small (13 cm) whitish finch with blackish wings and red crown spot and black chin. Similar in all age classes to Common Redpoll but much whiter with less streaking and only limited amount of pink on breast, side of face and rump. Tail forked; rump almost white. Iris—dark brown; bill—yellow; feet—black.

Voice: Calls similar to Common Redpoll but slightly higher-pitched. Song as Common Redpoll.

Range: Holarctic in Arctic tundra zone, some wintering to south.

Distribution and status: Uncommon. Race *exilipes* winters in Hulun Nur region of E Nei Mongol, Gansu (Mulin River) and E Tianshan range of NW Xinjiang.

Habits: Inhabits dwarf birch and willow thickets, sometimes in large winter flocks. Habits like Common Redpoll.

1253. Common Redpoll *Carduelis flammea* Baiyao zhudingque   PLATE 120

Description: Small (14 cm) grey-brown finch with red crown spot. Breeding male similar to Hoary Redpoll but browner with more streaking and broader pink area of breast extending to side of face. Rump is pale grey, with brown wash and black streaks compared to almost white rump of Hoary Redpoll. Female as male but lacks pink breast. Non-breeding male as female but pink scaling on breast. Tail forked. Iris—dark brown; bill—yellow; feet—black.

Voice: Song is short rippling trill interspersed with buzzing *errr errrr* in flight display. Call is distinctive metallic twitter, plaintive *teu-teu-teu-teu* or sharp grating *eeeeze*.

Range: N Holarctic. Breeding in northern conifer forest zone, wintering in temperate woodland zone. Introduced New Zealand.

Distribution and status: Common. Nominate race winters in W Tianshan in NW China and through NE provinces to Shandong and Jiangsu. Vagrant in NE Gansu.

Habits: Rapid bouncing flight. Lives in flocks feeding largely on ground and flying up to tree-top when disturbed.

1254. Twite *Carduelis flavirostris* Huangzui zhudingque   PLATE 120

Description: Small (13 cm) streaky brown finch with pink or whitish rump. Distinguished from redpolls by lack of red crown spot, browner, darker plumage, longer tail; also by calls. Distinguished from Eurasian Linnet by yellower, smaller bill; warmer brown head and more streaking on nape and mantle; less white on wing and base of tail. Some regional variation occurs. Race *korejevi* is rather pale and *montanella* even paler with whitish underparts and pale pink or buff rump. Race *rufostrigata* is browner and more heavily streaked, deep pink rump, blackish wings and tail with broad whitish wing bars and edges of tail feathers. Race *miniakensis* has buff edges to dark streaks on upperparts, white or pale pink rump. Iris—dark brown; bill—yellow; feet—blackish.

Voice: Call is nasal twittering in flight, like Eurasian Linnet but harsher. Race *rufostrigata* has distinctive *ditoo, didoowit* and twanging *twayee*. Song is extension of call with chattering trills.

Range: Discontinuous. NW Europe, C Asia, Tibetan Plateau, W and C Himalayas to C China.

Distribution and status: Fairly common. Race *korejevi* breeds in NW China; *montanella* is resident in CW China; *rufostrigata* is resident over most of Qinghai–Xizang Plateau; and *miniakensis* is resident from the eastern edge of the plateau and E Qinghai into Gansu and Sichuan.

Habits: Altitudinal migrant. Inhabits open hills, moors and conifer and mixed forests with clearings up to 4850 m in summer. Flight is fast and undulating but erratic. Feeds mostly on ground. Roosts communally.

1255. Eurasian Linnet *Carduelis cannabina*   PLATE 120
Chixiong zhudingque

Description: Small (13.5 cm) warm brown finch with pale belly and greyish head. Breeding male has crimson scaling on crown and breast and uniform grey head and nape contrasting with uniform warm brown mantle and wing coverts. Female is less richly coloured, lacks crimson and is more heavily streaked on crown, mantle, breast and flanks. Juvenile like female but browner on head. Distinguished from Twite by warmer brown colour, less streaking on nape, greyer head, larger and grey, rather than yellow, bill and more white edging on wing and base of tail; also paler feet. Iris—dark brown; bill—grey; feet—pinkish-brown.

Voice: Song is soft and varied musical warble with trills and twitters. Call is rapid trilling given in flight, also subdued *too tee* and alarm *isooeet*.

Range: Europe to N Africa and C Asia.

Distribution and status: Uncommon. Race *bella* is resident in NW China in Altai, Tianshan, Bogdo Olashan and Kashi regions.

Habits: Altitudinal migrant. Inhabits open rocky hillsides with scattered trees and bushes. Forms flocks in winter. Flight fast, undulating and direct. Feeds on ground and in trees.

1256. Plain Mountain Finch *Leucosticte nemoricola*   PLATE 121
[Hodgson's Mountain Finch, Hodgson's Rosy Finch] Lin lingque

Description: Medium-sized (15 cm) sparrow-like, brown finch with pale streaking. Has light supercilium and thin white or creamy wing bars. There is no white in notched tail. Sexes alike, but juvenile is warmer brown than adult. Race *altaica* is more rufous-brown with paler underparts. Distinguished from Brandt's Mountain Finch by paler head and lack of pink tips to feathers of rump. Iris—dark brown; bill—horn; feet—grey.

Voice: Song is sharp twitter *dui-dip-dip-dip* from rock perch. Call is soft twitter *chi-chi-chi-chi* or shrill double-noted whistle.

Range: C Asia and Himalayas, Tibetan Plateau to C China and Mongolia from 3600–5200 m.

Distribution and status: Common. Nominate race resident in N and E Xizang, E Qinghai, Gansu, Sichuan, S Shaanxi, NW Yunnan. Race *altaica* resident in NW China and extreme W China.

Habits: Inhabits stony hillsides and alpine meadows. Altitudinal migrant descending in winter to 1800 m at edge of cultivation. Often forms large flocks which wheel in fast twisting flight.

1257. Brandt's Mountain Finch *Leucosticte brandti*   PLATE 121
[Black-headed Mountain Finch] Gaoshan lingque

Description: Largish (18 cm) high altitude finch with dark head. Similar in shape and colour to Plain Mountain Finch but crown much darker, grey nape and mantle, distinctive pale wing coverts and pinkish rump. Much darker than any snowfinches. Seven Chinese races vary in intensity of dark plumage and shades of brown and grey. Clement *et al.* (1993) recognise only five of these. Iris—dark brown; bill—grey; feet—dark brown.

Voice: No song reported. Call is loud *twitt-twitt* often in flight; also has harsh *churr*.

Range: C Asia, W and C Himalayas to W China and Mongolia.

Distribution and status: Common. Race *margaritacea* in Tarbagatai and Altai ranges of NW Xinjiang; nominate race in Tianshan and Kashi areas of W Xinjiang; *pamirensis* confined to Pamirs; *pallidior* through Kunlun Mts; *intermedia* in Qinghai and Gansu; *haematopygia* in Karakoram range; *audreyana* over Xizang Plateau and Himalayas; and *walteri* on eastern edge of plateau across to Sichuan and N Yunnan.

Habits: Favours high crags and screes and wet boggy areas. Occurs higher than Plain Mountain Finch, from 4000–6000 m in summer, down to 3000 m in winter. Forms large flocks, sometimes with snowfinches.

1258. Sillem's Mountain Finch *Leucosticte sillemi*
Hongtou lingque

1258. Sillem's Mountain Finch
Leucosticte sillemi
红头岭雀
◼ Resident

Description: Largish (18 cm) plain greyish-brown mountain finch. Similar to Brandt's Mountain Finch but head is tawny-cinnamon, without black on forehead, mantle unstreaked, paler rump and underparts and complete absence of white fringes on flight feathers, drab grey instead of blackish ground colour of flight feathers, longer wings, shorter tail and more slender legs. Juvenile is more streaked on upperparts and chin and underparts are whiter than corresponding Brandt's Mountain Finch. Iris—brown; bill—grey; feet—dark brown.

Voice: Unknown.

Range: Known only from type-locality near Karakoram Pass in extreme SW Xinjiang (Kushku Maidam area controlled by China but claimed by India) at 5125 m.

Distribution and status: Uncertain. Listed by Collar *et al.* (1994) as Data Deficient species. One adult and one juvenile collected in 1929. No subsequent records, but there has been almost no ornithological work in this area since.

Habits: Collected in mixed flocks with other mountain finches.

Note: Remained unrecognised since collection in 1929 until its recent discovery and description (Roselaar 1992).

1259. Asian Rosy Finch *Leucosticte arctoa* [White-winged   PLATE 121
Mountain Finch] Fenhongfu lingque

Description: Medium-sized (17 cm) dark finch with rosy wings and underparts. Male *brunneonucha*: forehead crown and face grey; mantle cinnamon. Upperparts dark brown scaled sandy. Wings blackish with pink feather edges. Tail blackish with white edges. Underparts brown with pink feather centres. In winter crown, nape and nuchal collar buffy-rufous. Female: duller than male with pink on wings confined to coverts. Nominate race differs. Adult male with broad edges to wing feathers makes wing look white at rest. No pink in plumage and pale rump and vent. Distinguished from rosefinches by thicker yellow bill and lack of pink on head. Iris—brown; bill—yellow with black tip; feet—black.

Voice: Song is slow, descending series of *chew* notes given from ground or in descending circular undulating flight. Call is single repeated *chew*, dry *peut* or sparrow-like chirp.

Range: NE Asia from Altai Mts through Siberia and Japan.

Distribution and status: Nominate race probably breeds in Altai Mts. Race *brunneonucha* breeds in W Heilongjiang, migrating in winter to S Heilongjiang, Liaoning and Hebei. Locally common from 3600–5500 m. Occurs to sea-level on migration.

Habits: Pairs and small to large flocks feed on ground, or in low vegetation, on barren plateau and alpine tundra. Winters on bare hillsides with scattered trees.

[**1260. Spectacled Finch** *Callacanthis burtoni*　　　　　　　PLATE 121
Hongmei jinchique

Description: Largish (17.5 cm) big-headed finch with blackish crown and bright red (male) or yellow (female) eyestripe or 'spectacles'. Male is redder and blacker than female. Both sexes have black wings with white wing bars and spots. Could be confused with Spot-winged Rosefinch but easily distinguished on bill colour, lack of supercilium, dark crown and clean white spotting on wings. Juvenile like female but duller. Iris—dark brown; bill—yellow with black tip; feet—dull yellowish.
Voice: Song is loud melodious trill. Call is loud clear whistle followed by melodious descending *pweu*.
Range: W Himalayas to Sikkim.
Distribution and status: Not recorded in China but can be expected in those forested, south-flowing Himalayan valleys that cross the Chinese border.
Habits: Inhabits subalpine conifer and rhododendron forests from 2270–3330 m. Lives in pairs or small flocks. Eats seeds of deodars.]

1261. Crimson-winged Finch *Rhodopechys sanguinea*　　　　　PLATE 121
Chichi shaque

Description: Largish (17 cm) brown finch with heavy yellow bill, crimson wings and ocular region. Male: crown blackish-brown; back brown streaked black; rump brown washed pink. Ocular region crimson, cheeks brown, eyebrow, throat and side of neck sandy. Breast brown mottled with black; abdomen whitish. Wing coverts mostly pale crimson. Flight feathers black with crimson and white edges; tertials black with white tips. Notched tail is black and white with crimson edges. Female: as male but duller and less pink. Distinguished from other desert finches by darker, more mottled plumage, dark crown and heavier bill. Iris—brown; bill—yellow; feet—brown.
Voice: Song is repeated melodious *tchwili-tchwilichip*, also more rippling song in circling flight. Call is chat-like *wee-tll-wee* or soft *chee-rup* flight call.
Range: Spain, Morocco, Turkey, Iran and C Asia to NW China.
Distribution and status: Breeds in NW Xinjiang in Tianshan, Tarbagatai and Kashi S to Pamir. A rare bird from 2000–3000 m. Winters lower.
Habits: Inhabits montane scrub on rocky mountains. Nests in rock crevices and bushes.

1262. Mongolian Finch *Bucanetes mongolicus* [Mongolian　　　PLATE 121
Trumpeter Finch] Menggu shaque

Description: Medium-sized (15 cm) uniform sandy-brownish finch with heavy, dull horn-coloured bill. Pink edges to wing feathers usually visible. Breeding male pinker, with deeper crimson on greater coverts and pink wash on rump, breast and ocular region. Distinguished from other finches by uniform colour with pale bill. Iris—dark brown; bill—horn; feet—pinkish-brown.
Voice: Rather silent. Song is slow *do-mi-sol-mi* with repeated phrases and chirps. Call is soft *djudjuvu* or twittering when feeding in flocks.

Range: E Turkey to C Asia, Kashmir, N and W China and Mongolian Gobi.

Distribution and status: Widespread and locally common to 4200 m over most of W and N Xinjiang, Qinghai, Gansu, Ningxia and Nei Mongol.

Habits: Favours mountainous areas of dry stony desert and semi-arid scrub. Rather tame. Usually in flocks.

Note: Treated by some authors as a race of Trumpeter Finch *R. githagineus* (e.g. Cheng 1987).

1263. Desert Finch *Rhodospiza obsoleta* Jüzui shaque   PLATE 121

Description: Medium-sized (15 cm) sandy-coloured finch with pink wings and shiny black bill. Wing and tail feathers black edged white and pink. Male has black lores absent on female. Separated from all similar species by uniform sandy plumage and black bill. Iris—dark brown; bill—black; feet—dark brown.

Voice: Song is soft purring interspersed with trills, rolls and call notes; harsher and more nasal than Eurasian Linnet. Call is purring *r-r-r-r-ee* or harsh *turr*, also sharp *shreep* in flight.

Range: N Africa, Middle East to C Asia and NW China.

Distribution and status: Widespread but uncommon to locally common over most of W and N Xinjiang, Qinghai, Gansu and Nei Mongol.

Habits: Inhabits semi-arid areas with scattered bushes. Avoids dry stony or sandy desert. Also found in gardens and cultivated fields. Fast undulating flight.

1264. Long-tailed Rosefinch *Uragus sibiricus*   PLATE 122
Changwei que

Description: Medium-sized (17 cm) long-tailed rosefinch with very stout bill. Breeding male: face, rump and breast pink; forehead and nape frosted grey, wings mostly white; mantle brown streaked blackish and edged pink. Paler out of breeding season. Female: streaky grey with rufous rump and breast. Distinguished from Pink-tailed Bunting by stouter bill, white outer tail feathers, pale frosted brow and pink rump. Races *lepidus* and *henrici* have shorter tail. Iris—brown; bill—pale yellow; feet—grey-brown.

Voice: Call is melodious, liquid three-noted warble *pee you een* or rising *sit-it it*. Song resounds with Chaffinch-like trills.

Range: Kazakstan, S Siberia, N and C China, N Korea and N Japan.

Distribution and status: Race *sibiricus* in NW and NE China as far W as Shanxi (Pangquangou), winters in Tianshan; *ussuriensis* in southern NE China; *lepidus* is endemic to Qinling Mts of Shaanxi through Wushan in Gansu to E Xizang; and *henrici* is endemic to Sichuan, W Yunnan and SE Xizang. Locally common.

Habits: Adults as singles or pairs; Juveniles form flocks. Feeds like Agile European Goldfinch.

1265. Blanford's Rosefinch *Carpodacus rubescens*   PLATE 122
[Crimson Rosefinch] Chi zhuque

Description: Medium (15 cm) deeply-coloured rosefinch. Male: heavily suffused crimson with no supercilium, two red wing bars, maroon back and crown and lack of streaking on crown, mantle or breast. Female: warm uniform grey-brown with no streaking on underparts. Distinguished from all other rosefinches by unstreaked underparts. Iris—brown; bill—grey; feet—ashy-brown.

Voice: Call is short thin *sip* or series of short rising and falling notes. Song unknown.

Range: Himalayas, Xizang and SW China.

Distribution and status: Uncommon resident through Himalayas, S and SE Xizang, NW Yunnan, Sichuan and SE Gansu. From 1350–4500 m.

Habits: Breeds in scrub of high rocky valleys; wintering lower in conifer and birch forests.

1266. Dark-breasted Rosefinch *Carpodacus nipalensis* PLATE 122
[Dark/Nepal Rosefinch] Anxiong zhuque

Description: Smallish (15.5 cm) dark rosefinch. Nape and upperparts dark brown, washed with crimson. Male has bright pink forehead, eyebrow, cheeks and ear coverts with dark maroon breast. Distinguished from Dark-rumped and Vinaceous Rosefinches by pink forehead, more slender bill, and supercilium not extending in front of eye. Also separated from Vinaceous Rosefinch by dark breast. Female is rather uniform greyish-brown with two pale wing bars and distinguished from Dark-rumped by lack of pale supercilium and from Vinaceous by uniform underparts and lack of pale tips to tertials. Iris—brown; bill—greyish-horn; feet—pinkish-brown.

Voice: Song is monotonous chirping. Calls include plaintive, wailed double whistle, twitters and *cha-a-rrr* alarm note.

Range: Himalayas to S Tibetan Plateau and W China.

Distribution and status: Nominate race is not uncommon resident in S, E and SE Xizang. Race *intensicolor* is found in S Gansu, W Sichuan and NW Yunnan.

Habits: Shy, active rosefinch in mixed oak with conifers and rhododendrons at tree-line. Sometimes forms single sex flocks or mixes with Crimson-browed Finches.

1267. Common Rosefinch *Carpodacus erythrinus* [Scarlet/ PLATE 122
Hodgson's Rosefinch; Scarlet Grosbeak] Putong zhuque

Description: Smallish (15 cm) red-headed rosefinch with greyish-brown upperparts and white belly. Breeding male variously suffused bright red on head, breast, rump and wing-bars depending on race *roseatus* is mostly red; *grebnitskii* has paler pink underparts. Female lacks pink and is uniform grey-brown above with whitish underparts. Juvenile as female but browner and streaked. Male is distinguished from other rosefinches by brighter red colour. Lack of supercilium, white belly and dark cheeks and ear coverts eliminate most similar species. Females are less distinctive. Iris—Dark brown; bill—grey; feet—blackish.

Voice: Song is monotonously repeated, slowly-rising whistled *weeja-wu-weeja* or variants. Call is distinctive clear rising whistle *ooeet*. Alarm call is *chay-eeee*.

Range: Breeds across N Eurasian boreal zone plus high mountains of C Asia, Himalayas and W and NW China. Winters S to India, N Indochina and S China.

Distribution and status: Common resident and migrant, generally at 2000 m–2700 m but lower in NE China and higher on Qinghai–Xizang Plateau. Race *roseatus* is widespread in NW and W Xinjiang, the whole Qinghai–Xizang plateau and eastern outliers to Ningxia, Hubei and N Yunnan. Winters in mountains of tropical SW China. Race *grebnitskii* breeds in Hulun Nur and Greater Hinggan Mts of NE China, passing through E China to winter in coastal provinces and basins of S China.

Habits: Lives in subalpine forest zone but mostly in clearings, scrub and along streams. Found singly, in pairs or in small flocks. Has undulating flight. Less skulking than other rosefinches.

1268. Beautiful Rosefinch *Carpodacus pulcherrimus* PLATE 122

Hongmei zhuque

Description: Medium-sized (15 cm) rosefinch with mottled brown upperparts, lilac-pink supercilium, cheeks, breast and rump and whitish vent. Female lacks pink and has prominent buff eyebrow. Both sexes very similar to smaller Pink-rumped Rosefinch but bill is stouter and tail proportionally longer. Race *waltoni* is paler pink than other races. Iris—dark brown; bill—pale horn; feet—orange-brown.

Voice: Described calls include soft *trip* or *trillip*, tit-like twitter and harsh *chaaannn* given in flight. Song undescribed.

Range: Himalayas, Mongolia and SW China to C and N China.

Distribution and status: Common resident at 3600–4650 m: nominate race in Himalayas and S Xinjiang; race *waltoni* also in S and SE Xizang; *argyrophrys* from NE Xizang, Qinghai, Gansu, Ningxia, W Nei Mongol, Sichuan, Shaanxi and NW Yunnan; and *davidianus* in SE Nei Mongol, N Shaanxi, Hebei and Beijing.

Habits: Favours juniper scrub with dwarf oak and rhododendrons. Descends in winter. 'Freezes' in bush until danger has passed.

1269. Pink-rumped Rosefinch *Carpodacus eos* [Stresemann's PLATE 122

Rosefinch] Shuhong zhuque

Description: Small (12.5 cm) dark rosefinch with pink supercilium, cheeks, breast and rump. Very similar to Beautiful Rosefinch but smaller with more slender bill and shorter tail. Lacks buff-brown flanks of Beautiful Rosefinch. Forehead less bright than Pink-browed Rosefinch, mantle more heavily streaked and rump paler pink. Female lacks pink in plumage and coloured as Pink-rumped Rosefinch. Iris—dark brown; bill—brown-horn; feet—pale brown.

Voice: Song undescribed. Calls include assertive *pink*, bunting-like *tsip* and harsher double note *pitrit*.

Range: SE Xizang to SW China.

Distribution and status: Uncommon resident at 3900–4900 m in E Xizang, SE Qinghai and Sichuan. Some birds winter S to NW Yunnan where breeding may also occur.

Habits: Favours open alpine meadows and dry vlleys with bushes and shrubs. Forms flocks in winter and sometimes mixes with larger Beautiful Rosefinch.

1270. Pink-browed Rosefinch *Carpadacus rhodochrous* PLATE 122

Meihongmei zhuque

Description: Smallish (14.5) bright pink rosefinch with deep pink rump. Male has broad pink forehead and supercilium, broad dark reddish eye-stripe and generally unstreaked underparts. Distinguished from Beautiful and Pink-rumped Rosefinches by brighter forehead, deeper pink rump and pinker belly. Female lacks pink and is heavily streaked above and below with pale supercilium and abdomen. Iris—dark brown; bill—brownish with dark tip; feet—brownish-pink.

Voice: Song is sweet and lilting. Call is loud *per-lee* and canary-like *sweet*.

Range: Himalayas and S Tibetan Plateau.

Distribution and status: Rare resident in S Xizang from 2250–4500 m.

Habits: Found in undergrowth of subalpine forests, forest edge and alpine grassy slopes. Winters in forest at lower altitudes.

1271. Vinaceous Rosefinch *Carpodacus vinaceus*　　　PLATE 122
Jiuhong zhuque

Description: Smallish (15 cm) dark rosefinch. Male: all-dark crimson with paler rump, supercilium and tips to tertials pale pink. More uniformly dark than any other rosefinch. Smaller than Spot-winged; darker throat than Dark-breasted or Dark-rumped. Female: olive-brown with dark streaking; tips of tertials pale buff distinguish it from Dark-breasted or Blanford's Rosefinches. Iris—brown; bill—horn; feet—brown.

Voice: Call is sharp whiplash *pwit* or high-pitched *pink*. Song is simple two seconds *peedee, be do-do*.

Range: Himalayas, SE Xizang, C China and Taiwan.

Distribution and status: Race *vinaceous* is uncommon at 2000–3400 m in bamboo forest and scrubby hillsides. On Taiwan *formosana* is found from 2300–2900 m.

Habits: Lives singly or in small flocks, keeps close to ground. Sits still for long periods.

1272. Dark-rumped Rosefinch *Carpodacus edwardsii*　　　PLATE 122
Zong zhuque

Description: Medium (16 cm) dark rosefinch with prominent eyebrow. Male: dark mauve-brown with pale pink eyebrow and throat, chin and distal edges to tertials. Dark rump; lack of pink on forehead or underparts and lack of white on wings distinguishes from other dark rosefinches. Female dark brown above, buffy below with pale buff eyebrow and heavy dark streaking; lacks white on wings. Tail slightly notched. Males of Himalayan race *rubicunda* are washed crimson on upperparts, females are very dark. Iris—brown; bill—horn; feet—brown.

Voice: Usually silent; metallic *twink* and rasping *che-wee* calls.

Range: Himalayas to W China.

Distribution and status: Scarce or only locally common in upper forest strata and alpine scrub at 3000–4250 m. Race *rubicunda* occurs in Himalayas. Nominate race is found in S Gansu and W Sichuan mountains.

Habits: Skulks on or close to ground, singly or in small parties.

1273. Pale Rosefinch *Carpodacus synoicus* [Sinai Rosefinch]　　　PLATE 123
Shase zhuque

Description: Medium-sized (15 cm) pale unstreaked rosefinch. Male: pale brown upperparts, paler underparts with bright pink face, becoming paler on breast and pale pink rump. Race *beicki* has snowy brown. Female lacks pink. No other rosefinches live

in such arid habitat. Large buffy bill and uniform brown wing separates from it Mongolian and Desert Finches. Iris—brown; bill—buff; feet—buff.
Voice: High-pitched metallic chirps *tsweet* and *chig*; the latter given on ground or in flight. Song is musical jingle with buzzing notes.
Range: Negev and Sinai deserts, NE Afghanistan to W China.
Distribution and status: Locally common in arid mountains from 2000–3500 m. Race *stoliczkae* occurs from Qinghai Lake to Yarkand and W Kunlun Mts of SW Xinjiang. Race *beicki* is found from Langzhou in Gansu to E Qinghai.
Habits: Lives in small to large parties within range of water. Roosts in cliffs or crevices. Generally shy and quiet on ground.

1274. Pallas's Rosefinch *Carpodacus roseus* [Siberian Rosefinch]   PLATE 123
Bei zhuque

Description: Medium-sized (16 cm) stocky rosefinch with longish tail. Male: pinkish-crimson head, lower back and underparts. Crown is pale and forehead and chin are frosted white. Lacks contrasting eyebrow. Upperparts and wing coverts dark brown edged pinkish-white. Breast crimson, belly pink, wing with two pale wing bars. Female: duller with upperparts streaky brown, forecrown and rump pink. Underparts streaky buff, pinkish on breast with white vent. Iris—brown; bill—greyish; feet—brown.
Voice: Usually silent; short low whistle; soft song of rising and falling notes.
Range: C and E Siberia to N Mongolia. Migrates in winter to NE China, Japan, Korea and N Kazakhstan.
Distribution and status: Uncommon winter visitor in NE China as far S as Jiangsu and W to Gansu. Winters from 1500–2500 m. Summer breeding range higher.
Habits: Lives in coniferous forest but winters in cedar groves and brush-covered hillsides.

1275. Three-banded Rosefinch *Carpodacus trifasciatus*   PLATE 123
Banchizhuque

Description: Large (18 cm) rosefinch with two prominent pale wing bars. White edges to scapulars and outer webs of tertials create diagnostic third 'band'. Male has blackish face and deep crimson crown, nape, breast, rump and lower back. Female and juvenile have dark grey upperparts heavily streaked with black. Iris—brown; bill—horn; feet—dark brown.
Voice: Generally silent.
Range: Endemic to W and S China; some wintering to SE Tibet.
Distribution and status: Distributed from S Gansu through W Sichuan to Likiang range of NW Yunnan. Recorded in winter in SE Xizang. Generally scarce but locally common in winter.
Habits: Breeds in light conifer forest from 1800 to over 3000 m but descends to farmlands and orchards in winter.

1276. Spot-winged Rosefinch *Carpodacus rhodopeplus*   PLATE 123
Dianchi zhuque

Description: Medium-sized (15 cm) dark rosefinch. Breeding male has long pale pink supercilium, dull pink rump and underparts and diagnostic pale pink spots on tertials and wing coverts. Female lacks pink and is heavily streaked with buffish underparts and long pale supercilium. Pale tips to tertials distinguishes from female Pink-browed and Beautiful Rosefinches. Female *vinaceus* has pale tips to tertials but lacks pale supercilium. Race *verreauxii* is smaller and paler pink than nominate. Iris—dark brown; bill—greyish; feet—pinkish-brown.

Voice: Generally silent. Song unknown. Gives occasional loud canary-like chirp.

Range: Himalayas to W China.

Distribution and status: Rare. Nominate race resident in Nyalam region of S Xinjiang; *verreauxii* is resident in S and W Sichuan and NE Yunnan from 3000–4600 m in summer.

Habits: Inhabits tree-line scrub and alpine meadows in summer, descending to bamboo thickets in winter. Generally shy.

1277. White-browed Rosefinch *Carpodacus thura*   PLATE 123
Baimei zhuque

Description: Largish (17 cm) robust rosefinch. Male has pink rump and crown with long pale pink supercilium becoming diagnostically white posteriorly. White tips to median wing coverts give slight wing bar. Female is distinguished from other female rosefinches by deep yellowish rump, white posterior end to supercilium and in nominate race by warm brown wash on breast contrasting with white belly. All races have heavily streaked underparts. Races vary in tones; *blythi* males have pink wash on back; *dubius* males have brown back but crimson ear coverts without dark eye-stripe; male *femininus* is similar to *dubius* but deeper purple-pink on underparts; male *deserticolor* is paler brown above. Female *femininus* lacks any warm buff hue. Iris—dark brown; bill—horn; feet—brown.

Voice: Song is Eurasian Linnet-like twittering. Common call is sharp buzzing *deep-deep*, *deep-de-de-de-de* or fast piping *pupupipipipi*.

Range: Himalayas, E Tibet Plateau to NW China.

Distribution and status: Rather common resident at 3000–4600 m. Nominate race occurs in Himalayas and Chumbi valley; *femininus* in S and SE Xizang, SE Qinghai, Sichuan and NW Yunnan; *deserticolor* is confined to Qaidam Basin and Burhan Budai Mts of Qinghai; and *dubius* in NE and E Qinghai, Gansu, Ningxia and E Xizang (Qamdo). Specimens from W Himalayas may be attributable to *blythi*.

Habits: Altitudinal migrant using alpine and tree-line scrub in summer and scrub by hillsides in winter. Lives in pairs or small flocks, sometimes mixing with other rosefinches. Feeds mostly on ground.

1278. Red-mantled Rosefinch *Carpodacus rhodochlamys*   PLATE 123
Hongyao zhuque

Description: Large (18 cm) rosefinch with heavy bill. Breeding male has entire plumage washed pinkish with bright pink sides of neck, underparts, and unstreaked pink rump and supercilium and silvery flecking on side of face. Crown and eye-stripe are dark. Adult female pale grey-brown with dark streaks and no pink in plumage. Male like Pink-browed Rosefinth but larger with heavier bill and pinker underparts. Less bright red on face and underparts than Red-breasted Rosefinch. Female separated by large size, paleness of underparts, lack of pale supercilium or wing bars. Iris—dark brown; bill—horn; feet—brown.

Voice: Song not described. Call is plaintive wheezy, whistled *kwee* or sharp *wir*.

Range: C Asia, Afghanistan, NW India, NW China and Mongolia.

Distribution and status: Nominate race is rare resident in Tianshan, Kashi and Hami areas of NW Xinjiang.

Habits: Inhabits juniper and deciduous forests and alpine meadows from 2720–900 m in summer. Descends in winter. Generally in pairs or small flocks. Rather shy and secretive.

1279. Streaked Rosefinch *Carpodacus rubicilloides*   PLATE 123
Ni dazhuque

Description: Very large (19 cm) robust rosefinch with large bill and long wings and tail. Breeding male has deep red face, forehead and underparts with fine white streaks on crown and underparts. Nape and mantle grey-brown with dark streaks and only slight pink wash. Rump pink. Female grey-brown, heavily streaked. Male separated from Great Rosefinch by less intense red overall, browner, more streaked, nape and mantle. Female by streaking on nape, back and rump, also darker brown colour. Race *lucifer* is marginally larger than nominate form. Iris—dark brown; bill—pinkish-horn; feet—greyish.

Voice: Song is slowly descending *tsee-tsee-soo-soo-soo*. Call is loud *twink* note, softer *sip* or melancholy *dooid dooid*.

Range: Himalayas to C and N China.

Distribution and status: Uncommon resident from 3700–5150 m in W and S Xinjiang (*lucifer*) and from E Xizang to C China (nominate). Winters in S Sichuan and N Yunnan (Likiang).

Habits: Inhabits high altitude, rocky screes and plateaux with scattered bushes. In winter found in thickets near villages. Shy and secretive. Flight fast and bounding. Mixes with other rosefinches.

1280. Great Rosefinch *Carpodacus rubicilla* [Caucasian   PLATE 123
Great Rosefinch] Da zhuque

Description: Very large (19.5 cm) robust, very red, rosefinch with large bill and long wings and tail. Male has forehead to hindcrown and underparts deep red, frosted with white, cheeks, nape, mantle and rump uniform red or pink. Female lacks pink and is heavily streaked below but only finely streaked on mantle. Male is very similar to Streaked Rosefinch but generally much redder with bolder frosting; less streaking on upperparts. Female separated from Streaked by less streaking on nape, back and rump, greyish-buff on underparts. Race *severtzovi* much paler. Iris—dark brown; bill—yellowish-horn; feet—dark brown.

Voice: Song (*severtzovi*) is mournful low *weeep* and a series of soft chuckles. Call is loud *twink* or sharp *twit, ping*. Sometimes twitters in flight.

Range: Caucasus, C Asia, Himalayas to W and NW China.

Distribution and status: Uncommon resident from 3600–5000 m in Xinjiang and Qinghai–Xizang plateau (*severtzovi*). Race *kobdensis* is confined to a small area of Turpan depression in Xinjiang.

Habits: Inhabits rocky screes and alpine meadows above tree-line in summer, descending to village fields in winter. Mixes in flocks with other rosefinches.

1281. Red-fronted Rosefinch *Carpodacus puniceus* [Red-breasted   PLATE 123
Rosefinch] Hongxiong zhuque

Description: Very large (20 cm) stout rosefinch with rather long bill. Breeding male has red brow and short crimson supercilium, crimson from chin to breast, pink rump and darkish eye-stripe. Female lacks pink and is heavily streaked above and below. Male distinguished from similar-sized Great and Streaked Rosefinches by grey belly and from smaller but similarly-coloured Crimson-browed Finch by streaked belly and larger size. Female is more olive than Great or Streaked Rosefinches with yellow tinge to rump. Males of race *kilianensis* are paler than nominate and have narrow crimson brow band; *longirostris* is brighter, larger and longer-billed; *sikiangensis* is paler and females have whitish throat and breast. Clement *et al.* (1993) include *szetchuanus* in *sikiangensis*. Iris—dark brown; bill—brownish; feet—brown.

Voice: Song is short *twiddle-de-de* with snatches of warbling. Call is loud cheery whistle '*are-you-quite-ready*', also chips and has cat-like *maaau* call.

Range: C Asia, N Pakistan, N India and NW China.

Distribution and status: Common resident at 3900–5700 m. Nominate race in S Xizang; *kilianensis* in extreme W China; *sikiangensis* in SW Sichuan and NW Yunnan; *szetchuanus* in S Gansu and N and W Sichuan; and *longirostris* in NE Qinghai and N Gansu. Regarded as highest altitude breeding bird in Palearctic.

Habits: Inhabits alpine meadows and high altitude rocky screes and even glaciers at snowline. Hops on ground and does not fly far when disturbed. Descends to 3000–4600 m in winter.

1282. Tibetan Rosefinch *Corpodacus roborowskii*   PLATE 123
Zang que

Description: Large (18 cm) rosefinch with diagnostically long wings extending to end of tail and slender yellow bill. Male: glistening blackish-crimson head and no contrasting eyebrow; throat blackish-crimson with white spots. Rump, flanks and edges of tail pinkish. Mantle grey, scalloped with pink edges. Female buffy-brown, lacking any red; heavily streaked. Tail slightly notched. Iris—brown; bill—yellow; feet—dark brown.

Voice: Usually silent; short plaintive whistle or trill.

Range: Restricted endemic of NE Xizang and SW Qinghai in Buchan Boda and Amnemaqen ranges.

Distribution and status: Rare resident in barren rocky steppes at 4500–5400 m.

Habits: Feeds on ground, shuffling gait but with fast elegant flight.

Note: Sometimes placed in its own genus *Kozlowia* (e.g. Cheng 1987, 1994).

1283. Pine Grosbeak *Pinicola enucleator* Song que   PLATE 121

Description: Large (22 cm), long-tailed finch with thick, hooked bill and two conspicuous white wing bars contrasting with blackish wings. Adult male is deep pink with characteristic grey face pattern. Adult female is like male but olive-green instead of pink. Juvenile is uniform dull grey with buff wing bars. Similar pattern in both sexes to White-winged Crossbill but bill hooked rather than crossed, wing bars less pronounced, tail less forked and colour less intense. Iris—dark brown; bill—grey with pink base to lower mandible; feet—dark brown.

Voice: Song is loud musical warble with flute-like notes and trills. Call is fluty whistling *teu-teu-teu* with middle note highest. Also gives quiet twittering when feeding in flock. Alarm is rasped *caree*.

Range: Breeds in conifer forest of Holarctic, generally N of 65 °N. Migrates S in winter.

Distribution and status: Very rare: occasional birds of race *pacatus* winter in Heilongjiang and of race *kamtschatkensis* in W Heilongjiang, Liaoning and reported in S and E Sichuan.

Habits: Fairly tame. Forms flocks in winter; feeds on berries and seeds.

1284. Crimson-browed Finch *Propyrrhula subhimachala* PLATE 121
[Red-headed Rosefinch, Juniper Finch] Hongmei songque

Description: Large (19.5 cm) heavily-built finch with robust bill. Adult male has scarlet brow, malar area, chin and throat; reddish-brown upperparts with chestnut rump and grey underparts. Female has red areas of male olive-yellow; upperparts tinged green–olive and grey chin and throat. First-summer male is as adult male but orange rather than red. Grey rather than brownish belly distinguishes from Red-breasted Rosefinch; also less streaking on upperparts. Female distinguished from female Scarlet Finch by yellow forehead and side of breast. Iris—dark brown; bill—blackish-brown with paler base to lower mandible; feet—dark brown.

Voice: Rather silent. Song is varied warble; also a *ter ter tee* and occasional sparrow-like chirps.

Range: Himalayas from C Nepal to SE Tibetan Plateau and SW China.

Distribution and status: Uncommon resident in upper conifer forest from 3500–4200 m in S Xizang, NW Yunnan and Sichuan.

Habits: Descends to 2000–3000 m in winter months. Feeds in low canopy or on ground in small parties or pairs.

1285. Scarlet Finch *Haematospiza sipahi* Xue que PLATE 121

Description: Largish, distinctive (18.5 cm) red or olive-brown heavy-billed finch. Male unmistakable: all-uniform scarlet with blackish flight feathers edged red. Female: olive-brown above, grey below; mottled darkish with yellow rump. Juvenile male as female but rufous tinge on upperparts and more orange rump. Iris—dark brown; bill—pinkish-brown above, yellow-horn below; feet—pinkish-brown.

Voice: Song is clear liquid *par-ree-reeeeee*. Call is loud *too-eee, pleeau* or *chew-we-auh*.

Range: Himalayas to SE China and N Indochina.

Distribution and status: Uncommon resident in SE Xizang and W and S Yunnan from 1600–3400 m.

Habits: Prefers conifer or montane subtropical forests. Generally in clearings or at forest edge. Singly or in small single-sex flocks.

1286. Red Crossbill *Loxia curvirostra* [Common Crossbill] PLATE 121
Hong jiaozuique

Description: Medium-sized (16.5 cm) finch distinguished from all other finches except White-winged Crossbill by crossed mandibles. Breeding male brick-red varying from orange to rose and scarlet with race but generally a more yellow shade of red than any rosefinch. Red colour generally rather mottled. Bill more crossed than hooked bill of Pine Grosbeak. Female as male but dull olive-green rather than red. Juvenile as female but streaked. All classes separated from White-winged Crossbill by lack of bold white wing bars or white tips to tertials. Rare individuals of Red Crossbill do show some white wing bars but never as bold or complete as in White-winged Crossbill; head also less domed. Compared with nominate race, *transchanica* has slender bill and males are very yellow; *japonica* is bright and pale with vent often white; and *himalayana* is darkest with cherry-red males and brownish females. Iris—dark brown; bill—blackish; feet—blackish.

Voice: Call is hard explosive *jip jip*, given as series in alarm or subdued *chip* when feeding. Song is loud series of call notes interspersed with trills or warbles, sometimes given in circling display flight.

Range: Holarctic and Oriental in temperate conifer forest.

Distribution and status: Locally common in pine forest at moderate altitudes. Race *japonica* breeds in pine forest of NE China and in hills of Jiangsu, wintering in S Shaanxi, Henan, Shandong and Jiangsu; *tianschanica* breeds in Tianshan of NW Xinjiang, wintering in W Xinjiang, Qinghai and Liaoning and Hebei; and *himalayensis* breeds in S and E Xizang, and from NW Yunnan, W Sichuan and W Gansu to S Qinghai. Individuals of nominate race have been recorded wintering in Qinghai (Xining).

Habits: Nomadic and partially migratory, forming flocks in winter. Flight fast, undulating. Acrobatic feeder. Removes pine seeds with crossed mandibles.

1287. White-winged Crossbill *Loxia leucoptera* [Two-barred PLATE 121
Crossbill] Baichi jiaozuique

Description: Medium-sized (15 cm) finch with crossed bill. Very similar to Red Crossbill but smaller, more slender with more rounded and domed head. Distinguished from Red Crossbill by two bold white wing bars and white tips to tertials. Breeding male is dull rosy-crimson, brighter on rump. Female as male but dull olive-yellow with yellow rump. Juvenile grey and streaked but already with white wing bars. Iris—dark brown; bill—black with pinkish cutting edges; feet—blackish.

Voice: Call is soft *glib glib*, less hard than Red Crossbill also twittering when feeding. Song is rich Eurasian Siskin-like song with trills and harsh rattles from tree-top or in hovering display flight.

Range: Temperate forests of N America and Eurasia; winters to S.

Distribution and status: Rare. Race *bifasciata* probably breeds in Lesser Hinggan Mts of Heilongjiang, winters S as far as Liaoning and Hebei. May visit Altai Mts of N Xinjiang.

Habits: As Red Crossbill.

1288. Brown Bullfinch *Pyrrhula nipalensis* He huique   PLATE **124**

Description: Medium-sized (16.5 cm) grey finch with long notched tail, powerful bill, glossy dark greenish-purple tail and wings, pale wing patch and white rump. Male has mottled scaling on forehead and narrow black face mask. Female more uniform buffy-grey. Both sexes have small white patch below eye. Race *ricketti* has blackish lores and forehead. Iris—brown; bill—greenish-grey with black tip; feet—pinkish-brown.

Voice: Call is mellow *per-lee* similar to Red-headed Bullfinch. Song is repeated mellow *her-dee-a-duuee*.

Range: Himalayas to SW China, N Burma, SE China and Taiwan, also Malay Peninsula.

Distribution and status: Locally common in subalpine forests from 2000–3700 m; lower in winter. Nominate race resident in SE Xizang west of Zangpo (Brahmaputra) River, *ricketti* from SE Xizang through NW Yunnan and S China to SE China; and *uchidai* resident on Taiwan from 1300–2400 m.

Habits: Forms small flocks in winter. Flight fast and direct.

1289. Red-headed Bullfinch *Pyrrhula erythrocephala*   PLATE **124**
Hongtou huique

Description: Largish (17 cm) chunky finch with heavy, slightly hooked bill. Very similar to Grey-headed Bullfinch but male has orange head; female is greyer than female Grey-headed with yellow-olive crown and nape. Juvenile indistinguishable from juvenile Grey-headed Bullfinch but range overlap only limited. Iris—brown; bill—black; feet—pinkish-brown.

Voice: Call is plaintive whistle *pew pew* similar to Eurasian Bullfinch. Song is low mellow *terp-terp-tee*.

Range: Himalayas.

Distribution and status: Locally common in conifer forest of SE Xizang.

Habits: As other bullfinches.

1290. Grey-headed Bullfinch *Pyrrhula erythaca*   PLATE **124**
[Beavan's Bullfinch] Chixiong huique

Description: Largish (17 cm) bulky finch with heavy, slightly hooked bill. Similar to other bullfinches but adult has grey head. Male has deep orange breast and belly. Female has warm brown underparts and mantle and black band on back. Juvenile as female but entire head brown except minimal black mask. White rump and greyish-white wing bar conspicuous in flight. Iris—dark brown; bill—blackish; feet—pinkish-brown.

Voice: Call is slow *soo-ee* similar to Eurasian Bullfinch, sometimes as triple whistle. Race *owstoni* utters soft thin *yifu yifu*. Song undescribed.

Range: Himalayas to C China and Taiwan.

Distribution and status: Locally common from 2500–4100 m. Nominate race is resident from SE Xizang through C China to SW Shanxi and S to NW Yunnan. Race *wilderi* is localised in N Hebei to W Beijing and *owstoni* is endemic to Taiwan.

Habits: Inhabits subalpine conifer and mixed forests. Lives in small flocks in winter. Rather tame.

1291. Eurasian Bullfinch *Pyrrhula pyrrhula*   PLATE 124
[Common Bullfinch>Grey Bullfinch] Hongfu huique

Description: Medium-sized (14.5 cm) chunky finch with heavy slightly hooked bill, white rump and glossy black cap and mask. Male has grey mantle and white vent. Generally grey underparts with varying amounts of pink. Bold whitish bar contrasts with black wing. Race *cineracea* has no pink in plumage; *griseiventris* has the cheeks and part of throat area pink; *cassini* and nominate *pyrrhula* have cheeks, throat, breast and belly pink. Female patterned as male but pink replaced by warm brown. Juvenile as female but lacks black cap and mask and with buffy wing bar. Black crown separates from all other bullfinches. Iris—brown; bill—black; feet—blackish-brown.

Voice: Call is distinctive soft piping *teu*. Song is call note repeated with occasional interspersed trisyllabic piping note.

Range: Temperate Eurasia.

Distribution and status: Rare in China. Race *cassini* winters in Tianshan of NW Xinjiang. Lesser Hinggan Mts and Wusuli River area of N China, and once recorded in NE Hebei. Nominate race also recorded on passage through NE China. Race *griseiventris* winters in Tianshan of NW China and also S Heilongjiang, Liaoning and N Hebei; vagrant Jiangsu and Shanghai. Race *cineracea* is recorded and probably breeds in N Heilongjiang and NE Nei Mongol.

Habits: Favours woods, orchards and gardens. In winter usually in small flocks.

Note: Races *griseiventris* and *cineracea* are treated by some authors as races of a separate species, Oriental Bullfinch *Pyrrhula griseiventris*.

1292. Hawfinch *Coccothraustes coccothraustes* Xi zuique   PLATE 124

Description: Large (17 cm) dumpy, brownish finch with massive bill, shortish tail and broad white line on 'shoulder' of wing. Sexes almost alike. Adults have narrow black mask and chin; glossy blue-black wings (greyer in female) with unusual curved and pointed tips to higher primaries. Tail is warm brown, slightly notched and with narrow white tip and black subterminal edge to outer tail feathers. Black and white wing pattern distinctive from below and above. Juvenile, as adult but darker and with dark speckles and streaks on underparts. Iris—brown; bill—horn to blackish; feet—pinkish-brown.

Voice: Song starts with whistle and ends with musical liquid notes *deek-waree-ree-ree*. Call is abrupt *tzick*, also shrill *teee* or *tzeep*.

Range: Temperate Eurasia.

Distribution and status: Fairly common. Nominate race breeds in NE China, passing through E China to winter in Changjiang and Xinjiang catchments and SE coastal provinces; *japonicus* also winters in SE coastal provinces and is straggler to Taiwan.

Habits: Inhabits woods, gardens and orchards up to 3000 m in pairs or small flocks. Generally shy and quiet.

1293. Yellow-billed Grosbeak *Eophona migratoria*   PLATE 124
[Chinese Grosbeak; Black-tailed Hawfinch] Heiwei lazuique

Description: Largish (17 cm) stocky finch with massive black-tipped yellow bill. Breeding male superficially like large bullfinch with black hood, grey body and blackish wings. Distinguished from Japanese Grosbeak by black tip to bill, white tips to primaries, tertials and primary coverts and cinnamon vent. Female as male but black on head much reduced. Juvenile as female but browner. Iris—brown; bill—deep yellow with black tip; feet—pinkish-brown.

Voice: Song is series of whistles and trills similar to Eurasian Linnet. Call is loud harsh *tek-tek*.

Range: E Siberia, Korea, S Japan and E China, wintering to S China.

Distribution and status: Locally common. Nominate race breeds in deciduous and mixed forests in NE China passing S to winter in S China and Taiwan. Race *sowerbyi* breeds in C and E China, particularly in lower Changjiang catchment as far W as W Sichuan, wintering in SW China.

Habits: Uses woodlands and orchards, never found in dense forests.

1294. Japanese Grosbeak *Eophona personata* PLATE 124
Heitou lazuique

Description: Large (20 cm) chunky finch with massive yellow bill. Sexes alike. Similar to male Yellow-billed Grosbeak but bill larger and all-yellow, vent greyish, tertials brown and white pattern on wing different. Has small white panel near tip of primaries but does not have white tips to tertials, primary coverts and primaries. These differences are very obvious in flight. Juvenile browner with black of head reduced to narrow mask; also has two buffy wing bars. Race *magnirostris* is larger and paler than nominate race with larger bill and smaller white wing patch. Iris—dark brown; bill—yellow; feet—pinkish-brown.

Voice: Call is hard *tak-tak* given in flight. Song is 4–5 fluty whistled notes.

Range: Breeds E Siberia, NE China, N Korea and Japan; wintering to S China.

Distribution and status: Locally common: nominate race winters in S China, rarely on Taiwan. Race *magnirostris* breeds in NE China (Changbai and Lesser Hinggan Mts) passing through E China to winter in S China.

Habits: More lowland than other grosbeaks. Usually in small flocks. Rather shy and quiet.

1295. Collared Grosbeak *Mycerobas affinis* PLATE 124
Heichi nilazuique

Description: Large (22 cm) big-headed black and yellow finch with massive bill. Adult male: unmistakable with black head, throat, wings and tail otherwise yellow. Female has grey head and throat and dull greyish-yellow wing coverts, scapulars and mantle. Juvenile male like adult but duller. Distinguished from all other Chinese grosbeaks by yellow nape and collar. Iris—dark brown; bill—greenish-yellow; feet—orange.

Voice: Song is clear loud whistle of 5–7 rising notes *ti-di-li-ti-di-li-um*, also creaky song punctuated with bulbul-like notes. Call is rapid series of mellow *pip* notes, also alarm *kurr* sounding like stones being cracked together.

Range: Himalayas to C and SW China.

Distribution and status: Locally common in subalpine forest from 2700–4000 m; lower in winter. Resident in SE Xizang, NE Yunnan, W Sichuan and SW Gansu.

Habits: Inhabits conifer and mixed forests with dwarf oak and rhododendron and juniper scrub along tree-line. Forms flocks in winter. Flight fast and direct.

1296. Spot-winged Grosbeak *Mycerobas melanozanthos*   PLATE 124
Banchi nilazuique

Description: Large (22 cm) big-headed yellow and black finch with heavy bill. Breeding male unmistakable with black head, throat and upperparts and yellow breast, belly and vent. Separated from Collared Grosbeak by lack of yellow collar and back, and from White-winged Grosbeak by yellow breast and black rump. Wings have conspicuous yellowish-white spots on tips of tertials, greater coverts and secondaries. Female and juvenile are streaked black and yellow and quite distinctive. Both have same spot pattern on wing and juvenile is paler yellow than female. Iris—dark brown; bill—grey; feet—grey.

Voice: Song is loud melodious three-noted whistle *tew-tew-teeeu* or oriole-like whistle *tyop-tiu*. Call is rattling *krrr* like shaken, near-empty, matchbox.

Range: Himalayas to Burma, C and SW China.

Distribution and status: Uncommon in subalpine conifer and mixed forests from 2400–3600 m; lower in winter. Resident in SE Xizang, W and NW Yunnan and W Sichuan.

Habits: Similar to Collared Grosbeak but living at slightly lower altitudes. Feeding and roosting birds maintain cackling chatter.

1297. White-winged Grosbeak *Mycerobas carnipes*   PLATE 124
Baichi nilazuique

Description: Large (23 cm) big-headed, black and dull yellow finch with heavy bill. Breeding male: superficially like male Spot-winged Grosbeak but with yellow rump, black breast and yellow spots on tips of tertials and greater wing coverts. There is a white patch at base of primaries which is conspicuous in flight. Female is similar to male but duller; black being replaced by grey and with faint pale streaking on cheeks and breast. Juvenile as female but browner. Iris—dark brown; bill—grey; feet—pinkish-brown.

Voice: Song is repetitive harsh *add-a-dit-di-di-di-dit*. Call is nasal *shwenk* or *wet-et-et*.

Range: NE Iran, Himalayas to western edge of China in Tianshan and eastward to C and SW China.

Distribution and status: Locally common in fir, pine and dwarf juniper at tree-line from 2800–4600 m in W Xinjiang (Tianshan, Kashi), S, SE and E Xizang, Sichuan, NW Yunnan, Qinghai, Gansu, S Shaanxi, Ningxia and W Nei Mongol.

Habits: Forms flocks in winter and often mixes with rosefinches. Noisy feeder when cracking seeds. Quite tame.

1298. Gold-naped Finch *Pyrrhoplectes epauletta* [Gold-crowned   PLATE 124
Black Finch, Gold-headed Black Finch] Jinzhen heique

Description: Smallish (15 cm) finch with white tramlines formed by white edges to tertials. Male: unmistakable with black plumage, bright golden crown and nape and golden 'shoulder' flash. Female has warm brown wings and underparts, grey mantle and

olive-green and grey head. Behaviour and white tramlines distinguishes from any other species in habitat. Iris—dark brown; bill—black; feet—black.
Voice: Song is rapid, high *pi-pi-pi-pi* or softer piping. Call is whistled *teeu* or squeaky *plee-e-e*.
Range: Himalayas to SE Tibetan Plateau and SW China.
Distribution and status: Uncommon in subalpine forests of SE Xizang, W and NW Yunnan and SW Sichuan. Breeds from 2700–4000 m but lower in winter.
Habits: Keeps to undergrowth or ground in rhododendron and bamboos, sometimes in small flocks. Sometimes mixes with rosefinches.

BUNTINGS—Subfamily: Emberizinae

Buntings are a large, almost worldwide, subfamily of small, thick-billed seed-eating birds formerly treated as a family. They closely resemble the weavers but have longer, notched tails, slightly less massive bills and make open, cup-shaped, rather than covered, nests. They are flighty, flocking birds of open meadows and scrubland. Several northern breeding species migrate south into tropical Asia in winter. Thirty species are recorded in China of which 22 breed and two are endemic.

1299. Pink-tailed Bunting *Urocynchramus pylzowi* [Rose Bunting/ **PLATE 122** Przevalski's Rosefinch] Zhu wu

Description: Medium-sized (16 cm) rosefinch-like bunting with very long graduated tail and slender bill. Upperparts mottled brown. Breeding male has pink supercilium, throat, breast and edges to tail feathers. Female has buffy breast with dark streaks and pale pinkish-orange base to tail. The rosefinch with such a long tail is Long-tailed Rosefinch which has a stout bill, double wing bar and white outer tail feathers. Iris—dark brown; bill—horn with pinkish lower mandible in breeding male; feet—grey.
Voice: Generally quiet. Song is hurried *chitri-chitri-chitri-chitri*. Call is clear ringing *kvuit, kvuit* in flight or when alarmed.
Range: Endemic to C China.
Distribution and status: Uncommon resident in Qinghai, Gansu, N and W Sichuan and E Xizang from 3000–5000 m.
Habits: Inhabits scrub and alpine thickets near water. Lives singly, in pairs or small flocks. Flight is weak and fluttery.
Note: This monotypic genus is sometimes treated as a rosefinch (e.g. Meyer de Shauensee 1984).

1300. Crested Bunting *Melophus lathami* Fengtou wu **PLATE 125**

Description: Unmistakable, large (17 cm) dark bunting with characteristic long thin crest. Male is glossy black with chestnut wings and tail. Tip of tail is black. Female is dark olive-brown, heavily streaked on mantle and breast; has shorter crest than male and chestnut edges to dark wing feathers. Iris—dark brown; bill—grey-brown with pink base to lower mandible; feet—purplish-brown.
Voice: The call is a quite loud squeaky *pit-pit*; also has sweet song given from prominent

perch. This has monotonously repeated strophe falling in pitch. The first notes are grating and hesitant followed by clearer ending.

Range: India, Himalayas to SE China and N Indochina

Distribution and status: Common in grassy hillsides of C, SE and SW China; vagrant to Taiwan.

Habits: This is a bird of open ground and short grassland in hilly country over much of China. Lives and feeds mostly on ground where it is an active and conspicuous bird. In winter feeds among rice stubble.

1301. Slaty Bunting *Latoucheornis siemsseni* Lan wu   PLATE 126

Description: Small (13 cm) stocky bluish-grey bunting. Male: general plumage slaty-blue grey except white abdomen, vent and outer edge of tail. Tertials blackish. Female: unstreaked dull brown with two rusty wing bars, grey rump and rufous head and breast. Iris—dark brown; bill—black; feet—pinkish.

Voice: Song is high-pitched, metallic and variable similar to *Parus* tits. Call note is sharp repeated *zick*.

Range: C and SE China.

Distribution and status: Breeds in Qinling mountains of S Shaanxi, Min Mts of N Sichuan, S Sichuan and S Gansu. Winters eastward to Hubei, Anhui, Wuyi area of Fujian and N Guangdong. Spring records from Mangshan in S Hunan and Che Ba Ling in N Guangdong (Lewthwaite *et al.* 1996) suggest that breeding range extends much further S than previously known.

Habits: Inhabits secondary forest and scrub.

1302. Yellowhammer *Emberiza citrinella* Huang wu   PLATE 128

Description: Male: large (17 cm) bunting with yellow head lightly striped greyish-green and with chestnut moustachial streak. Underparts yellow with chestnut mottling on sides of breast forming breast band; dark streaks on flanks. Rump rufous and upperparts mottled rufous-brown with dark shaft streaks and yellow edges to most feathers. Distinguished from Pine Bunting by yellow hue to underparts. Female and non-breeding male similar but more streaked, duller and less yellow, outer tail feathers edged white. Iris—dark brown; bill—blue-grey; feet—pink-brown.

Voice: Call is single *steuf* or clicking *steelit* in flight. Song from prominent perch is characteristic repetitive buzz with penultimate note raised and last a lot longer and at lower pitch, *ze-ze-ze-ze-ze-ze-zoo-ziii*, '*little bit of bread and no cheese*'. Alarm is thin *dzee*.

Range: Europe to Siberia and N Mongolia, wintering in S of range.

Distribution and status: Vagrant. Race *erythrogenys* recorded in winter in Beijing, Heilongjiang and possibly Tianshan in W Xinjiang.

Habits: Found in heath and shrubland and cultivated fields in winter. Flicks notched tail when perched. Flight quick after sloping ascent.

1303. Pine Bunting *Emberiza leucocephalos* Baitou wu   PLATE 126

Description: Large (17 cm) bunting with unique head pattern and slight crest. Male unmistakable with white crown stripe enclosed by black lateral crown stripes and white central ear coverts enclosed by black edges to cheeks. Rest of head and throat chestnut contrasting with white breast band. Female paler and less distinctive, very similar to female Yellowhammer. Distinguished by two-toned bill, paler colour and pinkish tinge, lacking yellow tinge; submoustachial stripe is whiter in female Pine Bunting. Race *fronto* has more black on forehead and lateral crown stripes, also chestnut markings deeper. Iris—dark brown; bill—grey-blue with brown culmen; feet—pinkish-brown.

Voice: Song from tree or bush is very similar to Yellowhammer *ze-ze-ze-ze-ze-ze ziiii*. Call notes also similar to Yellowhammer.

Range: Siberian taiga to NW China and NC China.

Distribution and status: Race *leucocephala* breeds in NW China, Tianshan and Altai Mts and NE China (Hulun Nur) migrating S in winter to W Xinjiang, Heilongijiang, SE Nei Mongol, Hebei, Henan, S Shaanxi, S Gansu, SE Qinghai. Vagrant to Jiangsu and Hong Kong. A second resident race *fronto* occurs in E Qaidam Basin of Qinghai and adjacent Gansu.

Habits: Prefers forest edge and clearings or burnt and logged conifer or mixed forests. Winters in arable fields, wasteland and orchards.

Note: Closely related to Yellowhammer and hybrids regularly occur in W Siberia.

1304. Tibetan Bunting *Emberiza koslowi* Zang wu   PLATE 128

Description: Largish (16 cm) long-tailed bunting. Breeding male; black head with white supercilium extending from nares to nape; nuchal collar grey; back chestnut, rump grey, chin and lores chestnut. White bib and black gorget diagnostic. Underparts grey with whitish vent. White bar on wing. Flight feathers black with pale edges. Female and non-breeding male as breeding male but duller without black gorget. Back chestnut streaked black. Throat brown streaky with long pale supercilium. Iris—brown; bill—bluish-black; feet—orange-yellow.

Voice: Song is short twittering similar to Godlewski's Bunting. Flight call is *tsip tsip* and contact call is thin drawn-out *seee*.

Range: China endemic of E Tibetan Plateau valleys.

Distribution and status: Recorded from SE Qinghai and E Xizang only at 3600–4600 m. Not recorded S of 31 °N.

Habits: Favours open arid alpine scrub, dwarf junipers and rhododendrons and bare land above tree-line. Forms small flocks in winter.

1305. Rock Bunting *Emberiza cia* Huimei yanwu   PLATE 128

Description: Largish (16 cm) bunting with diagnostic grey and black striped head and warm brown underparts. Female as male but duller. Distinguished from Godlewski's Bunting by black rather than brown stripes on head and grey of head much whiter. Race *stracheyi* is smaller and has darker underparts than race *par* or nominate form; it also has darker, more rufous rump. Both *stracheyi* and *par* have buffy wing bars. Iris—dark rufous brown; bill—grey with blackish tip with yellow or pink basally on lower mandible; feet—orange-brown.

Voice: Song is quite long, accelerating into clear twittering phrases similar to Winter Wren or Reed Bunting. Call is sharp drawn out *tsii*; longer and repeated in alarm. Other calls include short *tiip*, twittering and rolling *trrr*.

Range: NW Africa, S Europe to C Asia and Himalayas.

Distribution and status: Locally common resident up to 4000 m. Race *par* in Altai and W Tianshan in NW Xinjiang and *stracheyi* in Zanda, Gar and Burang areas of SW Xizang.

Habits: Dry sparsely vegetated rocky hillsides and ravines, moving into open bushy habitat in winter.

1306. Godlewski's Bunting *Emberiza godlewskii* PLATE 126
Geshi yanwu

Description: Large (17 cm) bunting similar to Rock Bunting but with darker grey on head and lateral crown stripes chestnut and not black. Distinguished from Meadow Bunting by grey coronal stripe. Female similar to male but paler. Some racial variation: southern *yunnanensis* darker and more rufous than nominate race and westernmost race *decolorata* the palest of all. Juvenile has black streaked head, mantle and breast— undistinguishable in field from juvenile Meadow Bunting. Iris—dark brown; bill—blue-grey; feet—pinkish-brown.

Voice: Variable and similar to Rock Bunting but introduced by more high-pitched *tsitt* note. Call is thin, drawn-out *tzii* and hard *pett pett*.

Range: Distinct racial populations in Altai, Transbaikal Russia, Mongolia, N, C and SW China, NE India; winters NE Burma.

Distribution and status: Common in foothills of Tianshan and western rim of Tarim Basin in extreme W Xinjiang (*decolorata*); SE Xizang, S Qinghai and W Sichuan (*khamensis*); W Qinghai, Gansu, Ningxia and W Nei Mongol (*godlewskii*); N Yunnan and extreme SE Xizang to C Sichuan (*yunnanensis*); N and E Sichuan to S Heilongjiang (*omissa* including *styani*). Races *godlewskii* and *yunnanensis* show some southernly migration in winter.

Habits: Uses dry rocky hillsides and ravines with scrub close to forest, also cultivated lands.

Note: Regarded by some authors as race of Rock Bunting (e.g. Cheng 1994).

1307. Meadow Bunting *Emberiza cioides* PLATE 126
Sandaomei caowu

Description: Largish (16 cm) rufous bunting with bold head pattern and chestnut breast band with white eyebrow, upper moustachial stripe, chin and throat. Breeding male has unique brown, white and black face pattern with chestnut breast and rufous rump. Female is paler with buff supercilium and malar stripe and rich buffy breast. Both sexes similar to the rare Jankowski's Bunting found only in NE China. Meadow Bunting is separated from that species by sharp contrast between throat and breast, brown rather than grey ear coverts, less prominent white wing bars and less streaked mantle. Lacks chestnut belly patch. Juvenile is paler and more streaked very similar to juvenile Godlewski's and Rock Buntings but broader rufous edges to central tail feathers. Outer tail feathers edged white. Race *weigoldi* is brighter and more chestnut than nominate race; *tanbagataica* is the palest race with less rufous rump and narrower breast band. Race *castaneiceps* is smallest and darkest race with less streaked upperparts. Iris—dark brown; bill—two-toned, dark above, blue-grey below with dark tip; feet—pinkish-brown.

Voice: Song from prominent perch is short hurried phrase similar to Godlewski's Bunting but initial stressed note lower-pitched than high-pitched initial *tsitt* of Godlewski's. Call is sharp *zit-zit-zit* in rapid series of 3–4 notes.

Range: S Siberia, Mongolia, N and E China to Japan.

Distribution and status: Races intergrade and distributions not fully known. Race *tanbagataica* is resident in Tianshan region of NW China; *cioides* is resident in Altai Mts of NW China and in E Qinghai; *weigoldi* occurs through most of NE China; and *castaneiceps* resident in C and E China with wintering birds sometimes reaching Taiwan and S China coast.

Habits: Inhabits open scrub and forest edge in hills and mountains, descending to plains in winter.

1308. Jankowski's Bunting *Emberiza jankowskii* PLATE 128
[Rufous-backed Bunting] Libanfu wu

Description: Largish (16 cm) rufous bunting with white eyebrow and dark brown submoustachial stripe. Similar to Meadow Bunting but with grey ear coverts, more streaked mantle and white wing bars. Underparts lack contrast between throat and breast, with diagnostic dark chestnut patch on central belly. Distinguished by whitish breast if belly patch obscured. Female is paler version of male, pattern also similar to Meadow Bunting but distinguished by greyer ear coverts, heavier streaking on mantle, white wing bars and pale grey central breast. Iris—dark brown; bill—two-toned with dark upper mandible and bluish-grey below with dark tip; feet—pinkish-orange.

Voice: Simple and complex songs reported, with introduction similar to Rock Bunting but end of strophes similar to last note of Yellowhammer song. Other calls include contact *tsitt* and alarm *sstlitt* or thin *hsiu*.

Range: N Korea, SE Siberia and NE China.

Distribution and status: Breeds in SE Heilongjiang and Jilin migrating S in winter to Liaoning, Hebei and SE Nei Mongol. Has declined in numbers and range for unknown reasons (Collar *et al.* 1994) and is now listed as globally Vulnerable.

Habits: Inhabits scrub and grassland in low-lying hills and valleys especially evergreen sand dunes, also stunted forests on sand.

1309. Grey-necked Bunting *Emberiza buchanani* PLATE 126
Huijing wu

Description: Medium-sized (15 cm) bunting with plain grey head, pale eye-ring, pinkish underparts, yellow submoustachial stripe. Juvenile and non-breeding birds paler with black streaks on crown, breast and flanks. Distinguished from Ortolan Bunting by lack of clear division between breast and belly and by bluish-grey rather than greenish-grey head. Iris—dark brown; bill—pinkish; feet—pink.

Voice: Flight call is soft *tsip*. Also *tchcup* call. Song from elevated perch is similar to Ortolan Bunting, *ti-ti-ti tiu-tiu-tiuu u* with second strophe at lower pitch.

Range: Turkey, Iran, C Asia mountains to W China and W Mongolia, winters in Pakistan and W India.

Distribution and status: Race *neobscura* is locally quite common breeding in barren, mid-altitude areas of W Kashi, Wushi, mid-Turpan and Tianshan Mts in NW Xinjiang, migrates S in winter.

Habits: Forms flocks in autumn prior to migration, mixing with other buntings.

1310. Ortolan Bunting *Emberiza hortulana* PLATE 128
[Ortolan] Pu wu

Description: Largish (16 cm) bunting with uniform greenish-grey head and breast, prominent pale eye-ring and characteristic pattern of yellow submoustachial stripe and throat. Distinguished from Grey-necked Bunting by greyish breast clearly separated from rufous belly, also head much greener-grey. Wing bar usually white. Female and juvenile duller with black streaks on crown, nape and breast. Lack of supercilium, bold buff submoustachial stripe and greenish tinge to head distinguishes from other buntings. Iris—dark brown; bill—pink; feet—pink.

Voice: Sings from prominent perch. Song starts with 3–4 ringing units of similar pitch followed by 1–3, usually two, clear units of lower pitch, *dzii dzii huii huii*. Calls include short *tew*, dry *plet* and metallic *ziie*.

Range: W and C Europe, C Asia to Altai Mts and W Mongolia. Migrates to Africa for winter, rarely India.

Distribution and status: Uncommon in open dry country with scattered bushes in Altai and Tianshan Mts and W Kashi region of NW Xinjiang. Migrates S in winter.

Habits: Gregarious in small flocks; feeds in trees as well as on ground; generally hops.

1311. Tristram's Bunting *Emberiza tristrami* Baimei wu PLATE 125

Description: Medium-sized (15 cm) bunting with conspicuous striped head. Adult male unmistakable with bold black and white head pattern, black throat and rufous unstreaked rump. Female and non-breeding male duller with less contrast on head, but patterned as breeding male except pale chin. Lacks yellow brown of Yellow-browed Bunting and rusty nape of Rustic Bunting. Also differs from Yellow-browed in paler, more cinnamon tail, less streaking on breast and flanks and darker throat. Iris—dark chestnut-brown; bill—bluish-grey above, pinkish below; feet—pale brown.

Voice: Song given from canopy begins with clear, high-pitched unit followed by second unit of higher or lower pitch. Call ends with a variable but simple, rapid, repeated unit often with a short *chit* note at end.

Range: NE China and adjacent Siberia, wintering to S China and occasionally N Burma and N Tonkin.

Distribution and status: Breeds in forested areas of NE China, wintering in evergreen forest of S China, recorded on passage through eastern and coastal provinces.

Habits: Mostly keeps to forest floor in dense thickets on forested hillsides. Often forms small flocks.

1312. Chestnut-eared Bunting *Emberiza fucata* PLATE 126
[Grey-headed Bunting] Chixiong wu

Description: Largish (16 cm) bunting. Breeding male unmistakable with chestnut ear coverts contrasting with grey crown and side of neck. Neck pattern is unique: black malar stripe merging into black breast gorget of merged stripes contrasting with otherwise white throat and breast, backed by rufous breast band. Female and non-breeding male are similar but paler and more nondescript, approaching first-winter Ortolan Bunting but still distinguished by more rufous ear coverts and rump, also more white on side of tail. Male of race *arcuata* is darker and more richly-coloured than nominate with more black on gorget, less black streaking on mantle and broader rufous breast band. Similar race *kuatunensis* is darker and redder above with narrow breast band. Iris—dark brown; bill—black above with grey edge; bluish-grey below with pink base; feet—pink.

Voice: Song given from top of bush is more rapid and twittering than other buntings, with staccato *zwee* notes accelerating into twitter and ending with double *triip triip*. Call is explosive *pzick* similar to Rustic Bunting.

Range: W Himalayas to China, E Mongolia and E Siberia; wintering to Korea, S Japan and N Indochina.

Distribution and status: Common in NE China (*fucata*); C, SW China and SE Xizang (*arcuata*); and less common as breeder in S Jiangsu, Fujian and Jiangxi (*kuatunensis*). Winters in Taiwan and Hainan with passage migrants through most of E China.

Habits: Typical of genus. Forms flocks in winter.

1313. Little Bunting *Emberiza pusilla* Xiao wu PLATE 125

Description: Small (13 cm) streaked bunting with striped head. Sexes alike. Breeding adult unmistakable due to small size and striped black and chestnut head, also pale eye-ring. Wintering male and female both have ear coverts and crown stripe dull chestnut, malar stripe and edge to ear coverts greyish-black, eyebrow and second malar stripe dull buffy-rufous. Upperparts brown with dark streaks. Underparts whitish streaked with black on breast and flanks. Iris—dark red-brown; bill—grey; feet—reddish-brown.

Voice: High-pitched, quiet *pwick* or *tip tip*, also *tsew*.

Range: Breeds in extreme N Europe and N Asia; migrating in winter to NE India, China and SE Asia.

Distribution and status: Common on passage across NE China and wintering in extreme W Xinjiang and most of C, E, and S China and Taiwan.

Habits: Mixes with pipits. Hides in thick cover, reed beds.

1314. Yellow-browed Bunting *Emberiza chrysophrys* PLATE 125
Huangmei wu

Description: Smallish (15 cm) bunting with banded head. Similar to Tristram's Bunting but frontal half of supercilium yellow; underparts are whiter and more streaked; wing bars whiter; rump more mottled and tail darker. Black malar stripe of Yellow-browed is less pronounced than in Tristram's and breaks up into streaked pattern of breast. Distinguished from winter Black-faced Bunting by rufous rump and more striped and contrasting head pattern. Iris—dark brown; bill—pink with grey culmen and tip of lower mandible; feet—pink.

Voice: Song in breeding areas is similar to Tristram's but slower and less twittering and given from tree perch in dense forest. Contact call is short *ziit* similar to Black-faced Bunting.

Range: Breeds in Russia N of Lake Baikal, winters in S China.

Distribution and status: Uncommon. Wintering in open areas with scattered bushes and thickets in Changjiang valley and coastal provinces of S China.

Habits: Usually found at forest edges in secondary scrub. Often found in mixed flocks with other buntings.

1315. **Rustic Bunting** *Emberiza rustica* Tian wu   PLATE 126

Description: Smallish (14.5 cm) brightly coloured bunting with white belly. Adult male is unmistakable with black and white striped head and rufous nape, breast band, flank streaks and rump. Has slight crest. Female and non-breeding male similar but duller with white parts tinged buff and usually whitish spot behind cheeks. Juvenile less distinct and more heavily streaked. Race *latifascia* has crown blacker than nominate race and deeper red breast band and flank streaks. Iris—dark chestnut-brown; bill—dark grey with pinkish-grey base; feet—pinkish.

Voice: Song is melodious warble from elevated perch. Commonest call is sharp *tzip*. Alarm call is high-pitched *tsiee*.

Range: Breeds in N Eurasian taiga, wintering to China.

Distribution and status: Nominate race is common winter migrant in extreme W Xinjiang and in E provinces. Race *latifascia* is uncommon, wintering near E coast. Could breed in Taiga region of N Heilongjiang.

Habits: Inhabits taiga, heather and marshes, winters in open areas, cultivated woods and parkland.

1316. **Yellow-throated Bunting** *Emberiza elegans*   PLATE 125
Huanghou wu

Description: Medium-sized (15 cm) bunting with white belly and unmistakable black and yellow head pattern with short crest. Female similar to male but duller with brown replacing black and buff replacing yellow of male pattern. Distinguished from Rustic Bunting by uniform brown cheeks without black borders and lack of pale patch at rear of cheek. Race *ticehursti* is paler than nominate with narrower streaks on mantle; *elegantula* is darker than nominate with bolder darker streaking on mantle, breast and flanks. Iris—dark chestnut-brown; bill—blackish; feet—pale grey-brown.

Voice: Song is monotonous twittering given from tree perch. Similar to Rustic Bunting. Call is liquid, sharp, repeated *tzik*.

Range: Disjunct distribution. C and NE China, Korea and SE Siberia.

Distribution and status: Fairly common. Race *elegantula* is resident in C to SW China; *elegans* breeds in SE Siberia and N Heilongjiang, wintering in SE China and Taiwan; and *ticehursti* breeds in N Korea and adjacent NE China, wintering in coastal S and SE China.

Habits: Inhabits dry deciduous and mixed forests on hills and ridges. Winters in shady woodland, forest and secondary scrub.

1317. Yellowed-breasted Bunting *Emberiza aureola*   PLATE 127
[White-shouldered Bunting] Huangxiong wu

Description: Medium-sized (15 cm) colourful bunting. Breeding male unmistakable with chestnut crown and nape, black face and throat, yellow collar separated from yellow belly by chestnut breast band. Prominent white bar on shoulder of wing. Race *ornata* has more extensive black forehead and is darker than nominate. Non-breeding male much paler with yellow chin and throat and black confined to mottled ear coverts. Female and immature have pale sandy crown stripe flanked by dark-streaked lateral stripes; almost no malar stripe and long whitish-buff supercilium. All forms characterised by white shoulder stripe or patch as well as narrow white wing bar. White wing patch is conspicuous in flight. Iris—dark chestnut-brown; bill—grey above, pinkish-brown below; feet—pale brown.

Voice: Song from prominent perch slower and higher-pitched than Ortolan *djiiii-djiiii weee-weee ziii-ziii* and variants, mostly rising in pitch. Call is short, loud, metallic *tic*.

Range: Breeds in Siberia and Transbaikalia (Dauria) to NE China; wintering to S China and SE Asia.

Distribution and status: Common. Nominate race breeds in Altai Mts of N Xinjiang and *ornata* breeds in NE China. Both cross China in passage and winter in extreme S China, Taiwan and Hainan. Trapped in large numbers for food—sold as 'rice-birds'.

Habits: Lives in large rice fields, reeds or tall grass and moist thickets. Forms large flocks in winter and often mixes with other species.

1318. Chestnut Bunting *Emberiza rutila* Li wu   PLATE 127

Description: Smallish (15 cm) chestnut and yellow bunting. Breeding male unmistakable with entire chestnut head, upperparts and breast with yellow belly. Non-breeding male similar but duller with peppering of yellow over head and breast. Female is much less distinct with dark streaks on crown, mantle, breast and flanks. Distinguished from female Yellow-breasted and Black-faced Buntings by rufous rump and lack of white wing bar or white edge of tail. Juvenile is more heavily streaked. Iris—dark chestnut-brown; bill—brownish or bluish-horn; feet—pale brown.

Voice: Song is delivered from low tree perch. Variable and similar to Pallas's Leaf Warbler or Tree Pipit. Higher-pitched than Black-faced Bunting.

Range: Breeding in southern taiga of S Siberia and Transbaikalia (Dauria), wintering to S China and SE Asia.

Distribution and status: Breeds in extreme NE China and probably Changbai Mts. Fairly common winterer in southern provinces and Taiwan. On passage may be seen over eastern half of China.

Habits: Favours open conifer, mixed and deciduous forests with shrubs up to 2500 m. In winter, uses edges of woodlands and cultivated areas.

1319. Black-headed Bunting *Emberiza melanocephala*   PLATE 127
Heitou wu

Description: Largish (17 cm) mottled brown bunting with unstreaked yellowish underparts. Breeding male has black head but duller in winter and back brownish streaked black sometimes with a rufous wash on rump. Female and immature are buffy-brown streaked dark above. Both sexes have two whitish wing bars and unstreaked yellow underparts and vent. Immature are not safely separated in field from Red-headed Bunting. Female is distinguished from all buntings except Red-headed by uniform colour, yellow undertail coverts and lack of white in tail. Distinguished from Red-headed by larger, less conical bill. Iris—dark brown; bill—grey; feet—light brown.

Voice: Melodious, accelerating song similar to Corn Bunting, given from high perch. Flight call is deep, hard *tchip* like Yellowhammer or metallic *tzik*.

Range: Breeds in E Mediterranean to C Asia; wintering in India. Vagrants recorded in Thailand, China, Japan and Borneo.

Distribution and status: Rarely recorded on migration in Tianshan of W Xinjiang. Vagrant Fujian and Hong Kong.

Habits: A bird of open country with scattered bushes.

Note: Closely related to Red-headed Bunting and hybrids occur in overlap zone in N Iran.

1320. Red-headed Bunting *Emberiza bruniceps* PLATE 128
Hetou wu

Description: Largish (16 cm) yellowish bunting without any head stripes. Adult male unmistakable with chestnut head and breast contrasting with bright yellow nuchal collar and belly. Some males have reduced red area. Non-breeding male similar but paler and duller. Female is pale sandy-buff above and pale yellow below with blackish streaks on crown and mantle. Distinguished from female Black-headed Bunting by yellow rump and vent and buffy rather than white edges to wing feathers. Juvenile is greyer and heavier streaking extends to breast. Iris—dark brown; bill—greyish with dark tip; feet—pinkish-brown.

Voice: Song from prominent perch or in song flight is harsh monotonous phrase similar to but thinner than Black-headed Bunting, *twip*. Also metallic *ziff*, harsh *jiip* and *prrit*.

Range: C Asia, wintering to India.

Distribution and status: Locally common breeder in Altai, Tianshan and extreme W Xinjiang. Vagrant Beijing and Hong Kong.

Habits: Inhabits open arid steppes with scrub or bushes.

Note: Hybridises with Black-headed Bunting in N Iran.

1321. Japanese Yellow Bunting *Emberiza sulphurata* PLATE 127
[Yellow Bunting] Liuhuang wu

Description: Small (14 cm) bunting with uniformly greenish head, blackish lores and chin, prominent white eye-ring, bold double white wing bars and faint black streaking on flanks. Breeding male distinguished from male Black-faced Bunting of race *sordida* by paler head and lack of contrast between throat and breast. Female and non-breeding male distinguished from Black-faced Bunting by lack of supercilium, less streaking on breast, less prominent malar stripe and uniform bill colour. Distinguished from Eurasian Siskin by longer bill and tail; bunting-like jizz; dull rump and white on outer tail feathers. Iris—dark brown; bill—grey; feet—pinkish-brown.

Voice: Twittering song given from canopy, similar to Black-faced Bunting but shorter. Call is soft *tsip tsip*.

Range: Breeds in Japan, winters in Taiwan, S China and Philippines.

Distribution and status: Scarce, wintering in Taiwan and Fujian. Migrants occasionally seen on E coast from Jiangxi to Guangdong. Regularly winters in Hong Kong.

Habits: Prefers deciduous or mixed forests and secondary growth in foothills.

1322. Black-faced Bunting *Emberiza spodocephala*   PLATE 125

[Masked Bunting] Huitou wu

Description: Small (14 cm) black and yellow bunting. Nominate breeding male has grey head, nape and throat with black lores and chin. Upperparts are otherwise rich chestnut boldly streaked with black; pale yellow or whitish underparts. There is a white bar on 'shoulder' of wing and white edges to dark tail. Female and winter males have olive head with yellow eye-stripe and yellow crescent below ear coverts. Winter male is distinguished from Japanese Yellow Bunting by lack of black on lores. Race *sordida* and *personata* have more greenish-grey head than nominate and in *personata* the upper breast and throat are yellow. Iris—dark chestnut-brown; bill—blackish above with pale edge, pinkish below with dark tip; feet—pinkish-brown.

Voice: Song given from inconspicuous perch is lively succession of ringing chirps and trills, similar to Reed Bunting. Call is quiet sibilant *tsii-tsii*.

Range: Breeds in Siberia, Japan and NE and C China, winters to S China.

Distribution and status: Common. Nominate race breeds in NE China wintering in S China, Hainan and Taiwan. Japanese race *personata* occasionally found wintering close to E and S coasts. Chinese race *sordida* breeds in C China (E Qinghai, Gansu, S Shaanxi, Sichuan, N Yunnan, Guizhou and Hubei) winters to S and E provinces and Taiwan. Feeds on ground in forests, woodlands and scrub.

Habits: Continually twitches tail revealing white edges to outer feathers. Winters in reed beds and scrub and forest edge.

1323. Grey Bunting *Emberiza variabilis* Hui wu   PLATE 128

Description: Largish (17 cm) slaty-grey bunting. Breeding male: entire plumage slaty-grey with black streaks on mantle and a row of black spots, outlined pale grey, on scapulars. Winter male: feathers of upperparts and breast edged rufous; feathers of belly edged white. Female: brown and streaked similar to Black-faced Bunting but with no white pattern on outer tail feathers. Iris—dark chestnut; bill—greyish-black above, pinkish with dusky tip below; feet—pinkish-brown.

Voice: Song is simple phrase following soft drawn-out note. Call is sharp *zhii* similar to Black-faced Bunting.

Range: Breeds in N Japan and S Kamchatka, winters to S Japan and Ryukyus.

Distribution and status: Accidentals recorded along E coast of China, Taiwan and Helanshan in Ningxia.

Habits: Secretive bird of bamboo and shady undergrowth in mountain forests.

1324. Pallas's Bunting *Emberiza pallasi* Wei wu   PLATE 127

Description: Small (14 cm) bunting with black head. Breeding male: white submoustachial stripe contrasting with black head and throat; white nuchal collar, underparts grey, upperparts barred grey and black. Similar to Reed Bunting but slightly smaller with almost no brown or rufous on upperparts. Lesser wing coverts are blue-grey instead of rufous and white; wing bars are more marked. Female and non-breeding male and juveniles in all plumage forms are pale sandy-buff with dark streaking on crown, mantle, breast and flanks. Ear coverts not as dark as in Reed or Ochre-rumped Buntings. Grey lesser wing coverts distinguish from Reed Bunting; also straight rather than convex profile to upper bill. Tail is longer. Iris—dark chestnut; bill—greyish-black; feet—pinkish-brown.

Voice: Song from top of scrub or grass perch is uniform repetition of a single note. Common call is fine *chleep* like Eurasian Tree Sparrow. Also slurred *dziu* similar to Reed Bunting.

Range: Disjunct with northern alpine breeding range in tundra of Russia and Siberia (*polaris*) and southern breeding range in steppes of S Siberia and N Mongolia (nominate). Birds winter to south.

Distribution and status: Poorly understood. On passage through NW China to winter in Gansu, N Shaanxi and along the E coast from Liaoning to Guangdong. On passage Hong Kong. Breeding possible in W China—Altai Mts and Tekes (*polaris*) and NE China—Hulun Nur and N Heilongjiang (nominate). Breeding may also occur in Ordos Plateau.

Habits:

Note: The race *minor* of Reed Bunting has sometimes been assigned to this species (see Byers *et al.* 1995). A third subspecies, *lydiae* from Mongolia, probably also winters in China.

1325. Reed Bunting *Emberiza schoeniclus* [Common Reed Bunting] PLATE 127
Lu wu

Description: Smallish (15 cm) black-headed bunting with prominent white submoustachial stripe. Breeding male is similar to Pallas's Bunting but more rufous on upperparts. Female and non-breeding male lose much of the black of head and have mottled crown and ear coverts with a buffy supercilium. Again similar to Pallas's Bunting but distinguished by rufous rather than grey lesser wing coverts and convex profile to upper bill. Many races show minor variations. Of breeding races in China, *minor* is smallest and *pyrrhuloides* and *zaidamensis* are large and pale with bulbous bills. The latter is more buff and less grey than *pyrrhuloides*. Iris—chestnut-brown; bill—black; feet—dark brown to pinkish-brown.

Voice: Song from bush or reed stem is short series of hesitant tinkling units similar to House Sparrow. It is rather variable but usually ends with a trill. Common call is plaintive falling *seeoo* and harsh *brzee* contact call is given during migration.

Range: Palearctic.

Distribution and status: Locally common. Race *pyrrhuloides* breeds in extreme W Xinjiang (Kashi) and E Xinjiang (Hami) with wintering population in the upper Huanghe (Yellow River) valley and NW Gansu; *minor* breeds near Hulun Nur in E Nei Mongol, C and E Heilongjiang; wintering along E coast of China; *zaidamensis* is resident in the Zaidam basin of Qinghai; some *pallidior* winter in coastal SE China; and occasional wintering birds of races *passerina, parvirostris* and *incognita* have been found in NW China.

Habits: Lives in tall reed beds but wintering birds will feed in woods, fields and open countryside.

1326. Ochre-rumped Bunting *Emberiza yessoensis* [Chinese/ Japanese Reed Bunting] Hongjing weiwu   PLATE 127

Description: Smallish (15 cm) bunting. Breeding male black-headed, similar to Reed Bunting and Pallas's Bunting but lacks white submoustachial stripe and has rufous rump and nape. Breeding female as male except head which is like female Reed Bunting from which distinguished by less and paler streaking on underparts, pinkish-rufous nape and darker crown and ear coverts. Non-breeding male similar to female but with dark throat. Iris—dark chestnut; bill—blackish; feet—pinkish.

Voice: Song given from tall reed is brief twittering phrase often with short trill. Call is short *tick*. Also gives *bschet* call in flight, similar to Reed Bunting.

Range: Breeds in Japan, NE China and extreme SE Siberia (Ussuriland); wintering S to coasts of Japan, Korea and E China.

Distribution and status: Uncommon and globally rated as Near threatened (Collar *et al.* 1994). Race *continentalis* breeds in marshes of NE China (Harbin, Qiqihar and Xinghai (Khanka) Lake area); wintering near coasts of Jiangsu and Fujian; on passage through Liaoning, Hebei and Shandong and vagrant to Hong Kong.

Habits: Inhabits reed beds and shrubby marshland, also wet meadows in highlands. Winters in coastal marshes.

1327. Corn Bunting *Miliaria calandra* Shu wu   PLATE 128

Description: Large (19 cm) uniformly streaked, dull greyish-brown bunting. Sexes alike. Looks bulky with heavy bill. Flight heavier than larks and lacks pale trailing edge to wing. Iris—dark chestnut-brown; bill—pale horn; feet—yellow to pink/brown.

Voice: Persistent, distinctive, accelerating song like jingling of keys *tiik tiik*. Call is rapidly repeated dry clicking. In alarm gives *trrp* call.

Range: Temperate and Mediterranean zone of W Palearctic to Ukraine and Caspian Sea. Eastern population in N Afghanistan to S Kazakhstan and extreme W China.

Distribution and status: Race *buturlini* breeds in Tekes and Ili valleys of W Tianshan, wintering in Kashi area in W Xinjiang. Generally uncommon in scrub and grassland, winters on cultivated fields.

Habits: Sings from perch or in flight with dangling legs and uplifted wings. Males often polygamous. Forms flocks outside breeding season.

1328. Lapland Longspur *Calcarius lapponicus* Tiezhua wu   PLATE 126

Description: Medium-sized (16 cm) bulky bunting with large head, short tail and long hindtoe and claw. Breeding male unmistakable with black face and breast, rufous nape and white zig-zag pattern on side of head. Breeding female, less distinctive but has rufous nape and edges of greater wing coverts and blackish lateral crown stripes, pale supercilium and pale centre to ear coverts. Non-breeding birds and juvenile have streaked crown, buffy supercilium and bright rufous edges to greater coverts, secondaries and tertials. Race *kamtschaticus* is darker than nominate race with richer rufous tinge; *coloratus* is of similar colour but larger with large spur. Iris—chestnut-brown; bill—yellow with dark tip; feet—dark brown.

Voice: Common flight call is hard rattled *prrt* followed by short clear whistle *teuw*, similar to Snow Bunting but less melodious.

Range: Breeds on Arctic tundra, winters on prairies and coast to south.

Distribution and status: Winters in small numbers on bare meadows along E coast between 30° and 40° N and along Changjiang River; probably on Altai Mts. Passage birds found in open areas of NE China. Winters S to Gansu and E Qinghai. Races *lapponicus, kamtschaticus* and *coloratus* may all winter in China.

Habits: Gregarious; mixes with skylarks. Runs, walks and hops on ground. Perches on ground or boulders. Habits lark-like.

1329. Snow Bunting *Plectrophenax nivalis* Xue wu　　　　　　PLATE 128

Description: Large (17 cm) plump, black and white bunting with small black bill. Breeding male unmistakable with white head, underparts and wing bar contrasting with black of rest of plumage. Breeding female less contrasting with greyish streaking on crown, cheeks and nape and some orange-brown streaks on breast. First-winter birds less black and suffused with orange-brown especially on crown and breast. Juvenile has grey head and breast. Iris—dark; bill—black in adult, yellowish in juvenile and first-winter; feet—black.

Voice: Song from boulder or in quivering descending song flight, monotonous with alternating phrases. Common flight call is trilled *tiriririt* followed by liquid *tew*. Both calls similar to Lapland Longspur but more melodious. Also gives harsh *djeee* call.

Range: Breeds on Arctic tundra and sea cliffs. Winters S to about 50° N.

Distribution and status: Small numbers of race *vlasowae* winter in Tianshan, Altai Mts, E Nei Mongol and N Heilongjiang. Stragglers reach Hebei.

Habits: Keeps to bare ground. Gregarious in winter but generally not with other species. Usual gait is quick shuffling run but also hops. Birds from back of feeding flock leapfrog to front. Flocks rise up to 'dance' in undulating flight then drop suddenly to ground.

APPENDICES

Appendix 1

List of Endangered and Protected Species of Birds in China

B2W	CITES	RDB	PROT	NAME	ENGLISH
		R	P	*Lerwa lerwa*	Snow Partridge
			P	*Tetraogallus tibetanus*	Tibetan Snowcock
			P	*Tetraogallus altaicus*	Altai Snowcock
		R	P	*Tetraogallus himalayensis*	Himalayan Snowcock
NT		R	I	*Tetraophasis obscurus*	Chestnut-throated Partridge
NT		V		*Tetraogallus szechenyii*	Buff-throated Partridge
NT				*Alectoris magna*	Rusty-necklaced Partridge
		R		*Coturnix chinensis*	Blue-breasted Quail
		R		*Arborophila rufogularis*	Rufous-throated Partridge
NT		R		*Arborophila atrogularis*	White-cheeked Partridge
NT		I		*Arborophila crudigularis*	Taiwan Partridge
V		R		*Arborophila mandellii*	Chestnut-breasted Partridge
		R		*Arborophila brunneopectus*	Bar-backed Partridge
C		E	I	*Arborophila rufipectus*	Sichuan Partridge
V		I		*Arborophila gingica*	White-necklaced Partridge
E		E	I	*Arborophila ardens*	Hainan Partridge
V		R		*Arborophila charltonii*	Scaly-breasted Partridge
		V	P	*Ithaginis cruentus*	Blood Pheasant
	1	I	I	*Tragopan melanocephalus*	Western Tragopan
NT	3	R	I	*Tragopan satyra*	Satyr Tragopan
V	1	R	I	*Tragopan blythii*	Blyth's Tragopan
NT		V	P	*Tragopan temminckii*	Temminck's Tragopan
V	1	E	I	*Tragopan caboti*	Cabot's Tragopan
			P	*Pucrasia macrolopha*	Koklass Pheasant
	1	R	I	*Lophophorus impejanus*	Himalayan Monal
V	1	R	I	*Lophophorus sclateri*	Sclater's Monal
V	1	E	I	*Lophophorus lhuysii*	Chinese Monal
		V	P	*Gallus gallus*	Red Junglefowl
		R	P	*Lophura leucomelanos*	Kalij Pheasant
			P	*Lophura nycthemera*	Silver Pheasant
NT	1	I	I	*Lophura swinhoii*	Swinhoe's Pheasant
V	1	R	P	*Crossoptilon harmani*	Tibetan Eared Pheasant
V	1	V	P	*Crossoptilon crossoptilon*	White Eared Pheasant
V	1	E	I	*Crossoptilon mantchuricum*	Brown Eared Pheasant
NT		V	P	*Crossoptilon auritum*	Blue Eared Pheasant
V	1	V	I	*Syrmaticus ellioti*	Elliot's Pheasant
V	1	R	I	*Syrmaticus humiae*	Mrs Hume's Pheasant
NT	1	I	I	*Syrmaticus mikado*	Mikado Pheasant
V	1	E	P	*Syrmaticus reevesii*	Reeves's Pheasant
NT		V	P	*Crysolophus pictus*	Golden Pheasant
NT		V	P	*Chrysolophus amherstiae*	Lady Amherst's Pheasant
	2	R	I	*Polyplectron bicalcaratum*	Grey Peacock Pheasant
	2	R	I	*Polyplectron katsumatae*	Hainan Peacock Pheasant
V	1	E	I	*Pavo muticus*	Green Peafowl

B2W	CITES	RDB	PROT	NAME	ENGLISH
NT		E	I	*Dendragapus falcipennis*	Siberian Grouse
		I	P	*Lagopus lagopus*	Willow Ptarmigan
		I	P	*Lagopus mutus*	Rock Ptarmigan
		V	P	*Tetrao tetrix*	Black Grouse
			P	*Tetrao urogallus*	Western Capercaillie
		V	I	*Tetrao parvirostris*	Spotted Capercaillie
		E	P	*Tetrastes bonasia*	Hazel Grouse
NT			I	*Tetrastes sewerzowi*	Chinese Grouse
		V		*Dendrocygna javanica*	Lesser Whistling-duck
V	1	R		*Oxyura leucocephala*	White-headed Duck
		V	P	*Cygnus olor*	Mute Swan
		V	P	*Cygnus cygnus*	Whooper Swan
		V	P	*Cygnus columbianus*	Tundra Swan
V				*Anser cygnoides*	Swan Goose
			P	*Anser albifrons*	Greater White-fronted Goose
V				*Anser erythropus*	Lesser White-fronted Goose
			P	*Branta ruficollis*	Red-breasted Goose
		R		*Nettapus coromandelianus*	Cotton Pygmy-goose
NT		V	P	*Aix galericulata*	Mandarin Duck
NT				*Anas luzonica*	Philippine Duck
V				*Anas formosa*	Baikal Teal
V				*Marmaronetta augustirostris*	Marbled Duck
V				*Aythya nyroca*	Ferruginous Pochard
V				*Aythya baeri*	Baer's Pochard
T				*Polysticta stelleri*	Steller's Eider
V		R	I	*Mergus squamatus*	Scaly-sided Merganser
		I		*Turnix sylvatica*	Small Buttonquail
		I		*Turnix suscitator*	Barred Buttonquail
NT				*Indicator xanthonotus*	Yellow-rumped Honeyguide
NT				*Dendrocopos leucopterus*	White-winged Woodpecker
		R	P	*Dryocopus javensis*	White-bellied Woodpeck
V				*Picus rabieri*	Red-collared Woodpecker
	2	V	P	*Anthracoceros albirostris*	Oriental Pied Hornbill
	2	E	P	*Buceros bicornis*	Great Hornbill
NT		R	P	*Anorrhinus tickelli*	Brown Hornbill
V	1	R	P	*Aceros nipalensis*	Rufous-necked Hornbill
			P	*Aceros undulatus*	Wreathed Hornbill
		R	P	*Harpactes oreskios*	Orange-breasted Trogon
		V		*Harpactes erythrocephalus*	Red-headed Trogon
V				*Harpactes wardi*	Ward's Trogon
V				*Alcedo hercules*	Blyth's Kingfisher
			P	*Alcedo meninting*	Blue-eared Kingfisher
			P	*Halcyon capensis*	Stork-billed Kingfisher
			P	*Merops orientalis*	Green Bee-eater
			P	*Merops leschenaulti*	Chestnut-headed Bee-eater
		V	P	*Centropus sinstensis*	Greater Coucal
		V	P	*Centropus bengalensis*	Lesser Coucal
E				*Cacatua sulphurea*	Yellow-crested Cockatoo
		X	P	*Loriculus vernalis*	Vernal Hanging Parrot
	2	I	P	*Psittacula krameri*	Rose-ringed Parakeet
			P	*Psittacula finschii*	Grey-headed Parakeet
		R	P	*Psittacula roseata*	Blossom-headed Parakeet

B2W	CITES	RDB	PROT	NAME	ENGLISH
NT		V	P	*Psittacula derbiana*	Derbyan Parakeet
	2	V	P	*Psittacula alexandri*	Red-breasted Parakeet
			P	*Hirundapus cochinchinensis*	Silver-backed Needletail
			P	*Hemiprocne coronata*	Crested Treeswift
		R	P	*Tyto alba*	Barn Owl
			P	*Tyto capensis*	Grass Owl
	2	R	P	*Phodilus badius*	Oriental Bay Owl
	2	I	P	*Otus spilocephalus*	Mountain Scops Owl
			P	*Otus brucei*	Pallid Scops Owl
			P	*Otus scops*	Eurasian Scops Owl
			P	*Otus sunia*	Oriental Scops Owl
			P	*Otus elegans*	Elegant Scops Owl
			P	*Otus bakkamoena*	Collared Scops Owl
	2	R	P	*Bubo bubo*	Eurasian Eagle Owl
NT	2	I	P	*Bubo nipalensis*	Spot-bellied Eagle Owl
		I	P	*Bubo coromandus*	Dusky Eagle Owl
E			P	*Ketupa blakistoni*	Blakiston's Fish Owl
			P	*Ketupa zeylonensis*	Brown Fish Owl
NT	2	R	P	*Ketupa flavipes*	Tawny Fish Owl
			P	*Nyctea scandiaca*	Snowy Owl
			P	*Strix leptogrammica*	Brown Wood Owl
			P	*Strix aluco*	Tawny Owl
		R	P	*Strix uralensis*	Ural Owl
V		R	P	*Strix davidi*	Sichuan Wood Owl
	2	I	P	*Strix nebulosa*	Great Grey Owl
			P	*Surnia ulula*	Northern Hawk Owl
			P	*Glaucidium passerinum*	Eurasian Pygmy Owl
			P	*Glaucidium brodiei*	Collared Owlet
			P	*Glaucidium cuculoides*	Asian Barred Owlet
			P	*Athene noctua*	Little Owl
			P	*Athene brama*	Spotted Owlet
	2	I	P	*Aegolius funereus*	Boreal Owl
			P	*Ninox scutulata*	Brown Hawk Owl
			P	*Asio otus*	Long-earned Owl
			P	*Asio flammeus*	Short-eared Owl
		R		*Batrachostomus hodgsoni*	Hodgson's Frogmouth
V				*Caprimulgus centralasicus*	Vaurie's Nightjar
V				*Columba eversmanni*	Yellow-eyed Pigeon
			P	*Columba palumbus*	Common Wood Pigeon
V		R		*Columba punicea*	Pale-capped Pigeon
NT			P	*Columba janthina*	Japanese Wood Pigeon
		I		*Streptopelia senegalensis*	Laughing Dove
		R		*Macropygia unchall*	Barred Cuckoo Dove
			P	*Macropygia amboinensis*	Brown Cuckoo Dove
		R	P	*Macropygia ruficeps*	Little Cuckoo Dove
		V		*Chalcophaps indica*	Emerald Dove
		R	P	*Treron bicincta*	Orange-breasted Green Pigeon
		R	P	*Treron pompadora*	Pompadour Green Pigeon
		V	P	*Treron curvirostra*	Thick-billed Green Pigeon
		V	P	*Treron phoenicoptera*	Yellow-footed Green Pigeon
			P	*Treron apicauda*	Pin-tailed Green Pigeon
			P	*Treron sphenura*	Wedge-tailed Green Pigeon

B2W	CITES	RDB	PROT	NAME	ENGLISH
NT		R	P	*Treron sieboldii*	White-bellied Green Pigeon
NT		I	P	*Treron formosae*	Whistling Green Pigeon
			P	*Ptilinopus leclancheri*	Black-chinned Fruit Dove
		V	P	*Ducula aenea*	Green Imperial Pigeon
		V	P	*Ducula badia*	Mountain Imperial Pigeon
NT	2	I	I	*Tetrax tetrax*	Little Bustard
V	2	V	I	*Otis tarda*	Great Bustard
	1	I	I	*Chlamydotis macqueeni*	McQueen's Bustard
E	1	E	I	*Grus leucogeranus*	Siberian Crane
	2	R	I	*Grus antigone*	Sarus Crane
V	1	V	P	*Grus vipio*	White-naped Crane
			P	*Grus canadensis*	Sandhill Crane
	2	I	P	*Grus virgo*	Demoiselle Crane
			P	*Grus grus*	Common Crane
CD	1	E	I	*Grus monacha*	Hooded Crane
V	1	E	I	*Grus nigricollis*	Black-necked Crane
V	1	E	I	*Grus japonesis*	Red-crowned Crane
V			P	*Coturnicops exquisitus*	Swinhoe's Crake
		I		*Rallina fasciata*	Red-legged Crake
		I		*Rallina eurizonoides*	Slaty-legged Crake
		R		*Gallirallus striatus*	Slaty-breasted Rail
V			P	*Crex crex*	Corn Crake
		R	P	*Porzana bicolor*	Black-tailed Crake
			P	*Porzana parva*	Little Crake
NT				*Porzana paykullii*	Band-bellied Crake
		V		*Syrrhaptes tibetanus*	Tibetan Sandgrouse
		I	P	*Pterocles orientalis*	Black-bellied Sandgrouse
NT				*Gallinago hardwickii*	Latham's Snipe
V		I		*Gallinago nemoricola*	Wood Snipe
		I		*Limosa limosa*	Black-tailed Godwit
			P	*Numenius minutus*	Little Curlew
NT				*Numenius madagascariensis*	Eastern Curlew
E	1	I		*Tringa guttifer*	Nordmann's Greenshank
NT		R		*Limnodromus semipalmatus*	Asian Dowitcher
V				*Calidris pygmeus*	Spoon-billed Sandpiper
		R	P	*Metopidius indicus*	Bronze-winged Jacana
NT				*Charadrius placidus*	Long-billed Plover
NT				*Vanellus cinereus*	Grey-headed Lapwing
			P	*Glareola lactea*	Small Pratincole
E		V		*Larus saundersi*	Saunder's Gull
NT	1	V	I	*Larus relictus*	Relict Gull
			P	*Larus minutus*	Little Gull
			P	*Sterna caspia*	Caspian Tern
C		V	P	*Sterna bernsteini*	Chinese Crested Tern
V				*Sterna acuticauda*	Black-bellied Tern
			P	*Chlidonias niger*	Black Tern
NT				*Brachyramphus marmoratus*	Marbled Murrelet
		V		*Synthliboramphus antiquus*	Ancient Murrelet
V				*Synthliboramphus wumizusume*	Japanese Murrelet
	2	R		*Pandion haliaetus*	Osprey
NT	2	R		*Aviceda jerdoni*	Jerdon's Baza
			P	*Aviceda leuphotes*	Black Baza

B2W	CITES	RDB	PROT	NAME	ENGLISH
	2	V	P	*Pernis ptilorhynchus*	Oriental Honey-buzzard
	2	V	P	*Elanus caeruleus*	Black-shoulder Kite
			P	*Milvus migrans*	Black Kite
			P	*Milvus lineatus*	Black-eared Kite
	2	R	P	*Haliastur indus*	Brahminy Kite
	2	I	I	*Haliaeetus leucogaster*	White-bellied Sea Eagle
V	2	R	I	*Haliaeetus leucoryphus*	Pallas's Fish Eagle
NT	1	I	I	*Haliaeetus albicilla*	White-tailed Eagle
V	2	R	I	*Haliaeetus pelagicus*	Steller's Sea Eagle
NT			P	*Ichthyophaga humilis*	Lesser Fish Eagle
	2	V	I	*Gypaetus barbatus*	Lammergeier
NT		E	I	*Gyps bengalensis*	White-rumped Vulture
	2	R	P	*Gyps himalayensis*	Himalayan Griffon
			P	*Gyps fulvus*	Eurasian Griffon
NT	2	V	P	*Aegypius monachus*	Cinereous Vulture
NT	2	E	P	*Sarcogyps calvus*	Red-headed Vulture
	2	I	P	*Circaetus gallicus*	Short-toed Snake Eagle
	2	V	P	*Spilornis cheela*	Crested Serpent Eagle
			P	*Circus aeruginosus*	Eurasian Marsh Harrier
			P	*Circus spilonotus*	Eastern Marsh Harrier
			P	*Circus cyaneus*	Hen Harrier
NT			P	*Circus macrourus*	Pallid Harrier
			P	*Circus melanoleucos*	Pied Harrier
			P	*Circus pygargus*	Montagu's Harrier
	2	R	P	*Accipiter trivirgatus*	Crested Goshawk
	2	R	P	*Accipiter badius*	Shikra
			P	*Accipiter soloensis*	Chinese Sparrowhawk
			P	*Accipiter gularis*	Japanese Sparrowhawk
			P	*Accipiter virgatus*	Besra
			P	*Accipiter nisus*	Eurasian Sparrowhawk
			P	*Accipiter gentilis*	Northern Goshawk
			P	*Butastur teesa*	White-eyed Buzzard
NT	2	I	P	*Butastur liventer*	Rufous-winged Buzzard
	2	R	P	*Butastur indicus*	Grey-faced Buzzard
			P	*Buteo buteo*	Common Buzzard
	2	R	P	*Buteo rufinus*	Long-legged Buzzard
			P	*Buteo hemilasius*	Upland Buzzard
			P	*Buteo lagopus*	Rough-legged Buzzard
	2	R	P	*Ictinaetus malayensis*	Black Eagle
V	2	R	P	*Aquila clanga*	Greater Spotted Eagle
	2	V	P	*Aquila nipalensis*	Steppe Eagle
V	I	V	I	*Aquila heliaca*	Imperial Eagle
	2	V	I	*Aquila chrysaetos*	Golden Eagle
	2	R	P	*Hieraaetus fasciatus*	Bonelli's Eagle
			P	*Hieraaetus pennatus*	Booted Eagle
	2	R	P	*Hieraaetus kienerii*	Rufous-bellied Eagle
			P	*Spizaetus nipalensis*	Mountain Hawk Eagle
			P	*Microhierax caerulescens*	Collared Falconet
NT			P	*Microhierax melanoleucos*	Pied Falconet
V			P	*Falco naumanni*	Lesser Kestrel
			P	*Falco tinnunculus*	Common Kestrel
			P	*Falco vespertinus*	Red-footed Falcon

B2W	CITES	RDB	PROT	NAME	ENGLISH
			P	*Falco amurensis*	Amur Falcon
			P	*Falco columbarius*	Merlin
			P	*Falco subbuteo*	Eurasian Hobby
			P	*Falco severus*	Oriental Hobby
	2	R	I	*Falco cherrug*	Saker Falcon
			P	*Falco altaicus*	Altai Falcon
			P	*Falco rusticolus*	Gyrfalcon
	2	R	P	*Falco peregrinus*	Peregrine Falcon
			P	*Falco pelegrinoides*	Barbary Falcon
			P	*Podiceps cristatus*	Great Crested Grebe
			P	*Podiceps auritus*	Horned Grebe
		V	P	*Sula sula*	Red-footed Booby
		V	P	*Sula leucogaster*	Brown Booby
NT				*Anhinga melanogaster*	Darter
		V		*Phalacrocorax niger*	Little Cormorant
		R		*Phalacrocorax capillatus*	Japanese Cormorant
			P	*Phalacrocorax pelagicus*	Pelagic Cormorant
E		E	P	*Egretta eulophotes*	Chinese Egret
		R	P	*Egretta sacra*	Pacific Reef Egret
E				*Ardea insignis*	White-bellied Heron
C		E	P	*Gorsachius magnificus*	White-eared Night Heron
V				*Gorsachius goisagi*	Japanese Night Heron
		E		*Gorsachius melanolophus*	Malayan Night Heron
			P	*Ixobrychus minutus*	Little Bittern
NT				*Ixobrychus eurhythmus*	Von Schrenck's Bittern
		I	P	*Plegadis falcinellus*	Glossy Ibis
NT		R	P	*Threskiornis melanocephalus*	Black-headed Ibis
E			P	*Pseudibis davisoni*	White-shouldered Ibis
C	1	E	I	*Nipponia nippon*	Crested Ibis
	2	V	P	*Platalea leucorodia*	Eurasian Spoonbill
C		E	P	*Platalea minor*	Black-faced Spoonbill
		I	P	*Pelecanus onocrotalus*	Great White Pelican
V			P	*Pelecanus crispus*	Dalmatian Pelican
V			P	*Pelecanus philippensis*	Spot-billed Pelican
NT		X	P	*Mycteria leucocephala*	Painted Stork
	2	E	I	*Ciconia nigra*	Black Stork
			P	*Ciconia ciconia*	White Stork
E	1	E	I	*Ciconia boyciana*	Oriental Stork
V				*Leptoptilos javanicus*	Lesser Adjutant
V		I	I	*Fregata andrewsi*	Christmas Island Frigatebird
E	1		I	*Diomedea albatrus*	Short-tailed Albatross
NT				*Oceanodroma monorhis*	Swinhoe's Storm-petrel
DD				*Oceanodroma matsudairae*	Matsudaira's Storm-petrel
		R	P	*Pitta phayrei*	Eared Pitta
NT			P	*Pitta nipalensis*	Blue-naped Pitta
NT			P	*Pitta soror*	Blue-rumped Pitta
			P	*Pitta oatesi*	Rusty-naped Pitta
		R	P	*Pitta cyanea*	Blue Pitta
			P	*Pitta sordida*	Hooded Pitta
V		R	P	*Pitta nympha*	Fairy Pitta
			P	*Pitta moluccensis*	Blue-winged Pitta
			P	*Serilophus lunatus*	Silver-breasted Broadbill

B2W	CITES	RDB	PROT	NAME	ENGLISH
			P	*Psarisomus dalhousiae*	Long-tailed Broadbill
		R		*Lanius bucephalus*	Bull-headed Shrike
V				*Perisoreus internigrans*	Sichuan Jay
NT				*Urocissa whiteheadi*	White-winged Magpie
NT				*Cissa hypoleuca*	Indochinese Green Magpie
V				*Podoces biddulphi*	Xinjiang Ground-jay
V	V			*Oriolus mellianus*	Silver Oriole
NT				*Pericrocotus cantonensis*	Swinhoe's Minivet
		R		*Dicrurus annectans*	Crow-billed Drongo
		R		*Dicrurus remifer*	Lesser Racket-tailed Drongo
NT				*Dicrurus hottentottus*	Spangled Drongo
		R		*Dicrurus paradiseus*	Greater Racket-tailed Drongo
NT				*Terpsiphone atrocaudata*	Japanese Paradise-flycatcher
NT				*Bombycilla Japonica*	Japanese Waxwing
NT				*Turdus dissimilis*	Black-breasted Thrush
V				*Turdus feae*	Grey-sided Thrush
NT				*Turdus mupinensis*	Chinese Thrush
NT				*Brachypteryx stellata*	Gould's Shortwing
V				*Brachypteryx hyperythra*	Rusty-bellied Shortwing
V				*Rhinomyias brunneata*	Brown-chested Jungle Flycatcher
NT				*Muscicapa muttui*	Brown-breasted Flycatcher
NT				*Niltava davidi*	Fujian Niltava
NT				*Erithacus komadori*	Ryukyu Robin
V				*Luscinia ruficeps*	Rufous-headed Robin
V				*Luscinia obscura*	Blackthroat
NT				*Luscinia pectardens*	Firethroat
NT				*Tarsiger hyperythrus*	Rufous-breasted Bush Robin
NT				*Phoenicurus alaschanicus*	Ala Shan Redstart
NT				*Cinclidium frontale*	Blue-fronted Robin
NT				*Cochoa purpurea*	Purple Cochoa
NT				*Cochoa viridis*	Green Cochoa
V				*Saxicola insignis*	Hodgson's Bushchat
NT				*Saxicola jerdoni*	Jerdon's Bushchat
NT				*Sturnus sericeus*	Red-billed Starling
NT				*Sturnus philippensis*	Chestnut-cheeked Starling
NT				*Acridotheres albocinctus*	Collared Myna
	3		V	*Gracula religiosa*	Hill Myna
NT				*Sitta villosa*	Chinese Nuthatch
V				*Sitta yunnanensis*	Yunnan Nuthatch
V				*Sitta solangiae*	Yellow-billed Nuthatch
V				*Sitta magna*	Giant Nuthatch
V				*Sitta formosa*	Beautiful Nuthatch
NT				*Parus holsti*	Yellow Tit
NT				*Aegithalos fuliginosus*	Sooty Tit
NT				*Pycnonotus taivanus*	Styan's Bulbul
V				*Bradypterus major*	Long-billed Bush Warbler
V				*Locustella pryeri*	Japanese Swamp Warbler
V				*Acrocephalus sorghophilus*	Streaked Reed Warbler
NT				*Leptopoecile elegans*	Crested Tit Warbler
V				*Phylloscopus hainanus*	Hainan Leaf Warbler
NT				*Phylloscopus cantator*	Yellow-vented Warbler

B2W	CITES	RDB	PROT	NAME	ENGLISH
NT				*Tickellia hodgsoni*	Broad-billed Warbler
NT				*Garrulax maesi*	Grey Laughingthrush
NT				*Garrulax galbanus*	Yellow-throated Laughingthrush
V		R		*Garrulax sukatschewi*	Snowy-cheeked Laughingthrush
NT				*Garrulax lunulatus*	Barred Laughingthrush
V				*Garrulax bieti*	White-speckled Laughingthrush
NT				*Garrulax merulinus*	Spot-breasted Laughingthrush
NT				*Garrulax formosus*	Red-winged Laughingthrush
NT				*Garrulax milnei*	Red-tailed Laughingthrush
V				*Liocichla omeiensis*	Emei shan Liocichla
NT				*Xiphirhynchus superciliaris*	Slender-billed Scimitar Babbler
NT				*Rimator malacoptilus*	Long-billed Wren Babbler
V				*Spelaeornis caudatus*	Rufous-throated Wren Babbler
V				*Spelaeornis badeigularis*	Rusty-throated Wren Babbler
NT				*Spelaeornis formosus*	Spotted Wren Babbler
NT				*Sphenocichla humei*	Wedge-billed Wren Babbler
NT				*Chrysomma poecilotis*	Rufous-tailed Babbler
NT				*Babax waddelli*	Giant Babax
NT				*Babax koslowi*	Tibetan Babax
NT				*Pteruthius rufiventer*	Black-headed Shrike Babbler
NT				*Actinodura souliei*	Streaked Barwing
V				*Alcippe variegaticeps*	Gold-fronted Fulvetta
NT				*Alcippe cinerea*	Yellow-throated Fulvetta
NT				*Alcippe ruficapilla*	Spectacled Fulvetta
NT				*Alcippe rufogularis*	Rufous-throated Fulvetta
NT				*Heterophasia gracilis*	Grey Sibia
V				*Paradoxornis flavirostris*	Black-breasted Parrotbill
NT				*Paradoxornis conspicillatus*	Spectacled Parrotbill
NT				*Paradoxornis brunneus*	Brown-winged Parrotbill
V		R		*Paradoxornis zappeyi*	Grey-hooded Parrotbill
V		R		*Paradoxornis przewalskii*	Rusty-throated Parrotbill
V		R		*Paradoxornis davidianus*	Short-tailed Parrotbill
NT				*Paradoxornis atrosuperciliaris*	Lesser Rufous-headed Parrotbill
NT				*Paradoxornis ruficeps*	Greater Rufous-headed Parrotbill
NT		R		*Paradoxornis heudei*	Reed Parrotbill
V				*Lonchura oryzivora*	Java Sparrow
DD				*Leucosticte sillemi*	Sillem's Mountain-Finch
DD		R		*Carpodacus roborowskii*	Tibetan Rosefinch
NT				*Latoucheornis siemsseni*	Slaty Bunting
NT		R		*Emberiza koslowi*	Tibetan Bunting
V		R		*Emberiza jankowskii*	Jankowski's Bunting
V				*Emberiza sulphurata*	Japanese Yellow Bunting
NT				*Emberiza yessoensis*	Ochre-rumped Bunting

Codes: B2W = Birds to Watch list—V = Vulnerable, E = Endangered, C = Critical DD = Data Deficient, NT = Near Threatened. **CITES**—number refers to schedule. **RDB** = China Red Data Book—R = Rare, E = Endangered, V = Vulnerable, I = Indeterminate, X = Nationally extinct. **PROT** = listed under national protected species lists—P = listed as second-grade, I = listed as first-grade.

Note: Notice the severe bias towards protection of non-passerines under national protection lists and CITES. Bias is only partly improved in national Red Data Book compared to international Birds to Watch Listings. In fact it is the passerines that dominate in bird trade.

Appendix 2
List of endemic and limited distribution species of China's EBAs

D01 Taklimakan desert

Caprimulgus centralasicus
Podoces buddulphi
Leucosticte sillemi

D06 Eastern Tibet

Babax koslowi
Emberiza koslowi

D07 Southern Tibet

Crossoptilon harmani
Babax waddelli

D08 Eastern Himalayas

Arborophila mandellii
Tragopan blythii
Lophophorus sclateri
Harpactes wardi
Brachypteryx hyperythra
Phylloscopus cantator
Tickellia hodgsoni
Garrulax galbanus
Garrulax virgatus
Spelaeornis badeigularis
Spelaeornis caudatus
Spelaeornis longicaudatus
Sphenocichla humei
Stachyris oglei
Actinodura nipalensis
Actinodura waldeni
Alcippe cinerea
Alcippe ludlowi
Heterophasia gracilis
Heterophasia pulchella
Paradoxornis flavirostris

D11 Qinghai mountains

Alectoris magna
Phoenicurus alaschanicus

Phylloscopus kansuensis
Carpodacus roborowskii

D12 Central Sichuan mountains

Luscinia ruficeps
Luscinia obscura
Parus davidi
Garrulax sukatschewi
Garrulax lunulatus
Garrulax formosus
Actinodura souliei
Paradoxornis paradoxus
Paradoxornis przewalskii
Aegithalos fuliginosus
Latoucheornis siemsseni

D13 West Sichuan mountains

Lophophorus lhuysii
Strix davidi
Perisoreus internigrans

D14 Chinese subtropical forests

Arborophila rufipectus
Oriolus mellianus
Garrulax formosus
Liocichla omeiensis
Alcippe variegaticeps

D15 Yunnan mountains

Sitta yunnanensis
Garrulax bieti
Actinodura souliei
Paradoxornis brunneus

D20 Hainan

Arborophila ardens
Polyplectron katsumatae

Gorsachius magnificus
Sitta solangiae
Phylloscopus hainanus

D23 Shanxi mountains

Crossoptilon mantchuricum
Turdus feae

D24 South-east Chinese mountains

Arborophila gingica
Tragopan caboti
Syrmaticus ellioti
Gorsachius magnificus
Rhinomyias brunneata
Garrulax galbanus

D Taiwan

Arborophila crudigularis
Lophura swinhoii
Syrmaticus mikado
Treron formosae
Urocissa caerulea
Regulus goodfellowi
Pycnotous taivanus
Garrulax morrisonianus
Liocichla steerii
Actinodura morrisoniana
Yuhina bruneiceps
Heterophasia auricularis

Appendix 3

Birds expected or recorded from areas of Arunachal Pradesh claimed by China but not described in text

Cairina scutulata	White-winged Duck
Dinopium benghalense	Black-rumped Flameback
Hierococcyx varius	Common Hawk Cuckoo
Hirundapus giganteus	Brown-backed Needletail
Caprimulgus asiaticus	Indian Nightjar
Ichthyophaga ichthyaetus	Grey-headed Fish Eagle
Gyps indicus	Long-billed Vulture
Spizaetus cirrhatus	Changeable Hawk Eagle
Ardeola grayii	Indian Pond Heron
Tephrodornis pondicerianus	Common Woodshrike
Cyornis tickelliae	Tickell's Blue Flycatcher
Saroglossa spiloptera	Spot-winged Starling
Graminicola bengalensis	Rufous-rumped Grassbird
Garrulax nuchalis	Chestnut-backed Laughingthrush
Malacocincla abbotti	Abbott's Babbler
Turdoides earlei	Striated Babbler
Turdoides striatus	Jungle Babbler
Mirafra assamica	Rufous-winged Bushlark
Motacilla maderaspatensis	White-browed Wagtail

Appendix 4

List of bird species expected or confirmed from (Nansha) Spratly Islands but not described in text

Glossy Swiftlet	*Collocalia esculenta*
Silvery Pigeon	*Columba argentina*
Island Collared Dove	*Streptopelia bitorquata*
Peaceful Dove	*Geopelia striata*
Nicobar Pigeon	*Caloenas nicobarica*
Cinnamon-headed Green Pigeon	*Treron fulvicollis*
Little Green Pigeon	*Treron olax*
Pink-necked Green Pigeon	*Treron vernans*
Grey Imperial Pigeon	*Ducula pickeringii*
Pied Imperial Pigeon	*Ducula bicolor*
Beach thick-knee	*Esacus neglectus (=magnirostris)*

Appendix 5

List of clubs and societies for China area

1. Mainland China

Ornithological Society of China: many local chapters and a growing membership. Hosts symposia and publishes reports and collections of papers.
Contact: Dr. Zhang Zhengwang
Address: c/o Institute of Zoology, Chinese Academy of Sciences
7 Zhongguan-cun Lu, Haidian, Beijing, China.
E-mail: zzw@ns.buu.edu.cn

National Bird Banding Center: conducts a national bird ringing programme and publishes an annual report.
Address: c/o Forestry Academy of Science, Beijing.
E-mail: bird.hz@nicl.forestry.ac.cn

Beidaihe Birdwatching Society (BBS): actively monitors migrating birds passing through the key area of Beidaihe (Hebei) and publishes lists and a newsletter.
Contact: Dr Martin Williams (Vice-President)
Address: 1/F 15 Siu Kwai Wan, Ceung Chau Island, Hong Kong.

2. Hong Kong

The Hong Kong Bird Watching Society (HKBWS): founded in 1958 and aims to promote the study of birds in Hong Kong and the conservation of birds and their habitats in Hong Kong and Asia. The society runs regular outings in Hong Kong and members frequently visit sites in mainland China. The society has a growing membership and publishes a quarterly bulletin and an annual report—both of which contain colour illustrations. It also publishes occasional checklists—the next will appear in early 2000.
Address: Hong Kong Bird Watching Society, P.O. Box 12460, Hong Kong.
Website: http://www.hkbws.org.hk

Word-Wide Fund for Nature, Hong Kong (WWF-HK): manages Hong Kong's premier birding site—the Mai Po Marshes Wildlife Education Centre and Nature Reserve. For information about entry permits and on-site accommodation contact: WWF Hong Kong, GPO Box 12721, Hong Kong.
E-mail address: wwf@wwf.org.hk
Website: http://www.wwf.org.hk

3. Taiwan

Chinese Wild Bird Federation (CWBF): founded in 1988 and has over 3000 members in fifteen chapters in different towns on Taiwan. The federation has been active in conservation of wetlands and other bird habitats in Taiwan, serves as office of BirdLife International, runs an information site on the world wide web, conducts studies of rap-

tors and migrants, runs the bird banding programme of Taiwan and organises biannual symposia for ornithology of China and Taiwan.
Address: 1/F No. 34, Alley 119, Lane 30, Yung-chi Road, Taipei 110, Taiwan.
Telephone: 886-2-87874551; Fax: 886-2-87874547.
E-mail: cwbf@ms4.url.com.tw
Website: http://com5.iis.sinica.edu.tw/~cwbf

4. Japan

The Ornithological Society of Japan

Contact: Dr. Yuzo Fujimaki
c/o Laboratory of Wildlife Ecology,
Obihiro University of Agriculture and Veterinary Medicine.
Inada, Obihiro, 080–8555, Hokkaido, Japan.

Wild Bird Society of Japan: founded in 1934 with 86 local chapters and 50 000 members. Has an interest in Chinese birds and published a *Fieldguide to the Birds of Taiwan* with the Taiwan Wildbird Information Centre in 1991. WBSJ serves as the regional office of BirdLife International and coordinates their Red Data Book and Important Bird Areas in China programmes.
Contact: Simba Chan, Head International Cooperation Section.
Address: Wing, 2-35-2 Minamidaira, Hino, Tokyo 191-0041, Japan.
Tel: +81-42-593-6871; Fax: +81-42-593-6873.

Japanese Society for Preservation of Birds (JSPB)

Address: 2–5–5 Sinjuku Sinjukuku,
Tokyo, Japan 160-0022.
E-mail: jdy03271@nifty.he.jp

5. Korea

Institute of Ornithology

Address: Kyong Hee University,
Seoul, Korea.

6. International

BirdLife International (formerly the International Council for Bird Preservation, ICBP): studies and monitors bird populations worldwide, with a particular interest in their conservation. BirdLife publish many useful books, reports and lists including the *Birds to Watch 2: the world checklist of threatened birds*. BirdLife acts as the birds specialist group of IUCN (World Conservation Union).
Address: Wellbrook Court, Girton Road, Cambridge, CB3 ONA, UK.

The Oriental Bird Club (OBC): promotes interest in and conservation of the birds of the Oriental region (including all of China). The club has produced an important checklist of birds of the region and publishes an annual journal *Forktail* and a biannual bulletin.
Address: c/o The Lodge, Sandy, Bedforshire SG19 2DL, UK.

Wetlands International (WI): a consortium of bodies interested in the conservation of wetlands and includes the former Asian Wetland Bureau. The organisation has a special interest in wetland birds and has a China office.
Contact: Chen Keling, Coordinator.
Address: Wetlands International–China Programme
Room 501, Grand Forest Hotel
No 19a, Bei Sanhuan Zhonglu Road
Beijing, 100029.
E-mail: klchen@public.bta.net.cn

World Pheasant Association (WPA): has a special interest in China as the home of so many of the world's pheasants and has a series of projects sponsoring pheasant research and conservation in China.
Address: P.O. Box 5, Lower Basildon, Reading, RG8 9PF, Berkshire, UK.

International Crane Foundation (ICF): also has a special interest in China and provides assistance in monitoring and conserving crane populations in the region.
Address: International Crane Foundation, P.O. Box 447, Baraboo, WI 53913–0447, USA.
E-mail: gordon.icf@baraboo.com
Website: http://www.baraboo.com/bus/icf/whowhat.htm
Local Contact: Prof. Ma Yi Qing
Address: Haping Road, Harbin, 150040, China.

Appendix 6
New additions to China's Avifauna

Rufous-tailed Scrub Robin *Cercotrichas galactotes* has recently been added to Chinese lists, collected and recorded in extreme W Xinjiang.

A new monal pheasant *Lophophorus* sp. has been discovered in W Arunachal Pradesh, distinguished by an entirely white tail. The form should be looked for on the Chinese side of the border in SE Xizang.

References and Bibliography

Ali, S. and Ripley, S. D. (1948). The birds of the Mishmi Hills. *J. Bombay Nat. Hist. Soc.* 48: 1–37.

Ali, S. and Ripley, S. D. (1983). *Handbook of the birds of India and Pakistan.* Compact Edition. Oxford University Press.

Ali, S. and Ripley, S. D. (1987). *Compact edition of the handbook of the birds of India and Pakistan.* Second edition. Oxford University Press, Delhi.

Alström, P., Colston, P. R. and Olsson, U. (1990). Description of a possible new species of leaf warbler of the genus *Phylloscopus* from China. *Bull. Brit. Orn. Club* 110: 43–47.

Alström, P. and Olsson, U. (1995). A new *Phylloscopus* warbler from Sichuan Province, China. *Ibis* 137: 459–468.

Alstöm, P., Olsson, U. and Colston, P. R. (1992). A new species of *Phylloscopus* warbler from central China. *Ibis* 134: 329–334.

Alstöm, P., Olsson, U. and Round, P. D. (1991). The taxonomic status of *Acrocephalus agricola tangorum. Forktail* 6: 3–13.

Amadon, D. and Bull, J. (1988). Hawks and owls of the world. *Proc. W. Found. Vertebr. Zool..* 3: 293–357.

Baker, K. (1997). *Warblers of Europe and North Asia.* Helm, London.

Beaman, M. (1994). *Palearctic birds: a checklist of the birds of Europe, North Africa and Asia north of the foothills of the Himalayas.* Harrier Publications' Stonyhurst.

Beebe, W. (1918–1922). *A monograph of the pheasants.* Witherby, London.

Boonsong Lekagul and Round, P. D. (1991). *A guide to the birds of Thailand.* Saha Karn Bhaet, Bangkok.

Boswell, J. (1986a). Notes on the current status of ornithology in the People's Republic of China. *Forktail* 2: 43–51.

Boswell, J. (1986b). Some notes on bird markets, pigeon keeping and other bird/man relations in China. *Avicultural Magazine* 92 (3): 126–142.

Boswell, J. (1989). Ornithology in China: an update. *Forktail* 4: 55–61.

Brazil, M. A. (1991). *The birds of Japan.* Helm, London.

Byers, C., Olsson, U. and Curson, J. (1995). *Buntings and sparrows: a guide to the buntings and North American sparrows.* Pica Press, Mountfield.

Caldwell, H. R. and Caldwell, J. C. (1931). *South China birds.* Heyster May Vanderburgh, Shanghai.

Carey, G. J. (1993). The status and field identification of snipe in Hong Kong. *Hong Kong Bird Report 1992*: 139–152.

Chalmers, M. L. (1986). *Annotated checklist of the birds of Hong Kong.* Hong Kong Bird Watching Society.

Chan, S. (1991). The historical and current status of the Oriental White Stork. *Hong Kong Bird Report 1990*: 128–148.

Chantler, P. and Driessens, G. (1995). *Swifts: a guide to the swifts and treeswifts of the world.* Pica Press, Mountfield.

Cheng Tso-hsin (1976). [*Distribution list of Chinese birds.*] *Second (revised) edition. Science Press, Peking (In Chinese.)*

Cheng Tso-hsin (1987). A synopsis of the arifauna of China. *Science Press, Beijing.*

Cheng Tso-hsin (1994). *A complete checklist of species and subspecies of the Chinese birds.* Science Press, Beijing.

Chinese Bird Banding Office and Chinese bird banding Centre. (1986). *Chinese bird banding almanac.* Gansu Science and Technology Press.

Clement, P., Harris, A and Davis, J. (1993). *Finches and sparrows: an identification guide,* Helm, London.

Collar, N. J. and Andrew, P. (1988). *Birds to watch* International Council for Bird Preservation, Cambridge.

Collar, N. J., Crosby, M. J. and Stattersfield, A. J. (1994). *Birds to watch 2: the world list of threatened birds.* BirdLife International (Conservation Series 4), Cambridge, UK.

Cramp, S. (1992). *The birds of the Western Palearctic.* Volume 6. Oxford University Press.

Crosby, M. J. (1991). Little-known Oriental bird: Silver Oriole. *Bull. Oriental Bird Club* 14: 32–35.

Dai Bo, Dowell, S. D., Martins, R. P. and Williams, R. S. R. (1998). Conservation status of the Sichuan Hill-partridge *Arborophila rufipectus* in China. *Bird Conserv. Internatn.* 8: 349–360.

David, A. and Oustalet, E. M. (1877). *Les Oiseaux de Chine.* Two volumes. G. Masson, Paris.

Deignan, H. G. (1964). Subfamily Timalinae. In E. Mayr and Paynter, R. A. *Check-list of birds of the world.* Volume 10. Museum of Comparative Zoology, Cambridge, Massachusets.

Delacour, J. (1977). *The pheasants of the world.* Second edition. Saiga Publishing Co., Surrey.

Delacour, J. and Jabouille, P. (1931). *Les oiseaux de l'Indochine francaise.* Volumes 1–4. Exposition Coloniale Internationale, Paris.

Dickinson, E. C., Kennedy, R. S. and Parkes, K. C. (1991). *The birds of the Philippines.* British Ornithologists' Union (Checklist 12), Tring.

Étchécopar, R. D. and Hue, F. (1978–1983). *Les oiseaux de Chine, de Mongolie et de Coree.* Editions du Pacifique, Papeete, Tahiti.

Fan Zhongmin. (1990). *Essentials of China bird species.* Liaoning Science and Technology Press.

Fleming, R. L., Sir, Fleming, R. L. Jnr and Bangdel, L. S. (1984). *Birds of Nepal.* Third edition. Avalok Publishers, Kathmandu.

Forshaw, J. M. and Cooper, W. (1989). *Parrots of the world.* Third edition. Blandford Press, London.

Fry, C. H. (1984). *The bee-eaters.* T. and A. D. Poyser, Calton.

Fry, C. H., Fry, K. and Harris, A. (1992). *Kingfishers, bee-eaters and rollers.* Helm, London.

Gao Zhongxing *Zhalong birds.* Chinese Forestry Press.

Gaucher, P., Paillat, P., Chappuis, C., Saint Jalme, M., Lotfikah, F. and Wink, M. (1996). Taxonomy of the Houbara Bustard *Chlamydotis undulata* subspecies considered on the basis of sexual display and genetic divergence. *Ibis* 138: 273–282.

Goodwin, D. (1967). *Pigeons and doves of the world*. British Museum (Natural History), London.

Goodwin, D. (1976). *Crows of the world*. British Museum (Natural History), London.

Grimmett, R. (1991). Little-known Oriental bird: Biddulph's Ground Jay. *Bull. Oriental Bird Club* 13: 26–29.

Grimmett, R. and Taylor, H. (1992). Recent bird observations from Xinjiang Autonomous Region, China, 16 June to 5 July 1988. *Forktail* 7: 139–146.

Grimmett, R., Inskipp, G. and Inskipp, T. (1998). *Birds of the Indian Subcontinent*. Helm, London.

Hale, W. G. (1971). A revision of the taxonomy of the Redshank *Tringa totanus*. *Zool. J. Linnean Soc.* 53: 177–236.

Hancock, J. A., Kushlan, J. A. and Kahl, M. P. (1992). *Storks, ibises and spoobills of the world*. Academic Press, London.

Han Lianxian (1992). Wedge-billed Wren-Babbler *Sphenocichla humei*: a new species for China *Forktail* 7: 155–156.

Harrap, S. (1991). The Hainan Nuthatch. *Oriental Bird Club Bull.* 13: 35–36.

Harrap, S. and Quinn, D. (1996). *Tits, nuthatches and treecreepers: an identification guide*. Helm, London.

Harrison, P. (1983). *Seabirds: an identification guide*. Helm, Beckenham.

Hancock, J. and Kushlan, J. (1984). *The herons handbook*. Helm, London.

Hang Lianxian (1996). Notes on Slender-billed Scimitar-Babbler *Xiphirhynchus superciliaris* in Yunnan, China. *Forktail* 11: 168–169.

Harvey, W. G. (1986). Two additions to the avifauna of China. *Bull. Brit. Orn. Club* 106: 105.

He Fen-qi (1992). News on the Sichuan Hill Partridge. *W. P.A. News* 36: 33.

He Fen-qi and Lu Tai-chun (1991). Changes in status and distribution of China's pheasants. *W. P.A. News* 31: 19–24.

Howard, R. and Moore, A. (1991). *A complete checklist of the birds of the world*. Second edition. Academic Press, London.

Huang Zhengyi, Sun Zhenhua, Yikuai *et al*. *Bird resources and habitats in Shanghai*. Fudan University Press.

ICBP (1992). *Putting Biodiversity on the map: priority areas for global conservation*. International Council for Bird Preservation, Cambridge.

Inskipp, C. and Inskipp, T. (1991). *A guide to the birds of Nepal*. Helm, London.

Johansen, H. (1960). Die Vögelfauna Westsibiriens. III Teil (Non-Passeres). 9 Forsetzung: Alcidae, Laridae. *J. Orn.* 101: 316–339.

Johnsgard, P. A. (1986). *Pheasants of the world*. Oxford University Press.

Johnsgard, P. A. (1991). *Bustards, hemipodes and sandgrouse. Birds of dry places*. Oxford University Press.

Kennerley, P. R. (1987a). A survey of the birds of the Poyang Lake nature reserve. Jiangxi Province. China. 29 December 1985–4 January 1986. *Hong Kong Bird Report 1984–1985*: 97–111.

Kennerley, P. R. (1987b). Visit to Ba Bao Shan Nature Reserve 7–11 July 1987. Unpublished report.

Kennerley, P. R. (1989). A review of the status and distribution of the Black-faced Spoonbill. *Hong Kong Bird Report 1988*. 116–125.

Kennerley, P. R. and Bakewell, D. N. (1987). Nordmann's Greenshank in Hong Kong: a review of the identification and status. *Hong Kong Bird Report 1986*: 83–100.

Kennerley, P. R., Hoogendorn, W. and Chalmers, M. L. (1995). Identification and systematics of large white-headed gulls in Hong Kong. *Hong Kong Bird Report 1994>*: 127–156.

Kennerley, P. R. and Leader, P. J. (1992). The identification, status and distribution of small *Acrocephalus* warblers in eastern China. *Hong Kong Bird Report 1991*: 143–187.

King, B. (1984). Bird notes from the Anyemaqen Shan Range, Qinghai Province, China. *Le Gerfaut* 74: 227–241.

King, B. (1987a). Wild sighting of Brown Eared Pheasant. *W. P. A. News* 15: 14.

King, B. (1987b). Some bird observations at Pangquanguo Reserve in west central Shanxi Province in NE China. *Hong Kong Bird Report 1984–1985*: 112–114.

King, B. (1989a). Some bird observations at Kangwu Liangsi, southwest Sichuan province, China. *Hong Kong Bird Report 1988*: 102–110.

King, B. (1989b). Birds observed at Huang Nian Shan, Mabian County, southern Sichuan, China. *Forktail* 4: 63–68.

King, B. (1989c). Birds observed at Dafengding Panda Reserve, Mabian County, southern Sichuan, China. *Forktail* 4: 69–76.

King, B. F., Dickinson, E. C. and Woodcock, M. W. (1975). *A field guide to the birds of South East Asia*. Collins, London.

King, B. and Liao, W. P. (1989). Hainan Island bird notes. *Hong Kong Bird Report 1988*: 102–110.

Knystautas, A. (1993). *Birds of Russia*. HarperCollins, London.

Lambert, F. R. and Woodcock, M. (1996). *Pittas and broadbills*. Pica Press, Mountfield.

Lansdown, R. (1990). Little-known Oriental bird: Chinese Egret. *Bull. Oriental Bird Club* 11: 27–30.

La Touche., J. D. D. (1925–34). *A handbook of the birds of eastern China*. 1931–34>. Taylor and Francis, London.

Lewthwaite, R. W. (1996). Forest birds of southeast China: observations during 1984–1996. *Hong Kong Bird Report 1995*: 150–203.

Li, X. T. (1993). Surveys of the Brown Eared Pheasant in Dongling Mountain, Beijing. In Jenkins, D. *Pheasants in Asia 1992*. World Pheasant Association, Reading.

Long, A., Crosby, M. and Inskipp, T. (1994). A review of the taxonomic status of the Yellow-throated Laughingthrush *Garrulax galbanus*. *Bull. Oriental Bird Club*, 19: 41–47.

Ludlow, F. and Kinnear, N. B. (1933–34). A contribution to the ornithology of Chinese Turkestan. *Ibis* (13) 3: 240–259, 440–473, 658–694; 4: 95–125.

Ludlow, F. and Kinnear, N. (1944). The birds of south-eastern Tibet. *Ibis* (86) 14: 43–86, 176–208, 348–389.

Lu Tai-chun et al. (1991). *The rare and endangered gamebirds in China*. Fujian Science and Technology Press.

Ma, J. and Jin, L. (1987). The numerical distribution of the Red-crowned Crane in the Sanjiang Plain area of Heilongjiang province. *Acta. Zool. Sin.* 33: 82–87.

MacKinnon, J., Meng, S., Cheung, C., Carey, G., Zhu X. and Melville, D. (1996). *A biodiversity review of China*. WWF International China Programme, Hong Kong.

MacKinnon, J. and Philipps, K. (1993). *A field guide to the birds of Borneo, Sumatra, Java and Bali*. Oxford University Press.

Madge, S. and Burn, H. (1993). *Crows and jays: a guide to the crows, jays and magpies of the world*. Helm, London.

Melville, D. S. (1982). A preliminary survey of the bird trade in Hong Kong. *Hong Kong Bird Report 1980*: 55–102. *Hong Kong Bird Report 1980*: 55–102.

Meyer de Schauensee, R. (1984). *The birds of China*. Smithsonian, Washington DC.

National Environmental Protection Agency/Endangered Species Scientific Commission, P. R. C. (1998). *China Red Data Book of endangered animals—Aves*. Science Press, Beijing.

Olsson, U., Alström, P. and Colston, P. R. (1993). A new species of *Phylloscopus* warbler from Hainan Island, China. *Ibis* 135: 2–7.

Peters, J. L. (1934–1961). *Checklist of the birds of the world*. Volumes 1–15. Museum of Comparative Zoology, Cambridge, Massachusets.

Qian yanwen (1995). *Atlas of birds of China*. Henan Science and Technology Press.

Rand, M. (1994). More Saunder's Gull colonies found in north China. *Bull. Oriental Bird Club* 19: 19.

Ripley, S. D. (1977). *Rails of the world*. M. F. Feheley, Toronto.

Ripley, S. D., Saha, S. S. and Beehler, B. M. (1991). Notes on birds from the Upper Noa Dihang, Arunachal Pradesh, northeastern India. *Bull. Brit. Orn. Club* 111: 19–27.

Roberts, T. J. (1991–2). *The birds of Pakistan*. Two volumes. Oxford University Press, Karachi.

Robson, C. (1989). Birdwatching areas: Omei Shan, Sichuan, China. *Bull. Oriental Bird Club* 9: 16–21.

Robson, C. (1996). Russet Bush-Warbler *Bradypterus seebohmi*: a new species for Bhutan and the Indian subcontinent. *Forktail* 11: 161.

Roselaar, C. S. (1992). A new species of mountain finch *Leucosticte* from western Tibet. *Bull. Brit. Orn. Club*. 112: 225–231.

Round, P. D. (1992). The identification and status of the Russet Bush Warbler in China and continental southeast Asia. *Hong Kong Bird Report 1991*: 188–194.

Round, P. D. and Loskot, V. (1994). A reappraisal of the taxonomy of the Spotted Bush-Warbler *Bradypterus thoracicus*. *Forktail* 10: 159–172.

Scott, D. A. (1989). *A directory of Asian wetlands*, International Union for Conservation of Nature and Natural Resources, Gland, Switzerland and Cambridge.

Scott, D. A. (1993). The Black-necked Cranes *Grus nigricollis* of Ruoergai Marshes, Sichuan. China. *Bird Conserv. Internatn.* 3: 245–259.

Shaw, Tseng Hwang (1936). *The birds of Hopei Province*. Two volumes. Fan Memorial Institute of Biology, Peking.

Short, L. L. (1973). Habits of some Asian woodpeckers. (Aves, Picidae). *Bull. Amer. Mus. Nat. Hist.* 152: 253–364.

Short, L. L. (1982). *Woodpeckers of the world*. Delaware Museum of Natural History. Greenville, Delaware.

Sonobe, K. (1982). *A field guide to the birds of Japan.* Wild Bird Society of Japan, Tokyo.

Sibley, C. G. and Monroe, B. L., Jnr. (1990). *Distribution and taxonomy of birds of the world.* Yale University Press.

Sibley, C. G. and Monroe, B. L., Jnr. (1993). *Supplement to the distribution and taxonomy of birds of the world.* Yale University Press.

Singh, P. (1994). Recent bird records from Arunachal Pradesh, India. *Forktail* 10: 65–104.

Smythies, B. E. (1986). *The birds of Burma.* Third edition. Nimrod Press, Liss.

Stattersfield, A. J., Crosby, M. J., Long, A. J. and Wege, D. C. (1998). *Endemic Bird Areas of the world: priorities for biodiversity conservation.* Birdlife International, Cambridge (Conservation Series 7).

Stepanyan, L. S. (1990). [*Conspectus of the ornithological fauna of the USSR.*] Nauka, Moscow. (In Russian.)

Swinhoe, R. (1866). Ornithological notes from Formosa. *Ibis* 2 (2): 292–316, 392–406.

Thewlis, R. M., Duckworth, J. W., Anderson, G. Q. A., Dvorak, M., Evans, T. D., Nemeth, E., Timmins, R. J. and Wilkinson, R. J. (1996). Ornithological records from Laos, October 1992–August 1993. *Forktail* 11: 47–101.

Turner, A. and Rose, C. (1989). *A handbook to the swallows and martins of the world.* Helm, London.

Vaurie, C. (1951). Notes on some Asiastic swallows. *Amer. Mus. Novit.* 1529: 1–47.

Vaurie, C. (1959). *The birds of the Palearctic fauna. Volume 1 Witherby, London.*

Vaurie, C. (1962). The status of Larus relictus *and other hooded gulls from central Asia.* Auk *79: 303–309.*

Vaurie, C. (1972). *Tibet and its birds.* H. F. and G. Witherby, London.

Viney, C., Phillipps, K. and Lan Chiu Ying (1994). *Birds of Hong Kong and south China.* Sixth edition. Government Publications, Hong Kong.

Wang, J., Wu, C., Huang, G., Yang, X., Cai, Z., Cai, M. and Xiao, Q. (1991). [*Field guide: birds of Taiwan.*] Taipei, Taiwan. (In Chinese.)

Wang Zijiang (1991). *A manual for the identification of precious and rare and common birds in Yunnan province, China.* Yunnan University Press.

Watson, G. E., Traylor, M. A., Jnr. and Mayr, E. (1986). Family Muscicapidae. In Mayr, E. and Cotrell, G. W. (1986). *Check-list of birds of the world.* Volume 11. Museum of Comparative Zoology, Cambridge, Massachusetts.

Wei Tainhao (1989). [Brief history of modern ornithology in Yunnan]. *Zool. Research* 10 (2): 163–168. (In Chinese).

Williams, M. D., Carey, G. J., Duff, D. J. and Xu Weishu (1992). Autumn bird migration at Beidaihe, China, 1986–1990. *Forktail* 7: 3–56.

Wong, F. K. O. and Liang, Y. (1992). *Field survey of selected areas along the coast of Bohai Sea, People's Republic of China.* World Wide Fund for Nature, Gland Switzerland.

Wu Sen-hsiang and Yang Hsio-ying. (1991). [*A guide to the wild birds of Taiwan.*] Taiwan Wild Bird Information Centre and Wild Bird Society of Japan (In Chinese.)

Wu, Z., (1984). Birds in the Fanjingshan mountain area. In *Scientific investigations of Fanjingshan nature reserve, Guizhou.* Guizhou People's Publishing House.

Wu, Z., Lin, Q., Yang, J. and Wu, L. (1986). *The avifauna of Guizhou.* Guizhou People's Publishing House.

Wünderlich, K. (1991). *Aegithalos bonvaloti* Oustallet. In Stresemann, E. and Portenko, L. A. *Atlas der Verbreitung palaearktischer Vögel.* Volume 17. Akademie Verlag, Berlin.

Yang, Y. C. (1983). [Xishuangbanna bird distribution]. In *A comprehensive report on an investigation on Xishuangbanna Nature Reserve* (Eds. Xu, Y. C. *et al.*). Yunnan Technological Press. (In Chinese).

Zhang Yin-sun and He Fen-qi (1993). A study of the breeding ecology of the Relit Gull *Larus relictus* in Ordos, Inner Mongolia, China. *Forktail* 8: 125–132.

Zhao, Z. (1993). Is the Crested Shelduck extinct? *IWRB Threatened Waterfowl Research Group Newsletter* 3: 5.

Zheng, B. (1988). Discussion on survey of the birds from South Ailao mountains, Yunnan. *Zool. Res.* 9: 255–261 (In Chinese.)

Zheng Guangmei and Zhang Zhengwang (1992). The distribution and status of pheasants in China. In Jenkins, D. *Pheasants in Asia.* World Pheasant Association, Reading.

Zheng Zuoxin (1976). *Distributional list of Chinese birds* Scientific Press, Beijing (In Chinese).

Index

Upright numerals refer to species number; italics to plate number and in parenthesis to alternative names in common use in China.

Godwit
Bar-tailed 337; *37*
Black-tailed 336; *37*
Goldcrest, 889; *99*
Golden-Oriole, Eurasian 653; *69*
Goldeneye, Common 110; *11*
Goldfinch, European 1251; *120*
Goose
Bar-headed 74; *7*
Bean 70; *7*
Brent 77; *7*
Canada 76; *7*
Cotton Pygmy 83; *9*
Greater White-fronted 71; *7*
Greylag 73; *7*
Lesser White-fronted 72; *7*
Red-breasted 78; *7*
Snow 75; *7*
Swan 69; *7*
Gorsachius
goisagi 549; *61*
magnificus 548; *61*
melanolophus 550; *61*
Goshawk
Crested 482; *52*
Northern 488; *52*
Gracula religiosa 831; *85*
Graminicola bengalensis 1012; *95*
Grandala,795; *82*
Grandala coelicolor 795; *82*
Grassbird
Rufous-rumped 1012; *95*
Striated 1011; *100*
Grebe
Black-necked 522; *56*
Great Crested 520; *56*
Horned 521; *56*
Little 518; *56*
Red-necked 519; *56*
Green Pigeon
Orange-breasted 283; *30*
Pin-tailed 287; *30*
Pompadour 284; *30*
Thick-billed 285; *30*
Wedge-tailed 288; *30*
Whistling 290; *30*
White-bellied 289; *30*
Yellow-footed 286; *30*

Greenfinch
Black-headed 1248; *120*
Grey-capped 1246; *120*
Yellow-breasted 1247; *120*
Greenshank
Common 345; *38*
Nordmann's 346; *38*
Griffon
Eurasian 471; *50*
Himalayan 470; *50*
Grosbeak
Collared 1295; *124*
Japanese 1294; *124*
Pine 1283; *121*
Spot-winged 1296; *124*
White-winged 1297; *124*
Yellow-billed 1293; *124*
Ground-jay
Mongolian 637; *68*
Tibetan 639; *68*
Xinjiang 638; *68*
Grouse
Black 59; *6*
Chinese 63; *6*
Hazel 62; *6*
Siberian 56; *6*
Grus
antigone 298; *31*
canadensis 300; *31*
grus 302; *31*
japonensis 305; *31*
leucogeranus 297; *31*
monacha 303; *31*
nigricollis 304; *31*
vipio 299; *31*
virgo 301; *31*
Gull
Black-tailed 413; *43*
Brown-headed 423; *45*
Common Black-headed 424; *45*
Glaucous 416; *44*
Glaucous-winged 415; *44*
Herring 418; *44*
Heuglin's 419; *44*
Little 428; *43*
Mew 414; *44*
Pallas's 422; *45*
Relict 427; *45*